CourseMate

Engaging. Trackable. Affordable.

CourseMate brings course concepts to life with interactive learning, study, and exam preparation tools that support NUTR.

INCLUDES:

Integrated eBook, interactive teaching and learning tools, and **Engagement Tracker,** a first-of-its-kind tool that monitors student engagement in the course.

ON THE WEB

NUTR
Are you in?

ONLINE RESOURCES INCLUDED!

FOR INSTRUCTORS:

- First Day of Class Instructions
- Custom Options through 4LTR+ Program
- Instructor's Manual
- Test Bank
- PowerPoint® Slides
- Instructor Prep Cards
- Engagement Tracker
- Exam View
- JoinIn Questions

FOR STUDENTS:

- Interactive eBook
- Auto-Graded Quizzes
- Flashcards
- Crossword Puzzles
- Videos
- Animations
- Nutrition Tutorials
- Pop-up Tutors
- Nutrition Calculations

Students sign in at **login.cengagebrain.com**

WADSWORTH
CENGAGE Learning™

NUTR
Michelle Kay McGuire
Kathy A. Beerman

Publisher/Executive Editor: Yolanda Cossio

Nutrition Editor: Peggy Williams

Product Developmental Manager, 4LTR Press: Steven E. Joos

Associate Project Manager, 4LTR Press: Pierce Denny

Developmental Editor: Colin Grover, B-books, Ltd.

Assistant Editor: Elesha Feldman

Editorial Assistant: Shana Baldassari

Technology Project Manager: Miriam Myers

Marketing Manager: Laura McGinn

Marketing Manager, 4LTR Press: Courtney Sheldon

Marketing Assistant: Jing Hu

Sr. Marketing Communications Manager: Mary Anne Payumo

Production Director: Amy McGuire, B-books, Ltd.

Content Project Manager: Carol Samet

Frontlist Buyer, Manufacturing: Karen Hunt

Production Service: B-books, Ltd.

Art Director: John Walker

Rights Acquisitions Specialist: Thomas McDonough

Text Permissions Researcher: David Ferrell, B-books, Ltd.

Photo Researcher: Charlotte Goldman

Internal Designer: Ke Design, Mason, OH

Cover Designer: Joe Devise/Red Hanger Design

Cover Image: © Zac Macaulay/cultura/Corbis

Printed in the United States of America
3 4 5 6 7 16 15

Library of Congress Control Number: 2011940412

Student Edition ISBN-13: 978-1-111-57892-3
Student Edition ISBN-10: 1-111-57892-3

Wadsworth
20 Davis Drive
Belmont, CA 94002-3098
USA

Cengage Learning products are represented in Canada by Nelson Education, Ltd.

Design Photo Credits:
Chapter Opener: © Morgan Lane Photography/Shutterstock.com;
Box Graphic: © iStockphoto.com/Gergana Todorchovska

NUTR Brief Contents

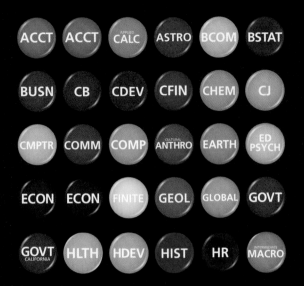

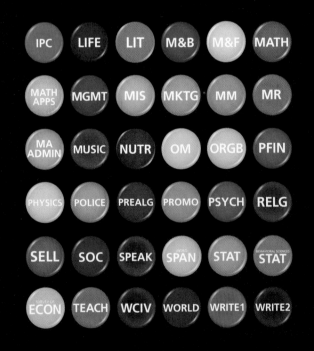

NUTR Contents

© Denise Kappa/Shutterstock.com

© Elena Elisseeva/Shutterstock.com

© iStockphoto.com/Danny Hooks

© iStockphoto.com/Bill Crawford

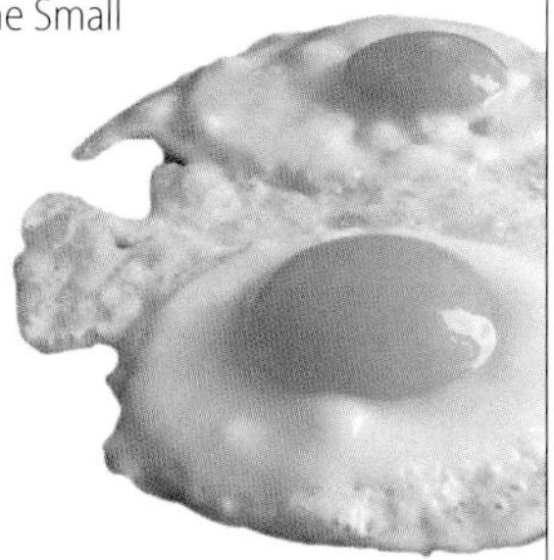

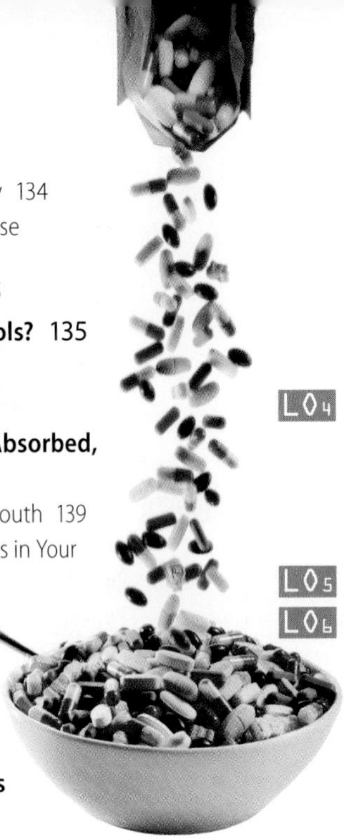

© pulp/Photodisc/Getty Images

7

THE VITAMINS 150

© iStockphoto.com/RedHelga

8

WATER AND THE MINERALS 182

© iStockphoto.com/bratan007

© Milos Luzanin/Shutterstock.com

© iStockphoto.com/R-O-M-A

© Vankad/Shutterstock.com

© diego cervo/Shutterstock.com

© Image Source/Jupiterimages

© iStockphoto.com/paul kline

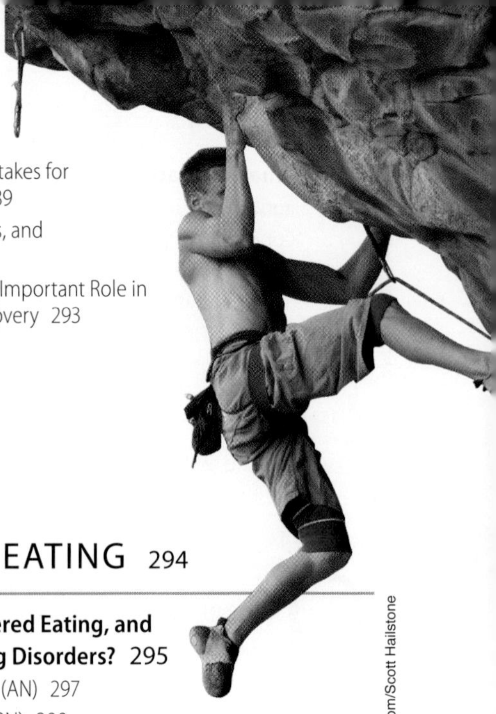

© iStockphoto.com/Scott Hailstone

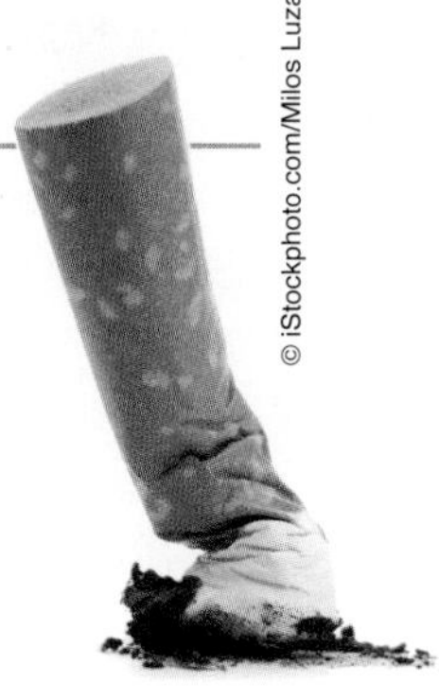

© iStockphoto.com/Milos Luzanin

NUTR

1 Why Does Nutrition Matter?

© iStockphoto.com/Stephanie Horrocks

LEARNING OUTCOMES:

- LO1 Define and understand the meaning of nutrition.
- LO2 Understand the purpose and classification of nutrients.
- LO3 Differentiate the major groups of nutrients.
- LO4 Define *calorie* and explain energy.
- LO5 Outline the scientific method.
- LO6 Evaluate the validity of a nutritional claim.
- LO7 Understand the connection between nutrition and health.
- LO8 Appreciate the importance of nutrition.

Chapter 1

LO1 What Is Nutrition?

Life would not be possible without the nourishment of food, and your quality of life depends greatly on which foods you choose to eat. Hopefully, you are interested in making sure that your nutrition is as good as possible. If you apply the information you learn in this course to your life, you may very well reach this goal. In this first chapter, you will learn many fundamental concepts necessary to understanding how good nutrition is essential to your health. You will also learn how scientists study nutrition, how national health is assessed, and how you can use scientific reason—not rumor—to select a healthy diet for years to come.

You have probably heard the terms *nutrition* and *nutrient*, but you may not know what they actually mean. The term **nutrition** refers to how living organisms obtain and use food to support all the processes required for their existence. Because this process is complex, the study of nutrition incorporates a wide variety of scientific fields. Scientists who study nutrition, *nutritional scientists*, work in many disciplines, such as immunology, medicine, genetics, biology, physiology, biochemistry, education, psychology, sociology, and of course nutrition. A **dietitian** is a nutrition professional who helps people make dietary changes and food choices to support a healthy lifestyle. Some dietitians are also involved in scientific research. Thus, the science of nutrition reflects a broad spectrum of academic and social disciplines. The credential *RD* stands for *registered dietitian*.

nutrition The science of how living organisms obtain and use food to support processes required for existence.

dietitian A nutrition professional who helps people make dietary changes and food choices to support a healthy lifestyle

nutrient A substance found in food that is used by the body for energy, maintenance of body structure, or regulation of chemical processes.

The foods you choose now will influence both your immediate and long-term health.

© Radius Images/Corbis

LO2 What Are Nutrients, and What Do They Do?

But what are nutrients, and why do you need them? In other words, why do people actually need to eat? A **nutrient** is a substance found in food that is required by the body and used for energy, maintenance of body structure, and/or regulation of chemical processes. For example, fats and carbohydrates provide the energy needed to fuel your body, calcium and phosphorus build and strengthen your teeth and bones, and many vitamins facilitate the chemical reactions that protect your cells from the damaging effects of excessive sunlight and pollution.

There are many ways to classify nutrients, foods, and food components. These classifications help nutritionists and other scientists distinguish the source, purpose, chemical composition, and importance to sustaining life of any given substance found in food. While a multitude of nutrient classification systems are important to scientific research, only three are needed to understand basic nutrition. Nutrients can be classified as essential, nonessential, or conditionally essential; as organic or inorganic; and as macronutrients or micronutrients. Each of these classification systems will be introduced in the following sections.

Essential, Nonessential, and Conditionally Essential Nutrients

One way nutritionists classify a nutrient is by whether it must be obtained from the foods you eat in order to support life. Although your body can theoretically use all the nutrients found in foods, you only *need* to consume some of them. A substance that must be obtained from the diet to sustain life is referred to as an **essential nutrient.** Your body needs essential nutrients, but it either cannot make them at all or cannot make them in adequate amounts. A **nonessential nutrient** is a substance that your body needs, but if necessary, can produce in amounts needed to satisfy its requirements. Therefore, you do not actually need to consume nonessential nutrients. Most foods contain a mixture of essential and nonessential nutrients. For example, milk contains a variety of essential vitamins and minerals (such as vitamin A and calcium) as well as several nonessential nutrients (such as cholesterol).

There are, however, circumstances under which a normally nonessential nutrient becomes essential. During these times, the nutrient is called a **conditionally essential nutrient.** For example, older children and adults must obtain two essential lipids through the diet, whereas babies require at least four that they are unable to produce because their physiologic systems are too immature. The additional lipids are therefore conditionally essential during early life. Certain diseases can also cause normally nonessential nutrients to become conditionally essential. You will learn about some of these in later chapters.

essential nutrient A substance that must be obtained from the diet to sustain life.

nonessential nutrient A substance that sustains life but is not necessarily obtained from the diet.

conditionally essential nutrient A normally nonessential nutrient that, under certain circumstances, becomes essential.

organic compound A substance that contains carbon and hydrogen atoms.

inorganic compound A substance that does not contain carbon.

Organic Nutrients, Inorganic Nutrients, and Organic Foods

Nutrients can also be distinguished as organic or inorganic. By definition, a substance that contains carbon and hydrogen atoms is an **organic compound.** Carbohydrates, proteins, lipids, and vitamins are therefore chemically organic nutrients because they all contain carbon and hydrogen atoms. An **inorganic compound** does not contain carbon. Because neither water nor minerals contain carbon, they are examples of inorganic compounds. In this way, all foods are considered organic—at least in the chemical sense of the term.

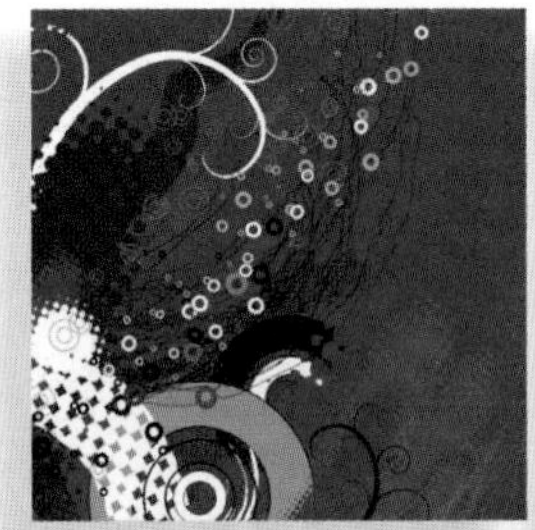

Are Organic Foods Healthier?

Because the USDA makes no claims that organically produced food is safer or more nutritious than conventionally produced food, the *organic* label is not meant to suggest superior nutritional quality or food safety. Furthermore, most scientific studies on the properties of organic food have not found that organic foods contain higher levels of nutrients than their nonorganic counterparts.[1] The only appreciable differences between organic and conventionally produced foods are the methods used to grow, handle, and process them. Whether these alternative agricultural practices promote enhanced environmental integrity and ecological balance is another area of active debate.

Figure 1.1 **The USDA Organic Foods Seal**

USDA ORGANIC

Certified organic foods can be identified by this seal.

United States Department of Agriculture

The term *organic* also has an additional and very different meaning when a person uses it to describe how a food is grown, harvested, or manufactured. When a food is labeled *certified organic*, it has been grown and processed according to U.S. Department of Agriculture (USDA) national organic standards. These foods are usually identified by the USDA's organic foods seal, which is displayed on the foods' packaging (see Figure 1.1). There are many rules and regulations that must be followed for a crop or food to be certified organic by the USDA. For example, a farmer cannot use conventional pesticides and herbicides on organically grown fruits and vegetables.

Not all organic foods are made entirely with organic ingredients. To find out the percentage of organic ingredients that a product contains, you can examine its food label (see Figure 1.2). Foods that carry the USDA organic seal and are labeled *100% organic* must have at least 95 percent organically produced ingredients. Foods labeled *organic* must have at least 70 percent organic ingredients. Products that contain less than 70 percent organic ingredients may list specific organically produced ingredients on the side panel of the package, but may not make any organic claims on the front of the package.

macronutrients A class of nutrients that humans need to consume in relatively large quantities (more than a gram per day).

micronutrients A class of nutrients that humans need to consume in relatively small quantities.

Macronutrients and Micronutrients

Finally, nutrients are classified based on how much of them a person must consume to maintain health. Nutrients consumed in relatively large amounts (more than a gram per day) are classified as **macronutrients.** The macronutrients include carbohydrates, proteins, lipids, and water. Because you need only very small amounts of vitamins and minerals (often micrograms or milligrams each day), these substances are called **micronutrients.** Over the course of a lifetime, a typical adult requires about 1,200 kilograms (2,700 pounds)

Figure 1.2 **Understanding Food Labels of Organic Products**

Must have 95–100% certified organic ingredients

Must have at least 70% certified organic ingredients

Organic ingredients can be listed on side panel

No organic claim is being made

When listed on food labels, the term *organic* can have different meanings.

phytochemical (or **phytonutrient**) A compound found in plants that likely benefits human health beyond the provision of essential nutrients and energy.

zoochemical (or **zoonutrient**) A compound found in animal-based foods that likely benefits human health beyond the provision of essential nutrients and energy.

functional food (or **super food**) A food that likely optimizes human health by providing a high concentration of nutrients, phytochemicals, or zoochemicals.

of protein, a macronutrient, but only about 0.14 kilograms (0.3 pounds) of iron, a micronutrient.

OTHER HEALTH-PROMOTING SUBSTANCES

As scientists research nutrition, they learn more and more about the relationship between diet and health. Not too long ago, scientists discovered that in addition to the traditional established nutrients, foods contain other substances that likely benefit your health. Scores of these compounds have only recently been uncovered and are therefore less understood than traditional nutrients. In fact, because scientific technology and nutritional knowledge have advanced so much during the last few decades, the definition of *nutrient* is evolving. The list of recognized nutrients will likely grow as researchers identify new substances and learn more about how the thousands of substances already identified promote health and well-being.

When a health-promoting compound is found in plants, it is called a **phytochemical** (or **phytonutrient**). When a health-promoting compound is found in animal-based food, it is called a **zoochemical** (or **zoonutrient**).[2] As scientists learn more about these compounds, some may be reclassified as nutrients.

You may have heard of functional foods or even seen them advertised. A **functional food** (or **super food**) is a food or product that likely promotes optimal health beyond simply helping the body meet its basic nutritional needs. Functional foods contain either (1) a high concentration of traditional nutrients, (2) phytonutrients, and/or (3) zoonutrients.[3] For example, soymilk is often referred to as a functional food because it contains phytochemicals that are believed to decrease the risk of some cancers. Other examples are cow's milk, which is rich in zoochemicals that may lower the risk of cancer and high blood pressure, and tomatoes, which may promote cardiovascular health. Although consuming functional foods may improve your health, the processes by which this occurs are often poorly understood.

LO3 How Are Macronutrients and Micronutrients Classified?

Scientists organize macronutrients and micronutrients into six general groups based on their chemical natures (see Table 1.1). Each major group or *class* of nutrients consists of many different compounds, and each contributes to the structure and/or function of your body in one way or another. In this section, each of the six macro- and micronutrient classes will be introduced. This is not the last time you will see them, however; each will be addressed in much greater detail in subsequent chapters.

Grains and cereals
are good sources of carbohydrates. Experts recommend that you choose whole-grain products at least half the time.

CARBOHYDRATES

Most dairy products are excellent sources of protein.

PROTEINS

Olives
are often recommended as sources of healthy oils.

LIPIDS

Table 1.1 **Grouping Macronutrients and Micronutrients**

Macronutrients	Micronutrients
Carbohydrates	Vitamins
Proteins	Minerals
Lipids	
Water	

Carbohydrates

Carbohydrates, consisting of carbon, hydrogen, and oxygen atoms, serve a variety of functions in the body. There are many different types of carbohydrates. For example, those found in starchy foods like rice and pasta are quite different from those found in fruits and sweet desserts. Of the many carbohydrates that exist, perhaps the most important is glucose, which you may have heard referred to as *blood sugar*. Glucose is so important because most of your body's cells use it as their primary source of energy. Your body uses other carbohydrates for many other purposes as well. For instance, some make up your genetic material (DNA). Others, such as dietary fiber, play roles in maintaining the health of your digestive tract and even decreasing the risk of certain conditions such as heart disease and type 2 diabetes.

Proteins

Protein is abundant in many foods, including meat, legumes (such as dried peas and beans), dairy foods, and some cereal products. Although most proteins consist primarily of carbon, oxygen, nitrogen, and hydrogen atoms, some also contain sulfur or selenium atoms. The thousands of proteins in your body have many roles, such as serving as hormones. Proteins also comprise the major structural materials of the body, including muscle, bone, and skin. Proteins allow you to move, support your complex internal communication systems, keep you healthy by fulfilling their roles in the immune system (which protects against infection and disease), and regulate many of the chemical reactions needed for life. When needed to do so, proteins can also serve as a source of energy.

Lipids

Lipids, liquid oils and solid fats, are the third major macronutrient class. Lipids generally consist of carbon, oxygen, and hydrogen atoms. They provide large amounts of energy, are important to the structure of cell membranes, and are necessary to the development and maintenance of your nervous and reproductive systems. Lipids also regulate a variety of processes that happen within cells. Many foods contain lipids, although the types of lipids found in plant-based foods such as nuts and corn oil are typically quite different from those found in animal-based foods such as meat, fish, eggs, and milk.

Water

Without water, the fourth macronutrient class, there

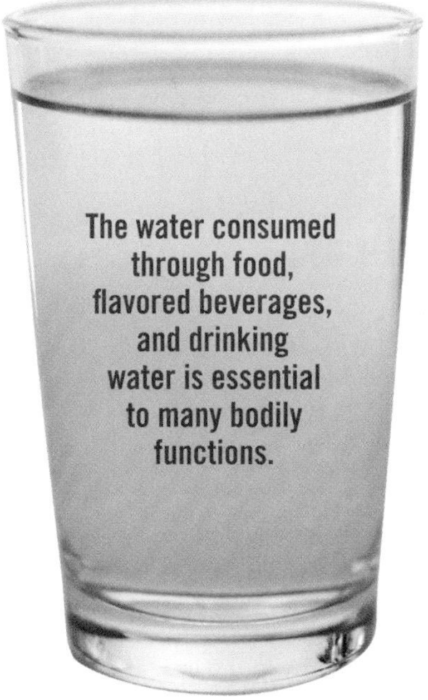
The water consumed through food, flavored beverages, and drinking water is essential to many bodily functions.

WATER

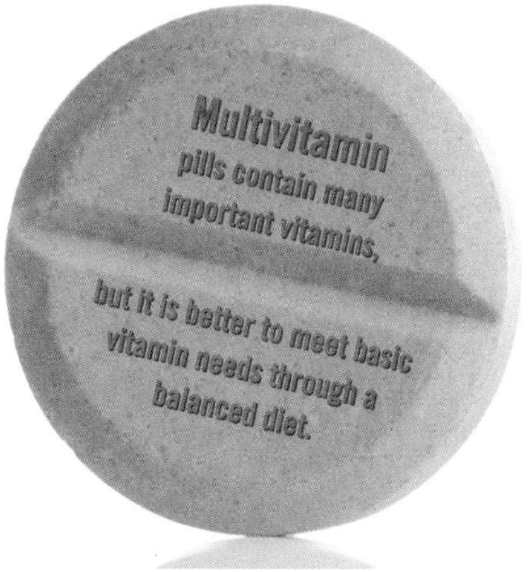
Multivitamin pills contain many important vitamins,
but it is better to meet basic vitamin needs through a balanced diet.

VITAMINS

The mineral sodium chloride (salt) is essential in small quantities, but harmful in excess.

MINERALS

energy The capacity to do work.

energy-yielding nutrient A nutrient that the body can use for energy.

adenosine triphosphate (ATP) A chemical that provides energy to cells in the body.

would be no life. Water, comprised of oxygen and hydrogen atoms, makes up approximately 60 percent of your total body weight. Humans typically consume water every day, whether as a beverage or in the foods they eat. The functions of water are varied and vital. Water functions as the transporter of nutrients, gases, and waste products; as the fluid in which chemical reactions occur; and as a partner in many chemical reactions needed for your body to function. Water is also important in regulating body temperature and protecting your internal organs from damage.

Vitamins

Vitamins, the first class of micronutrients, have a variety of chemical structures. Although they all contain carbon, oxygen, and hydrogen atoms, some vitamins also contain substances such as phosphorus and sulfur. Vitamins are abundant in most naturally occurring foods—especially fruits, vegetables, and grains. Your body requires vitamins to control hundreds of chemical reactions needed for its function. Vitamins also promote healthy and appropriate growth and development. Some vitamins, called antioxidants, protect your body from the damaging effects of harmful compounds such as air pollution. Unlike carbohydrates, proteins, and lipids, vitamins are not used directly for structure or energy. However, they play important roles in the chemical processes that build and maintain tissue and in the utilization of energy contained in macronutrients.

Vitamins can be subdivided based on how they interact with water. Water-soluble vitamins (vitamin C and the B vitamins) dissolve easily in water, while fat-soluble vitamins (vitamins A, D, E, and K) do not. Much contemporary nutritional research focuses on the role of vitamins in the prevention and management of diseases such as heart disease and certain types of cancer.

Minerals

At least 15 minerals, each of which serves a specific purpose, are considered to be essential nutrients. For example, calcium, abundant in dairy products, provides the matrix for various structural components in your body (such as bone). Other minerals, such as sodium, help regulate a variety of body processes (such as water balance). Still other minerals such as selenium, abundant in many seeds and nuts, facilitate chemical reactions. Like vitamins, minerals are not themselves used for energy, though many drive energy-producing reactions involving the macronutrients. Scientists are still discovering new ways that minerals prevent and perhaps even treat various diseases.

LO4 How Is the Energy in Food Measured?

As you have just learned, the macronutrients (with the exception of water) supply the body with energy. But what exactly is energy? And how can foods contain this important commodity? **Energy** is the capacity of a physical system to do work. In other words, if something has energy, it can cause something else to happen. Energy is not a nutrient, but in terms of nutrition, the body uses energy found in foods to grow, develop, move, and fuel the many chemical reactions required for life. Carbohydrates, proteins, and lipids all contain energy you can utilize. A nutrient that the body can use for energy is an **energy-yielding nutrient.** After you eat an energy-yielding nutrient, cells in your body transfer the energy that it contains into a special substance called **adenosine triphosphate (ATP),** which stores energy somewhat like a molecular battery. Your body can then use the energy stored in ATP to power its many processes (see Figure 1.3).

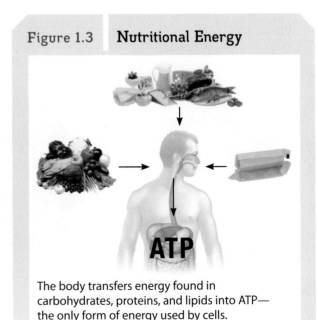

Figure 1.3 **Nutritional Energy**

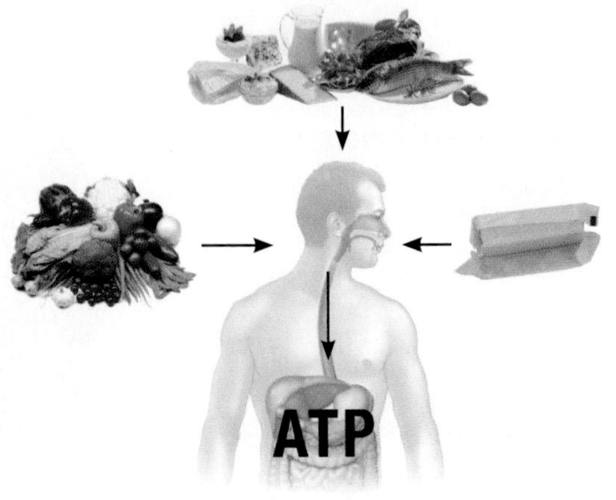

The body transfers energy found in carbohydrates, proteins, and lipids into ATP—the only form of energy used by cells.

Calories Represent the Amount of Energy in a Food

The amount of energy in the food you eat varies. The unit of measurement used to express the amount of energy in a food is called the **calorie.** The more calories a food has, the more ATP the body can make from it. Because a single calorie represents a very small amount of energy, the energy content of foods is typically expressed in a unit representing 1,000 calories—a *kilocalorie.* The kilocalorie is often abbreviated as *kcalorie* or *kcal.* Further, a kilocalorie is sometimes referred to as a Calorie (note the capital C) outside of scientific research, as on food labels. Therefore, 1 Calorie is equivalent to 1 kilocalorie, or 1,000 calories. A slice of cherry pie contains approximately 480 Calories, which is equivalent to 480 kilocalories, or 480,000 calories.

Is There Actually Energy in Your Energy Drink?

Have you ever looked at the ingredient list on the label of an "energy" drink or power shot? You might be surprised to learn that many of these products—especially those labeled *diet* or *sugar free*—do not actually contain energy-yielding nutrients. Instead, they are chock full of vitamins and other substances (like caffeine) loosely associated with increased mental energy and enhanced wakefulness.[4] You can easily tell if a product contains true energy by checking to see if it provides calories in the form of carbohydrates, protein, or fat.

Some energy drinks do not actually contain energy (calories).

© r19/r19/ZUMA Press/Newscom

Carbohydrates and proteins provide approximately 4 kcal/g. That is, they provide 4 kilocalories of energy for each gram (g) of substance consumed. Lipids provide approximately 9 kcal/g. Thus, 10 g of a pure carbohydrate or protein contain 40 kcal (4 kcal/g × 10 g), whereas 10 g of a pure lipid contain 90 kcal (9 kcal/g × 10 g). Although alcohol is not considered a nutrient, it provides 7 kcal/g. To practice figuring out how many calories are in a meal, try calculating the caloric content of a breakfast consisting of oatmeal, low-fat milk, brown sugar, raisins, and orange juice. The amount of each food's energy-yielding nutrients—carbohydrates, proteins, and lipids—can be found on the food's label or in any food composition table. By multiplying the weight of each energy-yielding nutrient by its caloric content and then adding up these values, you can easily determine the number of kilocalories provided by the meal. As confirmed in Table 1.2 on the next page, the total caloric content of this breakfast is 496 kcal, or 496 Calories. Because of rounding errors and other factors, the total number of kilocalories (or Calories) listed in a food composition table or on a label may differ slightly from the value obtained from calculations. However, these differences are usually very small.

calorie The unit of measurement used to express the amount of energy in a food.

scientific method A series of steps used by scientists to explain observations.

LO5 How Do Nutritional Scientists Conduct Their Research?

Few things are more significant or imperative to people than food. Food is also an important frontier of scientific research. Many questions about how food interacts with your bodies and ultimately your health are yet to be answered. But how do scientists carry out this research, and how can you distinguish reliable information from false or exaggerated claims? For centuries, scientists have explained observations (including those related to nutrition) using a series of steps collectively called the **scientific method.** There are three basic steps in the scientific method: making an observation, proposing a hypothesis, and collecting data.[5]

Table 1.2 **Calculating the Caloric Content of a Typical Breakfast**

Food	Carbohydrates (4 kcal/g)	Protein (4 kcal/g)	Lipids (9 kcal/g)	Total Kilocalories
	Kilocalories from Energy-Yielding Nutrients			
Oatmeal, 1 cup • Carbohydrates: 25 g • Protein: 6 g • Lipids: 2 g	25 g × 4 kcal/g = 100 kcal	6 g × 4 kcal/g = 24 kcal	2 g × 9 kcal/g = 18 kcal	142 kcal
Milk, 1 cup • Carbohydrates: 12 g • Protein: 8 g • Lipids: 2 g	12 g × 4 kcal/g = 48 kcal	8 g × 4 kcal/g = 32 kcal	2 g × 9 kcal/g = 18 kcal	98 kcal
Brown Sugar, 2 tablespoons • Carbohydrates: 24 g • Protein: 0 g • Lipids: 0 g	24 g × 4 kcal/g = 96 kcal	0 g × 4 kcal/g = 0 kcal	0 g × 9 kcal/g = 0 kcal	96 kcal
Raisins, 1/2 ounce • Carbohydrates: 11 g • Protein: 0 g • Lipids: 0 g	11 g × 4 kcal/g = 44 kcal	0 g × 4 kcal/g = 0 kcal	0 g × 9 kcal/g = 0 kcal	44 kcal
Orange Juice, 1 cup • Carbohydrates: 27 g • Protein: 2 g • Lipids: 0 g	27 g × 4 kcal/g = 108 kcal	2 g × 4 kcal/g = 8 kcal	0 g × 9 kcal/g = 0 kcal	116 kcal
Total	396 kcal	64 kcal	36 kcal	496 kcal

Step 1: The Observation Must Be Accurate

An appropriate and accurate observation about an event or phenomenon serves as both the framework and foundation for the rest of the scientific method. If the observation is flawed, any resulting conclusions will likely be flawed as well. For example, consider the observation that there has been an alarming rise in childhood obesity over the past few decades. Before a researcher can develop an explanation for this observation, she should first consider several questions about it. Are girls more likely to be obese than boys? At what age do the rates of obesity increase? Are adult obesity rates also rising? Answering these questions will help ensure that the observation is complete and accurate and provides a solid base on which the rest of the scientific method can be applied.

hypothesis A prediction about the relationship between variables.

Step 2: A Hypothesis Explains the Observation

Once the researcher makes an observation and understands the details associated with it, the next step is to explain why the event occurred. At this point, the scientist must propose a **hypothesis**, a prediction about the relationship between variables, to explain the observation. For example, the researcher might hypothesize that the increase in childhood obesity is due to a lack of exercise. Scientists make two general types of hypotheses: those that predict cause-and-effect relationships and those that predict correlations. A **cause-and-effect relationship** (or **causal relationship**) is a relationship by which an alteration to one variable causes a change in another variable. When two variables are clearly related, but one cannot be shown to cause the other, one can only say that there is a **correlation** (or **association**). For example, if your alarm clock goes off every morning around the time the sun rises, the two events are correlated, but neither actually causes the other.

Understanding the difference between cause-and-effect relationships and correlations is important in all scientific disciplines, not just nutrition. Although many studies are designed to test for correlations, their results are unfortunately interpreted or reported as proving causal relationships. Thus, it is important to remember one of the most frequently repeated scientific dictums: *correlation does not necessarily infer causation.*

Step 3: Experimentation

Although making an accurate observation (Step 1) and developing an appropriate hypothesis (Step 2) are critical to the scientific method, neither is nearly as complex as the final step—experimentation. Without supportive data generated through an experiment, a hypothesis is simply an unproven conjecture—not a scientific finding. Proper experimentation requires that the researcher designs an appropriate study, conducts the study carefully, and interprets the data correctly. If the study design is flawed, a good observation and/or hypothesis can be completely wasted. A basic understanding of experimentation and experimental practices will help you discern which nutrition claims are unfounded and which are valid.

EPIDEMIOLOGIC STUDIES

When a researcher wants to determine whether a variable is simply correlated with another variable, an **epidemiologic study** can be conducted. In this type of study, a researcher examines the relationship between variables in a group of people simply by making observations and recording information. In epidemiologic studies, researchers do not actually ask people to change their behaviors, alter their food intake patterns, or undergo any sort of treatment. Consequently, epidemiologic studies should not be used to test hypotheses predicting causal relationships—only correlations.

INTERVENTION STUDIES

In contrast to an epidemiologic study, an **intervention study** requires participants—regardless of whether they are humans, animals, or cell culture systems—to undergo a treatment or intervention (see Figure 1.4 on the next page). In most studies, some of the participants receive the treatment, while others do not.

cause-and-effect relationship (or **causal relationship**) A relationship whereby an alteration to one variable causes a change in another variable.

correlation (or **association**) A relationship whereby an alteration to one variable is related to a change in another variable.

epidemiologic study A study in which data are collected from a group of people who are not asked to change their behaviors in any way.

intervention study An experiment in which a variable is altered to determine its effect on another variable.

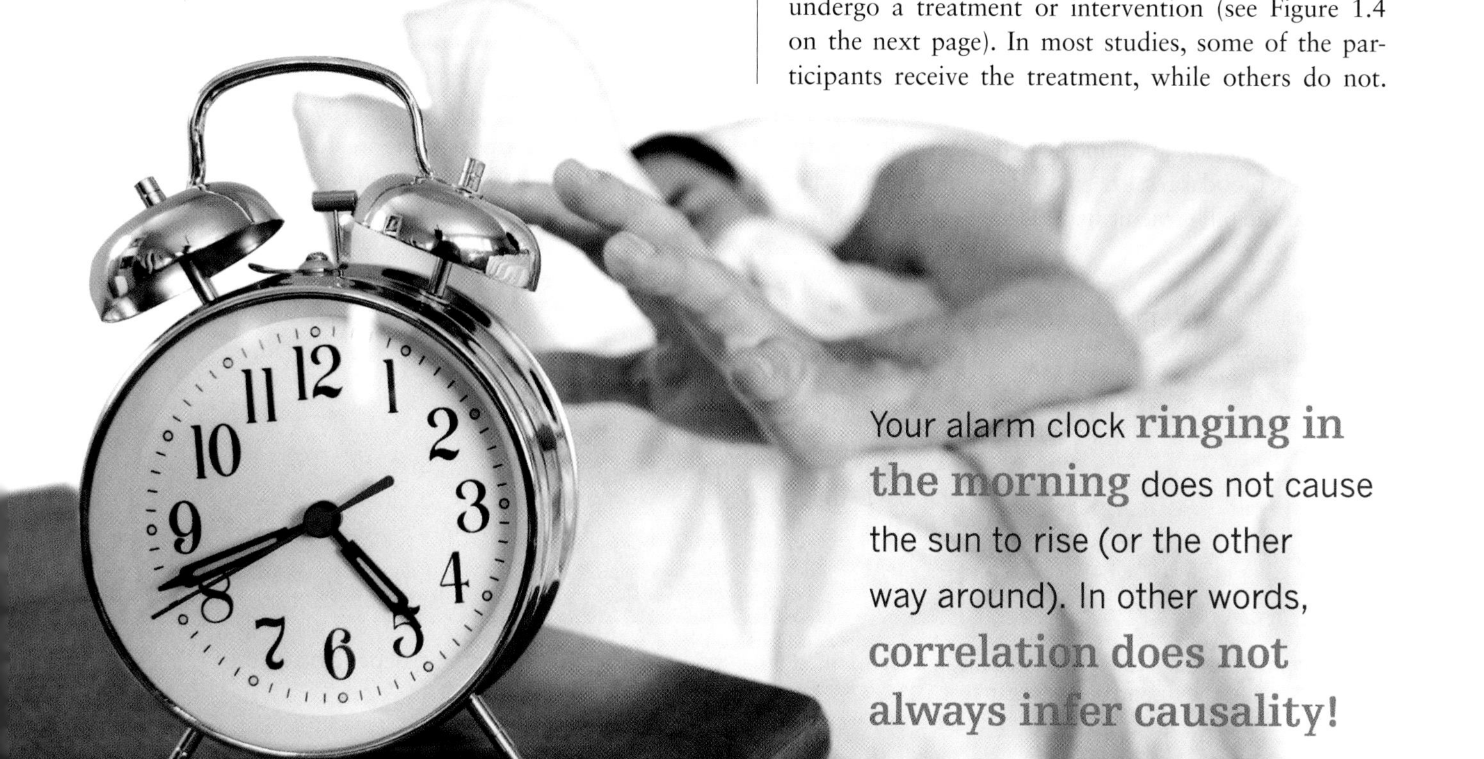

Your alarm clock ringing in the morning does not cause the sun to rise (or the other way around). In other words, correlation does not always infer causality!

control group Study participants that do not receive a treatment or intervention.

placebo effect A phenomenon whereby a study participant experiences an apparent effect of the treatment just because the participant believes that the treatment will work.

placebo An inert treatment given to the control group that cannot be distinguished from the actual treatment.

researcher bias A phenomenon by which the researcher influences the results of a study.

single-blind study A human experiment in which the participants do not know to which group they have been assigned.

double-blind study A human experiment in which neither the participants nor the scientists know to which group the participants have been assigned.

random assignment A condition by which study participants have equal chance of being assigned to the treatment and control group.

Participants that do not receive a treatment or intervention are considered to be in the **control group.** A control group is needed to determine whether the effects witnessed in the treatment group are caused by the treatment, are due to chance, or are a result of some other aspect of the study. For example, a researcher interested in understanding the childhood obesity epidemic might test the hypothesis that nutrition education can decrease obesity in children. To do so, the researcher might have some children attend a nutrition education class (the treatment group), and others not (the control group). The researcher could then measure whether the children who received the nutrition education gained less weight than those who received no extra education.

However, the presence of a control group does not necessarily ensure that the researcher's conclusions are accurate. This is because of a well-known phenomenon called the **placebo effect**, which occurs when a study participant experiences an apparent effect of the treatment just because he believes that the treatment will work. Indeed, many studies have shown that taking an unmedicated sugar pill actually influences blood pressure in some people if they believe that the pill is medicated.[6] How the placebo effect works remains mysterious to scientists, but it clearly connotes a strong mind–body connection. Researchers control for the placebo effect by requiring control group participants to consume or experience an inert treatment that looks, smells, tastes, and/or seems just like the actual treatment. This inert treatment is called a **placebo.**

It is also important to note that a researcher can inadvertently influence a study's outcome simply by knowing which subjects received the actual treatment and which received the placebo. This is a type of **researcher bias**. For example, if a researcher knows that certain participants received a treatment thought to improve memory, he might unconsciously score a memory test more favorably for those individuals. Because this type of bias may affect a study's final results, it should be avoided at all costs.

An important technique that minimizes the placebo effect and researcher bias is the blinding of the researchers and/or participants. When researchers know who is in the treatment group and who is in the placebo group but participants do not, the study is called a **single-blind study.** When neither the researchers nor the participants know, the study is called a **double-blind study.**

Another technique to avoid bias is the **random assignment** of participants to the treatment and control groups. For example, consider the study designed to test whether nutrition education influences childhood obesity. The random assignment of children to either the treatment or the control group would help ensure that important variables (such as usual intake of sweets and exercise) are distributed evenly between the two study groups. Every intervention study report should state whether participants were randomly assigned to study groups (see Figure 1.5).

As you now know, nutrition research takes many shapes and forms. Careful use of the scientific method

Figure 1.4 Intervention Studies

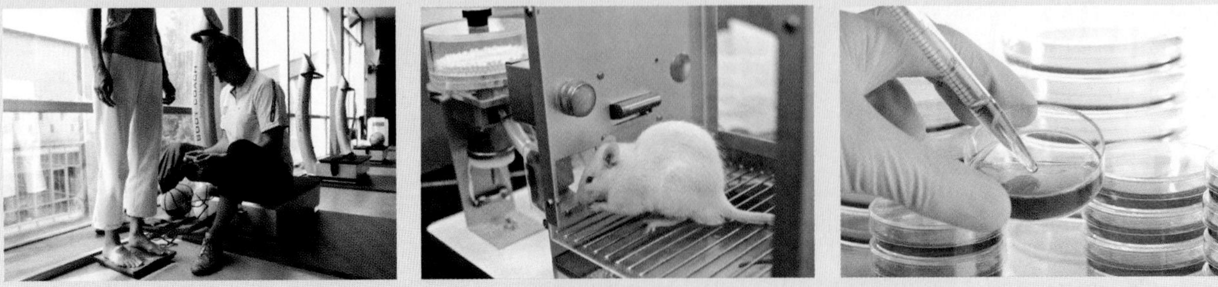

Nutrition intervention study participants can be humans, animal models, or cell culture systems, depending on the hypothesis that the researcher wants to test.

Figure 1.5 Components of a Nutrition Intervention Study

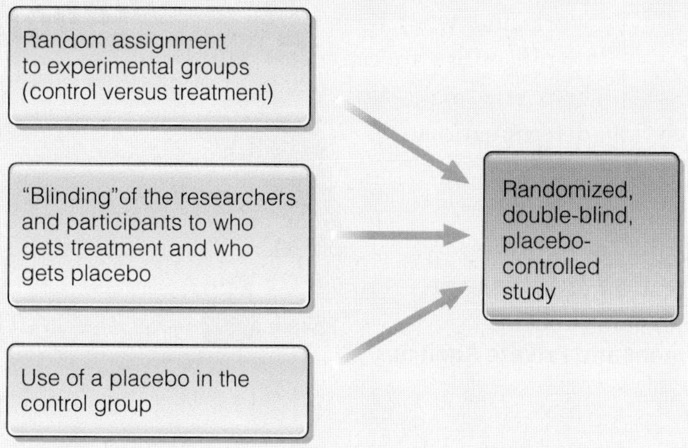
Random assignment to experimental groups (control versus treatment)
"Blinding"of the researchers and participants to who gets treatment and who gets placebo
Use of a placebo in the control group
Randomized, double-blind, placebo-controlled study

Randomized, double-blind, placebo-controlled intervention studies are considered the gold standards of nutrition research.

helps ensure that conclusions are valid and appropriate no matter the format of the research. Because nutrition research continues to reveal new findings, scientists' understanding of the relationship between diet and health is continually shifting. Therefore, you should not be surprised as dietary recommendations change over time.

LO6 Are All Nutrition Claims Believable?

Without a doubt, it is the nature of science to be a long and winding road of discovery. While the emergence of new theories, devaluation of previously accepted hypotheses, and shifts in popular opinion are to be expected, it is crucial that new nutritional claims are critically evaluated as they emerge. Sensational studies captivate public interest, but their findings are not always based on sound research. If unfounded nutrition claims go unchecked, they may come to be believed and communicated as fact. To help distinguish hearsay from science, keep the following questions in mind when you make nutrition-related decisions.

peer-reviewed journal A publication that requires a group of scientists to read and approve a study before it is published.

Where Was the Study Published?

The first important consideration when determining whether a nutrition claim is reputable is where the study was published. As you can imagine, not all sources of information are equally credible. This often makes it difficult to judge the validity of nutrition-related claims. It is important to determine the *primary source* of the information—the medium in which it was first reported or published. In general, a **peer-reviewed journal** is a trustable primary source of information. Studies published in peer-reviewed journals have been read and approved by a group of experts (peers) knowledgeable in the study's specific area of nutrition. Many private and governmental organizations, such as the National Institutes of Health, also publish credible studies and information concerning diet and health. These agencies are considered some of the most reliable and unbiased sources of information available to the public and to the scientific community. Table 1.3 on the next page lists both reputable peer-reviewed journals and organizations that publish nutrition-related articles. As a rule, you should question nutrition claims that have not first been published in a peer-reviewed journal or other highly regarded publication. Often, it is not necessary to go any further than this step to determine whether a nutrition claim is even worth considering.

Determining fact *from* fiction *when it comes to* nutrition claims *can sometimes be difficult.*

Who Conducted the Study?

Next, you should ask "Who conducted the research?" In general, reliable nutrition research is conducted by scientists at universities and medical schools. Researchers at private and public institutions and organizations also conduct sound nutrition research. It is important that the individuals conducting the research are qualified and knowledgeable. Finding out where the researchers conducting the study work and what their qualifications are can help you make this determination.

Annals of Nutrition and Metabolism Vol. 57, No. 3-4, 2010, reprinted with permission of S. Karger AG, Basel.

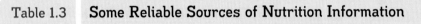

Table 1.3 Some Reliable Sources of Nutrition Information

Peer-Reviewed Journals	Government and Private Agencies
American Journal of Clinical Nutrition	American Cancer Society (http://www.cancer.org)
Annals of Nutrition and Metabolism	American Diabetes Association (http://www.diabetes.org)
Annual Review of Nutrition	American Dietetic Association (http://www.eatright.org)
Appetite	American Heart Association (http://www.americanheart.org)
British Journal of Nutrition	American Institute for Cancer Research (http://www.aicr.org)
Clinical Nutrition	American Medical Association (http://www.ama-assn.org)
European Journal of Nutrition	American Society for Nutrition (http://www.nutrition.org)
Journal of Human Nutrition and Dietetics	Centers for Disease Control and Prevention (http://www.cdc.gov)
Journal of Nutrition	Institute of Medicine (http://www.iom.edu)
Journal of the American College of Nutrition	Mayo Clinic (http://www.mayoclinic.org)
Journal of the American Dietetic Association	National Academy of Sciences (http://www.nas.edu)
Journal of the American Medical Association (JAMA)	National Institutes of Health (http://www.nih.gov)
Journal of Pediatrics Gastroenterology and Nutrition	NIH Office of Dietary Supplements (http://dietary-supplements.info.nih.gov/)
Lancet	NIH National Center for Complementary and Alternative Medicine (http://nccam.nih.gov)
Nature	U.S. Department of Agriculture and its Food and Nutrition Information Center (http://www.usda.gov and http://fnic.nal.usda.gov)
New England Journal of Medicine	U.S. Food and Drug Administration (http://www.fda.gov)
Nutrition	
Nutrition Research	
Public Health Nutrition	
Science	
Scientific American	

Who Paid for the Research?

Good research is often expensive. There are many ways that scientists obtain the money needed to fund their studies, but most acquire funding by applying for grants from private companies, foundations, and state and federal agencies. Researchers must take all necessary steps to ensure that their funding sources do not bias or influence the outcomes of their studies—especially if the funding agencies have something to gain or lose from the studies' results.

Did the Researchers Use the Right Study Design?

Once you have determined that a nutrition claim (1) has been published in a reputable journal or report, (2) was conducted by a qualified researcher, and (3) was likely not biased by its source of funding, you are ready to consider the research itself. Was the research conducted in a way that was appropriate to test the hypothesis? And do the conclusions fit the study design?

One of the best resources for finding information about a given research study is the U.S. National Library of Medicine, which hosts a searchable biomedical database called PubMed (http://www.ncbi.nlm.nih.gov/PubMed). The PubMed database allows easy access to more than 11 million biomedical journal citations. You can use details from PubMed to answer important questions about the study:

- Was it an epidemiologic study or an intervention trial?
- Do the results suggest an association or causal relationship?
- Was an appropriate control group used?
- Was the study double-blinded?
- Was a placebo used in the control group?

With your knowledge of experimental design, you can evaluate the research and determine whether the nutrition claim is likely to be valid.

Do Public Health Organizations Concur?

Even the best, most thorough experiment does not always provide conclusive evidence that a particular nutrient influences health in a certain way. Consequently, public health experts usually wait for several studies to produce similar results before they begin to make overall claims about the effect of a certain nutrient on health. Thus, before believing a nutritional claim to be true, determine whether it is supported by major public health organizations. For example, you might check whether the American Heart Association supports a study's claim that a certain phytochemical decreases one's risk of heart disease or whether the American Cancer Society supports a study's claim that a particular nutrient decreases one's risk of cancer.

LO7 Nutrition and Health: What Is the Connection?

Nutritional scientists continue to study the impact of nutrition because experts have very good reason to believe that it plays a dominant role in health and disease. Indeed, consuming either too little or too much of a nutrient can cause illness. But what do experts really know about nutrition and health? And how do scientists and organizations track the overall health of a nation? Although extensive answers to these questions are beyond the scope of this chapter, it is important to understand some basic concepts about the relationship between nutrition and health, how this relationship is assessed on a national scale, and how it evolves.

Public Health Agencies

To appreciate the complex relationship between nutrition and health, it is important to understand how researchers and public health experts assess the health of a community or nation. What do scientists and other health professionals measure when they want to determine if a population is becoming more or less healthy? And who keeps track of all this information? In the United States, the organization responsible for monitoring health trends is the U.S. Centers for Disease Control and Prevention (CDC). The CDC surveys and monitors national and international health to prevent disease outbreaks, implement disease prevention strategies, and maintain national health statistics.

Mortality and Morbidity Rates

The CDC monitors many aspects of societal health, but the most frequently cited are morbidity and mortality rates. A **rate** is a measure of some event, disease, or condition within a specific time span. For instance, speed is expressed as a rate (such as miles per hour). One example of a health-related rate is **mortality rate**, which assesses the number of deaths that occur in a certain population group in a given period of time. For example, cancer experts monitor cancer mortality rates to help determine whether deaths from this disease are increasing or decreasing over time.

An important health-related mortality rate is **infant mortality rate**, which is the number of infant deaths (occurring within the first year of life) per 1,000 live births in a given year. Infant mortality rate is often used to assess the well-being of a society, as it reflects a complex and interrelated web of environmental, social, economic, medical, and technological factors that influence overall health. The United States' infant mortality rate has declined dramatically in the past century, as illustrated in Figure 1.6.[7] Nonetheless, the most recent data suggest that the United States' current infant mortality rate is 6.7, placing the United States 28th internationally behind countries like Hong Kong (1.8) and Sweden (2.8). Many factors influence infant mortality rate: genetics, access to medical care, substance abuse, maternal nutrition, and weight gain during pregnancy all affect this statistic.

Whereas mortality rates assess the number of deaths in a given period of time, a **morbidity rate** reflects illness or disease in a given period of time. Like mortality rates, morbidity rates for certain diseases have changed drastically over the last few decades in the United States. For example, tuberculosis rates have continuously declined over the last century, whereas type 2 diabetes rates continue to increase.[8]

rate A measure of some event, disease, or condition within a specific time span.

mortality rate The number of deaths that occur in a certain population group in a given period of time.

infant mortality rate The number of infant deaths per 1,000 live births in a given year.

morbidity rate The number of illnesses or diseases in a given period of time.

life expectancy A statistical prediction of the average number of years of life remaining for a person at a particular age.

graying of America A phenomenon occurring in the United States by which an increasing proportion of the population is over the age of 65.

disease An abnormal condition of the body or mind that causes discomfort, dysfunction, or distress.

The Centers for Disease Control and Prevention (CDC) is a federal agency charged with monitoring health and disease rates among the U.S. population.

CDC/James Gathany

Life Expectancy

Another indicator of societal health is **life expectancy**, the average number of years of life remaining for a person at a particular age. Currently, male and female infants born in the United States can expect to live approximately 75 and 80 years, respectively. If they are still alive at the age of 65, they can expect to live to the ages of 82 and 85, respectively. As you might have guessed, average life expectancy in the United States has increased dramatically over the past century (see Figure 1.6).

There are many reasons why life expectancy has increased over the past century—one of which is better nutrition.[9] Regardless of its cause, an increasing life expectancy has resulted in a phenomenon often referred to as the graying of America. The **graying of America**, characterized by an increasing proportion of the population over the age of 65, continues to influence America's health significantly.[10] For example, the graying of America has contributed to a rise in the prevalence of diseases such as heart disease and cancer, which are now the nation's leading causes of disability and death.

Diseases Are Either Infectious or Noninfectious

You have just learned how scientists and public health experts use indices such as morbidity rates, mortality rates, and life expectancy to assess the health of a nation. But how do they categorize disease, and what is actually meant by this term? A **disease** is any abnormal condition of the body or mind that causes

Figure 1.6 **Changes in Life Expectancy and Infant Mortality Rate over the Past Century**

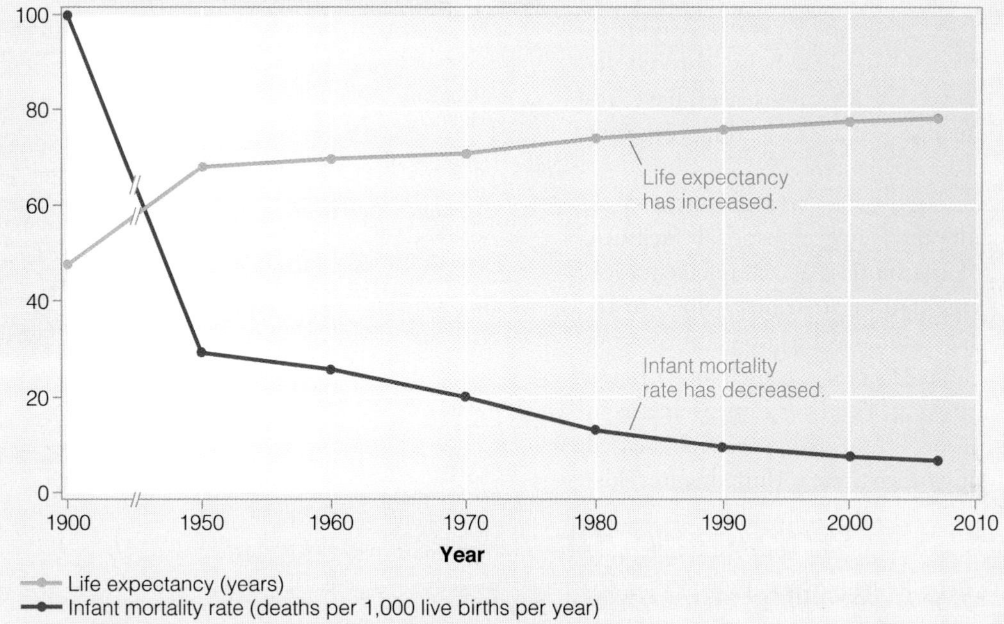

Since 1900, life expectancy has increased while infant mortality rate has decreased. These shifts indicate greater societal health.

Source: Xu JQ, Kochanek KD, Murphy SL, Tejada-Vera B. Deaths: Final data for 2007. National vital statistics reports. Hyattsville, MD: National Center for Health Statistics. 2010; 58:19. Available from: http://www.cdc.gov/NCHS/data/nvsr/nvsr58/nvsr58_19.pdf; National Center for Health Statistics. Health, United States, 2010. Available from: http://www.cdc.gov/nchs/data/hus/hus10.pdf.

discomfort, dysfunction, or distress. An **infectious disease** is caused by a pathogen (such as a bacterium, virus, fungus, parasite, or other microorganism), is contagious (meaning that it can be passed from one person to another person), and tends to be short-lived. An example of a common infectious childhood disease is chicken pox, which is caused by a virus. Conversely, a noninfectious disease is not spread from one person to another, does not involve an infectious agent, and tends to be long-term and chronic in nature. **Noninfectious diseases** tend to be treatable rather than curable, and are often preventable. An example of a noninfectious disease is type 2 diabetes.

A **chronic degenerative disease** is a noninfectious disease that develops slowly, persists over a long period of time (chronic), and tends to result in progressive breakdown of tissues and loss of function (degenerative). Type 2 diabetes, heart disease, osteoporosis, and cancer are chronic degenerative diseases that have been linked to poor nutrition. These diseases are among the most common and costly health problems that people face today—they are also among the most preventable. The adoption of healthy behaviors such as eating nutritious foods, being physically active, and avoiding tobacco use can prevent or control the devastating effects of these diseases. Scientists believe that most chronic degenerative diseases are caused by a combination of genetic and lifestyle factors. While poor nutrition is a leading cause, the exact processes by which chronic degenerative diseases develop are not often well understood. You will learn much more about this subject throughout this book.

infectious disease An illness that is contagious, caused by a pathogen, and tends to be short-lived.

noninfectious disease An illness that is not contagious, does not involve an infectious agent, and tends to be long-term and chronic.

chronic degenerative disease A noninfectious disease that develops slowly, persists over a long period of time, and tends to result in progressive breakdown of tissues and loss of function.

CHRONIC DEGENERATIVE DISEASES HAVE LARGELY REPLACED INFECTIOUS DISEASES

In the early 1900s, infectious diseases such as pneumonia, tuberculosis, and influenza were rampant throughout America. Infectious diseases were the leading causes of death, accounting for one-third of the nation's mortality rate. Severe

risk factor A lifestyle, environmental, or genetic factor related to a person's chances of developing a disease.

nutritional deficiencies contributed to the high number of deaths and illnesses that occurred throughout this period. It may surprise you to learn that today, chronic degenerative diseases have replaced infectious diseases as the leading causes of death in America (see Figure 1.7).[11]

Public health efforts such as investment in water treatment and sewage disposal facilities, development of antibiotics, and implementation of childhood vaccination programs helped to reduce the incidence of many infectious diseases. As a combined result of decreased infectious disease rates, better nutrition, and other improvements in health care, life expectancy rates have increased faster than at any other time in United States' history. While it seems like a testament to America's health, this trend is something of a double-edged sword because it has brought with it increases in the chronic degenerative diseases so common to today's society.

RISK FACTORS OF CHRONIC DISEASES

As previously mentioned, the specific causes (or *etiologies*) of today's most common chronic degenerative diseases are complex and often poorly understood. However, scientists do know that many lifestyle, environmental, and genetic factors are related to a person's risk of developing these diseases. Such a characteristic is called a **risk factor**. For example, the major lifestyle-related risk factors that are associated with heart disease, cancer, and stroke—the top three leading causes of death in the United States—include tobacco use, lack of physical activity, and a range of poor dietary habits. Cancer risk factors related to genetics and environment include a family history of cancer, asbestos- and lead-based housing, and exposure to pollution and excessive sunlight. Although some of these risk factors may actually play a role in causing chronic degenerative diseases,

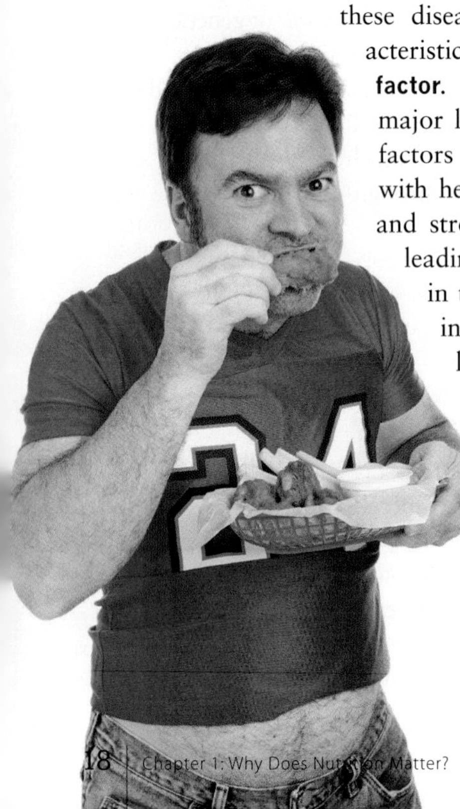

Many unhealthy eating patterns are considered risk factors for chronic degenerative diseases.

Figure 1.7 Five Leading Causes of Death over the Past Century

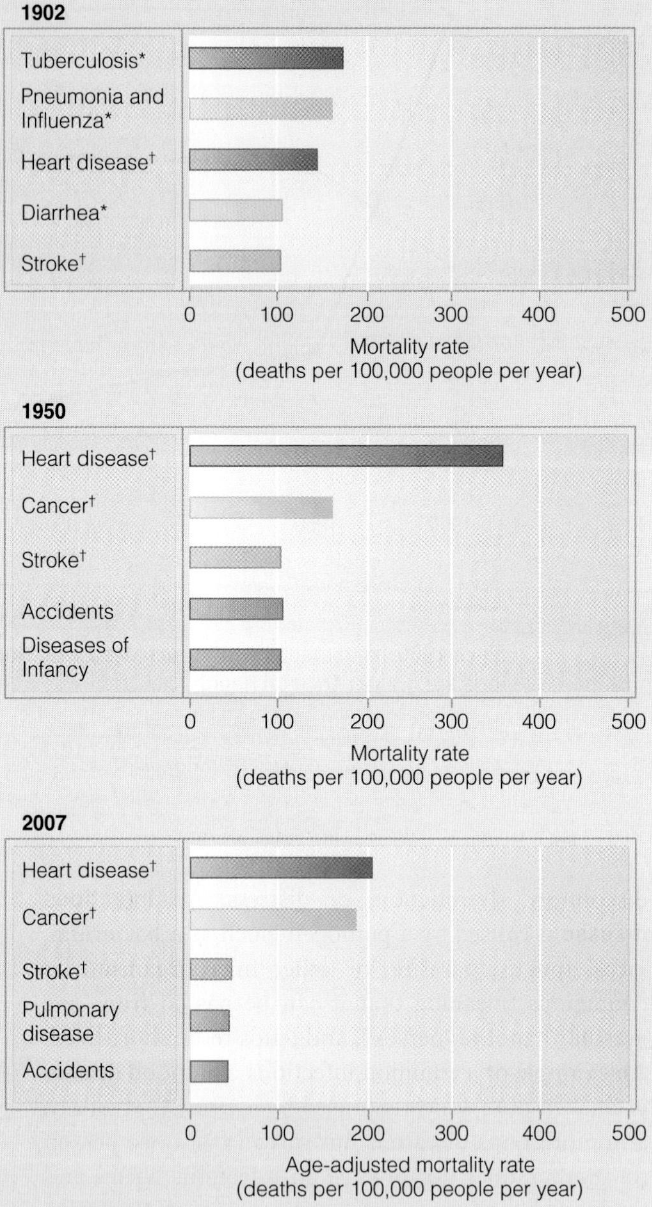

*Infectious disease
†Chronic disease

Source: Xu JQ, Kochanek KD, Murphy SL, Tejada-Vera B. Deaths: Final data for 2007. National vital statistics reports. Hyattsville, MD: National Center for Health Statistics. 2010; 58:19. Available from: http://www.cdc.gov/NCHS/data/nvsr/nvsr58/nvsr58_19.pdf; National Center for Health Statistics. Health, United States, 2010. Available from: http://www.cdc.gov/nchs/data/hus/hus10.pdf.

many are simply predictive—they help doctors and nutritionists know who is at the greatest risk. Remember that correlation does not always infer causality.

Obesity is a major risk factor for many chronic degenerative diseases. Although obesity itself is not a disease, it can predispose a person to life-threatening conditions such as heart disease, stroke, and type 2 diabetes. Obesity is becoming a health crisis of epidemic proportions both nationally and globally. As such, the dramatic shift from undernutrition to overnutrition or unbalanced nutrition that often occurs as a society transitions to a more industrialized economy—the **nutrition transition**—is strongly related to many of the chronic degenerative diseases facing humankind today. You will learn much more about obesity and its negative health consequences throughout this book.

Assessing the Nutritional Health of the Nation

As you have learned, nutritional concerns have changed dramatically over the past century. Specifically, Americans have transitioned from being vulnerable to micronutrient deficiencies to being over nourished and overweight. To help identify and tackle health challenges—including those related directly to nutrition—the U.S. Department of Health and Human Services (DHHS) has for three decades published a document called *Healthy People*. The most recent version, Healthy People 2020, outlines a set of overall health objectives for the nation to accomplish by 2020.[12]

Recognizing that chronic conditions such as obesity, type 2 diabetes, and osteoporosis are currently some of the most pressing health concerns throughout America, DHHS designed Healthy People 2020 to achieve four overarching health-related goals:

- Encourage long, high-quality lives that are free of preventable disease, disability, injury, and premature death.
- Achieve health equity, eliminate disparities, and improve the health of all groups.
- Create social and physical environments that promote good health for all.
- Promote quality of life, healthy development, and healthy behaviors across all life stages.

nutrition transition A shift from undernutrition to overnutrition or unbalanced nutrition that often occurs as a society transitions to a more industrialized economy.

These four broad goals are subdivided into 39 different topic areas. The topics of diabetes; food safety; nutrition/weight status; and maternal, infant, and child health each have corresponding nationwide goals. The goal for nutrition/weight status, for example, is the promotion of health and reduction of chronic disease risk through the consumption of healthful diets and achievement and maintenance of healthy body weights. Although discussing each of Healthy People 2020's topics and goals is beyond the scope of this book, you can learn more at the Healthy People website (http://www.healthypeople.gov).

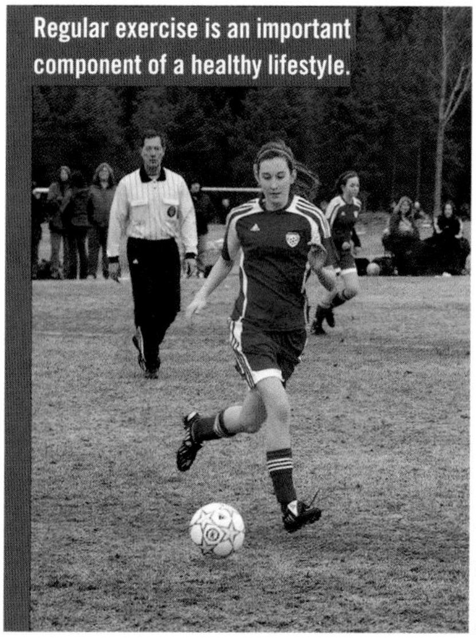
Regular exercise is an important component of a healthy lifestyle.
© Joan Rhoades

LO8 Why Study Nutrition?

Experts agree that chronic degenerative diseases and other health concerns caused at least in part by poor diet represent some of the United States' most serious and pressing public health issues. Researchers estimate that 70 percent of adults over the age of 20 are either overweight or obese, and that more than 280,000 deaths occur each year because of obesity alone.[13] More than 64 million Americans now have cardiovascular disease (the leading cause of death), 50 million have high blood pressure, and more than 1 million have type 2 diabetes. Cancer accounts for 25 percent of all deaths in the United States annually. Fortunately, consuming a healthy balance of nutrients, phytonutrients, and zoonutrients can decrease your risk of developing all these conditions. Indeed, as the occurrence of chronic diseases increases, it is ever more important to pay attention to what you eat throughout your entire life. As the old saying goes, "An ounce of prevention is worth a pound of cure." Or perhaps, stated most succinctly and accurately by Sir Francis Bacon more than 400 years ago, "Knowledge is power."

NUTR

2 Choosing Foods Wisely

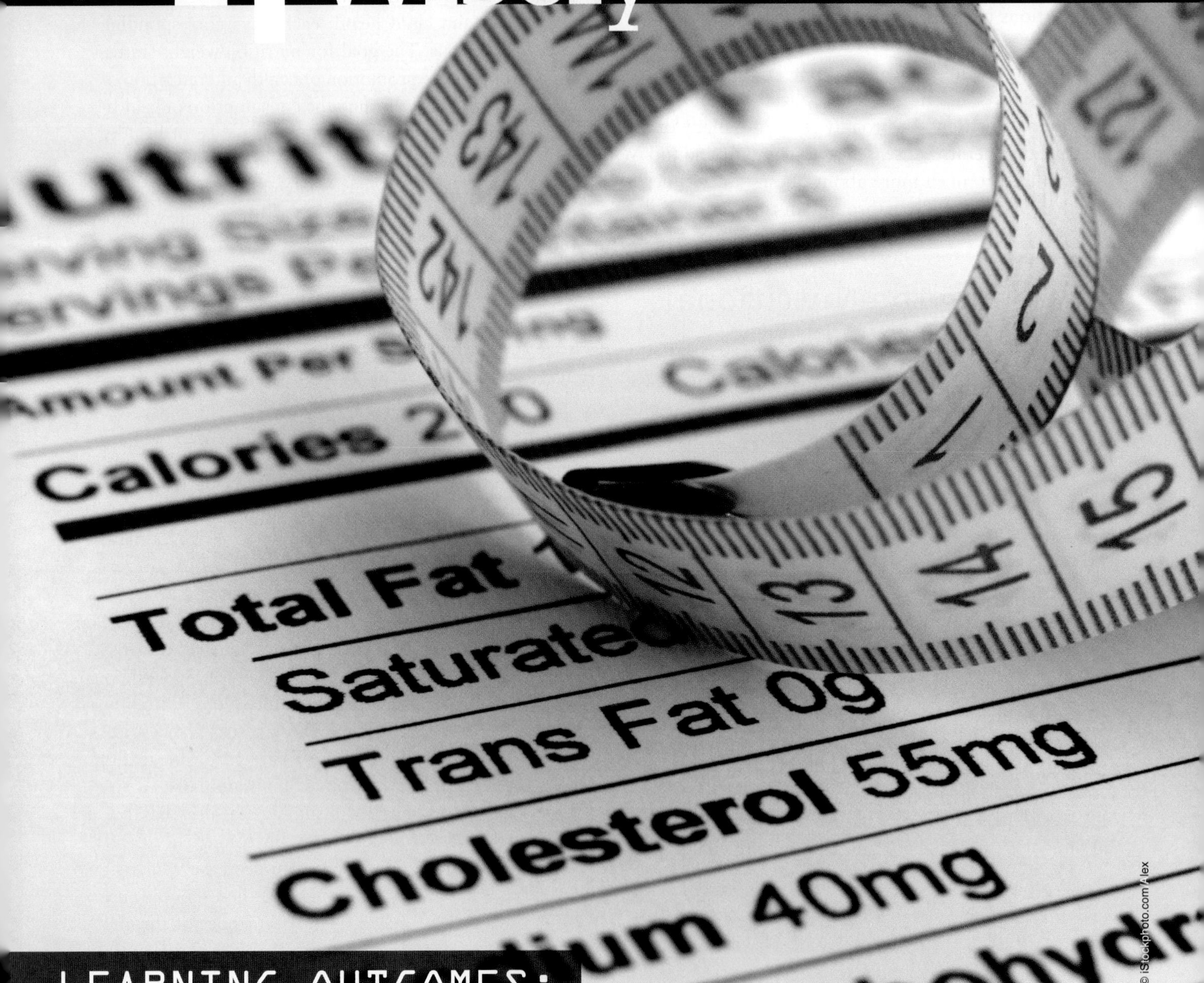

© iStockphoto.com/Alex

LEARNING OUTCOMES:

- LO1 Describe the continuum of *nutritional status.*
- LO2 Differentiate the methods by which nutritional status is assessed.
- LO3 Understand and estimate dietary intake.
- LO4 Utilize contemporary food guidance systems.
- LO5 Use food labels to plan a healthy diet.
- LO6 Put the chapter concepts into action.

LO1 What Is Nutritional Status?

Food and eating make up a large part of the American culture. On a more basic level, however, food is simply a vehicle for nutrients that provide the building blocks and energy for all your body's structures and functions. But how do you know if you are consuming enough—but not too much—of all the essential nutrients? And how can you choose foods better to meet your nutritional needs? To answer these questions, you need to understand several important concepts about the nutrients in foods and how much of them you need. In this chapter, you will learn the fundamentals of these concepts and how to apply them to your own food choices. You will also learn about important dietary regulations and guidelines that have been developed to help in this endeavor. This basic information and these helpful dietary tools will help you determine which foods—and how much of them—to choose to help optimize your health.

Most people know that not getting enough of the essential nutrients can lead to undesirable long-term consequences. Indeed, the inadequate intake of one or more nutrients and/or energy, **undernutrition**, can cause a **nutritional deficiency**. However, poor nutrition can also be caused by eating too much of a nutrient or food (overnutrition). For example, the consumption of too many fatty foods can lead to obesity and its related health consequences, and **nutritional toxicity**—overconsumption of a nutrient that results in dangerous toxic effects—can be fatal. Undernutrition and overnutrition make up the extreme ends of the nutritional status continuum. **Nutritional status** is the extent to which a person's diet meets her individual nutrient requirements. Both under- and overnutrition are examples of **malnutrition**, a state of poor nutritional status caused by an imbalance between the body's nutrient requirements and nutrient availability. Figure 2.1 on the next page illustrates the relationships among nutrient availability (or dietary intake), health, and nutritional status. As you can see, both undernutrition and overnutrition can increase the chance that you have suboptimal health. Thus, the center of the nutritional status continuum is where you want to be.

Primary and Secondary Malnutrition

Regardless of whether it causes undernutrition or overnutrition, malnutrition can be precipitated by a variety of underlying causes. **Primary malnutrition** is a condition by which poor nutritional status is caused strictly by inadequate diet, while **secondary malnutrition** is caused by factors other than diet, such as illness. For example, a person who is deficient in one of the B vitamins because his diet lacks vitamin-rich fruits and vegetables is experiencing primary malnutrition. If his vitamin B deficiency is instead caused by an illness that interferes with vitamin B absorption, he is experiencing secondary malnutrition. It is important to distinguish whether primary or secondary malnutrition is causing poor nutritional status because their treatments are very different. Nutritional deficiencies caused by primary malnutrition can often be cured by

undernutrition The inadequate intake of one or more nutrients and/or energy.

nutritional deficiency A condition caused by inadequate intake of one or more essential nutrients.

nutritional toxicity Overconsumption of a nutrient that results in dangerous toxic effects.

nutritional status The extent to which a person's diet meets his or her individual nutrient requirements.

malnutrition A state of poor nutritional status caused by an imbalance between the body's nutrient requirements and nutrient availability.

primary malnutrition A condition by which poor nutritional status is caused strictly by inadequate diet.

secondary malnutrition A condition by which poor nutritional status is caused by factors other than diet, such as illness.

nutritional adequacy A condition by which a person regularly consumes the required amount of a nutrient to meet physiological needs.

Both undernutrition (left) and overnutrition (right) are forms of malnutrition.

AP Images/Brennan Linsley / © Mark Richards/PhotoEdit

consuming certain foods, whereas secondary malnutrition requires addressing the underlying cause.

Adequate Nutrient Intake

How can a health care worker know if a person is malnourished? And how can you know the amount of each essential nutrient you need to be healthy? The answers to these questions are complex because different people need different amounts of nutrients for optimal health. Individual nutrient needs are determined by a host of factors, such as sex, age, physical activity, and genetics. For example, a man who plays for his university's soccer team is likely to have higher nutritional requirements than his retired grandmother. No matter the specific nutritional needs, a person who regularly consumes the required amount of a nutrient to meet her physiologic needs has achieved **nutritional adequacy** for that nutrient.

Dietitians and other health professionals use several tools to determine whether a person's nutritional status is optimal, or if it could benefit from different dietary choices. As you will learn in the following sections, you can use many of these same tools to assess your own nutritional status and dietary adequacy.

Figure 2.1 **Dietary Intake Influences Nutritional Status and Health**

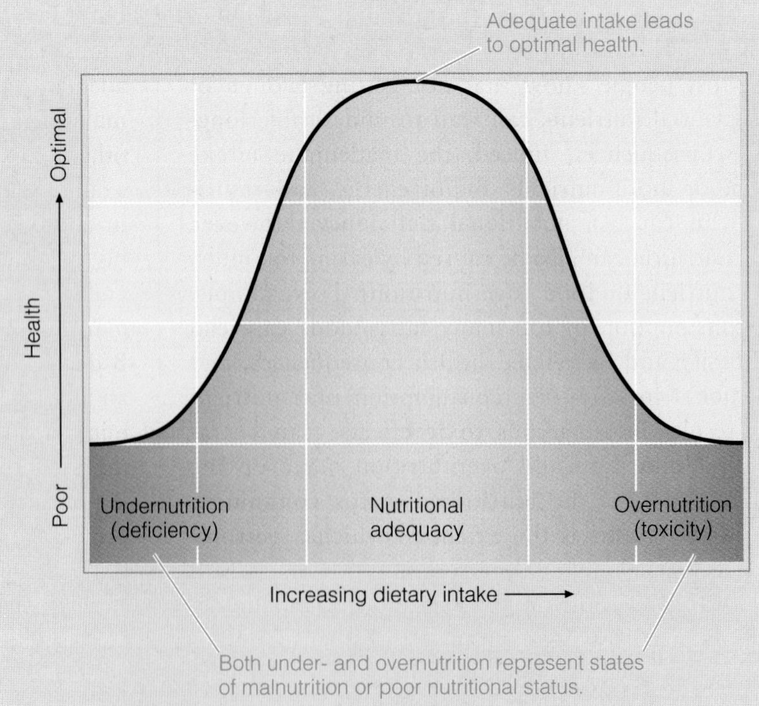

Both under- and overnutrition can result in malnutrition and poor health over time.

LO2 How Is Nutritional Status Assessed?

Because adequate nutrition is required for optimal health, it is important for health care providers to be able to assess a person's nutritional status. This is especially critical during periods of growth and development (such as infancy), when nutrient requirements are high and poor nutrition can have long-lasting consequences. In general, there are four ways by which nutritional status can be assessed. These tools are sometimes referred to as the *ABCD* methods of nutritional assessment:

- Anthropometric measurement
- Biochemical measurement
- Clinical assessment
- Dietary assessment

Although each of these nutritional assessment methods can provide some information about nutritional status, each one by itself cannot tell you everything. It is therefore important to use multiple approaches when determining nutritional adequacy.

Anthropometric Measurement

An **anthropometric measurement** assesses your body's physical dimensions (such as height) or composition (such as fat mass). The Greek term *anthropometry* means literally "to measure the human body." Because most anthropometric measurements are easy and inexpensive to conduct, anthropometry is routinely used in clinical and research environments. They are also common outside of scientific research: you take an anthropometric measurement every time you weigh yourself. Anthropometric measurement cannot confirm deficiency of any particular essential nutrient, but it can give a clinician clues that nutritional inadequacies might be present.

HEIGHT, WEIGHT, AND CIRCUMFERENCE

Because obesity can lead to chronic degenerative diseases such as heart disease and type 2 diabetes, height and body weight are often used to assess one's risk for such diseases. Changes in these anthropometric measures can also provide information about the progression of other diseases. For example, an elderly person's loss of height might indicate a decline in bone density, and a college student's significant loss of body weight might indicate an eating disorder. Height and weight are also commonly used to assess nutritional status during infancy, childhood, and pregnancy. Various circumferences, such as those of the waist, hips, and head, are also sometimes measured to assess health. Differences in waist and hip circumferences reflect variations in body fat distribution patterns, and head circumference is frequently measured to monitor brain growth during infancy.

anthropometric measurement
A measurement of a body's physical dimensions or composition.

body composition
The proportions of fat, water, lean tissue, and mineral (bone) mass that make up the body.

BODY COMPOSITION

Estimates of **body composition**—the proportions of fat, water, lean tissue, and mineral (bone) mass that make up the body—are also anthropometric measurements. The amount and distribution of these components can be an important indicator of one's nutritional status and overall health. For instance, adequate hydration status (water content) is important for optimal athletic performance; alterations in protein content can indicate advanced disease in cancer patients; too much body fat can lead to cardiovascular disease; and loss of bone mass is a major risk factor for osteoporosis. Many campus recreation centers and health clinics offer free body composition testing. Knowing this information about yourself might be useful as you learn more about nutrition and health.

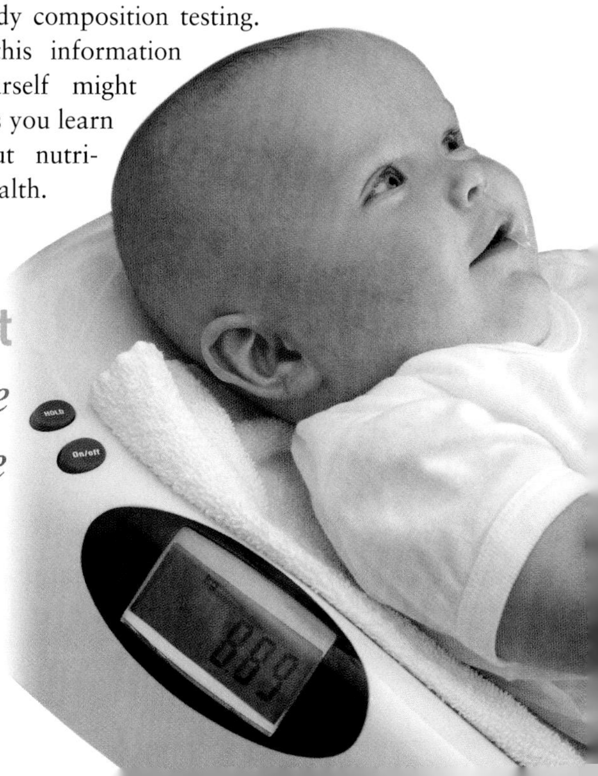

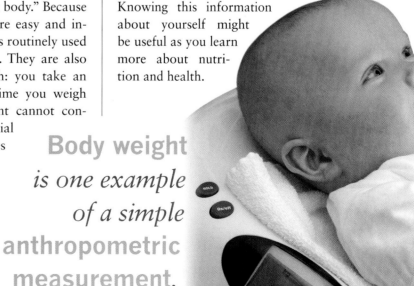

Body weight *is one example of a simple* **anthropometric measurement.**

biochemical measurement Laboratory analysis of a biological sample, such as blood or urine.

sign A physical indicator of disease that can be seen by others, such as pale skin and skin rashes.

symptom A subjective manifestation of disease that generally cannot be observed by other people.

Biochemical Measurement

Because anthropometric measurements are not diagnostic (that is, they cannot be used to diagnose a problem definitively), they must be supported by other measures of nutritional status. To assess health and nutritional status beyond anthropometric measurements, biochemical measurements are often used. A **biochemical measurement** is a laboratory analysis of a biological sample such as blood or urine. In some cases, the sample is analyzed for a specific nutrient or a substance related to the nutrient's function. For example, the selenium content of blood can be measured as an indicator of selenium status. Selenium status can also be assessed by measuring the activity of a protein that helps protect the body from damage in the blood. Biochemical measurements are integral to nutritional assessment because they can help diagnose a specific nutrient deficiency or excess. However, because the collection and analysis of biological samples often necessitate technical expertise and costly procedures, biochemical analyses are often done only when malnutrition is already suspected.

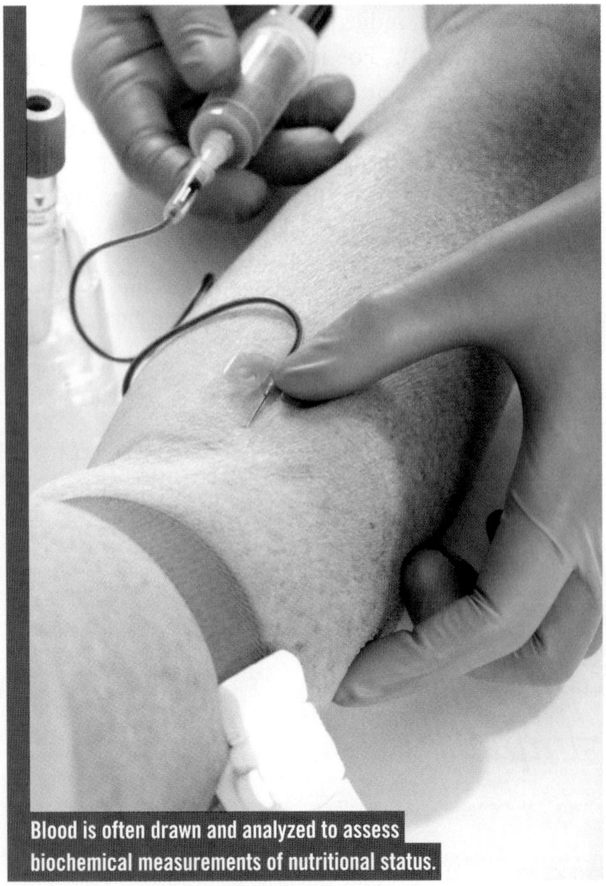

Blood is often drawn and analyzed to assess biochemical measurements of nutritional status.

© Alexander Raths/Shutterstock.com

Many nutrient deficiencies have distinct signs that can be seen by a trained eye. The disfiguration of this fingernail signals a vitamin B deficiency. Note, however, that fingernail disfiguration can also be caused by other nutrient deficiencies and conditions.

© Dr. P. Marazzi/Photo Researchers, Inc.

Clinical Assessment

Another way to assess nutritional status is to conduct a face-to-face evaluation. During a clinical assessment, a clinician may ask questions about whether a patient has had previous diseases, unusual weight loss or weight gain, surgeries, or has taken medications. The clinician may also ask about other relevant information such as family history. This process, called *taking a medical history*, can be helpful in determining one's overall health and risk for disease. Finally, the clinician may take anthropometric measurements. During a clinical assessment, the clinician will likely note each visible or measurable **sign** of illness—a physical indicator of disease that can be seen or assessed objectively by someone else. For example, signs of an iron deficiency might include pale skin and shortness of breath. Other observable signs that may indicate poor nutritional status include skin rashes and swollen ankles (edema). These signs may suggest vitamin B and protein deficiencies, respectively.

A patient may also be asked during a clinical assessment if she is experiencing anything unusual in her health—in other words—whether she has symptoms of disease or malnutrition. A symptom is different from a sign in that a **symptom** is a subjective manifestation of disease that generally cannot be observed by other people. For example, fatigue is a symptom commonly associated with iron deficiency, and loss of appetite is a symptom of zinc toxicity. Because they generally go unnoticed by clinicians, symptoms must be reported by the patient. Although nutrient deficiencies caused by primary malnutrition were once common in the United States, deficiencies are more likely to be caused by secondary malnutrition today. This is

© Denise Kappa/Shutterstock.com

why it is important for a clinician to gather as much information as possible about a person's health prior to treating a nutritional deficiency.

Clinical assessment has many advantages over other forms of nutritional assessment. For one, it is the only way that health care providers can determine if patients are experiencing symptoms of malnutrition. Furthermore, because signs of some extreme forms of malnutrition are very distinct, observing them clinically can make diagnosis of a particular nutrient deficiency or toxicity quite accurate.

Dietary Assessment

Although anthropometric, biochemical, and clinical assessments can be used to assess some aspects of nutritional status, it is also very important to evaluate the adequacy of one's dietary intake. This is called **dietary assessment**. There are three general types of dietary assessment. The first, **diet recall**, requires a person to record and evaluate every food and drink consumed over a given time span—typically 24 hours. Because diet recalls are generally based on a single day and require a person to remember information, they often fail to represent one's usual food intake. The second type of dietary assessment, the **food frequency questionnaire**, assesses food selection patterns over an extended period of time. For example, a food frequency questionnaire used to determine overall fruit intake might ask questions about which fruits the patient typically eats and how much she normally consume in a serving. Because it does not assess nutrient intake (only food intake patterns), information from a food frequency questionnaire is limited in accuracy and completeness.

Although diet recalls and food frequency questionnaires are relatively simple to conduct, their usefulness depends on a person's memory. To assess nutrient intake more accurately, it is best to record and analyze every food and drink as it is consumed over a given time span in a **diet record** (or **food record**). To keep a diet record, the patient must either estimate portion sizes using a standard household measurement (such as cup or ounce) or weigh the food before he eats it. To get an accurate assessment of nutrient intake, food records should be kept for at least three days—ideally two weekdays and

dietary assessment The evaluation of adequacy of a person's dietary intake.

diet recall A retrospective dietary assessment method by which a person records and analyzes every food and drink consumed over a given time span.

food frequency questionnaire A retrospective dietary assessment method by which food selection patterns are assessed over an extended period of time.

diet record (or **food record**) A prospective dietary assessment method by which a person records and analyzes every food and drink as it is consumed over a given time span.

Clinical assessments are valuable in **nutritional assessment** because they can uncover signs and symptoms of **malnutrition**.

© LWA/Dann Tardif/Blend Images/Jupiterimages

Dietary Reference Intakes (DRIs) A set of four dietary reference standards used to assess and plan dietary intake: Estimated Average Requirement, Recommended Dietary Allowance, Adequate Intake level, and Tolerable Upper Intake Level.

nutrient requirement The amount of a nutrient that a person must consume to promote optimal health.

one weekend day. When the diet record is complete, its information can be analyzed to estimate nutrient intake. Though the diet record method is usually more time- and labor-intensive than diet recall and the food frequency questionnaire, clinicians often prefer it for its superior accuracy.

FOOD COMPOSITION TABLES AND DIETARY ANALYSIS SOFTWARE

After completing a diet record, the next step is to determine the nutrient and energy (calorie) contents of one's diet. For example, if a person is interested in her vitamin C status, she will need to determine how much vitamin C she consumes. There are two ways to find information concerning the nutrient composition of foods: food composition tables and computerized nutrient databases. Food composition tables can be purchased or accessed free of charge on the website of the U.S. Department of Agriculture (USDA; http://www.ars.usda.gov/nutrientdata). Using printed food composition tables to calculate nutrient intake can be tedious. Fortunately, easy-to-use digital nutrient databases are also available. For example, an online dietary analysis tool accompanies the USDA's MyPlate food guidance system. The MyPlate system is described in more detail later in this chapter.

Although it should always be used in conjunction with anthropometric, biochemical, and clinical assessment, dietary assessment is one of the most effective methods of nutritional evaluation. However, after you have determined how much of each nutrient you consume on a regular basis, how can you know if your intake is adequate? To answer this question, you must be able to refer to a variety of dietary intake reference standards and recommendations, several of which are described next.

LO3 How Much of a Nutrient Is Adequate?

Recall that different people require different amounts of nutrients depending on their sex, age, and other variables. Because individuals' bodies and activities are so diverse, there is no simple and definitive way to know what a person's nutrient needs actually are. However, to help both medical professionals and interested individuals assess dietary adequacy, the Institute of Medicine developed a set of usable nutritional standards. These standards can help a person judge whether her typical dietary intake is likely to provide too little, too much, or the right amount of the essential micronutrients, macronutrients, and calories.

Dietary Reference Intakes

Because humans are all so different, establishing concrete dietary requirements proved an enormous hurdle for the scientific community. Indeed, the task required the input and analysis of an assembly of researchers organized by the Institute of Medicine, a division of the National Academy of Sciences. In 1994, the Institute of Medicine began developing the **Dietary Reference Intakes (DRIs)**, a set of four dietary assessment standards used to assess and plan dietary intake. The four reference values that comprise the DRIs are:

- Estimated Average Requirement (EAR)
- Recommended Dietary Allowance (RDA)
- Adequate Intake (AI) level
- Tolerable Upper Intake Level (UL)

These four dietary assessment standards are illustrated in Figure 2.2, outlined in Table 2.1 on page 30, and described below. Using all four types of DRI values allows individuals and health professionals to assess comprehensively the nutritional adequacy of individuals and populations. To utilize the DRI values, it is critical to understand some fundamental concepts underlying the DRIs. For example, because nutrient requirements differ by sex and stage of life, the DRIs provide different values for different groups of people. DRI values for the various life-stage groups take into account both age and physiologic condition, such as pregnancy and lactation. In all, there are 16 life-stage groups for females and 10 life-stage groups for males. When utilizing the DRIs, one must be careful to use the correct values for one's appropriate sex and life-stage group because using the incorrect ones may lead to misinterpretation of intake adequacy.

It is important to understand the term *nutrient requirement*. A **nutrient requirement** is the amount of a nutrient that a person must consume to promote optimal health. Individuals each have their own unique nutrient requirements. In general, the nutrient

Figure 2.2 Dietary Reference Intake Standards

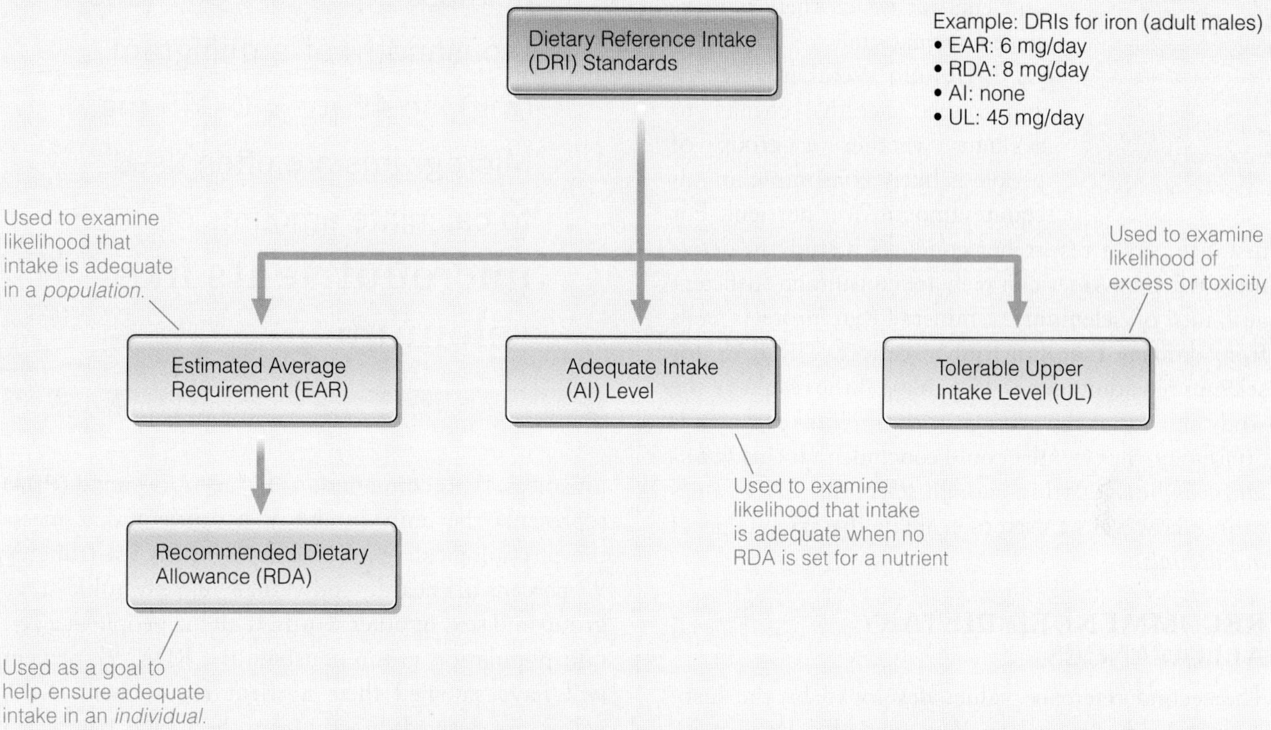

There are four sets of DRI reference values: Estimated Average Requirement, Recommended Dietary Allowance, Adequate Intake Level, and Tolerable Upper Intake Level.

requirements of a population are distributed in a bell shape, meaning that the vast majority of people have requirements at some mid-level amount, while equally few require much less and much more. Many factors influence a person's nutrient requirements—these factors are partially accounted for in the DRI life-stage groups. However, there are many other factors, such as genetics, certain diseases, medications, lifestyle choices, and other environmental influences that can also influence nutrient requirements. For instance, tobacco use can increase one's vitamin C requirement.[1] It is virtually impossible to know how much of each essential nutrient a person really needs; the DRIs are only *estimates* of nutrient requirements and intake goals in a healthy population. Your personal level may actually be less or more.

Nonetheless, the DRI values can be powerful tools in assessing your nutritional status and helping you plan a healthy diet. How can you use the DRIs for these purposes? The answer to this question requires a basic understanding of the standards that comprise the DRIs. In the following sections, you will learn about each of these four main dietary reference standards in detail and how to use them to determine if you are getting too little or too much of each important nutrient.

ESTIMATED AVERAGE REQUIREMENTS

When the Institute of Medicine assembly first set out to establish nutrient requirements, it developed Estimated Average Requirements for each essential micronutrient and macronutrient. An **Estimated Average Requirement (EAR)** value represents the daily intake of a nutrient that meets the physiological requirements of half the healthy individuals in a given life-stage group and sex (see Figure 2.3 on the next page). For example, the EAR for iron in breastfeeding women is 6.5 mg/day, meaning that half of the women in this life-stage group require less than 6.5 mg/day iron, and half need more. It is important to remember that an EAR value meets or exceeds the needs of half the population, but is inadequate for the

Estimated Average Requirement (EAR) The daily intake of a nutrient that meets the physiological requirements of half the healthy individuals in a given life-stage and sex.

Recommended Dietary Allowance (RDA) The daily intake of a nutrient that meets the physiological requirements of nearly all (roughly 97 percent) healthy individuals in a given life-stage and sex.

other half. Also, remember that EAR values are only available for nutrients for which there is sufficient information.

EARs are useful in research and public health settings to evaluate whether a group of people is likely consuming an adequate amount of a nutrient. For instance, say a researcher conducts a study to determine whether American men are consuming sufficient amounts of selenium, a mineral that protects cells from damage that can lead to cancer. The EAR for selenium in adult men is 45 μg/day. If the results of the study show that the average selenium intake in men is 50 μg/day, the scientist could conclude that this *population* is likely getting enough selenium. EAR values cannot however be used to evaluate dietary intakes of *individuals*.

RECOMMENDED DIETARY ALLOWANCES

The second reference values developed by the Institute of Medicine are the Recommended Dietary Allowances. A **Recommended Dietary Allowance (RDA)** represents the daily intake of a nutrient that meets the physiological requirements of nearly all (roughly 97 percent of) healthy individuals in a given life-stage group and sex. In other words, if all the people in a certain population group consume the RDA, 97 percent will have satisfied their nutrient requirement. RDA values are derived directly from the EARs; Figure 2.3 illustrates the relationship between these two sets of standards. In contrast to the population-oriented EARs however, the RDAs can be utilized as nutrient-intake goals for individuals. This is because each RDA has a built-in safety margin that ensures adequate intake by 97 percent of the population if recommended levels are consumed.

Unless specifically noted, an RDA value does not distinguish whether the nutrient is found naturally in foods, is added to foods, or is consumed in supplement form. Also, like EARs, RDAs are available only for nutrients for which there is sufficient information. Because of this, another set of standards was developed to

> A **microgram (μg)** is a unit of mass equal to 0.001 (one-thousandth) of a milligram (mg), so 45 μg = 0.045 mg. Micrograms are often used to calculate amounts of **micronutrients like selenium**.

Figure 2.3 Comparing EARs to RDAs

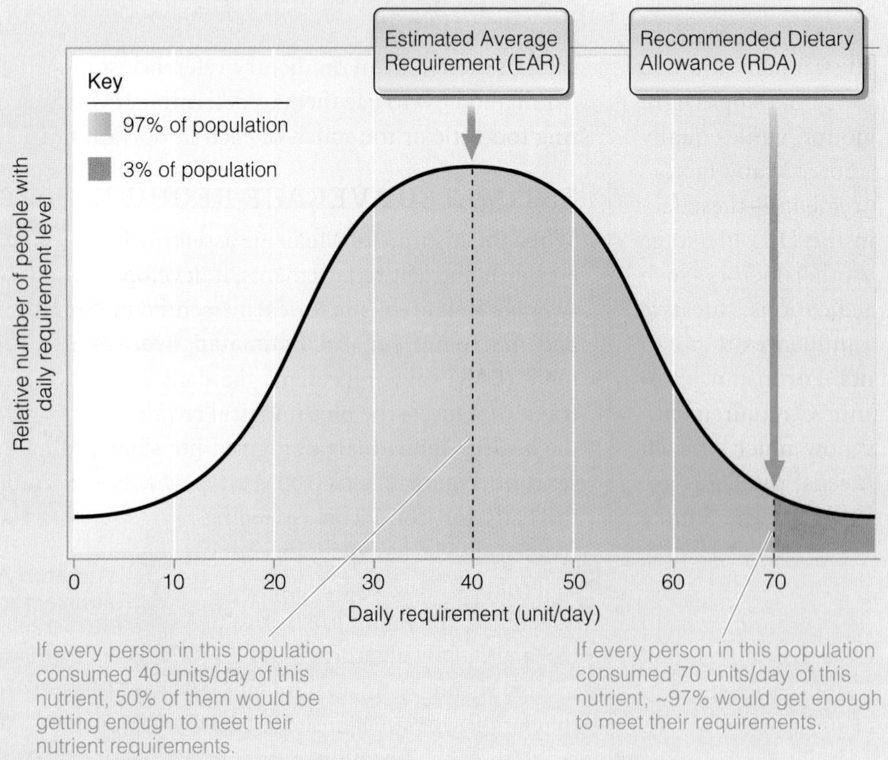

EAR values represent average requirements for a given population, whereas RDAs are intake goals for individuals.

provide provisional intake goals for nutrients without EARs or RDAs until more rigorous scientific studies are available. These standards are called Adequate Intake levels.

ADEQUATE INTAKE LEVELS

When scientific evidence was insufficient to establish an EAR (and thus accurately set an RDA), the DRI committee developed an **Adequate Intake (AI)** level instead. As such, if a nutrient has an AI instead of an RDA, you might safely assume that more research is needed before a conclusive dietary intake goal can be set. An example of a nutrient with an AI instead of an RDA is potassium. This may surprise you, as adequate potassium intake is essential to heart, brain, kidney, and muscle function. More research is required before EARs and RDAs can be established for this mineral. Sometimes further research is impossible. For example, because rigorous studies cannot ethically be conducted on infants younger than six months of age, there are no RDAs for this life-stage group—only AIs.

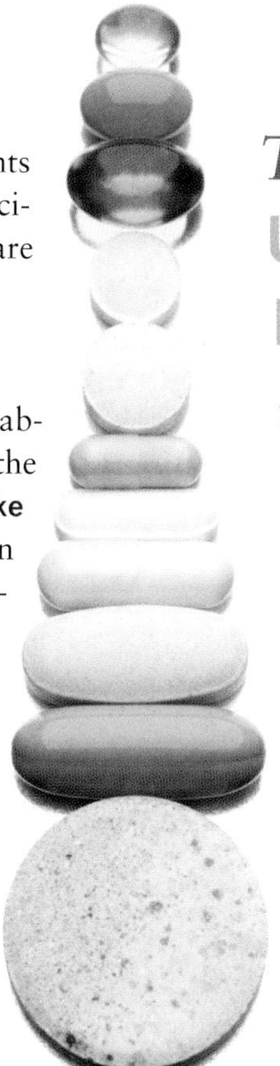

The Tolerable Upper Intake Levels (ULs) were developed to help consumers know how much of a nutrient supplement is safe to consume.

Adequate Intake (AI) The daily intake of a nutrient that appears to support adequate nutritional status; established when RDAs cannot be determined.

Tolerable Upper Intake Level (UL) The highest level of usual daily nutrient intake likely to be safe.

Like an RDA, an AI represents a daily nutrient intake goal for healthy individuals. Unlike the RDAs, which are based on rigorous scientific studies, the AIs are based on intake levels that appear to support adequate nutritional status. Continuing the above example, AIs for infants were developed by measuring average daily nutrient intakes by breastfed babies. Because breastfeeding is considered the ideal way to feed a baby, these nutrient intakes are assumed to be adequate. You can learn which nutrients have AI values and which have RDA values for each life-stage group by examining the DRI tables on the Dietary Reference Intakes card located at the back of the book as well as in Table 2.1 on the next page.

Together, RDA and AI values can be used to set goals to help you consume nutrients in sufficient quantities, allowing you to maintain health. Conversely, another set of values can help you avoid the consumption of nutrients in such large quantities that they actually do you harm. These values are called Tolerable Upper Intake Levels.

The Adequate Intake (AI) values for babies are based on average nutrient intakes of breastfed infants.

TOLERABLE UPPER INTAKE LEVELS

The RDA and AI values have been established to prevent deficiencies and optimize health. However, avoiding nutrient overconsumption and toxicity is also important. To help avoid such excesses, a **Tolerable Upper Intake Level (UL)** is established as the highest level of usual daily nutrient intake likely to be safe. For instance, the UL for selenium in men is 400 μg/day. This means that for a typical college-aged male, consuming 400 μg of this mineral a day is not likely to cause problems, but exceeding this amount might lead to selenium toxicity. The ULs are not meant to be used as goals for dietary intake. Instead, they provide limits for those who take supplements or consume large amounts of fortified foods.

Table 2.1 **Available Dietary Reference Intake (DRI) Standards for Adults**

Nutrient	EAR	RDA	AI	UL
Macronutrients[a]				
Water			■	
Linoleic acid			■	
Linolenic acid			■	
Carbohydrates	■	■		
Fats				
Protein	■	■		
Vitamins				
Thiamin	■	■		
Riboflavin	■	■		
Niacin	■	■		■
Biotin			■	
Pantothenic Acid			■	
Vitamin B_6	■	■		■
Folate	■	■		■
Vitamin B_{12}	■	■		
Vitamin C	■	■		■
Choline			■	■
Vitamin A	■	■		■
Vitamin D	■	■		■
Vitamin E	■	■		■
Vitamin K			■	
Minerals				
Sodium			■	■
Chloride			■	■
Potassium			■	
Calcium	■	■		■
Phosphorus	■	■		■
Magnesium	■	■		■
Iron	■	■		■
Zinc	■	■		■
Iodine	■	■		■
Selenium	■	■		■
Copper	■	■		■
Manganese			■	■
Fluoride			■	■
Chromium			■	
Molybdenum	■	■		■

[a] Note that there are no DRIs for energy, *per se*. Instead, you can estimate your caloric needs by using the Estimated Energy Requirement (EER) calculations.

Using EARs, RDAs, AIs, and ULs to Assess Nutrient Intake

The four sets of standards that comprise the DRIs should be used coordinately to assess nutrient intake. Although this process may seem complex at first, relatively simple guidelines can be used to make inferences about your diet—simply compare the results of your dietary assessment to your EAR, RDA, AI, and UL values.[2] These concepts are illustrated in Figure 2.4. When EARs, RDAs, and ULs have been established:

- If your intake of a nutrient is much less than the EAR, then it is likely to be inadequate, increasing your risk of nutrient deficiency.
- If your intake is between the EAR and the RDA, then you should probably increase your intake.
- If your intake is between the RDA and the UL, then it is probably adequate.
- If your intake is above the UL, then it is probably too high.

When only AIs are available:

- If your intake of a nutrient falls between the AI and UL, then it is probably adequate.
- If your intake is below your AI, no conclusion can be made about the adequacy of your intake.

Consider a 20-year-old female who upon completing a food record and dietary assessment learns that her vitamin A intake is 1,500 μg/day. Because this value falls between the RDA (700 μg/day) and the UL (3,000 μg/day), her vitamin A intake is probably adequate. Had her vitamin A intake been 600 μg/day, it would have been between her EAR (500 μg/day) and RDA (700 μg/day), indicating that she should probably consume more. Had it been 3,500 μg/day (above her UL), then she could conclude that her intake of vitamin A is too high.

Energy Intake Can Also Be Assessed

So far, you have learned about using the DRIs to assess your nutritional status in terms of micronutrients and macronutrients. However, it is also important to determine whether you are consuming the right amount of calories and whether these calories are coming from the right mix of foods. To address this issue, the Institute of Medicine developed two types of standards that you can use to assess your energy intake: Estimated Energy Requirements and the Acceptable Macronutrient Distribution Ranges.[3]

ESTIMATED ENERGY REQUIREMENTS

The Estimated Energy Requirements are similar in theory and application to the EAR. An **Estimated Energy Requirement (EER)** value represents the average energy intakes needed for a person to maintain a healthy weight. The EER values vary by age, sex, weight, height, and physical activity level. Note that this is different from the other DRI reference values, which only vary by lifestage and sex.

Estimated Energy Requirement (EER) The average energy intake needed for a healthy person to maintain weight.

EERs are calculated using relatively simple mathematical equations. Using the EER equations for adult men and women of healthy weight below, you can calculate your own EER in kilocalories per day (kcal/day). Additional EER equations for other lifestage groups can be found on the Estimated Energy Requirement (EER) Calculations and Physical Activity (PA) Values card at the back of the book.

Adult man: $\text{EER} = 662 - (9.53 \times \text{age}) + \text{PA} \times (15.91 \times \text{weight} + 539.6 \times \text{height})$

Adult woman: $\text{EER} = 354 - (6.91 \times \text{age}) + \text{PA} \times (9.36 \times \text{weight} + 726 \times \text{height})$

To solve these equations, you must insert your age in years, your weight in kilograms (kg), and your height in

Figure 2.4 Using DRI Values to Assess Nutritional Status

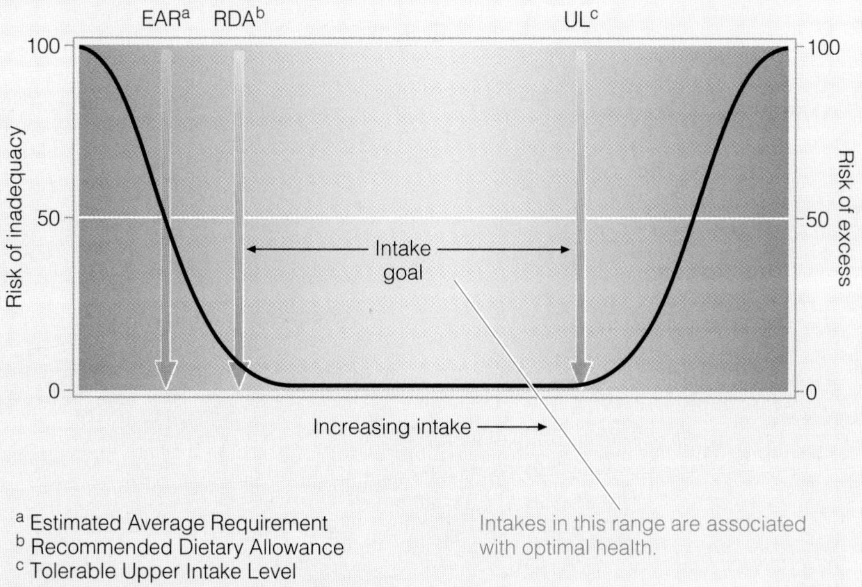

By determining your nutrient intake values and comparing them to the appropriate DRI values, you can easily determine whether you are likely getting too little, too much, or just the right amount of a given nutrient.

Calculating your **Estimated Energy Requirement (EER)** is a simple way for you to approximate how many **calories to consume daily** to maintain your current weight.

© iStockphoto.com/webphotographeer

Acceptable Macronutrient Distribution Range (AMDR) The recommended range of intake for a given energy-yielding nutrient, expressed as a percentage of total daily caloric intake.

meters (m). *PA* indicates a physical activity value that is based on whether one is sedentary, low active, active, or very active. The lower your PA value is, the less active you are, and consequently the lower your EER is. Table 2.2 provides examples of these activity categories and their corresponding PA values. Note that at every age, active individuals need more energy than do their sedentary counterparts.

Calculating your EER is not difficult, as it requires only that you insert your age, PA value, weight, and height into the correct equation. You can convert pounds to kilograms by dividing your weight in pounds by 2.2. You can convert feet to meters by dividing your height in feet by 3.3. Here is an example of an EER calculation:

Kyung-Soon is a 38-year-old woman who weighs 115 pounds (52.3 kg), is 5 feet 4 inches (5.3 feet, or 1.6 m) tall, and has a low activity level.

Kyung-Soon's EER is calculated as follows.

$$\begin{aligned} \text{EER} &= 354 - (6.91 \times \text{age}) + \text{PA} \times (9.36 \times \text{weight} + 726 \times \text{height}) \\ &= 354 - (6.91 \times 38) + 1.12 \times (9.36 \times 52.3 + 726 \times 1.6) \\ &= 354 - 262.6 + 1.12 \times (489.5 + 1161.6) \\ &= 91.4 + 1.12 \times 1651.1 \\ &= 1{,}941 \text{ kcal} \end{aligned}$$

Take the time right now to determine your own EER. Knowing how much energy you require on a daily basis can serve as a guide to help establish the right balance of energy in your diet.

ACCEPTABLE MACRONUTRIENT DISTRIBUTION RANGES

After you establish whether you are consuming the right amount of calories, you must determine whether your distribution of energy sources—carbohydrates, proteins, and fats—is healthy. To answer this question, the Acceptable Macronutrient Distribution Ranges were developed. An **Acceptable Macronutrient Distribution Range (AMDR)**, expressed as a percentage of total energy, represents the ideal range of intake for a given class of energy-yielding nutrient:

Carbohydrates: 45 to 65 percent of total energy
Protein: 10 to 35 percent of total energy
Fat: 20 to 35 percent of total energy

For example, if your EER is 2,400 kcal/day, you should be getting from 1,080 (45 percent of 2,400) to 1,560 (65 percent of 2,400) of those daily kilocalories from carbohydrates. It is important for you to determine how many calories you currently get and should get from each group of energy-yielding macronutrients. Maintaining a diet that adheres to the AMDRs both decreases your risk of chronic disease and promotes the adequate intake of essential micronutrients.

Table 2.2 Physical Activity (PA) Categories and Values[a]

Activity Level Category	Physical Activity (PA) Value: Men	Physical Activity (PA) Value: Women	Description
Sedentary	1.00	1.00	No physical activity aside from that needed for independent living.
Low Active	1.11	1.12	1.5 to 3 miles/day at 2 to 4 miles/hour in addition to the light activity associated with typical day-to-day life.
Active	1.25	1.27	3 to 10 miles/day at 2 to 4 miles/hour in addition to the light activity associated with typical day-to-day life.
Very Active	1.48	1.45	10 or more miles/day at 2 to 4 miles/hour in addition to the light activity associated with typical day-to-day life.

[a] These values only apply to normal-weight, nonpregnant, nonlactating adults. Values for children, pregnant or lactating women, and overweight or obese individuals are different.

Source: Institute of Medicine. Dietary Reference Intakes for energy, carbohydrate, fiber, fat, fatty acids, cholesterol, protein, and amino acids. Washington, DC: National Academies Press; 2005.

LO4 How Can You Assess and Plan Your Diet?

If you complete a dietary assessment using a food record and investigate your DRI, EER, and AMDR values, you can determine whether your nutrient and energy intakes are likely adequate. However, because this process can be rather cumbersome, it is usually only done in a research environment or when nutrient inadequacies are suspected. Luckily for the lay nutritionist, several additional process-simplifying tools are available. A number of reputable organizations have streamlined dietary assessment even further than the Institute of Medicine by summarizing what a healthy diet generally "looks like." In other words, they have attempted to answer the question "What kinds of foods should I consume, and how much of them should I eat?" by formulating general nutritional guidelines such as the Dietary Guidelines for Americans and the MyPlate food guidance system.

Food Guidance Systems

Several government agencies are involved in assuring the United States' health. In fact, some of these agencies have been collecting and providing information about nutrition and health for more than a century. One such agency is the United States Department of Agriculture (USDA), which published its first set of nutritional recommendations for Americans in 1894. The USDA has continued to publish recommendations for the American public since that time. In fact, over the past 100 years, a succession of federally supported recommendations—all designed to provide guidelines for dietary planning—have been published.[4] Two of these publications are shown in Figure 2.5. Each such **USDA Food Patterns** publication (formerly called USDA Food Guide) categorizes nutritionally similar foods into *food groups* and makes recommendations regarding the number of servings of each food group that should be consumed daily.

USDA Food Patterns A USDA publication that categorizes nutritionally similar foods into food groups and makes recommendations regarding the number of servings of each food group that should be consumed daily.

You may already be aware of some of the USDA's food groups. For example, poultry, eggs, meat, and seafood have been grouped together because they are all excellent sources of protein. However, the composition and number of food groups has changed over the years as science has evolved and socioeconomic times have changed. For instance, during the Great Depression, food was scarce and malnutrition was common. In response, the USDA shifted the focus of the Food Guide to recommend relatively inexpensive, high-fat

Figure 2.5 Dietary Guidance Has Evolved Tremendously over the Years

USDA (http://www.usda.gov)

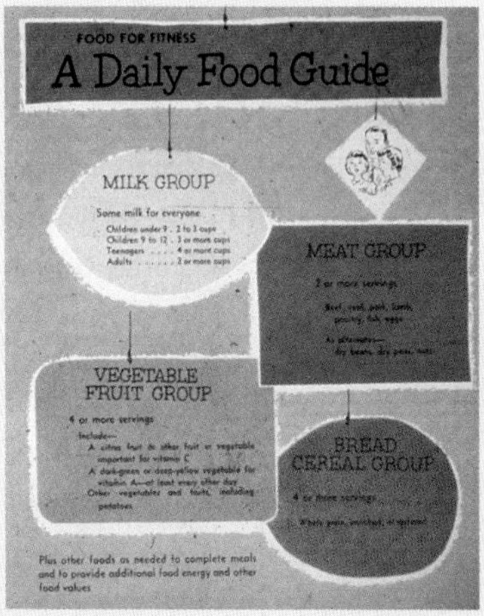

USDA (http://www.usda.gov)

The U.S. Department of Agriculture has published numerous food-related recommendations over the past century. Here, Depression Era (left) and "Basic 4" (right) recommendations are expressed visually.

Dietary Guidelines for Americans A series of recommendations that provide specific nutritional guidance and advice about physical activity, alcohol intake, and food safety.

foods such as milk, peanuts, and cheese. These foods provide maximal energy and nutrients at minimal cost. By contrast, when the *Basic 4* food groups were unveiled in 1956, the USDA Food Guide highlighted a bread/cereal group and a vegetable/fruit group, reflecting both a positive economic climate and growing evidence that these types of foods provided important vitamins and minerals to the diet. As you will soon learn, the current Food Patterns report has five food groups, each contributing an important set of nutrients.

In 1980, the U.S. Department of Health and Human Services (DHHS) and the USDA published an expanded dietary recommendation called the **Dietary Guidelines for Americans.** This new series of guidelines provided specific nutritional guidance and advice about physical activity, alcohol intake, and food safety. Since its introduction, the Dietary Guidelines for Americans has been revised every five years to reflect new scientific information. In addition to overarching advice concerning healthy eating, the Dietary Guidelines for Americans now incorporates the USDA Food Patterns. Indeed, such guidelines are designed to evolve as the understanding of nutrition and health does. You can follow these changes on the USDA website for years to come (http://www.usda.gov).

2010 Dietary Guidelines for Americans

The 2010 Dietary Guidelines for Americans, the nation's current nutritional recommendations, were developed by a highly esteemed committee of nutrition scientists. These experts systematically evaluated published scientific literature to formulate dietary and physical activity recommendations to optimize the health of the U.S. population. The committee had to keep two important and sometimes competing facts in mind while developing these recommendations. First, poor diet and physical inactivity are the most important factors contributing to overweight and obesity epidemics in all segments of U.S. society. Second, nearly 15 percent of American households are currently unable to acquire adequate food to meet their dietary needs.[5] To complicate matters, the committee recognized that while obesity has become the nation's leading health threat, many Americans (even those who are overweight or obese) still consume less than optimal amounts of certain nutrients. This is true even among people with adequate financial resources.

The 2010 Dietary Guidelines for Americans encompasses two overarching goals: (1) to help individuals maintain energy balance over time and (2) to help them choose nutrient-dense foods and beverages. Underlying these broad goals is the proposition that nutrient needs should be met primarily through the consumption of healthy foods. As such, the Dietary Guidelines establishes four groups of key recommendations, which are listed in Table 2.3 and discussed in the following sections.

The 2010 edition of the Dietary Guidelines goes a step further than previous editions by acknowledging that "everyone has a role in the movement to make American healthy." By working together to enact policies, programs, and partnerships that strengthen America's overall health, organizations and individuals can improve the health of the current generation as they ensure better health for generations yet to come.

BALANCE CALORIES TO MANAGE WEIGHT

It is important to balance the number of calories consumed with those expended. Otherwise, weight gain and its associated health risks are inevitable. In response to rising rates of overweight and obesity across America, the 2010 Dietary Guidelines includes several recommendations related to *calorie balance* (energy consumed relative to energy spent). For instance, recognizing that all people—regardless of age—can become obese, the Dietary Guidelines recommends maintaining appropriate calorie balance during every stage of life. To help actualize this recommendation, the following strategies were proposed:

- Focus especially on the total number of calories consumed.
- Monitor food intake, bodyweight, and physical activity so that you can detect changes that might lead to poor health.
- When eating out, choose smaller portions or lower-calorie options.
- Prepare, serve, and consume smaller portions of foods and beverages—especially those high in calories.
- Eat a nutrient-dense breakfast. In other words, eat foods that maximize nutrients for each calorie.

Table 2.3 **2010 Dietary Guidelines for Americans' Major Concepts and Key Recommendations**

Major Concept	Key Recommendations
Balance Calories to Manage Weight	• Prevent and/or reduce overweight and obesity by improving behaviors related to eating and physical activity. • Control total caloric intake to manage body weight. For people who are overweight or obese, this means consuming fewer calories from foods and beverages. • Increase physical activity and reduce sedentary activity. • Maintain appropriate calorie balance during each stage of life.
Reduce Consumption of Certain Foods and Food Components	• Reduce daily sodium intake to less than 2,300 mg/day. Individuals at increased risk of hypertension should further reduce intake to 1,500 mg/day. • Consume less than 10 percent of calories from saturated fats by replacing them with monounsaturated and polyunsaturated fats. • Consume less than 300 mg/day of dietary cholesterol. • Keep *trans* fatty acid consumption as low as possible. • Reduce intake of calories from solid fats (saturated and *trans* fats) and added sugars. • Limit consumption of foods that contain refined grains. • If alcohol is consumed, it should be consumed in moderation (up to one drink per day for women and two drinks per day for men) and only by adults of legal drinking age.
Increase Consumption of Certain Foods and Nutrients*	• Increase vegetable and fruit intake. • Eat a variety of vegetables, especially peas, beans, and vegetables that are dark green, red, or orange. • Consume at least half of all grains as whole grains. • Increase intake of fat-free or low-fat milk, milk-based products, and/or fortified soy beverages. • Choose a variety of protein foods, such as seafood, lean meat, poultry, eggs, beans, peas, soy products, and unsalted nuts and seeds. • Replace protein foods that are high in solid fats with choices that are lower in solid fats and calories. Choose protein foods that are sources of oils. • Choose foods that contain more potassium, dietary fiber, calcium, and vitamin D, such as vegetables, fruits, whole grains, and dairy products.
Build Healthy Eating Patterns	• Select an eating pattern that meets nutrient needs over time and allows for an appropriate caloric intake. • Account for all foods and beverages consumed and assess how they fit within a healthy eating pattern. • Follow food safety recommendations when preparing and eating foods to reduce the risk of foodborne illnesses.

* Additional intake recommendations were also made for specific population groups. For example, women capable of becoming pregnant should choose foods rich in heme iron and folate, and should consume additional iron sources, enhancers of iron absorption, and synthetic folic acid. Women who are pregnant or breastfeeding should strive to consume 227 to 340 g (8 to 12 oz) of seafood per week, but should be careful to avoid seafood high in mercury. Individuals who are age 50 or older should consume foods fortified with vitamin B_{12} or should take a dietary supplement.

Source: U.S. Department of Agriculture and U.S. Department of Health and Human Services. Dietary Guidelines for Americans, 2010. 7th edition, Washington, DC: US Government Printing Office, December 2010.

- Limit sedentary activities (such as watching television and playing computer games) to no more than one to two hours each day.

Clearly, reversing current obesity trends is a primary focus of the 2010 Dietary Guidelines. To determine your personal calorie balance, compare your daily caloric consumption to your daily caloric need. The EER equations are designed perfectly for this very purpose. Once you have determined your estimated energy requirement using the appropriate EER equation, either strive to eat that amount, or if you are overweight, consume fewer calories (such as 500 kcal/day less) to reach a healthy body weight. The Dietary Guidelines recommends that individuals control their caloric intakes by increasing consumption of whole grains, vegetables, and fruits, and by decreasing consumption of sugar-sweetened beverages. Monitoring calorie intake from alcoholic beverages is also important.

REDUCE CONSUMPTION OF CERTAIN FOODS AND FOOD COMPONENTS

Whereas the first group of recommendations promotes weight maintenance, the second focuses on reducing the intake of foods and food components known to increase one's risk of chronic degenerative disease. Following this group of recommendations also increases the likelihood that you will meet your nutritional needs without exceeding your energy requirements. In particular, the Dietary Guidelines recommends that Americans reduce their intakes of sodium, saturated fats, cholesterol, *trans* fatty acids, solid fats (including both saturated fats and *trans* fats), added sugars, and refined grains. To help reduce intake of these components, the following strategies were developed:

- Read Nutrition Facts labels for information about the sodium contents of foods and purchase foods that are low in sodium.
- Consume more fresh foods and fewer processed foods (which are often high in sodium).
- Eat more home-cooked foods and use little or no salt when preparing and eating foods.
- When eating at restaurants, ask that salt not be added to your food or order low sodium options.
- Focus on eating the most nutrient-dense forms of foods from all food groups.
- Limit the amount of solid fats and added sugars when cooking or eating out. Trimming fat from meat, using less butter and stick margarine, and using less table sugar are all good strategies to limit these food components.
- Consume fewer and smaller portions of foods and beverages that contain solid fats and/or added sugars, such as grain-based desserts, sodas, and other sugar-sweetened beverages.

INCREASE CONSUMPTION OF CERTAIN FOODS AND NUTRIENTS

The Dietary Guidelines identifies four "nutrients of concern" that are somewhat lacking in the American diet: these are potassium, dietary fiber, calcium, and vitamin D. Certain populations' intakes of iron, folate, and vitamin B_{12} are also of concern. In addition to these specific nutrients, Americans would benefit from eating more especially wholesome and healthy foods such as seafood, legumes, whole grains, and low-fat dairy products. The Dietary Guidelines makes several key recommendations aimed at increasing the nation's consumption of these important foods and food components. Of course, these goals should be met in conjunction with a stable calorie balance and a healthy eating pattern.

> "**A Long Island Iced Tea** cocktail can contain as many as **800 calories**."

BUILD HEALTHY EATING PATTERNS

Finally, the 2010 Dietary Guidelines emphasizes that there are many ways that individuals and families can incorporate these recommendations into their lives. In other words, there is more than one healthful eating pattern—especially considering the incredible diversity of cultural, ethnic, traditional, and personal preferences in the United States. Food costs and availabilities are critical pieces of the puzzle for many people when determining what to eat. To help consumers of all traditions, preferences, and socioeconomic statuses choose healthy dietary patterns, the Dietary Guidelines include three key recommendations related to overall meal planning and preparation:

- Select an eating pattern that both meets nutrient needs over time and allows for an appropriate caloric intake.
- Account for every food and beverage you consume. Asses how well your dietary choices fit within a complete and healthy eating pattern.
- Follow food safety recommendations when preparing and eating foods to reduce the risk of foodborne illnesses.

USDA Food Patterns

Knowing about the overall food intake patterns that tend to be healthy can help a person plan his own eating pattern. As such, the USDA Food Patterns, part of the 2010 Dietary Guidelines, describes an array of 12 eating patterns that have been scientifically shown to impart optimal health, supply all of the essential nutrients, and limit calories. These *food patterns*, equivalent to the *food guides* published in previous editions of the Dietary Guidelines, provide specific recommendations as to the kinds of foods a person should consume and the proportions in which she should consume them. One of the key recommendations, to eat a variety of vegetables, especially peas and beans (legumes) and vegetables that are dark green, red, or orange, speaks to these established food patterns.

But how many servings of these different types of vegetables does a person actually need? This question is answered by the USDA Food Patterns list, which currently considers five food groups: vegetables, fruits, grains, dairy products, and protein foods. Some of these food groups are further separated into subcategories. For example, the vegetables food group is separated into five important subgroups: dark green (such as broccoli and spinach), red and orange (such as tomatoes and pumpkins), beans and peas (such as kidney beans and lentils), starchy (such as white potatoes and corn), and other (such as iceberg lettuce and onions). Beyond the five food groups, the food patterns also address recommended allowances of oil, solid fats, and added sugars.

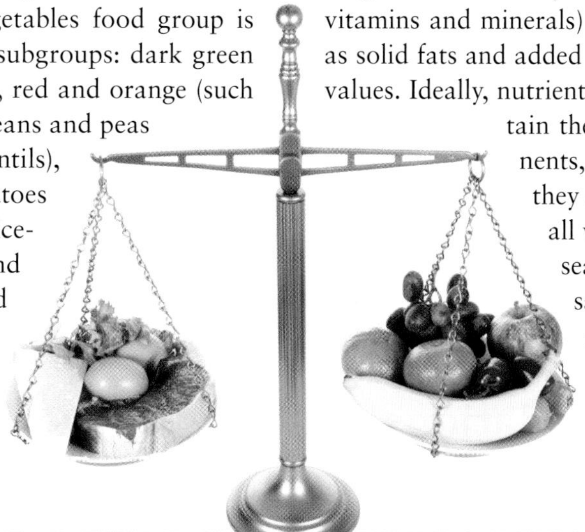

The 12 food patterns developed by the USDA are based on caloric needs. As you might assume, the more calories a person needs, the more food she should eat (see Table 2.4). For instance, a person who requires 2,000 kcal/day should eat 2½ cups of vegetables every day, whereas a person who requires 2,400 kcal/day should strive to eat 3 cups every day.

nutrient density
The relative ratio of a food's amount of nutrients to its total calories.

FOCUS ON NUTRIENT DENSITY

The USDA Food Patterns recommends that individuals consider **nutrient density**, the relative ratio of a food's amount of nutrients to its total calories, when selecting foods and beverages. Nutrient-dense foods and beverages contain many beneficial compounds (such as vitamins and minerals) and few unhealthy ones (such as solid fats and added sugars) relative to their caloric values. Ideally, nutrient-dense foods are those that retain their naturally occurring components, such as dietary fiber. Unless they are altered during preparation, all vegetables, fruits, whole grains, seafood, eggs, beans, peas, unsalted nuts, seeds, lean meats, fat-free and low-fat milk and milk products, and poultry are nutrient-dense foods.

Table 2.4 Amounts of Each Food Group Recommended by the USDA Food Patterns and MyPlate

Food Category	Amounts Recommended (per day)[a]	Dietary Significance
Grains	3 to 10 oz	Grains are a major source of B vitamins, iron, magnesium, selenium, energy, and dietary fiber.
Vegetables	1 to 4 cups	Vegetables are rich sources of potassium, vitamin C, folate, dietary fiber, and vitamins A and E.
Fruits	1 to 2½ cups	Fruits are good sources of folate, vitamin C, vitamin A, and fiber.
Dairy products	2 to 3 cups	Dairy products are major sources of calcium, potassium, vitamin D, and protein.
Protein foods	2 to 7 oz	Protein foods are rich sources of protein, magnesium, iron, zinc, B vitamins, vitamin D, energy, and potassium.

[a] Recommended amounts depend on age, sex, and physical activity level. Personalized recommendations can be generated at the MyPlate website (http://www.choosemyplate.gov).

REMEMBER THAT BEVERAGES COUNT

The 2010 Dietary Guidelines notes that individuals need to think about not only everything they eat—but also everything they drink—when considering how close their diets match the recommended dietary patterns. Indeed, beverages contribute substantially to most Americans' caloric intakes. Although caloric beverages provide the body with water (and sometimes important nutrients), many are not nutrient-dense. In fact, most flavored beverages contain empty calories and high amounts of added sugars. With the exception of recommended amounts of low-fat or fat-free milk and 100 percent fruit juices, the Dietary Guidelines recommends that individuals drink water and other beverages with few or no calories (such as tea).

Figure 2.6 **The USDA MyPlate Graphic**

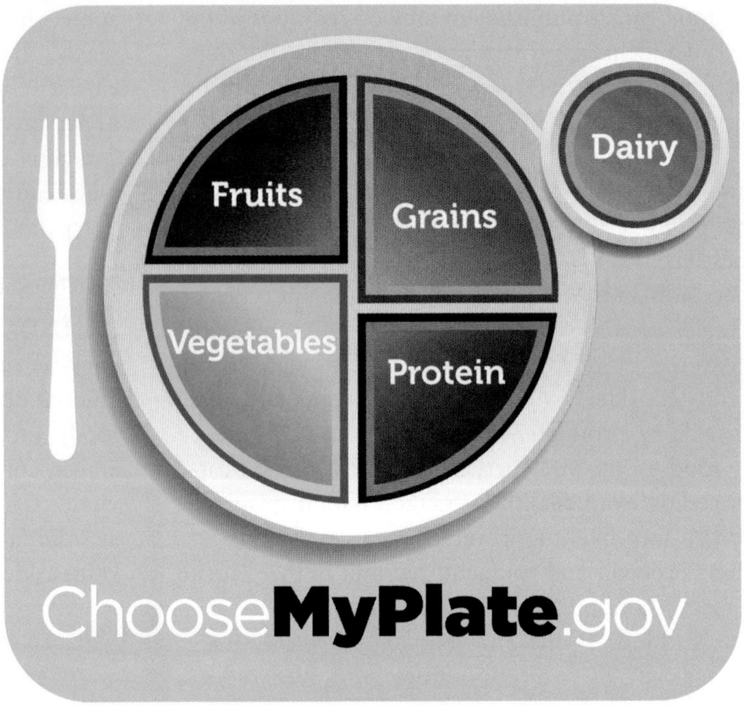

MyPlate illustrates the USDA's recommended food consumption pattern as a consumer-friendly graphic.

USDA (http://www.choosemyplate.gov)

MyPlate Illustrates How to Put Recommendations into Practice

Wading through the Dietary Guidelines and accompanying Food Patterns may be interesting to some, but it can also be time consuming and complex. To help convey the major messages of its reports to the public, the USDA developed **MyPlate**, a visual food guide that illustrates the most important food pattern intake recommendations put forth in the 2010 Dietary Guidelines for Americans. In 2011, MyPlate replaced MyPyramid as the federal government's official food guidance system graphic. Whereas some found MyPyramid to be obtuse and confusing, MyPlate was designed to be elegant and adaptable. The MyPlate graphic simply reminds Americans to eat healthful amounts of the five food groups using a familiar mealtime visual: a place setting. As you can see in Figure 2.6, recommended daily intakes of four food groups (fruits, vegetables, grains, and protein foods) are illustrated proportionately on the plate, whereas the fifth food group, dairy, is represented by the round glass (or bowl) on the upper-right periphery of the plate.

MyPlate A visual food guide developed by the USDA to illustrate the most important food intake pattern recommendations of the 2010 Dietary Guidelines for Americans.

Note that the MyPlate graphic does not specify the numbers of servings recommended by the USDA Food Patterns. The recommended amount of each food group depends on a person's age, sex, and physical activity level. The creators of MyPlate wanted to encourage each person to determine how much food is needed on an individual basis. To determine your own food needs, you must log on to the MyPlate website (http://www.choosemyplate.gov) and provide personal information (age, sex, and physical activity level) under the *Daily Food Plan* section, which can be found under *Interactive Tools*. Once you have provided your personal information, the MyPlate website generates individualized recommendations regarding food-intake patterns, serving sizes, and menu selection. In all, there are 12 different sets of recommended dietary patterns for the general population. This corresponds to the 12 different energy-intake levels listed in the USDA Food Patterns, on which MyPlate was based. An example of a personalized food plan, created for a 20-year-old woman with low physical activity, is illustrated in Figure 2.7. Food plans can also be generated for pregnant and breastfeeding women by clicking the *For Moms* section.

In addition to personalized food plans, the MyPlate website provides in-depth information about the types and amounts of foods that fit into each food group, assistance with meal planning, and material for special subgroups (such as vegetarians). For example, using the *Menu Planner* option, you can see how a meal or entire day's menu contributes to your overall MyPlate recommendations. Nutrient compositions and serving sizes of individual foods can be found in *MyFoodapedia*, while information about physical activity recommendations and Let's Move!, a federally-funded initiative designed to combat childhood obesity, can be found throughout other sections of the MyPlate website. At the time this textbook was revised, the USDA had not fully updated the MyPlate website. Consequently, there may be slight differences between what is written in this book and what is available on the MyPlate website.

BASIC THEMES AND KEY MESSAGES OF THE MYPLATE SYMBOL

The MyPlate graphic and website represent four basic themes:

1. Build a healthy plate.
2. Cut back on foods high in solid fats, added sugars, and salt.
3. Eat the right amount of calories for you.
4. Be physically active your way.

In this way, MyPlate is much like MyPyramid, which emphasized similar overarching goals. Unlike previous food guidance graphics, however, MyPlate is a singular part of a much larger communications initiative developed to help U.S. consumers make better food choices. This initiative involves strategic partnering between the USDA and dozens of public and private-sector groups to help promote and amplify seven key consumer messages based on MyPlate's four basic themes. These user-friendly messages are:

- Make half your plate fruits and vegetables.
- Enjoy your food, but eat less.
- Drink water instead of sugary drinks.
- Make at least half your grains whole grains.
- Avoid oversized portions.
- Compare sodium in foods and choose the foods with lower numbers.
- Switch to fat-free or low-fat (1%) milk.

Each of these messages will be granted a span of national focus in the popular media. For instance, in the spring and summer of 2012, you may notice (or may have noticed) numerous advertisements and other forms of communications encouraging people to drink water instead of sugary drinks. In the winter of 2013, the national focus will shift to choosing

Figure 2.7 An Example of a Daily Food Plan

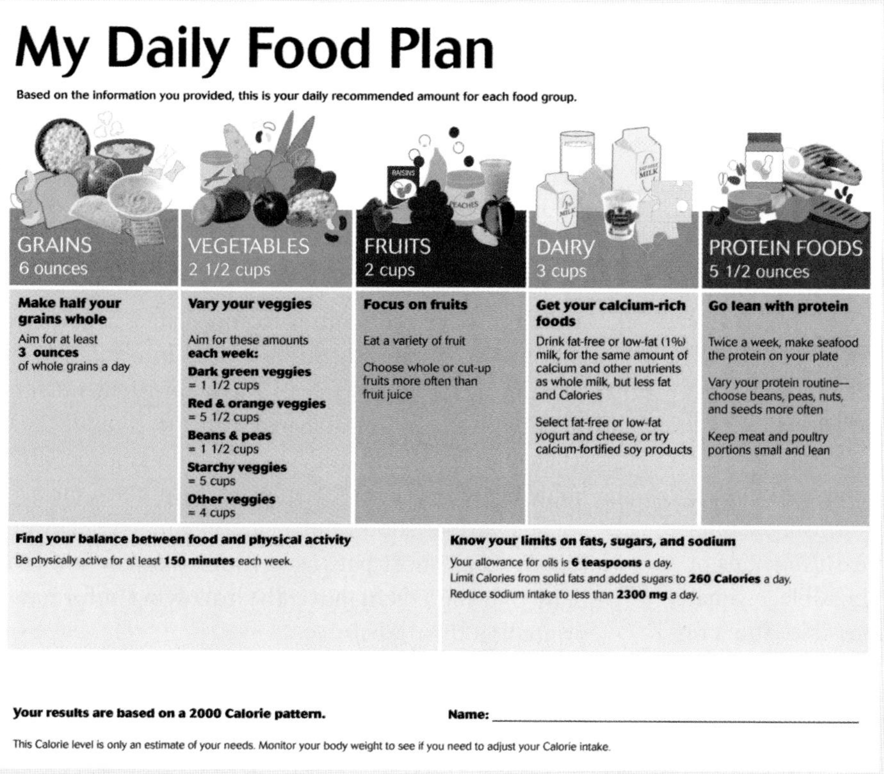

My Daily Food Plan
Based on the information you provided, this is your daily recommended amount for each food group.
GRAINS
6 ounces
VEGETABLES
2 1/2 cups
FRUITS
2 cups
DAIRY
3 cups
PROTEIN FOODS
5 1/2 ounces
Make half your grains whole
Aim for at least 3 ounces of whole grains a day
Vary your veggies
Aim for these amounts each week:
Dark green veggies = 1 1/2 cups
Red & orange veggies = 5 1/2 cups
Beans & peas = 1 1/2 cups
Starchy veggies = 5 cups
Other veggies = 4 cups
Focus on fruits
Eat a variety of fruit
Choose whole or cut-up fruits more often than fruit juice
Get your calcium-rich foods
Drink fat-free or low-fat (1%) milk, for the same amount of calcium and other nutrients as whole milk, but less fat and Calories
Select fat-free or low-fat yogurt and cheese, or try calcium-fortified soy products
Go lean with protein
Twice a week, make seafood the protein on your plate
Vary your protein routine—choose beans, peas, nuts, and seeds more often
Keep meat and poultry portions small and lean
Find your balance between food and physical activity
Be physically active for at least 150 minutes each week.
Know your limits on fats, sugars, and sodium
Your allowance for oils is 6 teaspoons a day.
Limit Calories from solid fats and added sugars to 260 Calories a day.
Reduce sodium intake to less than 2300 mg a day.
Your results are based on a 2000 Calorie pattern.
Name:
This Calorie level is only an estimate of your needs. Monitor your body weight to see if you need to adjust your Calorie intake.

There are 12 different daily food plans in the MyPlate food guidance system. Each food plan is based on a particular energy requirement. The food plan shown here is for a 2,000 kcal/day diet.

Source: U. S. Department of Agriculture - http://www.choosemyplate.gov/myplate/index.aspx

Food Tracker A component of the MyPlate website that allows individuals to conduct dietary self-assessments.

fat-free or low-fat milk. Throughout this unprecedented nationwide initiative, an overarching theme of "be[ing] physically active your way" will be integrated into each key message. It is the hope of the USDA that this targeted communications campaign will systematically translate the basic themes illustrated by the MyPlate graphic onto the American dinner plate.

USING THE MYPLATE FOOD TRACKER TO CONDUCT A DIETARY SELF-ASSESSMENT

The MyPlate website offers an excellent opportunity for you to conduct a self-assessment of your own dietary intake using the free **Food Tracker**. (To access this tool, click *Analyze my diet* on the MyPlate homepage.) Food Tracker allows you to assess whether your diet meets your recommended USDA Food Pattern (for example, whether you are consuming enough servings of milk) as well as recommendations put forth in the DRIs (for example, whether you are consuming your RDA for calcium). To use Food Tracker, you must keep track of everything you eat and drink for at least one day and then enter the information into the website. Ideally, however, you should keep a food record for three days—one of which should be a weekend day. Other tips for keeping an accurate food record include the following:

- *Detail is important.* Include as much detail as possible for all foods and beverages consumed. Brand names and preparation methods will help improve accuracy. For *mixed dishes* (or *composite foods*), estimate the amount of each ingredient. For example, "salad" might include lettuce, tomatoes, eggs, and so on. The more detail included in the diet record, the more accurate the analysis can be.
- *Estimate or measure serving sizes accurately.* Ideally, weigh or measure foods using standard household devices such as measuring cups or a kitchen scale. If this is not possible, estimate serving sizes as carefully as possible. Some restaurants provide detailed information about the amounts of food served. Such detail can be very helpful.
- *Choose representative, normal days.* Choose days that are representative of your typical eating patterns, and avoid special days such as holidays and birthdays. Because being sick or under unusual stress can influence food preferences and overall intake, it is best to avoid these circumstances as well.
- *Do not change your normal eating patterns.* When keeping diet records, people often alter their eating patterns to be more convenient or healthful. Resist this temptation, as it is especially important that diet records reflect normal intake.

When these guidelines are followed, the MyPlate Food Tracker can be a powerful tool in helping you determine whether the foods and beverages you typically choose are likely to encourage or deter you from being well.

LO5 How Can You Use Food Labels to Plan a Healthy Diet?

Understanding the recommendations put forth by the Dietary Guidelines and MyPlate is an excellent starting place for healthful eating. However, other resources are available to help you choose the appropriate amounts of nutritious foods and make the right food-related decisions every day. One of the most important resources is the food label, which contains a vast amount of information that can help you optimize your health.

Food Labels

Although paying attention to food labels is an excellent way to obtain nutrition-related information, this was not always the case. In fact, consistent nutrition labeling was not established for foods until 1973, when the U.S. Food and Drug Administration (FDA) implemented a series of rules to help consumers become aware of food's nutrient content. The FDA now requires that most packaged foods that contain more than one ingredient have the following information printed on their labels:

- Product name and place of business
- Product net weight
- Product ingredient content (from most abundant to least abundant ingredient)
- Company name and address
- Country of origin

- Product code (UPC bar code)
- Product dating, if applicable
- Religious symbols, if applicable (such as kosher)
- Safe-handling instructions, if applicable (such as for raw meats)
- Special warning instructions, if applicable (such as for aspartame and peanuts)
- Nutrition Facts panel outlining specified nutrient information

Of special interest to many people is the **Nutrition Facts panel**, a required component of most food labels that provides information about the nutrient content of the food. The FDA mandates that several critical elements be listed on every Nutrition Facts panel. For example, the food's serving size must be noted. Serving sizes have been standardized so that a person can easily evaluate the nutrient contents of similar foods. For instance, one can compare the amount of iron in a single serving of two similar breakfast cereals by looking at the amounts listed on their Nutrition Facts panels. Because the two cereals' serving sizes are the same, their iron contents can be compared directly. In addition to serving size, the label must list the total energy (Calories), total carbohydrates (including dietary fiber), sugar, and protein per serving. An example of a food label that includes a Nutrition Facts panel is shown in Figure 2.8 on the next page.

Nutrition Facts panel A required component of most food labels that provides information about the nutrient content of the food.

Nutrition Facts panels must also provide information concerning specific nutrients that individuals should try to limit. These include total fat, saturated fat, *trans* fat, cholesterol, and sodium. Conversely, because individuals are encouraged to consume more dietary fiber, vitamin C, vitamin A, calcium, and iron, the FDA requires that information concerning these "shortfall" nutrients be included on Nutrition Facts panels as well. In this way, food labels can help consumers choose foods that specifically meet current dietary intake goals. The federal government continually evaluates and updates its food labeling requirements. Exemplifying this ever-evolving process is a new requirement that nutrient information is included on some raw meat packaging. This new law is slated to take effect in 2012.[6] You can learn

The *Buy Fresh Buy Local* Campaign

Today, many people are interested in purchasing and consuming foods grown and manufactured in their local areas. This trend has developed in response to a variety of factors including economic downturn, concerns about energy costs and the safety of foods produced in other countries, and a belief that locally grown foods might be more nutritious or better tasting than those that have been transported across the country. Although there are no federal guidelines defining what is meant by *local*, many cities and individual neighborhoods now host food cooperatives and other venues (such as farmers' markets) whereby local produce is labelled and sold. Even if you do not see the appeal of buying local, you might be interested in a food's country of origin. You can usually find this information by reading its food label.

© Food Routes

The **Buy Fresh Buy Local campaign** encourages consumers to purchase **locally produced** and manufactured goods.

Figure 2.8 Understanding Food Labels and Nutrition Facts Panels

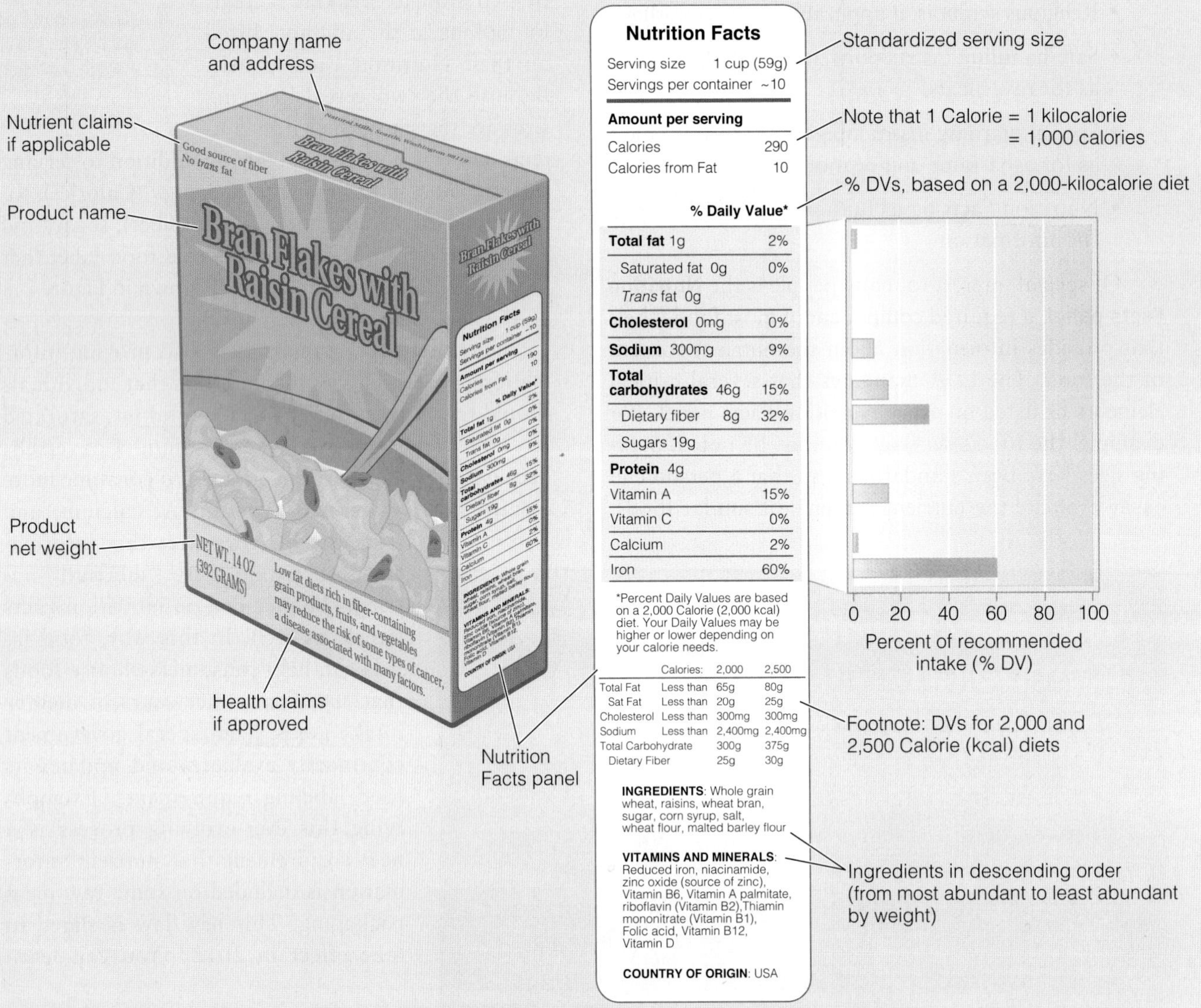

The FDA mandates that food labels provide specific information. A Nutrition Facts panel, required on most food labels, helps consumers choose their foods wisely.

Reading **food labels** *can* **help you choose** *the most nutritious products.*

more about food labels by visiting the FDA's website (http://www.cfsan.fda.gov/~dms/foodlab.html).

DAILY VALUES

Although Nutrition Facts panels provide an impressive amount of nutrition information, actually using this information to plan a diet can be somewhat challenging. For instance, a cereal box's Nutrition Facts panel may tell you how much vitamin C is in a serving of the cereal, but how do you know if that amount is a little or a lot? To help answer this question, Nutrition Facts panels incorporate an element called the

Daily Value (DV), which gives consumers a benchmark as to whether a food is a good source of a particular nutrient. Most importantly, Daily Values allow consumers to compare one food to another quickly and easily.

There are two basic types of DVs. The first type, used for select vitamins and minerals, represents a nutrient's recommended daily intake for a person who requires approximately 2,000 kcal/day. For example, vitamin C's DV is 60 mg/day. Thus, a cereal that provides 30 mg of vitamin C per serving fulfills half of the DV for vitamin C. Another example of this type of DV is evident in Figure 2.8. By reading the Nutrition Facts label, you can see that one serving of this cereal provides 15 percent of an average person's DV for vitamin A.

The second type of DV represents a nutrient's upper limit: it presents a daily amount that you should try *not to exceed.* This type of DV is used for total fat, saturated fat, cholesterol, sodium, and total carbohydrates. For example, one serving of the cereal illustrated in Figure 2.8 provides 2 percent of an average person's daily upper limit of total fat. If a food's package is big enough, actual upper limit DV amounts are provided in addition to percentages. As you can see in the lower portion of the food label in Figure 2.8, the saturated fat DV of a person who requires 2,000 kcal/day is 20 g. This amount increases to 25 g for a person who requires 2,500 kcal/day.

In addition to understanding the daily requirements and upper limits of nutrients in a food, you can use DVs to determine whether a food is a good source of a particular nutrient. A food that provides less than 5 percent of a nutrient's DV is considered low in that nutrient. Conversely, a food that provides at least 20 percent of a nutrient's DV is high in that nutrient. Thus, the cereal illustrated in Figure 2.8 is considered high in vitamin A and low in saturated fat.

Daily Value (DV) A benchmark as to whether a food is a good source of a particular nutrient. May represent a nutrient's recommended daily intake or upper limit.

nutrient content claim An FDA-regulated word or phrase that describes how much of a nutrient (or its content) is in a food.

health claim An FDA-approved statement that describes a specific health benefit of a food or food component.

Nutrient Content Claims and Health Claims

Food packaging often provides additional nutrition-related information that can help you plan a healthy diet. For example, a **nutrient content claim** describes in a very consumer-friendly way how much of a nutrient (or its *content*) is in a food. Nutrient content claims include phrases like "sugar free," "low sodium," and "good source of." Thus, if a person wanted to increase his fiber intake, he could consistently choose foods labeled "good source of fiber." The use of these claims is regulated by the FDA; a selection of approved definitions is provided in Table 2.5 on page 46.

Some food manufacturers also include information about potential health benefits of foods or food components on their products' packaging. This type of information is called a **health claim**. Health claims are quite different from nutrient content claims, which simply state that a food contains or does not contain a particular nutrient. Health claims must be supported by sufficient scientific evidence that increased consumption of a nutrient or food component impacts health positively. For example, in order to print "eating oatmeal significantly decreases

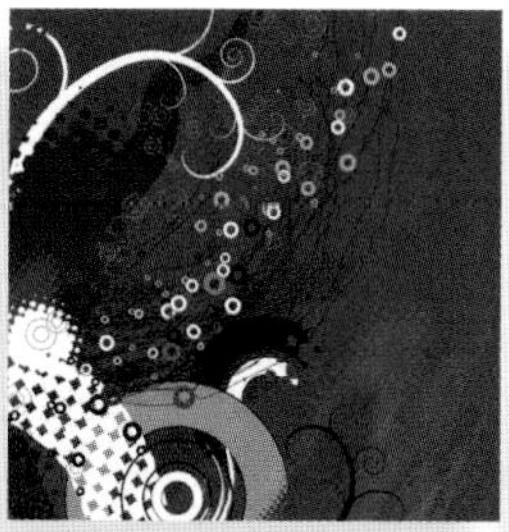

Why Don't All Foods Have Food Labels?

Although most purchased foods have food labels, you may have noticed that some do not. For instance, fresh fruits and vegetables, fish, meats, and poultry are currently not required to carry Nutrition Facts panels (although some packaged meats will be required to carry such labels beginning in 2012). Nonetheless, many grocers provide their customers with nutrition-related information voluntarily, often in the form of a poster or pamphlet. Other products such as gums and candies that come in very small packages are not required to carry nutrition labels either. To obtain nutrition information for these products, you frequently must contact the manufacturer or check the product's website.

regular health claim A health claim that is supported by considerable scientific research.

qualified health claim A health claim that has less scientific backing and must be accompanied by a disclaimer (or qualifier) statement.

your risk of heart disease"on a box of oatmeal, the manufacturer must first confirm that this health claim is supported by a series of conclusive experimental studies. Like other parts of a food's package, all health claims must be approved by the FDA.

Manufacturers can make two kinds of health claims: regular health claims and qualified health claims. Both kinds of claims concern the relationship between a specific food (or food component) and a health-related condition. However, while a **regular health claim** is supported by considerable research, a **qualified health claim** has less scientific backing and must be accompanied by a disclaimer (or qualifier) statement. You can usually tell that a health claim is qualified because it contains a statement such as, "However, the FDA has determined that this evidence is limited and not conclusive." As new research emerges, some new health claims are approved while others are disapproved.

LO6 Can You Put These Concepts into Action?

One benefit of studying nutrition is that much of what you learn can be applied directly to your own life. Remember that though dietary assessment is only one component of a complete nutritional assessment, it is an important first step toward a lifetime of health and nutritional awareness. You now have the information needed to assess your own diet and begin to choose the right foods to improve it. Consider the example of Jodi, a 21-year-old college student. In the following vignettes, Jodi uses the same basic knowledge you just learned to assess her diet and make changes that affect her health for the better.

Step 1: Set the Stage and Set Your Goals

After reading about dietary assessment, Jodi decides to make sure her own diet is adequate. Jodi is generally in good health, but has lately felt overly tired, even when she gets plenty of sleep. After doing some research, Jodi learns that iron and vitamin deficiencies might be responsible for her condition. As such, she wants to know whether her diet lacks adequate amounts of these or other nutrients.

Step 2: Assess Your Nutritional Status

First, Jodi examines her anthropometric measurements to determine whether there is any evidence of overall malnutrition. Jodi is 1.6 meters (5 feet, 3 inches) tall and weighs 61.4 kilograms (135 pounds). Because she is likely at a healthy body weight, Jodi concludes that she is probably not consuming too few or too many calories. However, because weight and height are not good indicators of overall nutritional adequacy, Jodi decides to conduct a dietary self-assessment using a three-day diet record and the MyPlate Food Tracker. She records everything she eats and drinks for three days, paying close attention to portion sizes. Jodi carefully notes every component of more complex foods such as the lasagna served in the cafeteria.

Next, Jodi logs on to the MyPlate website and enters her information into its database. Using this free

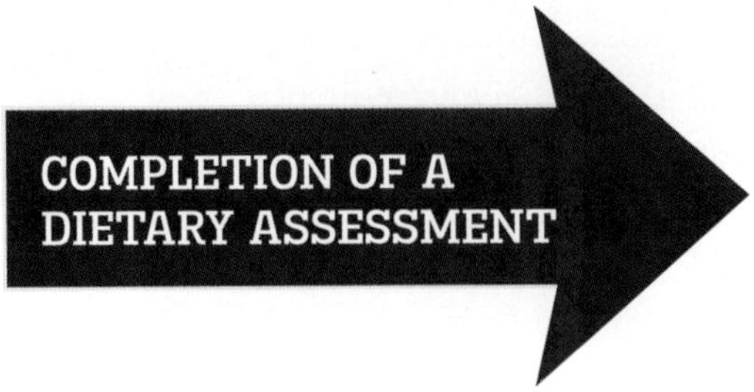

software, she is able to compare her dietary intake to the DRI values of all the required vitamins, minerals, and macronutrients, as well as to her EER. In addition, Jodi compares her dietary intakes of certain food groups to those recommended by the 2010 Dietary Guidelines for Americans, USDA Food Patterns, and MyPlate.

The results of Jodi's dietary analysis indicate that she consumes almost all of the necessary nutrients at levels above their AIs or between their RDAs and ULs. Further, Jodi's total energy intake is acceptable. However, her percentage of calories coming from fat is 40 percent—higher than recommended—and her daily fiber intake is below the recommended 20 to 25 g. Jodi's intake of iron is only 13 mg/day (72 percent of her RDA), and her vitamin B_{12} intake is only 1.9 μg/day (79 percent of her RDA). This suggests that Jodi may have inadequate intakes of fiber, iron, and vitamin B_{12}. Jodi does some research on why the body requires iron and vitamin B_{12} and is surprised to learn that both are needed for energy production—this might explain why she has been so tired. With this new knowledge, Jodi decides to increase her intakes of iron and vitamin B_{12}, decrease her fat intake, and increase her fiber consumption.

Doing what you can right now to improve your health will help you live a longer and healthier life.

© Sam Edwards/OJO Images/Getty Images

Step 3: Set the Table to Meet Your Goals

Using USDA nutrient composition databases, information on the MyPlate website, and a food composition table, Jodi learns that lean meats and fortified breakfast cereals are good sources of iron that do not supply high amounts of fat. She also learns that meat and dairy products contain vitamin B_{12}, and that fiber is found in whole-grain products, peas and lentils, and in some fruits and vegetables. Jodi begins to eat more of these foods and limit her total fat intake. She also begins to read Nutrition Facts panels on packaged foods and, when possible, choose foods with lower total fat contents. In the cafeteria, Jodi looks at cereal box labels and begins to eat high-fiber, fortified breakfast cereals instead of low-fiber, unfortified ones. (Fortified foods are those to which nutrients have been added during manufacturing.) When cooking for herself, Jodi chooses foods containing at least 20 percent of the DV for iron, looks for products that contain relatively large amounts of whole-grain components, and makes sure she eats enough lean meat and low-fat dairy products.

Step 4: Compare Your Plan and Your Assessment: Did You Succeed?

After a few weeks, Jodi wants to know whether the changes she made to her diet have improved her nutrient intake. She does another dietary self-assessment and learns that these simple dietary changes have resulted in adequate intakes of iron, vitamin B_{12}, and fiber and a reduced total fat intake. Jodi also notes that she has felt less tired and has been able to concentrate on her studies for longer periods of time. It appears that Jodi has succeeded, but just to make sure that she has not overlooked anything and that her overall health is good, she makes an appointment with her campus health care provider.

There Is No Time Like the Present

As evidenced by this example, dietary self-assessment and planning can be applied easily to anyone's life—including your own. Now that you know more about measuring your nutritional status and completing a dietary assessment, you can use the many reference standards (such as DRI values), dietary recommendations (such as the 2010 Dietary Guidelines), and nutrition tools (such as Nutrition Facts panels) to conduct your own assessment and choose a diet that fits your personal nutritional needs. Some professors require students to conduct a dietary self-assessment as a class assignment. If this is the case for you, then you will garner firsthand experience with and understanding of these procedures. If not, you are enthusiastically advised to do so on your own.

Whether required or optional, a dietary self-assessment will make the rest of this course more meaningful because you will be able to apply the insights you gain to your own diet and overall health. Experts agree that the food habits you establish now will not only affect your success in college but also influence your eating patterns and health for years to come. Take the time to set yourself up for success by eating right and being as healthy as possible.

Table 2.5 Selected FDA-Approved Nutrient Content Claims

Wording	Description
"Light" or "Lite"	If 50 percent or more of the regular product's calories are from fat, fat must be reduced by at least 50 percent, as compared to the regular product. If less than 50 percent of the regular product's calories are from fat, fat must be reduced by at least 50 percent, or calories must be reduced by at least one-third, as compared to the regular product.
"Reduced Calories"	Product contains at least 25 percent fewer calories per serving than a regular product.
"Calorie Free"	Product contains less than 5 kcal (Calories) per serving.
"Fat Free"	Product contains less than 0.5 g fat per serving.
"Low Fat"	Product contains 3 g or less fat per serving.
"Saturated Fat Free"	Product contains less than 0.5 g saturated fat and less than 0.5 g *trans* fatty acids per serving.
"Low in Saturated Fat"	Product contains 1 g saturated fat per serving and derives 15 percent or fewer calories from saturated fat.
"Cholesterol Free"	Product contains less than 2 mg cholesterol per serving. Cholesterol claims are only allowed if a food contains 2 g or less saturated fat per serving.
"Low in Cholesterol"	Product contains 20 mg or less cholesterol per serving.
"Sodium Free"	Product contains less than 5 mg sodium per serving.
"Low in Sodium"	Product contains 140 mg or less sodium per serving.
"Sugar Free"	Product contains less than 0.5 g sugar per serving. This does not include alcohol sugars.
"High," "Rich in," or "Excellent Source of"	Product contains 20 percent or more of the Daily Value per serving. This wording may describe protein, vitamins, minerals, dietary fiber, or potassium.
"Good Source of," "Contains," or "Provides"	Product contains 10 to 19 percent of the Daily Value per serving.
"More," "Added," "Extra," or "Plus"	Product contains 10 percent or more of the Daily Value per serving. This wording may describe protein, vitamins, minerals, dietary fiber, or potassium.
"Fresh"	A raw food that has not been frozen, heat processed, or otherwise preserved.
"Fresh Frozen"	Food was quickly frozen while still fresh.

Source: Adapted from U.S. Department of Health and Human Services and U.S. Food and Drug Administration. A food labeling guide—Appendix B. Available from: http://www.cfsan.fda.gov/~dms/flg-6b.html.

3 Body Basics

© Richard Bradbury/Taxi/Getty Images

LEARNING OUTCOMES:

LO1 Internalize the connection between chemistry and nutrition.

LO2 Compare and contrast cells, tissues, organs, and organ systems.

LO3 Describe the process of digestion.

LO4 Describe the process of absorption.

LO5 Understand the circulation of nutrients throughout the body.

LO6 Define *metabolism* and understand its role.

LO1 Why Learn about Chemistry When Studying Nutrition?

To satisfy its nutritional needs, your body extracts nutrients from thousands of complex foods. The first step in this process takes place in the gastrointestinal tract, where food is systematically broken down into its most basic components. Only once they are broken down can nutrients leave the gastrointestinal tract and circulate through the extensive network of blood and lymph vessels that make up your circulatory systems. Nutrients absorbed by the cells in your body undergo amazing chemical transformations that ultimately sustain your life. To ensure that these activities take place under optimal conditions, nonstop communication networks—your endocrine and nervous systems—orchestrate these events. In this chapter, you will learn about the chemical and physiological events that take place every time you eat, and you will gain an appreciation for the intricate and varied tasks required to nourish your body.

Chemistry is fundamental to the study of nutrition. Not only are nutrients chemicals themselves, but the body's utilization of nutrients involves a vast number of chemical reactions. The organization of atoms into molecules, molecules into macromolecules, macromolecules into cells, cells into tissues, tissues into organs, and organs into organ systems is indeed remarkable (see Figure 3.1 on the next page). This entire system is made of and fueled by the nutrients contained in food. In order to appreciate your body's life-sustaining functions, it is important first to have a basic understanding of chemistry—the science of matter.

Atoms Make Up the World around You

It is difficult to imagine the existence of something that you cannot see, taste, touch, or hear. Yet the **atom**, though invisible to the human eye, is the fundamental unit that makes up the world around you. Many atoms have an equal number of positively and negatively charged particles, and therefore are neutral. This is not the case for all atoms, however. Depending on the type and number of particles it contains, an atom could have a positive electrical charge or a negative electrical charge. A charged atom is called an **ion**. Ions serve many vital functions in the body. For example, ions such as calcium (Ca^{2+}) and magnesium (Mg^{2+}) are required for muscle contraction and nerve function. Note the + signs—these denote positively charged ions. Other important ions found in the human body include sodium (Na^{+}), potassium (K^{+}), chloride (Cl^{-}), iodide (I^{-}), and fluoride (F^{-}). The − signs of the latter three ions denote that these are negatively charged (while the former two are positively charged).

atom The fundamental unit that makes up the world around us.

ion An atom that has a positive or negative electrical charge.

electrolyte A molecule that when submerged in water separates into individual ions.

Ions with opposite charges are often attracted to one another, a phenomenon that can result in a type of chemical bonding. For example, the ions sodium (Na^{+}) and chloride (Cl^{-}) are always found together in food: because they are oppositely charged, they form a chemical bond, creating sodium chloride (NaCl), which you know better as table salt. You may have heard the term *electrolyte* before. Although the terms *ion* and *electrolyte* are often used interchangeably, they do not actually mean the same thing. An **electrolyte** is a

AP Images/Jon Elswick

Figure 3.1 Levels of Organization in the Body

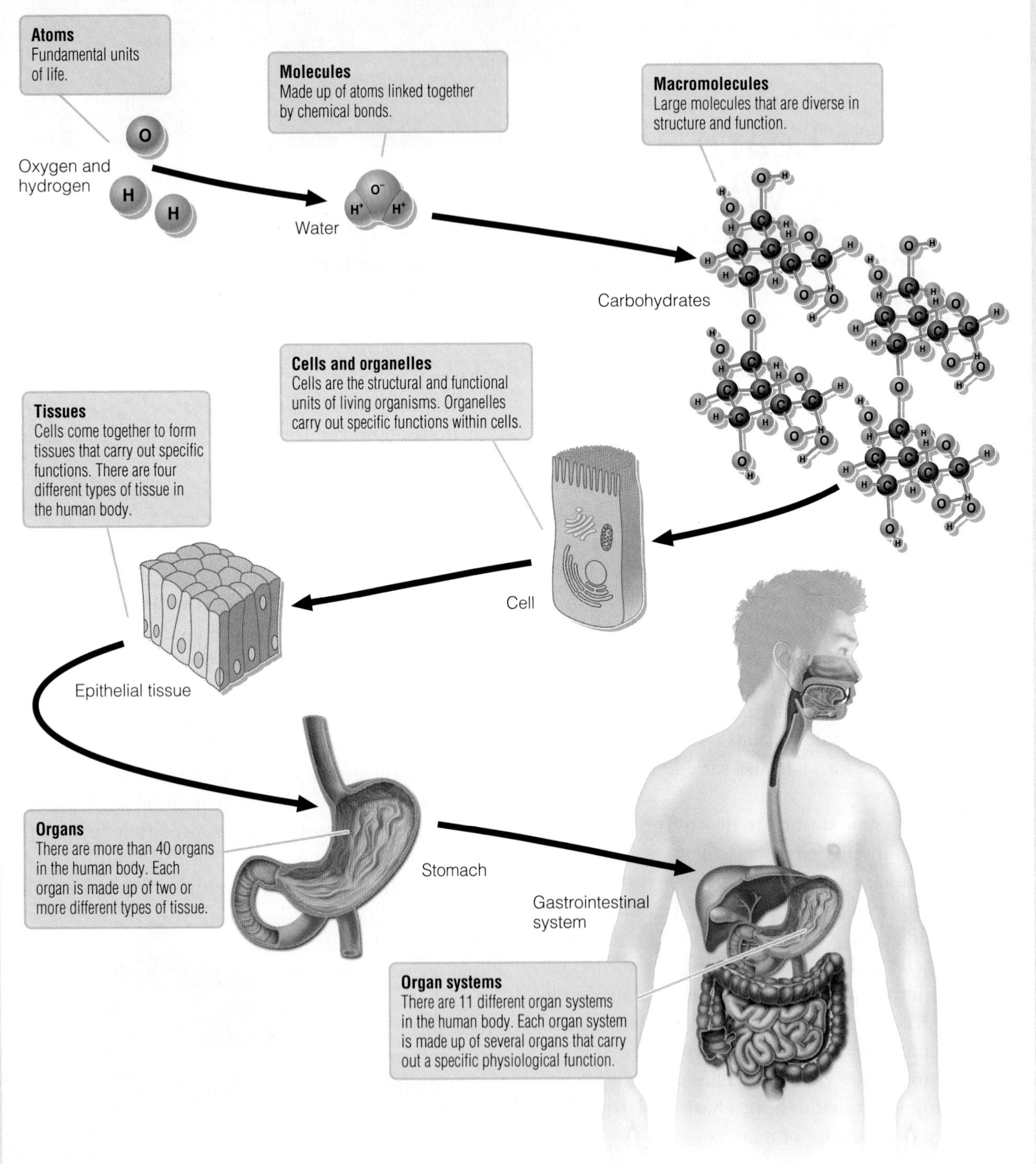

molecule that when submerged in water separates into individual ions. When this happens, the molecule is *ionized*. Because excessive sweating can cause a depletion of water and electrolytes from the body, athletes are often encouraged to drink beverages that contain both of these important components. In fact, the sports drink Gatorade® was developed by researchers at the University of Florida to help the school's football players—the Gators—sustain healthy fluid–electrolyte balances while exercising.

Elements and Molecules

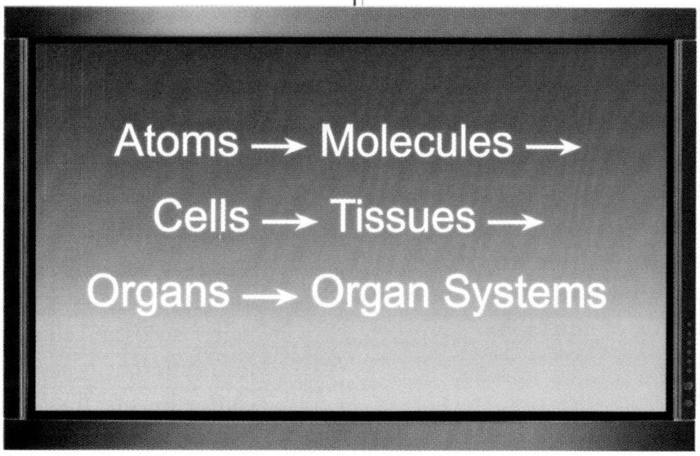

An **element** is a pure substance made up of only one type of atom. There are approximately 92 naturally occurring elements, 20 of which are essential to human health. Perhaps surprisingly, just six elements—carbon, oxygen, hydrogen, nitrogen, calcium, and phosphorus—account for 99 percent of your total body weight. These important elements provide the raw materials needed to form the molecules essential to living systems.

A **molecule** is formed when two or more atoms join together by one or more chemical bonds. Amazingly, such bonds enable a relatively small number of atoms to form millions of different molecules. Without chemical bonds, the molecular world would fall apart. In fact, you can think of chemical bonds as the glue that holds atoms together within a molecule. Many molecules are small and simple. However, molecules can also be very large: some consist of thousands of atoms bonded together. Carbohydrates, lipids, proteins, and nucleic acids (DNA and RNA) are large, complex molecules (or *macromolecules*) that are vital to the functions of cells. The elements that comprise these large molecules come from the nutrients in foods that you eat.

element A pure substance made up of only one type of atom.

molecule A unit of two or more atoms joined together by chemical bonds.

molecular formula A representation of the number and type of atoms present in a molecule.

UNDERSTANDING A MOLECULAR FORMULA

A **molecular formula** is a representation of the number and type of atoms present in a molecule. For example, glucose, an important source of energy in your body, has a molecular formula of $C_6H_{12}O_6$. This formula tells you that one molecule of glucose consists of six carbon (C_6), 12 hydrogen (H_{12}), and six oxygen (O_6) atoms. You may recognize the molecular formula for water, H_2O. Water is comprised of two hydrogen atoms (H_2) and one oxygen (O) atom. Note that when no number follows an element's symbol, it means that there is only one atom of that type of element present. When more than one molecule of a substance is present, the total number of molecules is placed before the molecular formula. For example, the molecular formula for three molecules of water is written $3H_2O$ (see Figure 3.2).

LO2 How are Cells, Tissues, Organs, and Organ Systems Related?

Molecules such as carbohydrates, proteins, lipids, water, and nucleic acids are basic to life as we know it. However, these molecules in and of themselves do not always function in useful ways. Rather, some molecules

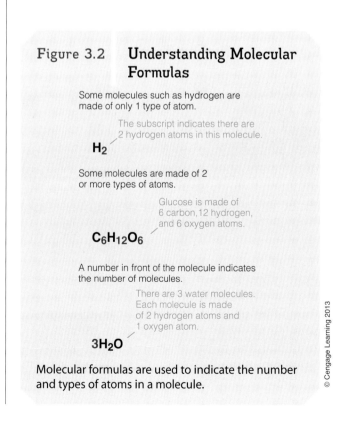

Figure 3.2 Understanding Molecular Formulas

Some molecules such as hydrogen are made of only 1 type of atom.

H_2 — The subscript indicates there are 2 hydrogen atoms in this molecule.

Some molecules are made of 2 or more types of atoms.

$C_6H_{12}O_6$ — Glucose is made of 6 carbon, 12 hydrogen, and 6 oxygen atoms.

A number in front of the molecule indicates the number of molecules.

$3H_2O$ — There are 3 water molecules. Each molecule is made of 2 hydrogen atoms and 1 oxygen atom.

Molecular formulas are used to indicate the number and types of atoms in a molecule.

cell A structural and functional unit that makes up body tissues.

passive transport mechanism A transport mechanism that does not require energy to move a substance across a cell membrane.

active transport mechanism A transport mechanism that requires energy to move a substance across a cell membrane.

osmosis The diffusion of water across a cell membrane.

simple diffusion A passive transport mechanism whereby a substance moves from a region of higher concentration to a region of lower concentration without using energy or the assistance of a transport protein.

facilitated diffusion A passive transport mechanism whereby a substance moves from a region of higher concentration to a region of lower concentration with the assistance of a carrier molecule.

function as building blocks, forming a structural and functional unit called a **cell.** Cells make up tissues, which in turn function as building blocks for organs. Finally, organs work together as components of organ systems. The human body has 11 organ systems, all of which are pertinent to the study of nutrition.

Passive and Active Transport Mechanisms

Cells are like microscopic cities—full of activity. Cell structures subsist in a gel-like substance called cytoplasm, which is surrounded by a protective membrane that regulates the movement of substances into and out of the cell. The movement of nutrients and other substances across a cell membrane occurs through a variety of processes, which are referred to collectively as *transport mechanisms*. A transport mechanism that does not require energy (ATP) is called a **passive transport mechanism**, whereas one that does require energy is called an **active transport mechanism.**

PASSIVE TRANSPORT MECHANISMS

As illustrated in Figure 3.3, there are three types of passive transport mechanisms: osmosis, simple diffusion, and facilitated diffusion. **Osmosis** is the diffusion of water across a cell membrane. Fluids in the body contain different types of dissolved substances. The concentrations of dissolved substances inside and outside the cell determine whether water crosses into or out of the cell. Specifically, water moves toward a higher concentration of dissolved substances, effectively equalizing a given substance's concentration on both sides of the cell membrane.

Simple diffusion enables a substance to cross a cell membrane from a region of higher concentration to a region of lower concentration without using energy in the process. In simple diffusion, a substance moves without assistance, like a raft floating downstream. In **facilitated diffusion**, a substance moves passively from a higher to a lower concentration, but only with the assistance of a carrier molecule. As a motorboat pulls a raft downstream, the carrier molecule escorts the substances across the cell membrane.

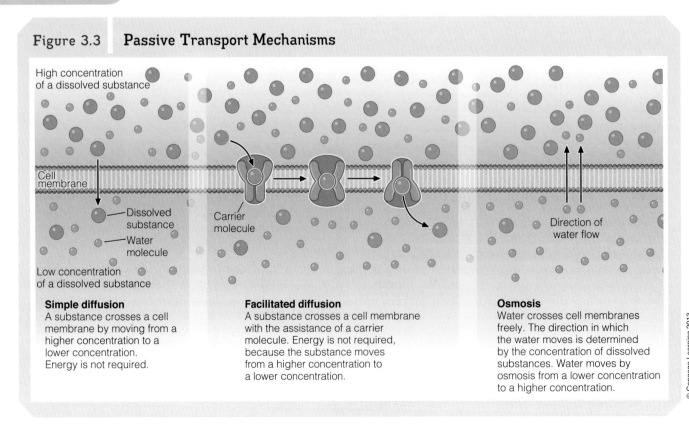

Figure 3.3 Passive Transport Mechanisms

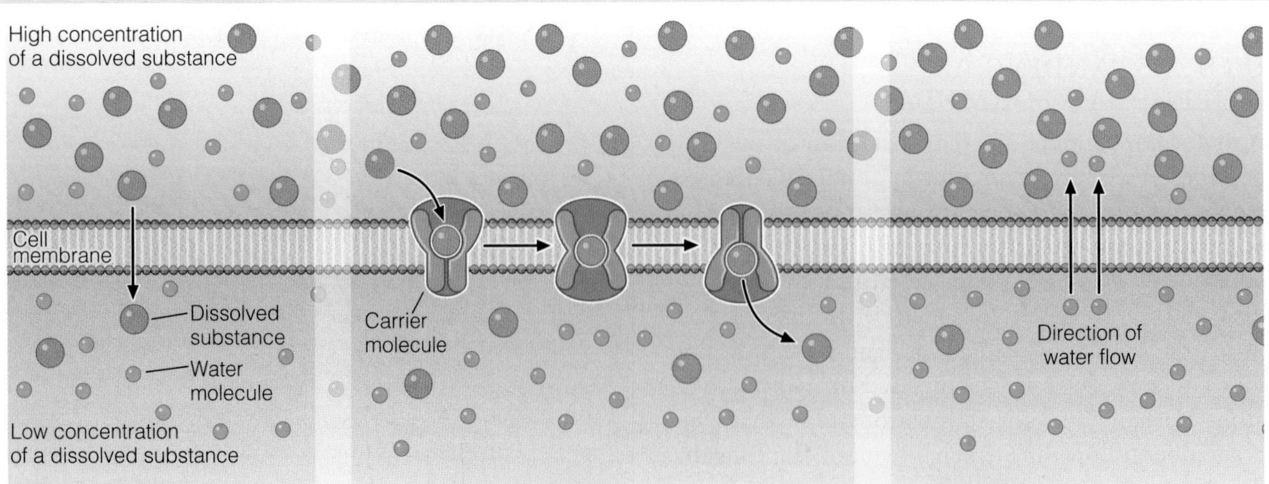

Simple diffusion
A substance crosses a cell membrane by moving from a higher concentration to a lower concentration. Energy is not required.

Facilitated diffusion
A substance crosses a cell membrane with the assistance of a carrier molecule. Energy is not required, because the substance moves from a higher concentration to a lower concentration.

Osmosis
Water crosses cell membranes freely. The direction in which the water moves is determined by the concentration of dissolved substances. Water moves by osmosis from a lower concentration to a higher concentration.

ACTIVE TRANSPORT MECHANISMS

Some substances must cross cell membranes against the prevailing concentration gradient, moving from a region of lower concentration to one of a higher concentration. Because energy is needed to pump the molecule across a cell membrane, this type of transport mechanism is considered active. **Carrier-mediated active transport** is a transport mechanism whereby a substance uses energy to move from a region of lower concentration to a region of higher concentration with the assistance of a carrier protein. This process is similar to a motorboat towing a raft upstream—because it is fighting against the current, it has to use energy (see Figure 3.4).

Cell Organelles

Within every cell, small structures carry out specialized functions that are critical to life. Each structure, or **organelle**, is responsible for a specific function within the cell. Some organelles produce substances necessary to cellular activity, while others function as waste-disposal systems, assisting the degrading and recycling of worn-out cellular components. Organelles called *mitochondria* serve as power stations, converting the chemical energy of energy-yielding nutrients (carbohydrates, lipids, and proteins) into energy that is usable by cells (ATP). Another organelle, the *nucleus*, houses the genetic material (DNA) that provides the blueprint for protein synthesis. Figure 3.5 on the next page provides an overview of a cell and its components.

Figure 3.4 Carrier-Mediated Active Transport

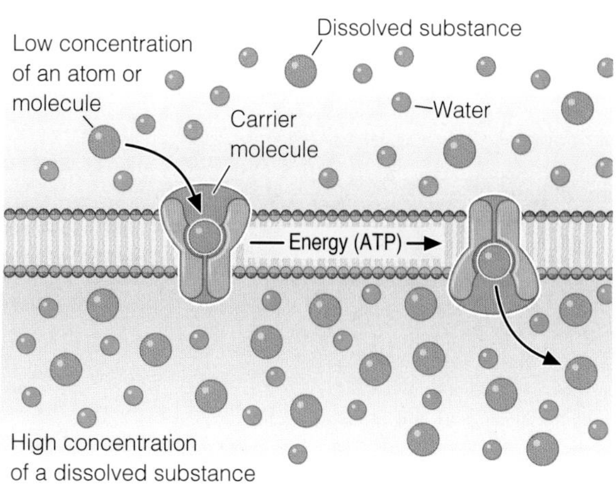

A substance crosses a cell membrane with the assistance of a carrier protein. Energy (ATP) is required, because the substance moves from a lower concentration to a higher concentration.

The Four Types of Tissue

Following the molecule and the cell, the next level of complexity is the tissue. A **tissue** is formed when cells of similar structure and function group together to accomplish a common task. The human body contains four different types of tissue:

- Epithelial
- Connective
- Muscle
- Neural

Your skin and the inner lining of your organs are made of **epithelial tissue**, which helps to protect the body. **Connective tissue** supports, connects, and anchors body structures—it is the glue that holds the body together. Tendons, cartilage, and some parts of bones are examples of connective tissue. The body contains several types of **muscle tissue**, all of which are used for movement. *Skeletal muscle*, a type of muscle tissue that you can control voluntarily, allows you to move the various parts of your body. There are approximately 640 different skeletal muscles in the human body that are under your active control. *Smooth muscle* tissue, which you cannot control voluntarily, is embedded within the epithelial lining of organs such as the esophagus, stomach, and small intestine. Smooth muscle is even found in the interior lining of blood vessels; it is this layer of smooth muscle that allows vessels to constrict (become narrower) and relax (become wider). Because smooth muscles are not under our conscious control, they are referred to as involuntary muscles—their movement is regulated in part by the nervous system. The final type of tissue is **neural tissue**, which makes up the brain, spinal cord, and nerves. Neural tissue plays an important communicative role in the body. The four types of tissue are illustrated in Figure 3.6 on the next page.

carrier-mediated active transport An active transport mechanism whereby a substance moves from a region of lower concentration to a region of higher concentration with the assistance of a carrier protein and energy.

organelle A structure that is responsible for a specific function within a cell.

tissue An aggregation of similarly structured and functioning cells that have grouped together to accomplish a common task.

epithelial tissue Tissue that helps protect the body.

connective tissue Tissue that supports, connects, and anchors body structures.

muscle tissue Tissue that is used for movement.

neural tissue Tissue that facilitates communication throughout the body.

organ Two or more different types of tissue functioning together to perform a variety of related tasks.

Organs and Organ Systems

In the human body, the four types of tissue comprise more than 40 organs, which in turn make up 11 unique organ systems. An **organ** consists of two or more different types of tissue functioning together to perform a variety of related tasks. An *organ system* is formed when several organs work together: each organ carries out its own

Figure 3.5 A Typical Cell

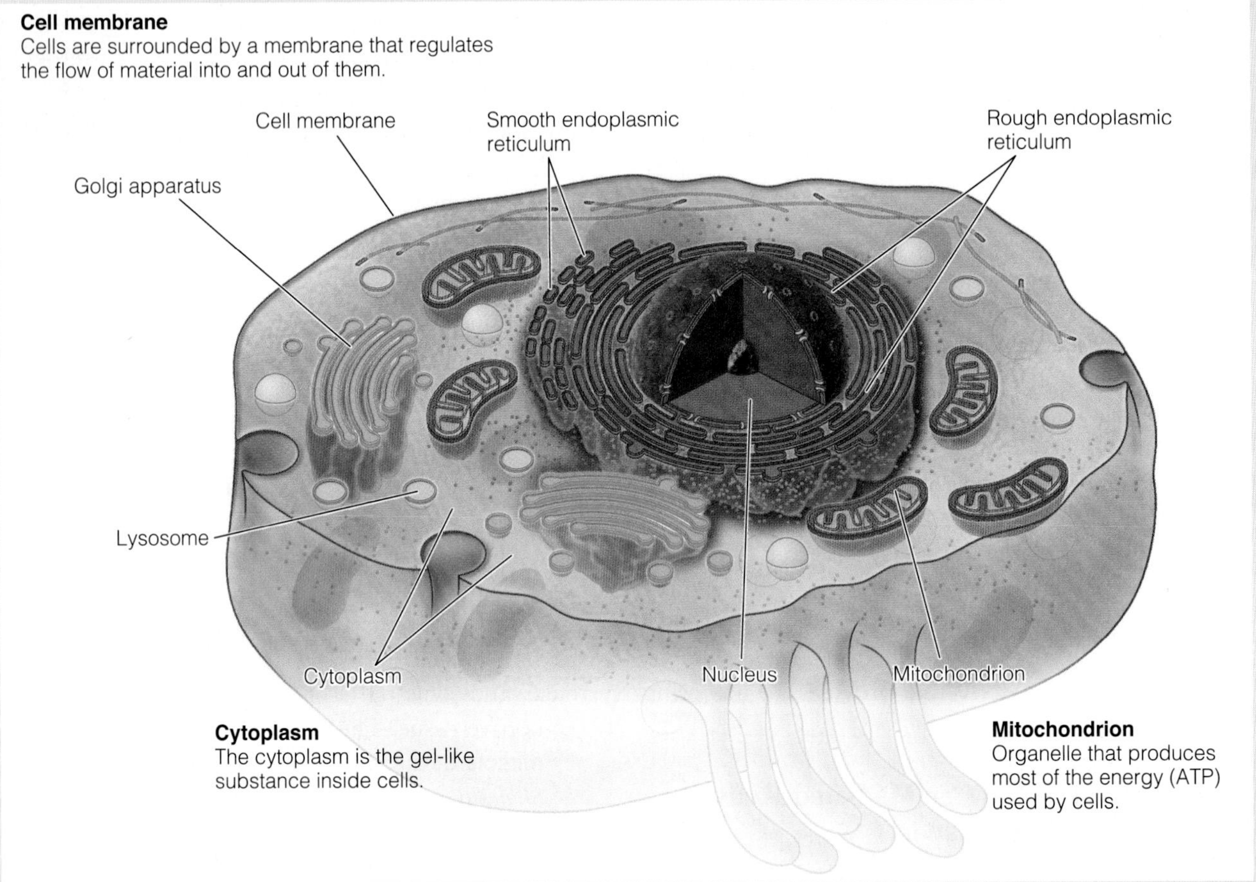

Figure 3.6 Four Basic Types of Tissue

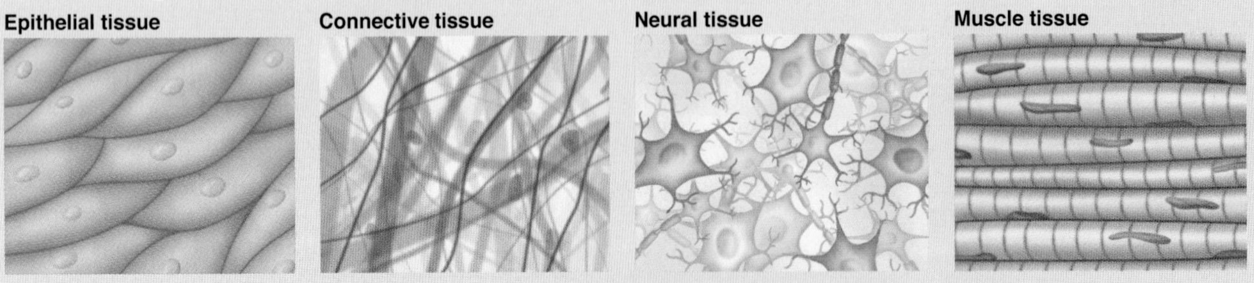

Epithelial tissue covers and lines body surfaces, organs, and cavities.

Connective tissue provides structure to the body by binding and anchoring body parts.

Neural tissue plays a role in communication by receiving and responding to stimuli.

Muscle tissue contracts and shortens when stimulated, playing an important role in movement.

Epithelial, connective, neural, and muscle tissue make up all the organs in the human body.

important physiological function, but each aids in the system's greater purpose. For example, the digestive system is composed of several organs that work together to break down food. The major organ systems and their basic functions are summarized in Table 3.1.

ORGAN SYSTEMS WORK TOGETHER TO MAINTAIN BALANCE

The ability of organs and organ systems to work together requires constant communication within the body. The right hand must know what the left hand is doing, so to speak. The body has two well-developed communication systems that coordinate physiologic functions: the nervous system and the endocrine system. The nervous system communicates information via nerve cells, whereas the endocrine system communicates by releasing chemical messengers into the blood. A chemical messenger released by the endocrine system is called a **hormone**. Together, the nervous and endocrine systems monitor our internal environment continuously, respond to change, and restore balance when necessary. These mechanisms allow you to adapt in an ever-changing, complex environment so that you can maintain homeostasis. **Homeostasis**, a state of balance or equilibrium, is an important

hormone A chemical messenger released into the blood by the endocrine system.

homeostasis A state of balance or equilibrium.

Table 3.1 Organ Systems and Related Major Functions

Organ System	Major Organs and Structures	Major Functions
Integumentary	Skin, hair, nails, and sweat glands	Protects against pathogens and helps regulate body temperature.
Skeletal	Bones, cartilage, and joints	Provides support and structure to the body. The bone marrow of some bones produces blood cells. Also provides a storage site for certain minerals.
Muscular	Smooth, cardiac, and skeletal muscle	Assists in voluntary and involuntary body movements.
Nervous	Brain, spinal cord, nerves, and sensory receptors	Interprets and responds to information. Controls the basic senses, movement, and intellectual functions.
Endocrine	Endocrine glands such as pituitary gland, thyroid gland, and adrenal glands	Produces and releases hormones that control physiological functions such as reproduction, hunger, satiety, blood glucose regulation, metabolism, and stress response.
Respiratory	Lungs, nose, mouth, throat, and trachea	Governs gas exchange between the blood and air.
Circulatory	Heart, blood vessels, blood, lymph vessels, lymph nodes, and lymph organs	Transports nutrients, waste products, gases, and hormones.
Digestive	Mouth, esophagus, stomach, small intestine, large intestine, liver, gallbladder, pancreas, and salivary glands	Governs the physical and chemical breakdown of food into chemicals that can be absorbed into the blood. Eliminates solid waste.
Reproductive	Gonads and genitals	Carries out reproductive functions and is associated with sexual characteristics, sexual function, and sexual behaviors.
Urinary	Kidneys, bladder, and ureters	Removes metabolic waste products from the blood and regulates water balance.
Immune	Lymph vessels, bone marrow, and lymphatic tissue	Protects against foreign bodies (such as bacteria and viruses) and helps regulate cell division and cell death.

lumen The cavity that spans the entire length of the GI tract.

transit time The amount of time it takes for food to travel the entire length of the GI tract.

accessory organ An organ that is not part of the digestive tract but nonetheless plays an important role in digestion and absorption.

digestion The physical and chemical breakdown of food into a form that allows nutrients to be absorbed.

absorption The transfer of nutrients from the GI tract into the blood or lymph.

elimination The process whereby solid waste is removed from the body.

peristalsis A vigorous, wave-like muscular contraction that propels food from one region of the GI tract to the next.

concept in nutrition. Even everyday events such as eating can disrupt the body's delicate, internal balance. For example, when a carbohydrate-rich meal causes the level of glucose in one's blood to rise, the pancreas releases the hormone insulin. This response lowers blood glucose to normal levels and restores homeostasis.

LO3 What Happens during Digestion?

In order to nourish your body, your digestive system methodically disassembles the complex molecules of food into simpler basic nutrient components. While this arduous task requires many organ systems, it is primarily the digestive system that gets the job done. After it is consumed, food travels through various regions of the digestive tract, undergoing physical and chemical transformations—a journey that takes 24 to 72 hours. By the time it reaches the small intestine, food barely resembles the substance you consumed. Each region of the GI tract makes its own unique contribution to the digestive process, which comes to completion when nutrients are able to move out of the digestive tract and into the blood or lymph. This complex process is explained in detail throughout the following sections.

The Digestive System

Your digestive system comprises both the digestive tract and its accessory organs. The digestive or gastrointestinal (GI) tract can be thought of as a hollow tube that runs from the mouth to the anus (see Figure 3.7). The inner cavity that spans the entire length of the GI tract is called the **lumen**. Organs that make up the GI tract include the mouth, esophagus, stomach, small intestine, and large intestine. The amount of time it takes for food to travel the entire length of the GI tract is called **transit time**.

An organ that is not part of the digestive tract but nonetheless plays an important role in the digestive process is referred to as an **accessory organ**. Accessory organs include the salivary glands, pancreas, liver, and gallbladder. Together, the GI tract and accessory organs carry out three important functions:

1. **Digestion**, the physical and chemical breakdown of food into a form that allows nutrients to be absorbed.
2. **Absorption**, the transfer of nutrients from the GI tract into the blood or lymph.
3. **Elimination**, the process whereby solid waste is removed from the body.

MOVEMENT AND SECRETIONS AID THE PROCESS OF DIGESTION

The process of digestion takes place in a highly regulated and organized manner. Like the conductor of an orchestra, neural and hormonal signals coordinate the movement of food through the GI tract. A vigorous, wave-like muscular contraction called **peristalsis** propels the food from one region of the GI tract to the next (see Figure 3.8 on page 58). The rate of peristalsis increases

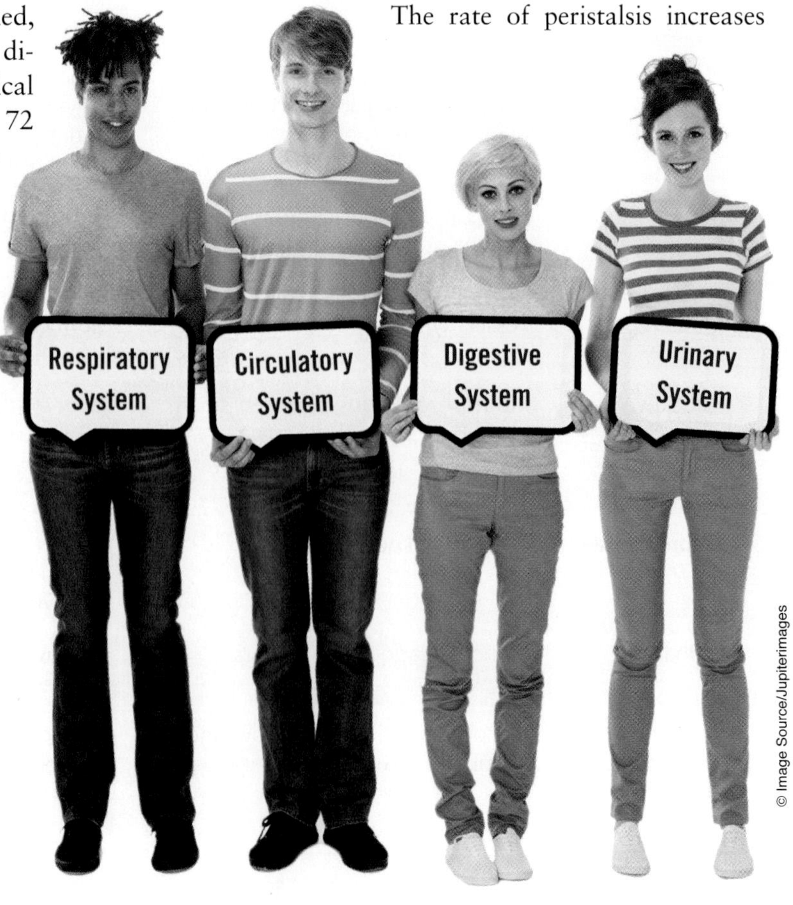

and decreases to ensure that the food mass moves along the GI tract at the appropriate rate. At certain points throughout the GI tract, a circular band of muscle called a **sphincter** regulates the flow of food. Sphincters act like one-way valves, opening and closing in response to neural and hormonal signals. Relaxation of the sphincter muscle allows the food to flow forward. Once the food reaches the next organ, the sphincter closes to prevent the food from flowing backward (see Figure 3.9 on the next page).

The innermost lining of the digestive tract, the **mucosa**, consists mainly of epithelial tissue that

sphincter A circular band of muscle that regulates the flow of food through the GI tract.

mucosa The innermost lining of the gastrointestinal tract.

Figure 3.7 **Organs of the Digestive System**

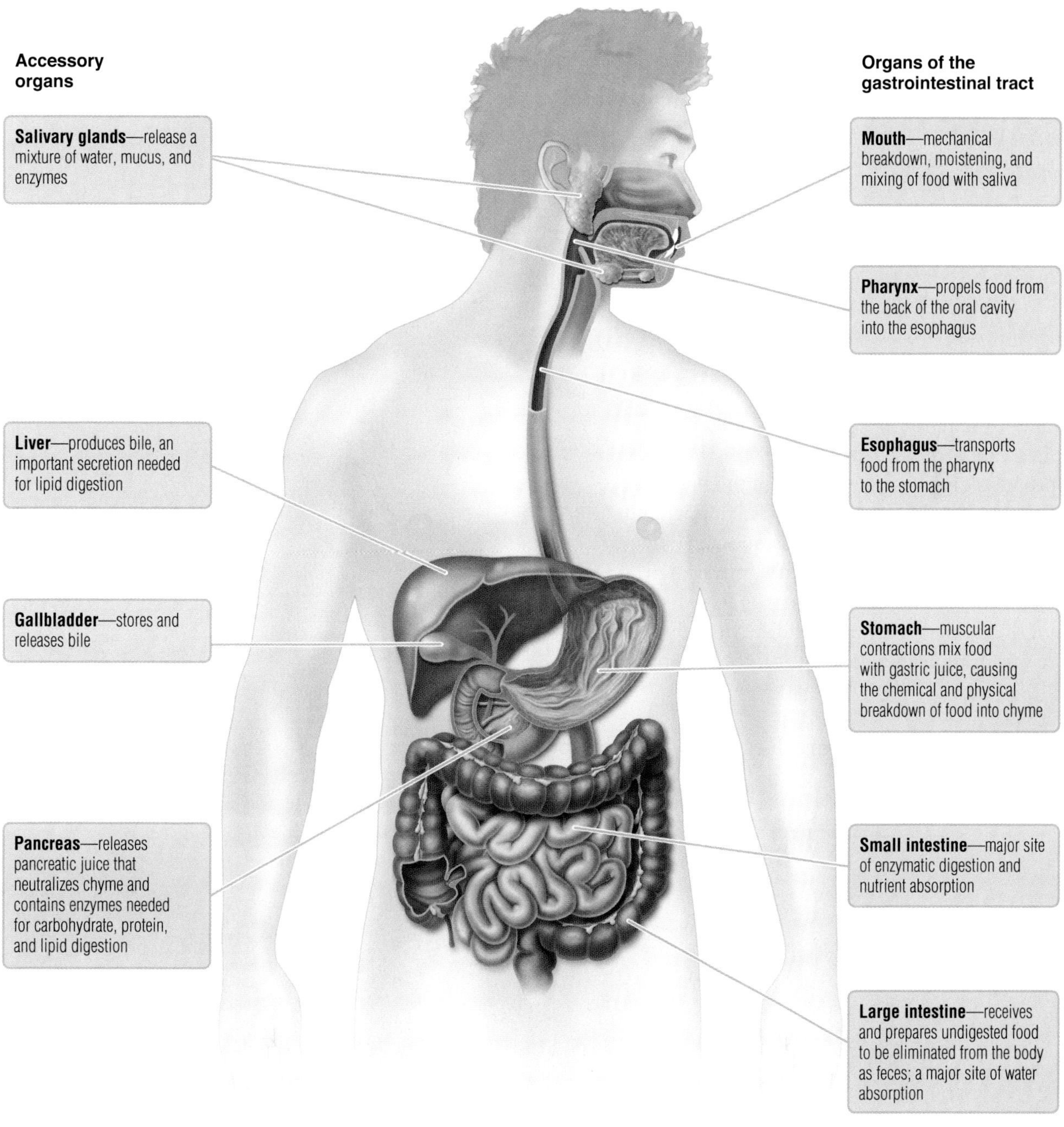

Accessory organs
Salivary glands—release a mixture of water, mucus, and enzymes
Liver—produces bile, an important secretion needed for lipid digestion
Gallbladder—stores and releases bile
Pancreas—releases pancreatic juice that neutralizes chyme and contains enzymes needed for carbohydrate, protein, and lipid digestion
Organs of the gastrointestinal tract
Mouth—mechanical breakdown, moistening, and mixing of food with saliva
Pharynx—propels food from the back of the oral cavity into the esophagus
Esophagus—transports food from the pharynx to the stomach
Stomach—muscular contractions mix food with gastric juice, causing the chemical and physical breakdown of food into chyme
Small intestine—major site of enzymatic digestion and nutrient absorption
Large intestine—receives and prepares undigested food to be eliminated from the body as feces; a major site of water absorption

The digestive system consists of the gastrointestinal tract and the accessory organs.

enzyme A biological catalyst that accelerates a chemical reaction.

saliva A secretion released into the mouth by the salivary glands that moistens food and starts the physical process of digestion.

produces and releases a variety of substances that facilitate the process of digestion. An **enzyme**, for example, is a biological catalyst that accelerates chemical reactions. Enzymes released into the GI tract facilitate reactions that aid in the chemical breakdown of nutrients. In addition to various enzymes, some epithelial cells in the mucosal lining release hormones that regulate the rate at which food moves through the GI tract. Another important substance that aids digestion is mucus, which keeps the inner lining of the GI tract moist and provides protection from irritating (and sometimes harmful) substances.

Digestion Begins in the Mouth

In truth, the process of digestion begins even before food enters your mouth—the mere thought, smell, and sight of food serve as a wake-up call to the digestive system. Sometimes called the *cephalic phase*, this response optimizes digestion by stimulating the muscles that move food through the digestive tract and by triggering the release of substances that help break food down chemically. The physical process of digestion begins when food enters the mouth and the forceful grinding action of teeth breaks food into manageable pieces. The presence of food in the mouth also stimulates the release of **saliva** from the salivary glands. As food is broken apart, it mixes with saliva. This makes food moist and easier to swallow and allows the digestive enzymes in

Figure 3.8 Peristalsis

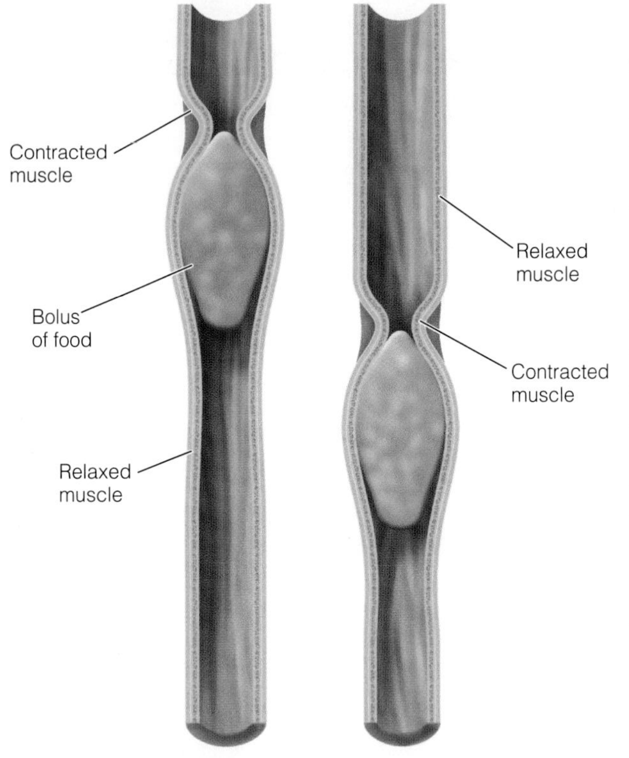

Peristalsis consists of a series of wavelike rhythmic muscular contractions and relaxation. This action propels food forward through the GI tract.

Figure 3.9 Sphincters Regulate the Flow of Food

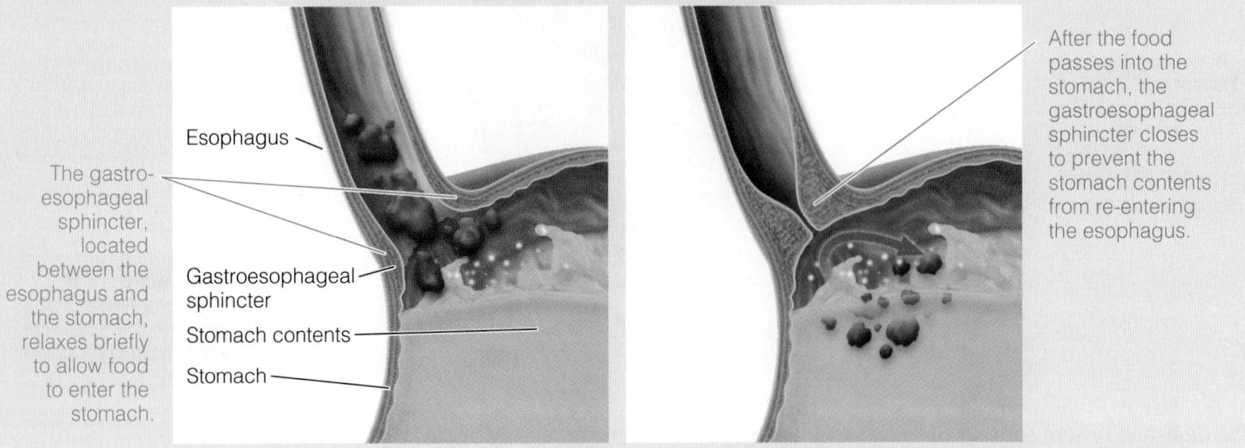

Sphincters are circular bands of muscles located throughout the gastrointestinal tract that regulate the flow of food.

saliva to begin the process of chemical digestion. Saliva serves another useful purpose—it helps you to taste your food. Some flavor-containing components of food must be dissolved in saliva before they can be detected by taste buds. The dissolved flavor components stimulate both taste buds on the tongue's surface and cells that line the nasal cavity, sending signals to the brain that enable it to distinguish the thousands of different flavors you enjoy in your food. This is why food can be difficult to taste when you are congested.

> "The **salivary glands** produce as **much as a liter** (roughly a quart) of saliva **per day**."

Food Moves from the Mouth to the Esophagus

The tongue is a powerful muscle that assists in chewing and swallowing. As food mixes with saliva, the tongue manipulates the food mass and pushes it up against the hard, bony palate that makes up the roof of your mouth. As illustrated in Figure 3.10 on the next page, swallowing takes place in two phases. First, as you prepare to swallow, your tongue directs the soft, moist mass of chewed food, now referred to as a **bolus**, to the region toward the back of your mouth, an area known as the **pharynx**. The pharynx is the shared space between the oral and nasal cavities. This first phase of swallowing occurs under voluntary control, but once the bolus reaches the pharynx, the second (involuntary) phase of swallowing begins. At this point, the bolus is ready to enter the **esophagus**, a narrow muscular tube that ends at the stomach.

bolus A soft, moist mass of chewed food.

pharnyx A region toward the back of the mouth that serves as the shared space between the oral and nasal cavities.

esophagus A narrow muscular tube that begins at the pharynx and ends at the stomach.

Cleft Palate: A Common Birth Defect

An infant born with a cleft palate often has difficulty swallowing because the bones that make up the hard palate (the roof of the mouth) are not formed properly, resulting in an opening between the oral and nasal cavities. Babies born with cleft palates often cannot suck and swallow normally, and therefore have problems feeding. Fortunately, a cleft palate can be corrected surgically shortly after birth. Until surgery can be performed however, special care must be given to ensure that the infant is adequately nourished.

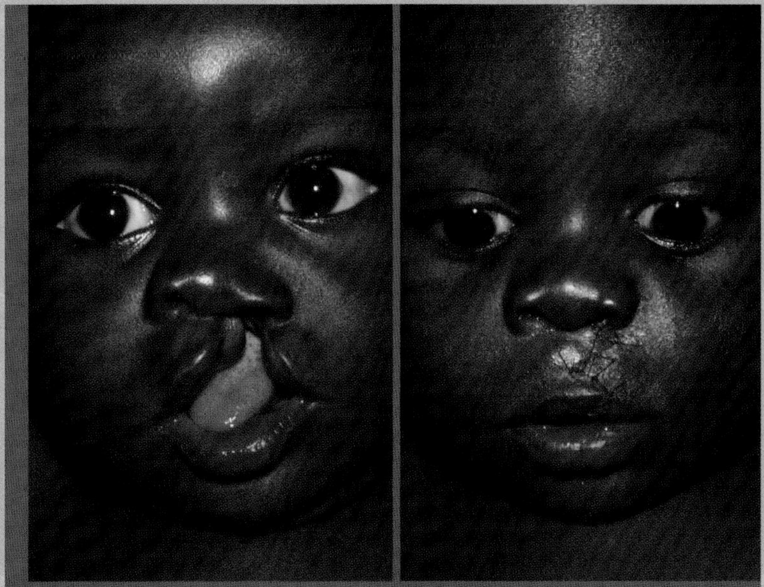

A cleft palate (left) occurs when the bones that make up the hard palate do not completely fuse together. This birth defect can be corrected surgically, often leaving minimal visible signs (right).

Figure 3.10 **Voluntary and Involuntary Phases of Swallowing**

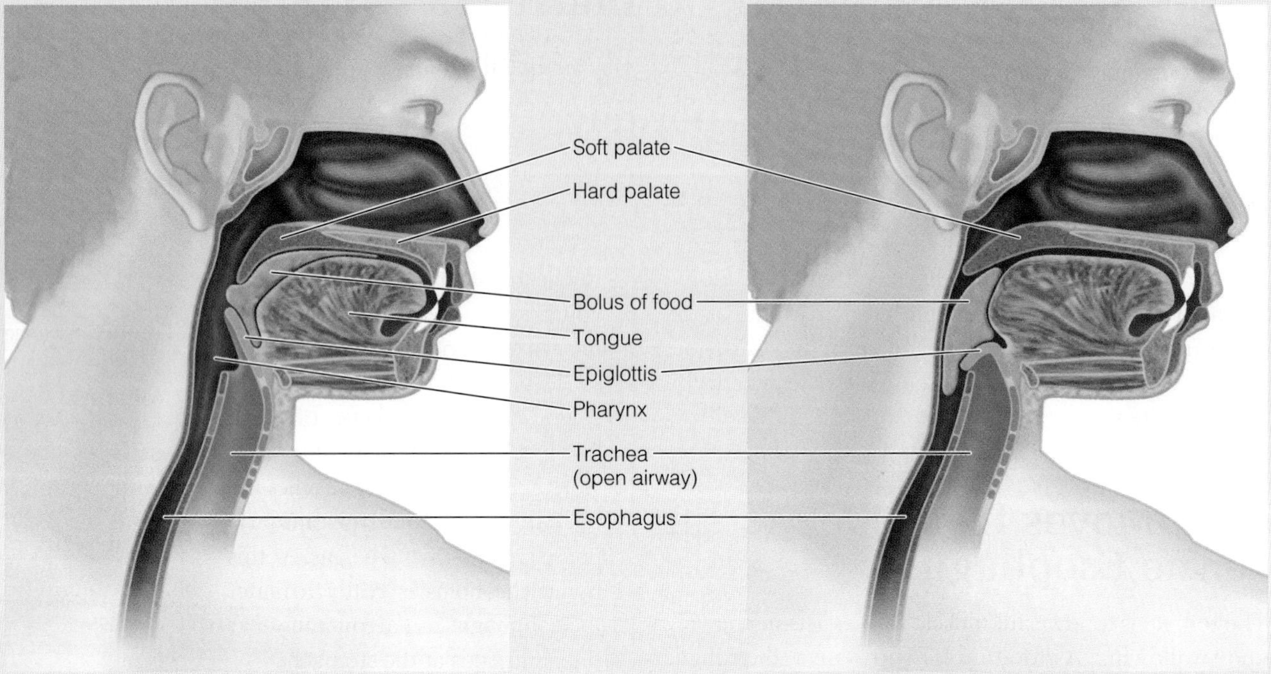

A. Voluntary phase
The tongue pushes the bolus of food against the hard palate. Next, the tongue pushes the bolus against the soft palate, which triggers the swallowing response.

B. Involuntary phase
The soft palate rises, which prevents the bolus from entering the nasal cavity. The epiglottis covers the trachea, blocking the opening to the lungs. The bolus enters the esophagus and is propelled toward the stomach by peristalsis.

The first phase of swallowing is under our control (voluntary) whereas the second phase is not (involuntary).

The Esophagus Delivers Food to the Stomach

During the involuntary phase of swallowing, the upper-back portion of the mouth lifts upward, blocking the entrance to the nasal cavity. This helps guide the bolus into the correct passageway—the esophagus. This movement also causes the *epiglottis*, a cartilage flap, to cover the *trachea*, the airway leading to the lungs, so that the bolus cannot enter the lungs. Once the bolus moves past this dangerous intersection, the muscles relax and prepare for the next swallow.

gastroesophageal sphincter A circular muscle that regulates the flow of food from the esophagus to the stomach.

gastroesophageal reflux disease (GERD) A condition caused by chronic reflux of the stomach contents into the esophagus, irritating the lining.

The esophagus is lubricated and protected by a thin layer of mucus, which facilitates the passage of food. Peristalsis propels the food through the esophagus toward the stomach, where the bolus encounters the first of several sphincters in the GI tract, the **gastroesophageal sphincter.** As the bolus moves toward the end of the esophagus, the gastroesophageal sphincter relaxes long enough for it to pass into the stomach. Once this occurs, the sphincter closes, preventing the contents of the stomach from reentering the esophagus. The entire trip from the pharynx to the stomach takes only 6 to 10 seconds.

GASTROESOPHAGEAL REFLUX DISEASE

The gastroesophageal sphincter is sometimes unable to prevent the contents of the stomach from reentering the esophagus; this phenomenon is referred to as *gastric reflux* (see Figure 3.11). Over time, chronic reflux can result in a condition called **gastroesophageal reflux disease (GERD),** whereby the lining of the esophagus becomes irritated, causing a burning sensation in the chest (commonly referred to as heartburn).

Figure 3.11 **Gastroesophageal Reflux Disease (GERD)**

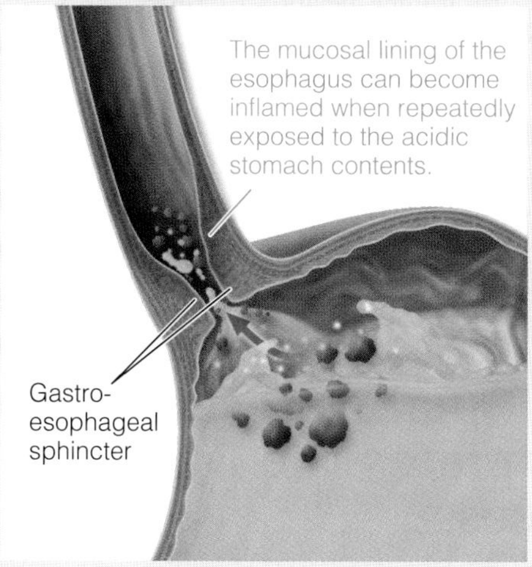

Gastroesophageal reflux disease (GERD) is caused by dysfunction of the gastroesophageal sphincter. If the gastroesophageal sphincter weakens, the stomach contents flow back into the esophagus. The lining of the esophagus can become inflamed when it is exposed to the contents of the stomach.

Fortunately, most people are able to manage heartburn by changing their lifestyles. Avoiding large meals and certain types of foods (such as caffeinated beverages, mint, and fried foods) are often effective prevention and management strategies. If left untreated, GERD can lead to serious conditions such as inflammation of the esophagus. For this reason, it is important for anyone experiencing GERD to seek medical attention and treatment.[1]

The Stomach Mixes and Stores Food

The stomach is uniquely equipped to carry out two important functions: (1) mixing food with the gastric secretions that aid in chemical digestion and (2) temporarily storing food. Specialized cells in the lining of the stomach produce a variety of substances that aid in the process of digestion. These digestive secretions, collectively referred to as **gastric juice**, consist mainly of water, hydrochloric acid, digestive enzymes, mucus, and intrinsic factor, a substance essential to vitamin B_{12} absorption.

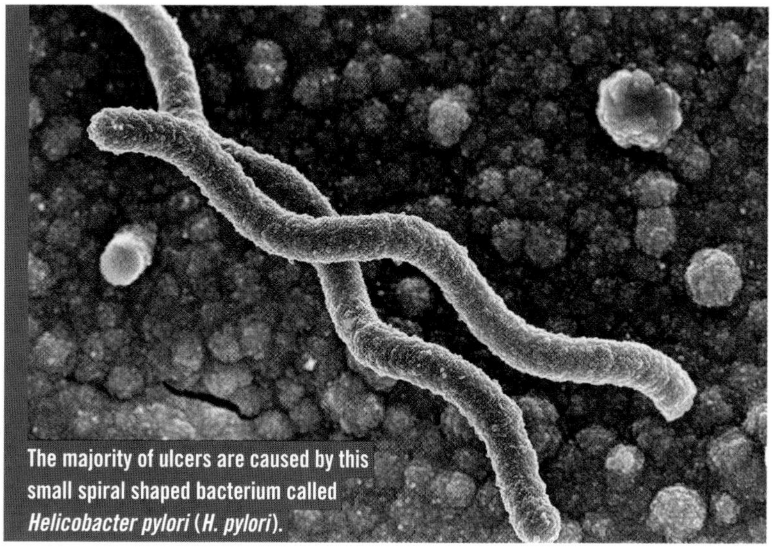

The majority of ulcers are caused by this small spiral shaped bacterium called *Helicobacter pylori* (*H. pylori*).

Your stomach produces more than 2 liters (roughly 2 quarts) of gastric juice daily. While the hydrochloric acid and digestive enzymes in gastric juice would normally damage the tissue that forms the stomach, a thick mucus layer protects the stomach lining from irritation. Damage to this protective mucus layer can result in inflammation or *gastritis*, which can subsequently lead to the formation of ulcers. Contrary to popular belief, the majority of ulcers are not caused by stress or by eating spicy foods. Most ulcers are caused by a small, spiral-shaped bacterium called *Helicobacter pylori* (*H. pylori*), but some medications (such as aspirin) can cause them as well.[2]

gastric juice Digestive secretions that consist mainly of water, hydrochloric acid, digestive enzymes, mucus, and intrinsic factor.

gastrin A hormone that stimulates release of gastric juice and causes the muscular wall of the stomach to contract vigorously.

chyme A semi-liquid paste resulting from the mixing of partially digested food with gastric juice in the stomach.

The presence of food in the stomach stimulates the release of a hormone called **gastrin**, which is produced by special cells in the stomach's lining. Gastrin stimulates the release of gastric juice and causes the muscular wall of the stomach to contract vigorously. These powerful muscular contractions, much like the action of kneading bread, force the bolus to mix with the acidic gastric juice. Within three to five hours after eating a meal, the partially digested food is mixed thoroughly with the gastric juice. By the time the bolus leaves the stomach, it has been transformed into a semi-liquid paste called **chyme**.

It is important that all the digestive events that are supposed to occur in the stomach do so before chyme enters the small intestine. To make sure that this happens, the stomach serves as a temporary storage

pyloric sphincter A circular muscle that regulates the flow of chyme from the stomach into the small intestine.

gastric emptying The process by which food leaves the stomach and enters the small intestine.

duodenum The first segment of the small intestine.

facility. As your stomach fills with food, its walls expand (much like an accordion). This stretching triggers a neural response, signaling the brain that the stomach is becoming full. In turn, the brain sends a signal that hunger has been satisfied, causing a person to stop eating. Feeling this sensation of fullness is an important aspect of regulating food intake.

The **pyloric sphincter**, located at the base of the stomach, regulates the flow of chyme into the small intestine. With each peristaltic wave, a few teaspoons of chyme squeeze through the sphincter as it opens briefly. The remaining chyme tumbles back and forth in the stomach, allowing for even more mixing. This slow release of chyme allows the small intestine to prepare for the processes that occur therein.

A number of factors influence the rate of **gastric emptying**—the process by which food leaves the stomach and enters the small intestine. The consistency of a food, for example, is an important factor. Because solid foods take more time to liquefy than soft foods, they must remain in the stomach longer. Even the nutrient composition of a food impacts the rate of gastric emptying. In general, high-fat foods make a person feel full for longer periods of time.[3] This is because high-fat foods take more time to digest, and therefore occupy the stomach longer than low-fat foods do. Beyond consistency and nutrient composition, volume, or how much a person eats, can also influence the rate of gastric emptying. Large meals increase the strength of peristaltic contractions, which in turn increase the rate of gastric emptying.

The Small Intestine Completes the Digestion Process

The small intestine, a narrow, 20-foot-long tube with a diameter of about 1 inch, is the primary site of chemical digestion and nutrient absorption. After leaving the stomach, chyme passes into the first segment of the small intestine, the **duodenum**. The duodenum also receives secretions from the gallbladder and the pancreas—more on those later (see Figure 3.12).

Figure 3.12 **Overview of the Small Intestine and Accessory Organs**

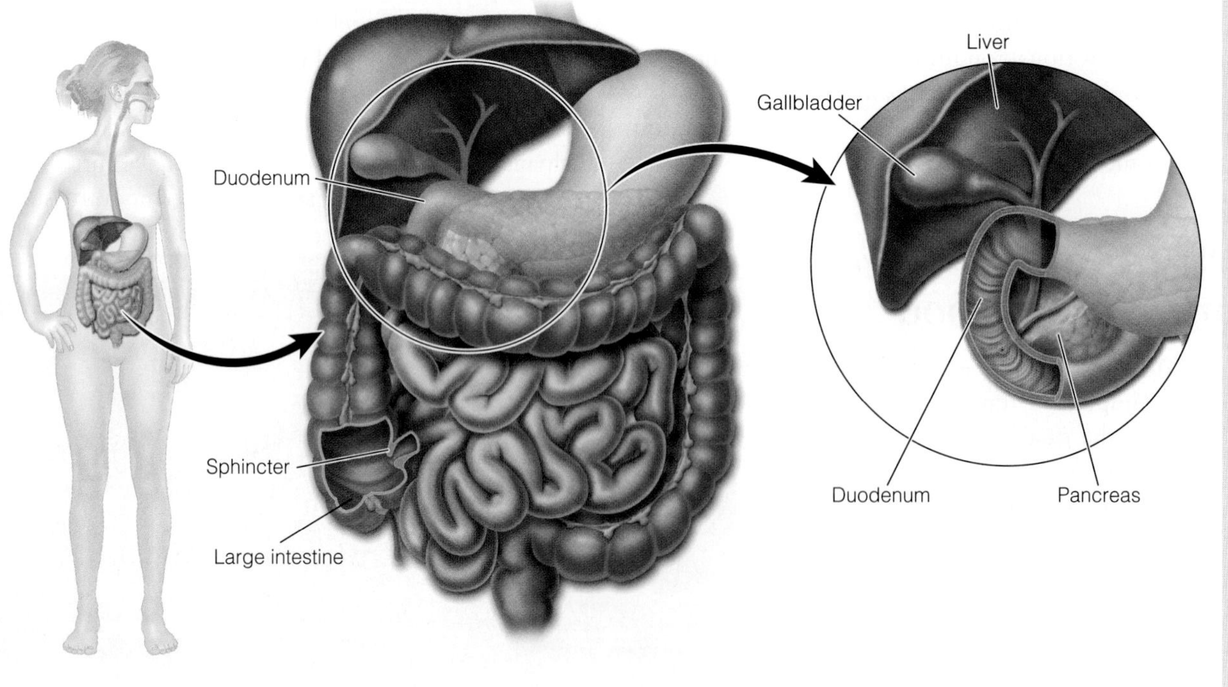

The accessory organs (liver, gallbladder, and pancreas) work closely with the small intestine to facilitate the process of nutrient digestion.

The lining of the small intestine has a large surface area, which is aptly suited for the processes of digestion and nutrient absorption. As illustrated in Figure 3.13, the inner lining of the small intestine forms large rounded folds that face inward (toward the lumen of the small intestine). These rounded folds are covered with small finger-like projections. Blood and lymph flow into each such **villus** through a vast network of blood capillaries and lymphatic vessels.

Another way to think about the inner lining of the small intestine is to imagine a bathroom rug folded like an accordion. The folds in the rug represent the large, rounded folds, whereas each tiny loop that covers the surface of the rug represents a villus. Villi (the plural form of villus) consist of specialized epithelial cells that are covered on their lumenal

villus A small finger-like projection. Villi cover the inner surface of the small intestine.

Figure 3.13 Absorptive Surface of the Small Intestine

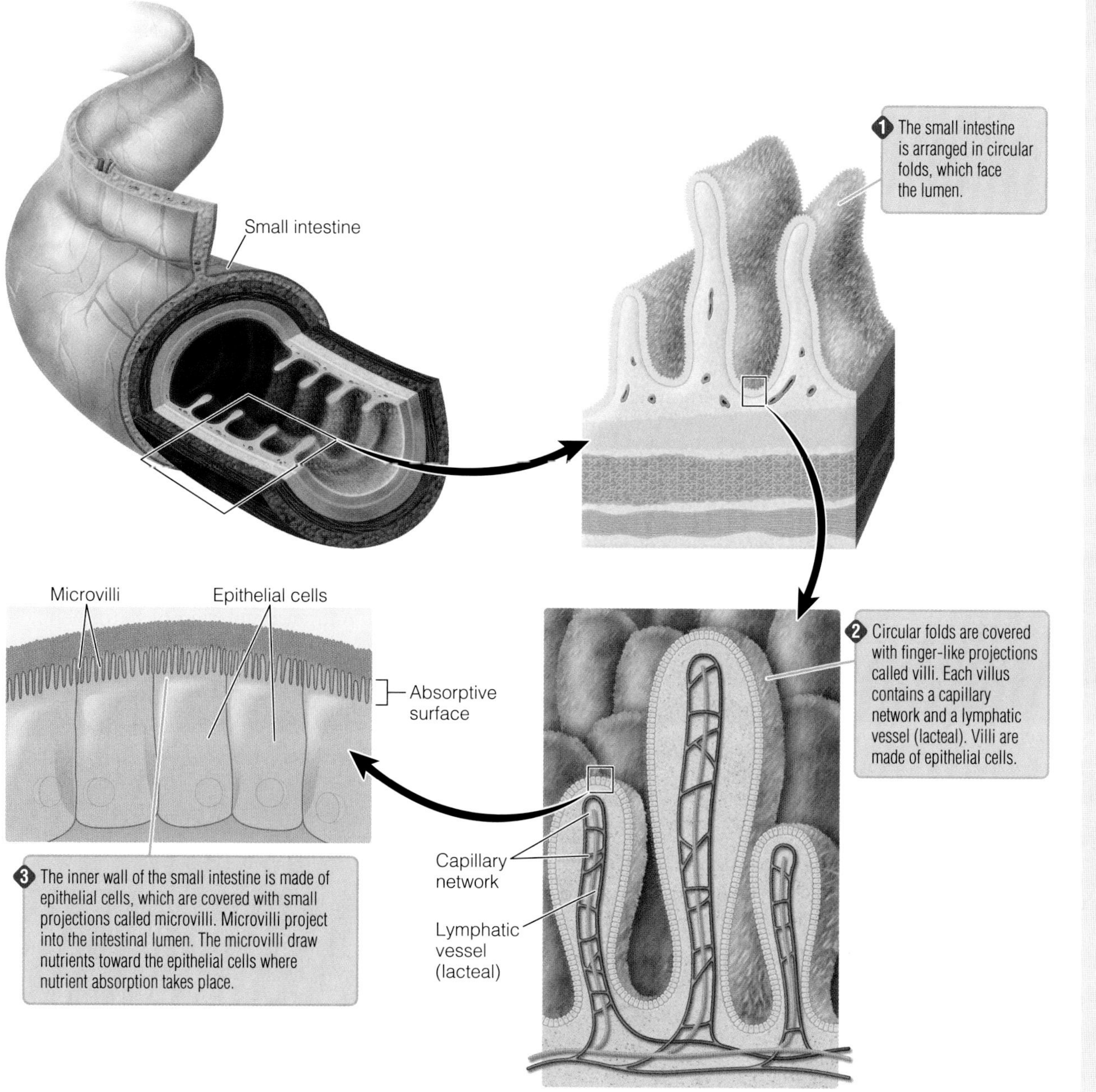

(outward) surfaces with thousands of minute projections, each of which is called a **microvillus**. It is here, on the microvilli (the plural form of microvillus), where the final stages of nutrient digestion take place.

The small intestine is a source of digestive enzymes. These enzymes facilitate the chemical reactions that break food particles down into the smallest components yet. To aid in this process, the small intestine releases hormones that help regulate digestion by coordinating the release of secretions from the pancreas and gallbladder with the relaxation of sphincters and peristalsis—all with great precision. These actions ensure that nutrient digestion and absorption in the small intestine are rapid and efficient. Indeed, within 30 minutes of the arrival of chyme in the small intestine, the final stages of digestion are usually complete.

Figure 3.14 The Pancreas Has an Important Role in Nutrient Digestion

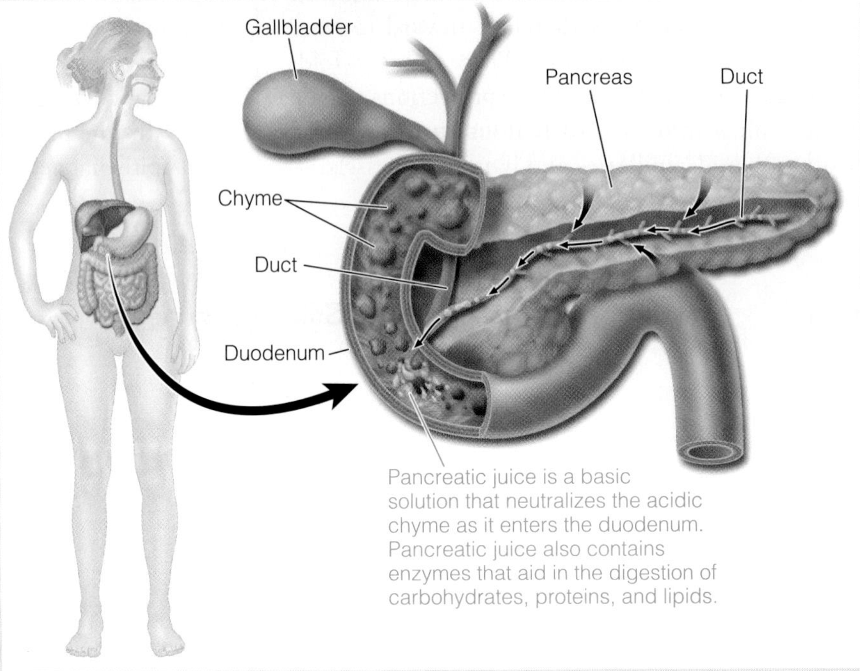
Gallbladder
Pancreas
Duct
Chyme
Duct
Duodenum
Pancreatic juice is a basic solution that neutralizes the acidic chyme as it enters the duodenum. Pancreatic juice also contains enzymes that aid in the digestion of carbohydrates, proteins, and lipids.

The Pancreas and Gallbladder Play Important Roles in Digestion

microvillus A tiny finger-like projection. Microvilli cover the lumenal surfaces of the epithelial cells that line the villi in the small intestine.

pancreatic juice A mixture of water, bicarbonate, and various enzymes released by the pancreas.

bile A fluid produced in the liver and released by the gallbladder that disperses large globules of fat into smaller droplets that are easier to digest.

The pancreas plays an important role: it protects the small intestine from the acidity of chyme (see Figure 3.14). The arrival of chyme in the small intestine stimulates the pancreas to release **pancreatic juice**, a mixture of water, bicarbonate, and various enzymes necessary to digestion. Being a particularly strong base, the bicarbonate quickly neutralizes the acidic chyme as it enters the duodenum. Another fluid that plays an important role in digestion—especially for those who consume a lot of fatty foods—is **bile**. Although bile is produced in the liver, it is stored in the gallbladder for quick release into the small intestine. When high-fat foods are consumed, the gallbladder releases bile, which acts like a detergent by dispersing large globules of fat into smaller droplets that are easier to digest.

GALLSTONES

The formation and accumulation of hard, pebble-like deposits inside the gallbladder, referred to as *gallstones*, can interfere with the normal flow of bile. Some people with gallstones have no symptoms, while others experience extreme pain. When gallstones block the

> "The combined surface area of the **villi** and **microvilli** in one human body is approximately the **size of a standard tennis court**."

passage of bile on its way from the gallbladder to the small intestine, the gallbladder can become enlarged and inflamed, causing pain. Although the exact cause of gallstone formation is not clear, it is more common in women than men, and the risk of developing gallstones increases with age. Other risk factors associated with gallstone formation include obesity, rapid weight loss, and pregnancy. Often, surgical removal of the gallbladder is the only way to treat the painful accumulation of gallstones. Since bile is necessary to fat digestion, you may wonder how people get by without their gallbladders. Although a person may experience difficulty with fat digestion after his gallbladder is removed, the liver continues to produce bile, which is still released into the small intestine.

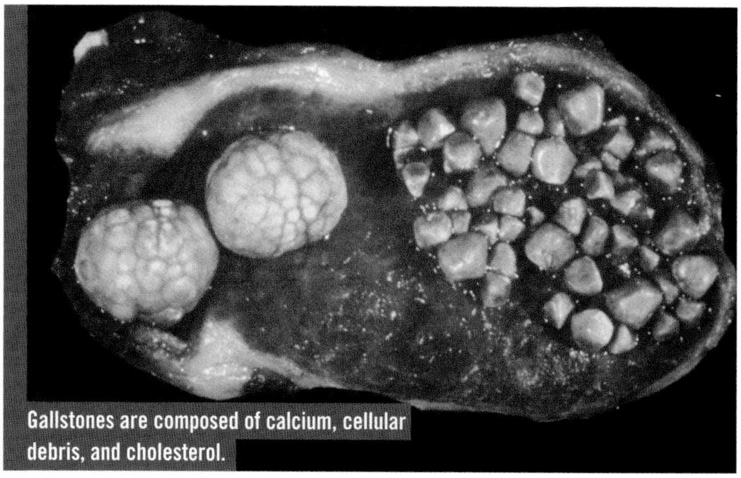

Gallstones are composed of calcium, cellular debris, and cholesterol.

LO4 Nutrient Absorption: What Happens after Digestion?

When the process of digestion is complete, nutrients are ready to be absorbed. Nutrient absorption is the transfer of nutrients from the lumen of the GI tract into either the blood or the lymph (see Figure 3.15 on the next page). Although some nutrient absorption takes place through the lining of the stomach, the vast majority of nutrients are absorbed in the small intestine. This is one reason why diseases that affect the absorptive surface of the small intestine such as celiac disease can lead to nutritional deficiencies. In the small intestine, microvilli

Celiac Disease

Celiac disease is an inflammatory response to gluten, a specific protein found in a variety of cereal grains such as wheat, rye, barley, and possibly oats. When a person with celiac disease consumes gluten-containing foods, the lining of the small intestine becomes damaged. Researchers only came to understand celiac disease and how to treat it within the last 50 years. They now know that people with celiac disease (or gluten-sensitive enteropathy) experience an immunological response to gluten. More specifically, consumption of gluten triggers the production of antibodies that attack the intestinal microvilli, causing them to flatten. As a result, nutrient absorption is impaired. Because of the progressive damage done to the inner lining of the small intestine, people with celiac disease often experience severe signs and symptoms such as diarrhea, weight loss, and malnutrition. Because its signs and symptoms are similar to other common GI disorders, celiac disease is often misdiagnosed. When it is suspected, a blood test may be performed to confirm. Fortunately, once diagnosed, a person with celiac disease can live symptom-free by eliminating gluten from the diet.[4]

For people with celiac disease, it is important to buy foods that are gluten-free. A variety of products are available in most grocery stores.

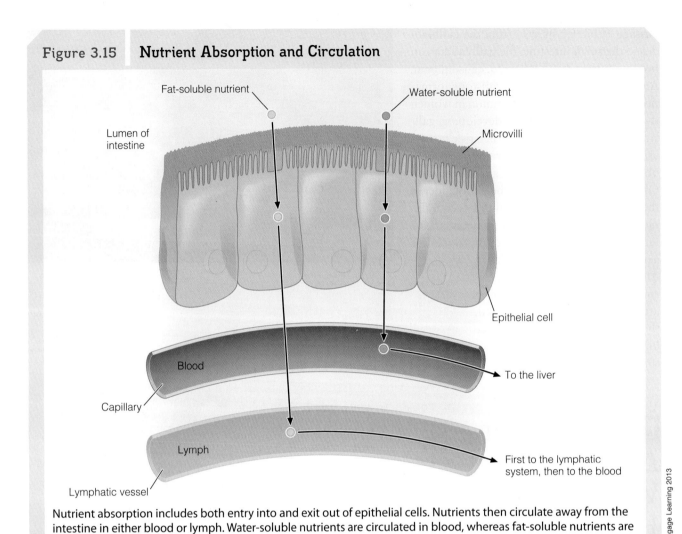

Figure 3.15 **Nutrient Absorption and Circulation**

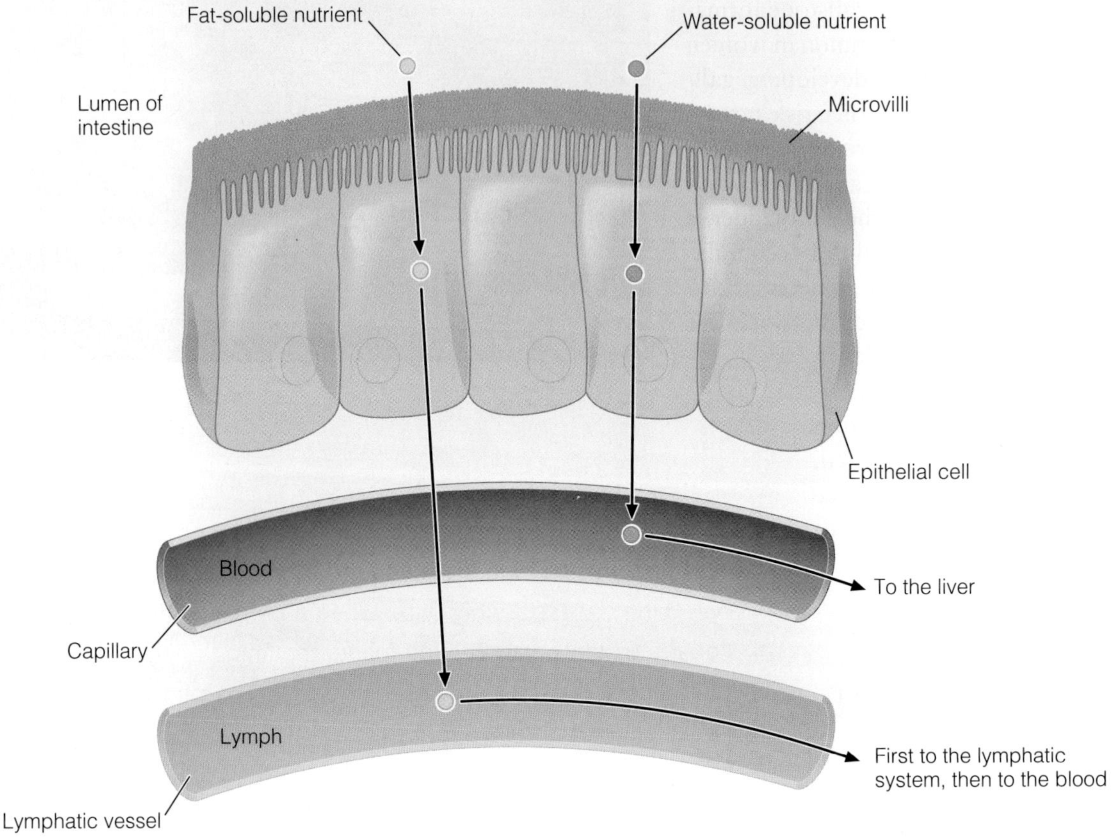

Nutrient absorption includes both entry into and exit out of epithelial cells. Nutrients then circulate away from the intestine in either blood or lymph. Water-soluble nutrients are circulated in blood, whereas fat-soluble nutrients are circulated in lymph. Eventually, the lymph empties into the blood.

move in a sweeping action to trap and pull nutrients toward the intestinal epithelial cells. During absorption, nutrients move into and out of these cells by both passive and active transport mechanisms. Once absorbed, nutrients enter either the blood or lymphatic circulatory system.

The extent to which a nutrient is absorbed into the blood or lymphatic system is called its **bioavailability**. The bioavailability of a particular nutrient can be influenced by physiological conditions, other dietary components, and certain medications. For example, the physiological needs of the body can influence the amount of iron absorbed, a function that protects you from iron toxicity. Some nutrients affect the bioavailability of other nutrients markedly. For example, the presence of vitamin C can enhance the absorption of certain forms of iron, and vitamin D can increase the bioavailability of calcium. It is important to be aware of the impact that certain nutrients and drugs have on nutrient bioavailability.

bioavailability The extent to which a nutrient is absorbed into the circulatory system.

colon The first portion of the large intestine.

The Large Intestine Eliminates Solid Waste

The large intestine is shaped like an inverted letter U (∩) and is approximately five feet long. The first portion of the large intestine, the **colon**, receives approximately two liters of undigested material from the small intestine every day. A sphincter separating the small from the large intestine serves two main functions: (1) to prevent the contents of the large intestine from flowing backward into the small intestine and (2) to prevent the premature flow of undigested material from the small intestine into the colon. Following the colon is the **rectum**, the segment of the GI tract that leads to the anal canal, which opens to the outside of the body (see Figure 3.16).

Materials entering the large intestine consist mostly of undigested remains from plant-based foods, water, bile, and electrolytes. Muscles embedded within the intestinal wall squeeze the undigested food residue, slowly propelling the material forward and exposing it to the inner (mucosal) lining of the colon. As material moves through the various regions of the colon, water and electrolytes are absorbed and returned to the blood for reuse by the body. This exemplifies the body's ability to reclaim its important resources.

rectum The segment of the GI tract that leads to the anal canal.

microbiota The natural microbial population that resides in the large intestine.

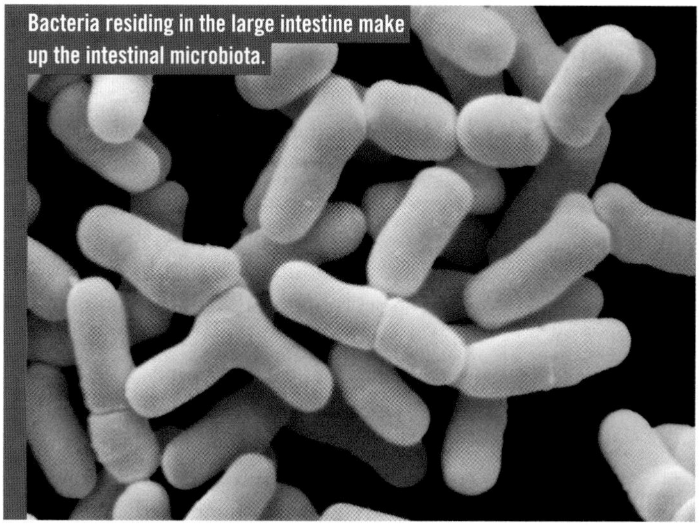
Bacteria residing in the large intestine make up the intestinal microbiota.

The slow, propulsive movement of the large intestine also provides an ideal environment for bacteria to grow and flourish. The number and variety of bacteria residing in the large intestine is astronomical—more than 400 different species call your large intestine home. This natural microbial population, also referred to as the intestinal **microbiota**, helps maintain a healthy environment in your colon. Some intestinal bacteria break down undigested food residue; others produce nutrients such as vitamin K, certain B vitamins, and some lipids. The presence

Figure 3.16 **Overview of the Large Intestine**

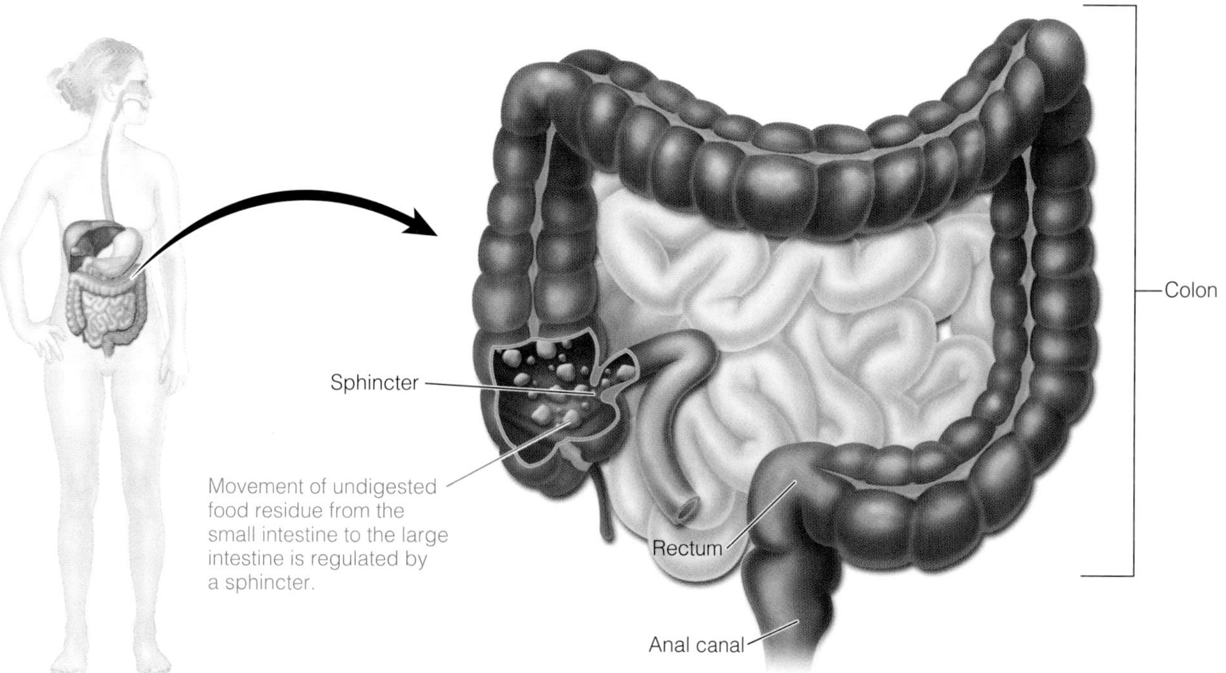

The large intestine has three general regions: the colon, the rectum, and the anal canal.

probiotic food A food that contains live bacteria, some of which thrive in the colon.

prebiotic food A typically fiber-rich food that may stimulate the growth of the microbial population in the large intestine.

feces Solid waste consisting mainly of undigested and unabsorbed matter, dead cells, secretions from the GI tract, water, and bacteria.

of healthy intestinal microbiota also inhibits the growth of other disease-causing bacteria.

To help establish and maintain a healthy intestinal microbiota, it may be helpful to consume substantial amounts of probiotic and prebiotic foods. A **probiotic food** is one that contains live bacterial cultures, some of which thrive in the colon. Yogurt and other fermented dairy products are examples of probiotic foods. Some dietary supplements also supply probiotic bacteria. A **prebiotic food** is typically fiber-rich and may stimulate the growth of the microbial population in the large intestine. Some health care professionals believe that consuming both probiotic and prebiotic foods provides a powerful defense against disease-causing bacteria.[5]

Once the remaining material completes its journey, it is ready to be eliminated from the body. This solid waste, now called **feces**, consists mainly of undigested and unabsorbed matter, dead cells, secretions from the GI tract, water, and bacteria. As the feces approaches the end of the colon, it passes into the rectum, which serves as a holding chamber. An accumulation of feces causes the walls of the rectum to stretch, signaling the need to defecate. Unlike other sphincters in the GI tract, the sphincter located between the rectum and the anal canal is under voluntary control. This enables a person to determine whether the time is right for waste elimination. When the sphincter relaxes, the feces move into the anal canal and are expelled from the body.

The consistency of feces depends mainly on its water content. If chyme moves too quickly

Nutrient + Pharmaceutical = Nutraceutical

The term *nutraceutical* is used to describe foods that are consumed to prevent and/or treat certain diseases. Though the term is a recent invention, the perception of certain foods as medicine is not new: many ancient civilizations used foods to treat a variety of ailments. Today, the use of foods to treat certain disorders is viewed by some as a type of alternative medicine. However, this practice has gained considerable recognition in the medical community as a viable way to manage and treat some health problems. The consumption of a probiotic food to promote intestinal health is a good example of how nutraceuticals might contribute to overall well-being. Researchers are currently exploring whether probiotics can relieve conditions such as infectious diarrhea, irritable bowel syndrome, inflammatory bowel disease, tooth decay, and vaginal infections. Other foods believed to have therapeutic purposes include cranberry juice (to help prevent urinary tract infections), soy (for the management of menopausal symptoms), garlic (to lower blood cholesterol levels), and certain types of fish (to prevent heart disease). Although consumption of neutraceuticals is a growing trend, it is important to recognize that some of the health benefits attributed to medicinal foods may be due to the placebo effect.[6]

Phytochemicals found in cranberries may help **protect against** the recurrence of **urinary tract infections**. This is an example of a nutraceutical food.

PRNewsFoto/Pierre's Ice Cream Company

through the colon, a sufficient amount of water cannot be extracted from it. This results in loose, watery feces referred to as *diarrhea*. Prolonged diarrhea can cause excessive water loss from the body, which can lead to dehydration. Conversely, when chyme moves too slowly through the colon, too much water may be absorbed. This can cause the fecal matter to become hard and dry, resulting in a condition called *constipation*. Constipation can make elimination difficult and can put excessive strain on the muscles in the colon wall.

INTESTINAL DISORDERS

Diarrhea and constipation are not the only conditions that can cause intestinal discomfort. **Irritable bowel syndrome (IBS)** is a disorder that typically affects the lower GI tract, causing bouts of cramping, bloating, diarrhea, and constipation. Because IBS is not associated with any known structural abnormalities, it is considered a functional (as opposed to anatomical) disorder. Although the underlying cause of IBS is yet to be determined, some believe that the severity of this disorder is associated with emotional stress. That is, emotional stress itself is not a cause of IBS, but it can make it worse. People with IBS are encouraged to keep a food diary to help identify the foods that trigger the disorder. Elimination of these foods from the diet is an important step in managing IBS.

Inflammatory bowel disease (IBD) is a category of chronic conditions characterized by inflammation of the lining of the GI tract. While the terms *irritable bowel syndrome* and *inflammatory bowel disease* may sound similar, they refer to very different conditions. Many researchers believe that IBD develops when exposure to something in the environment causes an immunologic response. Factors believed to trigger this response include certain proteins, viruses, and bacteria. When exposure triggers the production of antibodies, the intestinal lining becomes inflamed. IBD's signs and symptoms include diarrhea, fatigue, weight loss, abdominal pain, diminished appetite, and on occasion rectal bleeding.[7] Prolonged periods of inflammation can permanently damage the delicate lining of the large intestine. Ulcerative colitis and Crohn's disease are examples of inflammatory bowel disease.

LO5 How Does the Body Circulate Nutrients and Excrete Waste Products?

Once food has been digested and nutrients have been absorbed, the body's next task is to transport the nutrients throughout the body. This is accomplished by the veins and arteries that comprise the blood circulatory system and the lymphatic vessels that comprise the lymphatic circulatory system. These extensive systems both deliver the nutrients needed to operate and aid in the elimination of some types of waste products.

The Cardiovascular System Circulates Nutrients and Gases

Upon absorption, water-soluble nutrients enter the bloodstream through the capillaries contained within each villus. Once they reach the bloodstream, water-soluble nutrients circulate directly to the liver, giving it first access to the nutrient-rich blood as it leaves the small intestine. The liver then regulates the use of the nutrients to suit the body's needs. Small amounts of some nutrients are stored in the liver, but most either undergo chemical modification or are released directly into blood, whereby they are delivered to other parts of the body.

irritable bowel syndrome (IBS) A disorder that typically affects the lower GI tract, causing bouts of cramping, bloating, diarrhea, and constipation.

inflammatory bowel disease (IBD) A category of chronic conditions characterized by inflammation of the lining of the GI tract.

lacteal A lymphatic vessel found in an intestinal villus into which nutrients are absorbed.

excretion The removal of waste products produced in cells.

metabolism The sum of chemical processes that occur within a living cell to maintain life.

energy metabolism Chemical reactions that enable cells to use and store energy.

The Lymphatic System Circulates Fat-Soluble Nutrients

The lymphatic system plays an important role in the circulation of fat-soluble nutrients (mostly lipids and some vitamins) away from the GI tract. Each villus contains a lymphatic vessel—a **lacteal**—through which the nutrients are absorbed. Each lacteal connects to a larger network of lymphatic vessels that circulate a translucent liquid called lymph. Though the circulatory route of the lymphatic system initially bypasses the liver, it eventually empties into the bloodstream where nutrients can be taken up and used by cells.

Excretion of Wastes

Not only do cells require a continuous supply of essential nutrients and oxygen, they must also be able to rid themselves of waste products. While *elimination* indicates the removal of solid waste (feces) from the body, **excretion** indicates the removal of waste products produced in cells. Cellular waste includes water, carbon dioxide, and nitrogen-containing compounds. The excretion of these natural by-products generated by the cellular breakdown of nutrients is vital to your health.

Excretory organs involved in the removal of waste products include those in the circulatory system, liver, kidneys, lungs, and skin. Collectively, these organs help prevent the accumulation of toxic waste products and aid in their removal. For example, the liver converts ammonia—a nitrogen-containing by-product of protein breakdown—into a less toxic substance called urea, which is released into the blood. The kidneys then filter the urea out of the blood so it can be excreted from the body in urine. Other examples of the excretory process include the exhalation of air through the nose and mouth (to rid the body of carbon dioxide), and the active process of sweating (to rid the body of water, salt, and urea).

LO6 What Is Metabolism?

After they are circulated away from the GI tract, nutrients are delivered to and taken up by cells. The work of the digestive system is complete—at least until the next meal. Once a nutrient enters a cell, the complex process of nutrient metabolism begins. **Metabolism** is the sum of chemical processes that occur within a living cell to maintain life. In short, when a cell uses a nutrient, it metabolizes it. Metabolic waste products such as ammonia and carbon dioxide are removed from cells through the excretory process. Some metabolic processes utilize nutrients to build and support the body's structures (such as bones and muscle). Others transfer the energy stored in nutrients to the powerful chemical bonds of ATP, the major energy currency of cells. ATP is a perfectly designed, intricate molecule that fuels thousands of chemical reactions required to sustain life. The processes by which cells use and store energy involve a series of chemical reactions referred to collectively as **energy metabolism**. Energy metabolism enables cells to transform the chemical energy stored in certain nutrients, which provides cells with adequate ATP.

Metabolic Pathways

Metabolic processes are generally complex, requiring many steps. Recall that cells contain mitochondria, which are organelles that play an important role in converting the chemical energy found in certain nutrients into ATP.

This transformation occurs through a series of interrelated chemical reactions collectively called a **metabolic pathway**. Each chemical reaction in a metabolic pathway requires the help of at least one enzyme. To function properly, enzymes often require the assistance of certain vitamins and minerals. Thus, although they themselves do not provide cells with a source of energy, vitamins and minerals play an important role in energy metabolism.

Some metabolic pathways break down complex molecules into simpler ones, releasing energy in the process. This type of metabolic pathway is referred to as a **catabolic pathway**. Conversely, a series of metabolic reactions that uses energy to construct a complex molecule from simpler ones is referred to as an **anabolic pathway**. Thus, catabolic and anabolic pathways can be thought of respectively as series of "break it" and "make it" chemical reactions.

CATABOLISM AND THE RELEASE OF ENERGY

Energy-yielding nutrients (carbohydrates, proteins, and lipids) store energy in their chemical bonds. However, for mitochondria to synthesize ATP from these bonds, nutrients must undergo the chemical reactions that make up a catabolic pathway. Many catabolic pathways require oxygen in order to function. These are referred to as **aerobic**, or oxygen requiring. Catabolic pathways that can function under conditions of low oxygen availability are referred to as **anaerobic**. Both aerobic and anaerobic pathways play important roles in making energy available to cells. For example, when you perform a moderate-intensity physical exercise (such as jogging) for an extended period of time, your lungs and heart work harder to pump adequate amounts of oxygen into your body so that ATP can be generated through aerobic pathways. During more intense exercise (such as strength training), it can be difficult for your lungs and heart to deliver adequate amounts of oxygen to active muscle cells. When this occurs, anaerobic catabolic pathways help sustain you by generating enough ATP to prevent muscle fatigue. There is a limit to how much energy can be generated under conditions of limited oxygen availability, however. This is why muscles tire more quickly during high-intensity workouts.

When there is a need to generate ATP rapidly, muscles rely primarily on anaerobic metabolic pathways.

Courtesy of Kathy Beerman

metabolic pathway A series of interrelated chemical reactions that require the help of enzymes.

catabolic pathway A series of metabolic reactions that breaks down a complex molecule into simpler ones, releasing energy in the process.

anabolic pathway A series of metabolic reactions that uses energy to construct a complex molecule from simpler ones.

aerobic A metabolic pathway that requires oxygen in order to function.

anaerobic A metabolic pathway that can function under conditions of low oxygen availability.

ANABOLISM AND MAKING NEW MOLECULES

Whereas catabolic pathways usually result in the release of energy, anabolic pathways serve a different function—the synthesis of new molecules and/or the storage of energy. An example of an anabolic process at work is the construction of the many proteins needed to support growth, facilitate reproduction, and synthesize complex molecules. When a person has more energy than he needs, cells use anabolic pathways to store energy-yielding nutrients for later use—sort of like depositing money in the bank. For example, glucose molecules not immediately needed for energy can be converted to the glucose-storage molecule glycogen. When glycogen stores are full, cells turn the remaining glucose into fatty acids, which are subsequently stored as body fat. These anabolic reactions provide cells with an energy reserve, a critical component to survival when food is limited and/or energy requirements are high.

NUTR

4 Carbohydrates

LEARNING OUTCOMES:

LO1 Define and distinguish simple carbohydrates.

LO2 Define and distinguish complex carbohydrates.

LO3 Explain how the body digests, absorbs, and circulates carbohydrates.

LO4 Describe how the body stores, uses, and regulates glucose.

LO5 Differentiate between the two types of diabetes.

LO6 Summarize carbohydrate intake recommendations.

LO1 What Are Simple Carbohydrates?

Carbohydrates comprise a diverse group of compounds produced primarily by plants. The sugars in fruit, the fibers in celery, and the starch in potatoes are all different forms of carbohydrate. Although they are all made of sugar molecules, carbohydrates differ from each other in a variety of ways. Some carbohydrates contain only one or two sugar molecules, whereas others are comprised of hundreds of them. As the number of sugar molecules increases, so does the size and complexity of the carbohydrate. Carbohydrates, plentiful in a variety of foods, serve many essential functions within the body. As such, they are an important part of a healthy, well-balanced diet. In this chapter, carbohydrate chemistry, dietary sources of carbohydrates, the functions of carbohydrates in the body, and dietary guidelines for carbohydrate intake will be examined.

The simplest type of carbohydrate is a *sugar*—a molecule that cannot be broken down further to produce other sugars. Thus, a **carbohydrate** is an organic compound made up of one or more sugar molecules (see Figure 4.1 on the next page). Most people think of sugar as a substance used to sweeten their foods. Although this is true, the uses and functions of sugars extend far beyond sweetening. For example, cells use a special type of sugar (which you will soon read about) as an important source of energy. A carbohydrate consisting of a single sugar molecule is called a **monosaccharide**, and a carbohydrate consisting of two sugar molecules bonded together is called a **disaccharide**. Because of the small sizes of these molecules, every monosaccharide and disaccharide is referred to as a **simple carbohydrate** (or **simple sugar**).

Monosaccharides

Monosaccharides are single-sugar molecules comprised entirely of carbon, hydrogen, and oxygen atoms. There are hundreds of different naturally occurring monosaccharides, but the three that are most plentiful in food are glucose, fructose, and galactose. Although the structures of these sugars differ from one other, they all have one thing in common: each contains six carbon atoms (see Figure 4.2 on the next page).

> From the Greek words ***photo*** ("light") and ***synthesis*** ("putting together"), photosynthesis means literally, **"putting together from light."**

GLUCOSE

Glucose, the most abundant monosaccharide in the human body, is produced when chlorophyll-containing plants combine carbon dioxide (CO_2) and water (H_2O) using energy harvested from sunlight, a process called **photosynthesis** (see Figure 4.3 on page 75). The primary function of glucose in the body is to provide cells a source of energy. Plants use glucose both as an important energy source and to construct larger, more complex carbohydrates. When you consume plant-based foods, your body breaks down (or in other words, digests) these large carbohydrates into glucose, providing you with an important source of energy. In this way, plants and animals are interconnected in the delicate balance that exists in nature.

carbohydrate An organic compound made up of one or more sugar molecules.

monosaccharide A carbohydrate consisting of a single sugar molecule.

disaccharide A carbohydrate consisting of two monosaccharides bonded together.

simple carbohydrate (or **simple sugar**) A category of carbohydrates comprised of monosaccharides and disaccharides.

glucose The most abundant monosaccharide in the human body; used extensively for energy.

photosynthesis A process whereby chlorophyll-containing plants produce glucose by combining carbon dioxide (CO_2) and water (H_2O) using energy harvested from sunlight.

Figure 4.1 Classification of Carbohydrates

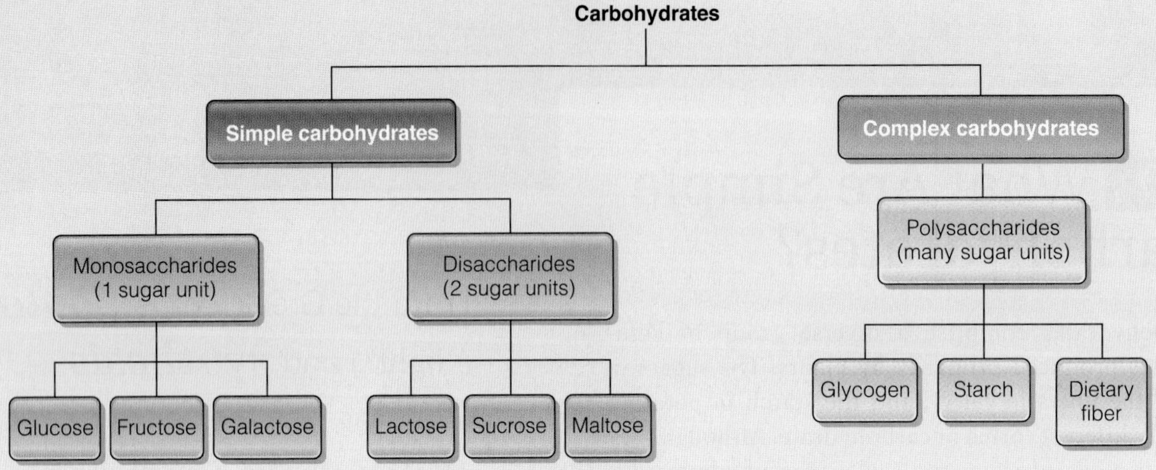

Carbohydrates are classified as either simple or complex. Simple carbohydrates include monosaccharides and disaccharides, whereas complex carbohydrates include polysaccharides such as glycogen, starch, and fiber.

Figure 4.2 The Structures of Monosaccharides

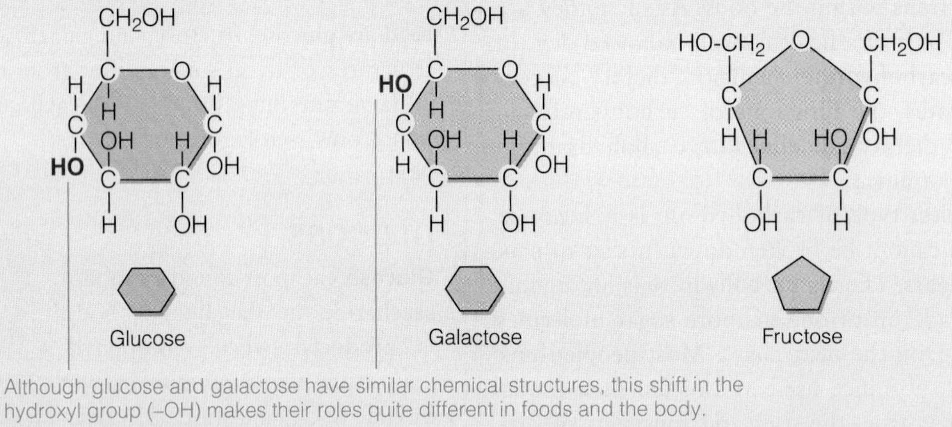

Glucose
Galactose
Fructose
Although glucose and galactose have similar chemical structures, this shift in the hydroxyl group (–OH) makes their roles quite different in foods and the body.

Although glucose, galactose, and fructose are all monosaccharides that contain 6 carbon atoms, each has a slightly different chemical structure.

fructose A naturally occurring monosaccharide found primarily in honey, fruits, and vegetables.

high-fructose corn syrup (HFCS) A widely used sweetener consisting of glucose and fructose that is manufactured from cornstarch.

FRUCTOSE

Fructose is a naturally occurring monosaccharide found primarily in honey, fruits, and vegetables. While it is the most abundant sugar in fruits and vegetables, the majority of fructose consumed in the Western diet comes from foods and beverages made with high-fructose corn syrup. **High-fructose corn syrup (HFCS)** is a widely used sweetener consisting of glucose and fructose that is manufactured from cornstarch. It is found in soft drinks, cereals, fruit juice beverages, soups, and a multitude of other foods. HFCS is used so extensively by food manufacturers that it now accounts for approximately 7 percent of total energy intake in the United States.[1] The U.S. Department of Agriculture estimates that Americans consume approximately 35 pounds of high-fructose corn syrup (per capita) each year.[2] In fact, the consumption of foods and beverages sweetened with high-fructose corn syrup now

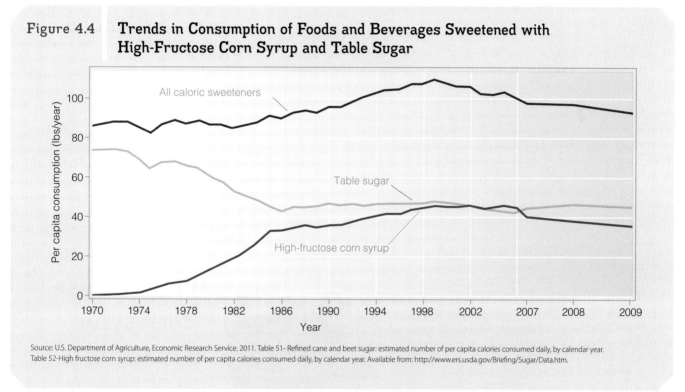

Figure 4.3 Photosynthesis

The process of photosynthesis combines carbon dioxide and water in the presence of sunlight to produce glucose and oxygen.

Figure 4.4 Trends in Consumption of Foods and Beverages Sweetened with High-Fructose Corn Syrup and Table Sugar

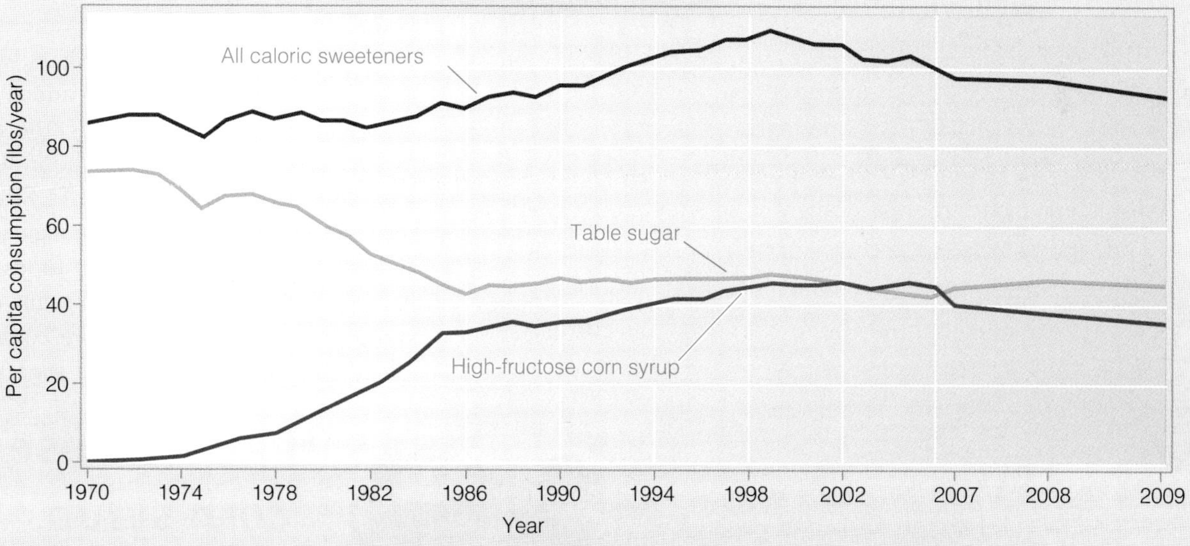

Source: U.S. Department of Agriculture, Economic Research Service. 2011. Table 51- Refined cane and beet sugar: estimated number of per capita calories consumed daily, by calendar year. Table 52-High fructose corn syrup: estimated number of per capita calories consumed daily, by calendar year. Available from: http://www.ers.usda.gov/Briefing/Sugar/Data.htm.

exceeds that of those sweetened with table sugar (see Figure 4.4). In the body, the majority of fructose is converted to glucose and used as a source of energy.

GALACTOSE

Glucose and **galactose** look rather similar. However, slight differences in their chemical structures result in important differences in their physiological functions. Few foods contain galactose in its free state. Rather, the majority of dietary galactose comes from a naturally occurring disaccharide in dairy products. Like fructose, the

galactose A monosaccharide that exists primarily as part of a naturally occurring disaccharide found in dairy products.

lactose A disaccharide comprised of galactose joined with glucose. It is the most abundant carbohydrate in milk and many other dairy products.

maltose A disaccharide comprised of glucose joined with glucose. It is not found in many foods.

sucrose A disaccharide comprised of fructose joined with glucose. It is found in many plants.

majority of galactose in the body is converted to glucose and used for energy.

Disaccharides

Disaccharides consist of two monosaccharides bonded together. Present in a wide variety of foods, the most common disaccharides are lactose, maltose, and sucrose. Note that in all of these disaccharides, at least one of the monosaccharides in the pair is glucose (see Figure 4.5).

- **Lactose:** Comprised of galactose joined with glucose, lactose is the most abundant carbohydrate in milk and many other dairy products (such as yogurt, cheese, and ice cream).
- **Maltose:** Comprised of glucose joined with glucose, maltose is formed during starch digestion. Maltose is not found in many foods, but the enzymatic breakdown of starches into maltose is an important step in the production of beer.
- **Sucrose:** Comprised of fructose joined with glucose, sucrose is found in many plants (and is especially abundant in sugar cane and sugar beets). Crushing these plants produces a juice that can be processed to make a thick brown liquid called molasses. Further treatment and purification of molasses forms pure crystallized sucrose, otherwise known as refined table sugar. Because most people enjoy the intense sweetness of sucrose, it is often added to processed foods.

High-Fructose Corn Syrup: Are There Concerns?

Fructose is a naturally occurring sugar that makes fruits deliciously sweet. However, most of the fructose Americans consume comes not from fruit but instead from foods and beverages sweetened with high-fructose corn syrup (HFCS). HFCS is made from cornstarch that has been treated with enzymes to convert some of its natural glucose to fructose. Recently, scientists observed that an increase in HFCS consumption has paralleled an increasing rate of obesity.[3] Some claim that HFCS has unique properties that may be to blame for this correlation, but studies have not shown a definite causal relationship between the consumption of HFCS and weight gain. The evident relationship between HFCS intake and obesity is likely a product of excess consumption of energy-dense foods and beverages in general, not HFCS *per se*. Until scientists learn more about the effects of this sweetener on health, we can all agree that consuming too many calories in any form can contribute to unwanted weight gain.

© iStockphoto.com/Danny Hooks

Naturally Occurring Sugars and Added Sugars

Foods like fruits, vegetables, and milk contain naturally occurring sugars, while many processed foods contain added sugars or syrups. On food labels, the term "sugar" usually accounts for both added and naturally occurring sugar. As you can see in Table 4.1, many types of sugars and syrups can be used to sweeten foods and beverages. White sugar, brown sugar, corn syrup, honey, and molasses are common added sugars. Although added sugars are

High-fructose corn syrup, *a widely used sweetener, is found in a* **variety of foods** *and beverages—such as soft drinks.*

chemically identical to naturally occurring sugars, foods with naturally occurring sugars usually contain other important nutrients in addition to their carbohydrates. For example, besides sucrose, fruits are naturally rich in vitamins, minerals, and fiber. By contrast, foods with large amounts of added sugars (such as soft drinks, cakes, and candy) often have little nutritional value beyond the calories they contain.

In recent years, consumption of added sugars has decreased slightly throughout the United States. Today, the average American consumes about 89 grams (22 teaspoons) of added sugars—a total of 355 kcal—every day.[4] Although this may not sound like much, the calories can add up quickly. Nearly half of the added sugars in the American diet come from regular soft drinks.[5] Substantial evidence suggests that a greater consumption of sugar-sweetened beverages is associated with increased body weight in adults.[6] This is not surprising when you consider that the average 12-ounce can of sweetened soft drink contains eight teaspoons of added sugar—the equivalent of 130 kcal.

ALTERNATIVE SWEETENERS

Some people believe that natural sweeteners (such as honey) present a healthier alternative to processed sweeteners (such as refined white sugar). However, like refined sugars, natural sweeteners possess limited nutritional value beyond the energy they contain. To add sweetness without increasing the caloric contents of foods, food manufacturers often use artificial low-calorie sweeteners such as saccharin, aspartame, and acesulfame K. In addition to these artificial sweeteners, food manufactures use natural substitutes such as sugar alcohols and stevia, a sweet-tasting herbal dietary product. None of these substances are actual

> "The average American consumes 22 teaspoons (355 kcal) of added sugars every day."

Figure 4.5 Disaccharides

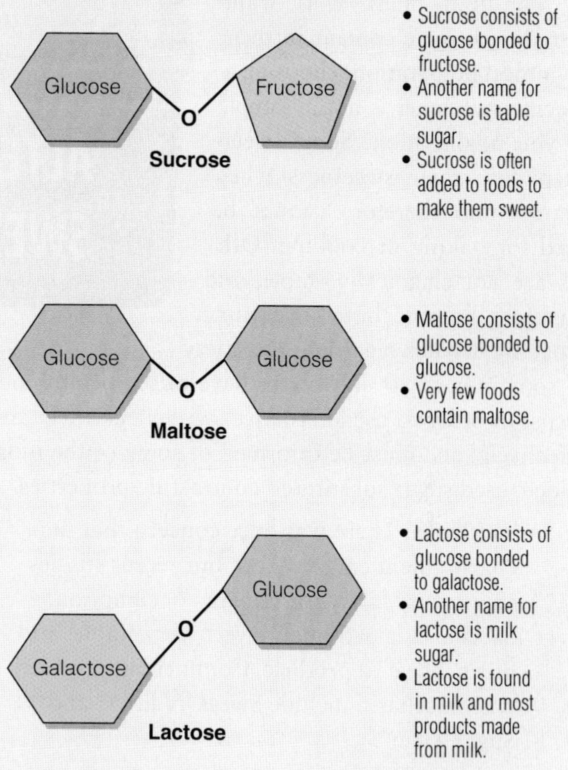

Glucose
O
Fructose
Sucrose
• Sucrose consists of glucose bonded to fructose.
• Another name for sucrose is table sugar.
• Sucrose is often added to foods to make them sweet.
Glucose
O
Glucose
Maltose
• Maltose consists of glucose bonded to glucose.
• Very few foods contain maltose.
Galactose
O
Glucose
Lactose
• Lactose consists of glucose bonded to galactose.
• Another name for lactose is milk sugar.
• Lactose is found in milk and most products made from milk.

The disaccharides sucrose, maltose, and lactose consist of two monosaccharides bonded together. Notice that each disaccharide has at least one glucose molecule.

Table 4.1 Added Sweeteners Commonly Listed on Food Labels

Brown rice syrup	Brown sugar
Concentrated fruit juice sweetener	Confectioner's sugar
Corn syrup	Dextrose
Fructose	Glucose
Granulated sugar	High-fructose corn syrup
Honey	Invert sugar
Lactose	Levulose
Maltose	Maple sugar
Molasses	Natural sweeteners
Raw sugar	Sucrose
Turbinado sugar	White sugar

© iStockphoto.com/John Sigler

sugars, but each replicates sugar's sweetness to some extent.

For individuals who want to reduce caloric content in their own food preparation, choosing a specific sweetener is not as simple as you might think. Some sweeteners lose their sweetness when heated, and therefore cannot be used for baking or cooking. Others are not chemically stable, and therefore become bitter over time. Ongoing debates regarding the safety of some non-sugar substitutes have raised health concerns. In order to use alternative sweeteners safely and effectively, one must be informed of some of the most widely used sugar substitutes' contrasting properties.

- **Saccharin:** There was once concern that saccharin might cause cancer, but recent studies support its safety for human consumption. Saccharin is extremely sweet, very stable, and inexpensive to produce. Commercial products with saccharin include Sweet 'N Low® and Sugar Twin®.
- **Aspartame:** Although it has the same energy content as sucrose (4 kcal per gram), aspartame is almost 200 times sweeter. The food industry uses aspartame in many sugar-free beverages, but because it is not heat stable, aspartame cannot be used in products that require cooking. Although the U.S. Food and Drug Administration (FDA) has ruled aspartame safe for most people, some individuals claim that it causes adverse effects such as headaches, dizziness, nausea, and seizures. While several studies have reported that aspartame increases the occurrence of cancer in laboratory animals, hundreds of studies have failed to find similar effects in humans.[7] People with the genetic disorder phenylketonuria (or PKU) should not consume products sweetened with aspartame because they are not able to metabolize it properly. Commercial products with aspartame include NutraSweet® and Equal®.
- **Acesulfame K:** Used extensively throughout Europe, acesulfame K was approved in 1998 for use in the United States, where it is sold under the trade name Sweet One®. Unlike aspartame, this artificial sweetener is heat stable and is used in a wide variety of prebaked commercial products.
- **Sucralose:** Even though sucralose is roughly 600 times sweeter than sucrose, it provides minimal calories because it is difficult for the body to digest and absorb. Because it is water soluble and stable, sucralose is used in a broad range of foods and beverages. Commercial products with sucralose include Splenda® and Altern®.
- **Sugar alcohols:** Sugar alcohols occur naturally in plants (particularly fruits) and have half the sweetness and calories of sucrose (roughly 2.5 kcal per gram). Sorbitol, mannitol, and xylitol, the most common sugar alcohols, are often found in sugar-free products such as chewing gums, breath mints, candies, toothpastes, mouthwashes, and cough syrups. One advantage of sugar alcohols is that unlike sucrose, they do not promote tooth decay. When consumed in excessive amounts however, sugar alcohols can cause diarrhea.
- **Stevia:** The term *stevia* is used to describe a group of related plants that grow in some regions of Central and South America, as well as in the southwestern region of the United States. The leaves of stevia plants are particularly sweet, a fact long recognized by the native peoples of these semi-arid regions. Stevia is essentially free of calories and is considerably sweeter than table sugar. Commercial products with stevia include Truvia® and NuStevia™.

Stevia, *an herb that grows in semi-arid climates, is essentially* **calorie free** *and is considerably* **sweeter than table sugar.**

PRNewsFoto/Cumberland Packing Corp.

LO2 What Are Complex Carbohydrates?

In contrast to the simple carbohydrates that contain one or two monosaccharides, a **complex carbohydrate** (or **polysaccharide**) is comprised of many monosaccharides bonded together. The types and arrangements of bonded sugar molecules determine the shape and form of the polysaccharide. For example, some sugar molecules bond in a way that gives a polysaccharide an orderly linear appearance. Others bond in a way that gives a polysaccharide a shape similar to the branches of a tree. Three of the most common polysaccharides—starch, glycogen, and dietary fiber—are discussed next.

Starch

Recall that plants generate glucose through the process of photosynthesis. To store this important source of energy, plants convert the glucose to starch. Thus, starch is made entirely of glucose molecules bonded together. The glucose molecules in starch are arranged in either an orderly unbranched linear chain or a highly branched configuration (see Figure 4.6). Plants typically contain a mixture of these two types of starch. Examples of starchy foods include grains (such as corn, rice, and wheat), products made from them (such as pasta and bread), and legumes (such as lentils and split peas). Potatoes and hard winter squashes are also good sources of starch.

Glycogen

Whereas plants store extra glucose as starch, the human body stores small amounts of excess glucose in the form of **glycogen**. Glycogen is a polysaccharide found primarily in liver and skeletal muscle. Like some forms of starch, the glucose molecules in glycogen are arranged in a highly branched configuration (see Figure 4.7 on the next page). The numerous branches of glucose molecules can be broken down quickly, providing cells a quick source of energy when one is needed.

Fiber

The term **fiber** (or **dietary fiber**) refers to a diverse group of plant polysaccharides found in a variety of plant-based foods such as whole grains, legumes, vegetables, and fruits. Different foods contain different types of dietary fiber. The type of chemical bond found in fiber differs from that found in starch. Because humans lack the enzymes needed to break these chemical bonds, fiber is virtually indigestible in the human small intestine (see Figure 4.8 on the next page). As a result, undigested fiber passes from the

complex carbohydrate (or **polysaccharide**) A category of carbohydrate comprised of many monosaccharides bonded together.

glycogen A polysaccharide found primarily in liver and skeletal muscle that is comprised of glucose molecules.

fiber (or **dietary fiber**) A diverse group of plant polysaccharides that are not digestible by human enzymes.

Figure 4.6 Plants Store Glucose as Starch

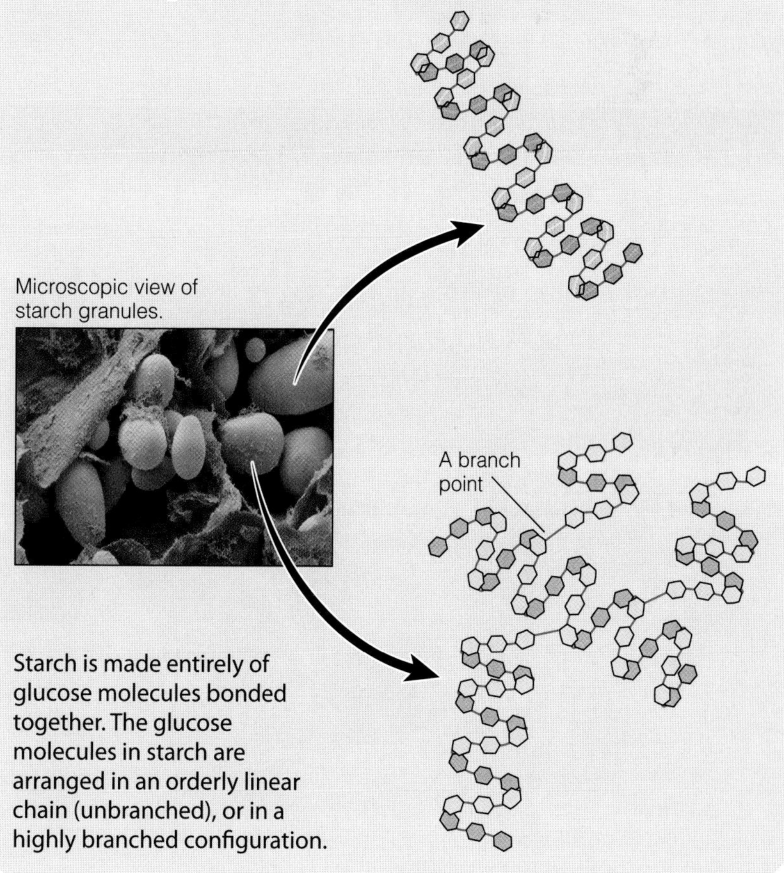
Microscopic view of starch granules.
A branch point
Starch is made entirely of glucose molecules bonded together. The glucose molecules in starch are arranged in an orderly linear chain (unbranched), or in a highly branched configuration.

Figure 4.7 **Animals Store Glucose in the Form of Glycogen**

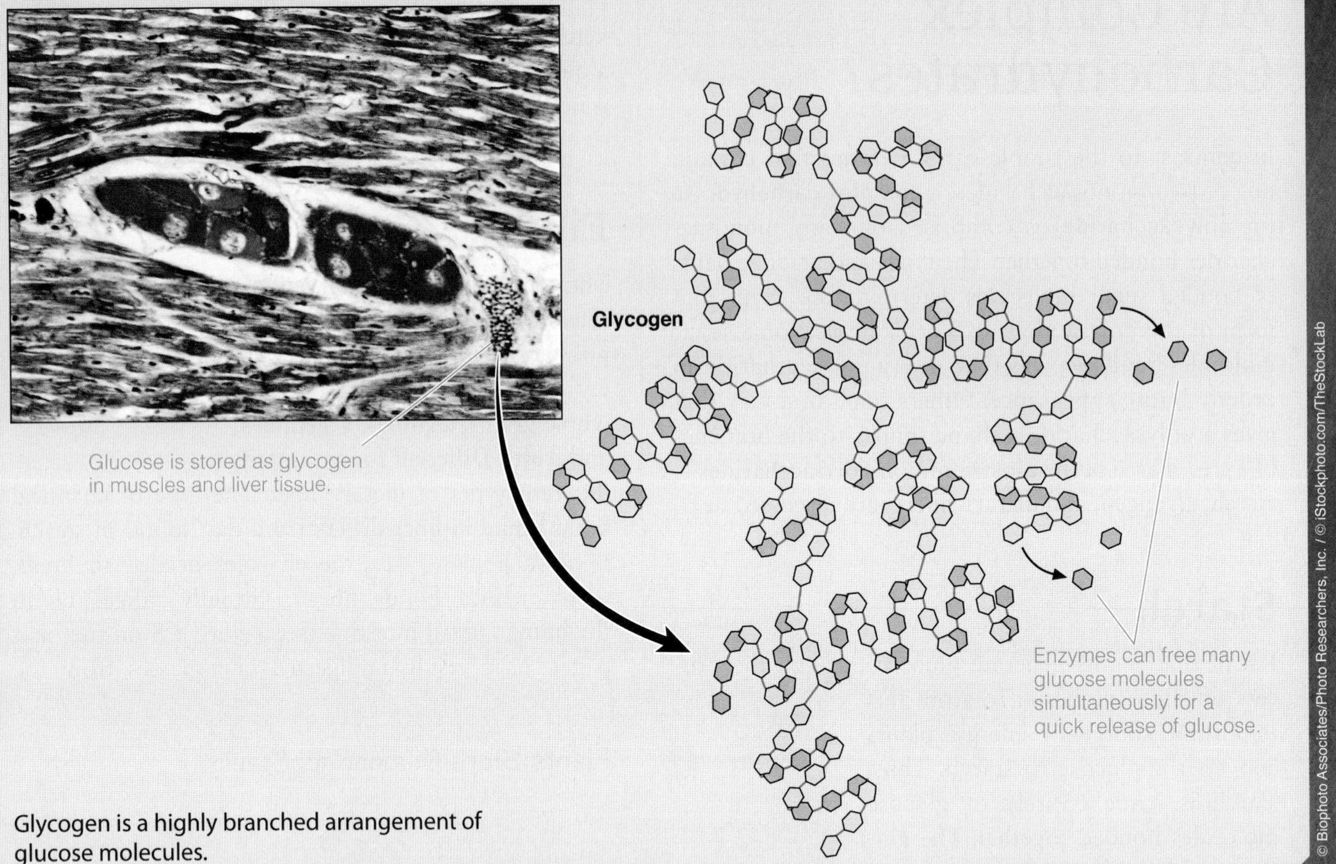
Glycogen
Glucose is stored as glycogen in muscles and liver tissue.
Enzymes can free many glucose molecules simultaneously for a quick release of glucose.
© Biophoto Associates/Photo Researchers, Inc. / © iStockphoto.com/TheStockLab

Glycogen is a highly branched arrangement of glucose molecules.

Figure 4.8 **Humans Are Unable to Digest Dietary Fiber**

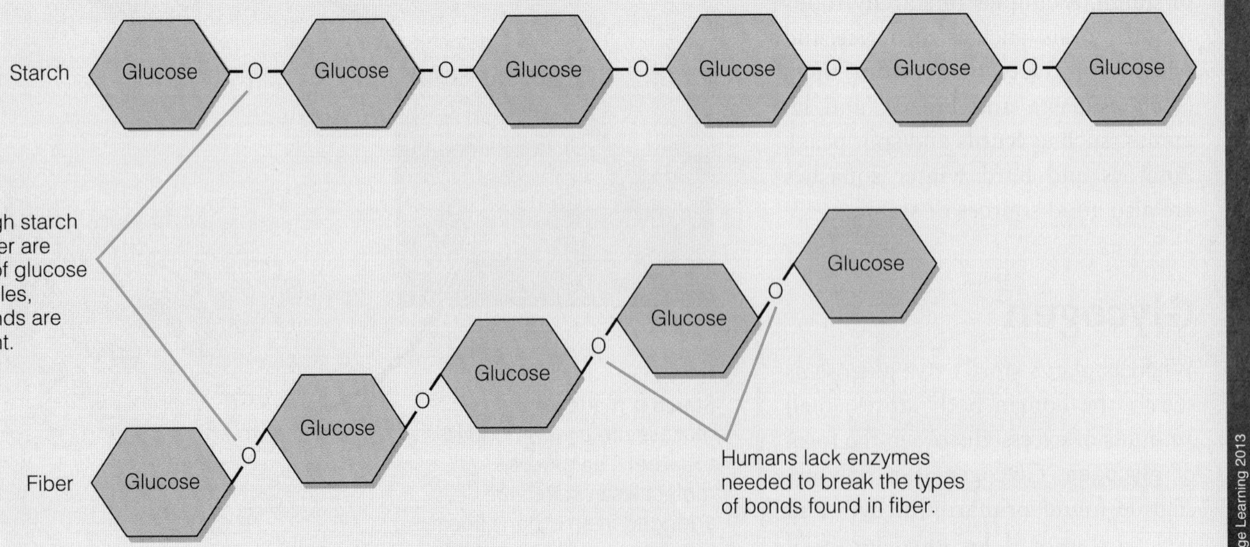
Starch
Glucose
O
Although starch and fiber are made of glucose molecules, the bonds are different.
Fiber
Humans lack enzymes needed to break the types of bonds found in fiber.
© Cengage Learning 2013

When humans consume dietary fiber, it remains intact in the gastrointestinal tract. This is because humans lack digestive enzymes that can break the bonds in fiber.

small to the large intestine relatively intact. Once in the large intestine, fiber is broken down by intestinal bacteria. There is evidence that dietary fiber helps promote the growth of beneficial intestinal bacteria, which in turn helps inhibit the growth of disease-causing (pathogenic) bacteria.[8] Dietary fiber is commonly classified on the basis of its solubility in water: **soluble fiber** tends to dissolve or swell in water, while **insoluble fiber** remains relatively unchanged.

POSSIBLE HEALTH BENEFITS OF FIBER

Although fiber is not an essential dietary component, it is clearly an important part of a healthful diet. Some studies suggest that increased fiber consumption is associated with lower risk of certain diseases. The evidence supporting the health benefits of fiber-rich foods is so substantive that the FDA has approved several fiber-related health claims (see Table 4.2).[9] Though fiber is not a magic bullet, its credentials are impressive. Consumption of soluble fiber can help lower blood cholesterol levels, promote satiety, and lower blood glucose levels.[10] In addition to these health-promoting benefits, soluble fiber may also function as a prebiotic by promoting the growth of beneficial bacteria in the colon.[11]

Another health benefit of dietary fiber pertains to the maintenance of intestinal health. Dietary fiber passes through the GI tract relatively intact, increasing fecal mass. Because larger amounts of feces move through the colon more quickly, eating foods that contain dietary fiber may help prevent or alleviate constipation. Not only can hard, dry feces make elimination more difficult, but it can also contribute to **diverticular disease** (or **diverticulosis**), a condition whereby straining (which is associated with persistent constipation) leads to the formation of pouches called *diverticula* that protrude along the colon wall (see Figure 4.9 on the next page). **Diverticulitis** occurs when the diverticula become infected or inflamed (note that the suffix *–itis* refers to inflammation). Symptoms of diverticulitis include cramping, diarrhea, fever, and occasionally, bleeding from the rectum. Preventing conditions such as diverticular disease is another reason why a diet high in fiber may be beneficial to one's health.

Some **manufacturers add fiber** to their products, making them **functional foods**. This extra fiber may provide additional **health benefits**.

© JONATHAN KIRN Feature Photo Service/Newscom

Table 4.2 FDA-Approved Health Claims Concerning Fiber

• Low-fat diets rich in fiber-containing grain products, fruits, and vegetables may reduce the risk of some types of cancer, a disease associated with many factors.
• Diets low in saturated fat and cholesterol and rich in fruits, vegetables, and grain products that contain some types of dietary fiber, particularly soluble fiber, may reduce the risk of heart disease, a disease associated with many factors.
• As part of a diet low in saturated fat and cholesterol, soluble fiber from foods such as oats may reduce the risk of heart disease.
• Diets high in whole-grain foods and other plant-based foods and low in total fat, saturated fat, and cholesterol may help reduce the risk of heart disease and certain cancers.

Source: U.S. Department of Health and Human Services. Health Claims Meeting Significant Scientific Agreement. Available from: http://www.fda.gov/food/labelingnutrition/labelclaims/healthclaimsmeetingsignificantscientificagreementssa/default.htm.

soluble fiber Dietary fiber that tends to dissolve or swell in water.

insoluble fiber Dietary fiber that remains relatively unchanged in water.

diverticular disease (or **diverticulosis)** A condition whereby straining leads to the formation of pouches called *diverticula* that protrude along the colon wall.

diverticulitis A condition whereby the diverticula become infected or inflamed.

Figure 4.9 **Diverticular Disease**

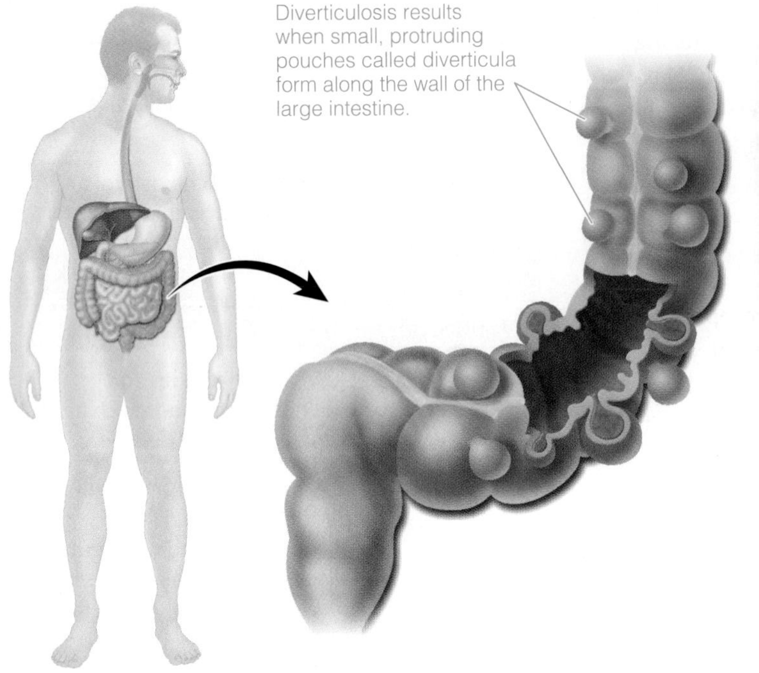

Diverticulosis results when small, protruding pouches called diverticula form along the wall of the large intestine.

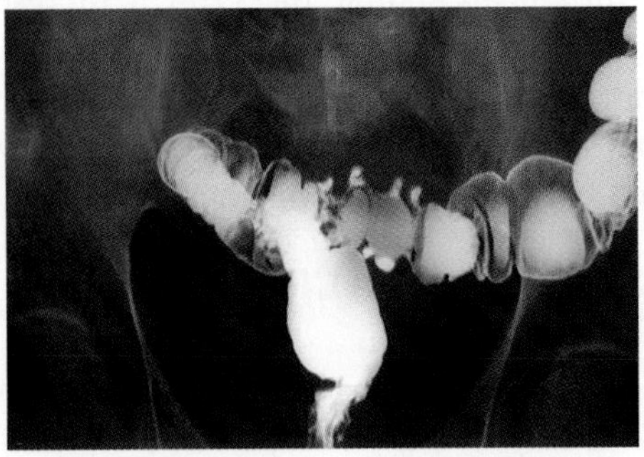

Diverticulosis can be detected by a special type of x-ray that uses contrast material.

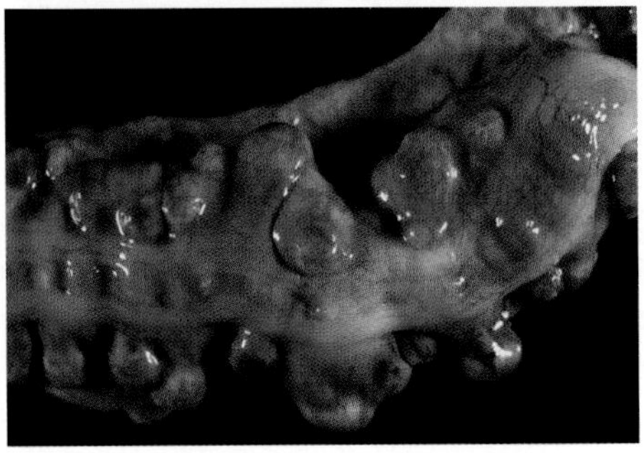

Section of a large intestine with diverticula.

Diverticular disease is most common in older adults who do not consume adequate amounts of fiber-rich foods. Diverticulitis is the inflammation of diverticula.

WHOLE-GRAIN FOODS

To ensure adequate fiber intake, it is important to eat a variety of fruits and vegetables every day. It is likewise important to eat **whole-grain foods.** The nutritional value of grains is greatest when all three components of the grain—bran, germ, and endosperm—are present in the same proportions as exist naturally (see Figure 4.10). Whereas the **bran** contains most of the grain's fiber, the **germ** contains much of its vitamins and minerals. The **endosperm** contains mostly starch. Because refinement removes the bran, products made with refined flour often contain very little fiber. When reading food labels, it is important to look for the phrases "whole-grain cereals"

whole-grain foods Cereal grains that contain bran, endosperm, and the germ in the same relative proportions as exist naturally.

bran The outer portion of a grain that contains most of the fiber.

germ The portion of a grain that contains most of the vitamins and minerals.

endosperm The portion of a grain that contains mostly starch.

Figure 4.10 **Anatomy of a Wheat Kernel**

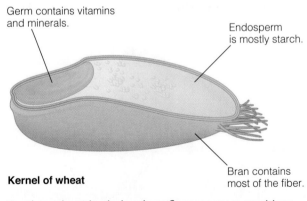

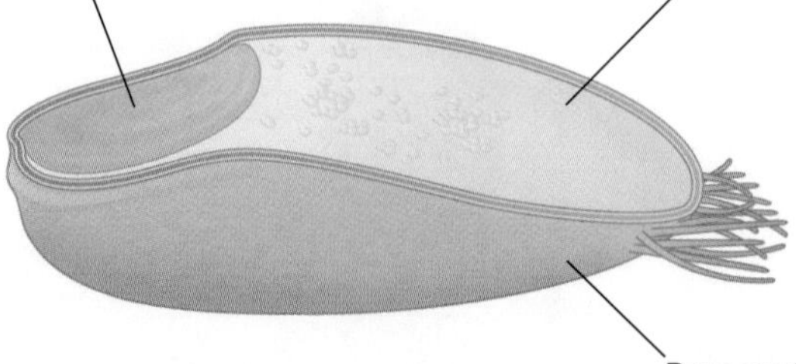

Foods made with whole-wheat flour are more nutritious than foods made with refined wheat flour because the components of the wheat kernel—the bran, germ, and endosperm—have not been removed. Each component contributes important nutrients needed for good health.

and "whole-wheat flour" because foods made with simply "wheat flour" are not necessarily high in fiber.

LO3 How Are Carbohydrates Digested, Absorbed, and Circulated?

The ultimate goal of carbohydrate digestion is to break down large, complex molecules such as starch into small, absorbable monosaccharides. Carbohydrates undergo extensive chemical transformations as they move through the GI tract during digestion. The bonds that hold disaccharides and starches together are broken with the help of digestive enzymes, and the resulting simple sugars are absorbed into the bloodstream.

Starch Digestion

Recall that starch is made entirely of glucose molecules. The chemical digestion of starch begins in the mouth when the salivary glands release the enzyme salivary amylase. **Salivary amylase** breaks the chemical bonds in starch, resulting in shorter chains of glucose molecules. However, because food stays in the mouth only a short time, very little starch digestion actually takes place there.

Once the partially digested starch enters the stomach, the acidic environment stops the enzymatic activity of salivary amylase. The short chains of glucose enter the small intestine, where they encounter *pancreatic amylase*, an enzyme made in the pancreas and delivered to the small intestine as part of the pancreatic juice. Pancreatic amylase picks up where salivary amylase left off, breaking the chemical bonds that hold the glucose molecules together. The glucose chains get shorter and shorter, eventually forming molecules of the disaccharide maltose. Lastly, **maltase**, an enzyme made in the cells that line the small intestine, is released into the intestinal lumen. Maltase finishes the job of digestion by breaking the last remaining chemical bond in maltose, resulting in two free (unbound) glucose molecules. These glucose molecules are now ready to be absorbed across the cells of the small intestine lining and into the blood (see Figure 4.11 on the next page).

Disaccharide Digestion

The digestion of disaccharides (maltose, sucrose, and lactose) takes place entirely in the small intestine. The small intestine is the lone source of enzymes needed for disaccharide digestion. Each disaccharide has its own specific digestive enzyme:

- Maltase (which you just learned about in the digestion of starch) digests maltose, releasing two glucose molecules.
- **Sucrase** digests sucrose, releasing glucose and fructose molecules.
- **Lactase** digests lactose, releasing glucose and galactose molecules.

Once disaccharides have been digested into their component monosaccharides, they can be absorbed across the cells of the small intestine and circulated in the blood (see Figure 4.12 on page 85).

salivary amylase An enzyme released from the salivary glands that breaks the chemical bonds in starch.

maltase An intestinal enzyme that digests maltose, releasing two glucose molecules.

sucrase An intestinal enzyme that digests sucrose, releasing glucose and fructose molecules.

lactase An intestinal enzyme that digests lactose, releasing glucose and galactose molecules.

lactose intolerance A condition whereby the body does not produce enough of the enzyme lactase, making it difficult to digest lactose.

LACTOSE INTOLERANCE

People who do not produce enough of the enzyme lactase have difficulty digesting lactose—a condition called **lactose intolerance.** Lactose intolerance is more common among certain ethnic and racial populations,

Lactose-reduced milk and other lactose-free dairy products are available at most grocery stores. These dairy products **contain all of the nutrients** found in regular milk products with the exception of lactose.

© ROSS HAILEY/MCT/Landov

Figure 4.11 **The Digestion of Starch**

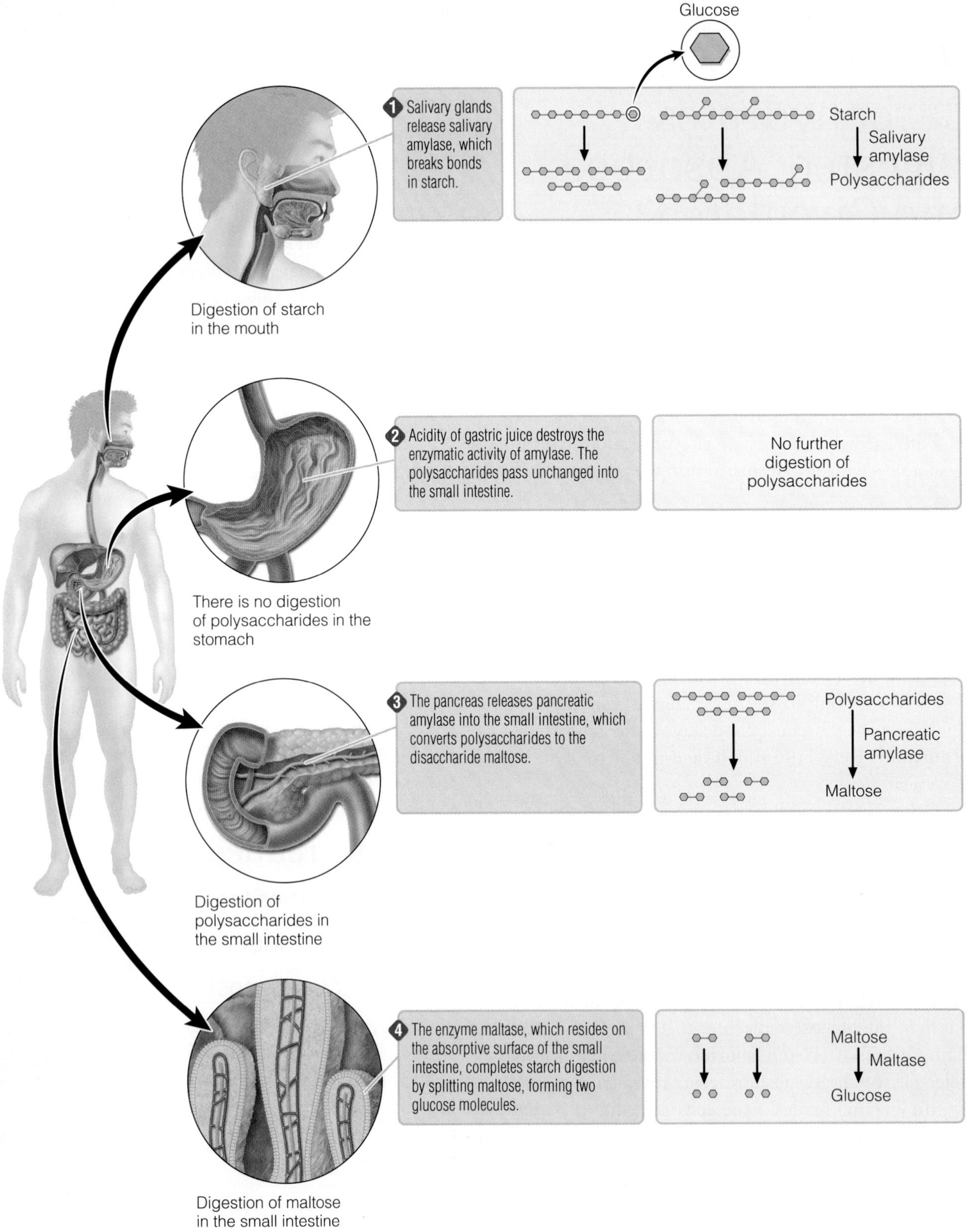

Glucose
1 Salivary glands release salivary amylase, which breaks bonds in starch.
Starch
Salivary amylase
Polysaccharides
Digestion of starch in the mouth
2 Acidity of gastric juice destroys the enzymatic activity of amylase. The polysaccharides pass unchanged into the small intestine.
No further digestion of polysaccharides
There is no digestion of polysaccharides in the stomach
3 The pancreas releases pancreatic amylase into the small intestine, which converts polysaccharides to the disaccharide maltose.
Polysaccharides
Pancreatic amylase
Maltose
Digestion of polysaccharides in the small intestine
4 The enzyme maltase, which resides on the absorptive surface of the small intestine, completes starch digestion by splitting maltose, forming two glucose molecules.
Maltose
Maltase
Glucose
Digestion of maltose in the small intestine

Through the process of digestion, starch is broken down into molecules of glucose. The majority of starch digestion takes place in the small intestine.

Figure 4.12 **Disaccharide Digestion**

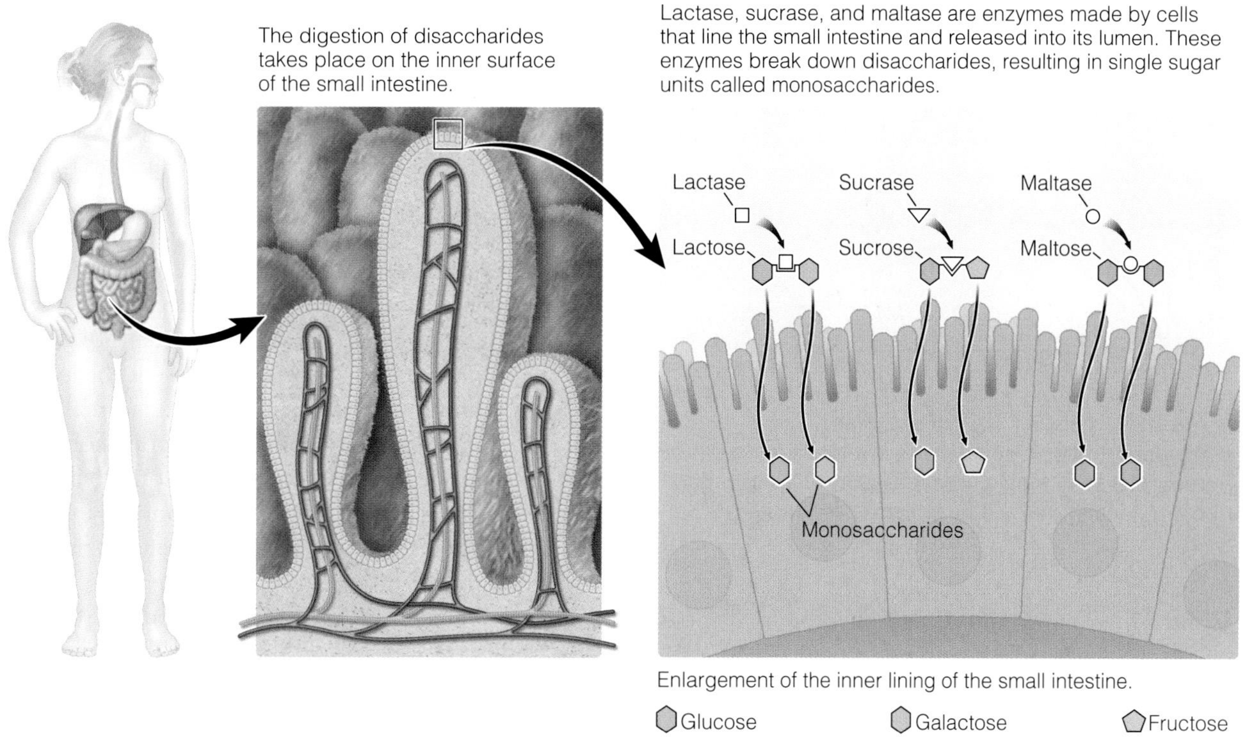

The digestion of disaccharides takes place on the inner surface of the small intestine.
Lactase, sucrase, and maltase are enzymes made by cells that line the small intestine and released into its lumen. These enzymes break down disaccharides, resulting in single sugar units called monosaccharides.
Lactase
Sucrase
Maltase
Lactose
Sucrose
Maltose
Monosaccharides
Enlargement of the inner lining of the small intestine.
Glucose
Galactose
Fructose

Enzymes made by the small intestine are needed for disaccharide digestion.

such as African Americans, Native Americans, and Asian Americans. In addition, the ability to produce lactase can decline with age, making lactose intolerance somewhat more common in older adults. When people with lactose intolerance consume lactose-containing foods, much of the lactose enters the large intestine undigested. Bacteria in the large intestine break down the lactose, which produces gas and sometimes induces other symptoms such as abdominal cramping and bloating. Most people with lactose intolerance can consume small amounts of low-lactose dairy products such as yogurt and cheese. The availability of lactose-free products and lactase-containing preparations makes it relatively easy for those with lactose intolerance to enjoy dairy products and obtain important nutrients such as calcium.

Monosaccharide Absorption

Once disaccharide and starch digestion is complete, the resulting monosaccharides (glucose, galactose, and fructose) are taken up by the cells lining the small intestine and subsequently released into the blood. The blood then carries the monosaccharides directly to the liver. Because monosaccharides enter the bloodstream relatively quickly after consumption, a rise in blood glucose levels can be detected shortly after you eat most carbohydrate-rich foods. However, not all carbohydrates have the same effect on blood glucose levels. Some foods cause blood glucose levels to rise quickly and remain elevated, while others elicit a more subdued or gradual increase.

The change in blood glucose following the ingestion of a food is called the **glycemic response** (see Figure 4.13 on the next page). Scientists have long believed that simple carbohydrates cause a greater glycemic response than complex carbohydrates. However, this is not always the case: starchy foods containing complex carbohydrates such as potatoes, refined cereal products, white bread, some whole-grain breads, and white rice tend to elicit similar glycemic

glycemic response The change in blood glucose following the ingestion of a food.

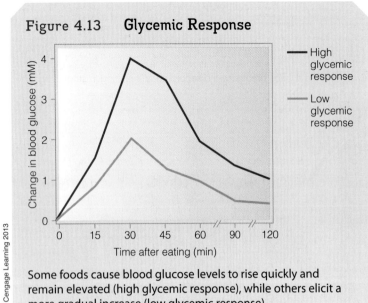

Figure 4.13 Glycemic Response

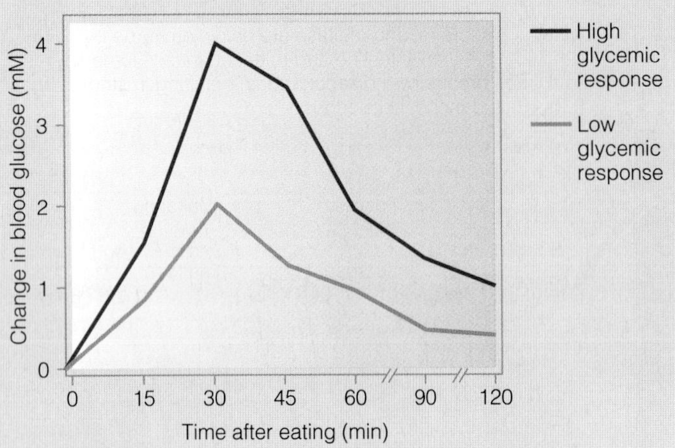

Some foods cause blood glucose levels to rise quickly and remain elevated (high glycemic response), while others elicit a more gradual increase (low glycemic response).

responses than many simple carbohydrate-rich foods such as soft drinks. Scientists now recognize that in addition to a carbohydrate's chemical complexity, other factors also influence a food's ability to raise blood glucose levels. For example, whether a food is raw or cooked, ground or whole, processed or unprocessed are all factors that must be considered.

GLYCEMIC INDEX AND GLYCEMIC LOAD

The **glycemic index (GI)**, a rating system based on a scale of 0 to 100, can be used to compare the glycemic responses elicited by different foods. In this system, the blood glucose response to the consumption of 50 grams of pure glucose is used as a reference point represented by the highest GI value, 100. The GI value of a given food is determined experimentally: subjects consume the amount of food that contains 50 grams of carbohydrate, and their blood glucose is measured. Using this standardized experiment, researchers can compare the glycemic response elicited by any food to that of eating 50 grams of pure glucose. Foods that elicit glycemic responses similar to that of pure glucose are considered high-GI foods (GI $\geq$ 70), whereas those that cause a lower and/or more gradual rise in blood glucose are considered low-GI foods (GI $\leq$ 55).

glycemic index (GI) A rating system based on a scale of 0 to 100 used to compare the glycemic responses elicited by different foods.

glycemic load (GL) A rating system used to compare the glycemic responses associated with different foods that takes into account the typical portion of food consumed.

One limitation of using GI values to compare the effects of foods on glycemic response is that the amount of carbohydrate found in a typical serving of a food is not taken into account. Rather, GI values are based on a standard amount of carbohydrate (50 grams), which may or may not represent the average amount that a person would normally eat. For example, to consume 50 grams of carbohydrates from carrots, a person would need to consume more than a pound of carrots, which is unrealistic. To counter this problem, another rating system is sometimes used to evaluate the glycemic response to foods. In this system, foods are assigned a value called a **glycemic load (GL)**. A food's GL value is different from its GI value in that the GL takes into account the typical portion of food consumed. Glycemic load is calculated by dividing the glycemic index of a food by 100 and multiplying that number by the food's available carbohydrate content (in grams). You can compare the GI and GL values of selected foods in Table 4.3.

There is considerable interest as to whether a diet emphasizing low-GL foods may be beneficial to health.[12] Although some health advocates believe that a diet based solely on foods with low GI and GL can greatly improve long-term health, the truth is that diets are significantly more complex than that. It is important to recognize that the GI and GL values associated with a food are determined experimentally and do not take into consideration other factors, such as the food's preparation method, its combination with other foods, and individual genetic differences. Until nutritional scientists know more, it is important to make healthy carbohydrate choices by increasing your consumption of whole grains, fruits, and vegetables, and decreasing your consumption of foods made with refined flour and added sugar.

Table 4.3 **Glycemic Index (GI) and Glycemic Load (GL) of Selected Foods**

Interpretation of value	Glycemic index (GI)	Glycemic load (GL)
High	≥70	≥20
Medium	56–69	11–19
Low	≤55	≤10

Food (serving size)	GI	Carbohydrates per serving (g)	GL	Food (serving size)	GI	Carbohydrates per serving (g)	GL
Pastas, grains, legumes, breads, starchy vegetables, and cereals							
Baked potato (150 g)	85	30	26	Corn Chex® (30 g)	83	30	25
Waffles (35 g)	76	13	10	Corn (80 g)	48	16	8
French fries (150 g)	75	29	22	Popcorn (20 g)	54	11	6
Bagel (70 g)	72	35	25	Cracked wheat (150 g)	48	26	12
Oat bran bread (30 g)	44	18	8	Pancakes (80 g)	67	58	39
White rice (150 g)	56	42	24	Apple muffins (60 g)	44	29	13
Angel food cake (50 g)	67	29	19	Lentils (50 g)	30	40	12
Raisin Bran (30 g)	61	19	12	Rye bread (30 g)	41	12	5
Fruits, beverages, and snack foods							
Apple (22 g)	38	22	8	Grapes (120 g)	43	17	7
Raisins (60 g)	64	44	28	Apple juice (250 ml)	39	25	10
Banana (120 g)	48	25	12	Tomato juice (250 ml)	38	9	3
Potato chips (28 g)	54	15	8	Plums (120 g)	24	14	3
Jelly beans (28 g)	80	26	21	Chocolate (28 g)	49	18	9
Cherries (120 g)	22	12	3	Ice cream (50 g)	61	13	8

Source: Foster-Powell K, Holt SH, Brand-Miller JC. International table of glycemic index and glycemic load values: 2002. American Journal of Clinical Nutrition. 2002;76:5–56.

LO4 How Does Your Body Regulate and Use Glucose?

Once absorbed, monosaccharides circulate directly to the liver, where the majority of galactose and fructose molecules are converted into other compounds—most notably glucose. Some molecules are converted to ribose, a constituent of many vital compounds including ATP, RNA, and DNA. While monosaccharides serve numerous functions within the body, the ability to convert the energy contained in glucose into ATP is probably the most notable role.

The concentration of glucose in your blood fluctuates throughout the day depending on when you eat, what you eat, and how much you eat. After several hours without eating, blood glucose decreases. An abnormally low level of glucose in the blood is referred to as **hypoglycemia**. Conversely, blood glucose increases

hypoglycemia A condition characterized by low blood glucose.

hyperglycemia A condition characterized by an excess of glucose in the blood.

insulin A hormone secreted by the pancreas in response to increased blood glucose.

glucagon A hormone secreted by the pancreas in response to decreased blood glucose.

insulin receptors Specialized proteins located on the outer membranes of certain types of cells that bind insulin.

after one eats a meal rich in carbohydrates. Elevated levels of glucose in the blood is referred to as **hyperglycemia**. Blood glucose levels are the lowest in the morning after an overnight fast, returning to normal shortly after eating. Because your cells need energy throughout the day, insulin and glucagon work vigilantly to keep blood glucose levels within an acceptable range at all times (see Figure 4.14). **Insulin** and **glucagon** are hormones secreted by the pancreas in response to increased and decreased blood glucose, respectively.

Insulin

After a person eats carbohydrate-rich foods, blood glucose levels quickly rise, leading to hyperglycemia. This in turn prompts the pancreas to increase its release of insulin. Insulin has several notable effects:

- It allows certain cells, like those in skeletal muscle and adipose tissue (body fat), to take up glucose from the blood.
- It increases the rate at which glucose is used as an energy source.
- It accelerates the formation of energy-storing glycogen molecules from glucose.
- It promotes the formation of protein in muscle.
- It promotes the conversion of glucose to fat stored in adipose tissue.

All of these actions work together to help bring blood glucose levels back down to normal—an excellent example of how the body maintains homeostasis. Insulin is capable of communicating only with tissues whose surfaces have built-in "receivers" for its message. These receivers, referred to as **insulin receptors**, are located on the outer membranes of certain types of cells, allowing the cells to bind insulin. The entire process of blood glucose regulation is illustrated in Figure 4.15.

EXCESS GLUCOSE IS CONVERTED TO GLYCOGEN AND FAT

After a meal, more glucose may be available than the amount that is needed to power the body. When this occurs, the body stores the excess energy supplied by the glucose for later use. The hormone insulin promotes the storage of this excess energy as glycogen, but the body can only store a limited amount of glycogen. Once this limit is reached, excess glucose is redirected to metabolic pathways that convert it to fat. Unlike glycogen storage, the body has a seemingly endless capacity to store body fat.

Glucagon

After several hours without food, blood glucose levels begin to decline. This shifts the hormonal balance away

Figure 4.14 Release of Insulin and Glucagon from the Pancreas

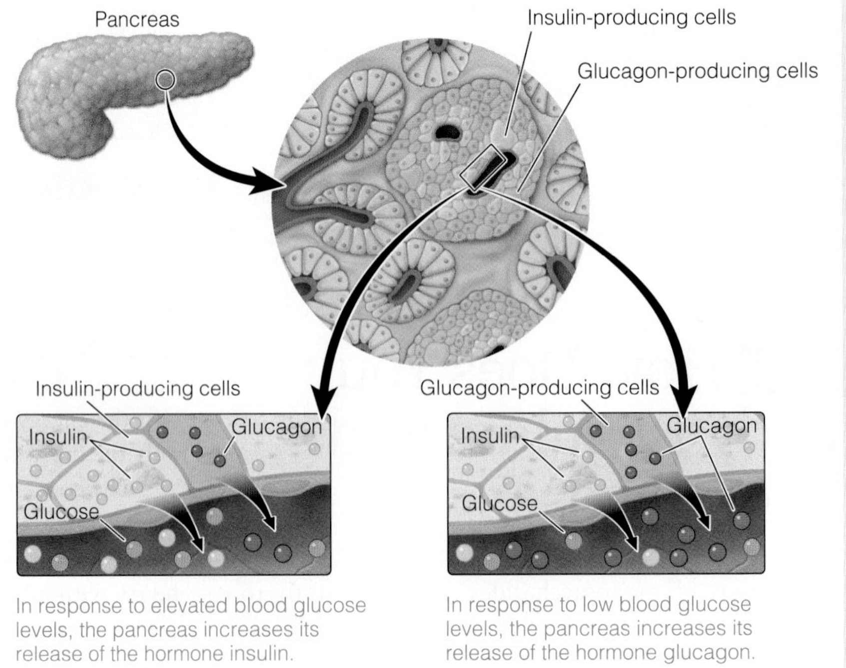

Pancreas
Insulin-producing cells
Glucagon-producing cells
Insulin-producing cells
Insulin
Glucagon
Glucose
Glucagon-producing cells
Insulin
Glucagon
Glucose
In response to elevated blood glucose levels, the pancreas increases its release of the hormone insulin.
In response to low blood glucose levels, the pancreas increases its release of the hormone glucagon.

The hormones insulin and glucagon are made and released by the pancreas. Both hormones play an important role in blood glucose regulation.

Figure 4.15 Hormonal Regulation of Blood Glucose

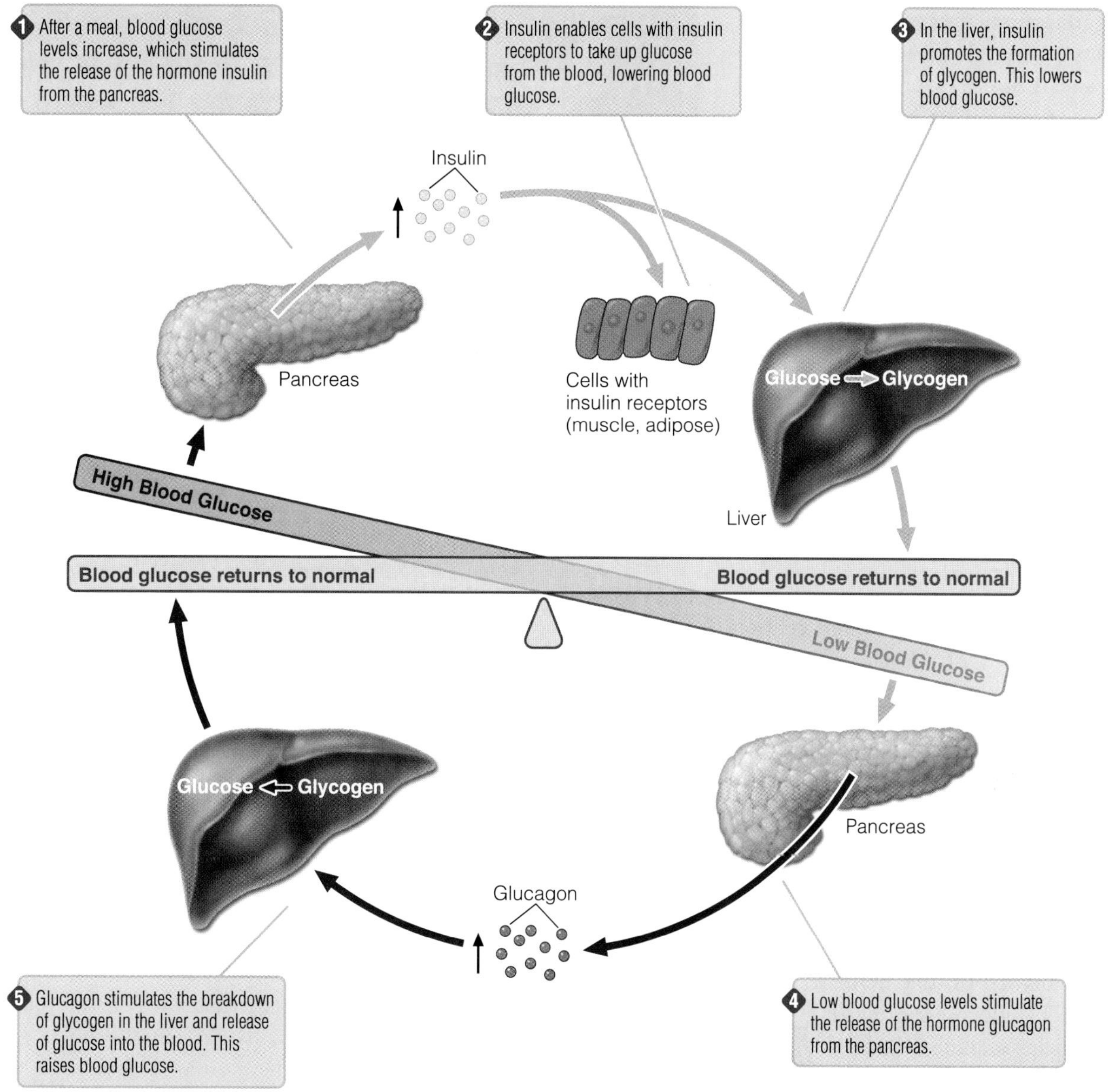

1 After a meal, blood glucose levels increase, which stimulates the release of the hormone insulin from the pancreas.
2 Insulin enables cells with insulin receptors to take up glucose from the blood, lowering blood glucose.
3 In the liver, insulin promotes the formation of glycogen. This lowers blood glucose.
Insulin
Pancreas
Cells with insulin receptors (muscle, adipose)
Glucose → Glycogen
Liver
High Blood Glucose
Blood glucose returns to normal
Blood glucose returns to normal
Low Blood Glucose
Glucose ← Glycogen
Pancreas
Glucagon
5 Glucagon stimulates the breakdown of glycogen in the liver and release of glucose into the blood. This raises blood glucose.
4 Low blood glucose levels stimulate the release of the hormone glucagon from the pancreas.

Insulin lowers blood glucose and promotes energy storage. Glucagon increases blood glucose by promoting the breakdown of liver glycogen.

from insulin and toward glucagon. Glucagon's primary function is to increase blood glucose. The brain is particularly sensitive to low blood glucose levels, and even a relatively minimal dip into hypoglycemia can make a person feel nauseous, dizzy, anxious, lethargic, and irritable. This is one reason why it is hard to concentrate when you have not eaten for a long time.

To increase glucose availability during these times, glucagon stimulates the breakdown of energy-storing glycogen molecules in the liver into glucose. These glucose molecules are released into the blood, increasing glucose availability to the rest of the body. The term for this metabolic process—**glycogenolysis**—literally means the breakdown ("lysis") of glycogen. Liver glycogen can supply glucose for approximately 24 hours before being depleted.

The breakdown of liver glycogen is an effective short-term solution for providing cells with glucose. However, because this

glycogenolysis The breakdown of glycogen into glucose.

reserve can be quickly depleted, the body must soon find an alternative glucose source. As glycogen stores dwindle, glucagon stimulates another metabolic process called gluconeogenesis. **Gluconeogenesis** is the synthesis of glucose from noncarbohydrate sources. Taking place mainly in the liver, gluconeogenesis synthesizes glucose mainly from amino acids derived from muscle protein, the body's major reservoir of amino acids. You may be wondering why cells do not convert fat into glucose. Although fat can provide cells with plentiful amounts of energy, the carbon atoms contained in fat cannot be utilized to make glucose.

KETONES SPARE THE USE OF AMINO ACIDS FOR GLUCOSE PRODUCTION

Gluconeogenesis increases glucose availability, but too much can have negative consequences; the body cannot rely on the breakdown of muscle protein for its source of glucose for very long. To minimize loss of muscle, the body reduces its dependency on glucose by using an alternative energy source called a **ketone**. These organic compounds are made from fatty acids under conditions of limited glucose availability. Ketones produced from the metabolism of body fat are released into the blood and used primarily by the brain for energy. This glucose-sparing response helps minimize the loss of muscle protein by lessening the body's demand for glucose. Ketone synthesis is not without its own consequences, however. When excessive ketones accumulate in the blood, a condition called **ketosis** can occur, causing a variety of complications including loss of appetite. In fact, this is one reason why many popular low-carbohydrate diets actually work: people who follow them often eat less.

Ketogenic Diets

Ketogenic diets are not new to the medical world; therapeutic ketogenic diets were once used to prevent seizures in children with severe epilepsy. Although physicians did not understand exactly why the diet worked in some cases, it appeared that ketones somehow altered metabolism in the brain. Today, ketogenic diets are employed primarily for weight loss. They mimic starvation by forcing the body to use fat rather than carbohydrates as its primary energy source. Some experts believe that chronic consumption of the carbohydrate-rich foods that cause insulin levels to rise can lead to weight gain. Thus, limiting starch and refined sugars in one's diet should theoretically help a person lose weight. Several studies evaluating the effectiveness and safety of low-carbohydrate weight-loss diets reported that these diets were effective alternatives to low-fat diets.[13] Although low-carbohydrate diets appear safe in the short term, some experts have raised doubts about their long-term effectiveness.

Some **popular weight-loss diets**, such as the Atkins Nutritional Approach, **promote ketosis** as the key to burning body fat, losing weight, and living a healthy life.

© Mario Ruiz/ZUMA Press/Newscom

gluconeogenesis The synthesis of glucose from noncarbohydrate sources.

ketone An organic compound used as an alternative energy source under conditions of limited glucose availability.

ketosis A condition characterized by excessive ketone accumulation in the blood.

EPINEPHRINE STIMULATES QUICK GLUCOSE RELEASE

Whereas glucagon is involved in the day-to-day regulation of blood glucose, another hormone, **epinephrine**, is released from the adrenal glands to stimulate glycogenolysis in emergency situations. Acting on the liver and muscle cells, epinephrine is released during times of emotional and physical duress. This immediate response, sometimes called the fight-or-flight reaction, signals the quick release of glucose into the blood. This helps ensure that glucose is available during extreme circumstances.

Glucose as an Energy Source

A rich source of energy, glucose can be used by every cell in the body to make ATP. This conversion is accomplished by means of a catabolic pathway (see Chapter 3). The first step in the metabolic breakdown of glucose is **glycolysis**, a series of chemical reactions that splits glucose, a six-carbon molecule, into two three-carbon **pyruvate** molecules. Because oxygen is not required for any of the steps in this pathway, glycolysis is considered to be an anaerobic metabolic pathway. The amount of energy released through glycolysis is small—enough to make two ATPs per glucose molecule—but there is much more energy yet to be harvested.

When conditions are right, the pyruvate molecules can be broken down further to yield additional energy. This requires another metabolic pathway that functions under oxygen-rich conditions. This series of oxygen-requiring chemical reactions, referred to as aerobic metabolism, releases considerably more energy: about 36 ATPs are produced through the complete breakdown of one glucose molecule. When ATP is produced this way, waste products such as water and carbon dioxide are also generated (see Figure 4.16). Recall that metabolic waste products such as these are eliminated from the body via expired air, urine, and sweat.

Figure 4.16 Glucose Metabolism

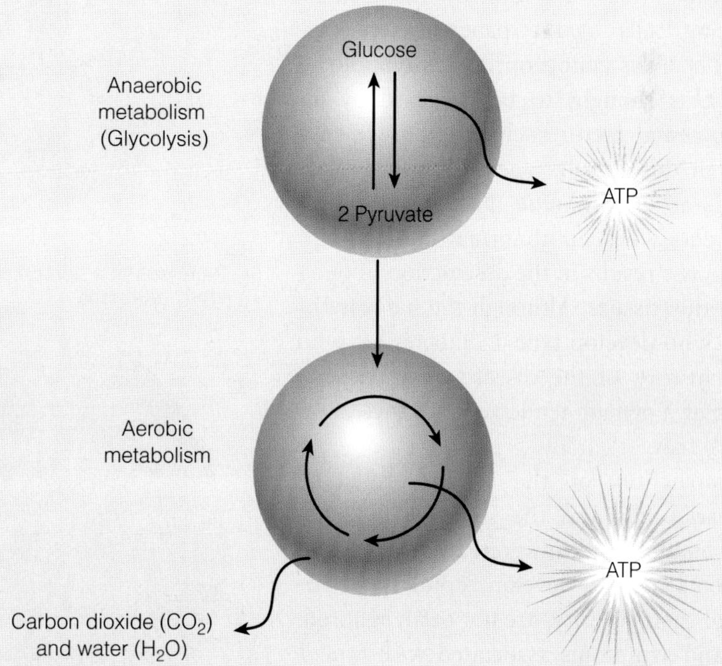

Glycolysis is an anaerobic pathway that splits glucose, a six-carbon molecule, into two three-carbon pyruvate molecules. The amount of energy released from glycolysis is small—enough to make just two ATPs. Under oxygen-rich conditions, pyruvate can be broken down further through aerobic metabolism. This results in more energy being released—enough to make about 36 ATPs. Metabolic waste products resulting from glucose metabolism include water and carbon dioxide.

LO5 What Is Diabetes?

Diabetes mellitus is a group of metabolic disorders characterized by elevated levels of glucose in the blood—otherwise known as hyperglycemia. There are different types of diabetes, which have been classified in a variety of ways since their discovery. Diabetes was once categorized according to the typical age of onset, but was later reclassified based on whether insulin injections were required as a treatment. Because there were many problems associated with both these classifications, the American Diabetes Association developed a new system of diabetes classification

epinephrine A hormone released from the adrenal glands that stimulates glycogenolysis in emergency situations.

glycolysis An anaerobic metabolic pathway made up of a series of chemical reactions that splits glucose into two three-carbon molecules.

pyruvate The end-product of glycolysis; formed by the breakdown of glucose.

diabetes mellitus A group of metabolic disorders characterized by elevated levels of glucose in the blood.

type 1 diabetes A form of diabetes whereby the pancreas is no longer able to produce insulin, causing blood glucose levels to become dangerously high.

autoimmune disease An illness that occurs when an abnormal immunological response results in the destruction of one's own bodily tissues.

glucometer A medical device used to monitor the concentration of glucose in the blood.

based on etiology, or underlying cause. Today, the two main types of diabetes are referred to as type 1 diabetes and type 2 diabetes. A third type, gestational diabetes, is discussed in Chapter 10.

Type 1 Diabetes

Type 1 diabetes occurs when the pancreas is no longer able to produce insulin. Without insulin, some types of cells cannot take up glucose, causing blood glucose levels to become dangerously high. Approximately 5 to 10 percent of all people with diabetes have type 1. Although type 1 diabetes can develop at any age, it most often does so during childhood and early adolescence.

Although the exact cause of type 1 diabetes is unknown, it develops when a person's immune system produces antibodies that mistakenly attack and destroy the insulin-producing cells of the pancreas (see Figure 4.17). This inappropriate immunologic response is thought to be triggered by an environmental factor such as exposure to a virus. For this reason, type 1 diabetes is classified as an **autoimmune disease**, an illness that occurs when an abnormal immunological response results in the destruction of one's own bodily tissues. Although the majority of people who develop type 1 diabetes have no family history of the disease, most experts agree that a genetic tendency does increase a person's risk.

Destruction of the insulin-producing pancreatic cells and the subsequent depletion of insulin result in severe hyperglycemia. Symptoms tend to develop rapidly, and because of their severity, are not easily ignored. Signs and symptoms associated with type 1 diabetes include rapid weight loss, extreme thirst, and frequent urination. Type 1 diabetes also causes a person to feel hungry and weak because cells are starved for energy. In fact, diabetes is often described as "starvation in the midst of plenty."

For people with type 1 diabetes, administration of insulin is necessary to control blood glucose levels. Insulin can be administered through a regimen of frequent injections or by an insulin pump. Without an exogenous source of insulin, type 1 diabetes is generally fatal. When diagnosed with type 1 diabetes, a person must learn to balance insulin treatments with a healthy diet and physical activity. A medical device called a **glucometer** helps those with type 1 diabetes monitor the concentration of glucose in the blood so they can know how much insulin to administer. Using a small sample of blood, these user-friendly devices provide immediate feedback regarding an individual's blood glucose levels.

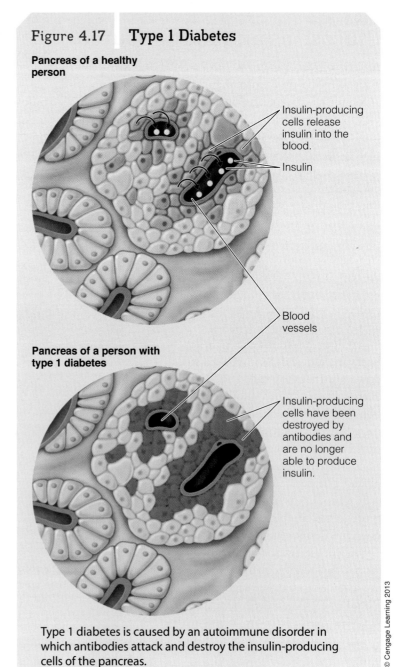

Figure 4.17 Type 1 Diabetes

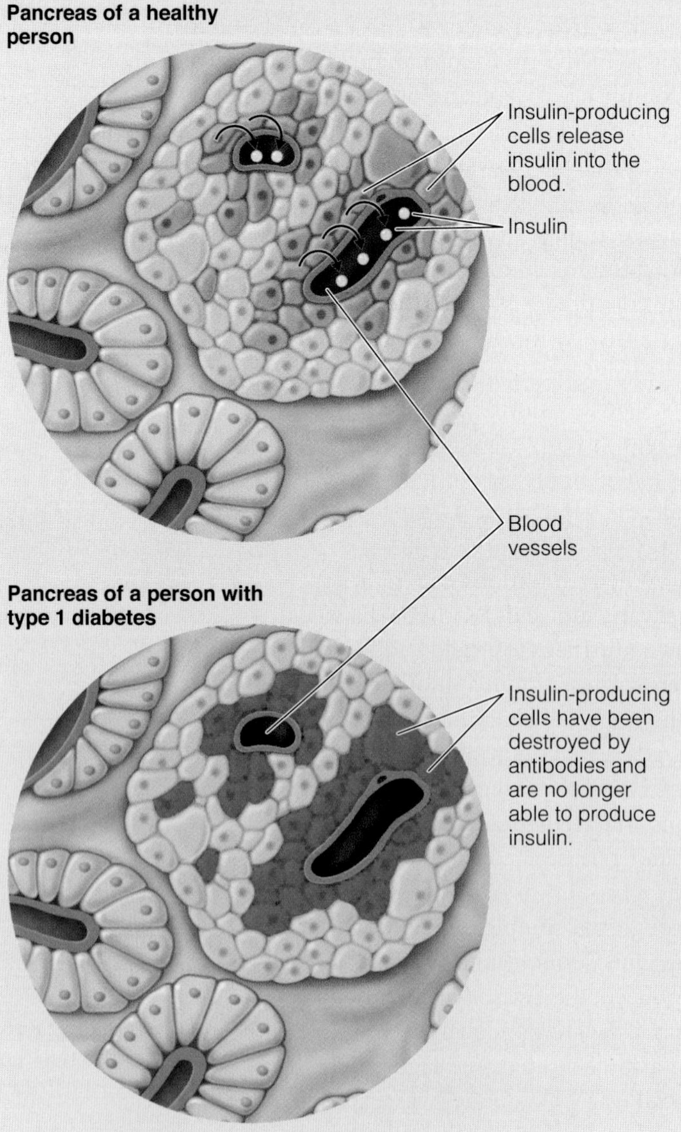

Type 1 diabetes is caused by an autoimmune disorder in which antibodies attack and destroy the insulin-producing cells of the pancreas.

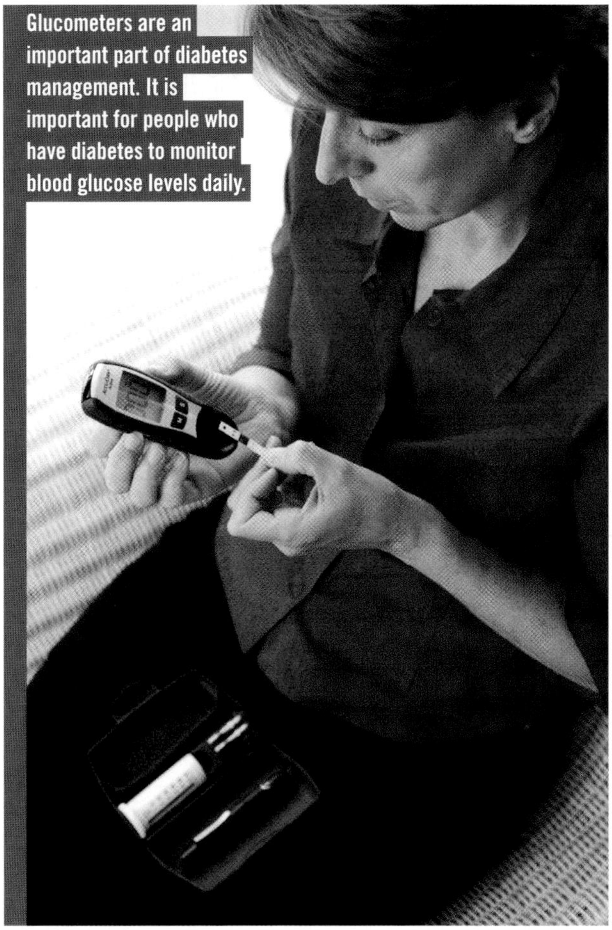

Glucometers are an important part of diabetes management. It is important for people who have diabetes to monitor blood glucose levels daily.

Type 2 Diabetes

Type 2 diabetes is by far the most common form of diabetes: 90 to 95 percent of people with diabetes fall into this category. In the United States, type 2 diabetes has become so widespread that an estimated one in four people has, will have, or has a family member who has this disease.[14] Although type 2 diabetes can occur at any age, it most frequently develops in adults middle-aged and older. However, the rising prevalence of obesity in children and teens throughout the United States has contributed to a new and alarming trend: an escalation in the number of children and teens diagnosed with type 2 diabetes.[15] Given this trend, type 2 diabetes can no longer be thought of as a condition that affects adults only.

Unlike type 1 diabetes, most people with type 2 diabetes have normal or even elevated levels of insulin in their blood. Type 2 diabetes is caused by **insulin resistance**, meaning that insulin receptors throughout the body are less responsive to insulin. Because the body's cells do not respond appropriately to insulin's signal, the amount of glucose they take up from the bloodstream is notably diminished. When blood glucose levels rise beyond the normal range, a person begins to experience symptoms associated with type 2 diabetes. Symptoms tend to develop gradually and are often ignored—this is why type 2 diabetes can go undiagnosed for many years. Some of the early symptoms associated with type 2 diabetes include fatigue, frequent urination, and excessive thirst.

type 2 diabetes A form of diabetes whereby insulin resistance prevents cells from taking up glucose from the blood.

insulin resistance A condition whereby insulin receptors throughout the body are less responsive to insulin.

There are numerous risk factors associated with the development of type 2 diabetes (see Table 4.4 on the next page). Although a person has little control over some of these risk factors (such as genetic makeup and ethnicity), one can make informed lifestyle choices to help prevent the disease. Studies show that even individuals who are genetically susceptible to the development of type 2 diabetes can significantly reduce their risk by eating a variety of healthy foods, staying physically fit, and maintaining a recommended body weight.[16] The last point is particularly important because obesity is a profoundly significant risk factor associated with type 2 diabetes. Approximately 80 percent of people diagnosed with type 2 diabetes are overweight or obese. Further, a particular distribution of body fat can pose additional risk for type 2 diabetes: body fat stored in the abdominal region of the body presents a greater risk than does that stored in the lower regions of the body.[17]

Preventing Complications Associated with Diabetes

As the prevalence of obesity in the United States continues to climb, so does the prevalence of type 2 diabetes. If these trends continue, researchers estimate that

> “A person diagnosed with **diabetes at age 30** spends about $305,000 on treatment over her lifetime, whereas a person **diagnosed at age 70** has a total cost of $67,000.”

Table 4.4 Risk Factors Associated with Type 2 Diabetes	
Having a parent, brother, or sister with diabetes	Being over 45 years of age
Delivering a baby weighing more than 9 pounds (for women)	Being of African, Hispanic, Native American, or Pacific Island descent
Maintaining a sedentary lifestyle (by exercising fewer than three times per week)	Having high blood pressure (greater than or equal to 140/90 mmHg)
Being obese, especially centrally obese	Having low HDL cholesterol (less than 35 milligrams per deciliter)
History of vascular disease and/or previous identification of impaired blood glucose regulation	Having high triglycerides in the blood (greater than 250 milligrams per deciliter)

Adapted from American Diabetes Association. Position statement from the American Diabetes Association: Screening for type 2 diabetes. Diabetes Care. 2004;27:s11-s14. Available from: http://care.diabetesjournals.org/content27/suppl_1/s11.full.pdf.

18 million Americans will have type 2 diabetes by the year 2020.[18] The greatest concern raised by these statistics is that diabetes can lead to other serious health problems. For example, diabetes increases one's risk of having a heart attack or stroke. It can also lead to a loss of feeling in the feet and hands, blindness, limb amputation, and impaired kidney function. These long-term complications are largely attributable to the harmful effects of excess glucose on blood vessels and nerves.

The good news is that maintaining a near-normal level of blood glucose can often prevent these complications. In the case of type 1 diabetes, one must take care to balance carbohydrate intake with insulin injections. In the case of type 2 diabetes, one can often control blood glucose through a combination of weight management, regular exercise, and a healthy diet. In fact, most of the nutritional guidelines developed to manage type 2

What the Pima Indians Have Taught Us about Type 2 Diabetes

Because of their unusually high rates of obesity and type 2 diabetes, researchers have studied the Pima Indians living in the American Southwest for over 30 years.[19] During this time, researchers have also studied a group of genetically-related Pima Indians still living in a remote area in Mexico. By contrast, this second group of Pima experiences very low rates of type 2 diabetes and obesity. Scientists have concluded that although both groups of Pima have similar genetics, only those living in an environment of abundant food and reduced physical activity tend to develop type 2 diabetes. Those who maintain traditional diets and physically active lifestyles are likely to remain free of diabetes. The Pima Indians' story exemplifies the fact that alterations in diet and physical activity can either lead to or help prevent type 2 diabetes—even in a genetically susceptible population.

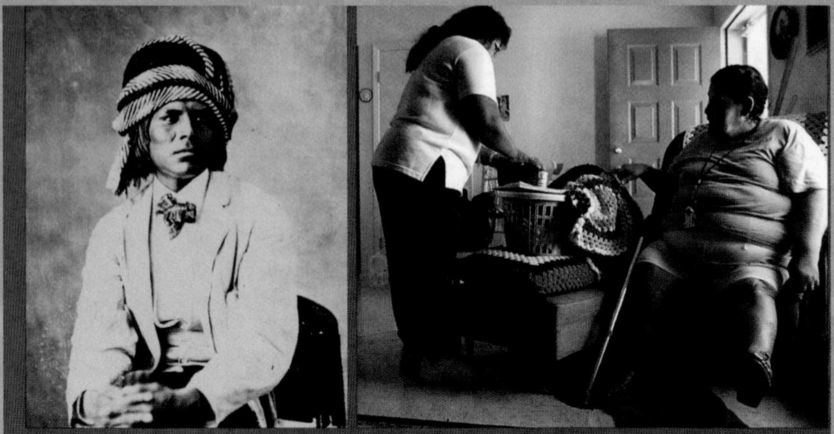

Genetics and environment both play a role in the development of type 2 diabetes. Research conducted on the Pima Indians has provided evidence that a healthy lifestyle can help prevent type 2 diabetes, even in a genetically susceptible population.

Photo by Alexander Gardner, Prints and Photographs Division, Library of Congress, LC-USZ62-126404 / AP Images/J. Pat Carter diabetes can also be applied by anyone who wants to maintain good health and eat a balanced diet that emphasizes fruits, vegetables, whole grains, legumes, low-fat dairy products, low-fat meat, and foods prepared with minimum amounts of added sugar.

LO6 What Are the Recommendations for Carbohydrate Intake?

Although carbohydrates are not technically essential nutrients, they are nonetheless an important part of your diet because they provide energy and dietary fiber. Some tissues, such as the brain, rely extensively on glucose for energy. Furthermore, there is evidence that consumption of certain carbohydrates may help prevent chronic diseases. Thus, to ensure that individuals make healthy carbohydrates a part of their eating patterns, the 2010 Dietary Guidelines for Americans emphasizes the importance of eating a variety of brightly colored vegetables (dark-green, red, and orange), and legumes such as dried peas and beans. Replacing at least half your intake of refined grains with whole-grain foods is also emphasized as an important step in improving your intake of healthy carbohydrates. This recommendation is particularly important because refined grain-based foods often contain high levels of solid fats, added sugars, and sodium. By making this dietary change, your overall health can benefit in many ways.

Dietary Reference Intakes for Carbohydrates

The Institute of Medicine's Dietary Reference Intake (DRI) values for carbohydrates were developed primarily to ensure that the brain has adequate glucose for its energy needs. The Recommended Dietary Allowance (RDA) for carbohydrates—the minimum amount of glucose needed by the brain each day—is 130 g for adults. For most adults, a well-balanced diet that includes two servings of fruit, three servings of vegetables, and six servings of grains every day easily provides the recommended amount of carbohydrates. In terms of overall energy distribution, the Acceptable Macronutrient Distribution Range (AMDR) suggests that 45 to 65 percent of your total caloric intake should come from carbohydrates.

In addition to ensuring that daily glucose needs are met, the Institute of Medicine's recommendations aim to minimize one's risk for chronic disease and promote optimal health. To this end, they suggest that in addition to the amount of carbohydrates you consume, it is important to pay attention to the types and sources of carbohydrates you choose. Although a candy bar and a serving of breakfast cereal contain roughly the same number of carbohydrate grams (24 g), a 2-ounce candy bar has approximately 275 kcal, whereas a 1-cup serving of breakfast cereal has approximately 150 kcal. Because some carbohydrate-rich foods are more nutrient-dense (and thus nutritional) than others, one must weigh a number of factors before deciding which carbohydrate-containing foods to consume.

Making the Right Choices

When it comes to carbohydrate-rich foods, there are many from which to choose. While the choice can

Although a **candy bar** *and a serving of* **whole-grain breakfast cereal** *contain roughly the same amount of carbohydrates (24 g), the breakfast cereal is more* **nutrient-dense** *and is high in* **dietary fiber.**

> Because **food labels' ingredients** lists are arranged in **descending order by weight**, such lists can give you a rough idea about the sugar content of a food.

be overwhelming, health experts agree that the best strategy is to maximize your intake of foods that are nutrient-dense and provide plenty of fiber, vitamins, and minerals. Conversely, it is important to minimize your intake of foods high in fat and sugar. Following the 2010 Dietary Guidelines for Americans can help eliminate much of the guesswork when it comes to determining which carbohydrate-rich foods to choose. One general guideline is to consume more fruits, vegetables, and whole-grain foods; these foods provide naturally occurring sugars, fiber, and plenty of vitamins and minerals. The MyPlate food guidance system, which emphasizes whole grains, fruit, and vegetables as the foundation of a healthy diet, is another useful tool to evaluate your carbohydrate food choices.

Another general guideline is to minimize highly processed foods that contain added sugars. Although there is no consensus as to how much total added sugar to consume (or more importantly, to avoid), health experts generally agree that one's consumption of foods with high amounts of added sugar (such as cookies, soda, sugary cereals, and heavy syrups) should be minimized. Table 4.5 lists the average amounts of added sugar in selected foods and beverages.

A BALANCED DIET NECESSITATES A HEALTHY INTAKE OF DIETARY FIBER

As you have learned, there are many health benefits associated with dietary fiber. One way to add more fiber to your diet is to read nutrition labels and choose those foods labeled "a good source of fiber." It is equally important when considering grain-based foods to look for terms such as *whole grain* and *whole wheat* on food packaging. The current recommendation for adults is to consume an equivalent of six to eight ounces of grain-based foods daily, half of which should be whole

Table 4.5 Added Sugars in Selected Food Items

Food	Serving Size	Added Sugar (g)[a]
Soft drink	12 oz	43
Milkshake	10 oz	36
Fruit punch	8 oz	38
Chocolate candy	1.5 oz	24
Sweetened breakfast cereal	1 cup	15
Yogurt with fruit	1 cup	33
Ice cream	1 cup	28
Cake with frosting	1 slice	28
Cookies	2 (medium)	14
Jam or jelly	1 tbsp	2

[a] 1 teaspoon equals 4.75 grams.

Source: United States Department of Agriculture database for the added sugars content of selected foods, Release 1 (2006). Available from: http://www.nal.usda.gov/fnic/foodcomp/Data/add_sug/addsug01.pdf.

grain. Examples of whole-grain foods are brown rice, pasta made with whole wheat, and whole-grain bread. Outside of grain-based products, fruits and vegetables should also contribute to one's daily intake of dietary fiber. The Institute of Medicine recommends that adults consume at least 14 g of dietary fiber per 1,000 kcal. The average daily intake of fiber in the United States is about half this amount.[20]

Most experts agree that a diet high in fiber is beneficial to one's health and carries minimal adverse effects. While some fibers may have a tendency to bind minerals such as calcium, zinc, iron, and magnesium, it is doubtful that the consumption of dietary fiber in recommended amounts affects mineral status in healthy adults. However, a few words of caution must be stated regarding the overconsumption of fiber: more is not always better. A sudden and/or large increase in fiber intake may cause a GI problem such as diarrhea or constipation. As such, a person should increase her fiber intake gradually and give the body time to adjust. It is also wise to increase fluid intake as one increases fiber—this can help alleviate common problems such as constipation.

THE
IN-
CROWD
Share your 4LTR Press story on Facebook at
www.facebook.com/4ltrpress for a chance to win.
To learn more about the In-Crowd opportunity 'like' us on Facebook.
© Go Media | © Cengage Learning 2011

NUTR

5 Protein

© iStockphoto.com/Ryan Fox

LEARNING OUTCOMES:

LO1 Define and differentiate various proteins and their components.

LO2 Describe the process by which cells make protein.

LO3 Explain the significance of a protein's shape.

LO4 Define and understand *genetics* and *epigenetics*.

LO5 Explain protein digestion, absorption, and circulation.

LO6 Describe the major functions of proteins in the body.

LO7 Explain how the body recycles and reuses amino acids.

LO8 Calculate the amount of protein you need every day.

LO9 Describe vegetarianism and vegetarian diets.

LO10 Know the consequences of protein deficiency and excess.

LO1 What Are Proteins?

The term *protein* was derived more than 170 years ago from the Greek *prota*, meaning "of first importance."[1] Indeed, proteins constitute the most abundant organic substances in the body, making up at least 50 percent of its dry weight.[2] But why exactly do you need proteins, and how do the proteins you eat influence your health?

In this chapter, you will learn what proteins are, what foods are good sources of them, how they are digested and absorbed, and how the body uses them to maintain health. You will also learn about eating patterns such as vegetarianism that can influence protein intake; how certain proteins in foods cause allergic reactions in some people; and how your dietary choices, genetics, and environment can interact to influence the proteins you make and your risk of various diseases. With this information, you can begin to make informed decisions about which protein sources to choose and how to optimize long-term health and well-being.

Perhaps the first questions you should ask when beginning your study of proteins are, "What are proteins, how does one protein differ from another, and what makes proteins different from other nutrients?" A **protein** (also called a peptide) is a nitrogen-containing macronutrient formed when smaller **amino acid** subunits join together via a type of chemical bond called a **peptide bond**. Chemically distinct from the other energy-yielding macronutrients (carbohydrates and lipids), all proteins contain not only carbon and hydrogen but also nitrogen. Proteins come in a variety of sizes: some are very simple, comprised of only a few amino acids, whereas others contain thousands of amino acid building blocks. Most proteins are of intermediate size, containing 250 to 300 amino acids. Because they vary so greatly in size, proteins can be classified based on their number of amino acids: *dipeptides* have two amino acids, *tripeptides* have three, and so forth. A protein with more than 12 amino acids is called a **polypeptide**.

Amino Acid Structure

The numerous proteins in the body are amazingly diverse. The key to this variety lies not only in the number and types of amino acids they contain but also in the order in which the amino acids are linked together. To understand protein diversity, you must first understand the basic components of an amino acid and what makes each one unique.

Amino acids have three *common* components:

- A central carbon atom bonded to a hydrogen atom
- A nitrogen-containing **amino group**
- A carboxylic acid group

In the body, the amino and carboxylic acid groups almost always exist in weakly charged states. These charges cause proteins to bend, ultimately giving rise to protein shape. As you will learn later in the chapter, it is the shape of a protein that imparts its function. In addition to the three common components, each amino acid contains a unique side-chain group called an **R-group**—it is the structure of the R-group that distinguishes each amino acid from the others. The subtle differences in the R-groups give each amino acid its distinctive chemical and physical nature. For example, some of the R-groups are negatively charged, some are positively charged, and

protein A nitrogen-containing macronutrient made from joined amino acids.

amino acid A nitrogen-containing subunit that combines to form proteins.

peptide bond A chemical bond that joins amino acids.

polypeptide A protein comprised of more than 12 amino acids.

amino group The nitrogen-containing component of an amino acid.

R-group The side-chain component of an amino acid that distinguishes it from other amino acids.

α-keto acid A compound similar to an amino acid that does not have an amino group; used to synthesize nonessential amino acids.

transamination The process by which nonessential amino acids are synthesized.

some have no charge at all. Figure 5.1 illustrates a generic amino acid and some examples of R-groups.

Essential, Nonessential, and Conditionally Essential Amino Acids

Just 20 different amino acids combine to construct all the proteins required by the human body. Amino acids can be categorized as essential (or indispensable), nonessential (or dispensable), or conditionally essential (or conditionally indispensable).[3] See Table 5.1. The nine essential amino acids are those you must consume in your diet because your body cannot make them or cannot make them in required amounts. For most people, the remaining 11 amino acids are nutritionally nonessential because they can be synthesized from other compounds. To do this, the body transfers an amino group from an essential amino acid to an **α-keto acid**, which is essentially an amino acid without an amino group. This process, called **transamination**, results in the synthesis of the nonessential amino acids.

Under some conditions, however, the body is unable to synthesize one or more of the nonessential amino acids. Some infants (especially those born prematurely) cannot construct several of the traditionally nonessential amino acids, for example. These amino acids are therefore considered conditionally essential because they must be obtained from the diet until the baby grows and matures.[4] Fortunately, sufficient amounts of the conditionally essential amino acids can typically be obtained from human milk or regular infant formula. Because the milk produced by women who deliver prematurely does not contain the essential amino acids in amounts sufficient to meet the needs of their infants,

Table 5.1 Essential, Nonessential, and Conditionally Essential Amino Acids

Essential	Nonessential
Histidine	Alanine
Isoleucine	Arginine[a]
Leucine	Asparagine
Lysine	Aspartic acid
Methionine	Cysteine[a]
Phenylalanine	Glutamic acid
Threonine	Glutamine[a]
Tryptophan	Glycine[a]
Valine	Proline[a]
	Serine
	Tyrosine[a]

[a] These amino acids are also classified as conditionally essential amino acids; for some people, these amino acids must be consumed in the diet.

Source: Institute of Medicine. Dietary Reference Intakes for energy, carbohydrate, fiber, fat, fatty acids, cholesterol, protein, and amino acids. Washington, DC: National Academies Press; 2005.

Figure 5.1 Components of an Amino Acid

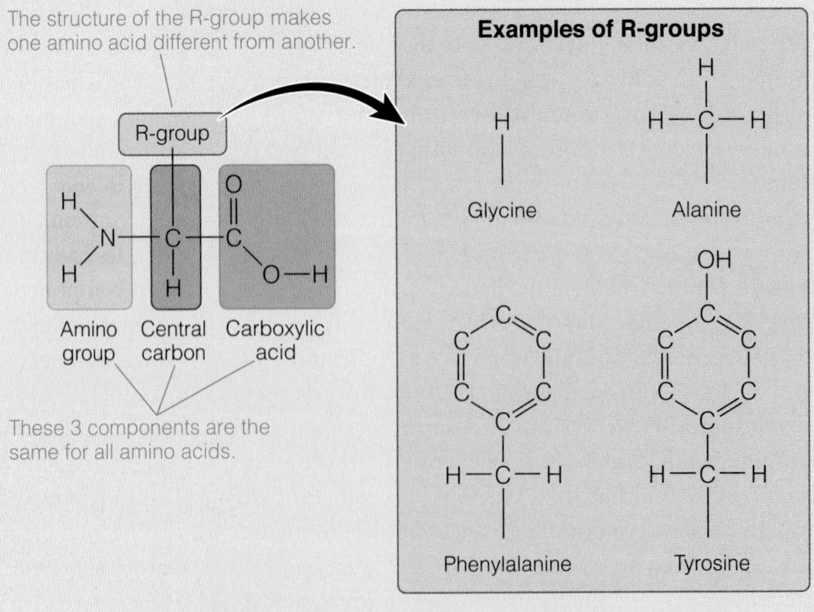

Amino acids have four parts: an amino group, a central carbon, a carboxylic acid group, and an R-group.

Animal-derived foods *such as meat, milk, fish, and eggs are considered* **complete protein sources** *because they contain all of the essential amino acids.*

premature infants' diets are usually supplemented with the proper amino acids.[5]

Certain diseases such as **phenylketonuria (PKU)** can also cause traditionally nonessential amino acids to become conditionally essential. People with PKU do not produce one of the enzymes required to convert the essential amino acid phenylalanine to the normally nonessential amino acid tyrosine. Thus, tyrosine becomes conditionally essential. In all, six amino acids are considered conditionally essential.

Not All Proteins in Food Are Created Equal

As with carbohydrates, some foods generally contain more protein than do others. Meat, poultry, fish, eggs, dairy products, and nuts contain more protein (per gram) than grains, fruits, and vegetables. Leguminous plants such as dried beans, lentils, peas, and peanuts are unique in that they are associated with bacteria that can take nitrogen from the air and incorporate it into amino acids, which legumes use to make their own proteins. This is why legumes tend to contain more protein than most other plants do.

Even foods with the same amounts of total protein can contain different combinations of amino acids. For example, both a cup of cottage cheese and a cup of cooked lima beans provide about 15 g of protein. The specific amino acids present in these two types of foods are quite different, however. Proteins of animal origin (like those found in cottage cheese) tend to be more valuable than those found in plant-derived foods (like lima beans) because the former proteins generally have greater amounts of essential amino acids than do the latter.

phenylketonuria (PKU) A disease whereby the body does not produce one of the enzymes required to convert the essential amino acid phenylalanine to the normally nonessential amino acid tyrosine.

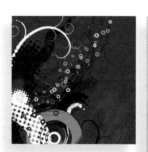

Living with PKU

People with phenylketonuria (PKU) are unable to make the amino acid tyrosine from the amino acid phenylalanine. This is a problem not only because tyrosine is a necessary component of many proteins, but also because high levels of phenylalanine in the blood can damage the brain. For this reason, almost all babies born in the United States are screened for PKU soon after birth. Children with PKU must follow a diet that restricts their intake of phenylalanine-rich foods such as milk, eggs, cheese, nuts, beef, fish, and chicken. They also need to avoid foods and medications made with aspartame (including many diet sodas) because this artificial sweetener contains phenylalanine. Adults with PKU must also be careful to not consume too much phenylalanine, although they do not usually need to be as vigilant as do children.

complete protein source A food that supplies an adequate and balanced amount of all essential amino acids.

incomplete protein source A food that lacks or supplies low amounts of one or more of the essential amino acids.

limiting amino acid An essential amino acid that is insufficient or absent in an incomplete protein source.

protein complementation The combining of diverse foods with different incomplete proteins to provide adequate amounts of all the essential amino acids.

cell signaling The process by which a cell is notified that it should make a particular protein.

A food that supplies an adequate and balanced amount of all essential amino acids is considered a **complete protein source**, whereas one that lacks or supplies low amounts of one or more of the essential amino acids is considered an **incomplete protein source**. An essential amino acid that is insufficient or absent in an incomplete protein source is called the **limiting amino acid**. When your body does not have access to a limiting amino acid, it cannot make any of the proteins that contain that particular amino acid—even if you have all of the other essential amino acids in full supply. As you might expect, meat, poultry, eggs, and dairy products are complete protein sources, whereas plant-based products are incomplete protein sources.

PROTEIN COMPLEMENTATION

You may be wondering how people who only eat plant-based foods (vegetarians) get all of their essential amino acids if these types of foods only contain incomplete proteins. The answer to this question is that diverse foods with different incomplete proteins can be combined to provide adequate amounts of all the essential amino acids. This dietary practice, called **protein complementation**, is customary around the world, especially in regions that traditionally rely heavily on plant-based foods for protein.[6] Examples of commonly consumed foods whose proteins complement each other are rice and beans, or corn and beans. Both rice and corn have several limiting amino acids (for example, lysine) but provide adequate amounts of others (for example, methionine). By contrast, beans and other legumes tend to be limiting in methionine but provide adequate amounts of lysine. In general, protein complementation allows diets containing a variety of plant-based protein sources to provide all of the necessary essential amino acids.

> "**Complete and incomplete proteins** are sometimes said to carry **high and low biological value**, respectively."

© iStockphoto.com/Elena Schweitzer

LO2 How Do Cells Make Proteins?

You now know the fundamental concepts related to what makes up a protein, how proteins differ from each other, and why some foods are considered better sources of protein than others. But to really understand why proteins are an essential part of your diet, you must also understand how your body converts the proteins you eat into the exact functional proteins that it needs. The following sections describe the process of protein synthesis, which involves three basic steps:

1. Cell signaling
2. Transcription
3. Translation

Figure 5.2 illustrates the three steps that comprise the process of protein synthesis.

Step 1: Cell Signaling

Virtually every cell in your body needs and makes proteins. However, different cells need different proteins, and the amounts needed at different times can be highly variable. As a result, protein synthesis within a particular cell is neither consistent nor random—it is tightly regulated by the amounts and types of proteins needed by the cell and the entire body at a given time. For example, synthesis of the proteins needed for calcium absorption in your small intestine is turned on or off depending on your body's need for calcium, ensuring that calcium availability is maintained at optimal levels. Because protein synthesis is not an ongoing process, a cell must be notified that it should make a particular protein. This communicative process, **cell signaling**, conveys physiological conditions or cellular needs to the nucleus of the cell, just like an indoor/outdoor

Figure 5.2 **Protein Synthesis**

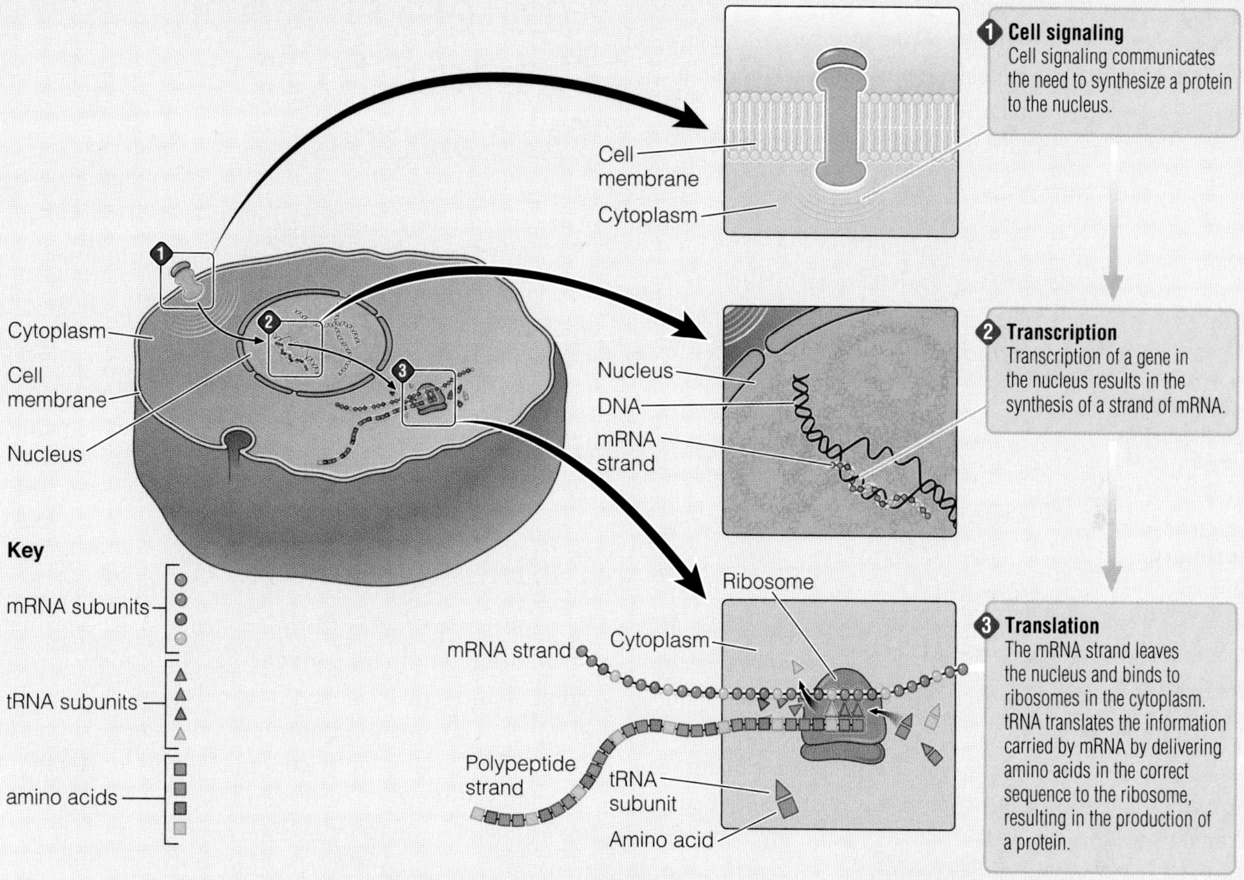

The steps involved in protein synthesis are three-fold: cell signaling, transcription, and translation.

thermometer conveys the outside temperature to the inside of a house.

Step 2: Transcription

Cell signaling initiates the second step of protein synthesis, **transcription**, whereby a specific type of ribonucleic acid (RNA), called messenger RNA (mRNA) is constructed using DNA as a template. One way to think about this process is that it is like reading a cookbook. A chemical called **deoxyribonucleic acid (DNA)** provides the fundamental instructions for protein synthesis. Found in a cell's nucleus, coiled strands of DNA combine with special proteins to form a **chromosome**, which serves as the complete, organized cookbook. Each chromosome is subdivided into thousands of units, each of which is like an individual recipe. Each recipe, or **gene**, provides the instructions and list of ingredients needed to make a protein. In other words, a gene (like a recipe) tells a cell (like a cook) which amino acids are needed and in what order they must be arranged to synthesize a protein (like a food).

In order for protein synthesis to take place, the information contained in the DNA's code must be communicated from inside the nucleus to the part of the

Cell signaling functions somewhat like an indoor/outdoor thermometer. Cues from outside the cell can "turn on" the protein synthetic machinery located in the cell's nucleus.

transcription The process by which mRNA is constructed using DNA as a template.

deoxyribonucleic acid (DNA) A chemical that provides the instructions for protein synthesis.

chromosome A substance comprised of coiled strands of DNA and special proteins found in a cell's nucleus.

gene A chromosome subunit that tells a cell which amino acids are needed and in what order they must be arranged to synthesize a protein.

messenger ribonucleic acid (mRNA) A chemical that carries the instructions contained in DNA outside of the nucleus.

ribosome A cellular organelle to which mRNA binds and on which proteins are made.

translation The process by which amino acids are joined via peptide bonds.

transfer ribonucleic acid (tRNA) A chemical that carries amino acids to a ribosome to be assembled into a protein.

primary structure (or **primary sequence**) The most basic level of protein structure; determined by the number and sequence of amino acids in a single peptide chain.

cell where proteins are made. To accomplish this process, the instructions contained in the DNA sequence are converted to a chemical called **messenger ribonucleic acid (mRNA).** A series of mRNA subunits bind to the particular gene coding for the protein that needs to be synthesized. These mRNA subunits then join, forming a strand of mRNA that is essentially a mirror image of the DNA. The newly formed strand of mRNA separates from the DNA, exits the nucleus, and enters the cytoplasm where it participates in the next step of protein synthesis: translation.

Step 3: Translation

Once the mRNA strand is outside the nucleus, it binds to a cytoplasmic organelle called a **ribosome,** on which the third step of protein synthesis occurs. **Translation**, the process by which amino acids are joined via peptide bonds, requires another form of RNA called **transfer ribonucleic acid (tRNA).** tRNA units carry amino acids to the ribosome to be assembled into a peptide chain. For translation to proceed, the ribosome moves along the mRNA strand reading its sequence. The sequence of the mRNA, in turn, instructs specific tRNAs to transfer the amino acids they are carrying to the ribosome. One by one, amino acids join together via peptide bonds to form a growing peptide chain. When translation is complete, the newly formed protein separates from the ribosome. This is not the final step in the formation of a new protein, however; the final structure and shape of the protein are yet to be determined.

LO3 Why Is a Protein's Shape Critical to Its Function?

After a peptide chain is released from the ribosome, it must fold (and sometimes combine with other peptide chains and substances) to form the complex shape and structure of the final protein. Because a protein's final shape is critical to its ultimate function within the body, it is important to understand the many levels of protein structure—and what can happen when something goes wrong.

Primary Structure

The most basic level of protein structure, the **primary structure** (or **primary sequence**), is determined by the number and sequence of amino acids in a single peptide chain (see Figure 5.3). A protein's primary structure is determined by the DNA code. Each peptide chain has a unique primary structure and is therefore a unique molecule. Although understanding the concept of primary structure is relatively simple, take a moment to contemplate the enormous number of primary structures that can be made from just 20 amino acids. Consider, as an analogy, the English alphabet, which has a similar number of letters. As you know, the number

Figure 5.3 **Primary Structure of a Protein**

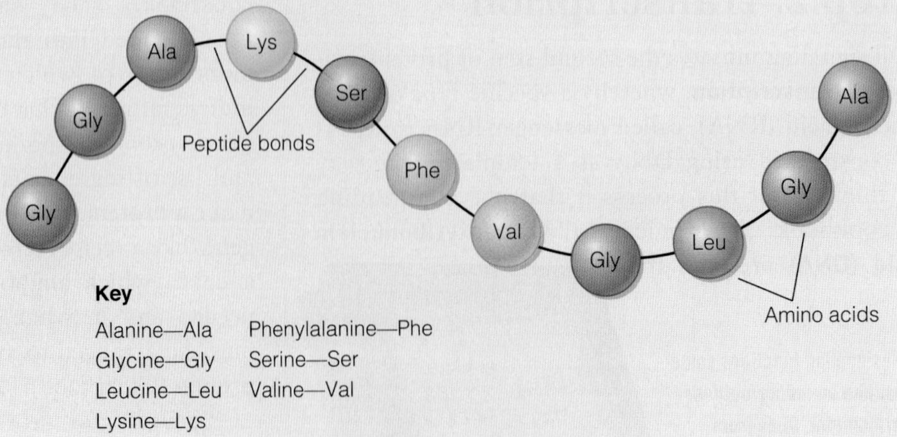

The number and sequence of amino acids in a single peptide chain is referred to as its primary structure.

and variety of words that can be constructed from just 26 letters is astonishing. Some words are short, some are long. Some words contain just a few different letters; others contain many different letters. The same holds true for proteins: some are short, some are long. Some contain a handful of different amino acids, whereas others contain all 20 amino acids. The possibilities are seemingly endless.

The primary structure of a protein is critical to its function because it determines the protein's most basic chemical and physical characteristics. Thus, a change in a protein's primary structure can profoundly affect the ability of the protein to do its job. Sometimes, alterations to a protein's primary structure can be caused by inherited genetic variations in the DNA sequence. An example of a disease caused by an inherited genetic variation is **sickle cell anemia** (or **sickle cell disease**). Sickle cell anemia is caused by a small alteration in the DNA that ultimately results in the production of defective, misshapen molecules of the protein hemoglobin within red blood cells.[7] Because hemoglobin is responsible for carrying oxygen and carbon dioxide in the blood, complications such as fatigue and increased risk of infections sometimes occur and can be serious.

sickle cell anemia (or **sickle cell disease**) A disease whereby a small alteration in the DNA results in the production of defective, misshapen molecules of the protein hemoglobin within red blood cells.

> **As opposed to humans** and other animals, most microorganisms and plants can **biosynthesize all** of the amino acids they need.

The number and variety of proteins that the body can make is as astonishing as the number of words that can be formed with the letters of the English alphabet.

© iStockphoto.com/Rick Rhay

Why Are Some People Especially Prone to Sickle Cell Anemia?

Scientists hypothesize that hundreds of years ago, the genetic alteration responsible for sickle cell anemia somehow protected people from the serious and sometimes deadly disease malaria.[8] As a result, individuals with the sickle cell anemia gene (or genes) were able to survive malaria epidemics, passing on their genetic code to their offspring. Today, millions of people have one or two genes for sickle cell anemia. Those with two genes have full-blown signs and symptoms of the disease, whereas those with one sickle cell gene are less affected, if at all. The genetic variation exists all over the world, especially in those with African, Mediterranean, Middle Eastern, or Indian ancestry. In the United States, sickle cell anemia is most common in people of black African heritage: one in 500 African Americans has the disease.

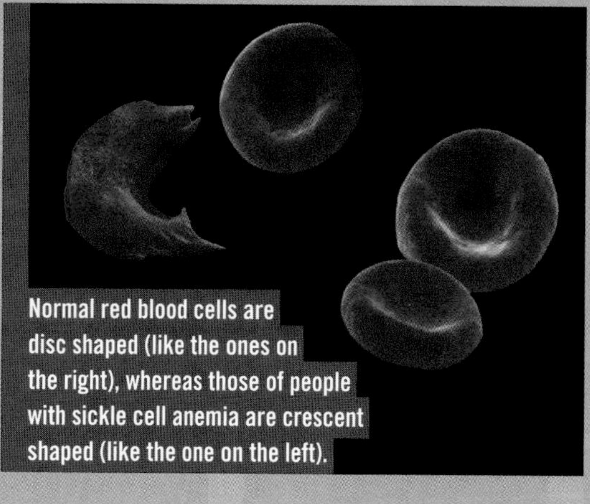

© Dr. Stanley Flegler/Visuals Unlimited/Getty Images

Normal red blood cells are disc shaped (like the ones on the right), whereas those of people with sickle cell anemia are crescent shaped (like the one on the left).

Secondary and Tertiary Structures

Peptide chains are relatively linear molecules. However, functional proteins are anything but linear—most have three-dimensional shapes. Because the backbone of the peptide chain is made of a series of weakly charged amino and carboxylic acid groups, these charges attract and repel each other like magnets. This causes portions of the polypeptide to fold into an organized and predictable pattern—the **secondary structure** of the protein. The two most common folding patterns are the **α-helix**, similar in shape to a spiral staircase, and the **β-folded sheet**, similar in shape to a folded paper fan. Both folding patterns are illustrated in Figure 5.4.

The next level of protein complexity, **tertiary structure**, is additional folding caused by interactions between the amino acids' R-groups. This folding transforms the entire protein into an even more complex, three-dimensional structure. Imagine for example what would happen to a folded paper fan if you were to crumple it gently in your hand. The folds (analogous to a protein's secondary structure) would remain, but they might be further bent and twisted in some regions (analogous to a protein's tertiary structure). This is similar to what happens in a protein when its tertiary structure is formed.

Quaternary Structure and Prosthetic Groups

The fourth level of protein structure is called **quaternary structure.** Quaternary structure occurs when two or more peptide chains join together, as shown in Figure 5.5. This level of complexity is somewhat like putting two or three crumpled paper fans together. Not all proteins have a quaternary structure—only those made from more than one peptide chain. In addition to the quaternary structure, a nonprotein component called a **prosthetic group** must sometimes be positioned precisely within a protein for it to function. Prosthetic

secondary structure Organized and predictable folds that develop in portions of a peptide chain because charged portions of the amino acid backbone attract and repel each other.

α-helix A common secondary structure folding pattern that resembles the shape of a spiral staircase.

β-folded sheet A common secondary structure folding pattern that resembles the shape of a folded paper fan.

tertiary structure Additional folding of a peptide chain that develops because of interactions between the amino acids' R-groups.

quaternary structure The most complex level of protein structure; occurs when two or more peptide chains join together.

prosthetic group A nonprotein component of a protein that often contains minerals needed for the protein to carry out its purpose.

Figure 5.4 Secondary Structure of a Protein

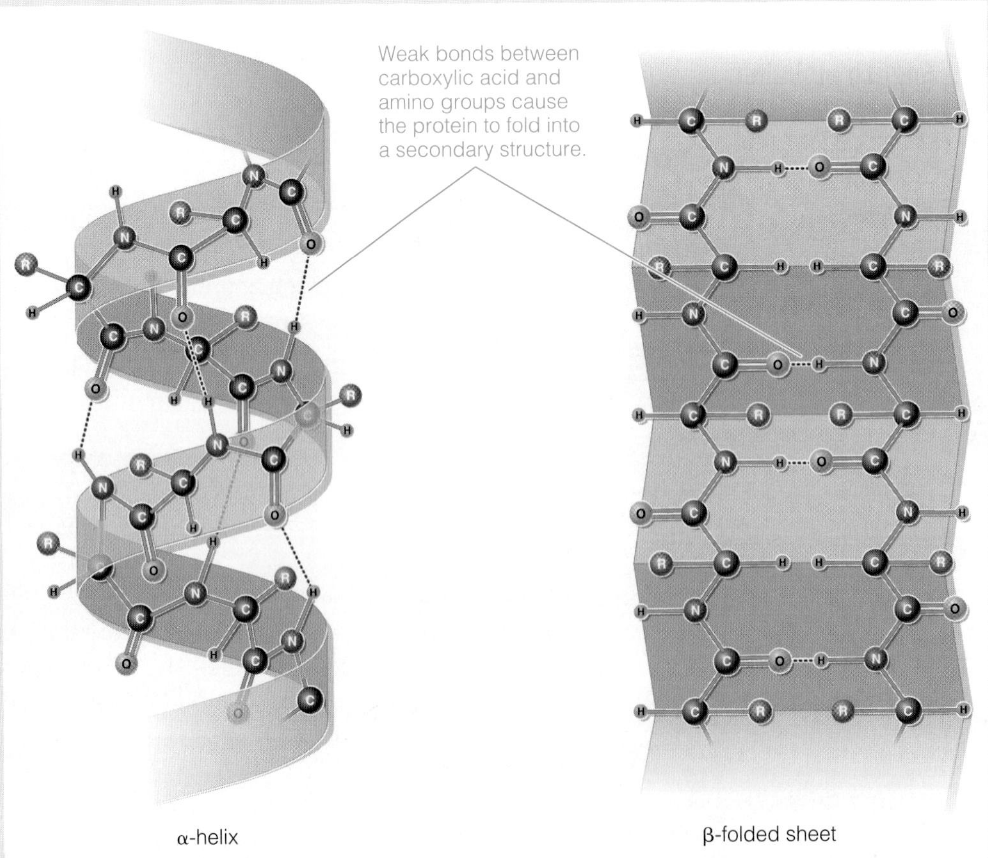

Weak bonds between carboxylic acid and amino groups cause the protein to fold into a secondary structure.
α-helix
β-folded sheet

Weak chemical bonds fold proteins into three-dimensional spiral- and fan-like shapes, resulting in secondary structures.

Figure 5.5 Quaternary Structure and Prosthetic Groups of Hemoglobin

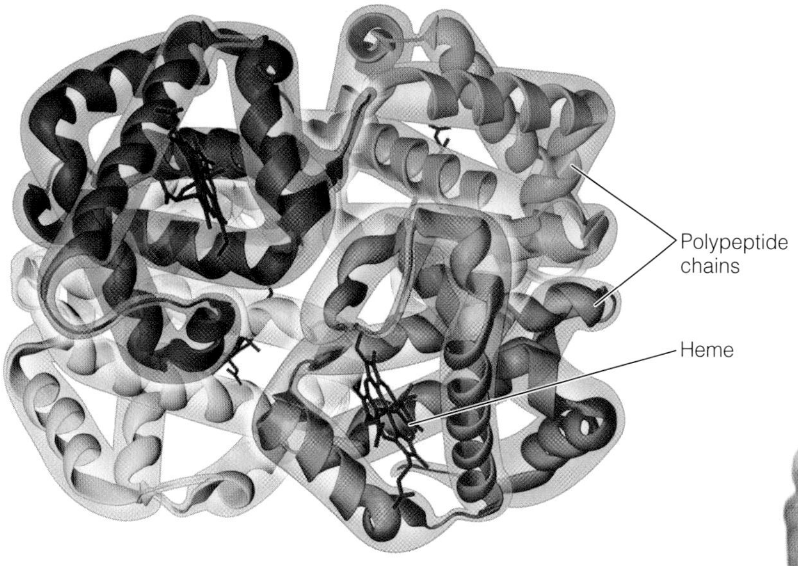

Hemoglobin is made from four polypeptide chains and four iron-containing prosthetic groups called heme.

Heat denatures the **proteins in egg white**, causing it to transform from a clear **gel-like substance** to a soft opaque solid.

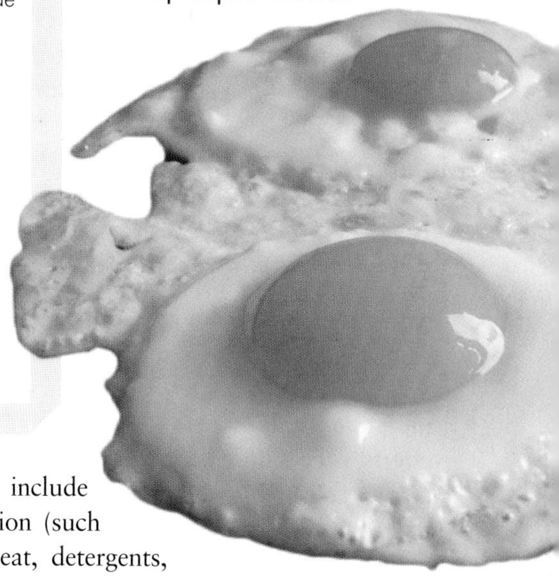

groups often contain minerals that are needed for the protein to carry out its purpose. Hemoglobin is an example of a protein with quaternary structure and prosthetic groups because it is made from four separate polypeptide chains, each of which contains an iron-containing prosthetic group called heme. Heme is the portion of hemoglobin that actually transports the oxygen and carbon dioxide gases in the blood.

A Protein's Shape Determines Its Function

You now know how a protein's primary, secondary, tertiary, and quaternary structures come to be, and how they determine the protein's final shape. Shape is critical to a protein's ability to carry out its function. When a protein's shape is disrupted, the protein can no longer complete the task for which it was designed.

DENATURATION

One way a protein's three-dimensional shape can be altered is by **denaturation.** Denaturation is akin to flattening out one of the pieces of paper that makes up a folded paper fan protein. Flattening out the paper results in a fan that probably does not work, just like denaturation can cause a protein to lose its function. Compounds and conditions that denature proteins are called denaturing agents. These include physical agitation (such as shaking), heat, detergents, acids, alkaline solutions, salts, alcohol, and heavy metals (such as lead and mercury).

LO4 What Is Meant by Genetics and Epigenetics?

As you now know, the DNA found in a cell's nucleus contains the instructions for all the proteins synthesized in your body. Each chromosome, consisting of double-stranded DNA, contains hundreds of genes. Each gene, in turn, contains the genetic code (information) to guide protein synthesis. Most chromosomes can code for many thousands of proteins. When a sperm cell fertilizes an egg cell, the chromosomes in each combine to become the particular DNA, or **genotype**, of the offspring. This is how an infant inherits his parents' genetic makeup. Except for identical twins, no two individuals have

denaturation The process by which a protein's three-dimensional structure is altered.

genotype The particular DNA inherited from one's parents.

phenotype The observable physical or biochemical characteristics of an organism.

mutation An alteration in a gene that occurs due to a chance genetic modification.

epigenetics Alterations in protein synthesis that do not involve changes in the DNA sequence.

the same genotype (which in part explains the vast diversity among humans). Some genetic differences impart individual physical characteristics, such as eye and hair color, which do not really affect your health. Such physical characteristics in part determine one's **phenotype**, the observable physical or biochemical characteristics of an organism. While alterations in genes affecting phenotype do not usually affect one's health, alterations in other genes can have more important consequences.

Mutations

An alteration in a gene sometimes results in a protein with an altered amino acid sequence—in other words, its primary structure is changed. When such an alteration occurs due to a genetic modification, it is called a **mutation.** While some mutations have no measurable effects, others do. Sickle cell anemia, for example, is a disease caused by a mutation that results in decreased oxygen transport. Other mutations, such as the one involved in PKU, can influence metabolism. Still others alter protein synthesis so that cells experience uncontrollable growth—a condition that can result in cancer. Mutations present in the DNA of egg and sperm cells can be passed on to offspring and are therefore considered to be *inherited*. PKU and sickle cell anemia are the result of inherited DNA mutations.

Epigenetics

A person's genetic makeup reflects the *sequence* of DNA that makes up her chromosomes. However, scientists recently learned that the connection between genes and physiology is actually much more complex than that. For example, DNA can be modified in ways that regulate whether a particular gene is expressed. The term **epigenetics** refers to alterations in protein synthesis that do not involve changes in the DNA sequence. Even if two people have exactly the same genetic sequences in their DNA (as is the case with identical twins), they may differ in terms of how those genes are expressed due to epigenetic variation. Interestingly, like mutations in DNA, some epigenetic differences can be passed on to the next generation. Scientists now think that epigenetic modifications may play important roles in the development of many chronic degenerative diseases such as cancer, type 2 diabetes, and cardiovascular disease.

NUTRITION AND EPIGENETICS

Growing evidence that nutritional status may affect long-term epigenetic modifications is of great importance to

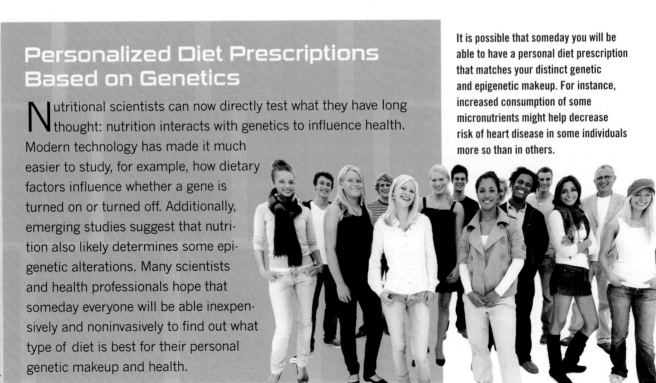

Personalized Diet Prescriptions Based on Genetics

Nutritional scientists can now directly test what they have long thought: nutrition interacts with genetics to influence health. Modern technology has made it much easier to study, for example, how dietary factors influence whether a gene is turned on or turned off. Additionally, emerging studies suggest that nutrition also likely determines some epigenetic alterations. Many scientists and health professionals hope that someday everyone will be able inexpensively and noninvasively to find out what type of diet is best for their personal genetic makeup and health.

It is possible that someday you will be able to have a personal diet prescription that matches your distinct genetic and epigenetic makeup. For instance, increased consumption of some micronutrients might help decrease risk of heart disease in some individuals more so than in others.

© iStockphoto.com/Jacob Wackerhausen

the field of nutrition. For example, there is strong evidence that babies who are malnourished during fetal life but then experience accelerated growth in childhood may be at increased risk for cardiovascular disease and type 2 diabetes as adults, partly due to epigenetic modifications in gene expression.[9] Whether later alterations in a person's environment (such as better nutrition) can reverse this effect remains to be discovered. Clearly, epigenetics is an exciting new area of nutrition research.

LO5 How Are Proteins Digested, Absorbed, and Circulated?

The process of digestion disassembles food proteins into amino acids that are then absorbed and carried to cells where they are used to make the proteins needed by the body. This is somewhat like disassembling someone else's house and then using the materials to build another house that perfectly fits your needs. The body also efficiently and systematically breaks down and recycles its own proteins when they become old and nonfunctional. In fact, you can think of your body as having its own protein-recycling center. The stages of protein digestion, absorption, and circulation are shown in Figure 5.6 and described next.

Protein Digestion Begins in the Stomach

Before you can use the proteins in the foods you eat, they must be broken down into their component amino acids. Although a small amount of mechanical digestion of protein occurs as you chew your food, chemical digestion of protein does not begin until it comes in contact with specialized cells in the stomach. The presence of food causes these stomach cells to release gastrin, which in turn triggers other cells in the stomach to release hydrochloric acid, mucus, and pepsinogen.

Figure 5.6 Overview of Protein Digestion

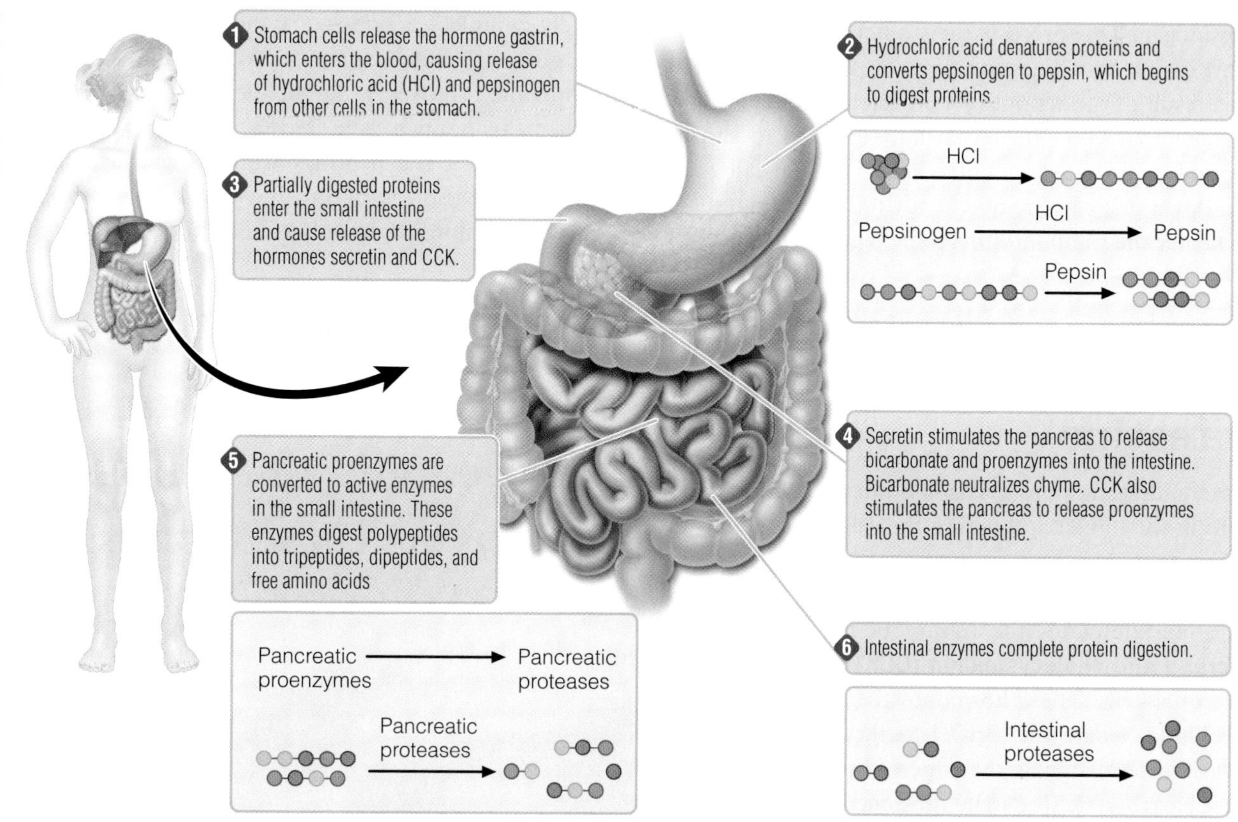

Protein digestion occurs in both the stomach and small intestine via pepsin, pancreatic proteases, and intestinal proteases.

pepsinogen The inactive form of the enzyme pepsin.

pepsin An enzyme needed for protein digestion.

proenzyme An inactive precursor of an enzyme.

protease A type of enzyme that breaks peptide bonds between amino acids.

secretin A hormone, secreted by intestinal cells, that signals the release of sodium bicarbonate and proteases from the pancreas.

cholecystokinin (CCK) A hormone, secreted by intestinal cells, that signals the release of bile from the gallbladder and proenzyme proteases from the pancreas.

food allergy A condition whereby the body's immune system responds to a food-derived peptide as if it is dangerous.

Pepsinogen is an inactive form of **pepsin**, an enzyme needed for protein digestion. In general, an inactive precursor of an enzyme is called a **proenzyme.** Once the components of the gastric juices are released, the chemical digestion of proteins can begin.

First, hydrochloric acid disrupts the chemical bonds responsible for the protein's secondary, tertiary, and quaternary structures. This process of denaturation straightens out the complex protein structure, helping to expose the peptide bonds to the digestive enzymes present in the stomach and small intestine. Second, hydrochloric acid converts the proenzyme pepsinogen into its active form, pepsin. Pepsin is an example of a **protease**, an enzyme that breaks peptide bonds between amino acids. Note that the stomach docs not produce the active protease enzyme pepsin. Instead, it produces and stores the inactive (or safe) proenzyme pepsinogen. This protects the stomach from the active enzyme's protein-digesting function until it is needed. As a result of the actions of stomach acid and enzymes, proteins that you eat are partially digested to shorter peptides and some free amino acids. The partially broken down proteins are now ready to leave the stomach and enter the small intestine to be digested further.

Protein Digestion Continues in the Small Intestine

Protein digestion in the small intestine takes place both in the lumen and within the cells that line it. Initiating this cascading series of digestive events, the arrival of amino acids and smaller peptides in the small intestine stimulates the release of the hormones **secretin** and **cholecystokinin (CCK)** from the intestinal cells. These hormones then enter the blood, where they travel to the pancreas. Secretin signals the pancreas to release bicarbonate (the same substance that makes up baking soda) into the lumen of the small intestine. Bicarbonate neutralizes the acid from the stomach and inactivates pepsin. Secretin and CCK also signal the release of proenzymes from the pancreas. Upon entering the small intestine, each of these proenzymes is activated and is then able to break peptide bonds holding specific sequences of amino acids together. The resulting di- and tripeptides are further broken down by a multitude of proteases produced in the cells that make up the absorptive surface of the small intestine. This usually results in the complete breakdown of proteins into their amino acid constituents.

Amino Acid Absorption and Circulation

When protein digestion is complete, some amino acids have already been taken up by the cells of the small intestine, but those remaining in the intestinal lumen must still be transported into the intestinal cells. Most amino acids are absorbed in the duodenum, where they enter the blood and circulate to the liver for further processing.

FOOD ALLERGIES AND INTOLERANCES

The breakdown of proteins into amino acids is a relatively complete process—it typically results in the absorption of amino acids (not proteins) into circulation. Sometimes, however, larger peptides are absorbed. When this happens, the body's immune system might respond as if the peptides were dangerous. In such cases, the person is said to have an *allergic response*, or what is more commonly called a **food allergy.**[10] The

More than **3 million people** *are* **allergic to peanuts** *or tree nuts in the United States.*

© Crystal Cartier/StockFood Creative/Getty Images

majority of food allergies are caused by proteins present in eggs, milk, peanuts, soy, and wheat. Researchers estimate that approximately 2 percent of adults and 5 percent of infants and young children in the United States have food allergies.[11] Note, however, that all adverse reactions to foods are not true food allergies. A negative physiological response to a substance in a food that does not trigger an immune response is called a **food intolerance** (or **food sensitivity**). An example of a food intolerance is lactose intolerance, which was discussed in Chapter 4.

An allergic reaction to a particular food protein typically causes minor physical discomfort such as skin rashes or gastrointestinal distress. For some people, however, an allergic food reaction can be frightening and even life threatening. Signs and symptoms usually develop within a few minutes to an hour after eating the food, depending on the type of food allergy a person has and how his immune system reacts to it. The most common signs and symptoms are:

- Tingling in the mouth
- Hives, itching, or eczema
- Swelling of the lips, face, tongue, throat, or other parts of the body
- Wheezing, nasal congestion, or trouble breathing
- Abdominal pain, diarrhea, nausea, or vomiting
- Dizziness, lightheadedness, or fainting

In a severe case, a person may experience more extreme symptoms, such as **anaphylaxis**. Anaphylaxis is a rapid immune response that causes a sudden drop in blood pressure, rapid pulse, dizziness, and a narrowing of the airways. This can block normal breathing almost immediately. When this occurs, emergency treatment is critical. Anaphylaxis due to food allergies is responsible for thousands of emergency room visits and as many as 200 deaths in the United States each year.[12]

WHAT TO DO IF YOU HAVE A FOOD ALLERGY

If you have a food allergy, it is important to know which foods to avoid. It is especially important to read food labels carefully; the U.S. Food and Drug Administration (FDA) requires that all foods containing the most common allergens be labeled as such. Wearing a medical alert bracelet or necklace may be advantageous in case an allergenic food is accidentally ingested. In the case of children with food allergies, parents should notify friends, family members, childcare providers, and school personnel so that they can help avoid exposure to offending foods. It is also important to make sure that the child knows which foods to avoid and to ask for help if needed.

food intolerance (or **food sensitivity**) A condition whereby the body reacts negatively to a food or food component but does not mount an immune response.

anaphylaxis A rapid immune response that causes a sudden drop in blood pressure, rapid pulse, dizziness, and a narrowing of the airways.

LO6 Why Do You Need Proteins and Amino Acids?

Once amino acids are circulated away from the GI tract, your body uses them to make the thousands of proteins it needs via protein synthesis. Using the previous analogy, this is the stage at which you would use all of the disassembled materials from someone else's house to build one of your own. The deconstructed materials could be used to build walls, cabinets, stairways, or furniture—each of these household items has its own function. Similarly, the proteins that your body makes can be classified into general categories based on their functions (see Table 5.2 on the next page). Some proteins, like those in your muscles, are used for movement. Others, like the hormone insulin, are used to regulate blood glucose. Some proteins can be broken down and used for energy, and some amino acids can be converted to glucose. In the following sections, you will learn more about the various types of proteins and amino acids your body needs, as well as other ways that they are used to promote functionality and health.

Proteins Provide Structure

Being constituents of muscles, skin, bones, hair, and fingernails, proteins comprise most of the structural materials in your body. Collagen, for instance, is a structural protein that forms a supporting matrix in bones, teeth, ligaments, and tendons. Proteins are also important structural components of cell membranes and organelles. The synthesis of structural proteins is especially important during periods of active growth and development such as infancy and adolescence.

Enzymes Catalyze Chemical Reactions

The myriad chemical reactions in your body are driven by a group of protein molecules called enzymes. These biological catalysts speed up chemical reactions

Table 5.2 Major Functions of Proteins in the Body

Function	Description	Selected Examples
Structure	Proteins that make up the basic structure of tissues such as bones, teeth, and skin	• Bone matrix proteins (such as hydroxyapetite) • Collagen in skin, teeth, ligaments, and tendons • Keratin in hair and fingernails
Catalysis	Proteins (enzymes) that facilitate chemical reactions	• Lingual lipase • Pancreatic amylase • Pepsin
Movement	Proteins found in muscles, ligaments, and tendons	• Actin and myosin in muscle • Elastin in ligaments
Transport	Proteins involved in the movement of substances across cell membranes and within the circulatory system	• Glucose and sodium transporters in cell membranes • Vitamin A-binding protein, which transports vitamin A in blood
Communication	Protein hormones, cell-signaling proteins, and neurotransmitters	• Insulin and glucagon • Cholecystokinin (CCK)
Protection	Proteins that constitute the skin and immune system	• Collagen in skin • Fibrinogen in blood clot • Antibodies
Regulation of fluid balance	Proteins that regulate the distribution of fluid in the body's various compartments via the process of osmosis	• Albumin
Regulation of pH	Proteins that take up and release hydrogen ions (H^+) to maintain appropriate pH of body fluids and tissues	• Hemoglobin

without being consumed or altered in the process. Without the catalytic functions of enzymes, the thousands of chemical reactions needed by your body to function would simply not occur, or at best, would occur at very slow rates. Examples of enzymes you have already learned about are amylase and pepsin, which catalyze reactions needed to digest carbohydrates and proteins, respectively.

Muscle Proteins Facilitate Movement

Protein is also necessary to movement, which results from the contraction and relaxation of the many muscle fibers in the body. Muscles are involved in both voluntary and involuntary movements such as those needed for cardiovascular function and physical activity, respectively. Nearly half of the body's protein is present in skeletal muscle, and adequate protein intake is required to form and maintain muscle mass and function throughout life. Although there are many proteins related to movement, perhaps the most important are actin and myosin, which make up much of the machinery needed for muscles to contract and relax. This is why protein deficiency can cause muscle wasting and weakness.

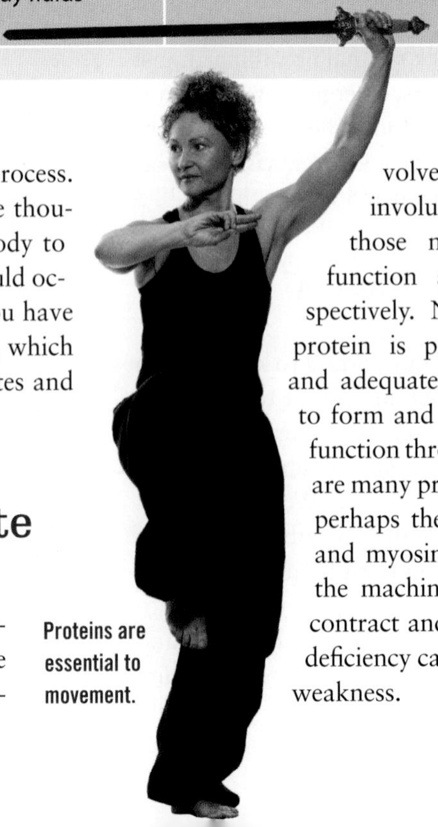

Proteins are essential to movement.

Some Proteins Serve as Transporters

Amino acids are used to make transport proteins, which are responsible for escorting substances into and around the body as well as across cell membranes. For example, absorption of many nutrients (such as calcium) requires one or more transport proteins to help the nutrients cross the lumen-facing cell membranes of intestinal cells. Protein deficiency can decrease the body's production of intestinal transport proteins, resulting in secondary malnutrition. In addition to facilitating the transport of substances across cell membranes, many proteins are critical for the transport of nutrients and other substances in the blood. Examples of circulating transport proteins include hemoglobin, which transports gases (oxygen and carbon dioxide), and a variety of binding proteins that transport hormones and fat-soluble vitamins in the blood.

> "Non-protein hormones, many of which are steroid hormones, include many of the reproductive hormones (such as estrogen and testosterone)."

Hormones and Cell-Signaling Proteins Are Critical Communicators

Tissues and organs have a variety of ways to communicate with each other, and most of these methods involve proteins. Although not all hormones are proteins, most are; secretin, gastrin, insulin, and glucagon are proteins, for example. Beyond hormonal communicators, there are also specialized proteins embedded in cell membranes that communicate information about the extracellular environment to the intracellular space. Some of these proteins are involved in the cell-signaling process that initiates protein synthesis itself. Others regulate cellular metabolism. Together, hormones and cell-signaling proteins make up part of the body's critical communication network. Thus, protein deficiency can have profound effects on your body's ability to coordinate all of its functions.

Proteins Protect the Body

One of the most vital and basic functions of the proteins in your body is protecting it from physical danger and infection. For instance, skin is mainly made of proteins that form a barrier between the outside world and your internal environment. If your skin gets cut, blood clots (produced by a series of clotting proteins) close off the possible entry point to infection. And if a bacterium or other foreign substance does enter the body, your immune system fights back by producing **antibody** (or **immunoglobulin**) proteins to help fight the infection. Antibodies bind foreign substances so they can be destroyed. Protein deficiency can make it difficult for the body to prevent and fight certain diseases because its natural defense systems become weakened. This is why infection and illness often accompany protein deficiency.

Fluid Balance Is Regulated in Part by Proteins

Another function of proteins is regulating how fluids are distributed in the body. As you might know, most of your body is made of water. This important fluid is found both inside and outside of cells. The fluid outside of cells can be subdivided into that found in blood and lymph vessels and that found between cells. The amount of fluid in these spaces is tightly regulated by a variety of means, some of which involve proteins. **Albumin**, a protein present in the blood in relatively high concentrations, plays such a role. As the heart beats, blood is propelled out of the heart and into blood vessels that become increasingly narrower. As the pressure builds, the fluid portion of the blood is squeezed out of the tiny capillaries. Albumin, which remains in the blood vessels, becomes more concentrated as more fluid is lost. The high concentration of albumin draws the fluid that was once squeezed out of the blood vessel back inside (see Figure 5.7 on the next page). Severe protein deficiency can impair albumin synthesis, resulting in low levels of albumin in the blood and causing fluid to accumulate in the space surrounding tissues. This condition, called **edema**, can sometimes be observed as swelling in the hands, feet, and abdominal cavity.

antibody (or **immunoglobulin**) A protein, produced by the immune system, that helps fight infection.

albumin A protein present in the blood that plays an important role in regulating fluid balance.

edema A condition whereby low levels of albumin in the blood cause fluid to accumulate in body tissues or cavities.

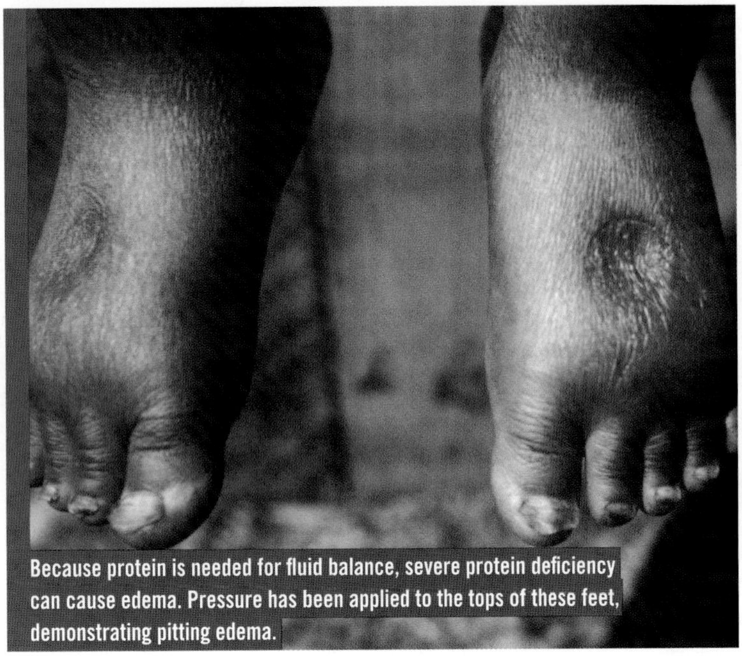

Because protein is needed for fluid balance, severe protein deficiency can cause edema. Pressure has been applied to the tops of these feet, demonstrating pitting edema.

Although edema is commonly seen in severely malnourished individuals, it can be caused by other factors as well.

Proteins Help Regulate pH

Proteins are involved in regulation of how acidic or alkaline your body fluids are, referred to as the fluids' pH. Your body must maintain certain pH levels in each of its fluids to maintain optimal health. One way that blood pH is maintained is through the action of certain proteins (such as hemoglobin), which act to increase and decrease the blood's pH as needed. As such, the body can have difficulty maintaining optimal pH balance during periods of severe protein deficiency.

Figure 5.7 Regulation of Fluid Balance by Albumin

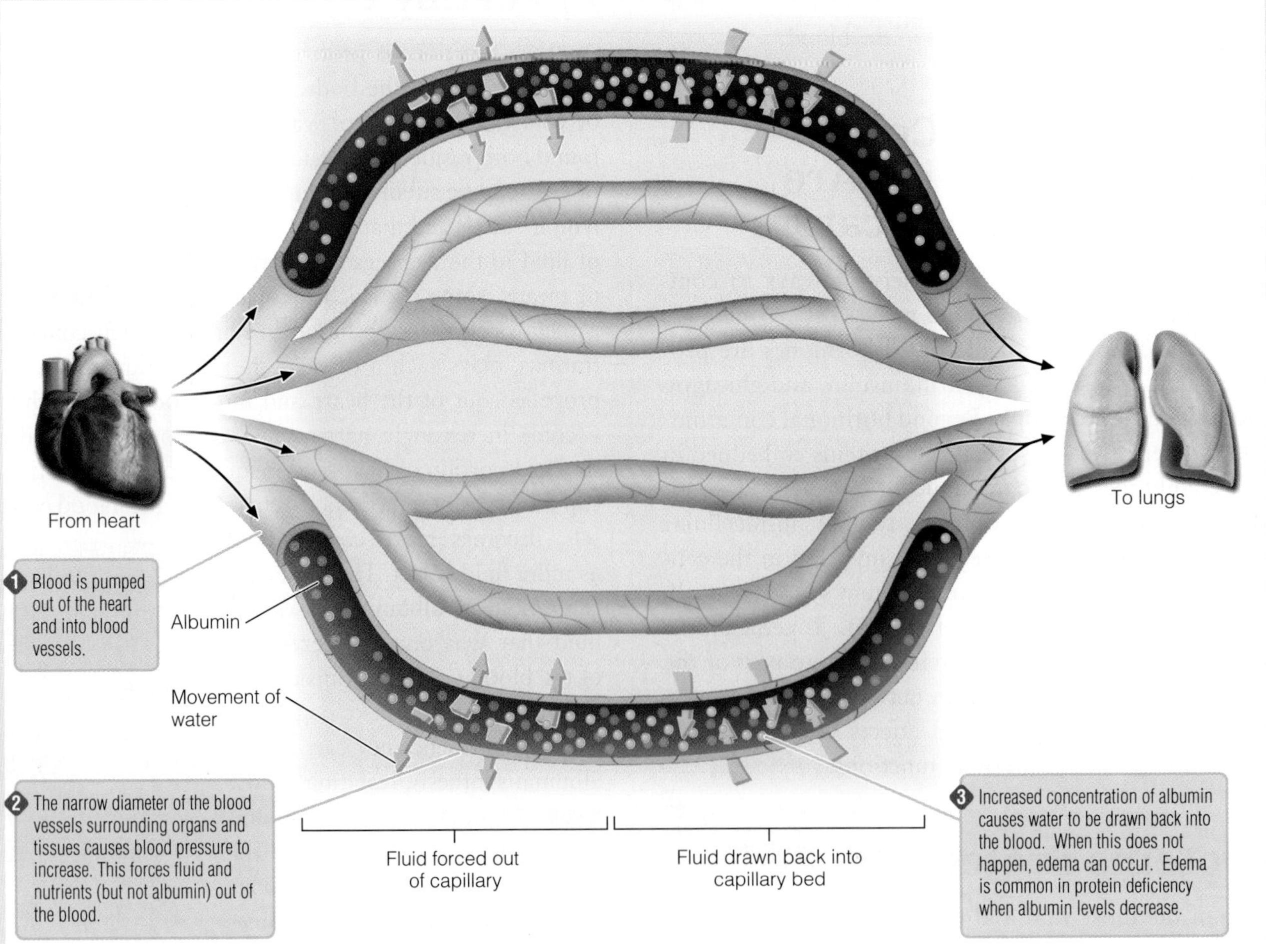

From heart
To lungs
1 Blood is pumped out of the heart and into blood vessels.
Albumin
Movement of water
2 The narrow diameter of the blood vessels surrounding organs and tissues causes blood pressure to increase. This forces fluid and nutrients (but not albumin) out of the blood.
Fluid forced out of capillary
Fluid drawn back into capillary bed
3 Increased concentration of albumin causes water to be drawn back into the blood. When this does not happen, edema can occur. Edema is common in protein deficiency when albumin levels decrease.

Albumin is a protein in the blood that helps regulate fluid balance within and around tissues.

Amino Acids Provide a Source of Glucose and Energy

Recall from Chapter 1 that proteins are categorized as energy-yielding macronutrients. This is because the body can:

- Break down proteins (a process called **proteolysis**) and use their constituent amino acids for energy (ATP) production
- Chemically transform some amino acids into glucose (a process called gluconeogenesis)
- Convert excess amino acids into fat (a process called lipogenesis)

Together, these processes help the body (1) generate ATP to power chemical reactions even when glucose and fat availability is limited, (2) maintain blood glucose at appropriate levels, and (3) store excess energy when dietary protein intake is more than adequate.

Amino Acids Serve Many Additional Purposes

In addition to serving as the building blocks for proteins and providing a source of energy, amino acids themselves have many unique purposes in your body. Some regulate protein breakdown, others are involved in cell communication, and still others are converted to neurotransmitters and other signaling molecules that function as messengers in the body. Thus, one needs amino acids not only for protein synthesis and energy production but for a multitude of other functions as well.

LO7 How Does the Body Recycle and Reuse Amino Acids?

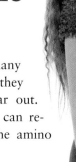

While proteins serve many functions within the body, they eventually—inevitably—wear out. Fortunately, human bodies can recycle and reuse most of the amino acids from retired proteins to synthesize new ones. The continual coordinated process of breaking down and resynthesizing protein is known as **protein turnover.** By regulating protein turnover, the body can adapt to periods of growth and development during childhood and maintain relatively stable amounts of protein during adulthood without requiring enormous amounts of protein from food.

proteolysis The breakdown of proteins.

protein turnover The continual coordinated process of protein breakdown and synthesis.

Nitrogen Excretion

As you have learned, amino acids can be converted to glucose or used as a source of energy. For this to occur, the nitrogen-containing amino group must first be removed. This process produces ammonia (NH_3), which is toxic to cells. In response to its production, the liver quickly converts ammonia to *urea*, a less toxic nitrogen-containing substance. The urea is then released into the blood, filtered out of the blood by the kidneys, and excreted in the urine.

Nitrogen Balance and Protein Status

Protein turnover results in a somewhat complex flux (or remodeling) of amino acids in your body every day. Measuring protein turnover can provide health professionals with important information about overall protein status. One's protein status can

Growing children are in a state of **positive nitrogen balance** *because they consume* **more nitrogen than they excrete.**

nitrogen balance A bodily state whereby nitrogen intake equals nitrogen loss; sometimes referred to as neutral nitrogen balance.

negative nitrogen balance A bodily state whereby nitrogen intake is less than nitrogen loss.

positive nitrogen balance A bodily state whereby nitrogen intake is greater than nitrogen loss.

be assessed by comparing protein *intake* to nitrogen *loss* in body secretions such as urine, sweat, and feces.[13] When nitrogen loss equals nitrogen intake, the body is in **nitrogen balance.** When nitrogen loss exceeds intake, as can occur during starvation, illness, or stress, a person is in **negative nitrogen balance.** When nitrogen intake exceeds loss, as can occur during childhood or recovery from an illness, a person is in **positive nitrogen balance.** Knowing whether a person is in neutral, positive, or negative nitrogen balance can help a clinician diagnose and treat certain disease states and physiologic conditions. For example, people who are on dialysis because of kidney failure often experience negative nitrogen balance and therefore require specialized nutritional support. Conversely, growing children should be in a state of positive nitrogen balance. If this is not the case, protein intake may need to be increased. Because the need for protein shifts throughout the course of one's life, dietary recommendations for proteins and amino acids have been developed to take into account both nitrogen balance and the need for specific amino acids during different life-stage periods.

LO8 How Much Protein Do You Need?

You need to consume dietary protein for two major reasons: (1) to supply adequate amounts of the essential amino acids and (2) for the additional nitrogen needed to make the nonessential amino acids and other nonprotein, nitrogen-containing compounds such as DNA. As such, recommendations for dietary amino acid and overall protein consumption reflect these needs.

Dietary Reference Intakes (DRIs) for Amino Acids

To begin with, consider how much of each essential amino acid you need to eat every day. Currently, the best estimates can be obtained from the Institute of Medicine's Recommended Dietary Allowances (RDAs), which are shown in Figure 5.8.[14] Note that these values are presented in units of milligrams per kilogram per day (mg/kg/day) because they represent requirements of the essential amino acids *relative to body size*: the larger you are, the more essential amino acids you need. For example, because the RDA for the essential amino acid valine in adults is 4 mg/kg/day, a woman weighing 64 kg (roughly 140 lb) would require 256 mg/d (4 mg × 64 kg) of this amino acid in her diet. A larger person weighing 91 kg (roughly 200 lb) would require 364 mg (4 mg × 91 kg) of valine each day.

Because the DRI committee concluded that there is no compelling evidence that high intake of any of the essential amino acids poses known health risks, the Institute of Medicine did not establish Tolerable Upper Intake Levels (ULs) for them.

Dietary Reference Intakes (DRIs) for Protein

Perhaps of greater interest are recommendations for overall protein consumption. Indeed, RDA for total protein intake have also been published and are found

Figure 5.8 RDAs for the Essential Amino Acids in Adults

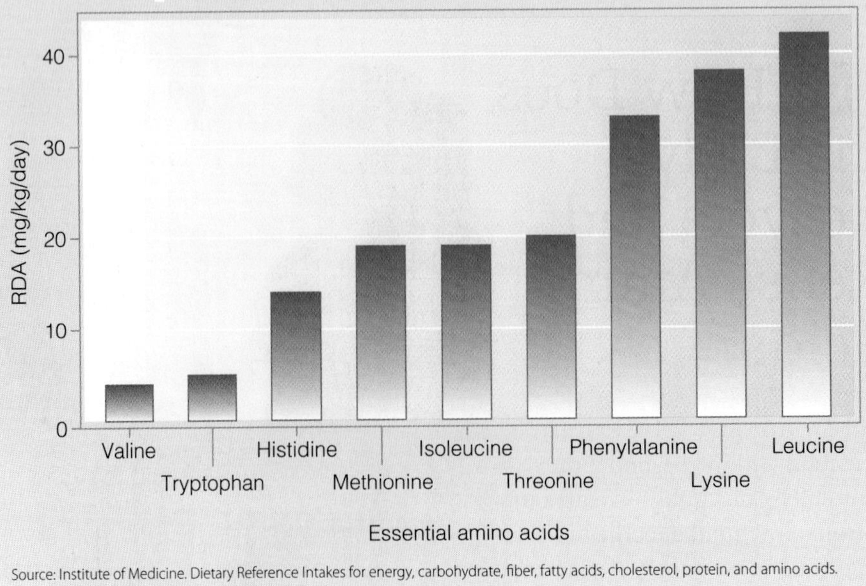

Source: Institute of Medicine. Dietary Reference Intakes for energy, carbohydrate, fiber, fatty acids, cholesterol, protein, and amino acids. Washington, DC: National Academies Press; 2005.

in the Dietary Reference Intakes card located at the back of this book. Although not all protein sources are created equal, researchers generally agree that most diets in affluent countries (such as the United States) provide a balanced mix of all the essential amino acids. Therefore, the dietary recommendations for protein intake do not distinguish between people who consume high-quality proteins and those who do not.

The RDA values for protein are expressed in two ways. The first, grams per day (g/day), reflects requirements for a typical person in a particular population group. These recommended protein intakes increase with age and are somewhat higher for men than women because, in general, men are larger than women and have more muscle mass. Using this set of values, a typical college-age man needs 56 g/day of protein, whereas a comparable woman needs 46 g/day.

The second way that an RDA value for protein can be expressed is as grams per kilogram body weight per day (g/kg/day). Like those for the essential amino acids, these recommendations adjust for body weight. The Institute of Medicine recommends that healthy adults consume 0.8 g/kg/day of protein. For example, an adult weighing 64 kg (roughly 140 lb) requires about 51 g of protein (0.8 g × 64 kg) every day regardless of whether the adult is a woman or a man. A person could easily get this much protein by eating a bowl of wheat flake cereal and low-fat milk (12 g protein) for breakfast, a hamburger (24 g protein) for lunch, and a bean burrito (15 g protein) for supper.

During infancy, the most rapid phase of growth in the life cycle, protein requirements (when adjusted for body weight) are relatively high. Protein requirements also increase during pregnancy and lactation, because additional protein is needed to support fetal growth and milk production.[15] Because healthy people show little evidence of harmful effects of high protein intake, no UL values are set for this macronutrient.

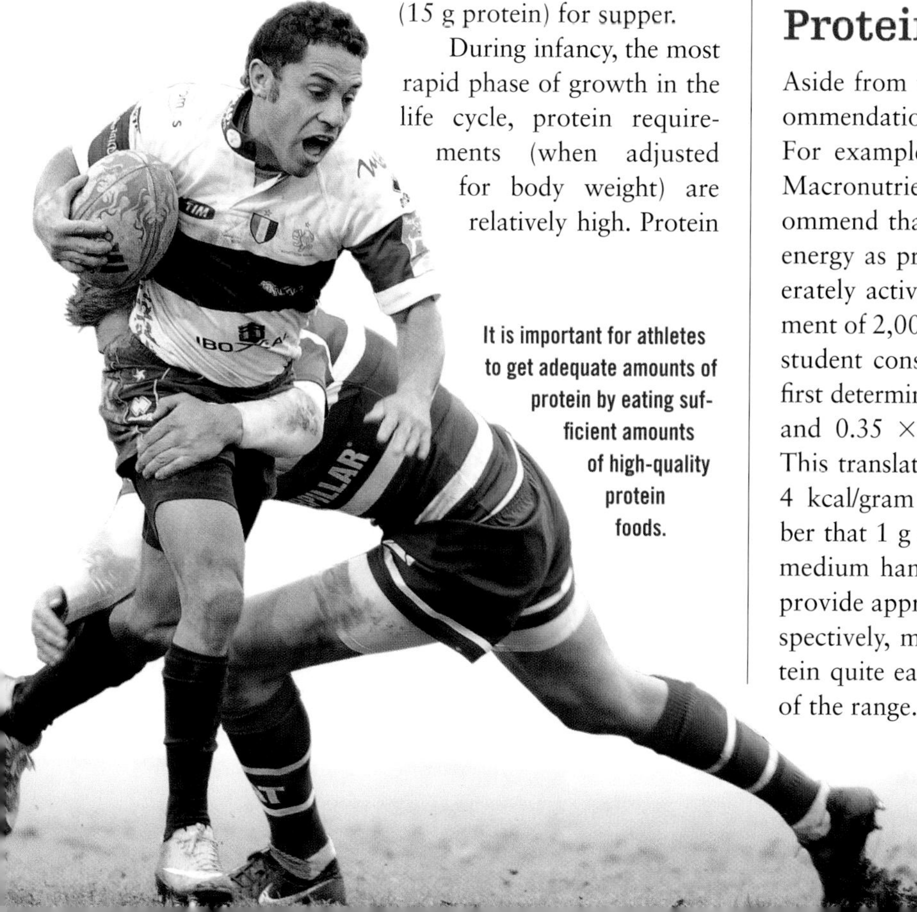

Press Association via AP Images

It is important for athletes to get adequate amounts of protein by eating sufficient amounts of high-quality protein foods.

Experts Debate Whether Athletes Need More Protein

Although many people believe that athletes have higher protein requirements than nonathletes, this is a topic of active debate. The DRI committee that established recommendations for amino acid and protein intake considered this question carefully. The committee concluded that physically active people likely require similar amounts of protein on a body-weight basis, and that adult athletes can generally estimate their protein requirements like other adults by using the same mathematical formula of 0.8 g/kg/day. On the other hand, in a position statement published in 2009, the American College of Sports Medicine concluded that protein intakes of 1.2 to 1.7 g/kg/day for endurance and strength-trained athletes may be beneficial.[16] Thus, scientists continue to grapple with this issue.

Additional Recommendations for Protein Intake

Aside from the RDA values, several other sets of recommendations for protein intake are also available. For example, the Institute of Medicine's Acceptable Macronutrient Distribution Ranges (AMDRs) recommend that you consume 10 to 35 percent of your energy as protein. Using this advice, consider a moderately active college student with an energy requirement of 2,000 kcal/day. How much protein should this student consume? To answer this question, you must first determine that 200 to 700 kcal (0.10 × 2,000 kcal and 0.35 × 2,000 kcal) should come from protein. This translates to 50 to 175 g of protein (200 kcal ÷ 4 kcal/gram and 700 kcal ÷ 4 kcal/gram). Remember that 1 g of protein supplies 4 kcal of energy. One medium hamburger patty and one cup of skim milk provide approximately 25 and 10 grams of protein respectively, making the recommended amount of protein quite easy to obtain—especially at the lower end of the range.

The 2010 Dietary Guidelines for Americans and the accompanying MyPlate food guidance system provide additional recommendations concerning intake of high-protein foods. Aside from supporting the AMDR for protein (10 to 35 percent of calories from protein), the Guidelines specifically encourage a range of intakes from fat-free or low-fat milk, lean meats, eggs, fish, nuts/seeds, and legumes to support optimal health. These food groups represent nutrient-dense, high-protein foods. More specifically, the Food Patterns and MyPlate recommend $1\frac{1}{2}$ to 5 ounces of lean meat and two to three cups of fat-free or low-fat dairy products daily, depending on caloric needs. Periodic consumption of legumes, such as dried beans and peas, is also encouraged. Note that three ounces of lean meat is generally equivalent to a small steak, lean hamburger patty, chicken breast, or a piece of fish. Portions of meat served in restaurants and cafeterias are often much larger than this. To determine precisely how many servings are recommended for you, visit the MyPlate website (http://www.choosemyplate.gov).

It is important to understand basic recommendations concerning how much of each essential amino acid you should consume. You can calculate your total protein requirement using the MyPlate food guidance system and can otherwise refer to MyPlate for suggestions that will help you meet your goals. But what if you were to decide not to eat meat or other high-quality protein sources? How would this affect your health, and what might you do to make sure that your diet was adequate?

Are Protein Supplements Helpful for Athletes?

Although there is very little evidence that protein supplementation is beneficial in terms of athletic performance, a handful of studies suggest that supplementation with certain amino acids—especially branched-chain amino acids (BCAA) such as valine and leucine—may help slow muscle breakdown during intensive training.[17] In response, the International Society of Sports Nutrition published a position paper on protein and exercise stating that, under certain circumstances, supplementation with branched-chain amino acids may improve exercise performance and recovery from exercise.[18] Nonetheless, the bottom line recommendations for athletes remain: (1) eat a well-balanced diet that provides sufficient energy and an appropriate mix of carbohydrates, fats, and protein, and (2) train long and hard.

© David Young-Wolff/PhotoEdit

LO9 Can Vegetarian Diets Be Healthy?

People have many different reasons for deciding which foods they will and will not eat. This seems especially true for meat and other animal-derived products. For example, some religious groups avoid some or all types of animal-based foods. Economic considerations and personal preference can also determine whether people eat meat and/or which types of meat they choose. In fact, it is very likely that you or someone you know avoids eating some or all animal-based products. Because these types of foods tend to provide high-quality protein as well as a multitude of other essential nutrients, it is important to consider the effect of animal-based food consumption (or lack thereof) on issues related to nutritional status.

> "**Su vegetarianism**, practiced by many Taiwanese Buddhists, **prohibits** the consumption of fetid (meaning smelly) vegetables such as **garlic**, **shallot**, **onion**, and **coriander** in addition to animal-based products."

There Are Several Forms of Vegetarianism

Most people have heard the term *vegetarian* (from the Latin *vegetus*, meaning "whole," "sound," "fresh," or "lively"). What does this word actually mean? The term was first used in 1847 by the Vegetarian Society of the United Kingdom to refer to a person who does not eat any meat, poultry, fish, or their related products such as milk and eggs. Today, the term **vegetarian** usually indicates a person who does not consume any, or consumes only some, foods and beverages made from animal-based products. Most vegetarians consume dairy products and eggs. Such a practitioner is called a **lacto-ovo-vegetarian.**[19] Alternatively, a **lactovegetarian** includes dairy products—but not eggs—in his diet. A vegetarian who avoids all animal-based products is referred to as a **vegan**. Thus, when people tell you they are vegetarian, you might want to ask what type they are.

© Beth Segal/StockFood Creative/Getty Images

Vegetarian diets can provide all the essential nutrients, but care must be taken to make sure sufficient protein, iron, calcium, zinc, and vitamin B_{12} are consumed.

Vegetarian Diets Sometimes Require Thoughtful Choices

vegetarian A person who does not consume or consumes only some foods and beverages made from animal products.

lacto-ovo-vegetarian A vegetarian who consumes dairy products and eggs in an otherwise plant-based diet.

lactovegetarian A vegetarian who consumes dairy products (but not eggs) in an otherwise plant-based diet.

vegan A vegetarian who consumes no animal products.

Do vegetarians run any special nutritional risks? The answer to this question depends on what kind of vegetarian a person is. In general, a well-balanced lacto-ovo- or lactovegetarian diet can easily provide adequate protein, energy, and micronutrients. Dairy products and eggs are convenient sources of high-quality protein and many vitamins and minerals. However, because meat is often the primary source of bioavailable iron, eliminating it can make it difficult to meet your iron requirements. Furthermore, vegans may be at increased risk of being deficient in several micronutrients, including calcium, zinc, iron, and vitamin B_{12}.[20] This risk is increased further during pregnancy, lactation, and periods of growth and development such as infancy and adolescence.[21] It is especially important that vegetarians consume sufficient amounts of plant-based foods rich in these micronutrients.

Special Dietary Recommendations for Vegetarians

Because some types of vegetarian diets pose certain nutritional risks, it is important to follow special dietary strategies if you decide to make this choice.

The MyPlate food guidance system specifically recognizes protein, iron, calcium, zinc, and vitamin B_{12} as nutrients that vegetarians should focus on and makes specific recommendations as to how to get adequate amounts of these substances. Special Food Patterns for lacto-ovo-vegetarians and vegans are also included in the 2010 Dietary Guidelines for Americans.

In addition, it is pointed out that some meat replacements (such as cheese) can be very high in calories, saturated fat, and cholesterol. Lower fat versions should

protein-energy malnutrition (PEM) A condition whereby protein deficiency is accompanied by a deficiency in energy, and usually, one or more micronutrients.

be chosen, and they should be consumed in moderation. The following comments and suggestions are provided to help ensure optimal health in individuals who choose to become vegetarians. Note that these recommendations are especially pertinent to vegans.

- Select protein sources that are naturally low in fat, such as skim milk and legumes.
- Minimize the use of high-fat cheese as a meat replacement.
- If you do not consume dairy products, consider drinking calcium-fortified, soy-based beverages. These can provide calcium in amounts similar to milk, are usually low in fat, and do not contain cholesterol.
- Add vegetarian meat substitutes (such as tofu) to soups and stews to boost protein without adding saturated fat and cholesterol.
- Recognize that most restaurants can accommodate vegetarian modifications to menu items by substituting meatless sauces, omitting meat from stir-fry recipes, and adding vegetables or pastas in place of meat.
- Consider eating out at Asian or Indian restaurants, as they often offer a varied selection of high-protein, nutrient-dense, vegetarian dishes.
- Be mindful of getting enough vitamin B_{12}. Because this vitamin is naturally found only in foods that come from animals, vitamin B_{12}-fortified foods or dietary supplements may be necessary for vegans.

The key to a healthy vegetarian diet, as with any diet, is to enjoy a wide assortment of foods and to consume them in moderation. Because no single food provides all the nutrients the body needs, eating a variety of foods can help ensure that vegetarians get the necessary nutrients and substances that promote good health.

Marasmus results from severe, chronic, overall malnutrition.

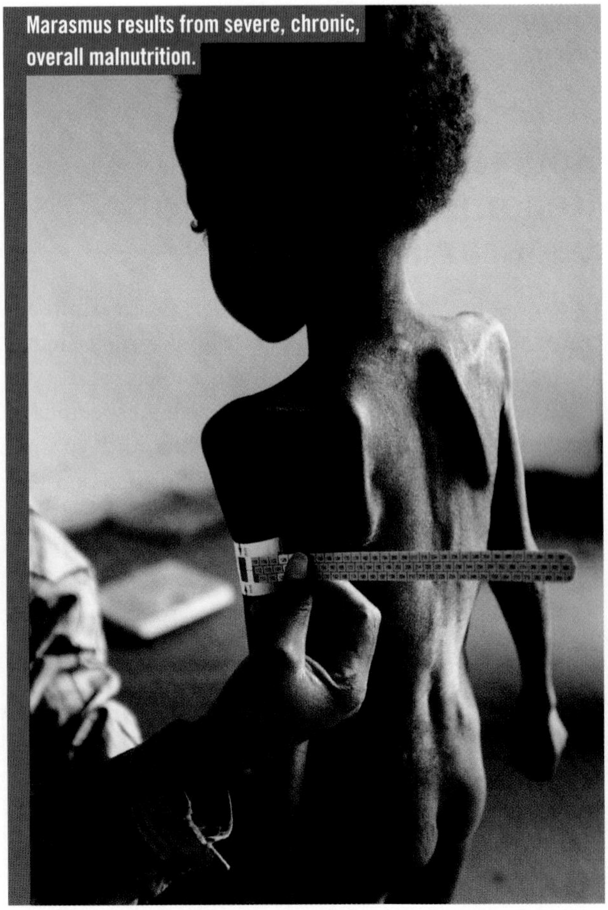

© Louise Gubb/CORBIS SABA

LO10 What Are the Consequences of Protein Deficiency and Excess?

Although generally not a concern in industrialized countries (even for those who consume a vegetarian diet), protein deficiency is more commonly seen in regions where the amount and variety of foods is limited. Protein deficiency is also seen in adults with some debilitating conditions such as acquired immune deficiency syndrome (AIDS) or cancer. Because proteins and amino acids are so important to optimal health, protein deficiency can have significant health implications—some of which are described next.

Protein Deficiency in Early Life

Protein deficiency is rare during the first months of life when infants consume most of their energy from human milk or infant formula. However, once weaned from these high-quality protein sources to foods that lack adequate protein, infants become at greater risk for protein deficiency. Because protein-deficient diets generally also lack energy, protein deficiency is often referred to as **protein-energy malnutrition (PEM).** Children with this condition are typically deficient in one or more micronutrients as well, so PEM is

> The word **marasmus** comes from Greek, meaning "starvation," whereas **kwashiorkor** is believed to come from the Ga language of **coastal Ghana**, meaning "the sickness the baby gets when the **new baby comes**."

Source: CDC/Dr. Lyle Conrad

Children with kwashiorkor often have distended abdomens (ascites), edema in their hands and feet, cracked and peeling skin, and an apathetic nature. These children are at increased risk for infections.

considered a condition of overall malnutrition. PEM has many implications for childhood health. For example, children with PEM are at great risk for infection and illness. Recall that protein is needed to make several components of your immune system as well as the skin and membranes that keep pathogens from entering the body. The World Health Organization estimates that PEM plays a role in at least 5 million child deaths each year, many of which are complicated by infection.[22]

Severe PEM actually encompasses a spectrum of malnutrition: at the extremes are two distinct types of severe PEM, and between them are conditions that combine features of both.[23] At one end of the spectrum is a condition called **marasmus**, which results from severe, chronic, overall malnutrition. In marasmus, fat and muscle tissue are depleted, and the skin hangs in loose folds, with the bones clearly visible beneath the skin. Children with marasmus tend at first to be alert and ravenously hungry, although with increasing severity they become apathetic and lose their appetites. Clinicians often say that marasmus represents the body's survival response to long-term, chronic dietary insufficiency.

The other extreme type of PEM, called **kwashiorkor**, is distinguished from marasmus by the presence of severe edema (swelling) in the extremities. Edema sometimes is present in children with marasmus, but those with kwashiorkor usually have more extensive edema, which typically starts in the legs but often occurs throughout the entire body. Remember that one of the bodily functions of protein is regulation of fluid balance. Children with kwashiorkor sometimes have large, distended abdomens due to fluid accumulation in the abdominal cavity. This condition is referred to as **ascites**. Because malnourished children often have intestinal parasites, worms sometimes contribute to this abdominal distension as well. Children with kwashiorkor are often apathetic and sometimes have cracked and peeling skin, enlarged fatty livers, and sparse unnaturally blond or red hair. Although many characteristics of kwashiorkor were once thought simply to be caused by protein deficiency, this does not appear to be the case.[24] Researchers now believe that many of the signs and symptoms of kwashiorkor are the result of micronutrient deficiencies (such as vitamin A deficiency) in combination with infection or other environmental stressors.

marasmus A form of PEM characterized by extreme wasting of muscle and loss of adipose tissue.

kwashiorkor A form of PEM characterized by severe edema in the extremities.

ascites A condition characterized by fluid accumulation in the abdominal cavity.

Protein Deficiency in Adults

PEM can also occur in adults. Unlike children, however, adults with PEM rarely experience kwashiorkor. Instead, they generally develop marasmus. There are many causes of PEM in adulthood, including inadequate dietary intake (such as sometimes occurs in alcoholics and those with eating disorders), protein malabsorption (such as occurs with some gastrointestinal disorders such as celiac disease), excessive and chronic blood loss, cancer, infection, and injury (especially burns).[25]

Adults with PEM sometimes experience extreme muscle loss because the body's muscles are broken down to provide glucose and energy. Fat accumulation in the liver and edema are common, and adults with severe PEM experience decreased function of many vital physiological systems, including the cardiovascular, renal (kidneys), digestive, endocrine, and immune systems. Treatment of PEM in adults is often long and difficult. For example, if the cause is infection, treatment may involve both dietary intervention and the use of antibiotics. By contrast, if protein deficiency is a result of an eating disorder, psychological counseling becomes a key component of the health care plan. Regardless of its cause, effective treatment of adult PEM presents a special challenge to any medical team.

Protein Excess

Protein deficiency can result in serious health concerns, but what about protein excess? Contrary to popular belief, high-protein diets do not appear to cause adverse health outcomes (such as osteoporosis, kidney problems, heart disease, obesity, and cancer) in most people. This conclusion was confirmed by the DRI committee, which carefully considered the peer-reviewed literature related to the potential health consequences of high-protein diets. In fact, the upper limit of the AMDR for protein intake (35 percent of energy from protein) was developed not because there was evidence that additional protein might pose a health risk, but solely to complement the recommendations for carbohydrate and fat intakes. Nonetheless, high intakes of protein are often accompanied by high intakes of fat, saturated fat, and cholesterol. Because these dietary components are risk factors for heart disease, it is important to choose a variety of lean and low-fat protein foods, such as those recommended by the MyPlate food guidance system.

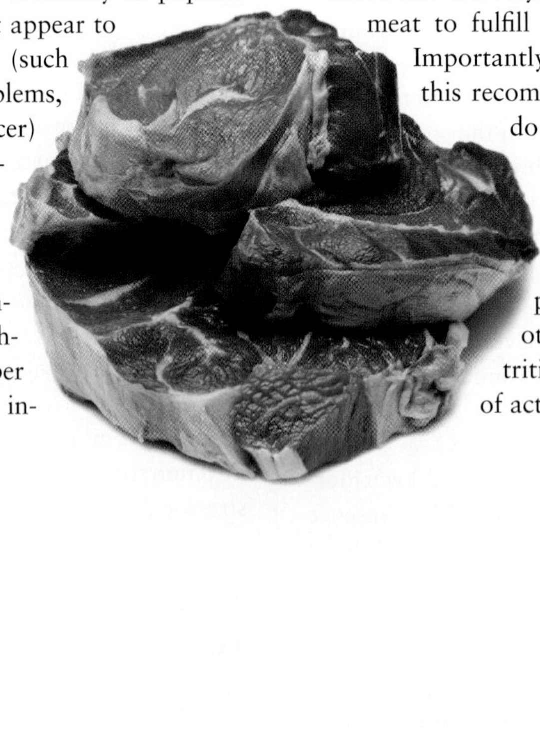

© iStockphoto.com/EddWestmacott

High Red Meat Consumption

Growing evidence suggests that chronically elevated intake of red meat (beef, lamb, and pork) or processed meats (bacon, sausage, hot dogs, ham, and cold cuts) is associated with increased risk for colorectal cancer.[26] As a result, the World Cancer Research Fund and the American Institute for Cancer Research recommended in 2007 that individuals limit their intakes of red meat to no more than 500 g (18 oz) each week and eat very little processed meat.[27] On average, this would be about 70 g (2.6 ounces) of meat each day—an amount less than that recommended by the 2010 Dietary Guidelines and MyPlate (if one were to eat only red meat to fulfill the protein foods requirement). Importantly, the panel of experts who made this recommendation emphasized that they do not suggest avoiding all meat or foods of animal origin. Clearly, these foods can be a valuable source of many essential nutrients and should be considered part of a healthy diet. Like several other issues related to protein nutrition, this topic continues to be one of active debate.

Complete the Speak Up survey in CourseMate at **www.cengagebrain.com**
Follow us at www.facebook.com/4ltrpress

NUTR

6 Lipids

© iStockphoto.com/Floortje

LEARNING OUTCOMES:

- LO1 Define and distinguish major lipid types.
- LO2 Differentiate essential, conditionally essential, and nonessential fatty acids.
- LO3 Describe the differences among mono-, di-, and triglycerides.
- LO4 Discuss the functions of phospholipids and sterols.
- LO5 Explain the processes of triglyceride digestion, absorption, and circulation.
- LO6 List and describe the types and functions of lipoproteins.
- LO7 Explain the impact of lipids on your health.
- LO8 Utilize dietary recommendations for lipids.

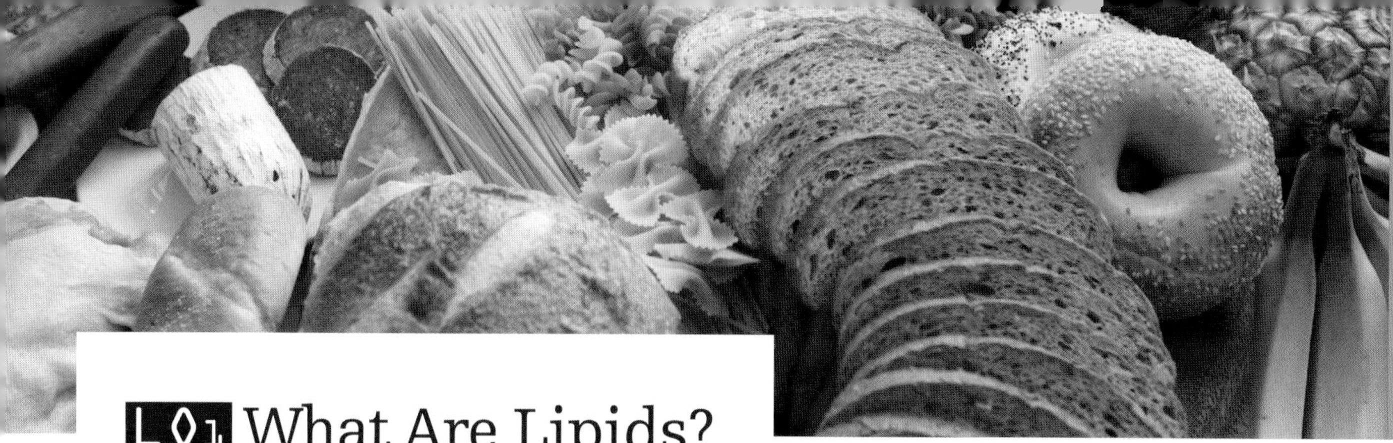

LO1 What Are Lipids?

Lipids (often referred to as fats and oils) are required for hundreds if not thousands of physiological functions in the body. Body fat (made primarily from lipids) protects the vital organs, and lipids surround each cell with a protective barrier. Lipids also make your favorite foods flavorful—butter, cream, olive oil, and well-marbled beef all taste the way they do largely because of lipids. The thought of fatty foods conjures up images of unhealthy living for many people. You may shop for "fat-free" foods and try to avoid fats altogether. Food manufacturers have even developed fat substitutes to replace some of the fats normally found in food. Although diets high in fat and its inherent calories can lead to health complications such as obesity and heart disease, getting enough of the right types of fat is just as essential to optimal health as avoiding excess fat and the wrong kinds of fat. In this chapter, you will learn about the variety of fats and oils in foods and how your body uses them to create substances vital to your health.

The term **lipid** refers to a diverse and important category of organic macronutrients that are relatively insoluble in water and relatively more soluble in organic solvents such as fingernail polish remover (acetone) and paint thinner (turpentine). Not only do lipids provide a major source of energy, they are also critical to optimal health. Lipids come in many varieties, and before you can truly understand their functions in foods and impact on health, you must first understand how lipids differ, both chemically and physiologically.

Fats and Oils

The first step to understanding lipids is learning some basic terminology. Most lipids are **hydrophobic**, or water "fearing." This means that they do not easily mix with water. A lipid that is liquid at room temperature is called an **oil**, and one that is solid at room temperature is called a **fat**. The major lipids in your body and the foods you eat include the fatty acids, triglycerides, phospholipids, sterols, and fat-soluble vitamins. While most of these substances are discussed in detail throughout this chapter, fat-soluble vitamins are addressed in Chapter 7.

Fatty Acids

Fatty acid molecules, made entirely of carbon, hydrogen, and oxygen atoms, comprise the most abundant type of lipid in your body and the foods you eat (see Figure 6.1 on the next page). A chain of carbon atoms forms the backbone of each fatty acid. One end of this carbon chain, called the **alpha (α) end**, contains a *carboxylic acid* group (–COOH). The other end, called the **omega (ω) end**, contains a *methyl* group ($-CH_3$). Both in the body and in foods, most fatty acids do not exist in their free (unbound) form. Instead, they are components of larger molecules, such as triglycerides and phospholipids (see Figure 6.2 on the next page).

There are hundreds of types of fatty acids. These substances are distinguished both by the number of carbons they contain and by the types and locations of chemical bonds holding the carbon atoms together. These variations influence the physical properties of the fatty acids and the roles they play.

Number of Carbons (Chain Length)

The number of carbon atoms in a fatty acid's backbone determines its **chain length**, as illustrated in Figure 6.3 on the next page. In fact, it may be helpful to think of a fatty acid as if it were a chain, with

lipid An organic macronutrient that is relatively insoluble in water and relatively soluble in organic solvents.

hydrophobic Water fearing; does not easily mix with water.

oil A lipid that is liquid at room temperature.

fat A lipid that is solid at room temperature.

fatty acid The most abundant type of lipid in the body and foods; comprised of a chain of carbons with a methyl ($-CH_3$) group on one end and a carboxylic acid (–COOH) group on the other.

alpha (α) end The end of a fatty acid that contains a carboxylic acid (–COOH) group.

omega (ω) end The end of a fatty acid that contains a methyl ($-CH_3$) group.

chain length The number of carbon atoms in a fatty acid's backbone.

short-chain fatty acid A fatty acid with fewer than eight carbon atoms in its backbone.

medium-chain fatty acid A fatty acid with eight to 12 carbon atoms in its backbone.

long-chain fatty acid A fatty acid with more than 12 carbon atoms in its backbone.

Figure 6.1 Fatty Acid Structure

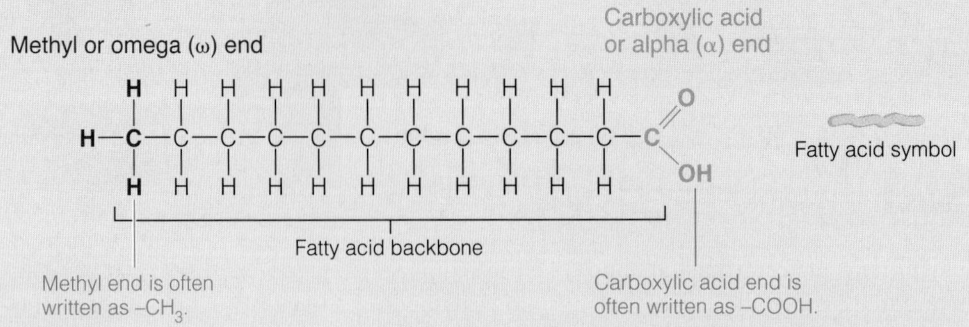

All fatty acids have three components: a carboxylic acid or α end (–COOH), a methyl or ω end ($-CH_3$), and a fatty acid backbone made from carbon and hydrogen atoms.

each link representing a carbon atom. Most naturally occurring fatty acids have an even number of carbon atoms: chains usually span from 12 to 22 carbons, although some may be as short as four or as long as 26 carbons. A fatty acid with fewer than eight carbon atoms in its backbone is called a **short-chain fatty acid**; one with 8 to 12 carbon atoms is called a **medium-chain fatty acid**; and one with more than 12 carbon atoms is called a **long-chain fatty acid.**

The chain length of a fatty acid affects its chemical properties and physiological functions. For example, chain length influences the temperature at which a fatty acid melts (its *melting point*), and lipids constructed predominantly of short-chain fatty acids are generally oils. Chain length also affects solubility in water: short-chain fatty acids are generally more water-soluble than long-chain fatty acids. Because the human body is mostly water, it is relatively easy for people to absorb and transport water-soluble substances such as short-chain fatty acids. Conversely, the body needs more complex processes to absorb, transport, and use dietary lipids that are more water insoluble, such as the long-chain fatty acids.

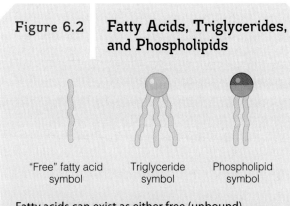

Figure 6.2 Fatty Acids, Triglycerides, and Phospholipids

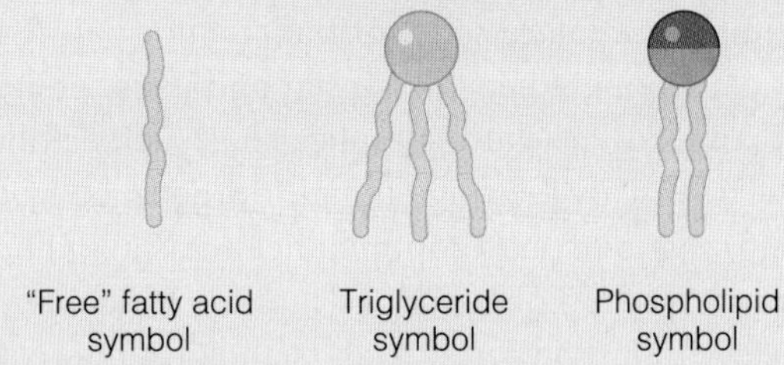

Fatty acids can exist as either free (unbound) molecules or components of larger lipids such as triglycerides and phospholipids.

Figure 6.3 Fatty Acids Can Have Different Chain Lengths

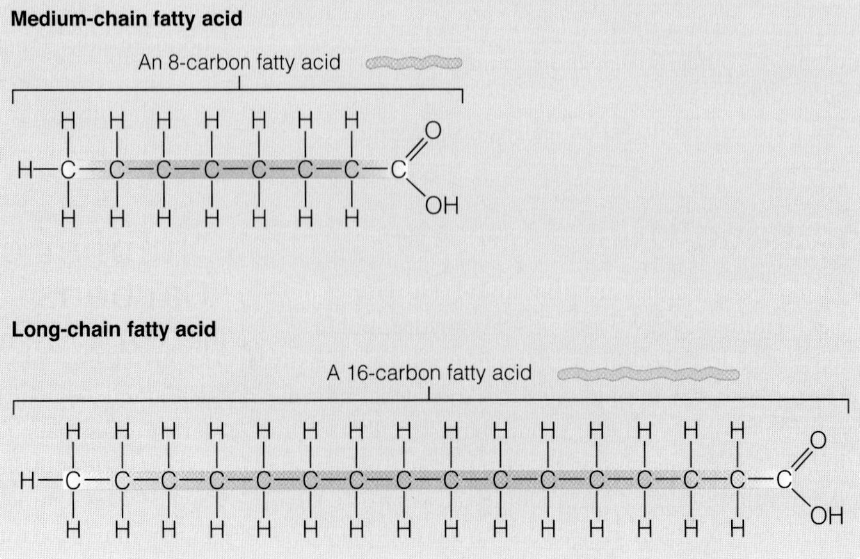

Number and Positions of Double Bonds

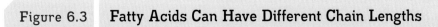

Aside from the number of carbon atoms they contain, fatty acids also differ in the types of chemical bonds between their carbon atoms (see Figure 6.4). A fatty acid's carbon–carbon bonds can be either single bonds or double bonds. If a fatty acid contains only carbon–carbon single bonds in its backbone, it is a **saturated fatty acid (SFA)**; if it contains at least one carbon–carbon double bond, it is an **unsaturated fatty acid.** A fatty acid with just one double bond in its backbone is a **monounsaturated fatty acid (MUFA),** while one with more than one double bond is a **polyunsaturated fatty acid (PUFA).**

As with chain length, the number of double bonds can influence the physical nature of a fatty acid. As you can see in Figure 6.4, each carbon atom in an SFA is surrounded (or saturated) by hydrogen atoms. Saturation prevents the fatty acid from bending. Because of this rigidity, SFAs are highly organized and dense, making them solid at room temperature. Foods containing large amounts of SFAs (such as butter) tend likewise to be solid at room temperature.

Compared with SFAs, unsaturated fatty acids (those with double bonds) have fewer hydrogen atoms flanking their carbon backbones—this allows them to bend. In fact, whenever there is a carbon–carbon double bond, there is a kink or bend in the fatty acid backbone. These bends cause

saturated fatty acid (SFA) A fatty acid that contains only carbon–carbon single bonds in its backbone.

unsaturated fatty acid A fatty acid that contains at least one carbon–carbon double bond in its backbone.

monounsaturated fatty acid (MUFA) A fatty acid that contains one carbon–carbon double bond in its backbone.

polyunsaturated fatty acid (PUFA) A fatty acid that contains more than one carbon–carbon double bond in its backbone.

Figure 6.4 Saturated and Unsaturated Fatty Acids

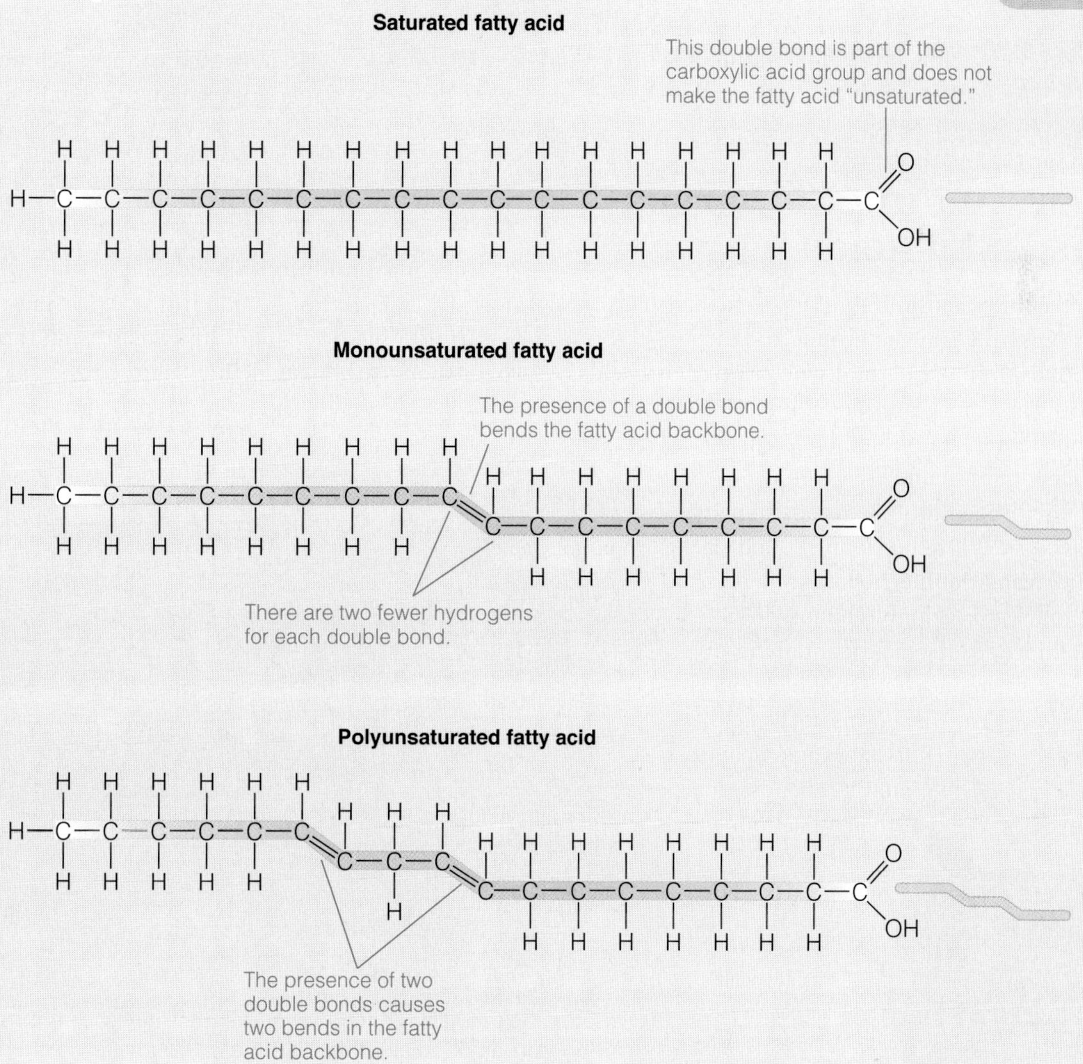

The type of carbon–carbon bonds in a fatty acid determines whether it is saturated or unsaturated.

cis double bond A carbon–carbon double bond in which the hydrogen atoms are positioned on the same side of the double bond.

trans double bond A carbon–carbon double bond in which the hydrogen atoms are positioned on opposite sides of the double bond.

trans fatty acid A fatty acid containing at least one *trans* double bond.

partial hydrogenation A process by which oil is converted into solid fat by changing many of the carbon–carbon double bonds into carbon–carbon single bonds; some *trans* fatty acids are also produced.

Some **animal products** *such as beef contain a* **relatively high** *amount of* **saturated fatty acids.**

unsaturated fatty acids to become disorganized, preventing them from becoming densely packed. In general, organized molecules like SFAs are solid at room temperature, while disorganized molecules like PUFAs are liquid. MUFAs have chemical characteristics that lie between those of SFAs and PUFAs, making them thick liquids or semi-solids at room temperature. You may have noticed that olive oil, which is high in MUFAs, is a thick oil at room temperature.

Understanding *Cis* versus *Trans* Fatty Acids

Unsaturated fatty acids differ in the ways that their hydrogen atoms are arranged around the carbon–carbon double bonds. In most naturally occurring fatty acids, the hydrogen atoms are positioned on the same side of the double bond. This is called a ***cis* double bond** (see Figure 6.5). When the hydrogen atoms are positioned on opposite sides of the double bond, it is called a ***trans* double bond**. Unlike *cis* double bonds, *trans* double bonds do not cause bending. A fatty acid containing at least one *trans* double bond is called a ***trans* fatty acid.** *Trans* fatty acids have fewer bends in their backbones than do their *cis* counterparts. For this reason, *trans* fatty acids are also more likely to be solid (fats) at room temperature.

Olive oil is a thick liquid because it contains a relatively high amount of MUFAs.

TRANS FATTY ACIDS IN FOOD

Trans fatty acids are found naturally in some foods, such as dairy and beef products. However, most dietary *trans* fatty acids are produced commercially via a process called **partial hydrogenation.** Partial hydrogenation converts oils (such as corn oil) into solid fats (such as margarine or shortening) by changing many of the carbon–carbon double bonds into carbon–carbon single bonds. This is achieved through the chemical addition of hydrogen atoms, hence the term *hydrogenation*. Aside from decreasing the number of double bonds (and thus increasing the number of carbon–carbon single bonds), the process of partial

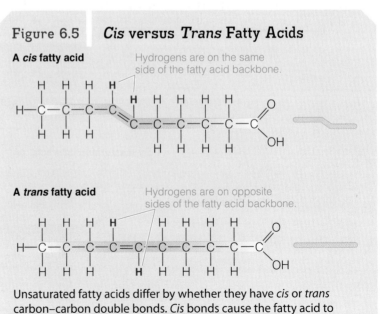

Figure 6.5 ***Cis* versus *Trans* Fatty Acids**

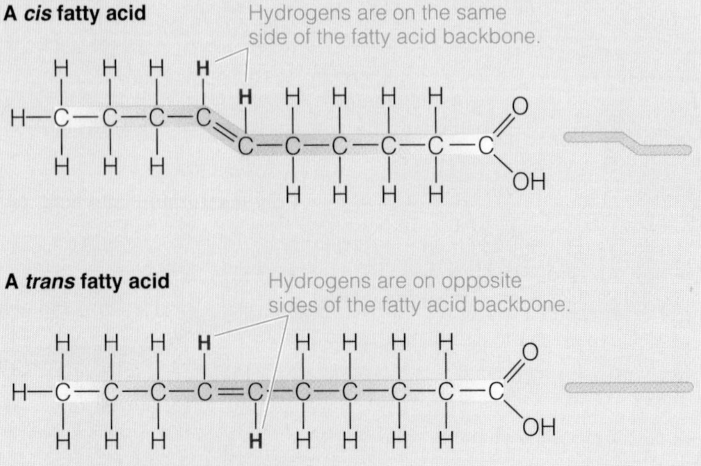

Unsaturated fatty acids differ by whether they have *cis* or *trans* carbon–carbon double bonds. *Cis* bonds cause the fatty acid to bend, whereas *trans* bonds do not.

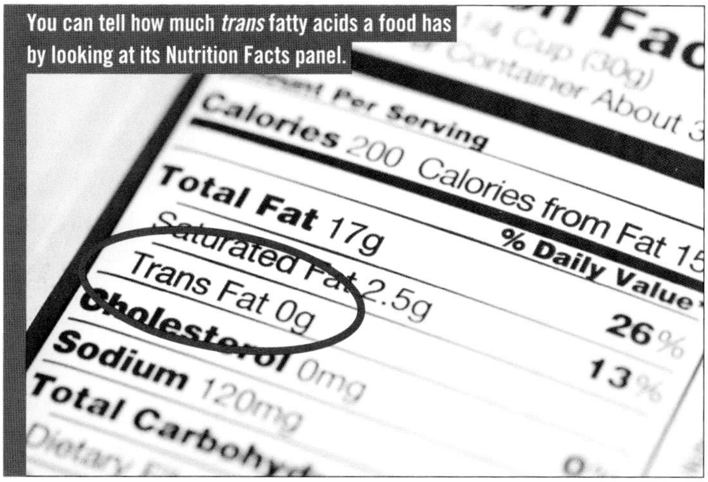

You can tell how much *trans* fatty acids a food has by looking at its Nutrition Facts panel.

hydrogenation converts some of the remaining *cis* double bonds to *trans* double bonds, causing the lipid to become high in *trans* fatty acids. Partial hydrogenation is used in food manufacturing because adding partially hydrogenated lipids imparts desirable food texture and reduces spoilage. Crackers, pastries, bakery products, shortening, and margarine are the main sources of the *trans* fatty acids in the diet.[1] However, a recent trend toward decreasing *trans* fatty acid intake has resulted in new food preparation and processing methods that reduce or eliminate *trans* fatty acids in many foods. For example, many fast-food chains have switched from frying their foods in high–*trans* fatty acid shortening to *trans* fat–free vegetable oils.

Many public health agencies (such as the USDA and the Institute of Medicine) suggest limiting your intake of *trans* fatty acids. These recommendations (described in more detail later on) were developed because some studies show that certain industrially produced *trans* fatty acids increase the risk for cardiovascular disease.[2] As of 2006, food manufacturers are required to state *trans* fatty acid content on their Nutrition Facts panels. Because of the considerable concern about *trans* fatty acids in the diet, some cities are even declaring themselves "*trans* fat–free zones."

Fatty Acids Are Named for Their Structures

There are several methods used to name or describe fatty acids. In general, names are based on the number of carbons, the number and types of double bonds, and the positions of the double bonds in a given fatty acid. Some fatty acids also have *common names* that are used to refer to substances informally.

ALPHA (α) NAMING SYSTEM

The *alpha* (α) naming system for fatty acids is based on the positions and types of double bonds relative to the carboxylic acid (α) end of the fatty acid. The basic formula for constructing an alpha (α) name is [*cis* or *trans*]*a*, [*cis* or *trans*]*b*–*x*:*y*, where *x* signifies the total number of carbon atoms and *y* signifies the total number of double bonds. In this formula, there are two

Trans Fat-Free Zones

Should your government have a say over the contents of your food? This has become an interesting issue in the case of *trans* fatty acids. For example, in 2008 the New York City Health Department banned restaurants from serving foods containing commercially produced *trans* fatty acids. It still, however, allows restaurants to serve *trans* fatty acid-containing foods that come in the manufacturer's original packaging. Thus, some but not all food sold in New York restaurants must be *trans* fatty acid free. While the New York City ban (like others across the nation) comes on the heels of significant scientific evidence that consumption of *trans* fatty acids found in partially hydrogenated oils can increase risk for heart disease, opponents believe that local municipalities have no business outlawing foods that the U.S. Food and Drug Administration (FDA) has deemed safe enough to allow in the public food system. What do you think?

double bonds. The type (*cis* or *trans*) and number of carbons from the carboxylic acid (α) end of each double bond (signified as *a* and *b*) are separated by a comma.

As an example, consider a fatty acid with the following characteristics: 18 carbons and two *cis* double bonds—one between the ninth and 10th carbons from the carboxylic acid (α) end and the other between the 12th and 13th carbons from the carboxylic acid (α) end. The fatty acid's name is constructed beginning with an 18, signifying that there are 18 carbons. The number 2 is added to form 18:2, signifying that there are two double bonds. Next, the two double bonds' locations are added to the front of the name: 9,12–18:2. Remember, the locations are determined by counting from the carboxylic acid (α) end. Finally, because both double bonds are *cis* double bonds, the name is modified to *cis*9,*cis*12–18:2. This fatty acid is illustrated in Figure 6.6.

Figure 6.6 The Alpha (α) Naming System

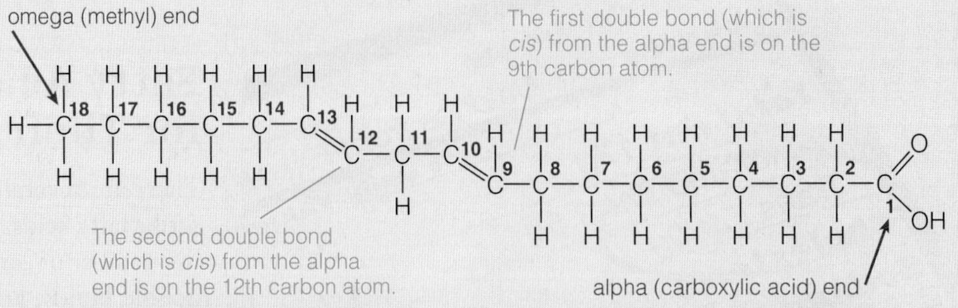

The alpha naming system for fatty acids requires that you count how far the double bonds are from the alpha (carboxylic acid) end of the carbon backbone. Using this system, this fatty acid's name is *cis*9, *cis*12–18:2.

OMEGA (ω) NAMING SYSTEM

An alternate system for naming a fatty acid is the omega (ω) system. In this system, the number of carbons and double bonds are again distinguished as *x*:*y* (for example, 18:2). However, in the omega naming system, fatty acids are expressed on the basis of the first double bond's distance from the methyl (ω; omega) end of the molecule. If the first double bond is located between the third and fourth carbons from the omega end, the fatty acid is an **omega-3 (ω-3) fatty acid**. If the first double bond is located between the sixth and seventh carbons from the omega end, it is an **omega-6 (ω-6) fatty acid**. There are also ω-7 and ω-9 fatty acids.

omega-3 (ω-3) fatty acid A fatty acid in which the first double bond is located between the third and fourth carbons from the omega (ω) end.

omega-6 (ω-6) fatty acid A fatty acid in which the first double bond is located between the sixth and seventh carbons from the omega (ω) end.

COMMON NAMES

In addition to the alpha and omega naming systems, fatty acids are often referred to by their common names, which usually reflect a prominent food source of the fatty acid. For example, *palmitic acid* (16:0) is found in palm oil, and *arachidonic acid* (*cis*5,*cis*8,*cis*11,*cis*14–20:4; from *arachis*, meaning legume or peanut) is found in peanut butter.

LO2 What Are Essential, Conditionally Essential, and Nonessential Fatty Acids?

There are potentially hundreds of different fatty acids, each with its own distinct structure, physiologic properties, and name. Although foods provide a variety of different fatty acids, many of which are needed for optimal health, only two are considered essential.

"According to some studies, the **consumption of** long-chain **ω-3 fatty acids** may help **lower** the risk of **heart disease**."

Because cells are unable to make these fatty acids, it is important that you get them from food.

The Essential Fatty Acids: Linoleic Acid and Linolenic Acid

The two essential fatty acids are **linoleic acid** and **linolenic acid** (or **α-linolenic acid**). Linoleic acid has 18 carbons, two double bonds, and is an ω-6 fatty acid. Linolenic acid has 18 carbons, three double bonds, and is an ω-3 fatty acid. Linoleic acid and linolenic acid are essential nutrients because the body cannot insert double bonds in the ω-3 and ω-6 positions of fatty acids. As such, you need linoleic acid and linolenic acid to make all of the other ω-6 and ω-3 fatty acids, respectively, and to make many related substances. For example, as illustrated in Figure 6.7, linoleic acid is used to make *arachidonic acid* (a 20-carbon, ω-6 fatty acid). Similarly, linolenic acid can be converted to *eicosapentaenoic acid* (EPA), a 20-carbon, ω-3 fatty acid, which can subsequently be converted to *docosahexaenoic acid* (DHA), a 22-carbon, ω-3 fatty acid. These important long-chain PUFAs have many functions in the body. For example, they are necessary to normal cell function and the regulation of gene expression. They also serve as precursors to hormone-like compounds called eicosanoids.

CONVERSION OF ESSENTIAL FATTY ACIDS TO EICOSANOIDS

The essential fatty acids and some other longer-chain fatty acids can be used to make other important compounds that are not themselves fatty acids, but are nonetheless vital for health. One example is the **eicosanoid family**, a collection of substances with diverse, hormone-like effects. Eicosanoids are involved in regulating the immune and cardiovascular systems and act as chemical messengers that direct a variety of physiologic functions.[3] Your body produces both ω-3 and ω-6 eicosanoids, which have somewhat opposing actions. The ω-6 eicosanoids (produced from linoleic acid) tend to cause inflammation and the constriction of blood vessels, whereas the ω-3 eicosanoids (produced from linolenic acid) tend to reduce inflammation, stimulate dilation (or relaxation) of the blood vessel walls, and inhibit blood clotting. Both ω-3 and ω-6 eicosanoids are important to health, and the body can shift its relative production in response to its needs.

linoleic acid An essential ω-6 fatty acid with 18 carbons and two double bonds.

linolenic acid (or **α-linolenic acid**) An essential ω-3 fatty acid with 18 carbons and three double bonds.

eicosanoid family A diverse group of hormone-like compounds that help regulate the immune and cardiovascular systems and act as chemical messengers.

Figure 6.7 Fatty Acid Metabolism and Eicosanoid Formation

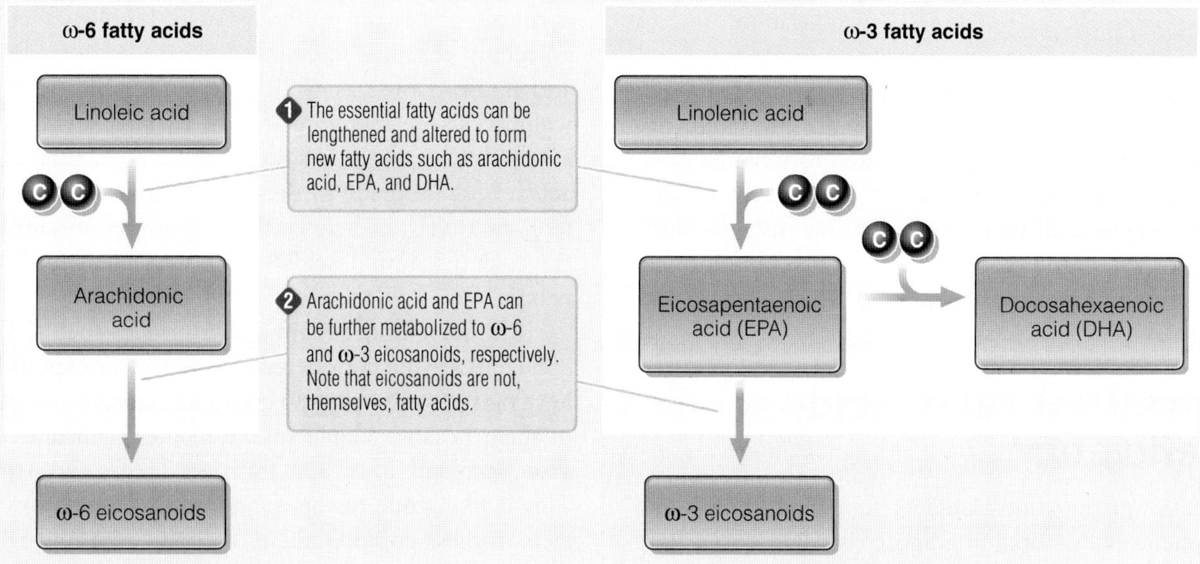

Linoleic acid and linolenic acid can be converted to longer-chain fatty acids, some of which are used to make hormone-like eicosanoids.

People who consume foods high in omega-3 fatty acids (such as many cold-water fish) may have lower risk for conditions related to inflammation, such as heart disease.

Dietary choices can influence the amount and types of eicosanoids that you make. For example, Alaska natives who consume high amounts of ω-3 fatty acids from fish and marine mammals (such as seals) have enhanced physiologic responses that are stimulated by ω-3 eicosanoids. Consequently, Alaska natives who consume traditional diets tend to take more time to form blood clots than people who consume fewer ω-3 fatty acids.[4] Many studies suggest that alterations in the balance of ω-3 to ω-6 eicosanoids may influence a person's risk of conditions related to inflammation, such as heart disease and cancer.[5] This is why experts often recommend that people consume fish regularly. These recommendations will be described in more detail later in this chapter.

Essential Fatty Acid Deficiency

Because of the almost endless supply of linoleic and linolenic acids stored in adipose tissue, essential fatty acid deficiencies do not generally occur in otherwise healthy people. Primary essential fatty acid deficiency is rare, generally occurring only in those with poor health. Secondary fatty acid deficiencies can occur in conjunction with diseases that disrupt lipid absorption or utilization such as cystic fibrosis. When an essential fatty acid deficiency is present, however, infections are common and wound healing may be slow. Children with essential fatty acid deficiencies also exhibit slow growth.

Conditionally Essential Fatty Acids in Infancy

In addition to linoleic and linolenic acids, other fatty acids such as arachidonic acid and DHA may be conditionally essential during infancy. While adults can synthesize these fatty acids from linoleic acid and linolenic acid respectively, a baby may not be able to make them because she cannot produce the needed enzymes in sufficient amounts. As might be expected, human milk contains ample amounts of these conditionally essential fatty acids, and many infant formulas are now fortified with arachidonic acid and DHA—making them more like human milk in this regard.

Dietary Sources of Fatty Acids

Although most foods contain a mixture of fatty acids, some are especially high in a particular fatty acid. For example, some nuts (such as walnuts), seeds, and oils (such as those made from soybean, safflower, and corn) are abundant in linoleic acid. Linolenic acid is especially plentiful in canola (or rapeseed), soybean, and flaxseed oils, as well as in some nuts (such as walnuts). Note that some foods (such as soybean oil and walnuts) are good sources of both essential fatty acids. Because many of these foods and oils are common to the U.S. diet, getting adequate amounts of the essential amino acids is not a problem for most people.

Although longer-chain ω-3 fatty acids such as EPA and DHA are not considered essential, some experts believe that it is important to consume adequate amounts of them because higher intake of these substances is associated with lower risk for heart disease and stroke. This is likely due to the potent anti-inflammatory effects of these compounds in the body. EPA and DHA are especially abundant in fatty fish and other seafood, while smaller amounts are found in meats and eggs. Longer-chain ω-6 fatty acids are also found in meat, poultry, and eggs.

Dietary Sources of Nonessential Saturated and Unsaturated Fatty Acids

Saturated fatty acids and most unsaturated fatty acids are not essential to your diet because your body can synthesize them when needed. Nevertheless, a healthy diet typically includes a variety of nonessential fatty acids, which serve as a source of energy, insulation, and protection. In general, animal-based foods supply the majority of dietary saturated fatty acids (SFAs), whereas plant-derived foods supply the majority of polyunsaturated fatty acids (PUFAs). There are exceptions, however: some tropical oils, such as coconut and palm oils, contain relatively high amounts of SFAs, and many oily fish have high levels of PUFAs. Monounsaturated fatty acids (MUFAs) are supplied by both plant- and animal-based foods. The relative amounts of SFAs, MUFAs, and PUFAs in commonly consumed fats and oils are summarized in Figure 6.8.

Essential Fatty Acids versus Essential Oils

An essential fatty acid is one that the body needs but must get from the diet because it is either unable to make it at all or cannot make it in sufficient amounts. This is very different from an *essential oil*, which is an oil derived from a natural substance, usually either for its purported healing properties or fragrance. Examples of essential oils include camphor and eucalyptus, which are believed to help relieve congestion. Thus, whereas the aroma or topical application of essential oils may provide valuable psychological or physical benefits, these oils are not for consumption.

Essential oils contain compounds that are added to medicinal preparations for their scents or putative health benefits. These essential oils are not the same as essential fatty acids, which are an important part of a healthy diet.

Figure 6.8 Distribution of Fatty Acid Types in Commonly Consumed Lipids

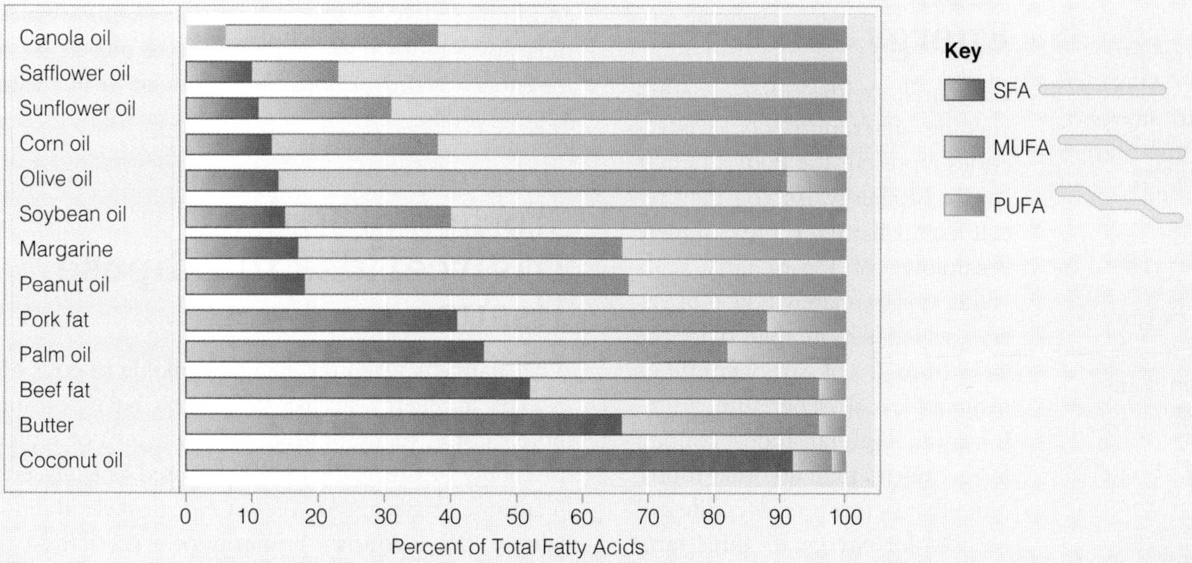

Dietary lipids contain different relative amounts of saturated fatty acids (SFAs), monounsaturated fatty acids (MUFAs), and polyunsaturated fatty acids (PUFAs).

LO3 What Is the Difference between Mono-, Di-, and Triglycerides?

As previously mentioned, most fatty acids do not exist in their free (unbound) form. Instead, they comprise various parts of larger, more complex molecules called monoglycerides, diglycerides, and triglycerides. What is the difference between these three types of molecules? As you might have guessed by their names, these substances are defined by the number of fatty acids present in their chemical structures. Whereas a **monoglyceride** molecule incorporates only one fatty acid, a **diglyceride** molecule incorporates two fatty acids, and a **triglyceride** molecule incorporates three fatty acids. The fatty acids can be saturated, monounsaturated, polyunsaturated, or a mixture thereof. In every case, the fatty acid molecule (or molecules) attach to a glycerol backbone (see Figure 6.9). The metabolic process by which fatty acids combine with glycerol to form triglycerides is called **lipogenesis**.

Figure 6.9 Triglyceride Structure

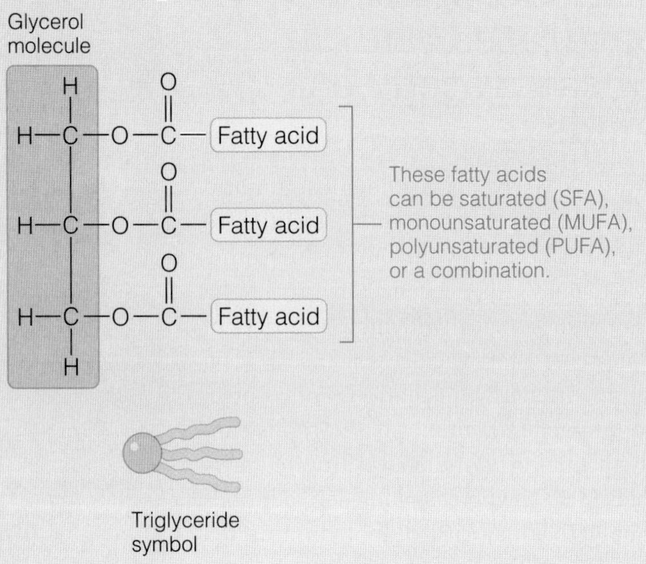

A triglyceride consists of a glycerol molecule and three fatty acids.

monoglyceride A lipid comprised of one fatty acid attached to a glycerol backbone.

diglyceride A lipid comprised of two fatty acids attached to a glycerol backbone.

triglyceride A lipid comprised of three fatty acids attached to a glycerol backbone.

lipogenesis The metabolic process by which fatty acids combine with glycerol to form triglycerides.

lipolysis The metabolic process by which a triglyceride's three fatty acids are removed from the glycerol backbone.

β-oxidation The metabolic breakdown of fatty acids to produce ATP.

adipocyte A specialized cell, found in adipose tissue, that can accumulate large amounts of triglycerides.

Triglycerides Are Rich Sources of Energy

Compared with other energy-yielding macronutrients, triglycerides represent the body's richest source of energy. As you may recall from Chapter 1, the complete breakdown of 1 g of fatty acids yields approximately 9 kcal of energy, which is more than twice the yield from 1 g of carbohydrate or protein (4 kcal). Therefore, gram-for-gram, high-fat foods contain more calories than do other foods. Note that before a triglyceride can be used for energy, its three fatty acids must be removed from the glycerol backbone. This metabolic process is called **lipolysis**. The subsequent breakdown of fatty acids to produce ATP is called **β-oxidation**.

In addition to using fatty acids as an immediate energy source, the body can convert them to other energy-yielding organic compounds called ketones. Recall from Chapter 4 that the production of ketones from fatty acids, called *ketogenesis*, occurs when glucose availability is low. Ketogenesis is important because some tissues (such as brain, heart, skeletal muscle, and kidney tissue) can use ketones to synthesize ATP, thus slowing down the mobilization of muscle protein for energy during periods of low calorie or low carbohydrate intake.

Storage of Excess Triglycerides in Adipose Tissue

What happens when the energy available to your body exceeds its energy needs? When you have consumed more calories than are needed, fatty acids not required for energy or other functions are stored as triglycerides in adipose tissue, and to a lesser extent, skeletal muscle. Adipose tissue consists primarily of a specialized type of cell called an **adipocyte**. Adipocytes can accumulate large amounts of triglycerides, thus serving as a reservoir

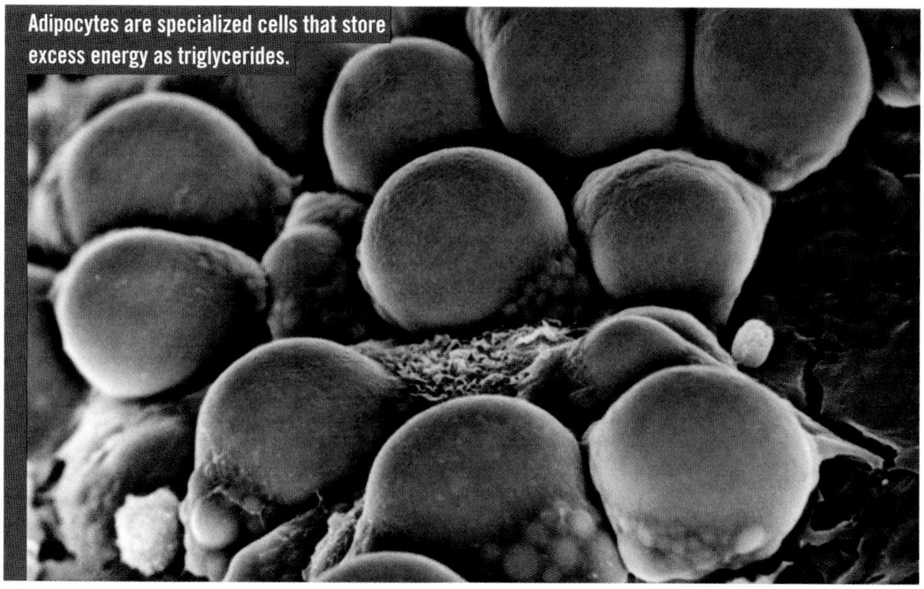

Adipocytes are specialized cells that store excess energy as triglycerides.

© SPL/Photo Researchers, Inc.

of energy. Adipose tissue is located in many parts of your body, including directly under your skin (**subcutaneous adipose tissue**) and around the vital organs in your abdomen (**visceral adipose tissue**). Many of your body's organs and tissues (such as the kidneys and breasts) also have adipose tissue associated with them. Not only does this help protect these organs, it also provides them with a readily available energy source.

Compared to glycogen, the storage of excess energy as triglycerides has two major advantages. First, because triglycerides are not stored with water which takes up space, a large amount can fit into a small space. Second, as previously discussed, a gram of pure triglyceride stores over twice the energy as does a gram of carbohydrate. Consequently, your body can store about six times more energy in one pound of adipose tissue than in one pound of liver glycogen. Indeed, the body has a seemingly infinite ability to store excess energy in adipose tissue, whereas its capacity to store glycogen is limited.

INSULIN STIMULATES LIPOGENESIS AND INHIBITS LIPOLYSIS

The hormone insulin stimulates the storage of triglycerides during times of energy excess—such as after a meal—by causing adipocytes (and to a lesser extent skeletal muscle) to take up glucose and fatty acids from the blood. Insulin also stimulates the conversion of excess glucose to fatty acids (another form of lipogenesis), which are in turn incorporated into triglycerides. Finally, insulin inhibits lipolysis. The increased lipogenesis and decreased lipolysis experienced after a meal help direct excess glucose and fatty acids to adipose tissue, where they are stored as triglycerides for later use.

Triglycerides Needed for Insulation

In addition to providing energy and protecting internal organs from injury, adipose tissue also insulates the body. Humans rely on subcutaneous adipose tissue to keep warm, and people with very little body fat often have difficulty regulating body temperature. In fact, a physiological response to insufficient body fat is the growth of very fine hair on the body. This hair, called **lanugo**, makes up for the absence of subcutaneous adipose tissue to some extent by providing a layer of external insulation. The presence of lanugo is common in very malnourished and underweight people, such as those with the eating disorder anorexia nervosa.[6]

LO4 What Are Phospholipids and Sterols?

Along with triglycerides, phospholipids and sterols are major types of lipids found throughout the body. These substances are essential components of cell membranes and are involved in the transport of lipids in your bloodstream. However, because the body can synthesize all the phospholipids and sterols it needs, there are no dietary requirements for either of them. Still, because they are commonly found in the foods you eat, an awareness of these two substances is important to a complete understanding of nutrition.

subcutaneous adipose tissue Adipose tissue located directly under the skin.

visceral adipose tissue Adipose tissue located around the vital organs in the abdomen.

lanugo Very fine hair that grows as a physiological response to insufficient body fat.

phospholipid A lipid composed of a glycerol molecule bonded to two hydrophobic fatty acids and a hydrophilic head group.

hydrophilic Water loving; mixes easily with water.

head group A phosphate-containing, hydrophilic chemical structure that serves as a component of a phospholipid.

amphipathic A characteristic of a substance that contains both hydrophilic and hydrophobic regions.

sterol A type of lipid with a multiple-ring chemical structure.

cholesterol An abundant and widely discussed sterol that is used to synthesize bile acids and a variety of hormones such as testosterone.

Phospholipids

A **phospholipid** is similar to a triglyceride in that both contain a glycerol molecule bonded to fatty acids (see Figure 6.10). However, instead of having three fatty acids (like a triglyceride), a phospholipid has only two. Replacing the third fatty acid is a phosphate-containing **hydrophilic** (or water-loving, meaning that it mixes easily with water) **head group**. There are many different types of head groups, but the most common are choline, ethanolamine, inositol, and serine.

Figure 6.10 Phospholipid Structure

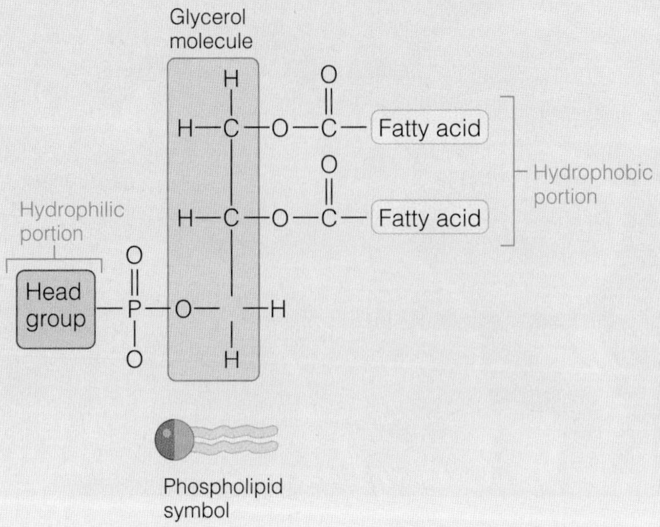

Phospholipids consist of a glycerol, a hydrophilic head group, and two hydrophobic fatty acids.

Phospholipids are **amphipathic**, meaning they contain both hydrophilic and hydrophobic regions. While the head group (the hydrophilic region) of each phospholipid molecule is attracted to water, the fatty acids (the hydrophobic region) repel water. This structure allows phospholipids both to act as major components of cell membranes and to play important roles in the digestion, absorption, and transport of lipids throughout your body. It is also helpful in terms of how phospholipids can be used in the food industry. For example, the phospholipid *lecithin* is often used in foods such as mayonnaise and salad dressings as a stabilizer. Lecithin prevents the lipid- and water-soluble components of these foods from separating from each other.

In the body, phospholipids are the main structural component of cell membranes (see Figure 6.11). Cell membranes consist of two layers (a bilayer) of phospholipids. The hydrophilic head groups of these back-to-back layers point to the inside and outside of the cell, both of which are areas comprised predominantly of water. The hydrophobic fatty acids face toward each other, forming a water-free, lipid-rich zone. The incorporation of two layers of amphipathic phospholipids allows cell membranes to carry out their functions effectively while remaining stable in a watery environment. Phospholipids serve other functions as well: some supply fatty acids for cellular metabolism, others activate enzymes important to blood clotting and cell replication, and still others donate their fatty acids for eicosanoid production or act as carriers of fatty substances throughout the body.

Phospholipids such as **lecithin** are added to some **salad dressings** to keep their components mixed.

Sterols

A **sterol** is different from other lipids in that it consists of a distinct multi-ring structure. There are many types of sterols, but the most abundant and widely discussed is **cholesterol** (see Figure 6.12). You may have heard of the potentially unhealthy relationship between elevated blood cholesterol and heart disease. You may

Figure 6.11 **Cell Membrane Made from Phospholipid Bilayer**

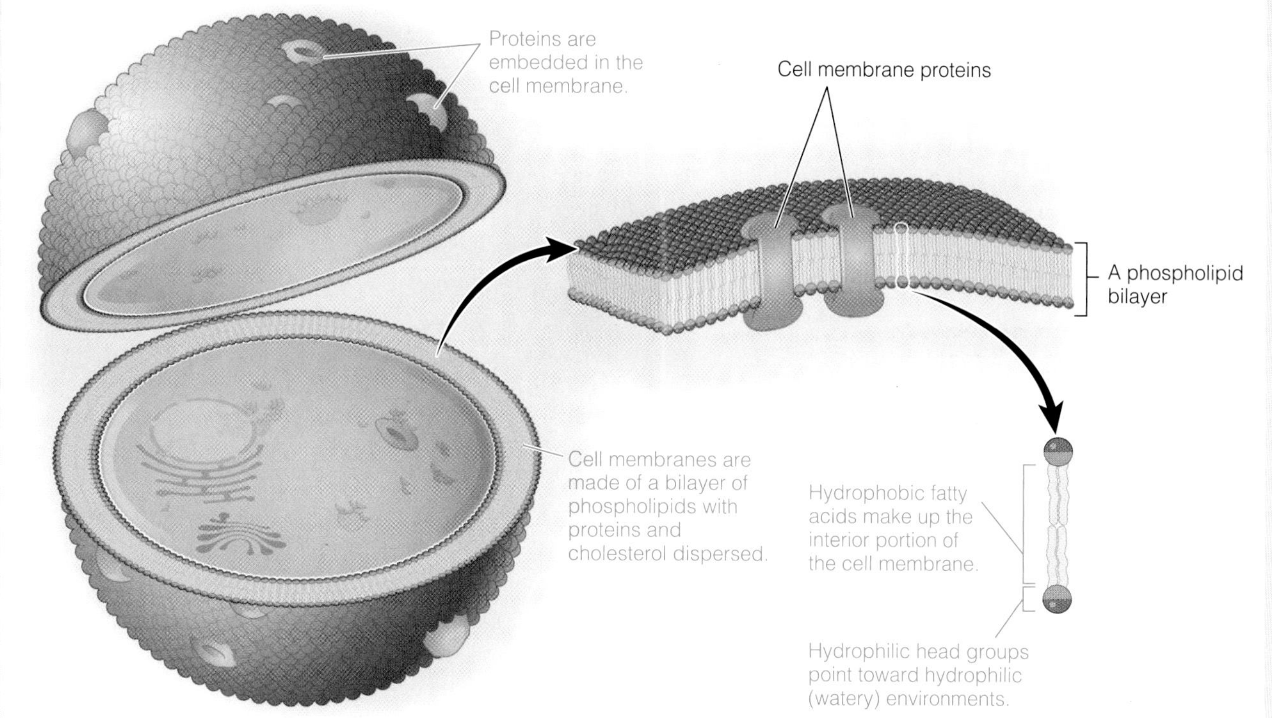

Proteins are embedded in the cell membrane.
Cell membrane proteins
A phospholipid bilayer
Cell membranes are made of a bilayer of phospholipids with proteins and cholesterol dispersed.
Hydrophobic fatty acids make up the interior portion of the cell membrane.
Hydrophilic head groups point toward hydrophilic (watery) environments.

The amphipathic nature of phospholipids allows cell membranes to carry out their functions.

not, however, be as familiar with the many important functions that cholesterol serves in your body. For instance, cholesterol is used to synthesize bile acids. A **bile acid**, which plays a critical role in digestion and absorption of lipids, consists of a cholesterol molecule attached to a very hydrophilic subunit, making a bile acid molecule amphipathic (much like a phospholipid). This structure allows bile acids to help break up dietary lipids in the intestine, preparing them for chemical digestion and subsequent absorption. Cholesterol is also a component of cell membranes, where it helps maintain flexibility. In addition to these functions, cholesterol is integral to the synthesis of many of the reproductive hormones (such as testosterone and estrogen), energy metabolism, calcium homeostasis, and electrolyte (salt) balance.

Figure 6.12 **Structure of Cholesterol**

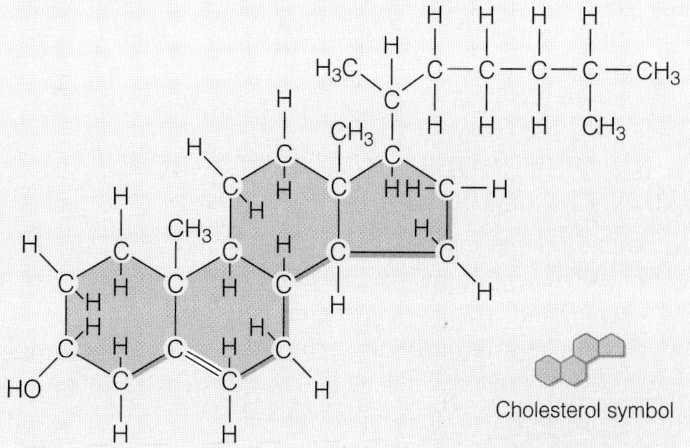

SOURCES OF CHOLESTEROL IN THE BODY

Almost every tissue in your body can make cholesterol—this is especially true for liver tissue. As such, cholesterol is not an essential nutrient. Many dietary factors influence how much cholesterol a person can make, however. For example, eating a low-calorie or low-carbohydrate diet decreases cholesterol synthesis in some people.[7] This is not the case for

bile acid An amphipathic substance, consisting of a cholesterol molecule attached to a very hydrophilic subunit, that plays a critical role in digestion and absorption of lipids.

> **Cholesterol** was first identified in solid form in 1769 by **François Poulletier de la Salle.**

everyone, however, because carbohydrate intake may only affect cholesterol synthesis in those with certain genetic make-ups.[8] Cholesterol production can also be lowered by certain medications, such as those taken by people at elevated risk for heart disease.

In addition to the cholesterol made by cells, the body obtains cholesterol from animal-derived foods such as shellfish, meat, butter, eggs, and liver (see Figure 6.13). Plants do not produce substantial amounts of cholesterol, so exclusively plant-based diets are relatively low in this substance. Because cholesterol is synthesized by the body, vegans who do not eat animal-based products are not at risk of cholesterol deficiency.

Figure 6.13 Cholesterol Content of Selected Foods

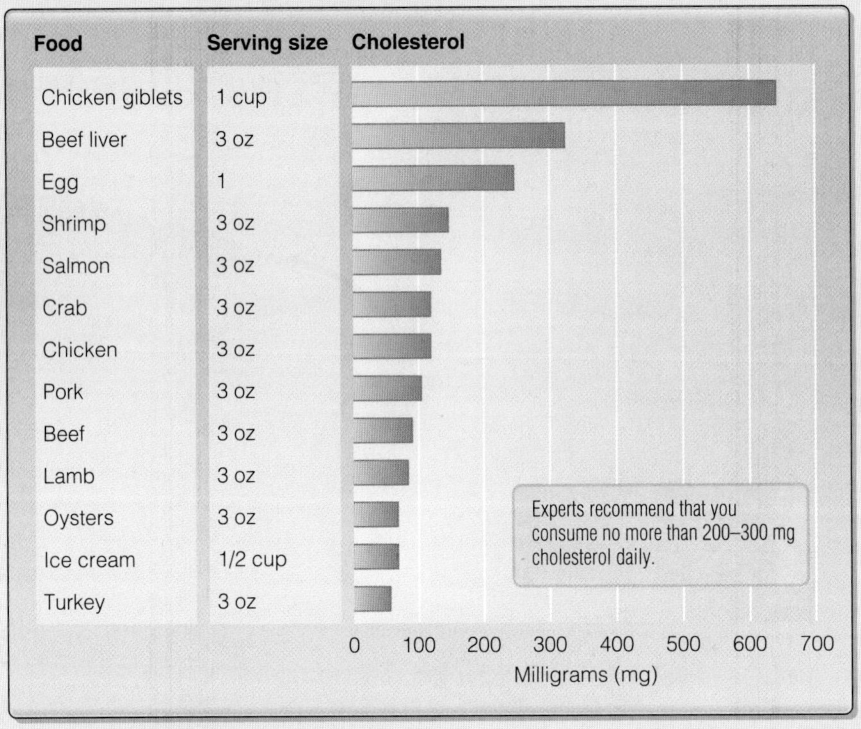

Source: USDA National Nutrition Database for Standard Reference, Release 18, 2005.

PHYTOSTEROLS ARE STEROL-LIKE PLANT COMPOUNDS

Although plants make very small amounts of cholesterol, some contain relatively large amounts of sterol-like compounds called **phytosterols**. An interesting group of phytosterols occurs naturally in corn, wheat, rye, and other plants. Similar sterols are produced commercially and marketed under various brand names, such as Benecol®. These products are often found in butter substitutes, yogurt drinks, salad dressings, and even dietary supplements. Some studies suggest that consuming roughly 500 to 2,000 milligrams of phytosterols a day may decrease blood cholesterol, lowering one's risk for cardiovascular disease.[9] Because a typical serving of a plant sterol–fortified table spread contains about 1.1 grams of the sterol, you would need to consume one or two servings daily to reach this goal. The mechanisms by which plant sterols decrease blood cholesterol are not well understood. However, most plant sterols are not easily absorbed and appear to bind cholesterol in the intestine. This may increase cholesterol elimination in the feces.

phytosterol A sterol-like compound made by plants.

LO5 How Are Triglycerides Digested, Absorbed, and Circulated?

Once ingested, lipids must be digested, absorbed, and circulated away from the small intestine. The basic goal of triglyceride digestion is to separate (or cleave) most of the fatty acids from their glycerol backbones—in other words, lipolysis. This process, which involves several enzymes and other secretions produced by the gastrointestinal tract and accessory organs, is relatively more complicated than digestion of the other macronutrients (carbohydrates and proteins). This is because it is more difficult for your body (which is made principally of water-soluble compounds) to digest water-insoluble molecules such as lipids than water-soluble molecules such as sugars and proteins (see Figure 6.14). Still, as you will soon understand, the digestion of dietary lipids occurs in your mouth, stomach, and small intestine every time you eat a meal containing fats or oils.

Triglyceride Digestion Begins in Your Mouth

A small portion of triglyceride digestion occurs in your mouth. As chewing breaks food apart, **lingual lipase**, an enzyme produced by your salivary glands, begins to remove fatty acids from the glycerol molecules. After the food is swallowed, lingual lipase accompanies the bolus into your stomach, where the enzyme continues to cleave additional fatty acids from glycerol.

lingual lipase An enzyme, produced by the salivary glands, that cleaves fatty acids from glycerol molecules.

gastric lipase An enzyme, produced in the stomach, that cleaves fatty acids from glycerol molecules.

Triglyceride Digestion Continues in Your Stomach

The second stage of triglyceride digestion begins when food enters your stomach, stimulating the release of gastrin from specialized stomach cells. Circulating in the blood, gastrin quickly stimulates the release of **gastric lipase**, another enzyme produced in the

Figure 6.14 Overview of Triglyceride Digestion

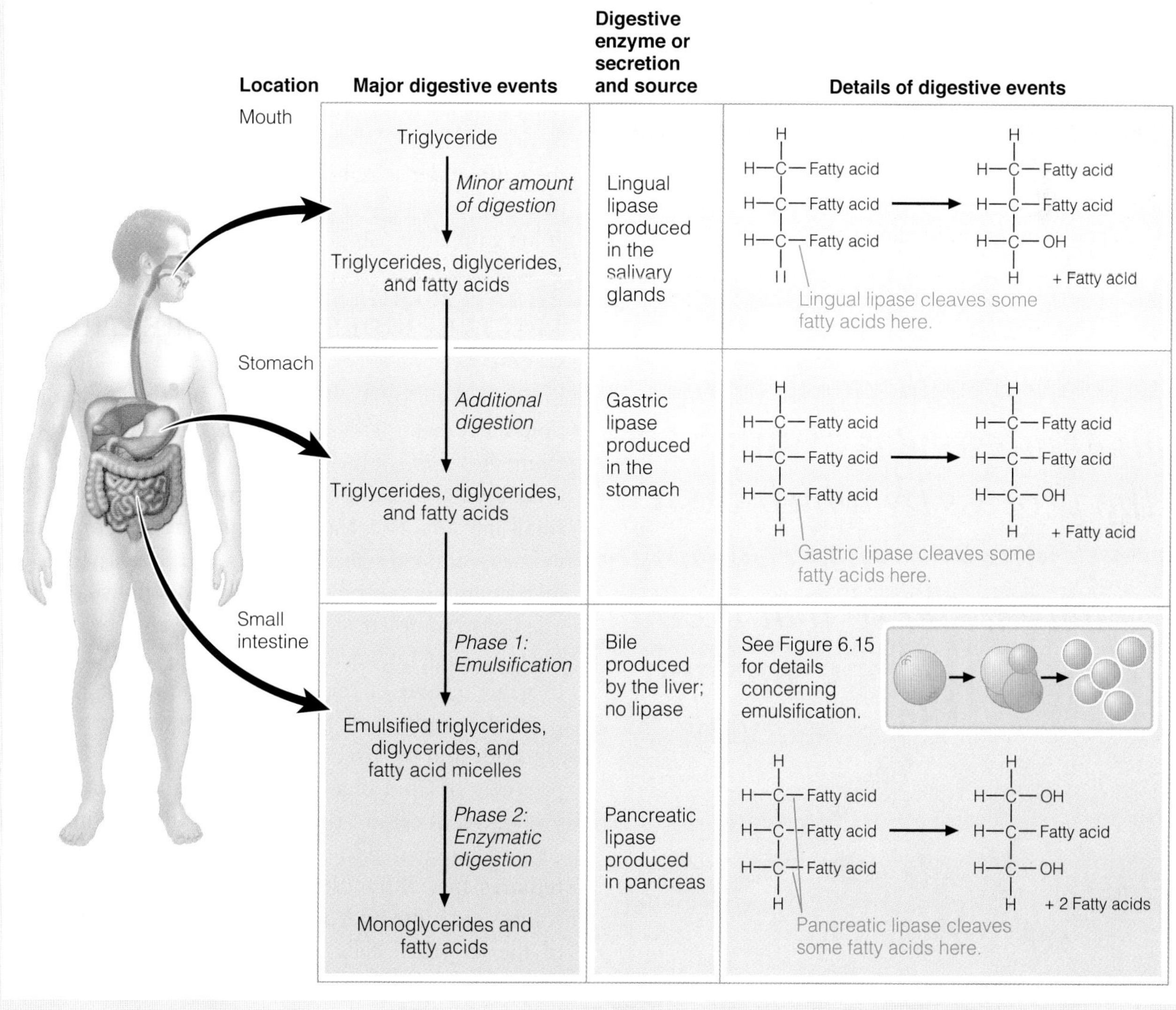

Location
Major digestive events
Digestive enzyme or secretion and source
Details of digestive events
Mouth
Triglyceride
Minor amount of digestion
Triglycerides, diglycerides, and fatty acids
Lingual lipase produced in the salivary glands
H–C–Fatty acid
H–C–OH
+ Fatty acid
Lingual lipase cleaves some fatty acids here.
Stomach
Additional digestion
Triglycerides, diglycerides, and fatty acids
Gastric lipase produced in the stomach
Gastric lipase cleaves some fatty acids here.
Small intestine
Phase 1: Emulsification
Emulsified triglycerides, diglycerides, and fatty acid micelles
Bile produced by the liver; no lipase
See Figure 6.15 for details concerning emulsification.
Phase 2: Enzymatic digestion
Monoglycerides and fatty acids
Pancreatic lipase produced in pancreas
+ 2 Fatty acids
Pancreatic lipase cleaves some fatty acids here.

micelle A small droplet of fat formed in the small intestine via emulsification.

emulsification The process by which large lipid globules are broken down and stabilized into smaller lipid droplets.

pancreatic lipase An enzyme, produced in the pancreas, that completes triglyceride digestion by cleaving fatty acids from glycerol molecules.

stomach. Gastric lipase, a component of the gastric juices, picks up where lingual lipase left off by cleaving further fatty acids from the glycerol molecules.

Triglyceride Digestion Is Completed in Your Small Intestine

Although some fatty acids are cleaved from the glycerol backbones in the mouth and stomach by lingual and gastric lipases, respectively, triglyceride digestion is not complete until the chyme interacts with bile and enzymes in your small intestine. This is in part because the watery environment of the gastrointestinal tract can cause lipids to clump together in large lipid globules that are difficult to digest. To overcome this problem, the final stage of triglyceride digestion occurs in two complementary and consecutive phases.

PHASE 1: EMULSIFICATION OF LIPIDS BY BILE–MICELLE FORMATION

In the first phase of intestinal triglyceride digestion, bile disperses large lipid globules into smaller lipid droplets, making the lipids more accessible to the intestine's digestive enzymes (see Figure 6.15). In response to the hormone *cholecystokin* (CCK) released by the small intestine in response to the arrival of chyme, the gallbladder contracts and releases bile into the duodenum. Recall from Chapter 3 that bile is made by the liver and stored in the gallbladder until needed. Bile is comprised of a mixture of bile acids, cholesterol, and phospholipids. Both bile acids and phospholipids are amphipathic. Because the hydrophilic and hydrophobic components of bile are attracted to water and lipids, respectively, the large lipid globules are pulled apart into smaller droplets when they mix with the bile and phospholipids. Bile remains with each newly formed droplet, or **micelle**, stabilizing it in the intestine. The process by which large lipid globules are broken down and stabilized into micelles is called **emulsification.**

When a person's bile contains an excess amount of cholesterol in relation to its other components, gallbladder disease may develop. Recall from Chapter 3 that the accumulation of calcium and cellular debris around the cholesterol can lead to the formation of gallstones. Surgical removal of the gallbladder is the most common treatment for persistent gallstone-related problems. Because some people may have difficulty emulsifying—and therefore digesting—fat after the gallbladder is removed, doctors often recommend that people initially avoid high-fat meals after undergoing gallbladder surgery.

The **emulsification of lipids** *by bile acids in the small intestine is* **similar to** *the process by which* **dish soap** *makes grease combine with water so that it* **can be washed** *away.*

© iStockphoto.com/Thomas Acop

PHASE 2: DIGESTION OF TRIGLYCERIDES BY PANCREATIC LIPASE

Emulsification by itself does not complete lipid digestion; fatty acids still need to be chemically cleaved from those still attached to glycerol. To accomplish this, lipid-containing chyme stimulates cells in the small intestine to release the hormone *secretin*, which in turn stimulates the pancreas to release pancreatic juice containing the enzyme **pancreatic lipase.** Pancreatic lipase completes triglyceride digestion by cleaving the remaining fatty acids from their glycerol molecules. In general, two of the three fatty acids are removed from the triglyceride molecules, producing a monoglyceride and two free (unbound) fatty acids. The final products of lipid digestion (fatty acids, glycerol, and monoglycerides) are then taken up into the intestinal cells and circulated to the rest of the body. This process requires special handling because many of these substances are quite hydrophobic, while both the interiors of the intestinal cells and the circulatory system are

Figure 6.15 Emulsification of Lipids to Form Micelles in the Small Intestine

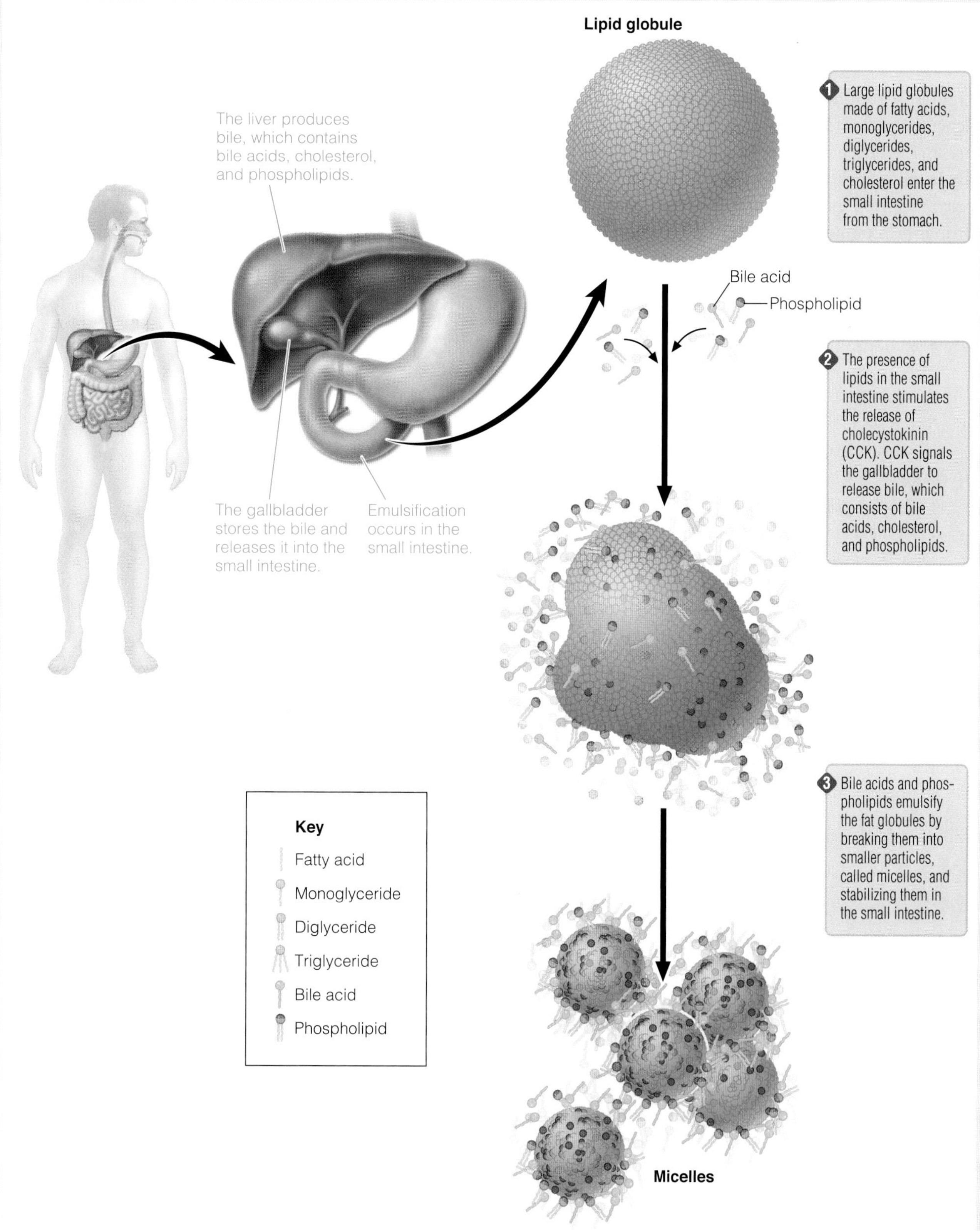

Intestinal emulsification of lipids requires bile, which is produced in the liver and stored in the gallbladder.

hydrophilic. Once again, it is the amphipathic property of phospholipids that makes absorption possible.

Lipid Absorption

Lipid absorption is accomplished in one of two ways, depending on how hydrophilic the lipid is (see Figure 6.16). Because they are relatively water-soluble (hydrophilic), short- and medium-chain fatty acids can be transported into intestinal cells unassisted. More hydrophobic compounds (such as long-chain fatty acids, monoglycerides, and cholesterol) must first be repackaged into micelles within the intestinal lumen before they can move into the intestinal cells. Once a lipid-containing micelle comes into contact with the lumenal surface of an intestinal cell, the micelle's contents are released into the cell's interior.

Lipid Circulation

Upon crossing the intestinal wall, dietary lipids are ready to be circulated away from the gastrointestinal tract for delivery to the body's tissues. Like absorption, the method by which a lipid is circulated depends on how hydrophilic it is.

MORE HYDROPHILIC LIPIDS ARE CIRCULATED IN THE BLOOD

The least complex method of dietary lipid circulation involves the short- and medium-chain fatty acids.

Figure 6.16 Absorption and Circulation of Lipids

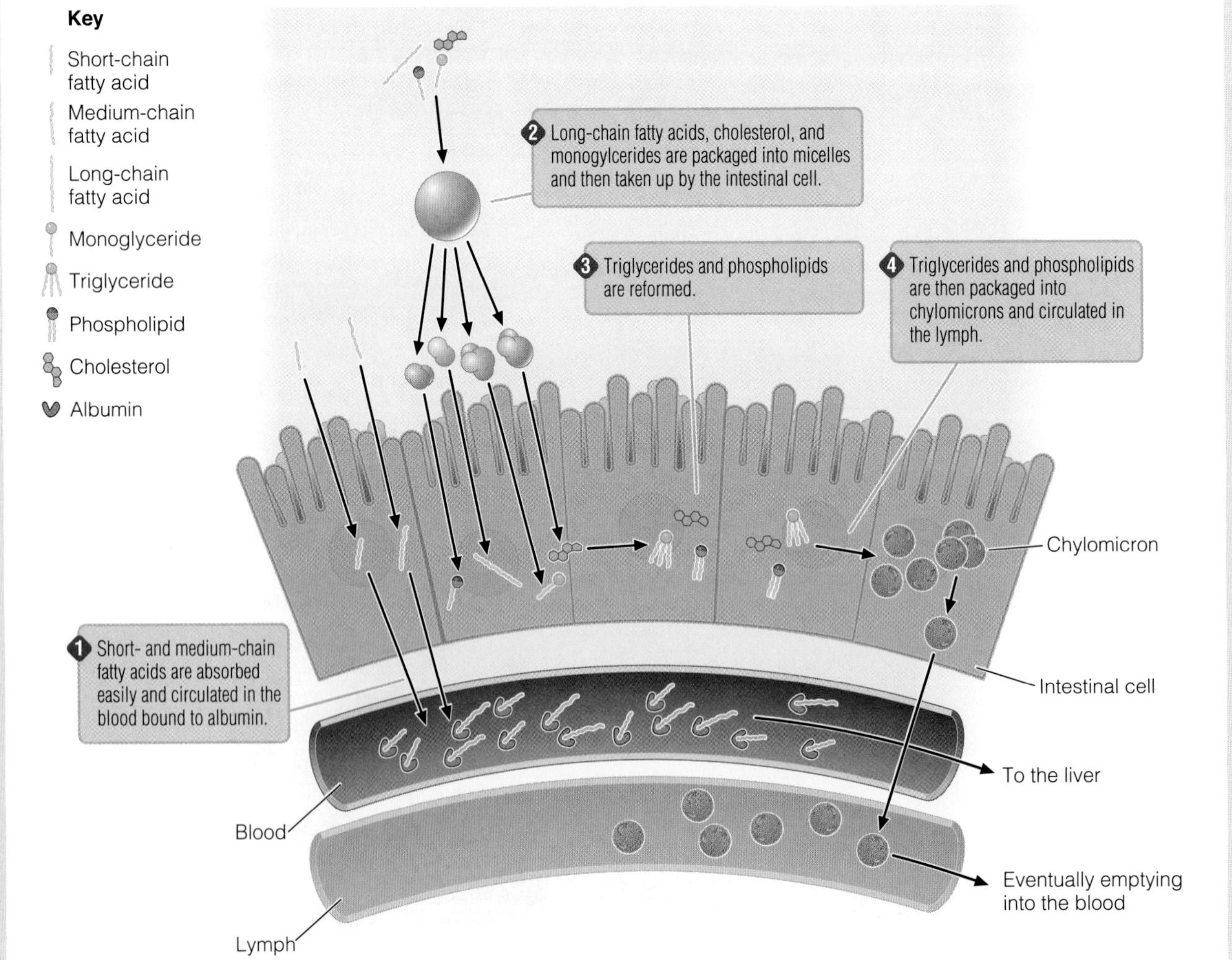

The process by which a lipid is absorbed and circulated depends on how hydrophilic it is. In general, it is easier for the body to absorb and circulate more hydrophilic substances than more hydrophobic ones.

Because they are somewhat soluble in water, these fatty acids can be circulated away from the small intestine by simply attaching to the protein albumin in the blood. This fatty acid-albumin complex circulates to the liver, where the fatty acid is either metabolized or rerouted for delivery to other cells in the body.

LESS HYDROPHILIC LIPIDS ARE CIRCULATED IN LYMPH

Compared to what happens with the short- and medium-chain fatty acids, circulation of larger, less hydrophilic lipids away from the gastrointestinal tract is more involved. First, long-chain fatty acids and monoglycerides must be reassembled into triglycerides inside the intestinal cell. These hydrophobic lipids, along with cholesterol and phospholipids, are then incorporated into **chylomicron** particles, which are released into the lymphatic system for initial circulation. Chylomicrons package their hydrophobic lipids (such as triglycerides) within a hydrophilic exterior shell formed mainly from phospholipids and proteins (see Figure 6.17). The lymphatic system eventually delivers the chylomicrons into the blood, where they travel to cells that take up their contents. The chylomicron is an example of a **lipoprotein**, a type of particle that your body makes to transport lipids. You will learn more about other lipoproteins in the following section.

The enzyme **lipoprotein lipase** enables chylomicrons to deliver their dietary fatty acids to cells. Lipoprotein lipase is produced in many tissues (especially adipose and muscle tissues) and resides within the lumens of the blood vessels surrounding the tissues that produce this specialized enzyme. As chylomicron-containing blood flows past, lipoprotein lipase attacks the chylomicrons, releasing fatty acids that can then be taken up by surrounding cells. After delivering dietary fatty acids to various tissues throughout the body, the residual

chylomicron A particle comprised of relatively hydrophobic lipids, cholesterol, and phospholipids that transports dietary lipids in the lymph for circulation.

lipoprotein A type of particle that transports lipids throughout the body.

lipoprotein lipase An enzyme that enables chylomicrons and other lipoproteins to deliver fatty acids to cells.

Figure 6.17 The Lipoproteins

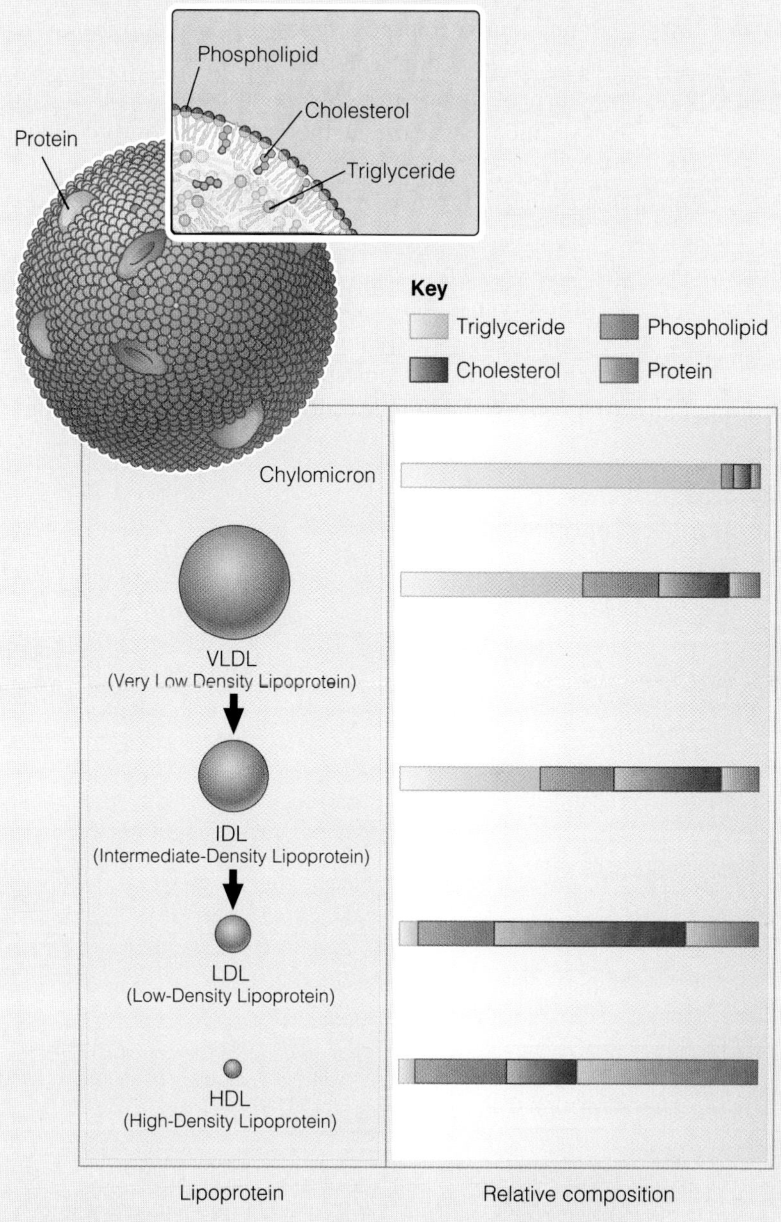

Lipoproteins are composed of varying amounts of triglycerides, phospholipids, cholesterol, and proteins. The ratio of lipids to proteins determines the lipoprotein's density.

Adapted from: Skipski VP. Blood lipids and lipoproteins: Quantitation, composition and metabolism. Nelson GJ, editor. 1972. Wiley-Interscience, New York. Christie W. American Oil Chemists' Society Lipid Library. Lipoproteins. Available from: http://lipidlibrary.aocs.org/lipids/lipoprot/index.htm.

chylomicron remnant A residual fragment of a chylomicron that remains in the blood until it is taken up by the liver.

fragments of the chylomicrons remain in the blood. These **chylomicron remnant** fragments are taken up by the liver where they are broken down and their contents are reused or recycled.

LO6 What Are the Types and Functions of Various Lipoproteins?

The liver serves as both the central command and recycling center for lipid metabolism. It receives and recycles dietary lipids (and as you have just learned, fatty acid-depleted chylomicron remnants), and synthesizes and metabolizes other lipids as needed. To deliver newly manufactured lipids and some dietary lipids throughout the body, the liver also produces a series of lipoproteins that circulate in the blood. A summary of the origins and functions of the various lipoproteins (including the chylomicrons) is presented in Figure 6.18.

Lipoproteins

Because lipids are relatively hydrophobic, their transport in the water-rich blood is somewhat complex. To aid in the process, the liver produces lipoproteins that function to transport lipids in the blood. These lipoproteins are different from the one you have already learned about—the chylomicron—which is the largest and least dense member of the lipoprotein family. Whereas chylomicrons are produced in the small intestine and transport *dietary* lipids in the bloodstream exclusively, other lipoproteins are produced in the liver and carry lipids originating from various other tissues and organs in addition to some coming from the diet. This includes both lipids manufactured in the liver and those released from storage in adipose tissue. Lipoproteins are complex globular structures that contain

Figure 6.18 **Origins and Major Functions of Lipoproteins**

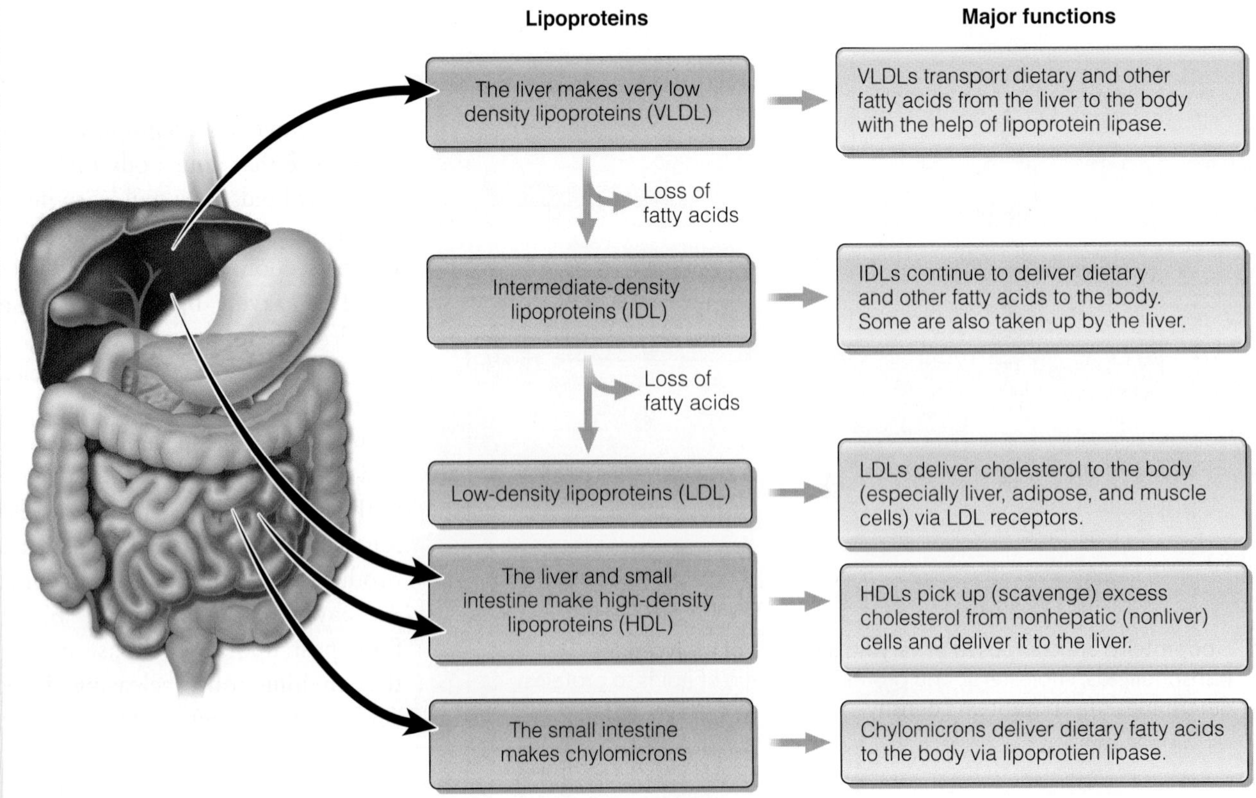

Both the liver and small intestine make lipoproteins that circulate lipids in the body.

varying amounts of triglycerides, phospholipids, cholesterol, and special proteins called apoproteins. Embedded within the outer shell of a lipoprotein, **apoprotein** (or **apolipoprotein**) molecules enable a lipoprotein to circulate in the blood and interact with the cells that require its contents. Like the chylomicrons made in the small intestine, the lipoproteins made in the liver are constructed so that their hydrophilic components (proteins and phospholipids) face outward and their hydrophobic components (such as triglycerides) face inward (recall Figure 6.17).

Because lipid is less dense than protein, the density of a lipoprotein depends on its relative amounts (or percentages) of lipids and proteins. Lipoproteins with more lipid and less protein have lower densities than those with more protein and less lipid. With the exception of chylomicrons, lipoproteins are named according to their densities, so the relative densities give each lipoprotein its name.

VERY LOW-DENSITY LIPOPROTEINS DELIVER FATTY ACIDS

One type of lipoprotein made by the liver is a **very low-density lipoprotein (VLDL).** VLDLs are similar to chylomicrons in that they contain triglycerides and cholesterol in their cores and are surrounded by phospholipids and proteins on their surfaces. However, VLDLs have a lower lipid-to-protein ratio than do chylomicrons, making them smaller and denser. Like chylomicrons, the primary function of VLDLs is to deliver fatty acids to cells. But unlike chylomicrons, which *only* transport dietary fatty acids, VLDLs also transport fatty acids derived from liver and adipose tissue. When VLDLs in the blood circulate past cells that produce lipoprotein lipase, the enzyme cleaves off fatty acids that are in turn taken up by the surrounding cells. In this way, tissues in need of fatty acids are able to obtain them directly from the bloodstream.

LOW-DENSITY LIPOPROTEINS DELIVER CHOLESTEROL

The removal of fatty acids from VLDLs causes them to become smaller and denser. The resulting lipoprotein that is smaller and less dense than a VLDL is referred to as an **intermediate-density lipoprotein (IDL).** Some IDLs are taken up by the liver, whereas others remain in circulation where they continue to be cleaved of additional fatty acids. Eventually, an IDL becomes a cholesterol-rich **low-density lipoprotein (LDL),** which is smaller and denser still. An **LDL receptor** is a specialized type of protein located on cell membranes—especially those of liver, adipose, and muscle tissue. LDL receptors bind to the proteins embedded in the surface of the LDLs, allowing the LDLs to be taken up and broken down by the cell. In this way, much of the cholesterol is removed from the blood and taken up by the many tissues that use it.

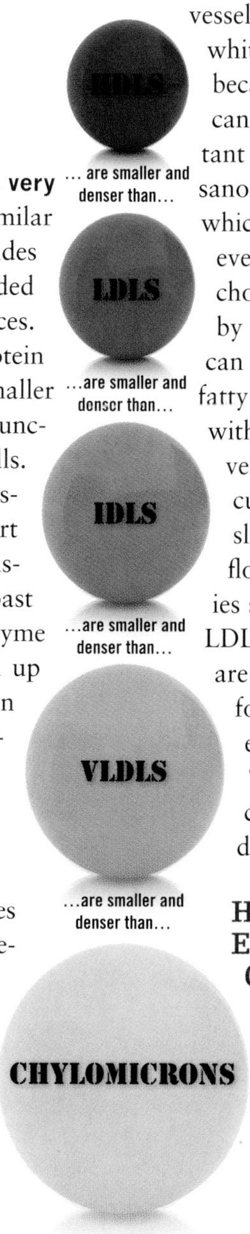

LDLs can also be taken up and used by white blood cells, some of which are located in major blood vessels. It is often necessary for white blood cells to do this because the contents of LDLs can be used to make important substances such as eicosanoids and immune factors, which keep you healthy. However, an uptake of too much cholesterol and other lipids by certain white blood cells can result in the buildup of a fatty substance called **plaque** within the walls of blood vessels. Eventually, the accumulation of plaque can slow or even block blood flow. Epidemiologic studies suggest that high levels of LDL cholesterol in the blood are related to increased risk for cardiovascular disease.[10] Thus LDL cholesterol is often deemed "bad cholesterol." The relationship between cholesterol and cardiovascular disease is addressed in greater detail later in this chapter.

HIGH-DENSITY LIPOPROTEINS ELIMINATE EXCESS CHOLESTEROL

The liver, and to a lesser extent the small intestine, also make a type of lipoprotein called **high-density lipoprotein (HDL).** Compared to other lipoproteins, HDLs have the lowest lipid-to-protein ratios, and therefore

apoprotein (or **apolipoprotein**) A protein embedded within the outer shell of a lipoprotein that enables it to circulate in the blood and interact with the cells that require its contents.

very low-density lipoprotein (VLDL) A lipoprotein, made by the liver, that has a lower lipid-to-protein ratio than do chylomicrons. It delivers fatty acids to cells.

intermediate-density lipoprotein (IDL) A lipoprotein that is slightly less dense than a VLDL.

low-density lipoprotein (LDL) A lipoprotein that delivers cholesterol to the body's cells.

LDL receptor A specialized type of protein, located on cell membranes, that binds to the proteins embedded in the surface of an LDL, allowing it to be taken up and broken down by the cell.

plaque A fatty substance that builds up within the walls of blood vessels.

high-density lipoprotein (HDL) A lipoprotein, made primarily by the liver, that has the lowest lipid-to-protein ratio. It circulates in the blood to collect excess cholesterol from cells and transport it back to the liver.

have the highest densities (hence, *high*-density lipoprotein). HDLs circulate in the blood to collect excess cholesterol from cells and transport it back to the liver. Because a high level of HDL cholesterol in the blood is generally associated with a lower risk of cardiovascular disease, HDL cholesterol is often referred to as "good cholesterol."[11] There are several types of HDL however, and not all forms are equally effective in removing excess cholesterol. Different HDLs have different proteins (apoproteins) on their surfaces, resulting in somewhat different functions. The presence of particular apoproteins makes some HDLs more or less efficient at cholesterol removal.

LO7 How Are Dietary Lipids Related to Health?

After lipids have been digested, absorbed, and circulated, they are ready to be used by your body for myriad purposes. As described previously, fatty acids and their by-products regulate metabolic processes within your cells and orchestrate a variety of physiological responses. Phospholipids are vital components of cell membranes, aid in lipid digestion and absorption, and are needed for many cellular functions. Cholesterol is incorporated into cell membranes, is a precursor for many hormones, and is involved in lipid digestion and absorption via its role in bile.

Although many lipids are vital to good health, too much of some and certain types of other lipids are associated with poor health. For example, an excessive dietary intake of some lipids is related to increased risks for cardiovascular disease and certain cancers. Of course, consuming too many lipid-rich foods can lead to excess calories, and subsequently, unwanted weight gain.

High-Fat Foods and Obesity

Obesity is defined as the overabundance of body fat. Although obesity is a complex condition, excess energy intake is a major factor in its development. Because fatty acids provide more than twice the calories per gram as carbohydrates and proteins, high fat intake is likely an important piece of the obesity puzzle. Whatever its causes, obesity is a major public health concern worldwide because of its association with increased risks for diseases such as cardiovascular disease, type 2 diabetes, and some forms of cancer. Because there is an intense interest in decreasing obesity rates, experts recommend that adults limit their energy intake—especially from high-fat foods. In response to consumer demand, many food manufacturers produce low-fat and fat-free items in addition to foods that contain fat substitutes. You will learn much more about lipid consumption and its relationship with obesity in Chapter 9.

CAN FAT SUBSTITUTES HELP?

Because many people want to consume low-fat diets, some food manufacturers have developed substances that have the desired characteristics of lipids but contain fewer or no calories. These fat substitutes are diverse in structure; some are made from complex carbohydrates, others are made from proteins, and still others are made from blends of carbohydrates and fatty acids. Olestra (trade name Olean®) is an example of a fat substitute made from sucrose with several fatty acids attached. Olestra cannot be digested by human lipases or intestinal bacteria, and therefore provides no usable energy to the body. Because olestra interferes with the absorption of fat-soluble vitamins in the gastrointestinal tract, the FDA requires that food manufacturers add vitamins A, D, E, and K to olestra-containing foods.

Fat substitutes made from carbohydrates and proteins are commonly used by food manufacturers. These substances, including cellulose, vegetable gum fiber (for example, guar gum), Maltrin®, and Stellar®, are used in many low-fat or fat-free foods such as salad dressings,

Although consumption of a high-fat diet can increase risk for obesity, there are many other factors (such as genetics and inadequate exercise) that contribute to this health problem.

table spreads, and frozen desserts. Because some of these ingredients are not heat-stable, however, they cannot all be used in foods that must be baked or fried. Although cellulose and vegetable gums are considered dietary fiber and therefore do not contribute calories to the diet, Maltrin® and Stellar® are glucose polymers—each provides 4 kcal per gram. Another fat substitute is Simplesse®, which is made from milk protein. Like Maltrin® and Stellar®, this substance contains 4 kcal per gram. Thus, foods containing these products may be *fat-free*, but they are certainly not *calorie-free*.

Dietary Lipids and Cardiovascular Disease

Like obesity, **cardiovascular disease**, defined as a disease of the heart or vascular system, is caused by a complex web of factors, including genetics, physical inactivity, and poor diet. The most common types of cardiovascular disease are **heart disease**, a slowing or complete obstruction of blood flow to the heart, and **stroke**, a slowing or complete obstruction of blood flow to the brain. When blood flow is restricted, cells do not receive adequate oxygen and nutrients, ultimately causing them to die. Restriction of blood flow is generally caused by a condition called **atherosclerosis**, characterized by a narrowing and stiffening of the blood vessels (often due to plaque buildup). A **blood clot**, a small, insoluble particle made of clotted blood and clotting factors, or **aneurysm**, the outward bulging of the vessel due to weakness, may also restrict blood flow. These conditions are illustrated in Figure 6.19.

cardiovascular disease A disease of the heart or vascular system.

heart disease A slowing or complete obstruction of blood flow to the heart.

stroke A slowing or complete obstruction of blood flow to the brain.

atherosclerosis A narrowing and stiffening of the blood vessels, causing the restriction of blood flow.

blood clot A small, insoluble particle made of clotted blood and clotting factors.

aneurysm An outward bulging of a blood vessel due to weakness in the vasculature.

Both total lipids and specific types of lipids can influence a person's risk for developing atherosclerosis, blood clots, and aneurysms—and therefore cardiovascular disease. For example, high intakes of SFAs, *trans* fatty acids, and cholesterol may increase risk for cardiovascular disease in some people, whereas MUFAs may have the opposite effect.

Figure 6.19 Causes of Cardiovascular Disease

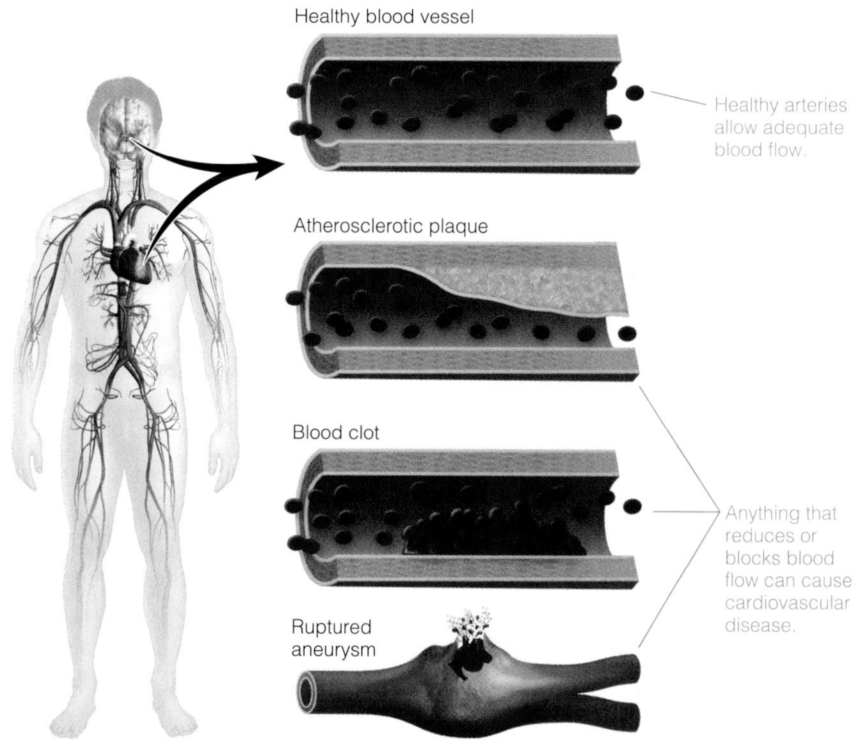

Atherosclerosis, blood clots, and aneurysms can all reduce or stop blood flow, leading to cardiovascular disease.

A person's genetic makeup often interacts with his or her diet to influence health—this holds especially true for lipids and cardiovascular disease. For instance, many studies have shown that the consumption of a diet high in cholesterol increases blood cholesterol and one's risk for cardiovascular disease.[12] However, some people can eat very high amounts of cholesterol without experiencing this effect. This inconsistency may arise because some people absorb dietary cholesterol efficiently while others may naturally excrete more cholesterol in their feces.[13] Genetic variation can also influence a person's HDL and LDL levels and functioning. For example, the ability of immune cells in blood vessel walls to take up LDL particles (and thus form plaques) may be influenced by variations in the genes that code for LDL receptor proteins.[14] As scientists learn more about how genetic makeup interacts with diet to influence health, health professionals will be better able to prescribe the most heart-healthful diet given a person's individual genetics.

NUTRITIONAL GUIDELINES FOR CARDIOVASCULAR HEALTH

The multifaceted and complex relationship between diet and cardiovascular risk is yet another example of how dietary variety, moderation, and balance are essential to health and well-being. To promote optimal cardiovascular health, agencies such as the American Heart Association, the U.S. Department of Agriculture (USDA), and the National Institutes of Health have set forth guidelines regarding dietary intakes that help lower risk for heart disease and stroke. Many of these recommendations do not relate to dietary fat *per se* but rather to the many different types of nutrients needed to keep the cardiovascular system healthy.

First and foremost, moderate your overall energy intake and balance your macronutrients. The primary goal set forth in many dietary recommendations related to energy consumption and cardiovascular disease is to consume only enough calories to maintain a healthy body weight. Being overweight or obese increases a person's risk for heart disease and stroke profoundly. A person who wants to lose weight is advised to decrease his energy intake and increase his energy expenditure to achieve a specific bodyweight goal. This is best accomplished by following the Institute of Medicine's recommendations for energy intake (recall the Estimated Energy Requirement formulas from Chapter 2) and Acceptable Macronutrient Distribution Range (AMDR) recommendations for balancing energy intake from the various macronutrient categories. In addition to following these recommendations, making sure that at least half of one's daily grain choices are whole grain helps increase intake of dietary fiber, which is known to lower risk for cardiovascular disease. Whole-grain foods are also rich sources of vitamins, minerals, and heart-healthy lipids.

Dietary Lipids and Cancer

Although some studies show a link between high-fat diets and a risk for cancer, the data are inconclusive.[15] Obesity, however, is a risk factor for several types of cancer, including breast cancer.[16] As obesity is often associated with consumption of high-fat diets, dietary lipids may play an indirect role. In 2007, the World Cancer Research Fund and the American Institute for Cancer Research jointly issued overall recommendations concerning diet and cancer prevention.[17] These recommendations advise that, to decrease your risk of cancer, you should both be as lean as possible within the normal range of body weight and avoid weight gain and increases in waist circumference as you age. They also recommended that you limit consumption of energy-dense foods and consume high-fat fast foods sparingly—if at all.

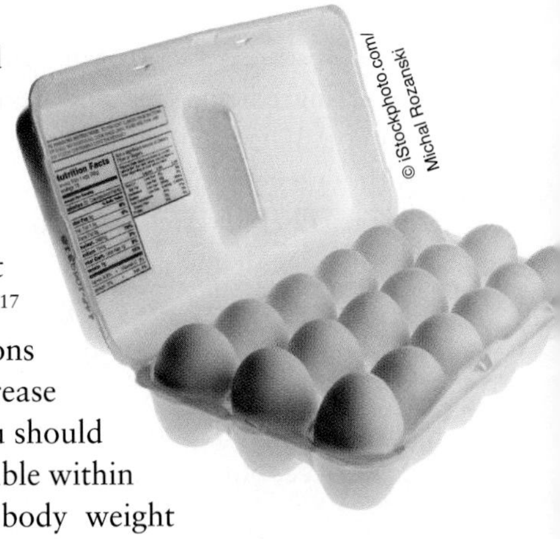

© iStockphoto.com/ Michal Rozanski

LO 8 What Are Some Overall Dietary Recommendations for Lipids?

It is important to maintain an optimal balance of healthy lipid intake throughout life. Here, a few of the many recommendations regarding dietary lipids are introduced. As you will see, many recommendations are related to the associations among dietary lipid intake, obesity, and cardiovascular disease. These recommendations will help you choose the right dietary lipids to help promote optimal health.

Consume Adequate Amounts of the Essential Fatty Acids

It is important to consume adequate amounts of the essential fatty acids; the Dietary Reference Intakes (DRIs) for these substances are found on the Dietary Reference Intakes card located at the back of this book. For example, Adequate Intake (AI) levels for linoleic acid are 17 and 12 g/day for adult men and women, respectively. This is the amount of linoleic acid contained in approximately 26 ml (or 1¾ tbsp) of corn oil. For linolenic acid, the AIs are 1.6 and 1.1 g/day for adult men and women, respectively. This is the amount contained in about 22 ml (or 1½ tbsp) of soybean oil.

Pay Particular Attention to the Long-Chain ω-3 Fatty Acids

In addition to specific recommendations for the essential fatty acids, the USDA and other health organizations suggest that you consume at least two servings of fish high in ω-3 fatty acids (such as salmon) each week, as well as other foods rich in linolenic acid (such as flaxseed and canola oils). People with cardiovascular disease are advised to consume about 1 g of ω-3 fatty acids daily from fish. This is the amount contained in approximately 57 g (2 oz) of Atlantic salmon. This recommendation has also been endorsed by the American Heart Association.

Limit Cholesterol, Saturated Fat, and *Trans* Fat

Scientists have long known that SFA intake is positively related to an increased risk of cardiovascular disease for some people. As such, many dietary guidelines suggest decreasing one's SFA intake. For example, the Institute of Medicine recommends that "intake of SFAs should be minimized while consuming a nutritionally adequate diet." The Dietary Guidelines suggests that SFA should constitute no more than 10 percent of a person's total caloric intake, and the American Heart Association recommends that SFA intake represent 7 percent or less of one's total calories. Reading food labels such as those illustrated in Figure 6.20 can help you keep track of your SFA intake.

Although there are no absolute guidelines as to how much *trans* fatty acids humans should consume, both the Dietary Guidelines and the Institute of Medicine recommend that people "minimize their intakes of *trans* fatty acids." Note that these guidelines apply to commercially produced *trans* fatty acids—not to naturally occurring ones.

No DRIs are established for cholesterol. However, the Dietary Guidelines and the American Heart Association recommend that you consume less than 300 mg of cholesterol daily. This is the amount found in one to two eggs, or two servings of beef. The Institute of Medicine recommends simply that "cholesterol intake should be minimized."

Figure 6.20 **Reading Nutrition Facts Panels**

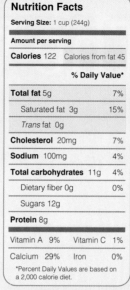

Nutrition Facts	
Serving Size: 1 cup (244g)	
Amount per serving	
Calories 122	Calories from fat 45
	% Daily Value*
Total fat 5g	7%
Saturated fat 3g	15%
Trans fat 0g	
Cholesterol 20mg	7%
Sodium 100mg	4%
Total carbohydrates 11g	4%
Dietary fiber 0g	0%
Sugars 12g	
Protein 8g	
Vitamin A 9%	Vitamin C 1%
Calcium 29%	Iron 0%
*Percent Daily Values are based on a 2,000 calorie diet.	

You can tell how much total fat, saturated fat, and *trans* fat is in a food by checking its Nutrition Facts panel.

Guidelines for Total Lipid Consumption

The Institute of Medicine has not established RDA or AI values for total lipid intake except during infancy, when AIs are set at approximately 30 g per day. However, the AMDRs recommend that healthy adults consume 20 to 35 percent of their energy from lipid. Based on a caloric requirement of 2,000 kcal per day, a person should consume between 400 to 700 kcal of lipid. Considering that 1 g of lipid contains about 9 kcal of energy, the daily lipid intake for most adults should be between 44 and 78 g. This amount is easy to obtain: a typical day's menu that includes three servings of low-fat milk, a bagel with cream cheese, a peanut butter sandwich, spaghetti with meatballs, and a salad with ranch dressing collectively contains about 56 g of lipid.

Even **high-fat foods** can be part of a **healthy diet** if **eaten in moderation**.

NUTR

7 The Vitamins

LEARNING OUTCOMES:

- LO1 Understand the history of vitamins.
- LO2 Discuss commonalities among the water-soluble vitamins.
- LO3 Understand the functions and sources of the B vitamins.
- LO4 Describe the functions and sources of vitamin C.
- LO5 Understand commonalities among the fat-soluble vitamins.
- LO6 Describe the functions and dietary sources of vitamin A and the carotenoids.
- LO7 Understand the functions and sources of vitamin D.
- LO8 Understand the functions and sources of vitamins E and K.
- LO9 Decide whether you should take dietary supplements.

LO1 What Do Scientists Know about Vitamins?

As described in previous chapters, your body uses carbohydrates, proteins, and lipids for a multitude of functions. Your body also requires micronutrients—vitamins and minerals—albeit in much smaller amounts. Both vitamins and minerals are needed to maintain a healthy body, but the two categories of micronutrients are quite different. Whereas minerals are simple inorganic elements, vitamins are complex organic compounds. You will first learn about vitamins in this chapter, and then about minerals in Chapter 8.

Over a century ago, scientists only classified vitamins by their solubility in water and fat. *Fat-soluble* vitamins were collectively called vitamin A, while *water-soluble* vitamins were collectively called vitamin B.[1] Researchers soon discovered, however, that vitamins A and B actually comprised several unique substances. Each newly discovered substance was designated with its own classification letter (such as C, D, and K). It was subsequently discovered that what was initially called vitamin B was not a single substance, but a group of several different interrelated ones. The B vitamins were not all renamed with new letters, but were differentiated by subscript numbers (such as B_1, B_6, and B_{12}). When grouped together, the B vitamins are often called *B-complex vitamins*.

Today, scientists know of at least 13 essential fat- and water-soluble vitamins. Beyond their letter designations, many vitamins have common names such as *thiamin* (B_1) and *riboflavin* (B_2). Most vitamins also have chemical names; vitamin C, for example, is referred to chemically as *ascorbic acid*. Because it is possible for a single vitamin to have numerous names, distinguishing one from the next can sometimes be confusing. In this chapter, you will learn about the essential vitamins' nomenclatures, dietary sources, functions, deficiencies, toxicities, and recommended intakes. You will also learn what dietary supplements are, where to find reliable information about them, and when you might want to consider taking them.

LO2 What Are Water-Soluble Vitamins?

Each nutrient class you have studied thus far has distinct characteristics that make it chemically unique from other classes. For instance, all proteins are made from nitrogen-containing amino acids, and all carbohydrates contain sugar units. This trend ends with vitamins, however. Unlike carbohydrates, proteins, and lipids, there is no single molecular characteristic that groups vitamins together and distinguishes them from other nutrients. In other words, aside from the fact that they are all organic molecules, each vitamin is chemically distinct from the next. Nonetheless, there are basic biological properties that define and characterize vitamins.

Commonalities among the Water-Soluble Vitamins

Even though they are not structurally similar, essential water-soluble vitamins tend to share three critical biological attributes. The first and perhaps most basic is that all water-soluble vitamins dissolve in water (and not in fat). This single attribute concretely distinguishes water-soluble vitamins from fat-soluble vitamins. Second, although there are exceptions, the body generally absorbs and transports water-soluble vitamins in a similar way. Water-soluble vitamins are absorbed in the small intestine, and to a lesser extent, the stomach. Bioavailability is influenced by many factors, such as nutritional status, other nutrients and substances in foods, medications, age, and health status. All of the water-soluble compounds are circulated in the blood (as opposed to the lymph), and are transported from the gastrointestinal tract directly to the liver. Finally, because the body cannot store most water-soluble vitamins, their overconsumption generally does not lead to toxic effects.

fortification The addition of nutrients to a food during processing.

enrichment The fortification of a select group of foods with FDA-specified levels of thiamin, niacin, riboflavin, folate, and iron.

thiamin (vitamin B_1) An essential water-soluble vitamin involved in energy metabolism and the synthesis of DNA and RNA.

coenzyme A vitamin that facilitates the function of an associated enzyme.

WATER-SOLUBLE VITAMINS CAN BE DESTROYED BY HEAT AND IMPROPER STORAGE

A balanced and varied diet can provide all of the nutrients needed to maintain health. Unfortunately, no matter how carefully you choose your foods, some preparation, cooking, and storage methods destroy nutrients—especially water-soluble vitamins. Whereas some are lost through exposure to water, air, heat, and/or light, others are unavoidably affected by acidity. Some vitamin loss is to be expected, but you can prevent excessive loss by properly preparing and storing foods. For example, overcooking some foods can diminish their vitamin B contents, and excessive sunlight can degrade vitamin C and certain B vitamins. Information about how each of the water-soluble vitamins can best be protected is addressed throughout the chapter.

Fortification and Enrichment of Food

Although all foods contain a variety of naturally occurring nutrients, some also have nutrients added during processing. This is called **fortification**. Because processing can destroy some nutrients, fortification is generally used to reintroduce nutrients and improve the nutritional value of processed foods. For example, because the portion of rice that contains most of the grain's thiamin is removed during the refinement process, this vitamin is often added to milled rice, making it a *fortified* food. Adding vitamins through fortification helps make many refined products such as white (polished) rice more nutritious. In some cases, a fortified food is even more nutritious than its original version.

Fortified food products that meet specified nutrient levels suggested by the U.S. Food and Drug Administration (FDA) can be labeled "enriched." **Enrichment** is the fortification of a select group of foods (rice, flour, bread or rolls, farina, pasta, cornmeal, and corn grits) with specified levels of B vitamins thiamin, niacin, riboflavin, folate, and the mineral iron.

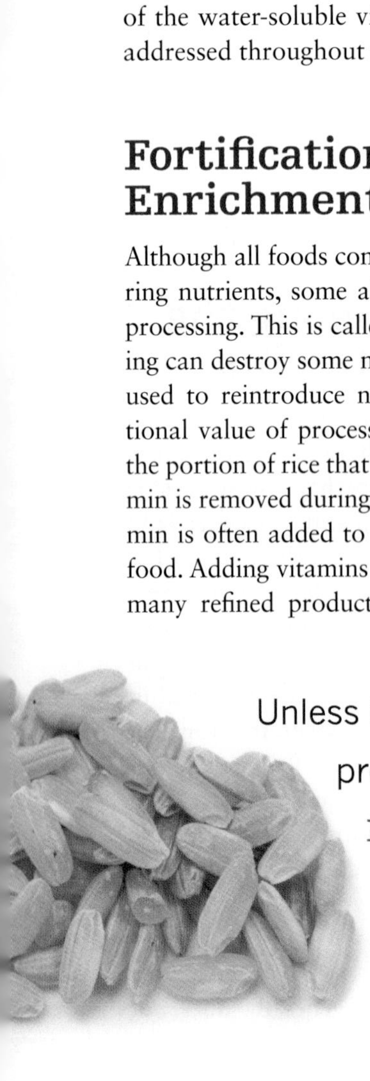

Unless it has been fortified, processed **white rice is not** a good source of **micronutrients** such as thiamin.

© iStockphoto.com/Elena Schweitzer

LO3 What Are the B Vitamins?

Although they differ chemically, the B-complex vitamins tend to serve similar biological functions within the body, especially in terms of energy metabolism. In this section, you will learn about each of these essential nutrients.

Thiamin

Thiamin (vitamin B_1) is an essential water-soluble vitamin involved in energy metabolism and the synthesis of DNA and RNA. It is found in a wide variety of foods, such as pork, peas, fish, legumes, soymilk, enriched cereal products, and whole-grain foods (see Figure 7.1). As you will note throughout this chapter, whole-grain foods tend to be good sources of many water-soluble vitamins. Thiamin is sensitive to heat and is easily destroyed during cooking. Shorter cooking times and lower temperatures can decrease this loss, although it is important to cook meat until it is done.

Several factors influence thiamin bioavailability. In general, its absorption increases when the body needs more of it, and decreases when thiamin status is adequate. Compounds found in raw fish, coffee, tea, berries, Brussels sprouts, cabbage, and alcohol can interfere with thiamin's absorption, and sulfites, often added to processed foods as preservatives, can destroy thiamin.[2] Conversely, vitamin C can increase thiamin bioavailability.

THIAMIN IS CRITICAL TO ATP, PROTEIN, AND TRIGLYCERIDE PRODUCTION

Although thiamin is not an energy-yielding nutrient per se, it is intimately involved in ATP production. To aid in energy synthesis, thiamin acts as a **coenzyme**, meaning

Figure 7.1 **Good Sources of Thiamin**

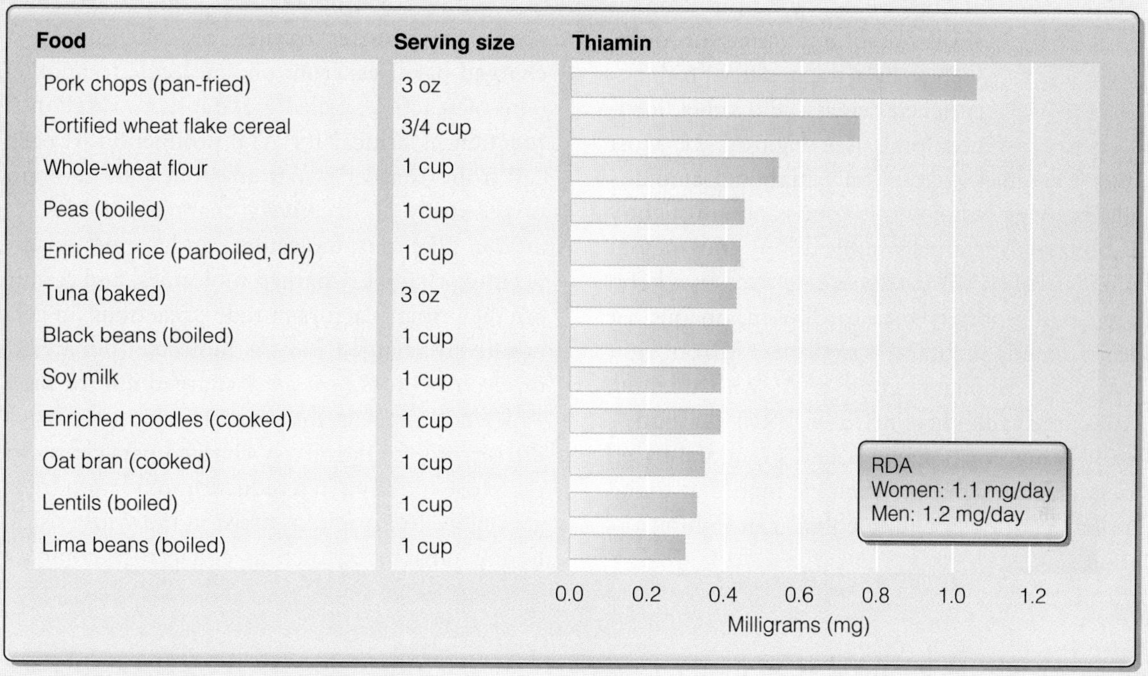

Source: USDA Nutrient Database for Standard Reference, Release 16-1.

that an associated enzyme will not function unless a specific vitamin—in this case thiamin—is present. As a component of a coenzyme, thiamin plays a role in the synthesis of DNA, RNA, and triglycerides. Recall from Chapter 5 that DNA and RNA are essential to protein synthesis. Without thiamin acting as a coenzyme, the enzymes that make DNA and RNA cannot work, and protein synthesis is halted. Further, cells cannot divide because they cannot replicate their genetic material.

THIAMIN DEFICIENCY CAUSES BERIBERI

Chinese physicians described **beriberi**, the life-threatening condition of thiamin deficiency, as early as 2600 B.C.E. Although now uncommon in the United States, beriberi is still prevalent in regions of the world that rely heavily on unfortified, refined rice as a major source of energy. There are four forms of beriberi. *Dry beriberi*, found mostly in adults, is characterized by severe muscle wasting, leg cramps, tenderness, and decreased feeling in the feet and toes. *Wet beriberi* is characterized by severe edema in the arms and legs. An enlargement of the heart and respiratory problems often result in heart failure. *Infantile beriberi*, found in babies breastfed by thiamin-deficient mothers, can also cause heart problems. *Cerebral beriberi* (or *Wernicke-Korsakoff syndrome*) is characterized by abnormal eye movements, poor muscle coordination, confusion, and short-term memory loss. Cerebral beriberi is typically associated with alcoholism both because alcohol decreases thiamin absorption and because alcoholics tend to maintain very poor diets.

RECOMMENDED THIAMIN INTAKE

The Recommended Dietary Allowances (RDAs) for thiamin are 1.2 and 1.1 mg/day for adult men and women, respectively. Because thiamin toxicity is very rare, there is no Tolerable Upper Intake Level (UL). A complete list of vitamin Dietary Reference Intakes (DRIs) is provided on the Dietary Reference Intakes card located at the back of this book.

Riboflavin

Riboflavin (vitamin B_2) is an essential water-soluble vitamin involved in energy metabolism, the synthesis of a variety of vitamins, nerve function, and protection of biological membranes. Riboflavin is named

beriberi A life-threatening condition caused by thiamin deficiency. There are four forms: wet, dry, infantile, and cerebral beriberi.

riboflavin (vitamin B_2) An essential water-soluble vitamin involved in energy metabolism, the synthesis of a variety of vitamins, nerve function, and protection of biological membranes.

reduction-oxidation (redox) reaction A transfer of oxygen or electrons from one molecule to another.

for its ribose structural component and its yellow color—*flavus* means "yellow" in Latin. Riboflavin is found in a variety of foods, such as liver, meat, dairy products, enriched cereals, and other fortified foods (see Figure 7.2). Most fruits and vegetables contain only marginal amounts of riboflavin, while whole-grain foods contain slightly more. Riboflavin is relatively stable during cooking but is quickly destroyed when exposed to excessive light. Milk is generally packaged in cardboard, opaque, or translucent (cloudy) containers to protect it from light and preserve its riboflavin content. As with thiamin, riboflavin's bioavailability increases when the body's need for it is high. Riboflavin found in animal-based foods is somewhat more bioavailable than that found in plant-based foods, and alcohol can inhibit its absorption.

RIBOFLAVIN ASSISTS IN REDUCTION-OXIDATION (REDOX) REACTIONS

Your body uses riboflavin to produce two coenzymes needed to transfer oxygen or electrons (negatively charged particles) from one molecule to another. This type of reaction, called a **reduction-oxidation (redox) reaction**, is critical for ATP production to occur. Recall from Chapter 3 that an atom with an abundance of positively or negatively charged particles is called an *ion*. Like ions, molecules can have net positive and negative charges. Charged molecules, and certain ions, are the primary actors in redox reactions. In general, a negatively charged ion (or molecule) donates oxygen or electrons to a positively charged ion (or molecule). When this happens, the positively charged particle is *reduced*, and the negatively charged particle is *oxidized*, therefore resulting in a reduction-oxidation (redox) reaction. Redox reactions assist in the synthesis of many bodily substances, such as collagen (a component of

Figure 7.2 Good Sources of Riboflavin

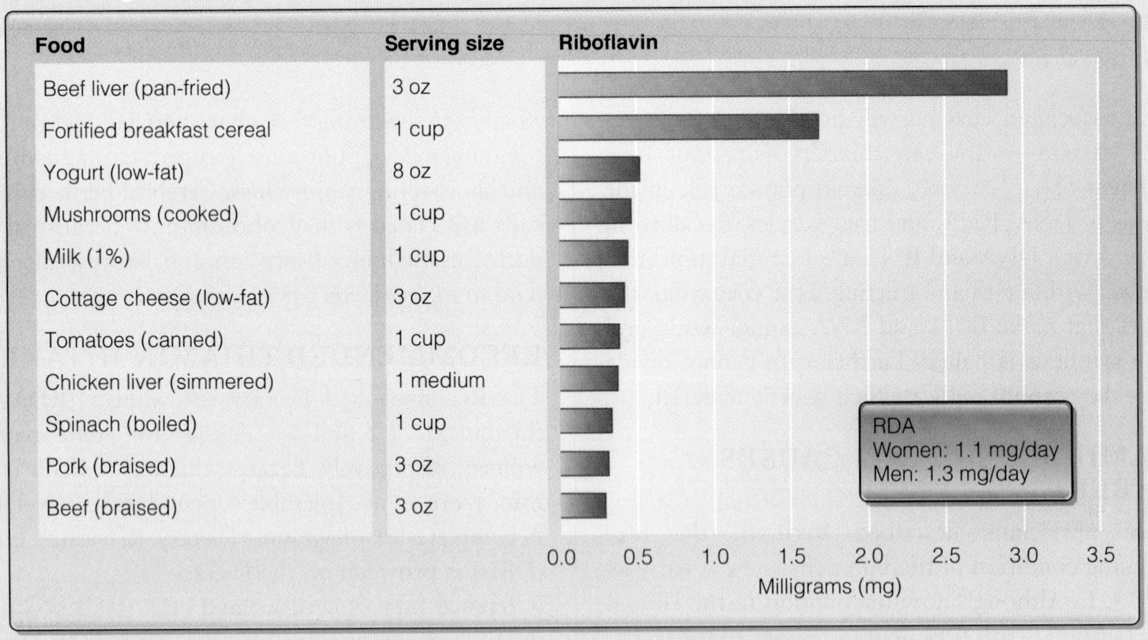

Source: USDA Nutrient Database for Standard Reference, Release 16-1.

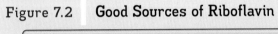

Riboflavin *is abundant in* **meat and dairy** *products.*

your skin, muscle tissue, and hair), carnitine (needed for fatty acid metabolism), and several neurotransmitters and hormones.

Beyond its role in redox reactions, riboflavin is also necessary to convert vitamin A and folate (a B vitamin) to their active forms, to synthesize niacin (a B vitamin) from tryptophan (an amino acid), and to form vitamins B_6 and K. Finally, riboflavin is critical to the metabolism of several neurotransmitters and is involved in reactions that protect biological membranes from oxidative damage.

Yellow Urine—A Cause for Concern?

Because changes in your body can indicate nutritional problems, it is important to pay close attention to how you look and feel. For example, rapid weight loss or gain might point toward a thyroid problem. Not all such signs and symptoms, however, are problematic. A change frequently reported by people taking vitamin B supplements is that their urine turns a bright, almost neon shade of yellow. What causes this color change (called *flavinuria*), and should it sound an alarm? The presence of excess riboflavin imparts an unusual color to the urine because riboflavin itself has a yellow hue. Although it may at first be startling, flavinuria is a harmless side effect of consuming more riboflavin than needed.[4]

RIBOFLAVIN DEFICIENCY CAUSES ARIBOFLAVINOSIS

Riboflavin deficiency causes a condition called **ariboflavinosis**, which does not usually occur on its own but is associated with general malnutrition. The signs and symptoms of ariboflavinosis include **cheilosis**, sores on the outside and corners of the lips, **stomatitis**, inflammation of the mouth, and **glossitis**, inflammation of the tongue, as well as muscle weakness and confusion. Riboflavin deficiency is rare in the United States but can occur in alcoholics who consume poor diets and in people with diseases that interfere with riboflavin utilization, such as thyroid disease.[3]

RECOMMENDED RIBOFLAVIN INTAKE

The RDAs for riboflavin are 1.3 and 1.1 mg/day for adult men and women, respectively. Because even at very high doses there are no known toxic effects of riboflavin consumption, no ULs have been established.

Niacin

Niacin (vitamin B_3) is an essential water-soluble vitamin involved in energy and vitamin C metabolism

Signs of ariboflavinosis include cheilosis and stomatitis (left), and glossitis (right).

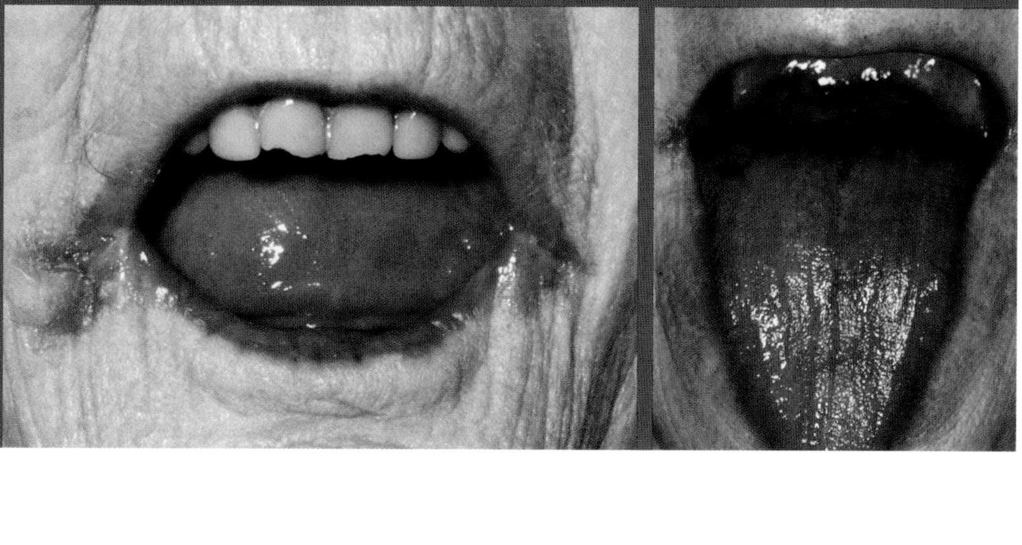

ariboflavinosis A condition caused by riboflavin deficiency whereby cheilosis, stomatitis, glossitis, muscle weakness, and confusion occur.

cheilosis A condition characterized by sores on the outside and corners of the lips.

stomatitis Inflammation of the mouth, often caused by dietary deficiencies.

glossitis Inflammation of the tongue.

niacin (vitamin B_3) An essential water-soluble vitamin involved in energy and vitamin C metabolism and the synthesis of fatty acids and proteins.

niacin equivalent (NE) A unit of measure for the combined amounts of niacin and tryptophan in food.

and synthesis of fatty acids and proteins. Niacin is usually obtained through the diet, but can also be synthesized in the body from tryptophan, an essential amino acid. As such, both niacin and tryptophan are considered dietary sources of niacin. The **niacin equivalent (NE)** is a unit of measure for the combined amounts of niacin and tryptophan in a given food. The NEs of selected foods are illustrated in Figure 7.3. Both niacin and tryptophan are found in a variety of foods, such as liver, poultry, fish, tomatoes, beef, and mushrooms. Whole-grain foods, enriched cereal products, and other fortified foods are also important sources of niacin, which is quite stable and not easily destroyed by preparation or exposure to light.

While the bioavailability of niacin is greater in animal- than plant-based products, treating plant-based foods with alkaline (basic) substances such as baking soda can increase bioavailability of the niacin they contain. In some traditional cultures, corn and cornmeal are soaked in lime water (an alkaline solution) to increase niacin availability. This practice is historically common in Mexico, where corn and corn tortillas have traditionally comprised a large part of many peoples' diets. As illustrated in Figure 7.4, lime is still added to some forms of cornmeal. Foods (such as tamales) prepared with these products are therefore good sources of niacin.

NIACIN IS INVOLVED IN REDUCTION-OXIDATION REACTIONS

Like thiamin and riboflavin, niacin is needed to synthesize ATP because niacin's coenzyme forms are critical to redox reactions related to energy metabolism. Niacin's coenzyme forms are also important to the synthesis of several important non-energy-related compounds, such as fatty acids, cholesterol, steroid hormones, and DNA. Finally, niacin's coenzyme forms are needed for the metabolism of vitamin C and folate.

> "Wondering why there is **no vitamin B_4**? Substances designated vitamins B_4, B_8, B_{10}, and B_{11} have been ***de*classified as vitamins** for various reasons, leaving gaps in the numbering system."

Figure 7.3 **Good Sources of Niacin**

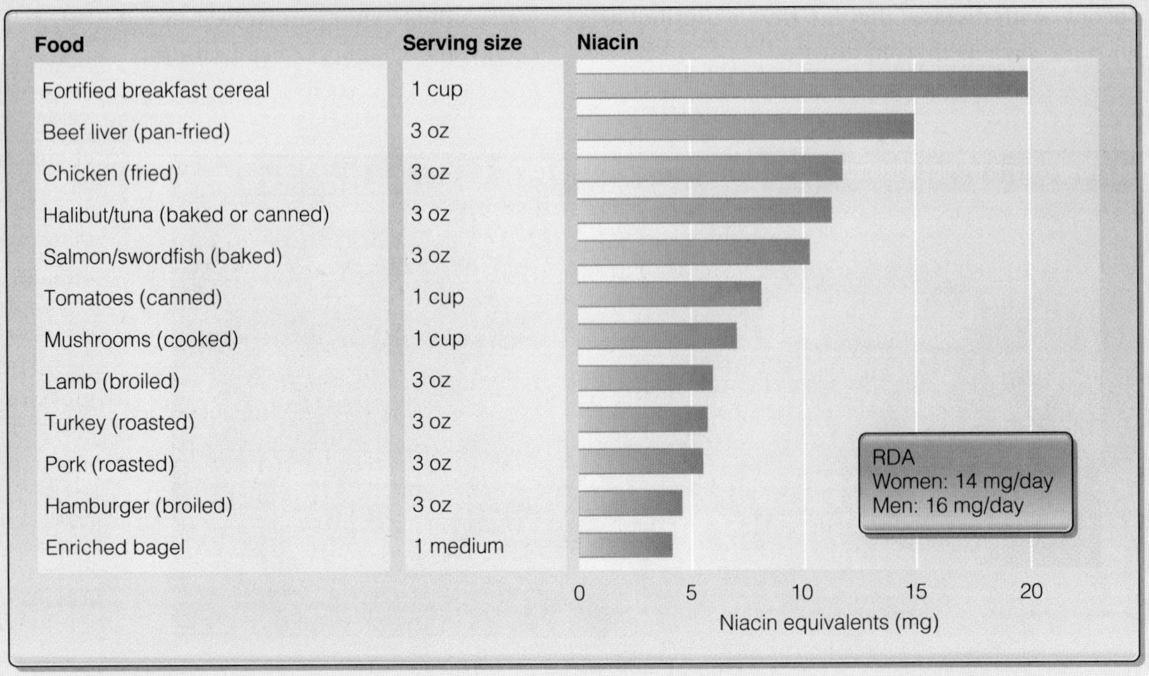
Food
Serving size
Niacin
Fortified breakfast cereal
1 cup
Beef liver (pan-fried)
3 oz
Chicken (fried)
3 oz
Halibut/tuna (baked or canned)
3 oz
Salmon/swordfish (baked)
3 oz
Tomatoes (canned)
1 cup
Mushrooms (cooked)
1 cup
Lamb (broiled)
3 oz
Turkey (roasted)
3 oz
Pork (roasted)
3 oz
Hamburger (broiled)
3 oz
Enriched bagel
1 medium
RDA
Women: 14 mg/day
Men: 16 mg/day
0
5
10
15
20
Niacin equivalents (mg)

Source: USDA Nutrient Database for Standard Reference, Release 16-1.

Figure 7.4 Increasing Niacin Bioavailability with Lime Water

SOMBRERO
INSTANT CORN MASA MIX
For tortillas, tamales, enchiladas, and atole
net wt. 4.4 lbs-2.0kg.
Ingredients: Specially ground and dehydrated whole kernel corn and lime. No preservatives added.
Tamales
Corn tortillas
Atole

Ground corn is often treated with lime (from limestone, not lime juice) to increase the bioavailability of niacin found naturally in corn.

In addition to its role as a coenzyme, niacin is important for the maintenance, replication, and repair of DNA, and may also play a role in protein synthesis, glucose homeostasis, and cholesterol metabolism. Little is known about how these effects occur, however.

NIACIN DEFICIENCY RESULTS IN PELLAGRA

Niacin deficiency causes **pellagra**, a condition that results in dermatitis, dementia, diarrhea, and death (sometimes referred to as the *four Ds*). Dermatitis—characterized by thick, rough, darkened, and sometimes red skin—is

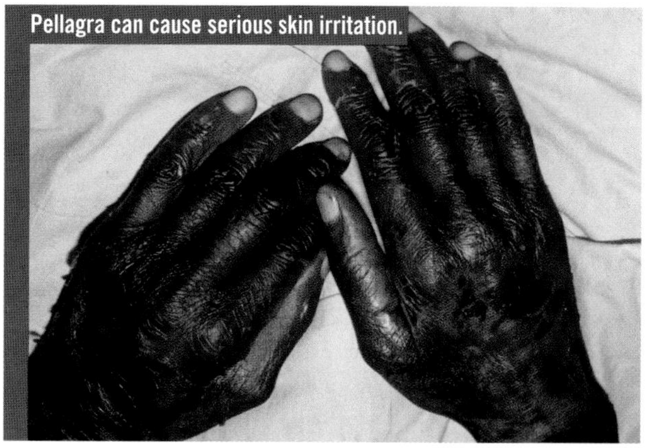

Pellagra can cause serious skin irritation.

often accompanied by neurological problems including depression, anxiety, irritability, dementia, and an inability to concentrate. Associated gastrointestinal disturbances cause loss of appetite, diarrhea, and a characteristically red and swollen tongue. If not treated, severe pellagra can cause death.

RECOMMENDED NIACIN INTAKE

RDAs for niacin are 16 and 14 mg/day NE for adult men and women, respectively. Because large doses of supplementary niacin can have ill effects (such as increased plasma glucose and liver damage), a UL of 35 mg/day NE has been set for niacin obtained from supplements and fortified foods. Note that this value does not apply to niacin that occurs naturally in foods.

Pantothenic Acid

Pantothenic acid (vitamin B_5) is a nitrogen-containing water-soluble vitamin involved in energy metabolism, hemoglobin synthesis, and phospholipid synthesis. Because it is found in almost every plant and animal tissue, pantothenic acid is named for the Greek word *pantos*, meaning "everywhere." Pantothenic acid is found in many foods, including mushrooms, organ meats (such as liver), and sunflower seeds (see Figure 7.5 on the next page). Dairy products, turkey, fish, and coffee are also good sources. High temperatures can destroy pantothenic acid, so cooking food at moderate temperatures is advisable to retain the pantothenic acid it contains. As with other B vitamins, pantothenic acid's bioavailability changes depending on how much the body needs.

pellagra A condition caused by niacin deficiency whereby dermatitis, dementia, diarrhea, and/or death may occur.

pantothenic acid (vitamin B_5) A nitrogen-containing water-soluble vitamin involved in energy metabolism, hemoglobin synthesis, and phospholipid synthesis.

Figure 7.5 **Good Sources of Pantothenic Acid**

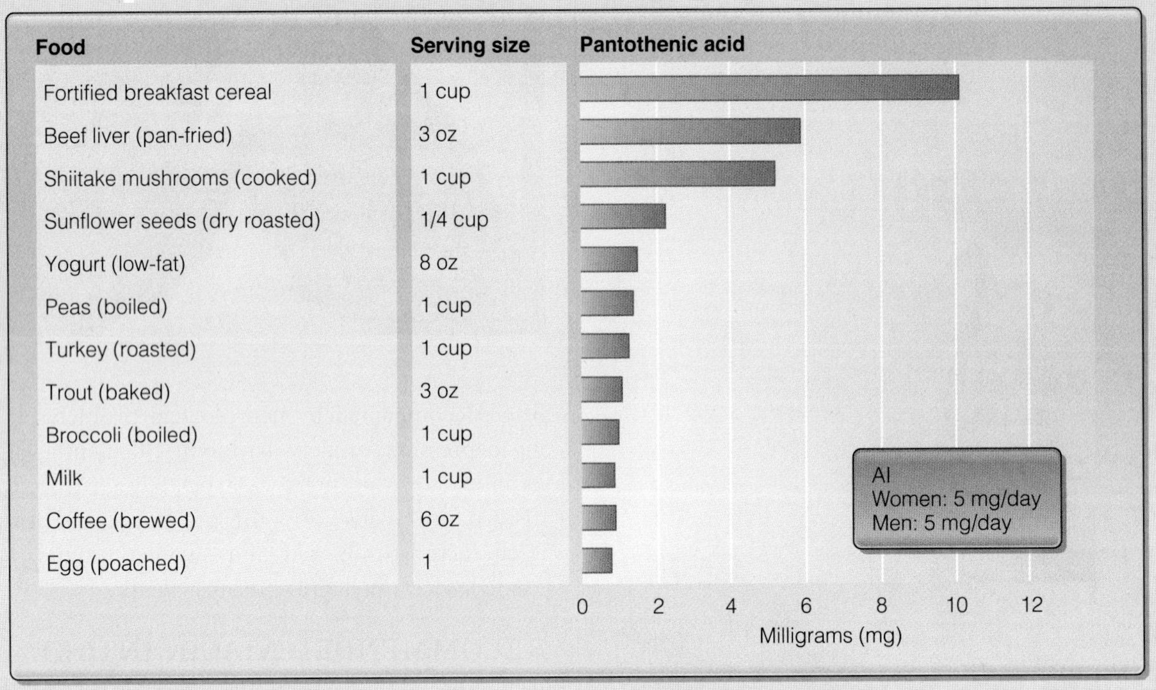

Source: USDA Nutrient Database for Standard Reference, Release 16-1.

burning feet syndrome A condition believed to be caused by pantothenic acid deficiency whereby tingling in the feet and legs, fatigue, weakness, and nausea may occur.

PANTOTHENIC ACID IS CRITICAL FOR CARBOHYDRATE, PROTEIN, AND LIPID METABOLISM

Pantothenic acid functions as a coenzyme in a variety of metabolic reactions—especially those involved in ATP production. The metabolism of glucose, amino acids, and fatty acids depends on pantothenic acid's coenzyme form. Beyond its coenzyme functions, this vitamin is critical for the synthesis of many important compounds, such as heme, cholesterol, bile, phospholipids, fatty acids, and some of the reproductive hormones.

Pantothenic acid *is found in a* diverse group *of foods, such as* mushrooms.

© iStockphoto.com/Adrian Assalve

PANTOTHENIC ACID DEFICIENCY MAY CAUSE BURNING FEET SYNDROME

Because pantothenic acid is found in almost every food, deficiencies are rare. Nonetheless, some nutrition researchers believe that a condition called **burning feet syndrome** may be caused by severe pantothenic acid deficiency. As its name implies, burning feet syndrome is characterized by a tingling sensation in the feet and legs, as well as fatigue, weakness, and nausea. Note that burning feet syndrome can be caused by factors other than pantothenic acid deficiency.

RECOMMENDED PANTOTHENIC ACID INTAKE

There is not enough information to establish RDAs for pantothenic acid, but an Adequate Intake (AI) level of 5 mg/day has been set for adults. Overconsumption of pantothenic acid is not fatal, but very high intakes have been associated with nausea and diarrhea. Because there is no evidence of pantothenic toxicity, a UL has not been set for this essential vitamin.

Vitamin B_6

Vitamin B_6 is a water-soluble vitamin involved in the metabolism of glucose, proteins, and amino acids; synthesis of neurotransmitters and hemoglobin; and regulation of steroid hormone function. Like the other B vitamins, vitamin B_6 is found in a variety of plant- and animal-based foods. Chickpeas (garbanzo beans), fish, liver, potatoes, and chicken are particularly good sources of this vitamin, as are fortified breakfast cereals and bakery products (see Figure 7.6). Unlike other B vitamins, vitamin B_6 does not have a widely used common name. The vitamin exists in seven known forms, the most common of which are *pyridoxine*, *pyridoxal*, *pyridoxamine*, and *pyridoxal phosphate*. Vitamin B_6 is somewhat unstable and is destroyed by extreme or prolonged heat and cold.

VITAMIN B_6 IS CRITICAL TO NONESSENTIAL AMINO ACID SYNTHESIS

Like most of the B vitamins, vitamin B_6 functions as a coenzyme—its pyridoxal phosphate form is involved in numerous chemical reactions related to the synthesis of nonessential amino acids from essential amino acids. Vitamin B_6 is also essential for the production of some nonprotein substances, such as heme and the neurotransmitters serotonin and dopamine.

VITAMIN B_6 DEFICIENCY AND TOXICITY

Vitamin B_6 deficiency impedes heme production, in turn lowering the concentrations of hemoglobin in red blood cells. This condition, called **microcytic, hypochromic anemia**, results in red blood cells that are small (*microcytic*) and light in color (*hypochromic*). Microcytic, hypochromic anemia can diminish red blood cells' ability to deliver oxygen to tissues, which in turn impairs ATP production. Vitamin B_6 deficiency also causes cheilosis, glossitis, stomatitis, and fatigue. Because these signs and symptoms are very similar to those of riboflavin deficiency, these two deficiencies are often difficult to distinguish without analysis of the blood.

Vitamin B_6 deficiency is rare, but cooking foods at high temperatures can destroy some of the vitamin B_6 they contain. This was not well understood until the

vitamin B_6 A water-soluble vitamin involved in the metabolism of proteins and amino acids, the synthesis of neurotransmitters and hemoglobin, glycogenolysis, and regulation of steroid hormone function.

microcytic, hypochromic anemia A condition characterized by low concentrations of hemoglobin, which causes red blood cells to be small and light in color.

Figure 7.6 Good Sources of Vitamin B_6

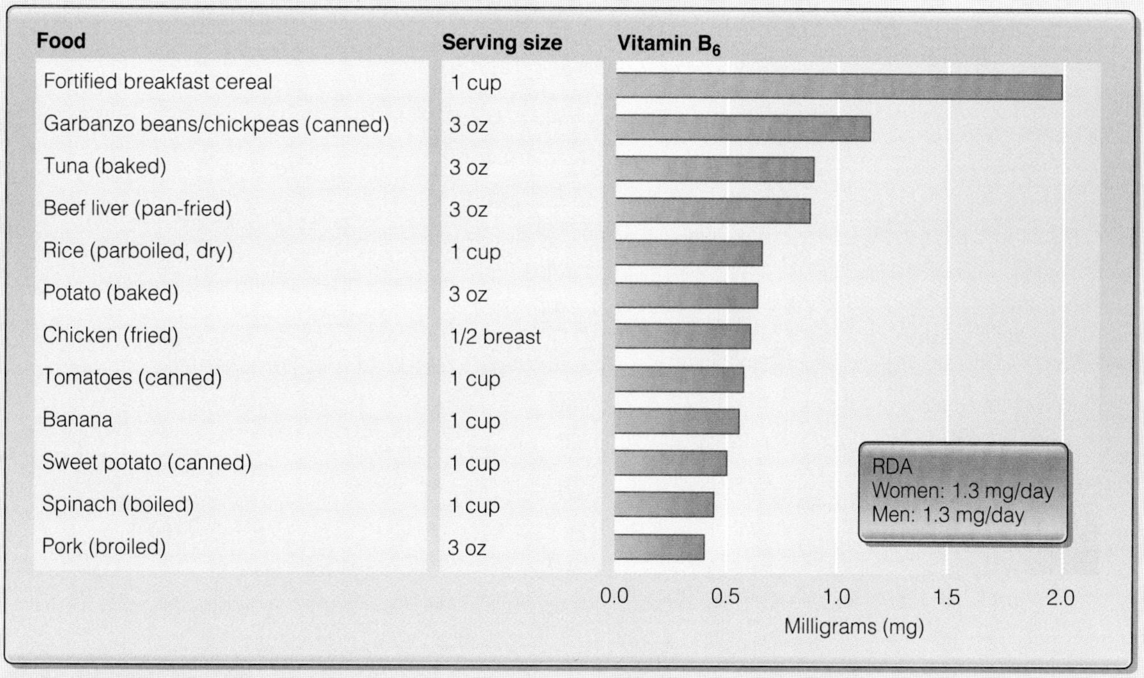

Source: USDA Nutrient Database for Standard Reference, Release 16-1.

biotin (vitamin B_7) A water-soluble vitamin involved in energy metabolism that is obtained both from the diet and bacteria in the large intestine.

1950s, when vitamin B_6 added to infant formula was often destroyed during heat processing. In those days, many formula-fed infants developed vitamin B_6 deficiencies, which caused serious complications such as seizures and convulsions. Manufacturing procedures have since changed so that infant formulas now contain sufficient amounts of this essential nutrient.

Relatively large amounts of pyridoxal phosphate are retained within muscle and liver cells. As such, vitamin B_6 toxicity occurs more frequently than does that of other water-soluble vitamins. Vitamin B_6 toxicity causes severe neurological problems, such as difficulty walking and numbness in the feet and hands. Vitamin B_6 toxicity is not likely to result from the overconsumption of vitamin B_6–rich foods, but may result from excessive intake of supplements.

RECOMMENDED VITAMIN B_6 INTAKE

Depending on age and sex, RDAs for vitamin B_6 vary from 1.3 to 1.7 mg/day for adults. To prevent the neurological problems associated with excess vitamin B_6, a UL of 100 mg/day has been established for this essential vitamin. Supplements containing 500 mg of vitamin B_6 are widely available, making it relatively easy to consume more than the recommended daily amount. If you supplement your diet with vitamin B_6 (or B-complex vitamins), be sure to read supplement packaging and labels carefully and make sure that your intake does not exceed the UL value.

Biotin

Biotin (vitamin B_7) is yet another water-soluble vitamin involved in energy metabolism. However, biotin is somewhat unique among the water-soluble vitamins because it is obtained from both the diet and biotin-producing bacteria in the large intestine. As illustrated in Figure 7.7, sources of dietary biotin include peanuts, tree nuts (such as almonds and cashews), mushrooms, eggs, and tomatoes. Biotin is often bound to proteins in food, but is released during digestion. Sometimes biotin is bound too tightly to food proteins, making it difficult to absorb. In fact, biotin was first discovered because of its role in *egg white injury*, a condition whereby biotin's bioavailability is reduced severely because it is consumed with *avidin*, a protein present in large quantities in raw egg whites. (Do not let the threat of egg white injury dissuade you from eating cooked eggs, however—heat denatures avidin, allowing biotin to be absorbed normally.) Alcohol can also

Figure 7.7 Good Sources of Biotin

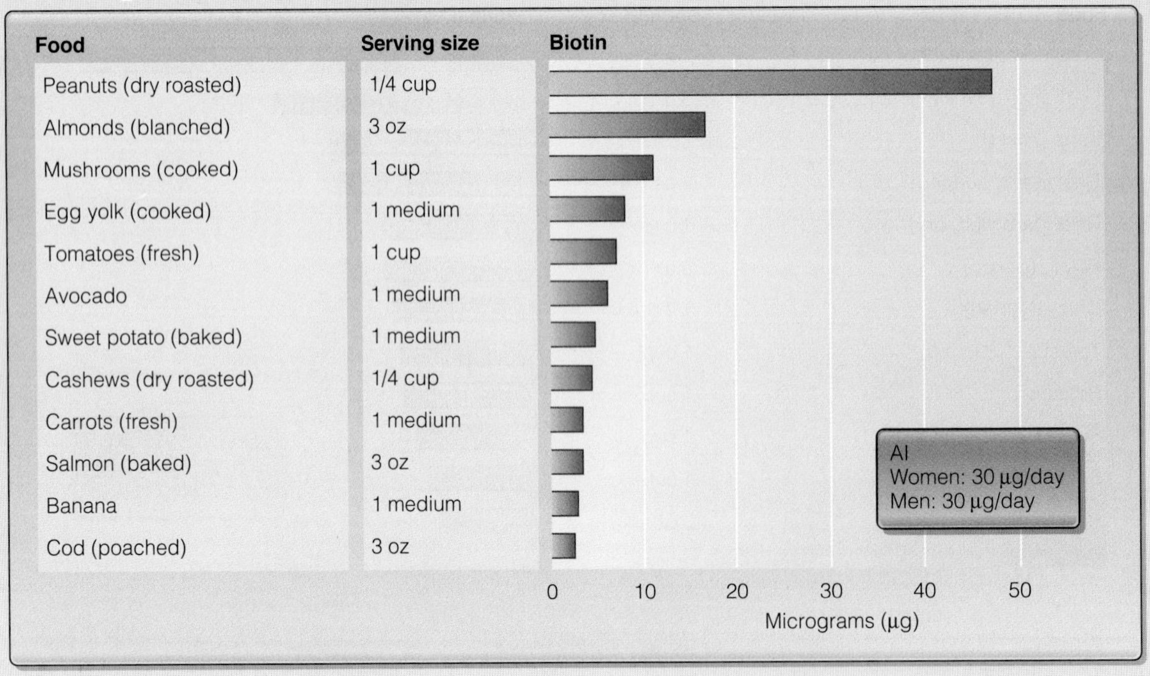

Source: Hands E. *Food Finder Vitamin and Mineral Source Guide*, 3rd ed. Salem, OR: ESHA Research.

Avidin, a protein found in raw egg whites, binds tightly to biotin and decreases its absorption.

decrease biotin absorption, and high preparation temperatures can destroy biotin in foods.

BIOTIN ADDS BICARBONATE (HCO_3^-) SUBUNITS

Biotin is a critical component of several coenzymes involved in energy metabolism pathways. Specifically, biotin-dependent reactions shift bicarbonate subunits (HCO_3^-) from one molecule to another. Beyond its role as a coenzyme, biotin serves several functions related to gene expression, cell growth, and development.

BIOTIN DEFICIENCY CAUSES DEPRESSION AND OTHER SIGNS AND SYMPTOMS

Though biotin deficiency is uncommon, it is sometimes caused by genetic disorders and other conditions that impair intestinal absorption (such as inflammatory bowel disease). Signs and symptoms of biotin deficiency are poorly understood, but can include depression, hallucinations, skin irritations, infections, hair loss, poor muscle control, seizures, and developmental delays.

RECOMMENDED BIOTIN INTAKE

Because information regarding biotin intake is still insufficient, RDAs have not yet been set for this important micronutrient. However, an AI of 30 μg/day has been set for adults. Because excessive biotin intake does not carry any known detrimental effects, a UL has not been established.

folate (vitamin B$_9$) A group of related water-soluble vitamins involved in single-carbon transfers, amino acid metabolism, and DNA synthesis.

Folate

Folate (vitamin B$_9$) refers to a group of related water-soluble vitamins involved in single-carbon transfers, amino acid metabolism, and DNA synthesis. Good sources of folate include organ meats, legumes (such as lentils and pinto beans), okra, and many green leafy vegetables such as spinach (see Figure 7.8 on the next page). A common form of folate called *folic acid* is rarely found in foods, but is often included in vitamin supplements and is added during the fortification of food. Because cereal grain enrichment has necessitated folic acid fortification since 1998, enriched foods are always very good sources of this vitamin. Folate is destroyed by excessive heat, so cooked foods often contain less folate than do their uncooked counterparts.

The bioavailability of folate depends on its form in food, genetic factors, and the use of certain medications. In general, the body absorbs folate found in supplements and fortified foods (folic acid) better than it absorbs naturally occurring folate. Because folate

Folate *(from an Italian word* meaning "foliage"*) is found in many* plant-based foods.

Figure 7.8 **Good Sources of Folate**

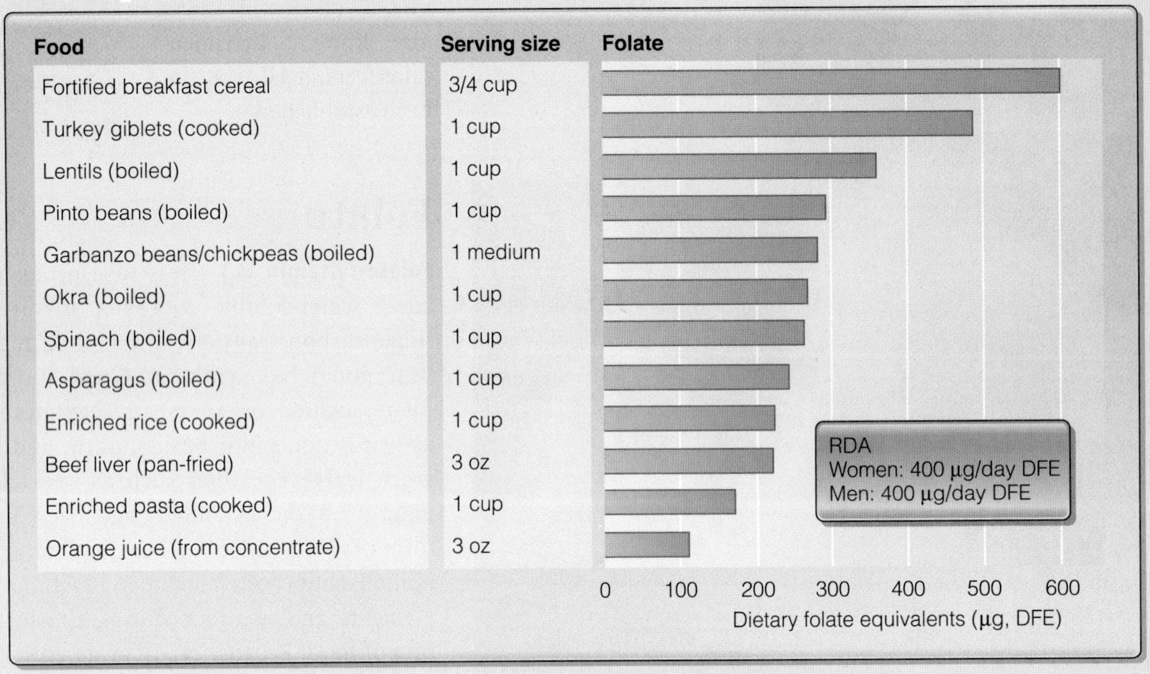

Source: USDA Nutrient Database for Standard Reference, Release 16-1.

absorption is so variable, the amount of folate present in a food is expressed in terms of **dietary folate equivalent (DFE)**, an approximation of the quantity actually absorbed by the body.

FOLATE FACILITATES SINGLE-CARBON TRANSFERS, AMINO ACID METABOLISM, AND FORMATION OF THE NEURAL TUBE

Once folate is absorbed by an intestinal cell, it bonds with four hydrogen atoms, converting it to its active form, **tetrahydrofolate (THF)**. In its active form, folate facilitates *single-carbon transfers*, which are needed to synthesize many important organic substances such as amino acids. In a single-carbon transfer, a carbon atom in the form of a methyl group ($-CH_3$) is pulled from a molecule and bound to THF, producing *5-methyltetrahydrofolate* (*5-methyl THF*). 5-methyl THF migrates to another molecule and allows it to bind to the methyl group, effectively transferring a single carbon from one molecule to another.

dietary folate equivalent (DFE) A unit of measure for the approximate amount of folate in a food that is absorbed by the body.

tetrahydrofolate (THF) The active form of folate.

neural tube defect A malformation whereby neural tissue does not form properly during fetal development.

Folate-requiring single-carbon transfers play a large role in amino acid metabolism. For example, to convert the amino acid homocysteine to the amino acid methionine, a methyl group must be transferred to homocysteine. This transfer results in the production of both THF (because 5-methyl THF loses its methyl group) and the essential amino acid methionine (see Figure 7.9). Because the production of methionine from homocysteine requires both folate and vitamin B_{12} (which you will learn about shortly), a deficiency in either can result in a buildup of homocysteine in the body. High levels of homocysteine are associated with an increased risk of heart disease and may have other serious health consequences.[5]

Because it is involved in single-carbon reactions, THF is critical to the formation of DNA and RNA, which are in turn essential to the growth, maintenance, and repair of every tissue in the body. The production of DNA and RNA is especially important during periods of rapid growth (such as embryonic and fetal growth) and in cells with very short life spans (such as those lining the gastrointestinal tract).

Folate is clearly an important vitamin, but it is particularly important during pregnancy because of its role in the formation of the neural tube, which later develops into the spinal cord and brain. Increased maternal intake of folate (both in the form of foods and supplements) has been shown to decrease the risk of a **neural tube defect** in some newborns.[6] A neural tube defect is a malformation whereby neural tissues do not form properly during very early development. When this happens, the spinal cord is not protected adequately, and in severe cases, the brain does not form properly. The most common form of neural tube defect is **spina bifida** (Latin for "split spine"), a failure of the neural tube to close properly during the first months of fetal life (see Figure 7.10).[7]

Figure 7.9 The Role of Folate and Vitamin B_{12} in the Conversion of Homocysteine to Methionine

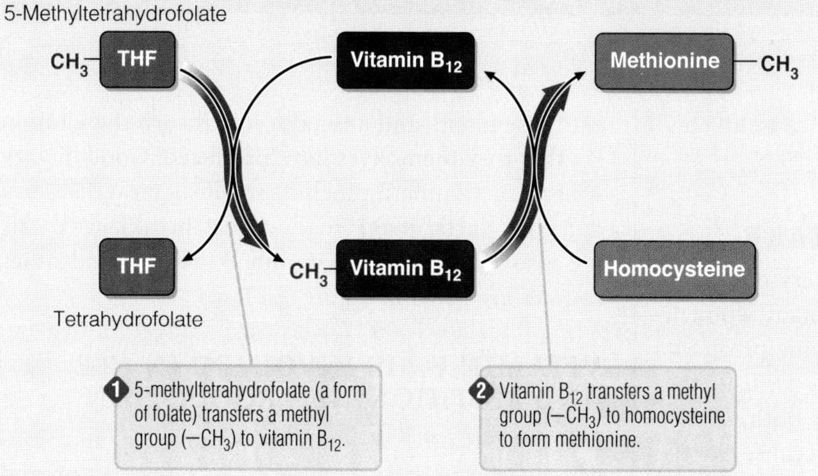

The conversion of homocysteine to methionine requires folate (in the form of 5-methyl THF) and vitamin B_{12}.

FOLATE DEFICIENCY CAUSES MEGALOBLASTIC, MACROCYTIC ANEMIA

Folate deficiency was once relatively common in the United States, but because enriched cereal products are now fortified with folate, this is no longer the case.[8] Today, folate deficiency occurs most often in alcoholics, people with intestinal diseases, people taking certain medications, and the elderly. Severe folate deficiency causes a condition called **megaloblastic, macrocytic anemia.** Because folate is integral to DNA synthesis and cell maturation, a dearth of dietary folate can cause cells (including red blood cells) to remain large (*macrocytic*) and immature (*megaloblastic*). Folate deficiency is only one cause of this form of anemia, however. Because several nutrient deficiencies can result in megaloblastic, macrocytic anemia, it is sometimes difficult to determine

spina bifida A failure of the neural tube to close properly during the first months of fetal life.

megaloblastic, macrocytic anemia A condition caused by folate deficiency whereby cells (including red blood cells) remain large and immature.

Figure 7.10 Spina Bifida

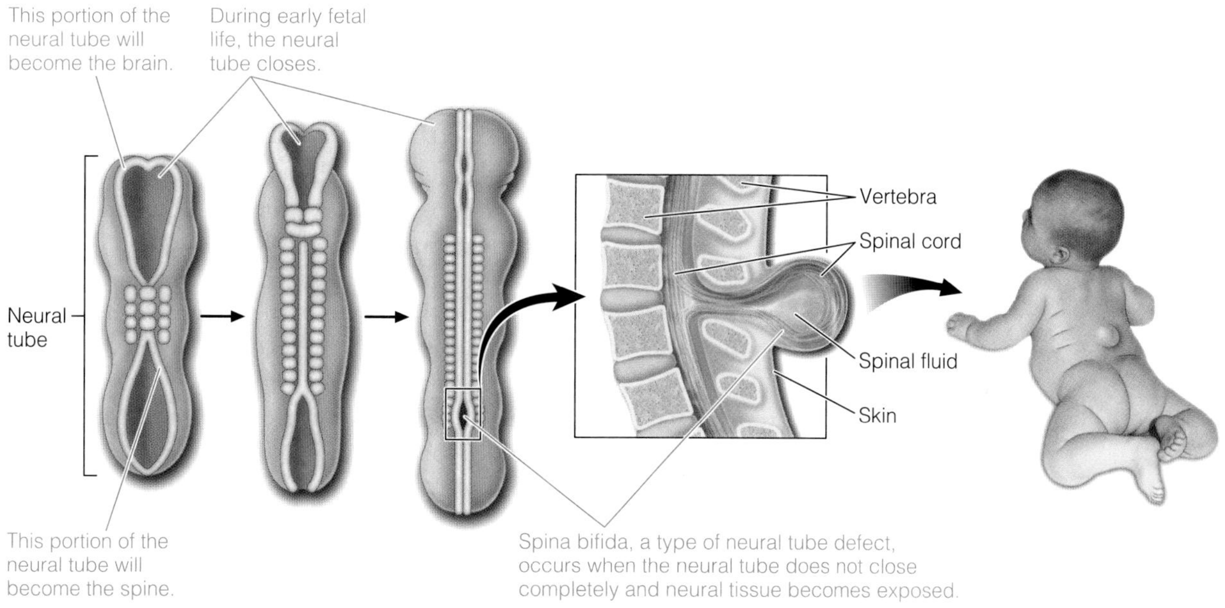

This portion of the neural tube will become the brain.
During early fetal life, the neural tube closes.
Neural tube
Vertebra
Spinal cord
Spinal fluid
Skin
This portion of the neural tube will become the spine.
Spina bifida, a type of neural tube defect, occurs when the neural tube does not close completely and neural tissue becomes exposed.

Spina bifida is a neural tube defect that occurs early in pregnancy.

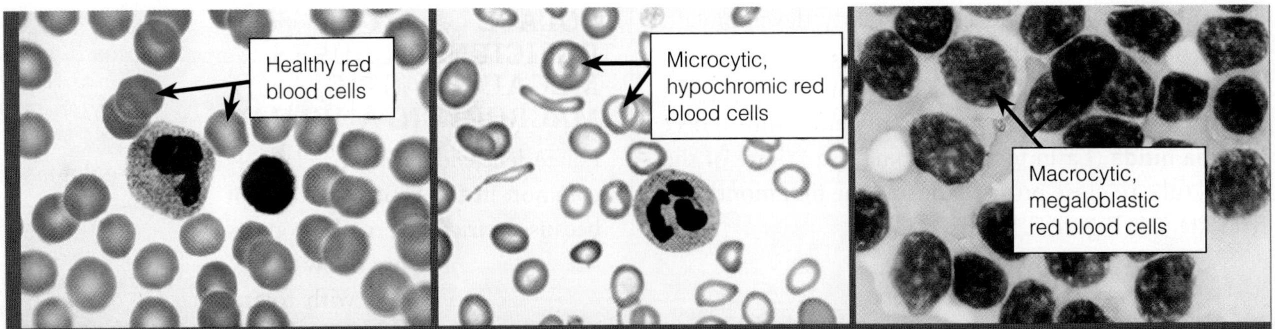

Under a microscope, healthy red blood cells (left) can be distinguished visually from cells affected by microcytic, hypochromic anemia (middle) and megaloblastic, macrocytic anemia (right).

whether suboptimal folate levels are the cause of this potentially dangerous condition.

RECOMMENDED FOLATE INTAKE

Adults should consume 400 μg DFE every day. This RDA increases to 600 μg/day for pregnant women. Women capable of or planning to become pregnant are encouraged to consume 400 μg/day DFE in the form of supplements, fortified foods, or both, in addition to a naturally folate-rich diet. Because an elevated folate intake may make it difficult to detect a vitamin B_{12} deficiency, a UL of 1,000 μg/day DFE has been set for folate derived from fortified foods and/or supplements. There is no evidence that a high intake of naturally occurring folate poses any risk of toxicity.

Vitamin B_{12}

Vitamin B_{12} (cobalamin) is a water-soluble vitamin involved in energy metabolism and methionine production. Cobalamin is named such because it contains the trace element *cobalt* and several nitrogen (or *amine*) groups. Vitamin B_{12} is a unique vitamin because it can *only* be synthesized by microorganisms such as bacteria and fungi. That is, plants and higher animals do not synthesize vitamin B_{12} themselves—they must obtain it from microorganisms living in either their environments or their gastrointestinal tracts. When you eat these plants and animals, you absorb the vitamin B_{12} that they themselves have produced. Good dietary sources of vitamin B_{12} include shellfish, meat, fish, and dairy products; many ready-to-eat breakfast cereals are also fortified with vitamin B_{12}. These and other sources are listed in Figure 7.11.

vitamin B_{12} (cobalamin) A water-soluble vitamin involved in energy metabolism and methionine production.

intrinsic factor A protein produced by the stomach that is needed for vitamin B_{12} absorption.

pernicious anemia An autoimmune disease caused by vitamin B_{12} deficiency whereby antibodies destroy the stomach cells that produce intrinsic factor.

VITAMIN B_{12} IS INVOLVED IN ATP AND METHIONINE PRODUCTION

Like several other B vitamins, vitamin B_{12} functions as a coenzyme. It is essential to ATP production and aids folate in amino acid metabolism. Recall that during the conversion of homocysteine to methionine, 5-methyl THF loses a methyl group and is converted to THF. Without adequate vitamin B_{12}, homocysteine blood levels rise, 5-methyl THF cannot be converted to THF, and folate deficiency symptoms appear. In this way, vitamin B_{12} deficiency can cause secondary folate deficiency.

VITAMIN B_{12} DEFICIENCY CAUSES PERNICIOUS ANEMIA

Vitamin B_{12} deficiency is caused by either inadequate dietary intake (primary vitamin B_{12} deficiency) or poor absorption (secondary vitamin B_{12} deficiency). Primary vitamin B_{12} deficiency occurs most often in vegans and infants breastfed by vitamin B_{12}–deficient mothers. Secondary vitamin B_{12} deficiency sometimes occurs when stomach cells fail to produce sufficient amounts of hydrochloric acid or **intrinsic factor**, a protein needed for vitamin B_{12} absorption. The production of either of these substances may slow or cease as a person ages, resulting in poor vitamin B_{12} absorption. Another cause of secondary vitamin B_{12} deficiency is **pernicious anemia**, an autoimmune disease whereby antibodies destroy the stomach cells that produce

Figure 7.11 **Good Sources of Vitamin B_{12}**

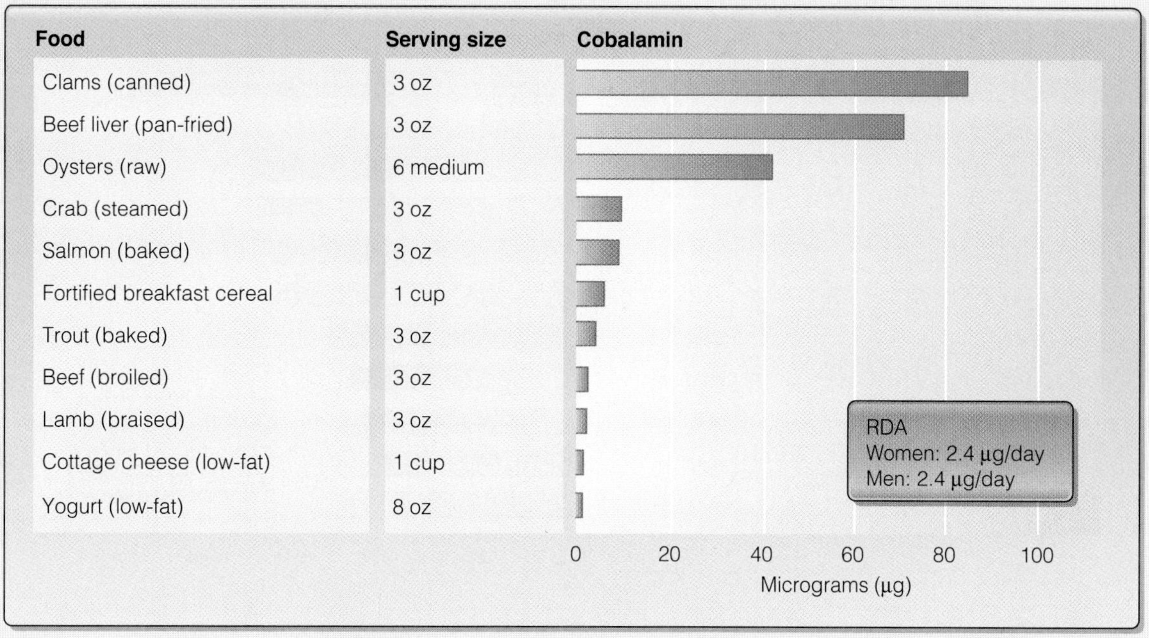

Source: USDA Nutrient Database for Standard Reference, Release 16-1.

intrinsic factor.[9] People with pernicious anemia or who do not otherwise produce adequate amounts of intrinsic factor cannot simply treat their vitamin B_{12} deficiencies with oral supplements—they must receive vitamin B_{12} by injection.

While megaloblastic, macrocytic anemia is often associated with both primary and secondary vitamin B_{12} deficiency, this condition is actually caused by an underlying folate deficiency. As you learned above, vitamin B_{12} deficiency can cause secondary THF deficiency, even when a seemingly adequate amount of folate is consumed. Because large doses of folate alleviate some of the symptoms of vitamin B_{12} deficiency (such as anemia), such doses can mask a vitamin B_{12} deficiency. However, while anemia may clear up with folate supplementation, other less evident complications associated with vitamin B_{12} deficiency (such as neurologic damage) may not. For this reason, misdiagnosis of a vitamin B_{12} deficiency as a folate deficiency can be dangerous.

RECOMMENDED VITAMIN B_{12} INTAKE

The RDA for vitamin B_{12} is 2.4 µg/day for adults. Vegans and vegetarians who do not eat any animal-based products should consider taking a supplement or eating foods that have been fortified with vitamin B_{12}. No ULs have been established for this vitamin.

LO4 What Is Vitamin C?

Vitamin C (ascorbic acid) is a water-soluble vitamin that serves antioxidant functions within the body. As illustrated in Figure 7.12 on the next page, vitamin C is found in many fruits and vegetables, such as citrus fruits, peppers, papayas, broccoli, strawberries, and peas. The bioavailability of vitamin C is generally high, although its structure can be destroyed by heat and exposure to oxygen; freshly peeled and/or prepared fruits and vegetables tend to provide more vitamin C than cooked, processed, and/or stored ones.

Vitamin C Is a Potent Antioxidant and May Benefit the Immune System

Unlike the B vitamins, vitamin C is not a coenzyme. Instead, it acts independently as an **antioxidant**, a compound that donates electrons

vitamin C (ascorbic acid) A water-soluble vitamin that serves antioxidant functions within the body.

antioxidant A compound that donates electrons or hydrogen ions to other substances, inhibiting oxidation.

Figure 7.12 Good Sources of Vitamin C

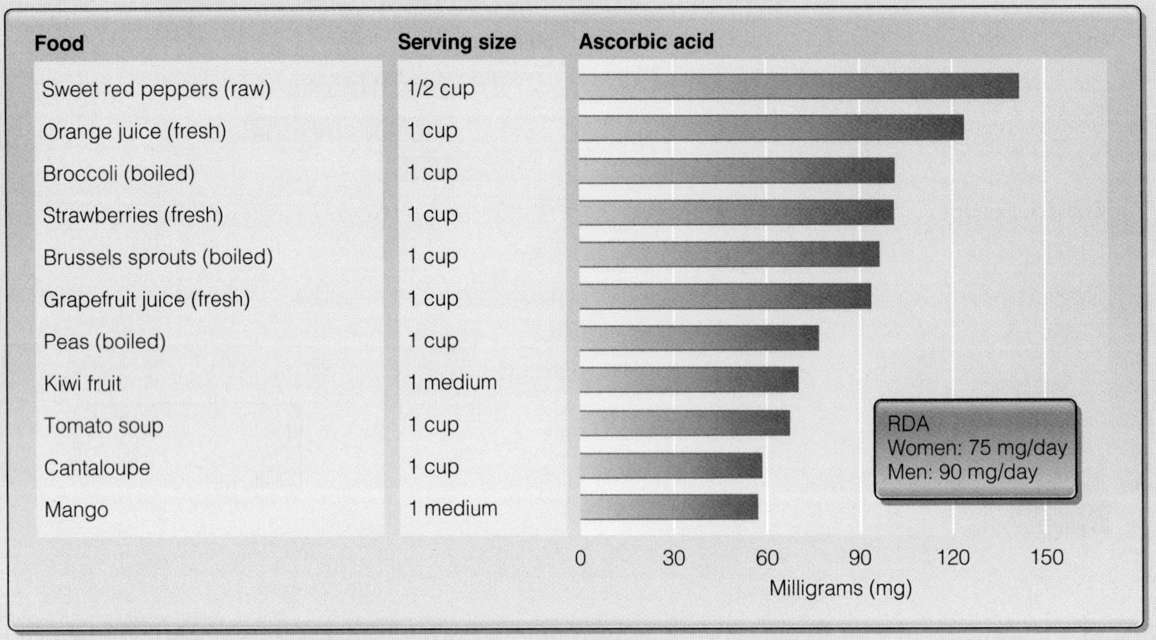

Food
Serving size
Ascorbic acid
Sweet red peppers (raw)
1/2 cup
Orange juice (fresh)
1 cup
Broccoli (boiled)
1 cup
Strawberries (fresh)
1 cup
Brussels sprouts (boiled)
1 cup
Grapefruit juice (fresh)
1 cup
Peas (boiled)
1 cup
Kiwi fruit
1 medium
Tomato soup
1 cup
Cantaloupe
1 cup
Mango
1 medium
RDA
Women: 75 mg/day
Men: 90 mg/day
0
30
60
90
120
150
Milligrams (mg)

Source: USDA Nutrient Database for Standard Reference, Release 16-1.

or hydrogen ions to other substances, inhibiting oxidation. Because vitamin C (like other antioxidants) can easily accept and donate electrons, it is involved in a wide variety of redox reactions. Vitamin C's antioxidant function also protects cells from **free radical** molecules, which have one or more unpaired electrons and are therefore highly reactive and unstable. In most cases, free radicals are produced in the body during normal cellular metabolism. Exposure to intense sunlight, some drugs, and toxic substances such as smog, cigarette smoke, and ozone can increase free radical production. Antioxidants are important because free radicals try to stabilize themselves by taking electrons from other molecules. Because antioxidants have electrons to give, they can effectively neutralize free radicals, in turn protecting other molecules from oxidative damage. If free radicals cannot take electrons from antioxidant molecules, they instead take them from—and are thus extremely destructive to—other molecules such as lipids in cell membranes, DNA, and proteins. Free radical damage leads to a variety of diseases. For example, damage done to DNA can lead to faulty protein synthesis. If the affected proteins regulate cell growth, their faulty production might lead to uncontrolled growth—otherwise known as cancer.

The ability of vitamin C to donate its electrons (and thereby reduce the charges of other molecules) also plays an important role in nutrient absorption in the gastrointestinal tract. Minerals such as iron, copper, and chromium are better absorbed in their reduced states. Consequently, the consumption of vitamin C can increase the bioavailability of these essential minerals. By drinking orange juice with iron-fortified cereal, you can significantly increase the amount of iron that is absorbed.

Some research suggests that increasing one's consumption of fruits and vegetables rich in vitamin C

free radical A highly reactive molecule with one or more unpaired electrons. Destructive to cell membranes, DNA, and proteins.

Vitamin C, found in oranges and many other fruits and vegetables, **protects the body** from the damaging **effects of free radicals**.

may decrease the risk of certain diseases, including the common cold, cancer, heart disease, and cataracts.[10] Controlled clinical intervention studies have not consistently demonstrated a protective effect of vitamin C.[11] Growing evidence, however, shows that large doses of vitamin C can benefit the immune system.[12] Because the immune system is directly involved in your body's defense against disease, it is possible that vitamin C may have an indirect effect.

Vitamin C Deficiency Causes Scurvy

Vitamin C deficiency causes **scurvy**, a deadly condition characterized by bleeding gums, skin irritations, bruising, and poor wound healing. In the 18th century, British medical doctor James Lind conducted what was perhaps the first controlled nutrition-related experiment in an attempt to determine scurvy's cause. Lind treated 12 sailors suffering from scurvy with lemons, limes, cider, nutmeg, seawater, or vinegar. He found that consumption of lemons or limes—but not the other treatments—prevented and cured scurvy. Although scurvy was once common, increased availabilities of fruits and vegetables have made it relatively rare. However, scurvy is still common in developing countries, alcoholics, and older adults on very restricted diets.[13]

Recommended Vitamin C Intake

RDAs for vitamin C are 90 and 75 mg/day for adult men and women, respectively. Because cigarette smoke increases exposure to free radicals (and consequently the body's need for antioxidants), smokers are advised to increase their vitamin C intake by an additional 35 mg/day. Whether this recommendation also applies to people exposed to secondhand smoke is not clear. To avoid possible gastrointestinal distress, a UL of 2,000 mg/day has been set for supplemental vitamin C.

scurvy A condition caused by vitamin C deficiency whereby bleeding gums, bruising, poor wound healing, and skin irritations occur.

retinoid (or **preformed vitamin A**) A vitamin A compound.

LO5 What Are Fat-Soluble Vitamins?

Like their water-soluble counterparts, fat-soluble vitamins share some common characteristics but are chemically unique. Fat-soluble vitamins are absorbed mostly in the small intestine, a process that requires the presence of both dietary lipids and bile. After they are absorbed, fat-soluble vitamins are circulated away from the small intestine by chylomicrons in the lymph. Fat-soluble vitamins eventually circulate to the blood as components of either lipoproteins (such as VLDLs) or transport proteins.

Like many of the water-soluble vitamins, each fat-soluble vitamin has several forms, and some forms are more biologically active than others. Unlike water-soluble vitamins, however, your body can store most of the fat-soluble vitamins. Consequently, consuming high doses of some fat-soluble vitamins (especially in supplement form) can result in toxicities—sometimes with serious consequences.

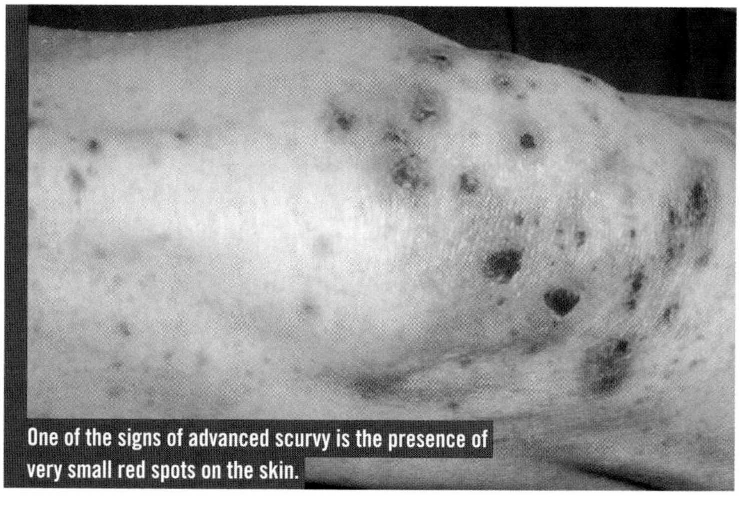

One of the signs of advanced scurvy is the presence of very small red spots on the skin.

LO6 What Are Vitamin A and the Carotenoids?

The term *vitamin A* refers to a series of three **retinoid** (or **preformed vitamin A**) forms: retinol, retinoic acid, and retinal. Although all three forms are important, retinol is the most biologically active and is synthesized in the body from retinal (see Figure 7.13 on the next page). Retinoic acid can also be synthesized

carotenoid A dietary compound with a similar structure to those of the retinoids. Some, but not all, can be converted to vitamin A.

provitamin A carotenoid A carotenoid that can be converted to vitamin A.

nonprovitamin A carotenoid A carotenoid that cannot be converted to vitamin A.

retinol activity equivalent (RAE) A unit of measure for the combined amounts of preformed vitamin A and provitamin A carotenoids in a food.

from retinal, but retinoic acid itself cannot be converted to any other retinoid.

Beyond the retinoids, the vitamin A family also includes several **carotenoid** compounds, which have structures similar to those of the retinoids. Some carotenoids can be converted to vitamin A; such a carotenoid is called a **provitamin A carotenoid.** One of the most common provitamin A carotenoids is *beta-carotene* (*β-carotene*), which the body converts to two retinal molecules. A carotenoid that cannot be converted to vitamin A is called a **nonprovitamin A carotenoid.** Lycopene, astaxanthin, zeaxanthin, and lutein are examples of nonprovitamin A carotenoids.

Good dietary sources of vitamin A or provitamin A carotenoids are listed in Figure 7.14. Because it exists in several forms (each of which has a unique biological identity), the amount of vitamin A in a given food can be difficult to determine. The **retinol activity equivalent (RAE)** is a unit of measure for the combined amounts of preformed vitamin A and provitamin A carotenoids in a food. In general, you consume preformed vitamin A when you eat animal-based foods (such as liver and other organ meats) and fatty fish. Whole-fat dairy products (such as whole milk), cheese, and butter are also good sources of vitamin A (especially when they are fortified). Reduced-fat dairy products are not good sources unless they are vitamin A-fortified.

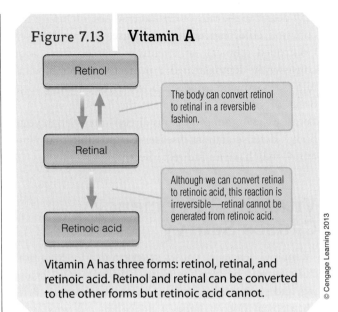

Figure 7.13 Vitamin A

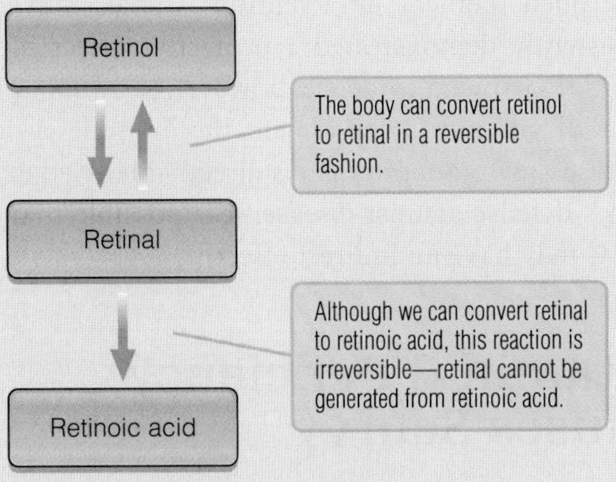

Vitamin A has three forms: retinol, retinal, and retinoic acid. Retinol and retinal can be converted to the other forms but retinoic acid cannot.

Figure 7.14 Good Sources of Vitamin A

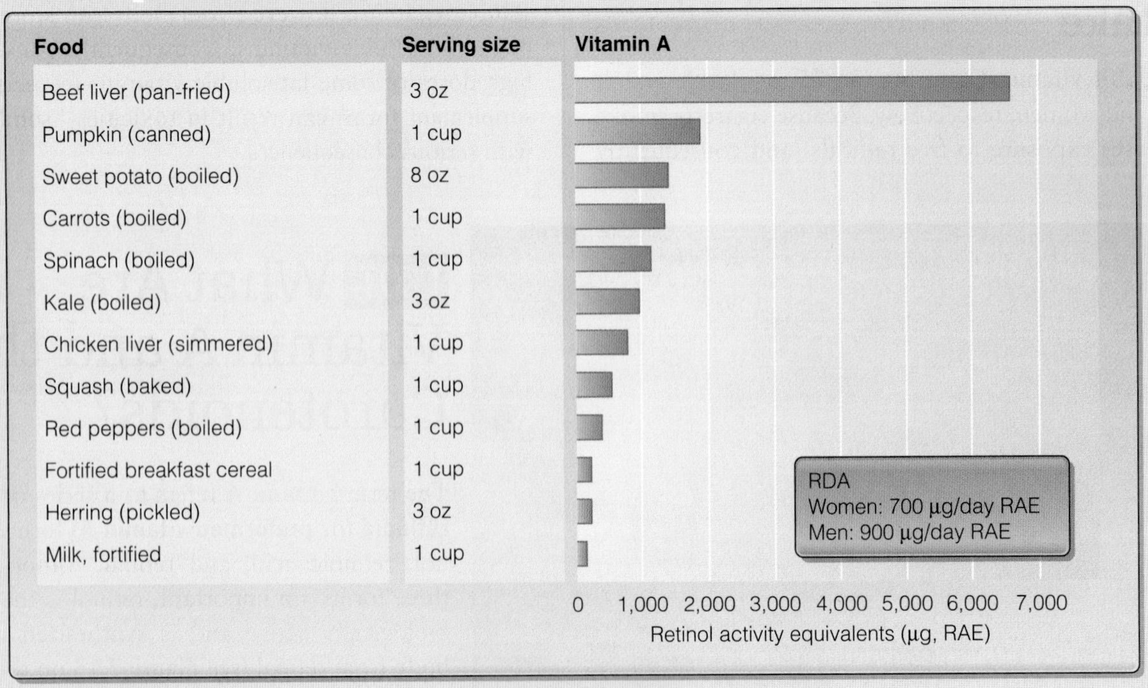

Source: USDA Nutrient Database for Standard Reference, Release 16-1.

Peppers are excellent sources of **provitamin A carotenoids.**

Whereas animal-based foods tend to contain preformed vitamin A, plant-based foods tend to contain provitamin A carotenoids. The yellow and red hues of many carotenoids make carotenoid-rich plant fibers and animal tissues brightly colored. As such, yellow, orange, and red fruits and vegetables (such as cantaloupe, carrots, and peppers) are particularly good sources of the carotenoids, as are brightly colored animal-based foods such as egg yolks, lobsters, crabs, and shrimp. Leafy greens are also good plant-based sources of carotenoids. Vitamin A compounds appear to be relatively stable in food. Processing and heating may actually increase the bioavailability of some carotenoids (such as β-carotene).

Vitamin A Is Critical to Vision, Growth, and Reproduction

Anecdotal evidence suggests that ancient Egyptian physicians prescribed vitamin A–rich liver to treat poor vision. Scientists did not begin to understand the mechanisms by which vitamin A–rich foods improve eyesight until the 20th century, however. As illustrated in Figure 7.15, when light enters your eyes, it passes

Figure 7.15 Vitamin A and Vision

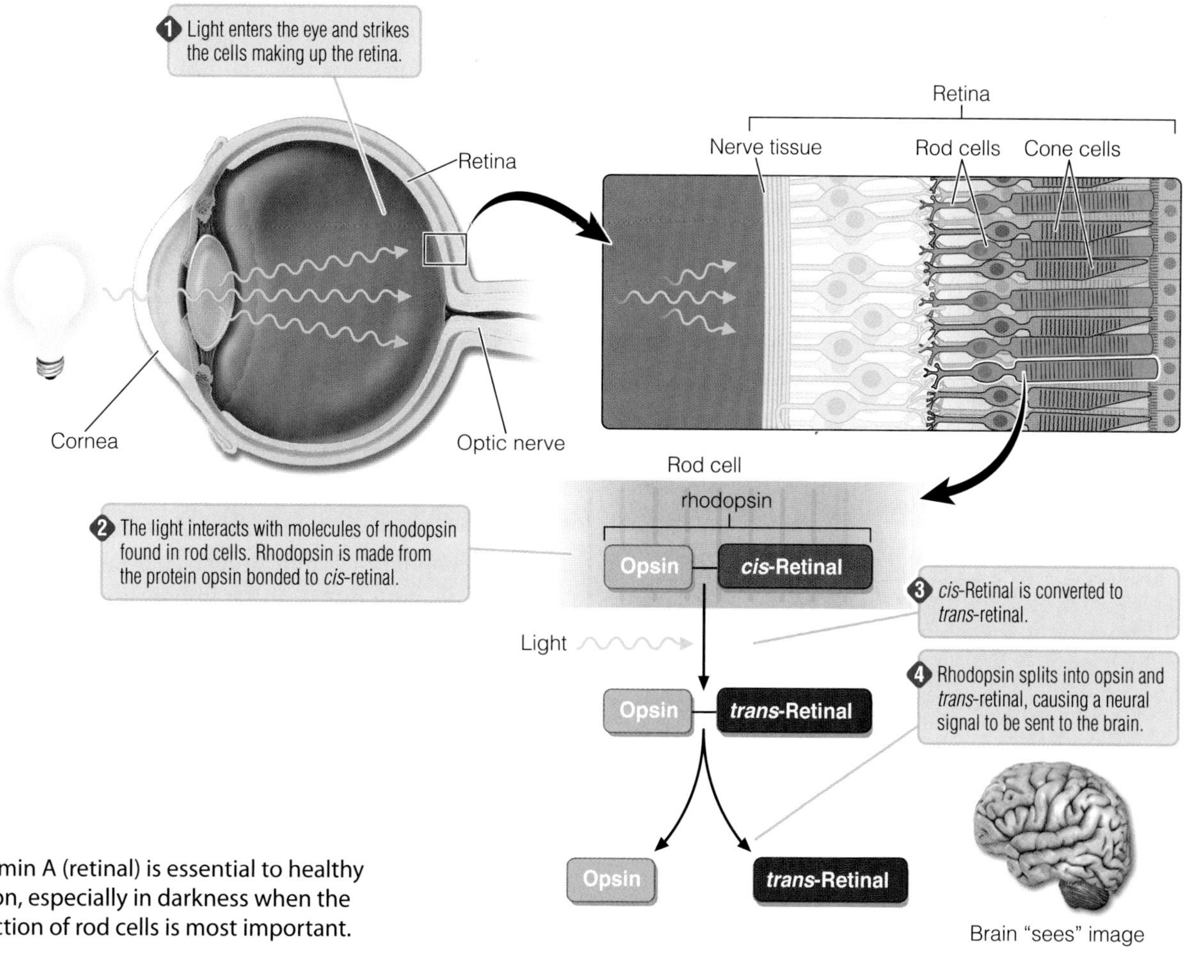
1 Light enters the eye and strikes the cells making up the retina.
Retina
Cornea
Optic nerve
Retina
Nerve tissue
Rod cells
Cone cells
Rod cell
rhodopsin
2 The light interacts with molecules of rhodopsin found in rod cells. Rhodopsin is made from the protein opsin bonded to *cis*-retinal.
Opsin
cis-Retinal
3 *cis*-Retinal is converted to *trans*-retinal.
Light
Opsin
trans-Retinal
4 Rhodopsin splits into opsin and *trans*-retinal, causing a neural signal to be sent to the brain.
Opsin
trans-Retinal
Brain "sees" image

Vitamin A (retinal) is essential to healthy vision, especially in darkness when the function of rod cells is most important.

night blindness A condition characterized by an impaired ability to see in low-light environments.

cell differentiation The process by which a nonspecialized, immature cell type becomes a specialized, mature cell type.

macular degeneration A chronic disease that causes deterioration of the retina.

vitamin A deficiency disorder (VADD) A spectrum of health-related consequences caused by vitamin A deficiency.

to an inner back lining called the *retina.* The retina consists of a layer of nerve tissue and millions of cells called *cones* and *rods.* Cones help you see color, whereas rods distinguish black from white (a critical aspect of night vision). Rods contain thousands of *rhodopsin* molecules, which are each comprised of bonded *cis*-retinal (a form of vitamin A) and *opsin* (a protein) molecules. When light strikes the rhodopsin, the *cis*-retinal is converted to *trans*-retinal and separates from the opsin. This reaction causes a neural signal to be sent to the brain. The light signal is then interpreted by the brain as a recognizable image. Because *cis*-retinal is a form of vitamin A, adequate consumption of vitamin A is needed for this cascade of events to occur. Vitamin A is also important to the health of the eye's outermost tissue layer, the *cornea*, and especially important to sight in low-light environments. This is why adequate vitamin A intake is needed to prevent **night blindness.**

Vitamin A has many important functions beyond vision. An example is its role in **cell differentiation**, the process by which a nonspecialized, immature cell type becomes a specialized, mature cell type.[14] Vitamin A is also critical to growth and reproduction, and it has a variety of immunological functions. These include the maintenance of protective barriers (such as skin and those that line organs) and the production of immune cells (such as those that produce antibodies). Preformed vitamin A is also essential to healthy bones.

> "Except for minor rearrangement of hydrogen atoms, ***cis*-retinal** molecules are **chemically similar** to ***trans*-retinal** molecules."

The Carotenoids Are Important Antioxidants

Adequate intake of the nonprovitamin A carotenoids is associated with reduced risks of heart disease, age-related eye disease, and cancer (in animal models).[15] Researchers believe that these effects may be attributable to the potent antioxidant functions of some carotenoids.[16] For instance, high circulating levels of lutein and zeaxanthin may decrease one's risk of **macular degeneration**—a chronic and often age-related disease that causes deterioration of the retina.[17] This progressive disease is the leading cause of visual impairment in older adults. It is not clear however whether increased nonprovitamin A carotenoid intake actually improves vision in older people.

Vitamin A Deficiency Causes Vitamin A Deficiency Disorder (VADD)

Primary vitamin A deficiency is uncommon in industrialized countries such as the United States. Nonetheless, secondary vitamin A deficiency sometimes occurs in people with diseases that impair lipid digestion and absorption, such as cystic fibrosis. Vitamin A deficiency is also prevalent in alcoholics with liver damage. Excessive alcohol consumption depletes the body's stores of vitamin A, but the mechanisms of this process are not well understood.

Vitamin A deficiency disorder (VADD) is a spectrum of health-related consequences caused by vitamin A deficiency. VADD is pervasive in developing countries, especially among children. In its mildest form, VADD causes night blindness. Severe forms of VADD can lead

> **Vitamin A deficiency** *can cause* **hyperkeratosis,** *a condition by which skin and nails become* **rough and scaly.**

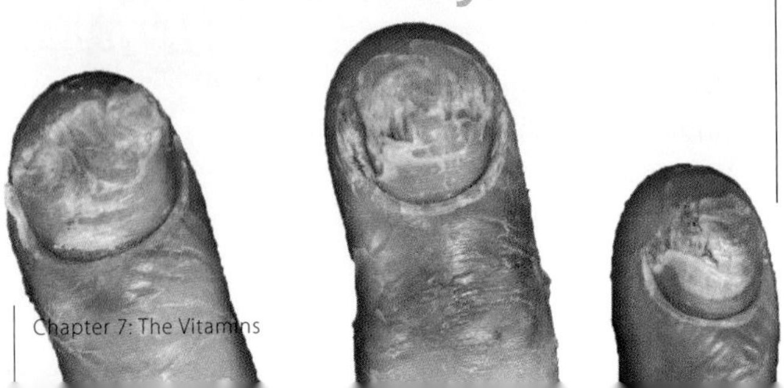

Vitamin A and International Child Health

Although vitamin A deficiency disorder (VADD) is not common in developed countries, it is still a serious health concern around the world—especially for children. Because vitamin A intake is essential to a healthy immune system, children with VADD are at an increased risk of infection. Protein-energy malnutrition, prevalent in vitamin A-deficient parts of the world, can make vitamin A deficiency even worse. In response to the prevalence of VADD in developing countries, researchers and public agencies have in recent years focused their efforts to combat this dangerous condition. For example, the World Health Organization urges healthcare professionals in the most at-risk countries to provide vitamin A supplements to all children six months of age and older. In other parts of the world, *biotechnology* (or *bioengineering*) is used to produce new kinds of rice that are naturally enriched with provitamin A carotenoids. With the aid of these and other promising health interventions, fewer children will suffer from vitamin A deficiency in years to come.

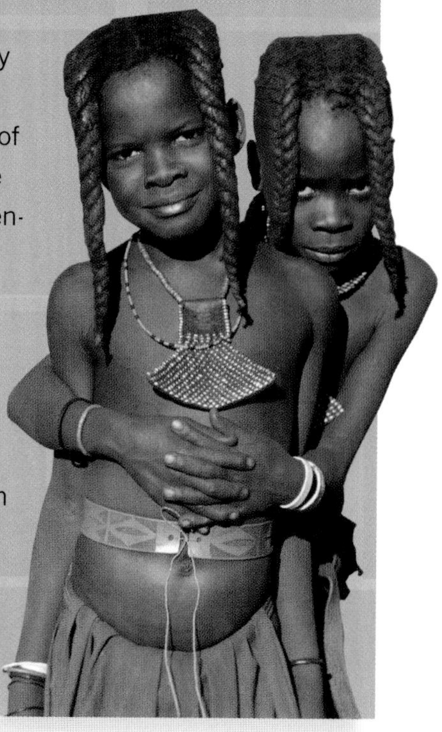

International efforts to decrease vitamin A deficiency have improved the health of children worldwide.

to **xerophthalmia**, a condition whereby the cornea and other portions of the eye are damaged, leading to dry eyes, scarring, and even blindness. This complex condition is often accompanied by *Bitot's spots*, white spots on the surface of the eye caused by accumulations of dead cells and secretions (see Figure 7.16 on the next page). Vitamin A deficiency also causes **hyperkeratosis**, a condition whereby skin and nail cells overproduce the protein keratin, causing them to become rough and scaly.

Vitamin A and Carotenoid Toxicities

Chronic consumption of large doses of preformed vitamin A (three to four times the RDA) can lead to vitamin A toxicity, which can cause **hypervitaminosis A**, a condition characterized by blurred vision, liver abnormalities, and reduced bone strength. Very high doses of naturally occurring and/or synthetic vitamin A can also lead to birth defects such as neurological damage and physical deformities. Isotretinoin (trade name Accutane®), a drug used to treat acne, is a form of vitamin A that can cause severe, life-threatening birth defects. Not only must pregnant women avoid this drug, but sexually active women of childbearing age must agree to use birth control before it can be prescribed. You can find out more about the dangers of isotretinoin at the National Center for Biotechnology Information's PubMed Health site (http://www.ncbi.nlm.nih.gov/pubmedhealth/PMH0000532/).

High-dose carotenoid supplementation has been found to increase the risk of lung cancer in some people.[18] Overconsumption of carotenoids can also cause them to accumulate in the skin, turning it a yellow-orange color—a benign condition called **hypercarotenodermia**.

xerophthalmia A condition caused by vitamin A deficiency whereby the cornea and other portions of the eye are damaged, leading to dry eyes, scarring, and even blindness.

hyperkeratosis A condition caused by vitamin A deficiency whereby skin and nail cells overproduce the protein keratin, causing them to become rough and scaly.

hypervitaminosis A A condition caused by vitamin A toxicity whereby blurred vision, liver abnormalities, and reduced bone strength occur.

hypercarotenodermia A condition whereby carotenoids accumulate in the skin, causing it to become yellow-orange.

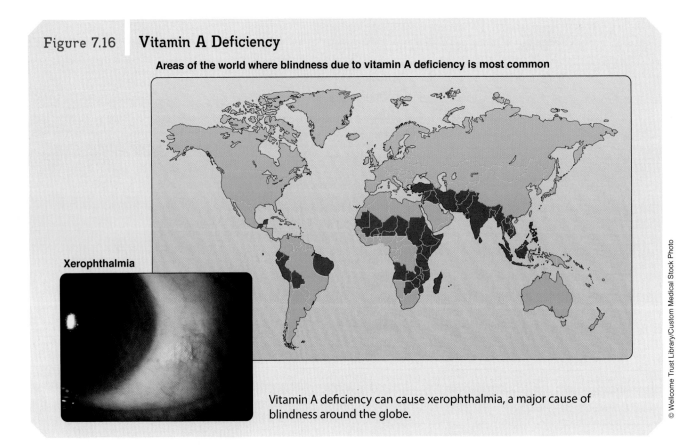

Figure 7.16 Vitamin A Deficiency

Areas of the world where blindness due to vitamin A deficiency is most common

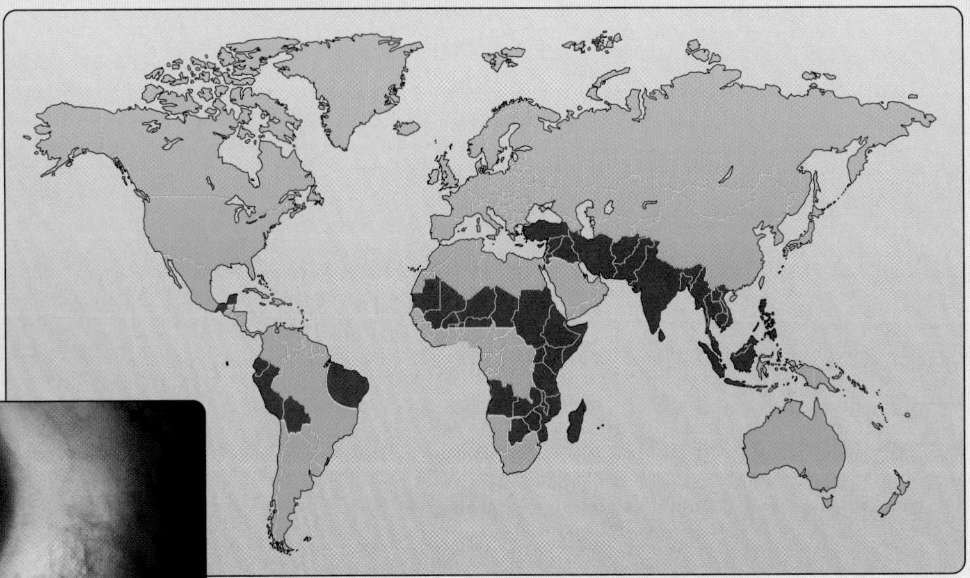

Xerophthalmia

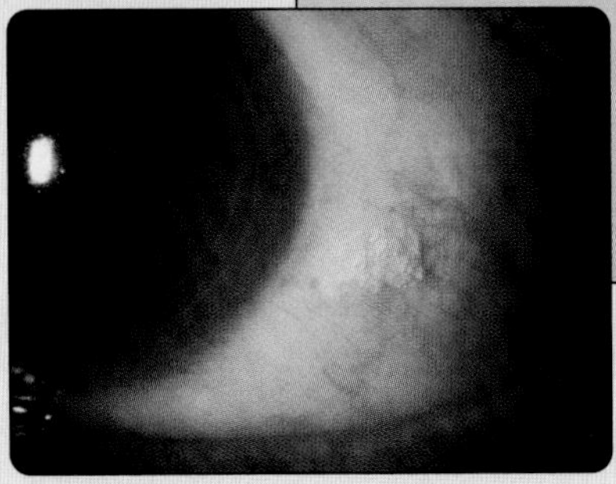

Vitamin A deficiency can cause xerophthalmia, a major cause of blindness around the globe.

Recommended Vitamin A and Carotenoid Intake

RDAs for vitamin A intake are 900 and 700 μg RAE/day for adult men and women, respectively. Vitamin A can be consumed either as preformed vitamin A or as provitamin A carotenoids. Because preformed vitamin A is found mostly in animal-based products, vitamin A supplementation may be necessary for vegans and vegetarians who do not consume sufficient amounts of provitamin A carotenoids or vitamin A–fortified foods. DRIs have not been set for the nonprovitamin A carotenoids because there is insufficient evidence to support their essentiality. To prevent the known toxic effects of excessive preformed vitamin A consumption, a UL of 3,000 μg RAE/day has been set for adults. The Institute of Medicine advises against dietary carotenoid supplementation for most people, although there are no UL values set for these compounds.

prohormone A compound converted to an active hormone in the body.

ergocalciferol (vitamin D_2) A form of vitamin D found in plant-based foods, fortified foods, and supplements.

cholecalciferol (vitamin D_3) A form of vitamin D found in animal-based foods, fortified foods, and supplements that is also synthesized in the body.

LO7 What Is Vitamin D?

Vitamin D is considered by many to be both a nutrient and a **prohormone**, a compound converted to an active hormone in the body. But is vitamin D essential, nonessential, or conditionally essential? Some health professionals consider vitamin D a nonessential nutrient because the body can synthesize it in sufficient amounts under normal conditions. However, because some people are not able to produce vitamin D in adequate amounts, other experts consider it to be conditionally essential.

There are two forms of vitamin D in foods: **ergocalciferol (vitamin D_2)** is found in plant-based foods, and **cholecalciferol (vitamin D_3)** is found in animal-based foods. Cholecalciferol is also synthesized in the body. Both forms of vitamin D are found in supplements and fortified foods. Good sources of vitamin D are listed in Figure 7.17. Egg yolks, butter, whole milk, fatty fish, and mushrooms are some of the few foods that contain naturally occurring vitamin D. Many dairy products and breakfast cereals are fortified

Figure 7.17 **Good Sources of Vitamin D**

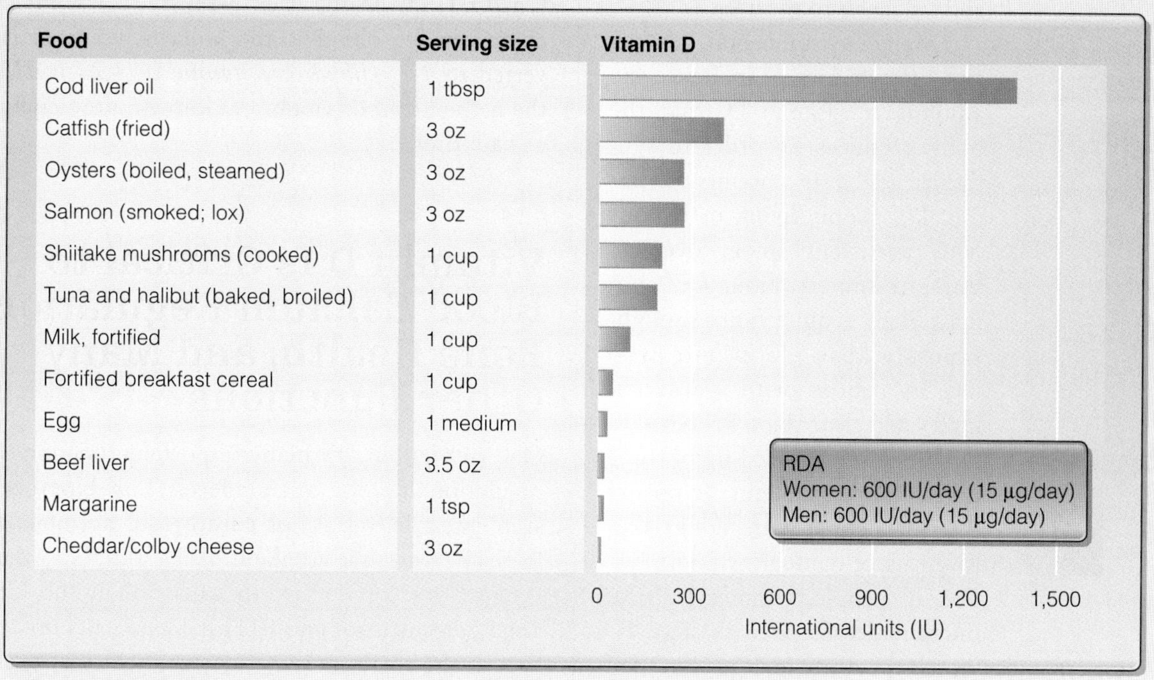
Food
Serving size
Vitamin D
Cod liver oil
1 tbsp
Catfish (fried)
3 oz
Oysters (boiled, steamed)
3 oz
Salmon (smoked; lox)
3 oz
Shiitake mushrooms (cooked)
1 cup
Tuna and halibut (baked, broiled)
1 cup
Milk, fortified
1 cup
Fortified breakfast cereal
1 cup
Egg
1 medium
Beef liver
3.5 oz
Margarine
1 tsp
Cheddar/colby cheese
3 oz
RDA
Women: 600 IU/day (15 µg/day)
Men: 600 IU/day (15 µg/day)
0
300
600
900
1,200
1,500
International units (IU)

Source: USDA Nutrient Database for Standard Reference, Release 16-1.

with vitamin D; most dietary vitamin D comes from these foods. Vitamin D is quite stable and is not easily destroyed by food preparation, processing, or storage.

Unlike other vitamins, vitamin D can be synthesized through exposure to sunlight; exposing your skin to sunlight for 10 to 15 minutes three times a week allows your body to make adequate amounts of vitamin D. This is why vitamin D is sometimes referred to as the "sunshine vitamin." This chemical conversion is not quite as simple as it sounds, however. The body's synthesis of vitamin D actually involves two steps, as illustrated in Figure 7.18 on the next page. First, a cholesterol derivative is converted by ultraviolet light to **previtamin D_3** (or **precalciferol**) in the skin. Tanning machines also emit the ultraviolet light that allows the first step to take place, but relying on tanning sessions to obtain enough vitamin D_3 is not recommended. This is because tanning can damage the skin and increase the risk of skin cancer.[19] After the cholesterol derivative is converted to previtamin D_3, this substance is converted in the skin

previtamin D_3 (precalciferol) An intermediate product made in the skin during the conversion of a cholesterol derivative to cholecalciferol.

Because the body can synthesize vitamin D when exposed to sufficient sunlight, this vitamin is sometimes called the *sunshine vitamin*.

vitamin D_3 (or **cholecalciferol**) The form of vitamin D that diffuses into the blood and circulates to the liver.

calcitrol The active form of vitamin D produced in the kidneys. Also known as 1,25-dihydroxyvitamin D (1,25-$[OH]_2$ D_3)

parathyroid hormone (PTH) A hormone released from the parathyroid gland that stimulates the conversion of 25-(OH) D_3 to calcitriol in the kidneys.

rickets A childhood condition caused by vitamin D deficiency whereby slow growth and bone deformation occur.

to **vitamin D_3** (or **cholecalciferol**), which then diffuses into the blood and circulates to the liver.

Many environmental, genetic, and lifestyle factors influence the amount of vitamin D_3 that the body produces. In order to synthesize adequate amounts of vitamin D_3, people living in regions with persistent smog, overcast skies, or limited amounts of sunlight likely require more sunlight exposure than do people who live in regions with warm climates and clear skies. People with darker skin may need up to three times more sunlight exposure than do people with lighter skin to produce enough vitamin D_3. For all people, vitamin D_3 production decreases with age. Finally, sunscreen can block the ultraviolet rays needed for vitamin D_3 formation.

Whether consumed through the diet or produced in the skin, vitamin D_3 must be further metabolized before it can be used by the body. This two-step process, illustrated in Figure 7.19, occurs in the liver and kidneys. First, vitamin D_3 is converted to *25-hydroxyvitamin D* (*25-[OH] D_3*) in the liver. Next, the 25-(OH) D_3 circulates in the blood to the kidneys, where it is converted to *1,25-dihydroxyvitamin D* (*1,25-$[OH]_2$ D_3*), the active form of vitamin D known more commonly as **calcitriol**.

Figure 7.18 **Synthesis of Cholecalciferol (Vitamin D_3) in the Skin**

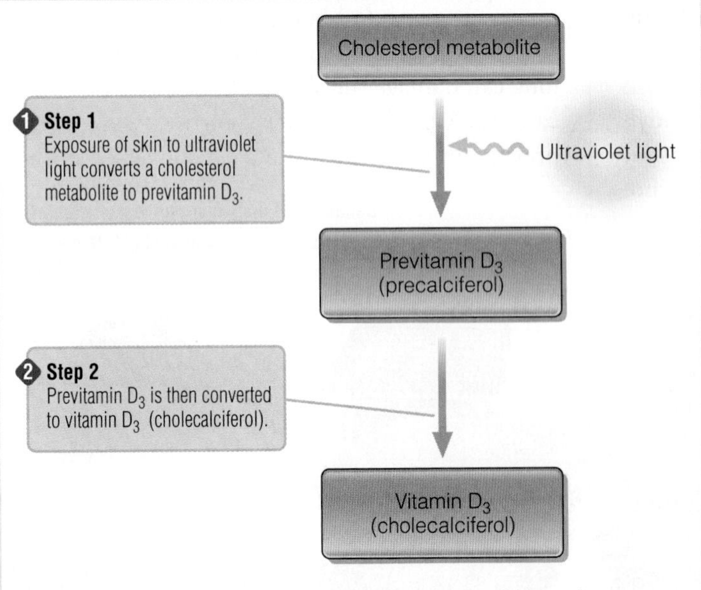

Cholesterol metabolite
Step 1
Exposure of skin to ultraviolet light converts a cholesterol metabolite to previtamin D3.
Ultraviolet light
Previtamin D3 (precalciferol)
Step 2
Previtamin D3 is then converted to vitamin D3 (cholecalciferol).
Vitamin D3 (cholecalciferol)

The body can make vitamin D_3 from a cholesterol-like substance when the skin is exposed to ultraviolet light.

Vitamin D Is Critical to Blood Calcium Regulation, Bone Health, and Many Other Functions

One of vitamin D's many important functions is the regulation of blood calcium levels. Low blood calcium stimulates the release of **parathyroid hormone (PTH)** from the parathyroid gland. PTH stimulates the conversion of 25-(OH) D_3 to calcitriol in the kidneys. Together, calcitriol and PTH increase calcium absorption in the small intestine, decrease calcium excretion in the urine, and facilitate the release of calcium from bones (see Figure 7.20 on page 176).

In addition to the regulation of blood calcium levels and bone health, vitamin D plays important roles in blood pressure regulation, muscle contraction, and nerve function. Vitamin D is also needed for cell differentiation and maturation. Like vitamin A, vitamin D activates selected genes, which in turn prompt cells to synthesize the specific proteins that allow the cells to become both specialized and functional. Although the connection warrants further investigation, some studies have suggested that vitamin D's role in cell regulation may help prevent certain types of cancers, such as those of the colon, breast, skin, and prostate.[20]

Vitamin D Deficiency Causes Rickets, Osteomalacia, and Osteoporosis

In infants and children, vitamin D deficiency can result in improper bone mineralization—a disease called **rickets**.

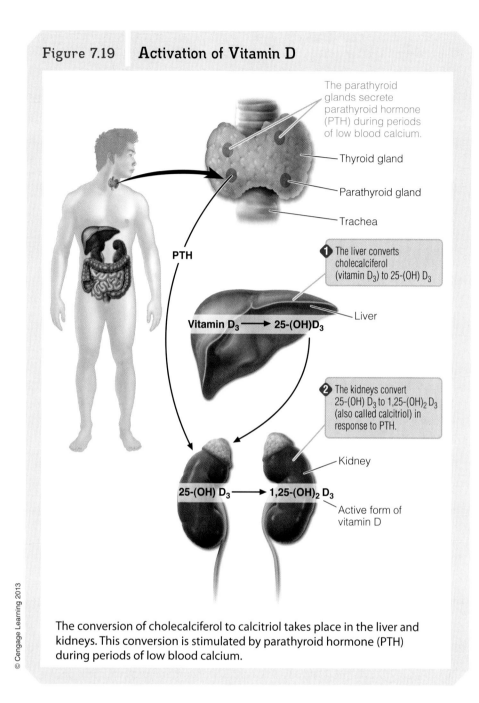

Figure 7.19 Activation of Vitamin D

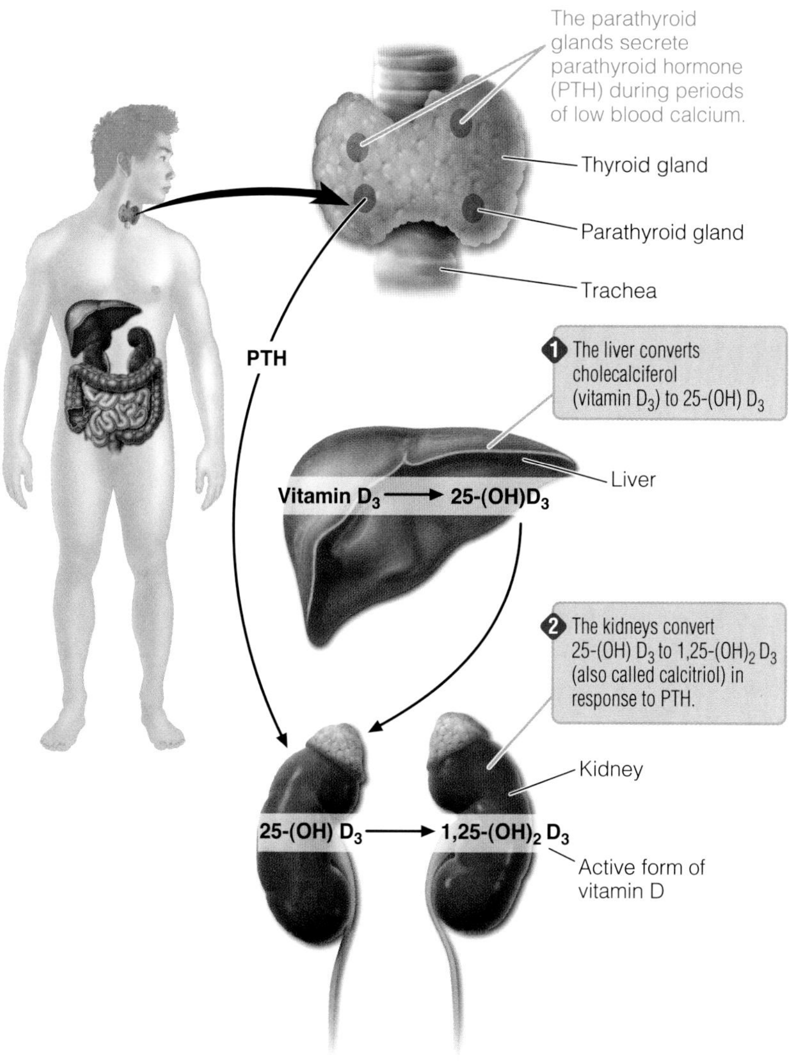

The conversion of cholecalciferol to calcitriol takes place in the liver and kidneys. This conversion is stimulated by parathyroid hormone (PTH) during periods of low blood calcium.

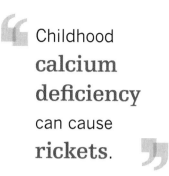

> Childhood **calcium deficiency** can cause **rickets.**

Rickets is a significant public health concern in some parts of the world.[21] Children with rickets experience slow growth and have characteristically bowed legs or knocked knees that develop because their long leg bones cannot support the stress of weight-bearing activities such as walking. Rickets can also cause the breastbone to protrude outward and the rib cage to become narrow, often resulting in cardiac and respiratory problems.

Rickets was common in the United States during the early 1900s. Vitamin D–fortified milk and infant formulas nearly eradicated rickets in this country—until recently, that is. In 2000, a medical research group published a disturbing report documenting a rise in the incidence of rickets, especially among dark-skinned, breastfed babies.[22] Further research suggested that some infants nourished entirely by human milk may be at increased risk of inadequate vitamin D intake. This phenomenon may be attributable to their mothers' insufficient exposure to sunlight, which results in diminished amounts of vitamin D in their milk. Infants' own insufficient exposure to sunlight may also contribute to the recent rise in rickets, especially among those with darker skin. To counter the potential development of rickets, the American Academy of Pediatrics recommends that all breastfed infants be supplemented with vitamin D until they can consume adequate amounts of vitamin D–rich foods (such as vitamin D–fortified cow milk).[23]

When vitamin D intake is inadequate in adults, bones can become soft and weak—a condition called **osteomalacia**. Symptoms of osteomalacia include diffuse bone pain and muscle weakness. People with osteomalacia are also at an increased risk of bone fractures. Inadequate vitamin D intake can also result in the demineralization

osteomalacia An adult condition caused by vitamin D deficiency whereby bones become soft and weak.

of previously healthy bone, ultimately leading to **osteoporosis.** Osteoporosis is a serious chronic disease—especially in the elderly—that affects more than 28 million Americans. To help prevent both osteomalacia and osteoporosis, people over 50 years of age are advised to get at least 15 minutes of sun exposure each day, consume adequate amounts of vitamin D–rich foods, and in some cases, take vitamin D supplements.

Recommended Vitamin D Intake

Until very recently, there were insufficient data to establish RDAs for vitamin D. However, in 2011 the Institute of Medicine established an RDA of 600 IU/day for adults.[24] This is approximately the amount of vitamin D in one liter (4.7 cups) of vitamin D–fortified milk. To help diminish bone loss in the elderly, the RDA increases to 800 IU/day at age 71.

Vitamin D toxicity is uncommon unless the diet is supplemented with large amounts of vitamin D. Excess vitamin D can cause calcium levels in the blood and urine to increase, which in turn can cause calcium to be deposited in soft tissues such as the heart and lungs. **Hypercalciuria,** a condition characterized by elevated urinary calcium levels, is also associated with kidney stone formation and renal (kidney) failure. Vitamin D toxicity promotes bone loss, and when severe enough, can be fatal. Because of these health concerns, the Institute of Medicine has established a UL of 4,000 IU/day for vitamin D in supplemental form.

Figure 7.20 **Regulation of Calcium**

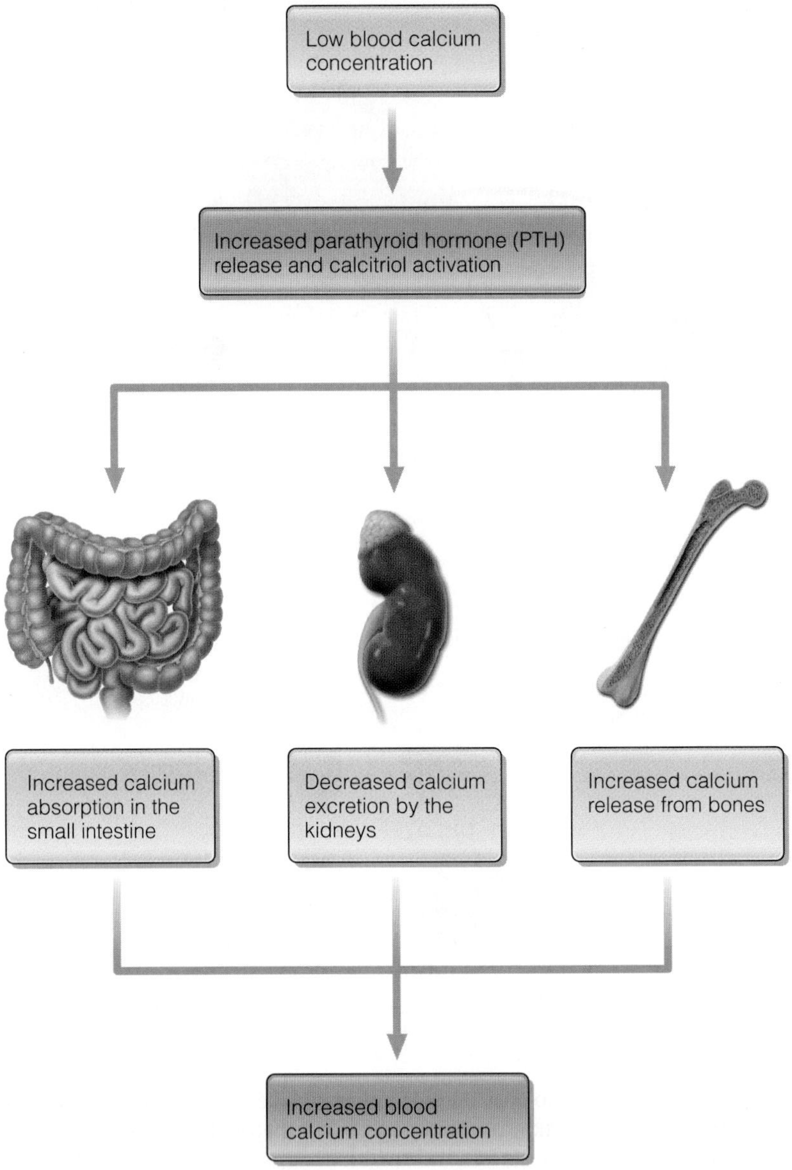

Parathyroid hormone (PTH) and calcitriol work together to increase blood calcium when it is low.

osteoporosis A serious disease caused by vitamin D deficiency whereby bones become weak and porous.

hypercalciuria A condition characterized by elevated urine calcium levels.

LO8 What Are Vitamins E and K?

Beyond vitamins A and D, adequate intakes of vitamins E and K are needed to meet your fat-soluble vitamin requirements. The functions and importance of these essential micronutrients are discussed next.

Vitamin E

Nuts and seeds *are good sources of* **vitamin E.**

© iStockphoto.com/Stefanie Timmermann

The term *vitamin E* refers collectively to eight different compounds that have similar chemical structures. Of these, **α-tocopherol** is the most biologically active. Vitamin E is found in both plant- and animal-based foods, but it is especially abundant in vegetable oils, nuts, and seeds. Some dark green vegetables such as broccoli and spinach contain vitamin E as well. Good dietary sources of vitamin E are listed in Figure 7.21. Vitamin E is easily destroyed during food preparation, processing, and storage. As such, expose vitamin E–rich foods to as little heat as possible during cooking to help preserve their vitamin E content. Proper storage in airtight containers is also important.

VITAMIN E IS A POTENT ANTIOXIDANT

As described in Chapter 6, biological membranes (such as cell membranes) are composed of a bilayer of phospholipid molecules. Naturally, maintenance of these membranes is vital to the stability and function of cells and their contents. Vitamin E, an antioxidant, plays a major role in this maintenance by protecting the fatty acids embedded in biological membranes from free radical-induced oxidative damage. This protection is especially important for cells that are exposed to high levels of oxygen, such as those in the lungs and red blood cells. Vitamin E's ability to act as an antioxidant is enhanced by the presence of other antioxidant micronutrients, such as vitamin C and selenium.

Because antioxidants protect DNA from cancer-causing free radical damage, some researchers believe that vitamin E might prevent or cure cancer. Indeed, diets high in vitamin E are associated with decreased cancer risk, but there is little experimental evidence that vitamin E itself is the cause of this.[25] In other words,

α-tocopherol The most biologically active form of vitamin E.

Figure 7.21 Good Sources of Vitamin E

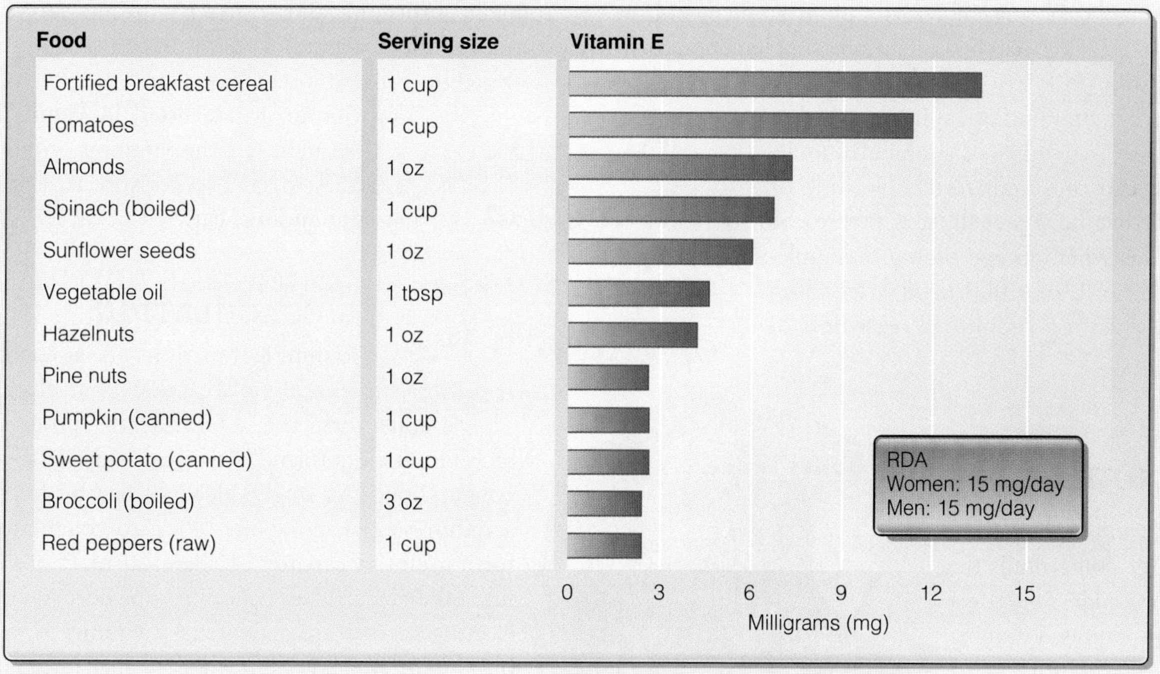

Food
Serving size
Vitamin E
Fortified breakfast cereal
1 cup
Tomatoes
1 cup
Almonds
1 oz
Spinach (boiled)
1 cup
Sunflower seeds
1 oz
Vegetable oil
1 tbsp
Hazelnuts
1 oz
Pine nuts
1 oz
Pumpkin (canned)
1 cup
Sweet potato (canned)
1 cup
Broccoli (boiled)
3 oz
Red peppers (raw)
1 cup
RDA
Women: 15 mg/day
Men: 15 mg/day
0
3
6
9
12
15
Milligrams (mg)

Source: USDA Nutrient Database for Standard Reference, Release 16-1.

hemolytic anemia A condition caused by vitamin E deficiency whereby red blood cell membranes weaken and rupture, reducing the blood's ability to transport oxygen.

phylloquinone A form of vitamin K found naturally in plant-based foods.

menaquinone A form of vitamin K produced by bacteria in the large intestine.

menadione A form of vitamin K produced commercially.

coagulation The process by which blood clots are formed.

eating foods rich in vitamin E is likely better for you than taking supplementary vitamin E.

VITAMIN E DEFICIENCY CAUSES HEMOLYTIC ANEMIA

Vitamin E deficiency is uncommon; cases have only been observed in infants fed formulas lacking sufficient amounts of vitamin E, people with genetic abnormalities, and people with diseases that cause fat malabsorption. Nonetheless, vitamin E deficiency is characterized by a variety of symptoms, such as neuromuscular problems, loss of coordination, and muscular pain. A condition called **hemolytic anemia** is also associated with vitamin E deficiency. Hemolytic anemia is characterized by the rupture of red blood cell membranes, reducing the blood's ability to transport oxygen. Hemolytic anemia often results in weakness and fatigue.

RECOMMENDED VITAMIN E INTAKE

The RDA for vitamin E is 15 mg/day—an amount easily obtained from a balanced diet. Unlike the other fat-soluble vitamins, vitamin E rarely causes toxicity, even when large amounts of vitamin E supplements are consumed. This may be because the supplemental form of vitamin E is less biologically active than naturally occurring vitamin E. In some people, however, very high doses of vitamin E supplements can cause dangerous bleeding or hemorrhaging. As such, it is best to be cautious when taking vitamin E supplements. A UL of 1,000 mg/day has been established for vitamin E, regardless of its source.

> “**Vitamin K** was discovered by Danish researcher Dr. Henrik Dam. Dam named the vitamin in honor of its **importance in coagulation** (*koagulation* in Danish). Dam received a **Nobel Prize in Physiology or Medicine** in 1943 for the discovery.”

Vitamin K

The term *vitamin K* refers to three compounds that have similar structures and functions. Vitamin K is found naturally in plant-based foods in a form called **phylloquinone**. This form is also found in some vitamin K supplements and is often given to infants at birth. Another form of vitamin K called **menaquinone** is produced by bacteria in the large intestine. Because this bacterial production does not produce sufficient amounts of vitamin K to sustain health, vitamin K is considered an essential nutrient. A third form of vitamin K called **menadione** is neither found naturally in food nor synthesized by intestinal bacteria, but is produced commercially.

In general, dark green vegetables such as kale, spinach, broccoli, and Brussels sprouts are good sources of dietary vitamin K. You can also obtain this vitamin from fish and legumes. Good dietary sources of vitamin K are listed in Figure 7.22. Like many of the vitamins, vitamin K is destroyed by exposure to excessive light and/or heat.

Broccoli *is a good dietary source of* **vitamin K.**

© iStockphoto.com/Liz Van Steenburgh

VITAMIN K IS CRITICAL TO COAGULATION

Vitamin K functions as a coenzyme in a variety of reactions that ultimately constitute the life-or-death process by which blood clots form. Without this process, called **coagulation**, you might bleed to death after even a minor scrape. For your blood to coagulate and form a clot, a cascade of chemical reactions must first take place. After a cut or scrape occurs, various clotting factors are activated by vitamin K, allowing the next reactions in the cascade to take place.

Figure 7.22 **Good Sources of Vitamin K**

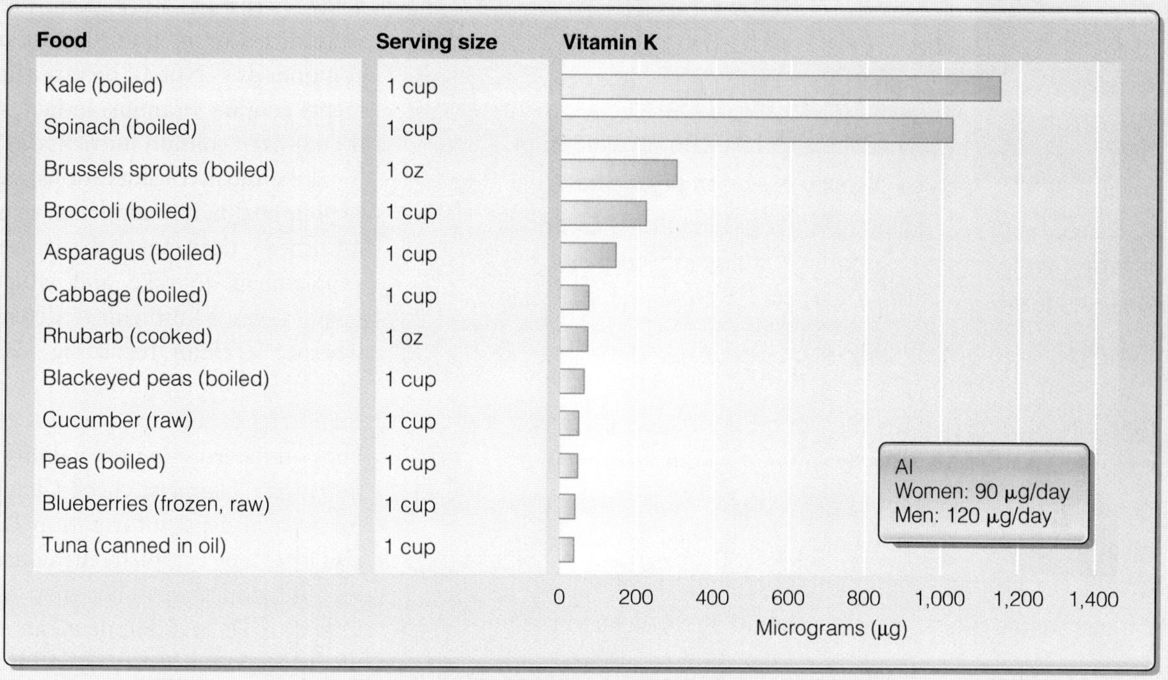
Food
Serving size
Vitamin K
Kale (boiled)
1 cup
Spinach (boiled)
1 cup
Brussels sprouts (boiled)
1 oz
Broccoli (boiled)
1 cup
Asparagus (boiled)
1 cup
Cabbage (boiled)
1 cup
Rhubarb (cooked)
1 oz
Blackeyed peas (boiled)
1 cup
Cucumber (raw)
1 cup
Peas (boiled)
1 cup
Blueberries (frozen, raw)
1 cup
Tuna (canned in oil)
1 cup
AI
Women: 90 μg/day
Men: 120 μg/day
0
200
400
600
800
1,000
1,200
1,400
Micrograms (μg)

Source: USDA Nutrient Database for Standard Reference, Release 16-1.

These reactions ultimately result in the production of *fibrin*, a protein that forms a web-like clot that stops the bleeding. Beyond its coagulation functions, vitamin K is also essential to the synthesis of proteins that aid in bone and tooth formation.

VITAMIN K DEFICIENCY CAUSES SEVERE BLEEDING

Although rare in healthy adults, vitamin K deficiency occurs in some infants. Because a newborn's large intestine completely lacks vitamin K–producing bacteria at birth and human milk often contains very low levels of this essential nutrient, babies receive only minimal amounts of vitamin K during their first few weeks of life. Although this does not present a problem to most infants, some develop severe vitamin K deficiency. This leads to a condition called **vitamin K deficiency bleeding**, originally called *hemorrhagic disease of the newborn*, which is characterized by uncontrollable internal bleeding. To counter the possibility of this condition, the American Academy of Pediatrics recommends that all newborns be given vitamin K injections.[26] Vitamin K deficiency also sometimes occurs in children and adults with diseases that impair lipid absorption, which again causes bleeding. Finally, prolonged use of antibiotics can kill bacteria residing in the large intestine, resulting in vitamin K deficiency in all age groups.

vitamin K deficiency bleeding A disease caused by vitamin K deficiency whereby uncontrollable internal bleeding occurs.

Vitamin K is essential to coagulation, which **helps heal your body and protect** it from infections.

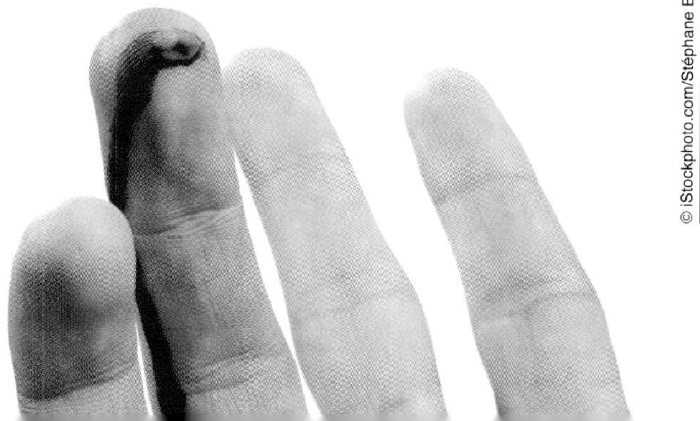

© iStockphoto.com/Stéphane Bidouze

RECOMMENDED VITAMIN K INTAKE

Although RDAs have not been established for vitamin K, AIs of 120 and 90 μg/day have been set for men and women, respectively. Because even very high amounts of vitamin K are rarely toxic, a UL has not been established for this vitamin.

Vitamin K and Blood-Thinning Drugs

Individuals with cardiovascular disease are often prescribed blood-thinning medication such as warfarin (trade name Coumadin®). Many of these drugs work by interfering with the actions of vitamin K. Because they decrease the formation of blood clots, these drugs can cause serious health problems if they are taken haphazardly. Because people taking blood-thinning drugs are at increased risk of excessive bleeding (even after sustaining a minor injury), they should take care to avoid engaging in hazardous activities. Also, because a sudden increase or decrease in vitamin K could interfere with the effectiveness of these medications, patients should strive to keep their vitamin K intakes constant. It is important to understand drug-nutrient interactions within your body, especially if you are taking potentially life-saving medications.

LO9 Should You Take Dietary Supplements?

Now that you know about the essential vitamins, you should be able to choose foods that contain adequate amounts of each of these important dietary components. Still, you may be wondering about vitamin supplements and the conditions under which they might be taken. Understanding some basic concepts about supplementation can help you make wise choices about what to take—and what not to take.

dietary supplement
Product intended to supplement the diet that contains vitamins, minerals, amino acids, herbs or other plant-derived substances, or a multitude of other compounds.

Dietary Supplements Can Contain Many Substances

The U.S. Food and Drug Administration (FDA) defines a **dietary supplement** as a product intended to supplement the diet that contains vitamins, minerals, amino acids, herbs or other plant-derived substances, or a multitude of other compounds.[27] Not all dietary supplements contain vitamins. In fact, some do not even contain nutrients.

Like most considerations related to your health, it is important to make informed decisions about which supplements to take and which to avoid. It can be difficult to determine whether a claim regarding vitamin supplementation and health is accurate. Refer to the next page for guidelines on determining the validity of a nutritional claim, or check Chapter 1 for more in-depth information. Before you take any supplement, consult a reliable scientific source such as the Office of Dietary Supplements (part of the National Institutes of Health). The Office of Dietary Supplements maintains an excellent user-friendly website (http://ods.od.nih.gov) that provides information about most dietary supplements and a wealth of news, studies, and recommendations. For example, before buying and/or using any supplement, contemplate the following tips developed by the Office of Dietary Supplements.

- **Consider safety first.** Some supplement ingredients can be toxic—especially in high doses. Do not hesitate to check with a health professional before taking any dietary supplement.
- **Think twice about chasing the latest headline.** Sound health advice is generally based on research over time, not a single study touted by the media. Be wary of results claiming a quick fix. If something sounds too good to be true, it probably is.
- **More may not be better.** Some products can be harmful when consumed in high amounts, and/or for a long time. Do not assume that more is better—it might be toxic.
- ***Natural* does not always mean *safe*.** The term *natural* simply means that something is not synthetic or human-made. Do not assume that this term ensures wholesomeness or safety.

When to Consider Taking a Supplement

There are no "hard and fast" rules about when to take a dietary supplement. However, if you have difficulty consuming a variety and balance of healthy foods in adequate amounts, a dietary supplement may help you achieve the essential nutrients' recommended intakes. The following situations often call for dietary supplementation:

- When your food availability or variety is limited by time or cooking constraints (as sometimes happens in college)
- When you decide not to consume certain foods (such as red meat or dairy products)
- During periods of rapid growth and development (such as infancy, childhood, adolescence, and pregnancy)
- When the consequences of normal aging (such as the loss of calcium from bone) make it difficult to consume adequate amounts of a nutrient from food
- When you consume a low-calorie diet for weight loss
- If you suffer from a health condition that increases your nutrient requirements or decreases the bioavailability of nutrients you eat

After deciding to take a dietary supplement, you must then determine which type of supplement to take. This can sometimes be a long and difficult process. Again, one of the best resources in this regard is the Office of Dietary Supplements' website, which contains dozens of *Supplement Fact* sheets. Carefully reading these sheets and consulting with your health care provider should provide valuable information to help you determine which type of supplement best fits your needs.

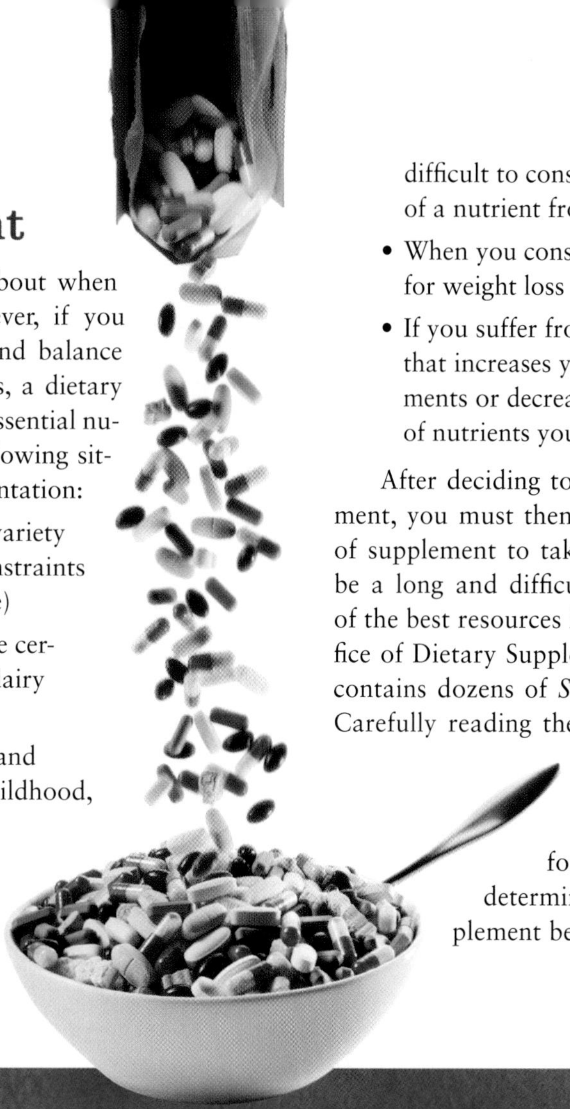

© pulp/Photodisc/Getty Images

DETERMINING THE VALIDITY OF A NUTRITIONAL CLAIM:

1. **WHERE WAS THE STUDY PUBLISHED?**
   As a rule, you should question nutrition claims that have not first been published in a peer-reviewed journal or other highly regarded publication.
2. **WHO CONDUCTED THE STUDY?**
   It is important that the individuals conducting the research are qualified and knowledgeable.
3. **WHO PAID FOR THE RESEARCH?**
   Researchers must take all necessary steps to ensure that their funding sources do not bias or influence the outcomes of their studies.
4. **DID THE RESEARCHERS USE THE RIGHT DESIGN?**
   Was the research conducted in a way that was appropriate to test the hypothesis? And do the conclusions fit the study design?
5. **DO PUBLIC HEALTH ORGANIZATIONS CONCUR?**
   Before believing a nutritional claim to be true, determine whether it is supported by major public health organizations.

© iStockphoto.com/TheStockLab

NUTR

8 Water and the Minerals

© iStockphoto.com/Håkan Karlsson

LEARNING OUTCOMES:

LO1 Appreciate the importance of water.

LO2 Differentiate and understand the major minerals.

LO3 Differentiate and understand the trace minerals.

LO1 Why Is Water Essential to Life?

"Water, water, everywhere, Nor any drop to drink." This famous passage from Samuel Taylor Coleridge's *The Rime of the Ancient Mariner* relates the lamentation of an experienced seafarer tormented by thirst during a long ocean voyage. Although he is surrounded by water, not a drop is fit for the mariner to consume. Like the ancient mariner, your very existence depends on a plentiful supply of drinkable water. But have you ever wondered why the human body needs water to stay alive? Because water is the most abundant molecule in the human body, a constant supply of it is essential to replenish the liquids lost through normal activity and physiological processes. Humans can survive for weeks—even months—without food, but this is not the case for water. Water is so vital to the human body that a person cannot survive without it for more than one or two weeks. Truly, water is the essence of life.

For its seemingly simple elegance, water is rarely a pure substance. In fact, it has much greater chemical complexity than you might think. With the exception of distilled water, the water you drink contains not only hydrogen and oxygen atoms, but also a variety of dissolved inorganic minerals that are also vital to your health. Although drinking water provides some of the life-sustaining minerals in your diet, the vast majority of these important nutrients are consumed in food. In this chapter, you will learn about the role of water in the body and how some minerals play a role in regulating water balance. You will also learn about the dietary sources, regulation, functions, deficiencies, toxicities, and requirements of the essential minerals in your body.

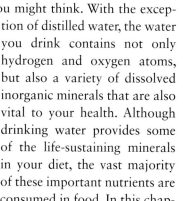

The water molecule is comprised of two hydrogen atoms and one oxygen atom.

Water is an indispensable macronutrient that is involved in myriad physiologic functions. Scientists have long marveled at water's unique physical and chemical properties, such as its ability to dissolve a large variety of chemical compounds. Because it can dissolve so many substances, water is sometimes referred to as the *universal solvent*. Not only does water act as a biological solvent, it facilitates chemical reactions and helps regulate body temperature.

Distribution of Water in the Body

In adults, approximately 50 to 70 percent of total body weight is comprised of water—about 42 liters (11 gallons) total. Water's percentage of body weight is even higher in newborn infants: water may account for as much as 75 percent of an infant's total body weight. Because lean tissue (muscle) retains more water than does fat, men generally have greater water volumes than women. For the same reason, physically fit individuals (with more muscle mass) tend to have greater water volumes than their sedentary counterparts.

As the main component of biological fluid, water is distributed both inside and outside of cells. As illustrated in Figure 8.1 on the next page, fluid located inside of a cell is referred to as **intracellular fluid**, whereas **extracellular fluid** is located outside of a cell. Extracellular fluid that fills spaces between or surrounding cells is referred to as **intercellular fluid**, whereas the extracellular fluid component of your blood and lymph is referred

intracellular fluid Fluid located inside of a cell.

extracellular fluid Fluid located outside of a cell.

intercellular fluid Extracellular fluid that fills spaces between or surrounding cells.

to as **intravascular fluid.** You may be surprised to learn that the majority of bodily fluid (23 liters, or roughly 6 gallons) is located inside cells.

Figure 8.1 Fluid Compartments

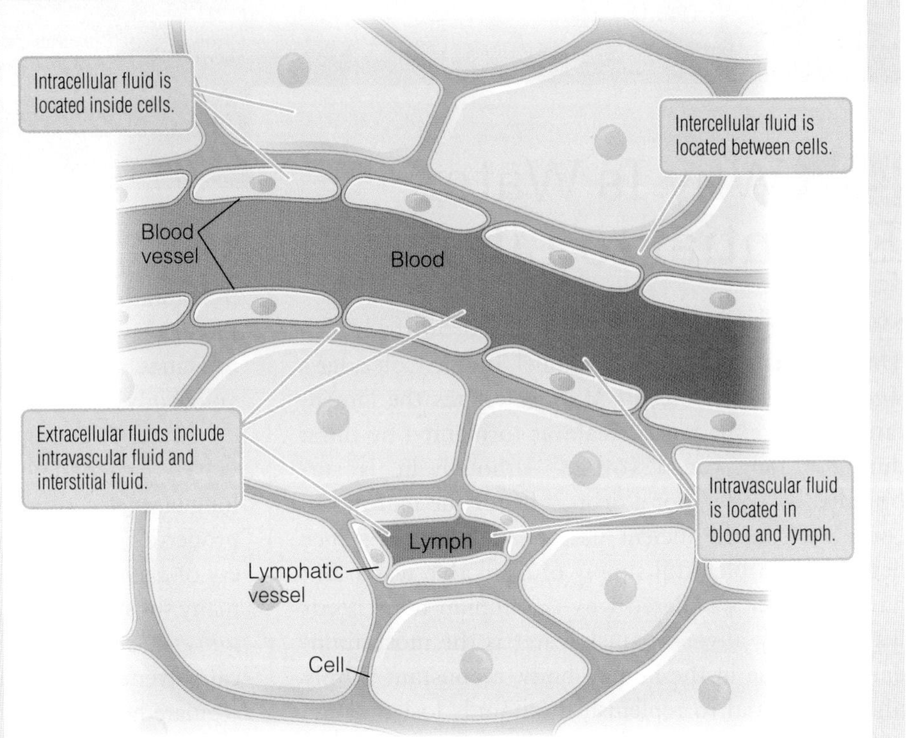

Water is distributed both inside and outside of cells. Extracellular fluids include fluids surrounding cells and fluid in blood and lymph. Intracellular fluid is located inside cells.

WATER BALANCE

Cells rupture if they contain too much water and collapse if they contain too little water. To ensure stable fluid balance, the movement of water across cell membranes is carefully regulated. Recall from Chapter 3 that osmosis is the process by which water molecules move from a region of low solute concentration through a cell membrane to a region of high solute concentration. A **solute** is a substance (such as a protein or electrolyte) that is dissolved in a fluid. While water molecules can move freely across most cell membranes, solutes cannot. A high concentration of solutes generates a force that attracts water, diluting the higher concentration of solutes. Thus, the driving force behind osmosis is the difference in intracellular and extracellular fluids' solute concentrations. The movement of water ceases when the concentrations of solutes are the same on both sides of the cell membrane (see Figure 8.2). As an example of osmosis in action, consider what happens when vegetables are pickled. When a cucumber is placed in a salty brine solution, water moves out of the cucumber in the direction of the greater solute (salt) concentration. As water leaves the cucumber, it shrivels and pickles. When the right spices are added to the brine, the pickle takes on a unique and delicious flavor.

The pickling of fruits and vegetables depends on osmosis. When a cucumber is placed in a salty brine solution, water is forced out of the cucumber, leaving behind a shriveled cucumber, more commonly referred to as a pickle.

Water's Functions Are Critical to Life

Although water does not provide the body with energy, it does play an active role in hundreds of chemical reactions involved in energy metabolism. Beyond its energy-related functions, this

intravascular fluid Extracellular fluid located in blood and lymph.

solute A substance dissolved in a fluid.

Figure 8.2 **Osmosis**

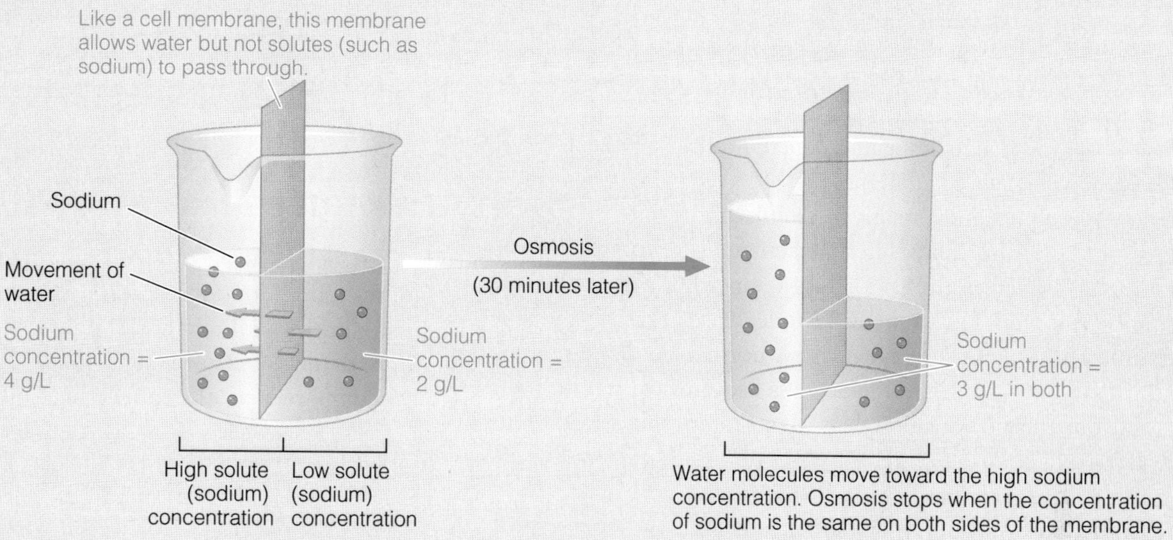

Osmosis is the movement of water molecules across a selective membrane. The direction in which water molecules move is determined by the concentration of dissolved substances (solutes).

versatile molecule provides protection and serves as an important solvent and lubricant. Finally, water helps maintain your internal body temperature, even when the surrounding environment is very cold or very hot.

HYDROLYSIS AND CONDENSATION REACTIONS

Water is essential to many chemical reactions. A **hydrolysis reaction** occurs when a chemical bond is broken by the addition of a water molecule. You have already learned about many kinds of hydrolytic reactions (such as those required for carbohydrate and triglyceride digestion) in preceding chapters. In contrast to hydrolysis reactions, a chemical reaction that *produces* water is called a **condensation reaction.** Such reactions join molecules together through the formation of a chemical bond, releasing water in the process. The typical adult produces about 200 to 300 milliliters (approximately 1 cup) of water every day from condensation and other metabolic reactions. Condensation and hydrolysis can be thought of as opposite processes—*make it* and *break it* reactions, respectively.

WATER AS A SOLVENT, TRANSPORT MEDIUM, AND LUBRICANT

Water is the primary solvent in bodily fluids such as blood, saliva, and gastrointestinal secretions. Blood and lymph, for example, are solutions that consist of water, cells, and a variety of dissolved solutes such as nutrients and metabolic waste products. Such water-based solutions serve as the body's transport mediums. Substances dissolved in blood move from inside the blood vessel into the watery environment within and around tissues, delivering important nutrients to cells. Conversely, waste products produced by cells are dissolved in water, released into the blood, and subsequently eliminated from the body. Without water to dissolve these substances, such critical processes would not be possible.

Water also acts as a lubricant in the GI tract, respiratory tract, skin, and reproductive system, which produce digestive juices, mucus, sweat, and reproductive fluids, respectively. The body's ability to incorporate water into its secretions is vital to good health. As an example of its lubricating functions, water is an essential component of mucus produced in the lungs. Mucus both protects the lungs from harmful substances and lubricates the sensitive lung tissue so that it can remain moist and supple. When insufficient amounts of water are available or the body's ability to regulate water balance across membranes is impaired, the production of secretions (such as mucus) is compromised.

hydrolysis reaction A chemical reaction whereby a chemical bond is broken by the addition of a water molecule.

condensation reaction A chemical reaction whereby a chemical bond joins two molecules together, releasing water in the process.

evaporative cooling The process by which sweat evaporates from the skin, taking heat with it.

dehydration A condition whereby the body has an insufficient amount of water.

WATER REGULATES BODY TEMPERATURE

Body heat is generated when energy-yielding nutrients are metabolized, such as during physical activity. Despite metabolic and environmental fluctuations, the average human is able to maintain a relatively stable and comfortable internal temperature of 37 °C (98.6 °F). The stability of this temperature is critical because even a slight increase or decrease in the body's internal temperature can disrupt normal body functions. To prevent overheating, excess heat must be released from the body. This release is accomplished primarily through a process that involves both the skin and sweat glands.

Sweating is your body's most effective method of cooling itself. Because it can absorb and release large amounts of heat without itself changing appreciably in temperature, water is an ideal medium for heat transfer. In order to rid the body of excess heat, blood vessels located near the surface of the skin dilate, releasing sweat that beads on the surface of the skin. When sweat evaporates from the skin, it takes heat with it. This process, called **evaporative cooling**, is an effective and efficient method of dissipating heat from (and thus cooling) the body. Evaporative cooling's efficiency is one reason why **dehydration**, a condition whereby the body has an insufficient amount of water, impairs the body's ability to regulate body temperature. Without adequate hydration, the body is unable to produce enough sweat to eliminate excess heat. When a dehydrated person exercises in hot, humid conditions, evaporative cooling is further diminished, making it extremely difficult to dissipate excess heat. Therefore, it is particularly important for athletes to stay fully hydrated.

Because body heat is dissipated through the process of sweating, it is important to stay fully hydrated. The effectiveness of evaporative cooling is impaired when one exercises in hot, humid conditions.

Dehydration

In its early stages, dehydration often results in thirst and a dry mouth—conditions that prompt the dehydrated person to drink. Failure to respond to these initial symptoms of dehydration can lead to serious consequences. Compared to the other macronutrients, your body is least tolerant of water loss. Even a 2 percent loss in water weight can lead to serious complications:

- Mental confusion, reduced attention span, and impaired memory
- Muscle weakness and poor coordination
- Impaired body temperature regulation, especially when exercising
- Urinary tract infections
- Reduced blood pressure
- Seizures and coma

Such complications pose especially serious health concerns to children and the elderly, who have little tolerance for the negative consequences of dehydration. (Of particular concern in regard to the elderly is that signs and symptoms of dehydration can easily be mistaken for those of dementia.) To avoid complications, athletes must also be sensitive to the signs and symptoms of dehydration. Athletes who sweat a great deal lose large amounts of water, and are thus at particularly increased risk of dehydration. In short, ensuring adequate water intake is important for everyone, but is especially important for infants, children, the elderly, and athletes.

THE BODY'S RESPONSE TO DEHYDRATION

To meet your body's need for water, it is important to balance your daily intake of water with your daily water loss (see Figure 8.3). Adults excrete approximately 1,500 ml (1½ quarts) of fluid every day as urine. Water is also expelled as water

> **Amateur wrestlers** and other athletes sometimes **induce dehydration intentionally** to lose water weight and meet strict weight class limits. Nutrition **experts advise against** this form of weight reduction.

vapor in expired air (300 ml/day), as perspiration (500 ml/day), and in feces (200 ml/day). In total, the body loses approximately 2,500 ml (about 10.6 cups) of fluid every day. To counter this loss, adults typically need to consume 2,300 ml of fluid daily through a combination of foods and beverages. If you were wondering about the discrepancy between water loss and consumption, remember that cellular metabolic reactions generate approximately 200–300 ml of fluid every day.

When a person becomes dehydrated, several homeostatic control mechanisms work together to restore fluid balance (see Figure 8.4 on the next page). These mechanisms involve two hormones: antidiuretic hormone and aldosterone. A progressive loss of body fluids results in low blood volume. In response, the concentration of solutes in extracellular fluids (including blood)

Figure 8.3 Daily Water Balance

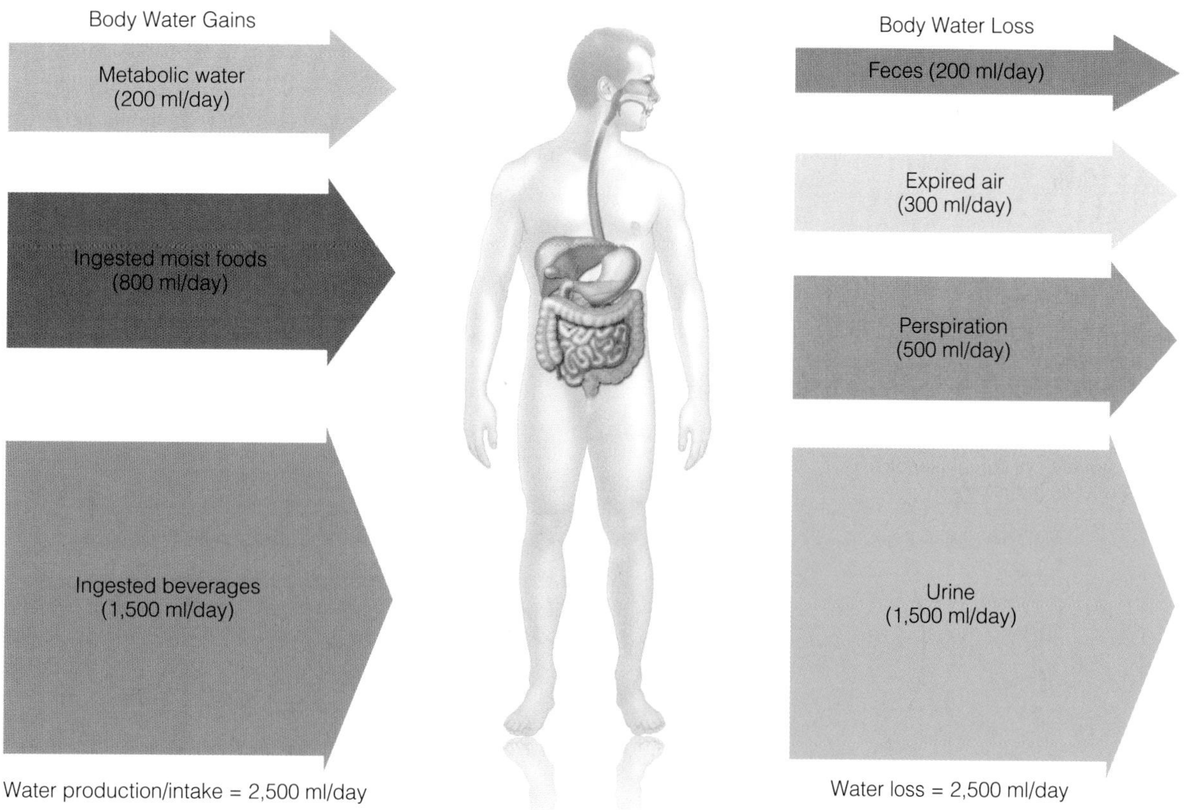

The body gains and loses water every day. For the body to function properly, daily water intake must equal average daily water loss.

antidiuretic hormone (ADH, or **vasopressin)** A hormone released by the pituitary gland during periods of low blood volume that decreases the amount of water excreted in the urine.

aldosterone A hormone produced by the adrenal glands in response to low blood pressure.

increases. This stimulates the pituitary gland to release **antidiuretic hormone** (**ADH**, or **vasopressin**) into the blood. ADH circulates to the kidneys, where it stimulates water conservation by decreasing the amount of water excreted in the urine. The increased water retention helps restore blood volume to healthy levels.

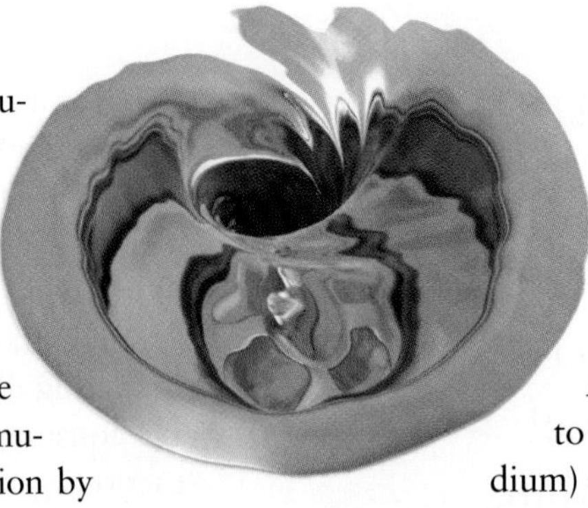

As the pituitary gland releases ADH in response to low blood volume, low blood pressure (also caused by dehydration) stimulates the adrenal glands to release the hormone **aldosterone.** Aldosterone signals the kidneys to reduce the amount of solutes (sodium) eliminated in the urine and instead return them to the blood. Remember: where solutes go, water will follow via osmosis. Together, the complementary actions of ADH and aldosterone restore blood volume to a healthy level.

Figure 8.4 Regulation of Blood Volume and Solute Concentration

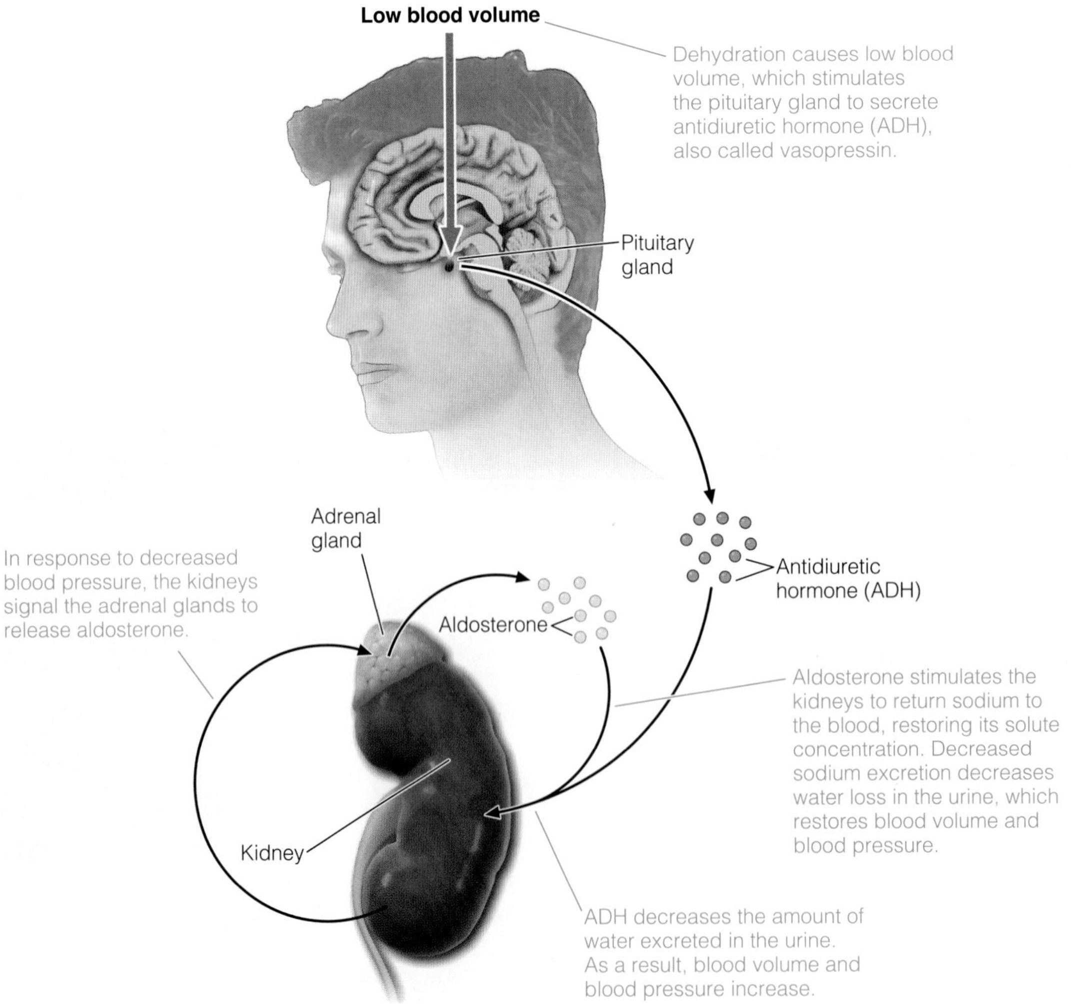

Low blood volume
Dehydration causes low blood volume, which stimulates the pituitary gland to secrete antidiuretic hormone (ADH), also called vasopressin.
Pituitary gland
Antidiuretic hormone (ADH)
Adrenal gland
Aldosterone
In response to decreased blood pressure, the kidneys signal the adrenal glands to release aldosterone.
Aldosterone stimulates the kidneys to return sodium to the blood, restoring its solute concentration. Decreased sodium excretion decreases water loss in the urine, which restores blood volume and blood pressure.
Kidney
ADH decreases the amount of water excreted in the urine. As a result, blood volume and blood pressure increase.

Blood volume and solute concentration are regulated by the brain and kidneys. This process involves two hormones: antidiuretic hormone and aldosterone.

Does Coffee Cause Dehydration?

Millions of people start the day with a steaming cup of coffee. Indeed, coffee is among the most widely consumed beverages in the world. But is there more to coffee than the pick-me-up that it provides? Aside from being a stimulant, coffee is also renowned as a diuretic. Although caffeine does stimulate urination, studies have shown that caffeine-containing beverages do not appear to cause a net loss of total body water.[1] That is, the amount of water consumed while drinking caffeinated beverages adequately compensates for the total amount of water excreted from the body.

Although caffeine increases urine production, studies have shown that caffeine-containing beverages do not appear to cause a net loss of total body water.

© iStockphoto.com/Graham Clarke

Recommendations for Water Intake

In 2004, the Institute of Medicine released its first recommendations for water intake. These guidelines advise women to consume about 2.7 liters (11 cups) of water every day. The recommended water intake for men is 3.7 liters (16 cups) a day. These amounts may sound exorbitant, but they include the water contents of all beverages and foods consumed. In general, about 80 percent of total water intake comes from beverages, while the remaining 20 percent comes from foods. Upon careful consideration, the DRI committee concluded that caffeinated beverages such as coffee and cola contribute to total water intake to the same degree as noncaffeinated beverages.[2] This conclusion stands in contrast to previous recommendations, which suggested that caffeinated beverages (which have diuretic properties) could not be relied upon to supply water to the body.

Physical activity increases a person's need for water, especially when performed in a warm environment. Studies show that very active people who live and/or work in warm climates may have daily water requirements of up to 7 liters (30 cups).[3] Specific guidelines have not been established for these individuals, but the Institute of Medicine recommends that very active people take special care to consume enough fluids every day. Consuming sports drinks instead of water during endurance training and competition can help athletes prevent serious dehydration-related complications and replenish electrolytes and energy stores. For less active people, however, water should be the drink of choice for simple hydration and rehydration.

LO2

What Are Minerals?

In nutrition, the term **mineral** describes an inorganic substance other than water. Because your body requires them in very small amounts, dietary minerals are considered micronutrients. All of the minerals needed to sustain good health are essential nutrients because the body cannot synthesize them from other compounds. Thus, you must rely on the foods you eat to provide adequate amounts of these important nutrients. Minerals can be neither created nor destroyed: even if you completely combust (burn) a food, all of its minerals will remain as ash. Although minerals comprise only a small fraction of total body weight, they play prominent roles in numerous bodily structures and functions.

Essential minerals are classified as either major or trace, depending on how much the body needs in a given day. A **major mineral** is one required in an amount greater than 100 mg/day. Your body requires seven major minerals: calcium, phosphorus, magnesium, sodium, chloride, potassium, and sulfur. A **trace mineral** is one required only in a minute amount—less than 100 mg/day. Humans require at least eight trace minerals: iron, copper, iodine, selenium,

mineral An inorganic substance other than water that is required by the body in small amounts.

major mineral An essential mineral required in an amount greater than 100 mg/day.

trace mineral An essential mineral that is required in an amount less than 100 mg/day.

chromium, manganese, molybdenum, and zinc. In addition to the major and trace minerals, several minerals are not currently considered essential nutrients, but may be reclassified as we learn more about them. These minerals include fluoride, arsenic, boron, lithium, nickel, silicon, and vanadium. The functions of these minerals in the body are not fully understood.

Is Bottled Water Better?

More than half of all Americans drink bottled water, and one-third drink it regularly.[4] Despite its popularity, there is no evidence that bottled water offers any health advantages over water that runs freely from the tap. In fact, because most municipalities add fluoride (which helps keep teeth strong) to their water supplies, drinking bottled water instead of tap water may increase a person's risk of tooth decay. Furthermore, by choosing tap water instead of bottled water, Americans could eliminate about 1.5 million tons of plastic waste per year. In fact, most experts agree that the environmental costs associated with bottled water far outweigh any potential nutritional benefits. While some developing countries' water supplies are contaminated (thus necessitating bottled water), American tap water is as wholesome as bottled varieties and is certainly healthier for the environment.

Americans buy nearly 30 billion bottles of water per year; 90 percent of bottles are not recycled and end up in landfills. It takes thousands of years for plastic to decompose.

Minerals Serve Diverse Roles

Minerals serve both structural and functional roles. In fact, many serve so many roles that they influence virtually every physiological system in the body. Most minerals serve multiple distinct physiologic functions, while some work with several other minerals to carry out interrelated tasks. Calcium, magnesium, and phosphorus function together to form and strengthen the structure of the skeleton, for example. Some enzymes need to be activated by a mineral to function. When a mineral such as copper, magnesium, or zinc combines with and activates an enzyme, the mineral is called a **cofactor** and the enzyme is called a **metalloenzyme.** Some minerals are involved in the metabolic breakdown of energy-yielding nutrients, whereas others are essential to nerve function, muscle contraction, and the regulation of bodily fluids. Although nutritional scientists understand much about the major and trace minerals, they do not yet know all of the functions of each one. Some of the known roles of minerals are summarized in Table 8.1.

> "**Mineral water**, promoted for its purity and fresh taste, contains **naturally occurring and/or added minerals** such as sulfur and calcium. Natural mineral water is both consumed and bathed in, and natural mineral springs are often popular **tourist attractions** because of their **curative properties**."

cofactor A mineral that activates an enzyme by combining with it.

metalloenzyme An enzyme that is activated when it combines with a mineral.

Minerals in Food

Minerals are abundant in both plant- and animal-based foods. In general, animal-based products have higher mineral contents than

Table 8.1 **Selected Roles of Important Minerals in the Body**

	Antioxidant function	Blood health	Bone health	Cellular metabolism	Fluid balance	Growth and development	Neurological function and muscle health
Major Minerals							
Calcium		✓	✓	✓		✓	✓
Chloride					✓	✓	✓
Magnesium			✓	✓	✓	✓	✓
Phosphorus		✓	✓	✓		✓	✓
Potassium				✓	✓	✓	✓
Sodium					✓		✓
Sulfur				✓		✓	
Trace Minerals							
Chromium				✓		✓	
Fluoride			✓				
Copper	✓	✓	✓	✓		✓	✓
Iodine				✓		✓	
Iron	✓	✓	✓	✓		✓	
Manganese	✓		✓	✓		✓	
Molybdenum	✓			✓		✓	
Selenium	✓			✓		✓	✓
Zinc	✓		✓	✓		✓	✓

do plant-based foods. For example, a one-cup serving of milk contains approximately 300 mg calcium, while a one-cup serving of broccoli contains approximately 43 mg calcium. The location where a plant grows and where an animal grazes can also influence mineral contents. For instance, a plant grown in selenium-deficient soil will have a lower selenium content than one grown in selenium-rich soil. Not surprisingly, selenium deficiency is more common in people living in geographic regions with low-selenium soil. The extent to which a food is processed can also influence its mineral content. For example, cereal grains contain minerals such as copper, selenium, and zinc, but these important nutrients are lost during processing. Food manufacturers sometimes fortify their products with minerals lost during processing, but most health care professionals recommend choosing foods made with whole grains whenever possible.

As with foods, the types and amounts of minerals found in drinking water vary by region. *Hard water* has appreciably high levels of dissolved minerals such as calcium and magnesium. The types of minerals present in water can give it a particular taste and smell. Also, some minerals can pose problems to homeowners because mineral deposits tend to collect on the hard surfaces of pipes, sinks, and bathtubs.

Mineral Availability in the Body

Like most nutrients in the body, optimal and safe mineral levels are maintained through adjustments in absorption and excretion. Changes in these regulatory processes optimize mineral availability and prevent

Healthy Plants Require Healthy Soil

The 1980 eruption of Mount St. Helens wreaked havoc throughout the United States Pacific Northwest. The volcano's eruption blanketed the region in a thick layer of volcanic ash and destroyed many of the area's agricultural crops. Apple farmers were particularly devastated as they watched their entire orchards wither away.[5] Although the eruption initially seemed catastrophic to the apple industry, something unexpected happened the following year. Apple trees were covered not only with blossoms in the spring, but also with plentiful clusters of large, succulent fruit in late summer. This bountiful harvest occurred because the fine powder ash that spewed from the eruption was extremely rich in minerals, which reinvigorated the soil.

toxicity. In addition to its absorption and excretion processes, the body is capable of storing small amounts of certain minerals in the liver, bones, and other tissues. Because these three processes (absorption, excretion, and storage) are tightly regulated, toxicity is rare for most minerals. Still, despite the body's effective mineral management system, genetic disorders and overconsumption of mineral-containing supplements or medications can sometimes have serious outcomes.

In some cases, humans can absorb nearly all of the minerals present in food. This is not the case for all minerals, however; some minerals' bioavailabilities are influenced by factors such as genetics, developmental maturity, nutritional status, and interactions with other compounds in food. During periods of growth, for instance, mineral requirements are high. To meet the demands of a growing body, the absorption of some minerals increases. Conversely, mineral absorption usually decreases as a person ages. In addition to these factors, a mineral's bioavailability is influenced by the body's need for it. For example, calcium absorption increases when circulating concentrations of calcium are low.

calcium (Ca) A major mineral located throughout the skeleton that is essential to coagulation, muscle and nerve function, and energy metabolism.

hydroxyapatite A large crystal-like molecule that combines with other minerals to form the structural matrix of bones and teeth.

In some instances, the presence of one mineral can interfere with the absorption of another. This is particularly true for minerals that have similar charges. For example, calcium, iron, copper, magnesium, and zinc all carry a positive (+2) charge. High intake of any one of these nutrients may reduce absorption of the others. Another dietary factor that can decrease the efficiency of mineral absorption is the presence of binding factors, such as those found in some nuts, grains, and vegetables. Some of these foods contain substances that bind to minerals in the GI tract, inhibiting their absorption. Despite these nutrient interactions, the Institute of Medicine does not recommend reducing the consumption of these otherwise nutritious foods.

Calcium

Calcium (Ca) is a major mineral located throughout the skeleton that is essential to coagulation, muscle and nerve function, and energy metabolism. Calcium is the most abundant mineral in your body, making up about 1 kg (2.2 pounds) of your weight. You may already know that calcium plays a critical structural role in bones and teeth. In fact, more than 99 percent of your body's calcium is located in your skeleton. The rest is located in the blood and other tissues, where it participates in vital functions such as muscle contraction, neural signaling, blood clot formation, and blood pressure regulation.

CALCIUM'S STRUCTURAL ROLE

The skeletal system makes up the basic architecture of your body. Although you might think of the skeleton as an inert structure, your bones and teeth are in fact comprised of living tissue. Skeletal calcium is a component of a large crystal-like molecule called **hydroxyapatite**, which combines with other minerals such as fluoride and magnesium to form the structural matrix of your bones and teeth. Hydroxyapatite also functions as a storage depot for calcium. Bone tissue is complex, and is composed of two different kinds of bone cells: osteoblasts and osteoclasts. An **osteoblast** promotes bone formation, whereas an **osteoclast** promotes the breakdown of older bone. These cells work in concert to keep your bones healthy and strong, largely by synthesizing and breaking down calcium-containing hydroxyapatite as needed. Every day, osteoclasts break down small pockets of old and damaged bone, releasing calcium, phosphorus, and other substances into the blood. In order to maintain stable bone mass, osteoblasts fill the dissolved pockets with new bone. This continuous process, referred to as **bone remodeling** (or **bone turnover**), is illustrated in Figure 8.5.

CALCIUM'S REGULATORY FUNCTIONS

Aside from its importance to structural maintenance, calcium is also required for many regulatory functions. Calcium works with vitamin K to stimulate blood clot formation and facilitates both muscle contraction and nerve impulse transmission.[6] Calcium supports healthy vision, the regulation of blood glucose and cell division.[7] Although it is not used directly as an energy source, calcium is a cofactor for several enzymes that aid in ATP production. In addition to the multitude of functions already discovered, scientists are just beginning to understand additional roles that calcium might play in the body. For example, there is growing (albeit controversial) evidence that adequate calcium consumption may help reduce the risk of cardiovascular disease, some forms of cancer, and obesity.

HORMONAL REGULATION OF BLOOD CALCIUM

Calcium deficiency affects many tissues in the body. For instance, because calcium plays a role in muscle contraction and nerve function, low blood calcium levels can cause muscle pain, muscle spasms, and a tingling sensation in the hands and feet. The amount of calcium in your blood is tightly regulated by hormonal mechanisms (see Figure 8.6 on the next page). This well-orchestrated system involves three hormones: calcitriol (vitamin D), parathyroid hormone (PTH), and calcitonin. These hormones work together to maintain healthy blood calcium levels at all times.

Recall from Chapter 7 that vitamin D is critical to the regulation of blood calcium levels. When blood

osteoblast A bone cell that promotes bone formation.

osteoclast A bone cell that promotes the breakdown of older bone.

bone remodeling (or **bone turnover**) The continuous process by which older and damaged bone is replaced by new bone.

Figure 8.5 Bone Remodeling

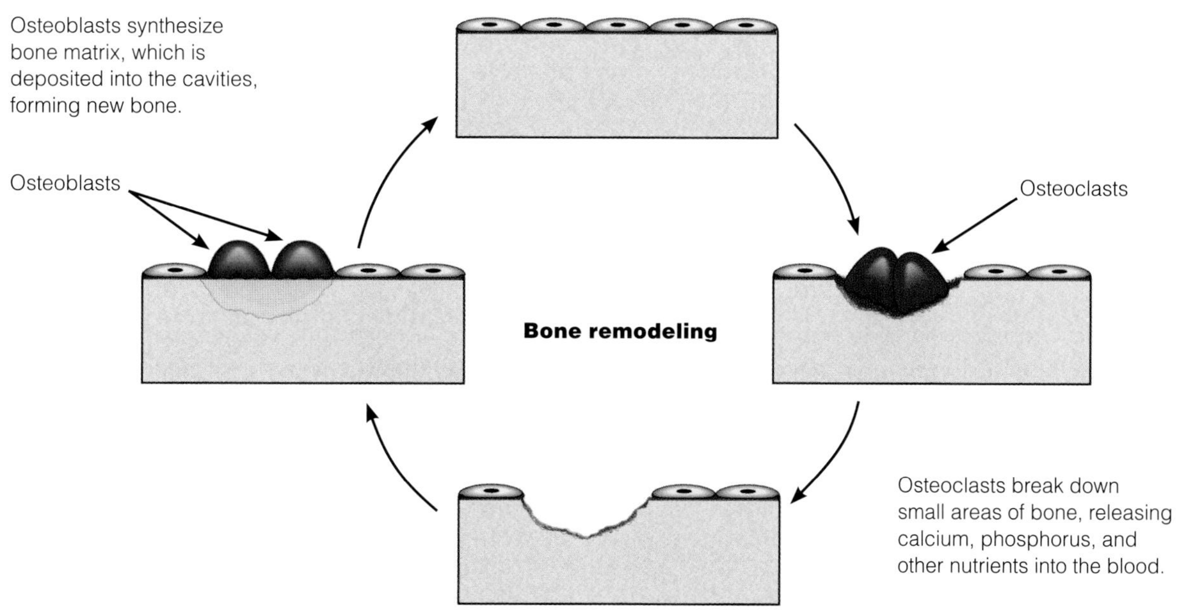

Osteoblasts synthesize bone matrix, which is deposited into the cavities, forming new bone.
Osteoblasts
Osteoclasts
Bone remodeling
Osteoclasts break down small areas of bone, releasing calcium, phosphorus, and other nutrients into the blood.

Bone mass is kept stable through a process called bone remodeling. This process involves cells that build bone (osteoblasts) and cells that break bone down (osteoclasts).

Figure 8.6 **Regulation of Blood Calcium**

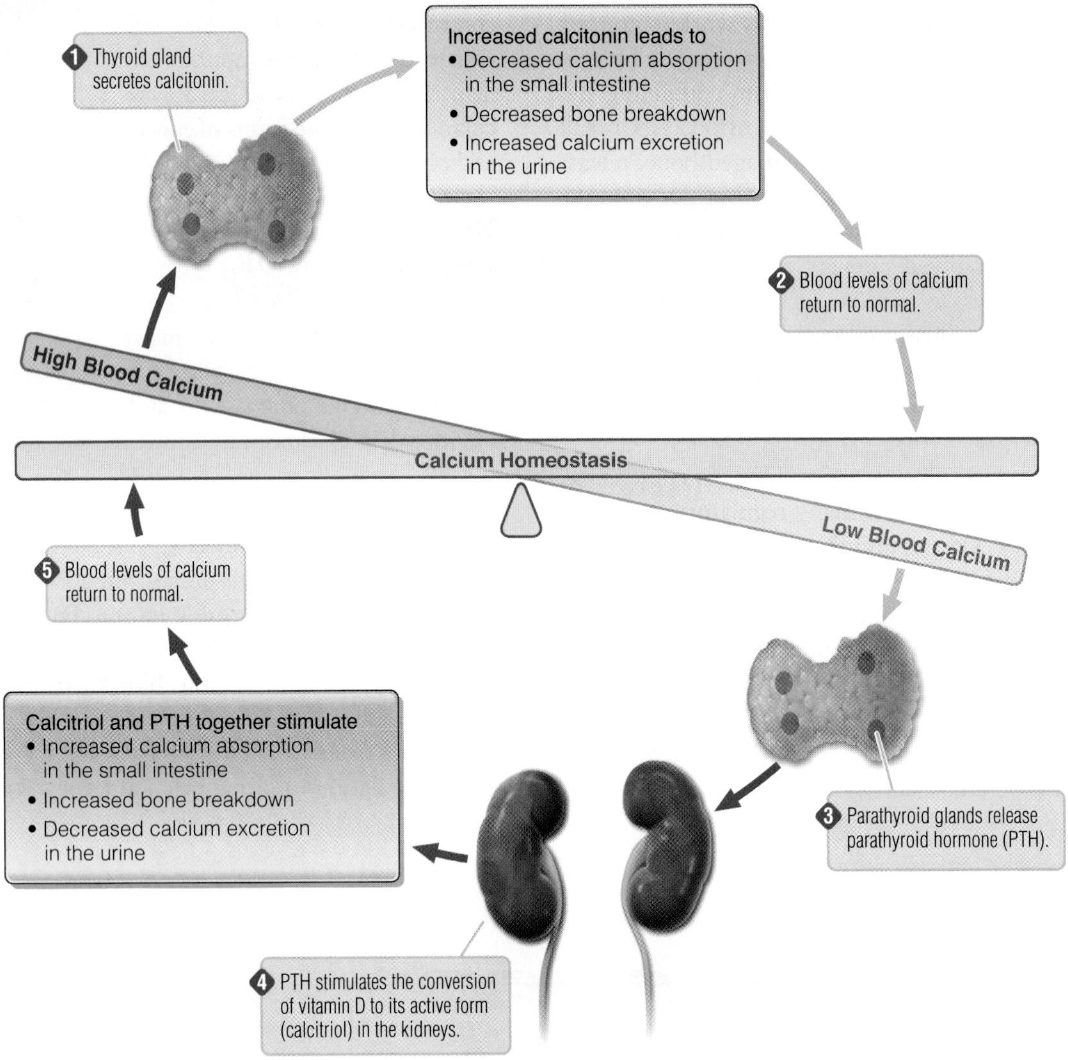

Blood calcium concentration is regulated by three hormones: parathyroid hormone (PTH), calcitriol (vitamin D), and calcitonin.

calcium levels are low, the parathyroid glands release PTH, which in turn circulates to the kidneys, where it stimulates the conversion of 25-hydroxy vitamin D_3 (25-[OH] D_3) to calcitriol (1,25-$[OH]_2$ D_3, the active form of vitamin D). Working together, PTH and calcitriol increase calcium absorption in the small intestine, reduce calcium excretion in the urine, and signal the release of calcium from bones into the blood. Collectively, these actions restore blood calcium levels to within a healthy range.

calcitonin A hormone released by the thyroid gland that decreases calcium loss from bone, decreases calcium absorption in the small intestine, and increases calcium loss in the urine.

While low calcium intake can cause health problems, excessive calcium intake is equally dangerous. Overconsumption can cause calcium to deposit in soft tissues such as muscle and kidney and can interfere with the bioavailability of other nutrients, such as iron and zinc. High calcium intake is also associated with impaired kidney function and the formation of certain types of kidney stones. When blood calcium levels are too high, the parathyroid glands produce less PTH. This decreases the formation of calcitriol, which in turn reduces calcium absorption in the small intestine. Elevated blood calcium also stimulates the thyroid gland to produce a hormone called **calcitonin**, which decreases calcium loss in bone, decreases calcium absorption in the small intestine, and increases calcium levels in the urine. Together, these processes help lower blood calcium levels back to normal.

BONE HEALTH AND OSTEOPOROSIS

Most bone growth is completed before age 20, but a small amount of bone (about 10 percent) is deposited between ages 20 and 30. The amount of mineral contained in a bone is referred to as **bone mass. Peak bone mass**, the greatest total amount of bone that a person has in her lifetime, is typically attained at about 30 years of age. To determine bone strength, clinicians test **bone density** by measuring the amount of bone tissue in a certain volume of bone. Bones with higher densities tend to be stronger and better able to withstand injury, while those with lower densities tend to break more easily.

Because low bone density can lead to negative health consequences later in life, it is important to develop and maintain healthy bones in childhood and even early adulthood. You may recall from Chapter 7 that rickets, typically caused by a dietary vitamin D deficiency in young children, can also be caused by a lack of calcium in the diet. In adults, moderate bone loss is referred to as **osteopenia**, whereas osteoporosis (meaning "porous bone") is reserved for severe bone loss. Osteopenia and osteoporosis are characterized by progressive bone loss, which over time causes bones to become weak and fragile. These conditions are especially prevalent in the elderly, for whom even minor falls can result in traumatic fractures that take great lengths of time to heal. Some people with osteoporosis lose significant height during old age and can develop a curvature in the upper spine, a condition called **kyphosis** (or **dowager's hump**). The National Osteoporosis Foundation estimates that osteoporosis is a major health concern for 44 million Americans, and 55 percent of people age 50 and older.[8]

bone mass The amount of minerals contained in bone.

peak bone mass The greatest total amount of bone that a person has in his or her lifetime.

bone density The amount of bone tissue in a certain volume of bone.

osteopenia A disease characterized by moderate bone loss in adults.

kyphosis (or **dowager's hump**) A curvature of the upper spine caused by osteoporosis.

A person's risk of developing bone disease depends in part on the peak bone mass achieved in early adulthood. After bone mass peaks, it begins the process of gradual decline. There are many reasons for this decline.[9] First, because of decreased production, absorption, and metabolism of vitamin D, calcium absorption often decreases with age. This prompts bones to release more calcium in order to maintain healthy blood calcium concentrations. Second, reproductive hormone concentrations naturally decline as humans age. For example, *estrogen*, a reproductive hormone produced in the ovaries, is important to the maintenance of bone strength in women. When women reach menopause, declining estrogen levels accelerate bone loss. Men also lose bone mass as they age, but the loss is more gradual (see Figure 8.7 on the next page).

Because bone fractures are a common cause of hospitalization and loss of independence in older adults, it is important to do all you can to prevent osteoporosis. Many factors are associated with a person's risk of osteoporosis. Biological risk factors, such as sex, are related to a person's genetic makeup. Although biological risk factors cannot be modified, it is important to be aware of which factors apply to you because they are believed to contribute about half of your risk of osteoporosis.[10] Conversely, lifestyle factors such as amount of exercise, diet, and smoking habits can be modified to decrease the risk of osteoporosis. See Table 8.2 on page 197 for common biological and lifestyle risk factors for bone disease.

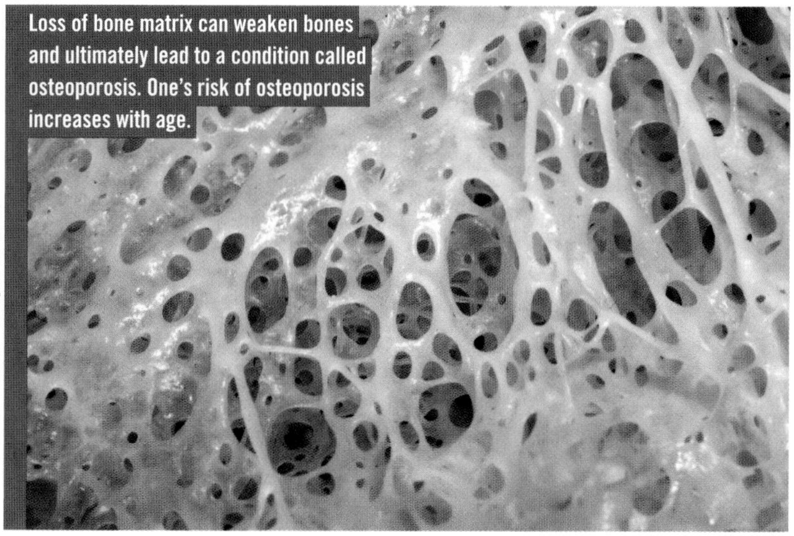

Loss of bone matrix can weaken bones and ultimately lead to a condition called osteoporosis. One's risk of osteoporosis increases with age.

DIETARY AND SUPPLEMENTAL SOURCES OF CALCIUM

Calcium is found naturally in both plant- and animal-based foods, but the best sources of calcium are dairy products (see Figure 8.8). Other good sources of calcium include dark green leafy vegetables such as collards and spinach, salmon and sardines (with bones), and some legumes. Calcium-fortified foods such as breakfast cereals, orange juice, and soy products also provide considerable amounts of calcium. It is relatively easy to assess the amount of calcium in most packaged foods because Nutrition Facts panels are required to list calcium content. Furthermore, foods fortified with calcium are frequently labeled as such.

If a person cannot obtain adequate amounts of calcium from his diet, many forms of calcium supplements are available. In general, calcium supplements are relatively well utilized by the body, but they are

Figure 8.7 Changes in Bone Mass during the Life Cycle

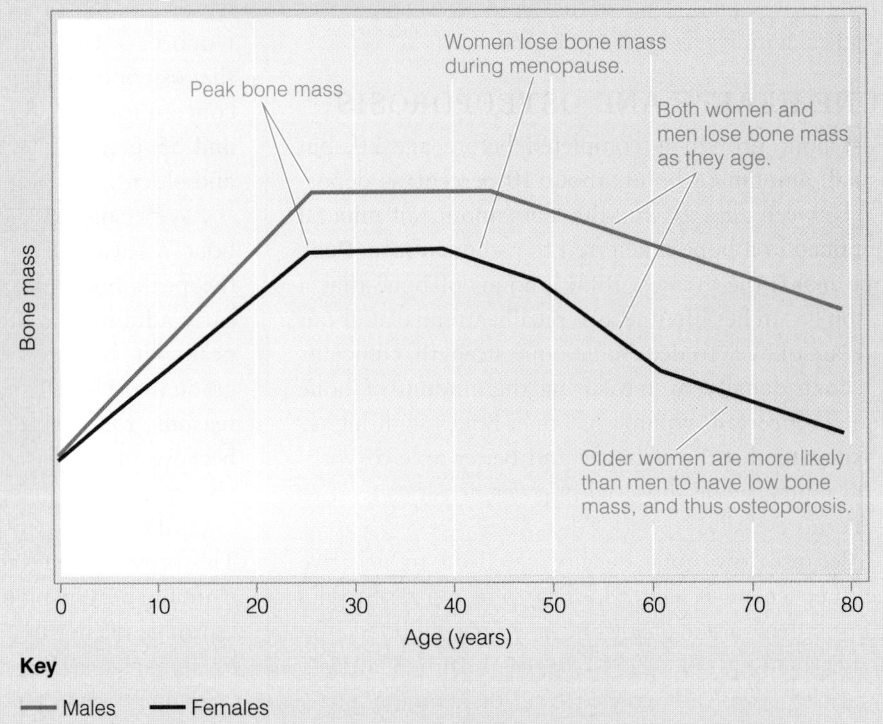

Women are at greater risk of osteoporosis than men because they tend to have lower peak bone mass and are likely to lose more bone mass as they age.

Adapted from Compston JE. Osteoporosis, corticosteroids and inflammatory bowel disease. Alimentary Pharmacology and Therapeutics. 1995;9:237–250.

Figure 8.8 Good Sources of Calcium

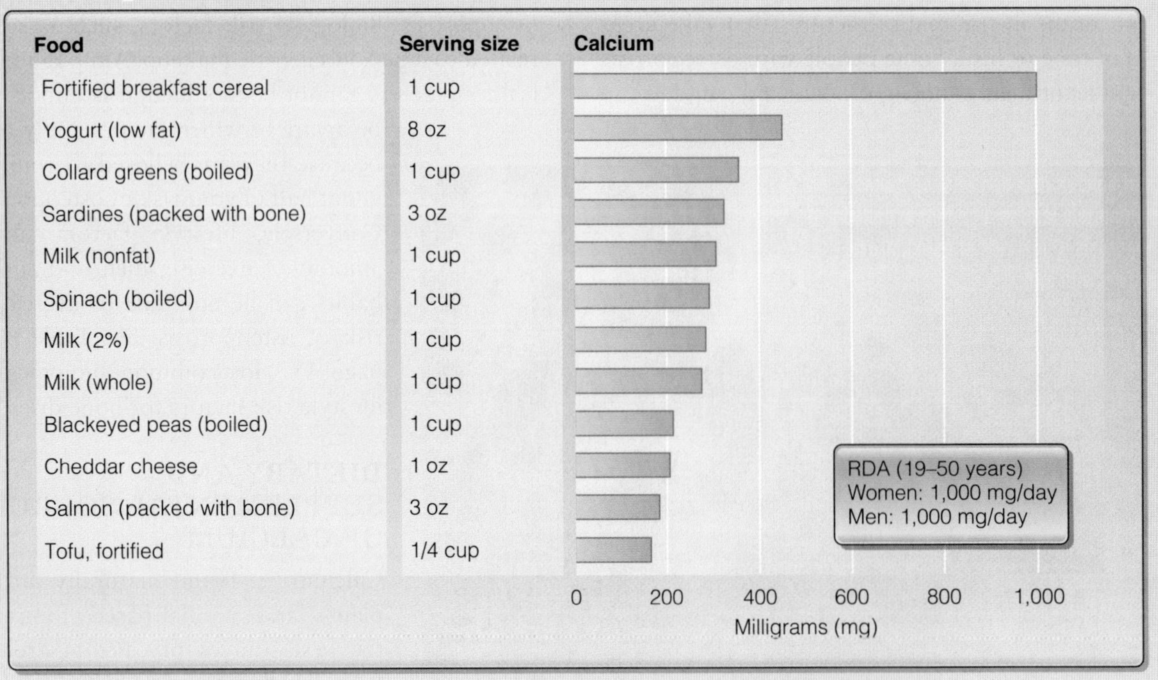

Source: USDA Nutrient Database for Standard Reference, Release 17.

Table 8.2 **Factors Related to Increased Risk of Osteoporosis**

Risk Factor	Effect on Risk of Osteoporosis
Unmodifiable (Biological) Factors	
Sex	Women are at greater risk for osteoporosis than are men because of loss of estrogen production after menopause and overall smaller body size.
Age	The rate of bone loss increases with age, especially in post-menopausal women.
Body size	Thinner people are at increased risk of osteoporosis because they tend to have lower bone mass.
Ethnicity/Race	People of northern European and Asian descent are at greatest risk, while those of African and Hispanic descent are at lowest risk.
Family history	Those with a family history of osteoporosis are at especially high risk.
Modifiable (Lifestyle) Factors	
Smoking	Smoking appears to weaken bones.
Medication	Chronic use of some medications (such as thyroid hormones, some diuretics, and corticosteroid therapy) can weaken bones and increase bone loss.
Low estrogen concentration	Low levels of estrogen increases bone loss in women.
History of eating disorder	A history of either anorexia nervosa or bulimia nervosa increases risk of lower bone density. This is partly because eating disorders frequently lead to very low body fat, which in turn can lower estrogen levels in women. People with eating disorders also often have multiple nutrient deficiencies.
Alcohol consumption	Chronic alcoholism increases risk of osteoporosis. This is mostly due to the negative influence of alcoholism on overall diet, especially in terms of calcium, vitamin D, and protein intake.
Physical inactivity	Regular physical activity, especially weight-bearing exercise, increases bone density and decreases risk of osteoporosis.
Nutritional factors	Chronically low calcium intake can decrease peak bone mass in young adults and increase bone loss in later life. Phosphorus, magnesium, vitamin D, vitamin K, vitamin C, fluoride, and protein are also important to bone health.

most effectively absorbed when taken with a meal and in doses of no more than 500 mg.[11] As such, if you choose to take a daily calcium supplement, you might want to consider taking half with breakfast and half with lunch or dinner. Although adequate calcium intake is certainly critical to the maintenance of healthy bones, studies suggest that taking calcium and vitamin D supplements produces only small improvements in bone health.[12] Furthermore, limited research suggests that calcium supplementation might actually increase risk of hip fractures in some people.[13] Clearly, calcium's impact on bone health is a complex subject—one that is convoluted by calcium's interaction with other dietary components.

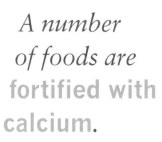

A number of foods are **fortified with calcium.**

renin An enzyme, secreted by the kidneys, that converts angiotensinogen to angiotensin I.

RECOMMENDED CALCIUM INTAKE AND BIOAVAILABILITY

The Recommended Dietary Allowance (RDA) for calcium has been set at 1,000 mg/day for adults. This is approximately the amount of calcium contained in three to four cups of milk. Because calcium absorption tends to decrease with age and women are especially prone to bone loss, the RDA increases to 1,200 mg/day at 51 years of age for women and at 71 years of age for men. To decrease the risk of calcium toxicity, a Tolerable Upper Intake Level (UL) of 2,500 mg/day has been established for calcium. At 51 years of age, this UL decreases to 2,000 mg/day.

Many factors influence calcium's bioavailability. For example, *oxalate* and *phytate* are compounds that bind calcium in the intestine, hindering its absorption. Additionally, nondietary factors such as age also affect how much calcium is absorbed from a meal. Because calcium absorption tends to decrease around age 50, it can be difficult for older people to satisfy their need for this important mineral by dietary means alone.[14]

Electrolytes: Sodium, Chloride, and Potassium

Recall from Chapter 3 that an electrolyte is a salt that separates into individual charged ions when dissolved in water. When table salt (sodium chloride), one of the most familiar electrolytes, is added to water, it separates into its charged components: sodium (Na^+) and chloride (Cl^-). These, along with potassium (K^+), are the three most abundant ions in the human body. Each of these charged ions is found in various fluids throughout the body, and each plays important roles in fluid balance regulation, nerve impulse conduction, and nutrient transport. While most people associate sodium with health problems such as hypertension (high blood pressure), it and other ions are essential for many physiological functions within the body.

ELECTROLYTE FUNCTION: NERVE IMPULSE AND MUSCLE CONTRACTION

Tissues, fluids, and cells all contain varying concentrations of electrolytes. As you previously learned, it is the concentrations of dissolved sodium, chloride, and potassium ions in fluids that direct the flow of water during osmosis. In this way, electrolytes play important roles in regulating fluid balance. A proper distribution of positively and negatively charged ions inside and outside of cells orchestrates many important physiological functions, including nerve cell transmission and muscle contraction.

Nerve impulses and muscle contractions occur when balances of ions shift across cell membranes. When a nerve cell is at rest, positively charged potassium ions are concentrated inside of the cell while positively charged sodium ions and negatively charged chloride ions are concentrated outside of the cell. Upon stimulation, sodium ions rush into the cell, and potassium ions rush out. This exchange of ions across the cell membrane produces an electrical signal, or a nerve impulse. Muscle fibers function in much the same manner, but their activation causes calcium (instead of sodium) to flow inside the cell, signaling the muscle to contract. Once calcium has been pumped back out of the muscle cell, it can again relax.

ELECTROLYTE FUNCTION: BLOOD PRESSURE REGULATION

Blood volume and circulating electrolyte concentrations are intimately linked. A change in blood volume can affect its electrolyte concentrations, and a change in electrolyte concentrations can affect blood volume. Although electrolyte concentrations and blood volume are clearly connected, their regulatory mechanisms are somewhat different. You may recall that a substantial loss of body fluid causes blood volume to decrease. When blood volume decreases, so too does blood pressure. This triggers a series of homeostatic responses that ultimately cause the kidneys to retain sodium and return it to the blood. Again—where electrolytes go, water will follow. Sodium retention increases blood volume, which subsequently restores blood pressure. While this homeostatic response involves ADH and aldosterone, a related homeostatic response called the renin-angiotensin-aldosterone system utilizes other substances to assist in blood volume and blood pressure regulation.

As illustrated in Figure 8.9, the kidneys respond to low blood pressure by releasing an enzyme called **renin**.

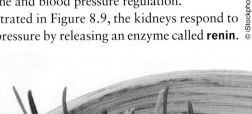

© iStockphoto.com/joanna wnuk

Figure 8.9 Regulation of Blood Volume and Blood Pressure

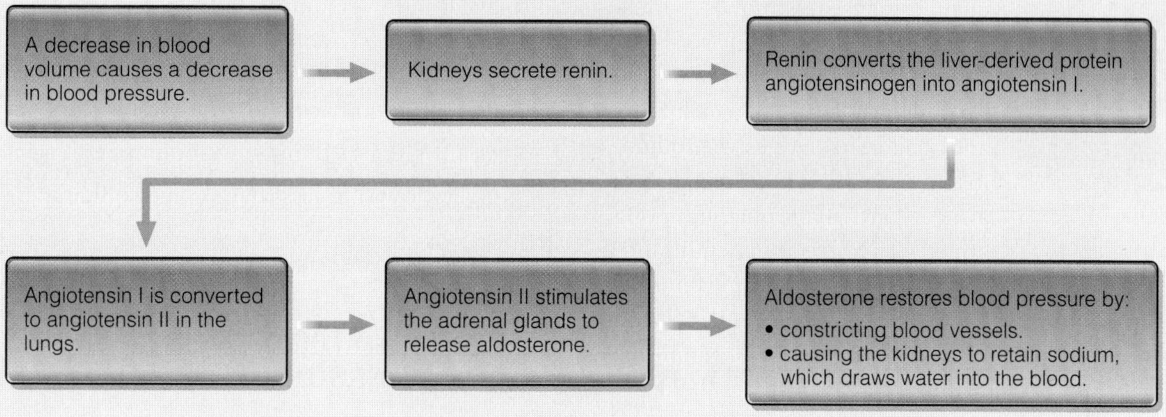

A loss of body fluids triggers a homeostatic response called the renin-angiotensin-aldosterone system, which helps restore blood volume and blood pressure.

Once in the blood, renin converts the liver-derived protein *angiotensinogen* into *angiotensin I*, which in turn is converted to *angiotensin II* in the lungs. Angiotensin II stimulates the adrenal glands to release aldosterone, whose principle effect is to signal the kidneys to retain sodium and return it to the blood. This process assists in the restoration of healthy blood volume by drawing water into the blood via osmosis. Aldosterone also causes blood vessels to constrict, which along with restored blood volume re-establishes normal blood pressure. This helps explain why some people with high blood pressure are advised to restrict their sodium intakes. In fact, some drugs control high blood pressure by disrupting this system.

ELECTROLYTE BALANCE

Sodium, chloride, and potassium are easily absorbed in the small intestine, and to a lesser extent, the colon. Once electrolytes are absorbed, their concentrations in the blood are regulated by the kidneys, which filter excess amounts into the urine. In other words, when blood electrolyte levels are elevated, the excretion of urinary sodium, potassium, and chloride increases. The opposite is true when blood electrolyte levels are low.

Low blood sodium is caused by inadequate sodium intake, loss of sodium through excessive sweating, diarrhea, vomiting, and/or excessive water consumption that is unaccompanied by increased sodium intake. When the concentration of blood sodium dips below a comfortable range, a condition

Not All Sports Drinks Are the Same

Some mechanisms associated with intestinal cells' absorption of sodium require the assistance of glucose. In fact, sports drinks designed to replace electrolytes lost during physical activity historically contained both salt and glucose.[15] Today however, many sports drinks are made with high-fructose corn syrup, a substance that typically contains nearly equal amounts of glucose and fructose. Although it is unclear if the type of sweetener used in a sports drink influences hydration status, sports drinks that contain both glucose and sodium have been found to effectively restore and maintain hydration status after a hard workout.[16] Most experts advise athletes to rely on water for rehydration unless their activity level is especially strenuous or the environment is very warm and/or humid, however.

hyponatremia A condition characterized by low blood sodium concentration.

diuretic A substance or drug that helps the body eliminate water.

hypokalemia A condition characterized by low blood potassium concentration.

called **hyponatremia** develops. Consequences of hyponatremia include nausea, dizziness, muscle cramps, and in severe cases, coma. Hyponatremia can be life threatening—especially in infants and children. Excessive sodium loss associated with sweating is also of particular concern among athletes involved in endurance sports such as marathon running.

Although potassium deficiency is rare, it too can occur because of prolonged diarrhea and/or vomiting. Heavy use of certain diuretics can also result in excessive urinary potassium loss. A **diuretic** is a substance or drug that helps the body eliminate water. Diuretics are often taken to lower blood pressure, but when the body excretes excessive amounts of fluids, it also unavoidably excretes electrolytes. This can lead to **hypokalemia**, a condition characterized by low blood potassium concentration. People with eating disorders that involve vomiting (such as bulimia nervosa) are at increased risk of hypokalemia. Potassium deficiency causes muscle weakness, constipation, irritability, and confusion. In severe cases, potassium deficiency can lead to irregular heart function, muscular paralysis, decreased blood pressure, and difficulty breathing.

Most experts recommend that you limit your sodium intake to less than 2,300 mg each day!

© iStockphoto.com/Suzannah Skelton

DIETARY SOURCES OF ELECTROLYTES

A wide variety of foods contain sodium and chloride. Table salt, meat, dairy products, and pickled foods such as sauerkraut are especially good sources of these electrolytes. Highly processed foods such as sandwich meats, chips, fast food, and condiments such as ketchup and soy sauce are often made with high amounts of salt. In general, manufactured foods often contain substantial amounts of sodium and chloride, whereas unprocessed foods such as fresh fruits and vegetables contain only small amounts. Dietary sources of potassium include legumes, potatoes, seafood, dairy products, meat, and a variety of fruits and vegetables, such as sweet potatoes and bananas. Good sources of sodium, chloride, and potassium are illustrated in Figure 8.10.

You can easily determine the salt content of packaged foods by reading Nutrition Fact panels. Because limiting dietary salt intake may decrease the risk of hypertension, the following health claim is sometimes displayed on low-sodium foods: "Diets low in sodium may reduce the risk of high blood pressure, a disease associated with many factors." The following terms are also used on food packaging to describe salt content:

- Salt free: less than 5 mg sodium per serving
- Very low salt: less than 35 mg sodium per serving
- Low salt: less than 140 mg sodium per serving

SODIUM CHLORIDE INCREASES BLOOD PRESSURE

Aside from making you feel thirsty, an elevated salt intake does not have any demonstrated toxic effects for most people. In some people, however, high intakes of sodium chloride are associated with increased blood pressure, a major risk factor of both heart disease and stroke.[17] Because blood sodium concentration is one of the major regulators of blood volume, higher levels of sodium cause blood volume, and thus blood pressure, to rise.

Epidemiologic studies have long suggested that high blood pressure, also called hypertension, is most likely to occur in people who consume large amounts of salt.[18] However, because hypertensive people are also more likely than nonhypertensive people to be overweight and smoke, it is somewhat difficult to discern which specific factors are *causally* related to blood pressure. Even though a definite causal relationship between salt intake and hypertension has not been established for all people, public health organizations have long recommended that individuals lower blood pressure by losing weight, quitting smoking, and restricting salt intake.

Not all people with hypertension benefit from low-salt diets; there is wide variation in how people respond to salt intake. This may be attributable to factors such as genetics, exercise, and the responsiveness of the renin-angiotensin-aldosterone system. Although there is no easy way to determine whether you are salt sensitive, certain populations experience higher incidence of salt sensitivity than do others. Salt sensitivity is most common among the elderly, women, African Americans, and people with hypertension, diabetes, or chronic kidney disease.[19] Regardless of whether you are salt sensitive, one of the best ways to control hypertension is to follow a diet plan that

Figure 8.10 **Good Sources of Sodium and Potassium**

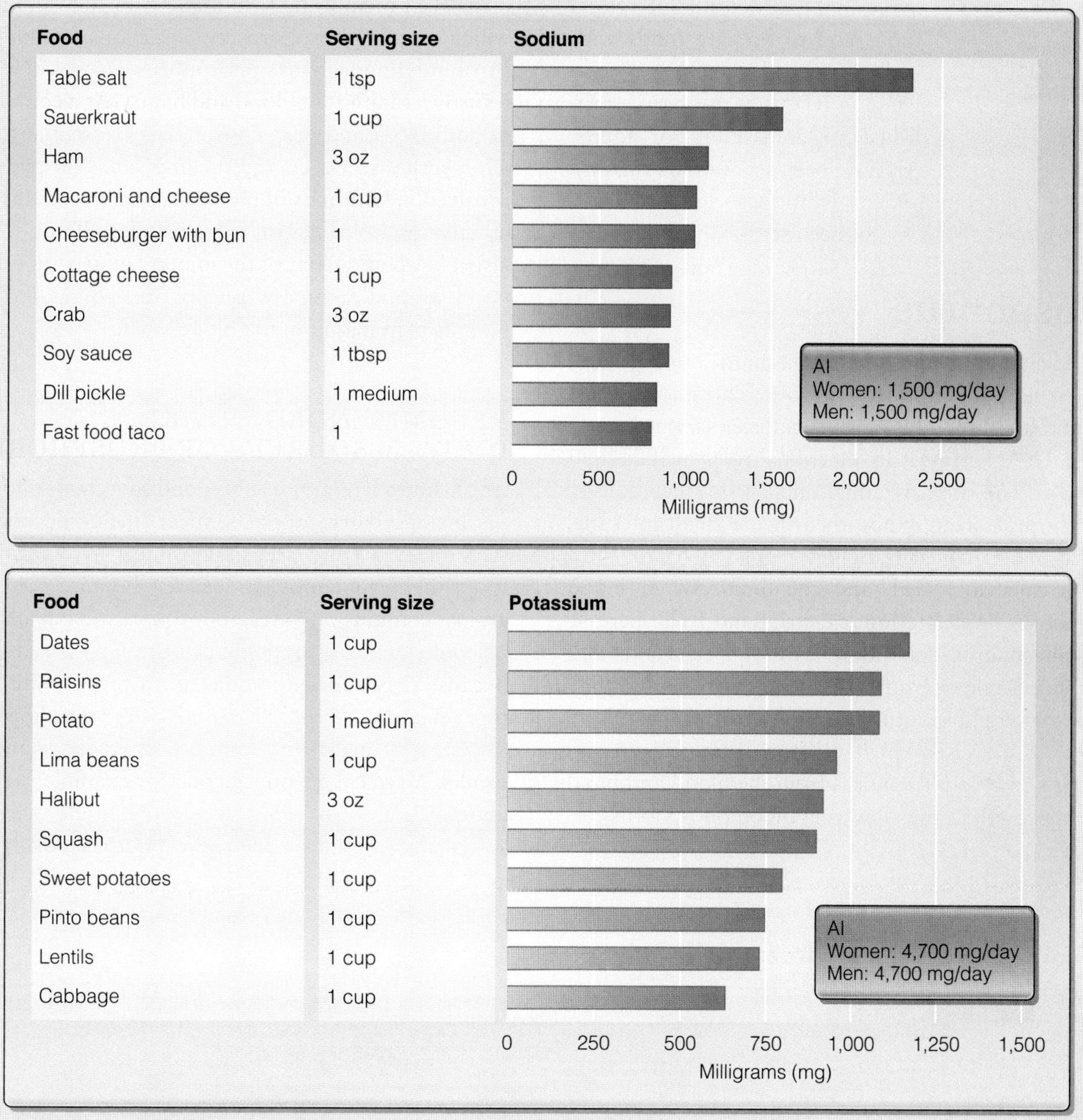

Source: USDA Nutrient Database for Standard Reference, Release 17.

emphasizes reasonable amounts of whole grains, fruits and vegetables, low-fat dairy products, and lean meats.

RECOMMENDED INTAKES FOR SODIUM, CHLORIDE, AND POTASSIUM

Although the AI for sodium is set at 1,500 mg/day for most adults (19–50 years), the 2010 Dietary Guidelines for Americans recommends a sodium intake of less than 2,300 mg/day. As the majority of dietary chloride is associated with sodium chloride and other salts, the AI for chloride is set at 2,300 mg/day. Adults age 51 and older, African Americans, and people with hypertension, diabetes, and/or chronic kidney disease should consume less than 1,500 mg/day of sodium. Highly active individuals (such as endurance athletes) and people living in very warm climates may require additional sodium in their diets. Recommended intake values have not been established for infants, but health professionals suggest that salt not be added to foods during the first year of life. Most people consume about 3,400 mg (1½ tsp) of sodium every day.

phosphorus (P) A major mineral that is essential to cell membranes, bone and tooth structure, DNA, RNA, ATP, lipid transport, and a variety of processes in the body.

Some of the best ways to prevent excessive sodium intake are to avoid highly-salted manufactured foods and to prepare foods with less salt. As with the other electrolytes, an RDA has not been established for potassium. For adults, AIs have been set at 4,700 mg/day. Because potassium toxicity is not associated with dietary sources, a UL has not been set for this mineral.

Phosphorus

Like calcium, **phosphorus (P)** is a major mineral with both structural and functional roles. Phosphorus is essential to the structures of cell membranes, bones, teeth, DNA, RNA, and ATP. Its functional roles include lipid transport, enzyme activation, and energy metabolism. Phosphorus is readily absorbed in the small intestine, and its blood concentration is regulated by the hormones calcitriol, PTH, and calcitonin. When blood phosphorus levels are low, calcitriol and PTH increase both phosphorus absorption in the small intestine and phosphorus release from bone tissue. These actions help return blood phosphorus levels to normal. When blood phosphorus levels are high, calcitonin stimulates osteoblasts to take up phosphorus from the blood and build new bone with it, thus lowering blood phosphorus levels back to normal.

The primary role of phosphorus in the body is as a component of cell membranes. Recall from Chapter 6 that cell membranes are made from phosophlipids, which contain phosphorus. Because phospholipids also surround lipoproteins, phosphorus is critical to the transport of lipids in blood and lymph. As a component of both DNA and ATP, phosphorus is essential to protein synthesis and energy metabolism. Phosphorus is involved in hundreds of metabolic reactions, for which it activates molecules (mostly enzymes). Phosphorus is also a key component of the mineral matrix that makes up bones and teeth, and phosphorus-containing compounds help maintain the blood's pH.

RECOMMENDED INTAKE AND DIETARY SOURCES OF PHOSPHORUS

As illustrated in Figure 8.11, good sources of phosphorus include dairy products, meat (including poultry), seafood, nuts, and seeds. Phosphorus's bioavailability is high from most foods, but because seeds and grains contain phytates, phosphorus bioavailability is lower for these foods than for animal-based foods. Because it is added to some foods to promote moisture retention, smoothness, and taste, phosphorus can also be obtained from some processed foods and soft drinks. A typical 12-ounce cola, for example, contains about 50 mg of phosphorus. To determine whether a soft drink contains added phosphorus, look for the

Figure 8.11 Good Sources of Phosphorus

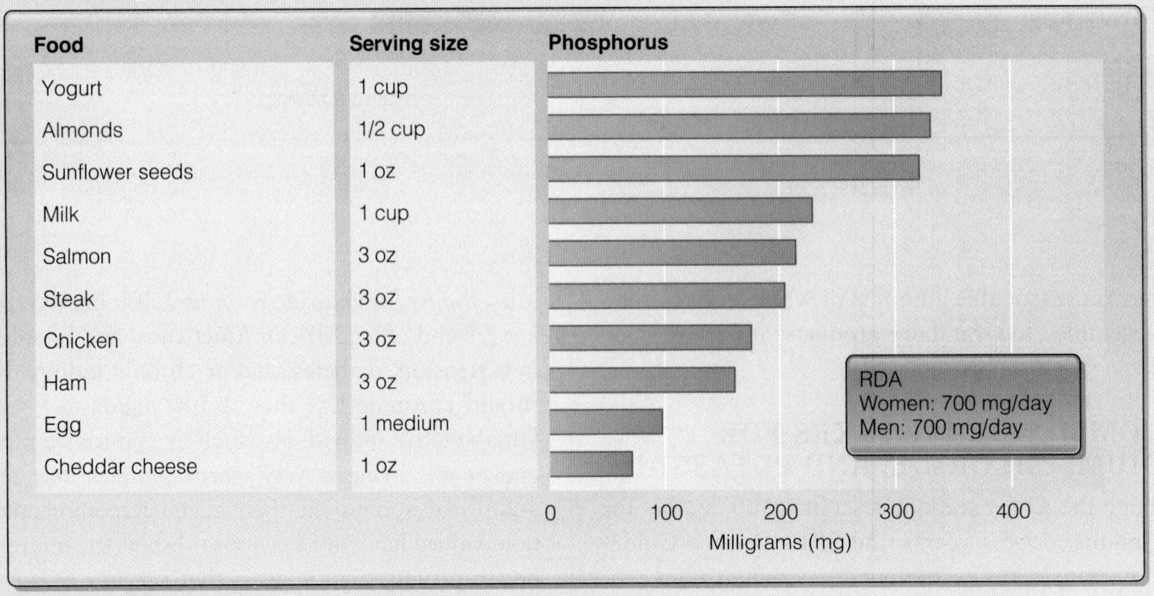

Source: USDA Nutrient Database for Standard Reference, Release 17.

ingredient *phosphoric acid* on the food label. Reliance on soft drinks to supply dietary phosphorus is not recommended because these beverages have low nutrient densities and contribute very little to nutrient requirements.

PHOSPHORUS DEFICIENCY, TOXICITY, AND RECOMMENDED INTAKES

Because of the abundance of phosphorus in most foods—especially those rich in protein—phosphorus deficiency is rare. When it does occur, individuals experience a loss of appetite, anemia, muscle weakness, poor bone development, and, in extreme cases, death. Phosphorus toxicity, on the other hand, is more common, resulting in mineralization of soft tissues such as the kidneys. The RDA for phosphorus is 700 mg/day—approximately the amount you would get from eating two servings of meat and two cups of milk. Because of the dangerous side effects associated with high amounts of phosphorus, a UL of 4,000 mg/day has been established for this mineral.

Magnesium

Magnesium (Mg) is a major mineral that is important to many physiological processes, such as energy metabolism and enzyme function. Blood magnesium concentrations are regulated primarily by the small intestine and to a lesser extent by the kidneys. As with most minerals, magnesium absorption occurs in the small intestine. Absorption increases when blood levels are low and decreases when blood levels are high.

magnesium (Mg) A major mineral that is important to many physiological processes, such as energy metabolism and enzyme function.

The majority of the body's magnesium content is associated with bones, for which it helps provide structure. Magnesium typically exists in the body as a positively charged ion (Mg^{2+}), which stabilizes enzymes and neutralizes negatively charged ions. For instance, magnesium helps stabilize high-energy compounds such as ATP, and is therefore vital to energy metabolism. All told, magnesium participates in more than 300 chemical reactions. Though perhaps most notable for its role in DNA and RNA metabolism, magnesium also influences nerve and muscle function—especially in heart tissue.

DIETARY SOURCES, DEFICIENCY, TOXICITY, AND RECOMMENDED INTAKE OF MAGNESIUM

Good sources of dietary magnesium include whole grains, green leafy vegetables, seafood, legumes, nuts, chocolate, and unprocessed (brown) rice (see Figure 8.12). In general, whole-grain foods contain more magnesium than do those made with refined grains. Magnesium's bioavailability is variable, and it is likely

Figure 8.12 **Good Sources of Magnesium**

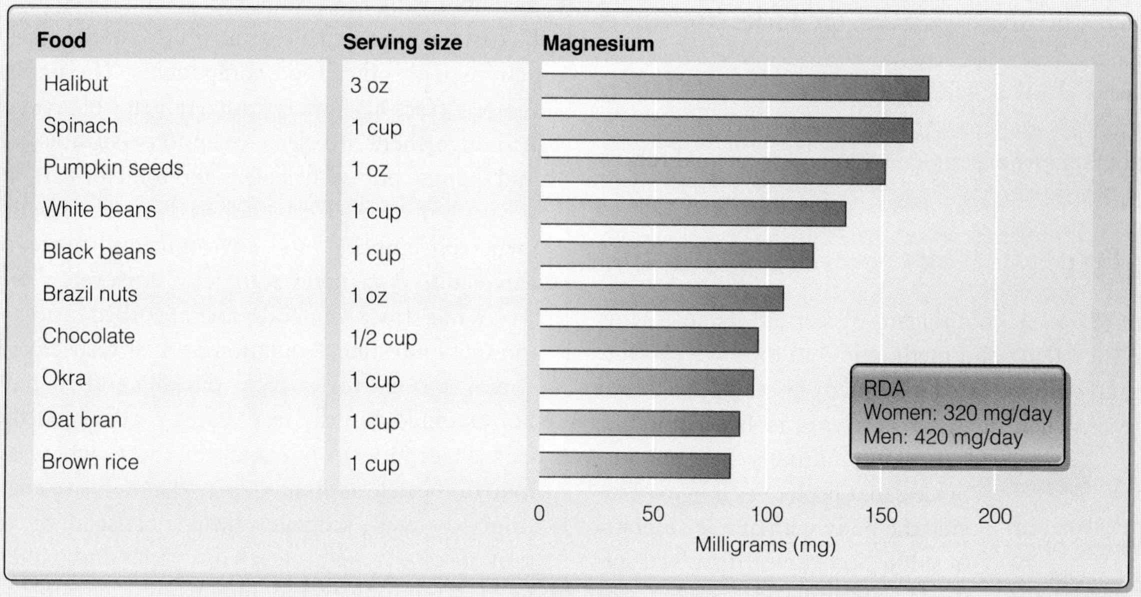

Source: USDA Nutrient Database for Standard Reference, Release 17.

Chocolate *is a* **good source** *of dietary* **magnesium.**

that several dietary factors influence its absorption. Some studies suggest that a diet high in calcium or phosphorus may decrease the bioavailability of magnesium.[20] Protein deficiency and/or high levels of dietary fiber may decrease the absorption of magnesium.[21]

Magnesium deficiency is rare, but is sometimes observed in alcoholics who suffer from dietary inadequacy and poor overall nutrient absorption. Severe magnesium deficiency causes abnormal nerve and muscle function. Magnesium toxicity can occur when large amounts of supplemental magnesium or medications that contain magnesium, such as milk of magnesia (magnesium hydroxide), are consumed. Such medications are often used to treat heartburn, indigestion, and constipation. Symptoms of magnesium toxicity include diarrhea, intestinal cramping, and nausea. Because severe magnesium toxicity can cause heart failure, the consumption of magnesium supplements should be carefully monitored.

RDAs for magnesium are 420 and 320 mg/day for men and women, respectively. You might consume this amount of magnesium by eating three ounces of halibut, a cup of spinach, and a serving of rice. The UL for magnesium is 350 mg/day for adults. This UL is unusual because it is actually similar to or less than the established RDA values. This is because the UL does not apply to magnesium obtained from foods—only that obtained from supplemental and medicinal forms.

Sulfur

Sulfur (S) is a component of certain amino acids (such as cysteine and methionine) in the body. Sulfur-containing amino acids give many proteins structural rigidity. As such, they are necessary to healthy joints, hair, skin, and nails. It is not surprising that protein-rich foods provide the body with ample amounts of sulfur. Sometimes, dissolved sulfur is detectable in drinking water by its egg-like (but harmless) odor and taste. Certain juices, beers, wines, and ciders contain sulfur, and sulfur-containing agents are added to some dried foods to preserve their freshness and color.

sulfur (S) A major mineral that is a component of certain amino acids (such as cysteine and methionine).

Adequate sulfur intake enables the synthesis of compounds that construct healthy connective tissue and facilitate nerve function. Sulfur is also an important component of the B vitamins thiamin and biotin, and is therefore essential to energy metabolism. Sulfur requirements are easily met when adequate amounts of sulfur-containing amino acids are consumed. The DRI committee took this into account when developing intake recommendations for protein and decided not to publish additional DRI values for sulfur.

LO3 What Are Trace Minerals?

Although the body only requires the essential trace minerals (iron, copper, iodine, selenium, chromium, manganese, molybdenum, and zinc) in minute amounts, a daily intake of each of these nutrients is vital to one's health. Another trace mineral, fluoride, is not technically an essential nutrient, but may nonetheless influence health. You obtain trace minerals from a wide variety of plant- and animal-based foods. Bioavailability of these minerals is influenced by many factors, such as genetics, nutritional status, and interactions with other food components. The natural aging processes also affects trace mineral bioavailability. Although there are clear exceptions (such as for iron and iodine), both deficiencies and toxicities are rare for the trace minerals. Genetic disorders, excessive supplement consumption, and environmental overexposure can lead to trace mineral toxicity, however.

While trace minerals are absorbed primarily in the small intestine, regulation of their respective blood concentrations varies from one mineral to the next. For example, healthy iron status is regulated through iron absorption in the small intestine, whereas other minerals (such as iodine and selenium) are regulated primarily by the kidneys. The liver assists in the regulation of some minerals (such as copper and manganese) by incorporating excess amounts of trace minerals into bile, which is subsequently eliminated in the feces.

Most trace minerals do not exist in their free forms within cells—they are instead bound to specific proteins. Some trace minerals act as cofactors (enzyme activators), while others function as components of larger nonenzymatic molecules such as hemoglobin. Like several of the major minerals, some trace minerals (such as fluoride) provide strength and structure to bones and teeth.

Iron

Iron (Fe) is a trace mineral that is necessary for oxygen and carbon dioxide transport, energy metabolism, the stabilization of free radicals, and the synthesis of DNA. Although iron is likely the most studied trace mineral, iron deficiency remains the most common micronutrient deficiency in the United States and around the world.[22] Scientists once believed that iron's only bodily function was to transport oxygen and carbon dioxide in the blood, but it is now clear that it serves many other important functions as well.

IRON'S FUNCTIONS IN THE BODY

The majority of the body's iron is incorporated into the oxygen-carrying molecules hemoglobin and myoglobin. **Hemoglobin**, found in red blood cells, is a complex protein that transports oxygen from the lungs to cells and carbon dioxide, a waste product of cellular metabolism, from cells to the lungs. **Myoglobin** serves as an oxygen-storage molecule within muscles. When circulating hemoglobin is not able to deliver enough oxygen to muscles (as is the case during strenuous physical exertion), the oxygen stored in myoglobin is released. This allows the muscle cells to continue to produce ATP, even though blood oxygen availability is low. Beyond its roles in hemoglobin and myoglobin, iron is a component of several metalloenzymes, is necessary for DNA synthesis, and plays a role in immunity.

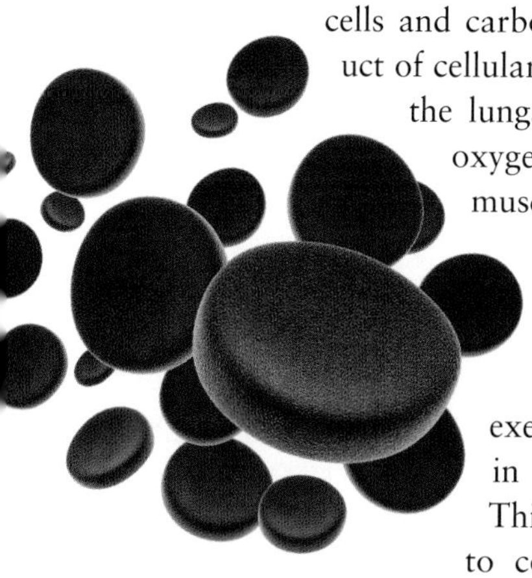

IRON BIOAVAILABILITY

Except for blood loss associated with menstruation, injury, and childbirth, the body loses very little iron once it has been absorbed. The body must therefore be careful not to absorb more iron than it needs. Iron bioavailability is influenced by several factors, such as the type of iron consumed, the presence or absence of other dietary components, and an individual's iron status.[23] Iron status greatly influences the amount of iron absorbed from a food. Specifically, absorption increases when iron stores are low and decreases when they are high. Because iron absorption is highest among those with the greatest need for iron, people with iron deficiency anemia, women with heavy menstrual losses, and pregnant women tend to absorb the most iron.

> "Iron's chemical symbol, **Fe**, was named for the **Latin word** for iron—***ferrum***."

Aside from an individual's iron status, another important factor that influences iron bioavailability is the type of iron consumed. There are two forms of dietary iron: heme iron and nonheme iron. The bioavailability of heme iron is two to three times greater than that of nonheme iron. Because **heme iron** is a part of hemoglobin and myoglobin, it is found only in meat such as beef, pork, poultry, and fish. **Nonheme iron** is not part of hemoglobin or myoglobin, and is found mainly in plant-based foods such as green leafy vegetables, mushrooms, and legumes.

Many dietary factors influence the bioavailability of nonheme iron. For instance, nonheme iron exists in two charged forms: *ferric iron* (Fe^{3+}) and *ferrous iron* (Fe^{2+}). The latter is more bioavailable than the former. The following factors can also influence the bioavailability of nonheme iron:

- Vitamin C (ascorbic acid) is one of the best-known enhancers of nonheme iron absorption. Because vitamin C converts ferric iron to ferrous iron in the small intestine, consuming it with nonheme iron in a meal enhances iron absorption.
- **Meat factor** is a compound found in meat that increases the bioavailability of

iron (Fe) A trace mineral needed for oxygen and carbon dioxide transport, energy metabolism, stabilization of free radicals, and synthesis of DNA.

hemoglobin A complex protein that transports oxygen from the lungs to cells and carbon dioxide from the cells to the lungs.

myoglobin An oxygen-storage molecule located within muscles.

heme iron Iron that is part of hemoglobin and myoglobin and is found only in meat.

nonheme iron Iron that is not part of hemoglobin and myoglobin and is found primarily in plant-based foods.

meat factor A compound found in meat that increases the bioavailability of nonheme iron.

chelator A compound that binds to nonheme iron in the intestine, making it unavailable for absorption.

transferrin A protein in the blood that binds to dietary iron once it is released from an intestinal cell.

ferritin A protein complex that stores iron and releases it only when it is needed.

nonheme iron. Consuming even a small amount of meat factor along with nonheme iron-containing grains or vegetables will increase the iron's bioavailability.

- **Chelator** compounds such as phytates and *polyphenols* bind to nonheme iron in the intestine, making the iron unavailable for absorption. Phytates are found in many vegetables, grains, and seeds, whereas polyphenols are found primarily in plant-based foods such as spinach, tea, coffee, and red wine. If you were to consume one cup of coffee or tea with a nonheme iron-containing meal, iron absorption might decrease by 40 to 70 percent.[24]

IRON ABSORPTION, TRANSPORT, AND STORAGE

Although iron deficiency causes many problems, iron toxicity is perhaps even more dangerous. In response, complex homeostatic mechanisms regulate the amount of iron absorbed and stored in the body, ensuring that there is neither too little nor too much. Transport proteins begin the process of absorption by escorting iron from the microvilli into the intestinal cells. Once iron is inside a cell, the first level of iron regulation begins. Iron-storing proteins acting as gatekeepers determine iron's initial fate by either releasing it into the blood or retaining it in the intestinal cell for subsequent elimination in the feces. This dynamic process optimizes iron bioavailability in relation to the body's iron status (see Figure 8.13).

Dietary iron released from the intestinal cell into the blood binds to a protein called **transferrin**, which delivers the iron to special receptors on cell membranes. The number of iron-binding receptors on a cell membrane changes in response to the cell's need for iron: cells increase the number of receptors when they need more iron, and decrease the number of receptors when they need less.

When more iron is consumed than is needed, small amounts of excess iron are stored in the body. Surplus iron is incorporated into **ferritin**, a protein found primarily in liver, skeletal muscle, and bone marrow cells. Ferritin forms a protein complex that only releases iron when it is needed. Blood ferritin levels can be used to assess a person's iron status.

IRON DEFICIENCY

When iron-containing red blood cells wear out, the body quickly reclaims the iron contained therein so

Figure 8.13 Effect of Iron Status on Iron Absorption

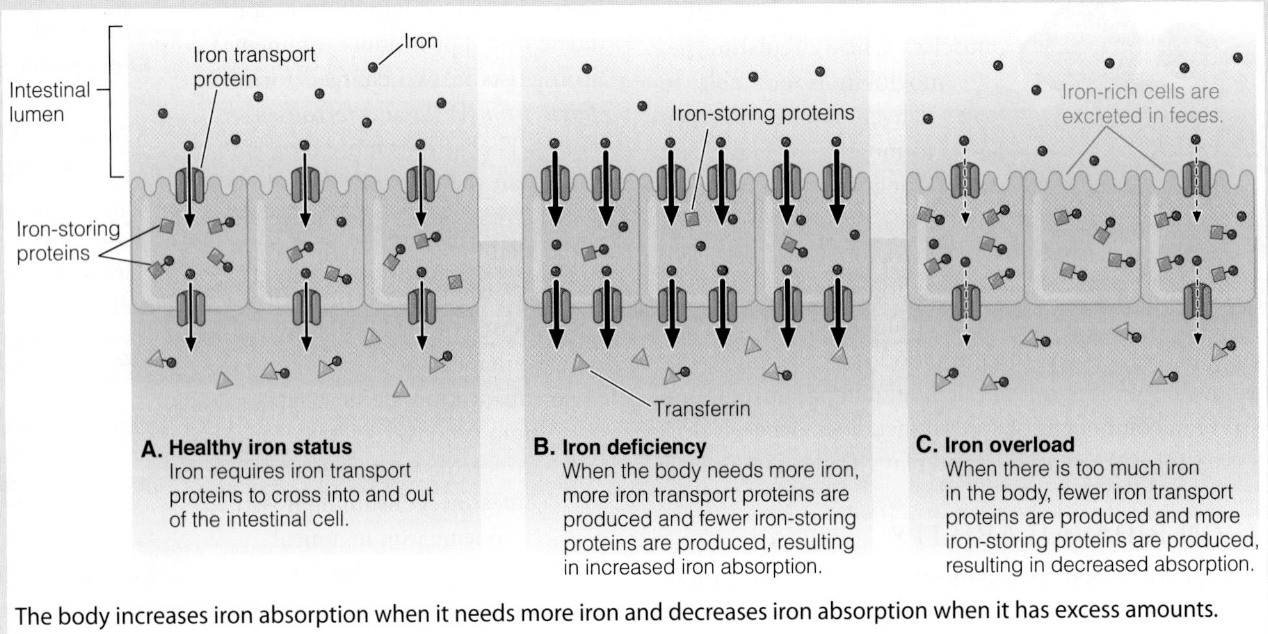

Intestinal lumen
Iron transport protein
Iron
Iron-storing proteins
Iron-storing proteins
Iron-rich cells are excreted in feces.
Transferrin
A. Healthy iron status
Iron requires iron transport proteins to cross into and out of the intestinal cell.
B. Iron deficiency
When the body needs more iron, more iron transport proteins are produced and fewer iron-storing proteins are produced, resulting in increased iron absorption.
C. Iron overload
When there is too much iron in the body, fewer iron transport proteins are produced and more iron-storing proteins are produced, resulting in decreased absorption.

The body increases iron absorption when it needs more iron and decreases iron absorption when it has excess amounts.

that it can be used to make new red blood cells. This system of recycling and reusing is not enough to meet the body's requirement for iron, however. If iron intake does not adequately replace iron lost from the body, iron deficiency can develop. Because iron requirements increase during periods of growth and development, iron deficiency is typically observed in infants, growing children and teenagers, and pregnant women. Women are especially at risk for iron deficiency because blood iron is lost each month during menstruation. Iron defi ciency was once believed only to cause iron deficiency anemia, a condition characterized by a decreased ability of the red blood cells to carry oxygen. Scientists now recognize that iron deficiency can influence many other aspects of health as well.[25]

The consequences of impaired iron status begin to occur before the noticeable onset of iron deficiency anemia. Early symptoms of iron deficiency are fatigue and impaired physical work performance. Mild iron deficiency impairs body temperature regulation—especially in cold conditions—and may negatively influence the immune system.[26] Even mild iron deficiency can affect behavior and intellectual abilities in children.[27] Unfortunately, some of these effects are irreversible, even after iron supplementation. Studies suggest that mild iron deficiency experienced during pregnancy increases the risk of premature delivery, low birth weight, and maternal death.[28] Because blood ferritin levels (which reflect iron storage) decrease long before anemia develops, measuring blood ferritin concentration allows detection of iron deficiency in its early stages.

Severe iron deficiency is characterized by the presence of microcytic (small), hypochromic (lightly colored) red blood cells. Microcytic, hypochromic anemia develops due to an inability to produce enough heme (and thus hemoglobin), which causes the concentration of hemoglobin in red blood cells to decrease. This impairs the delivery of oxygen to cells. Blood hemoglobin concentrations of less than 13 and 12 g/dL suggest the possibility of iron deficiency in men and women, respectively. Another indicator of severe iron deficiency is low **hematocrit**, the percentage of blood volume comprised of red blood cells. Hematocrit values decrease during iron deficiency because red blood cells are small and therefore take up less volume. Hematocrit values of less than 39 and 35 percent indicate anemia in men and women, respectively. Because microcytic, hypochromic anemia is also associated with other nutritional deficiencies (such as that of vitamin B_6), it can sometimes be difficult to determine whether iron deficiency is to blame.

hematocrit The percentage of blood volume comprised of red blood cells.

hereditary hemochromatosis A genetic abnormality whereby too much iron is absorbed, causing it to accumulate in the body.

IRON OVERLOAD AND TOXICITY

Because it is difficult for the body to excrete iron once it is absorbed, the potential for iron toxicity is high. The regulation of iron absorption generally protects healthy individuals from iron overload, even when an abundance of iron-rich foods are consumed. Iron regulation is not always absolute, however. Ingestion of large amounts of iron (especially in supplemental forms) can overload the body's regulatory mechanisms, resulting in the absorption of too much iron. Signs and symptoms associated with acute iron toxicity include nausea, vomiting, diarrhea, rapid heart rate, confusion, and damage to the liver and other organs. Because iron toxicity is a leading cause of accidental death in young children, parents and care-providers must be sure to store nutrient supplements out of their reach.

Iron toxicity can also be caused by a genetic abnormality called **hereditary hemochromatosis**, which involves a defect in or overproduction of one of the iron-regulating proteins. Too much iron is absorbed, causing it to accumulate in the body. As iron accumulates in the body, it eventually causes joint pain, bronze skin color, and organ damage. This condition is easily overlooked in the early states—most people are not properly diagnosed until middle age. One of the most effective treatments for hereditary hemochromatosis is regular removal of small amounts of blood to lower iron levels in the body. This disorder (which affects approximately one out of every 1,000 Americans) is more likely to develop in men than in women and tends to be more common in Caucasians than in other racial groups.

DIETARY SOURCES AND RECOMMENDED INTAKE OF IRON

Excellent sources of heme iron include shellfish, beef, poultry, and organ meats such as liver. Enriched foods and other iron-fortified foods are good sources of nonheme iron (see Figure 8.14). Iron RDAs are 8 and 18 mg/day for men and women, respectively. For women, the RDA increases to 27 mg/day during pregnancy. Because a three-ounce serving of beef contains only about 3 mg of iron, it can be difficult for some women to meet their RDAs—especially while pregnant. For this reason, iron supplements are often recommended during pregnancy. To prevent GI distress and other complications from iron excess, a UL of 45 mg/day has been set for iron.

SPECIAL RECOMMENDATIONS FOR VEGETARIANS AND ENDURANCE ATHLETES

Because only meat contains substantial amounts of highly bioavailable iron, vegetarians may have difficulty consuming adequate amounts of iron. As such, the Institute of Medicine estimates that vegans' dietary iron requirements are 80 percent higher than those of omnivores.[29] To take advantage of meat factor's positive effect on nonheme iron absorption, vegetarians who eat fish should try to consume some fish when they consume nonheme iron sources. Eating even small amounts of fish alongside iron-fortified foods can greatly increase the amount of nonheme iron absorbed. Some fish-eating vegetarians may still require iron supplements, however.

Athletes engaged in endurance sports such as long-distance running may also have increased dietary iron requirements. Increased blood loss in feces and urine and/or chronic rupture of red blood cells caused by the impact of running on hard surfaces may be why endurance athletes have increased iron requirements. The Institute of Medicine suggests that endurance athletes' iron requirements may increase by as much as 70 percent, although specific DRI values have not been established for this group.

Figure 8.14 Good Sources of Iron

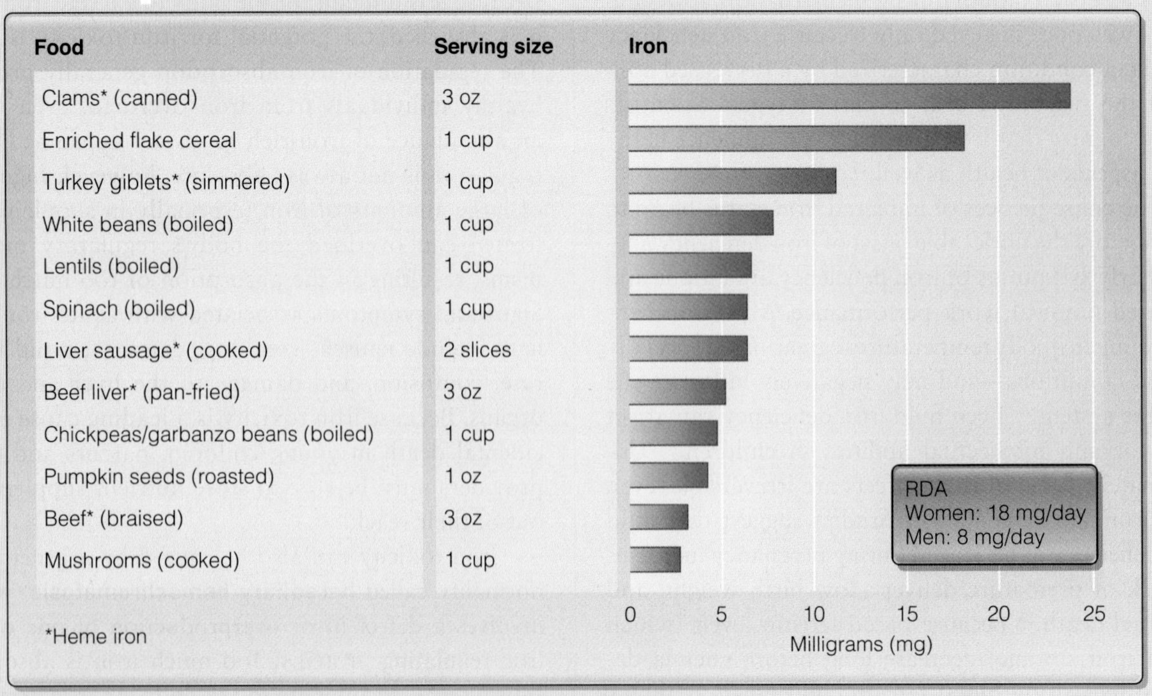

Source: USDA Nutrient Database for Standard Reference, Release 16-1.

IRON SUPPLEMENTATION

Iron supplementation is sometimes necessary when diet alone cannot sustain healthy iron status. Supplemental iron is available in both ferrous iron and ferric iron forms. Of these, the ferrous forms (ferrous fumarate, ferrous sulfate, and ferrous gluconate) are most easily absorbed.[30] The U.S. Centers for Disease Control and Prevention (CDC) recommends taking 50 to 60 mg of iron, the approximate amount in one 300 mg tablet of ferrous sulfate, twice daily for three months to treat iron deficiency anemia. Starting with half this dose and gradually increasing to the full dose may help minimize potential side effects such as gastrointestinal upset and dark stools.

> **Menkes disease** is a rare genetic condition that **impairs copper metabolism.** First discovered by Dr. John Menkes in 1962, this disease is **characterized by** poor muscle tone, subnormal body temperatures, and kinky, fair- or steel-colored hair that breaks easily.

Copper

Copper (Cu) is an essential trace mineral that acts as a cofactor for nine enzymes involved in reduction-oxidation (redox) reactions. Perhaps best known for its role as a cofactor, copper performs a host of other functions in the body as well. Similar to iron, copper is found in two distinct charged states: the *cupric form* (Cu^{2+}) and the *cuprous form* (Cu^{1+}). This is not the only commonality that copper and iron share, however. Copper and iron also have similar food sources, absorption mechanisms, and functions in the body.

As a cofactor for at least nine metalloenzymes, copper plays an important role in energy metabolism, iron metabolism, neural activity, antioxidant defense, and connective tissue synthesis. **Superoxide dismutase**, a particularly important copper-containing enzyme, acts as an antioxidant, stabilizing the free radical molecules that cause cell and tissue damage in the body. Copper is also required for the syntheses of collagen and norepinephrine, a neurotransmitter that is essential to brain function.

COPPER BIOAVAILABILITY, ABSORPTION, AND FUNCTIONS

Copper is absorbed primarily in the small intestine. As with iron, copper absorption increases when stores are low and decreases when stores are high. The mechanisms by which this occurs are poorly understood. Because they react with copper to form insoluble complexes that cannot be absorbed, antacids diminish copper bioavailability. High amounts of dietary iron can also decrease copper absorption because these two substances compete for the same transport proteins in the small intestine.

Once absorbed, copper circulates in the blood to the liver, where it is bound to its primary transport protein, **ceruloplasmin**. The majority of copper in the body does not circulate freely, but is bound to various transport proteins, storage proteins, and copper-containing enzymes. Excess copper is not stored in the body—it is instead incorporated into bile and eliminated in the feces.

COPPER DEFICIENCY AND TOXICITY

Copper deficiency is rare in individuals who eat balanced and varied diets. Still, it occurs occasionally in hospitalized patients and premature infants who receive improper nutritional support. Because antacids impair copper absorption, people who consume large amounts of them sometimes develop copper deficiencies. Signs and symptoms of copper deficiency include anemia, abdominal pain, nausea, diarrhea, and vomiting. In extreme cases, neurological problems may develop. Because copper toxicity is rare, very little is known about it. Nonetheless, people who have consumed large amounts of copper in drinking water and contaminated soft drinks have experienced cramping, nausea, and diarrhea.[31] Copper toxicity may also cause liver damage.

copper (Cu) An essential trace mineral that acts as a cofactor for nine enzymes involved in redox reactions.

ceruloplasmin A transport protein that binds to copper and transports it in the blood.

superoxide dismutase A copper-containing enzyme that helps stabilize highly reactive free radical molecules.

iodine (I) An essential trace mineral that serves as a component of the thyroid hormones.

thyroid-stimulating hormone (TSH) A hormone, produced by the pituitary gland, that regulates iodine uptake by the thyroid gland.

goitrogen A dietary compound, found in some plants, that decreases iodine utilization by the thyroid gland.

DIETARY SOURCES AND RECOMMENDED INTAKE OF COPPER

Although organ meats such as liver are likely the best sources of copper, this mineral is also found in shellfish, whole-grain products, mushrooms, nuts, and legumes (see Figure 8.15). An RDA of 900 μg/day has been established for adults. To prevent possible liver damage caused by copper toxicity, a UL of 10 mg/day has also been established.

Iodine

Iodine (I) is an essential trace mineral that serves as a component of the thyroid hormones. Although iodine affects the body in many ways, it is only essential to one function—the production of hormones by the thyroid gland. These hormones regulate growth, reproduction, and energy metabolism. Thyroid hormones also influence the immune system and neural development. Technically, most of the iodine in your body exists in the form of *iodide* (I^-). Consistent with much of the nutrition literature, however, this mineral is referred to herein and commonly as *iodine*.

IODINE BIOAVAILABILITY, ABSORPTION, AND FUNCTIONS

Iodine is highly bioavailable, and is absorbed mostly in the small intestine. Once absorbed, iodine is circulated to the thyroid gland and incorporated into the thyroid hormone *thyroxine* (T_4). Thyroxine, which contains four iodine atoms, can be converted to *triiodothyronine* (T_3), a more active form that contains three iodine atoms. T_3 and T_4 help regulate energy metabolism, growth, and development, and are critical to proper brain, spinal cord, and skeletal development during fetal growth. Because thyroid hormones are involved in the regulation of energy metabolism, low levels can cause severe fatigue.

Thyroid-stimulating hormone (TSH), a hormone produced by the pituitary gland, regulates iodine uptake by the thyroid gland (see Figure 8.16). During periods of iodine deficiency, TSH production increases, in turn increasing the percentage of iodine taken up by the thyroid gland. As you might expect, TSH production decreases when excess iodine is consumed. Iodine not needed by the body is excreted in the urine.

A type of dietary compound called a **goitrogen** can decrease the thyroid gland's utilization of iodine.

Figure 8.15 Good Sources of Copper

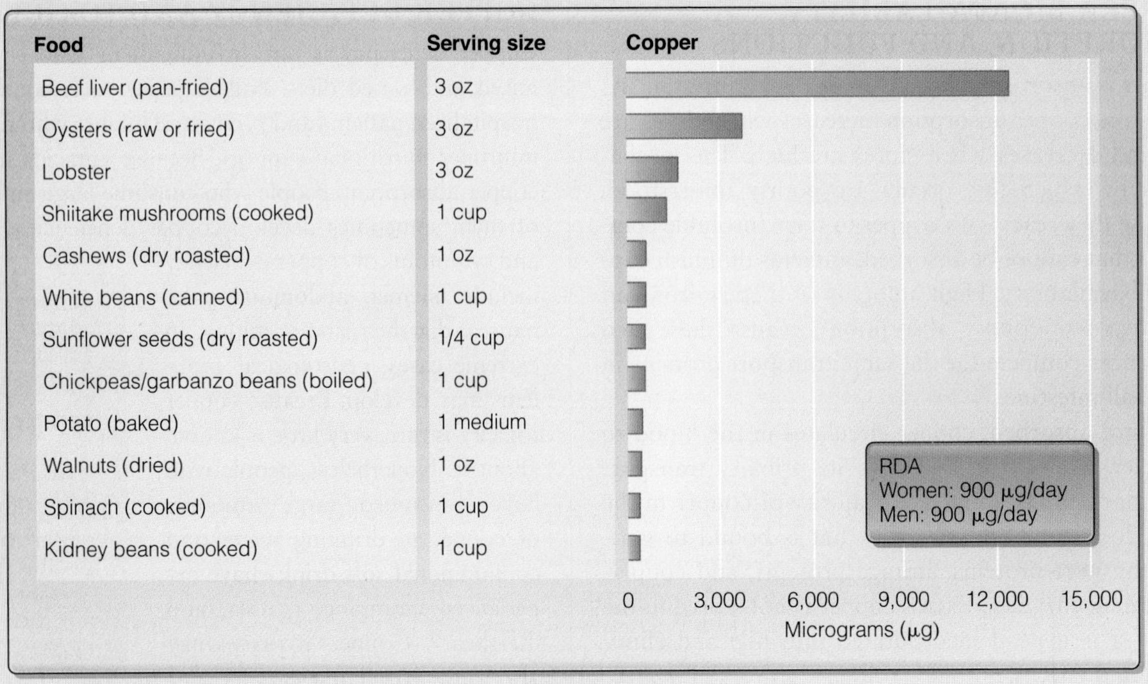
Food
Serving size
Copper
Beef liver (pan-fried)
3 oz
Oysters (raw or fried)
3 oz
Lobster
3 oz
Shiitake mushrooms (cooked)
1 cup
Cashews (dry roasted)
1 oz
White beans (canned)
1 cup
Sunflower seeds (dry roasted)
1/4 cup
Chickpeas/garbanzo beans (boiled)
1 cup
Potato (baked)
1 medium
Walnuts (dried)
1 oz
Spinach (cooked)
1 cup
Kidney beans (cooked)
1 cup
RDA
Women: 900 μg/day
Men: 900 μg/day
0
3,000
6,000
9,000
12,000
15,000
Micrograms (μg)

Source: USDA Nutrient Database for Standard Reference, Release 17.

Figure 8.16 Regulation of Iodine Uptake by the Thyroid Gland

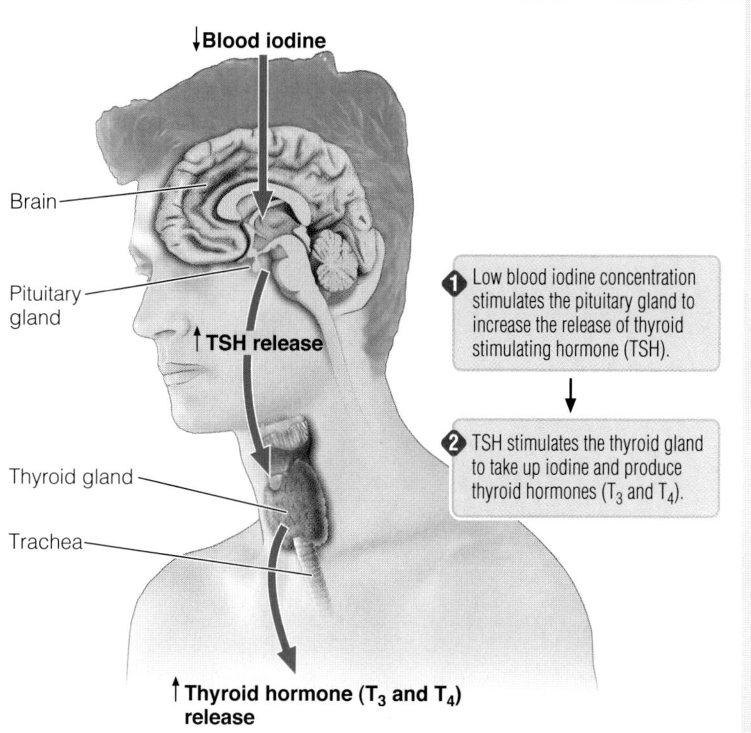
↓Blood iodine
Brain
Pituitary gland
↑TSH release
1 Low blood iodine concentration stimulates the pituitary gland to increase the release of thyroid stimulating hormone (TSH).
2 TSH stimulates the thyroid gland to take up iodine and produce thyroid hormones (T_3 and T_4).
Thyroid gland
Trachea
↑Thyroid hormone (T_3 and T_4) release

Produced in the pituitary gland, TSH stimulates iodine uptake in the thyroid gland and the production of thyroid hormones T_3 and T_4. The production of TSH increases during iodine deficiency.

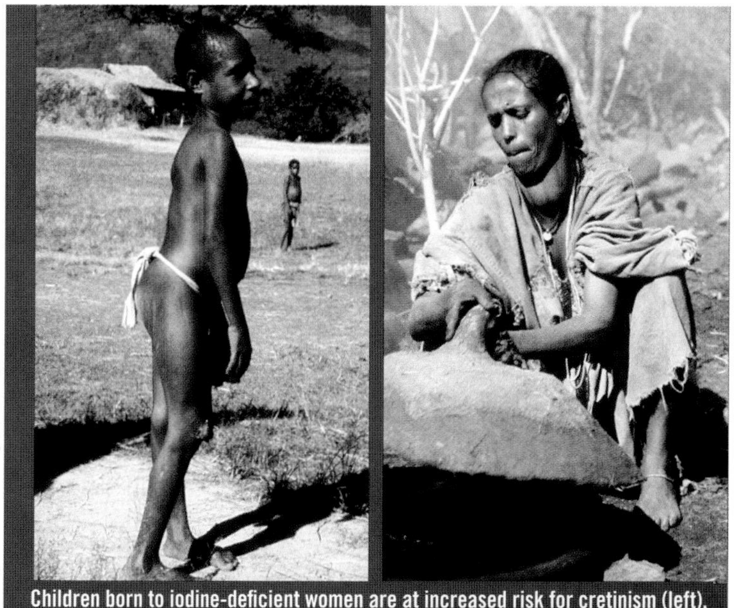

Children born to iodine-deficient women are at increased risk for cretinism (left), which causes delayed growth, developmental delay, and other abnormal features. Goiter (right), caused by enlargement of the thyroid gland in response to iodine deficiency, affects both children and adults.

Goitrogens are found in soybeans, cassava (a root eaten worldwide), and cruciferous vegetables such as cabbage, cauliflower, and Brussels sprouts. Goitrogens were named as such because they can potentially cause goiter, a disease described in the following section. The consumption of goitrogens does not typically pose a problem, except in conditions of very low iodine intake or in people who have thyroid dysfunction.

IODINE DEFICIENCY AND TOXICITY

Iodine deficiency is a significant public health problem worldwide. It is most prevalent in countries that do not have access to iodized salt and those that are not near an ocean or sea.[32] Iodine deficiency manifests in a broad spectrum of conditions collectively called **iodine deficiency disorders (IDDs)**. The type of IDD that develops in an iodine-deficient individual depends on factors such as genetics, severity of the deficiency, and age. The two most studied forms of IDD are cretinism and goiter.

The most severe form of iodine deficiency, **cretinism**, affects babies born to iodine-deficient mothers. During fetal growth, the baby relies on the mother's thyroid hormones for its own growth and development. If a mother is iodine-deficient, she does not produce sufficient thyroid hormones, which affects the baby's growth and development. Cretinism causes severe mental retardation, poor growth, infertility, and increased risk of death. This disease can be prevented in iodine-deficient women through consumption of iodine-rich foods or iodine supplements in early pregnancy.

When iodine deficiency occurs in children and adults, low levels of T_3 and T_4 signal the pituitary gland to secrete increased amounts of TSH in order to boost thyroid

iodine deficiency disorders (IDDs) A broad spectrum of conditions caused by iodine deficiency.

cretinism A severe form of iodine deficiency that affects babies born to iodine-deficient mothers.

hormone production. When chronically elevated levels of TSH stimulate the thyroid gland, they cause it to enlarge. An enlarged thyroid gland, known commonly as a **goiter**, is thus a classic sign of iodine deficiency. When a goiter grows very large, it can obstruct a person's trachea, making breathing and swallowing difficult. In some cases, portions of the thyroid gland are surgically removed to prevent these complications.

A high intake of iodine can cause iodine toxicity. This can take several forms, including both *hypothyroidism* (underactive thyroid activity) and *hyperthyroidism* (overactive thyroid activity). While iodine deficiency is the main cause of goiter, iodine toxicity can cause thyroid gland enlargement as well. The mechanisms by which this occurs are poorly understood. Further complicating the diagnosis of iodine deficiency, both hypothyroidism and hyperthyroidism can be caused by many factors other than diet. For example, the autoimmune disease Hashimoto's thyroiditis is caused by the production of antibodies that destroy the thyroid hormone-producing cells in the thyroid gland, resulting in hypothyroidism.

Iodine Deficiency and Iodization

Iodine deficiency is relatively rare in the United States, but this was not always the case. Prior to salt iodization in the 1920s, iodine deficiency was a major public health problem in some regions of the country. In fact, the region that spanned the Great Lakes and Rocky Mountains was once referred to as the *Goiter Belt.* When iodine deficiency was discovered to be the cause of goiter, state-wide campaigns promoting iodized salt as the cure for goiter were launched. A 75 percent reduction in childhood goiter was observed in just four years, and by the 1950s, less than 1 percent of children were found to be affected by goiter. Clearly, iodized salt has drastically improved the health of many people—a prime example of how scientists can team up with public health officials and food manufacturers to improve the well-being of millions of people through better nutrition.

DIETARY SOURCES AND RECOMMENDED INTAKE OF IODINE

The amount of iodine contained in a food frequently depends on the iodine content of the soil or water in which the food was grown (see Figure 8.17). Ocean fish and mollusks tend to contain high amounts of iodine because they concentrate iodine from seawater into their tissues. Edible varieties of seaweed (such as nori) are excellent sources of iodine, as are many dairy products because iodine is used in milk processing. Most of the iodine you consume probably comes from iodized table salt. Approximately half of all salt consumed in America is iodized, meaning that it has been fortified with iodine. (Kosher salt and sea salt are not traditionally iodized.) Iodine's RDA is 150 μg/day. To prevent iodine toxicity, a UL of 1,100 μg/day has been established for this mineral.

goiter A sign of iodine deficiency characterized by an enlarged thyroid gland.

selenium (Se) An essential trace mineral that is critical to redox reactions, thyroid function, and the activation of vitamin C.

selenoprotein A protein comprised of selenium-containing amino acids.

Selenium

Selenium (Se) is an essential trace mineral that is critical to redox reactions, thyroid function, and the activation of vitamin C. The essentiality of selenium was only discovered in the 1950s. Since that time, much has been learned about how the body uses this trace mineral. Optimal selenium intake is believed to decrease the risk of cancer, protect the body from toxins and free radicals, activate vitamin C, slow the aging process, and enhance immunity.

SELENIUM BIOAVAILABILITY, ABSORPTION, AND FUNCTIONS

Selenium's bioavailability is high, and its intestinal absorption is not regulated. Once absorbed, selenium circulates in the blood to the body's cells. Some selenium is incorporated into amino acids, which are subsequently used to make **selenoprotein** molecules. There are at least 14 selenoproteins in the body. One group of enzymatic selenoproteins helps protect against oxidative

Figure 8.17 Good Sources of Iodine

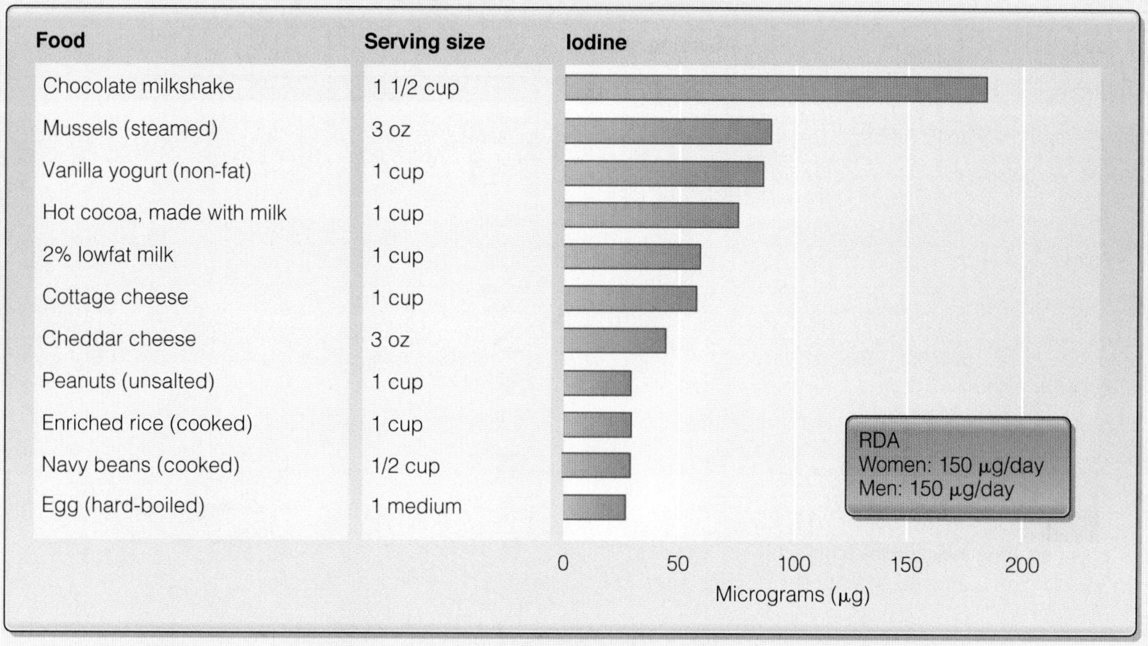

Source: Hands E. Food Finder Vitamin and Mineral Source Guide, 3rd ed. Salem, OR: ESHA Research. 1995.

damage and may protect against certain types of cancer, but more research is required to understand the role of selenium in disease prevention.[33] The kidneys maintain blood selenium concentrations, and excess amounts are simply excreted in the urine. When consumption is high, selenium can also be expelled in the breath, causing a garlicky odor.

DIETARY SOURCES, DEFICIENCY, TOXICITY, AND RECOMMENDED INTAKE OF SELENIUM

Good sources of selenium include Brazil nuts, meat (especially organ meat), and fish (see Figure 8.18 on the next page). Severe selenium deficiency results in **Keshan disease**, a potentially fatal disease that causes serious heart problems and mostly affects children. Keshan disease was first documented in the Keshan region of China, which has very low levels of selenium in its soil. Consuming high amounts of selenium can result in nausea, vomiting, diarrhea, and brittleness of the teeth and fingernails. Selenium's RDA is 55 µg/day, and its UL is 400 µg/day.

Chromium

Chromium (Cr) is a trace mineral that may be critical to proper insulin function. Chromium was first designated as an essential nutrient because scientists discovered that its deficiency caused a diabetic-like state in animals. It was discovered later that chromium might be critical to glucose regulation and insulin function in humans as well. While there is considerable debate as to whether chromium is a required nutrient for humans, it is still classified by some experts as essential.[34]

CHROMIUM BIOAVAILABILITY, ABSORPTION, AND FUNCTIONS

Chromium bioavailability is increased by the presence of vitamin C and acidic medications such as aspirin (acetylsalicylic acid). Conversely, chromium absorption is decreased by the presence of antacids. Regardless of how much is consumed, however, very little (less than 2 percent) chromium is absorbed—most is excreted in the feces. Absorbed chromium is circulated in the blood to the liver, and excess is excreted in the urine. When the body is unable to excrete excess chromium in the urine, it is deposited mainly in the liver. Although scientists do not know why, consuming large amounts of simple sugars can increase urinary chromium excretion.[35]

Keshan disease A disease caused by severe selenium deficiency that mostly affects children and causes serious heart problems.

chromium (Cr) A trace mineral that may be critical to proper insulin function.

Figure 8.18 Good Sources of Selenium

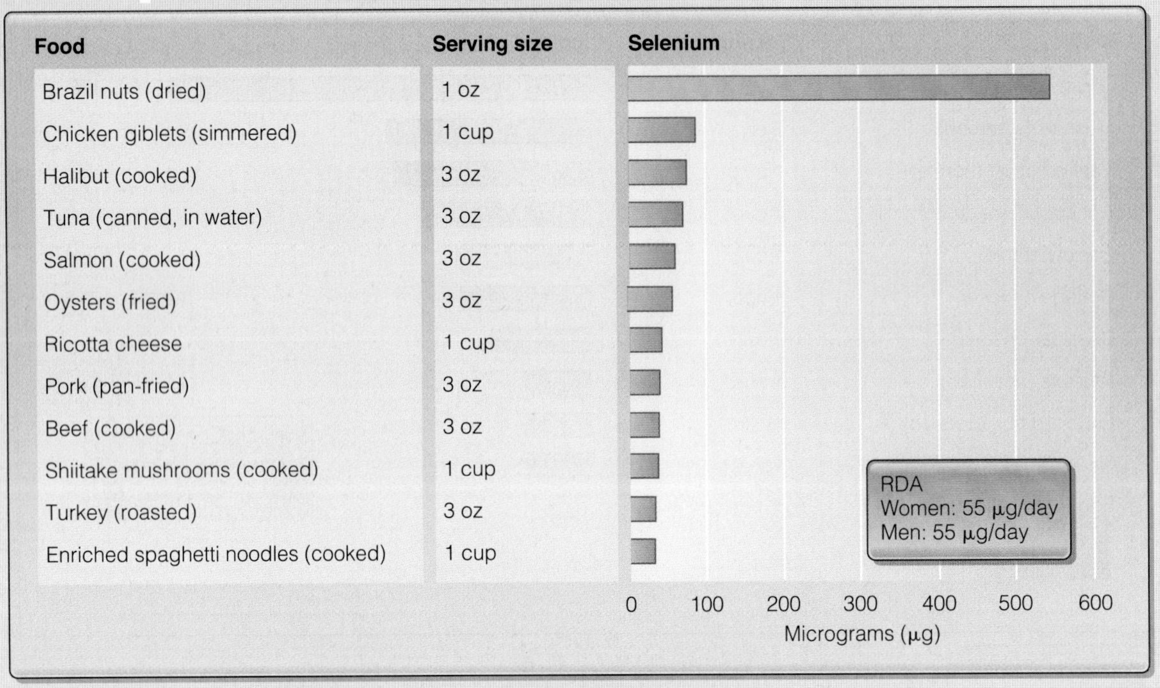
Food
Serving size
Selenium
Brazil nuts (dried)
1 oz
Chicken giblets (simmered)
1 cup
Halibut (cooked)
3 oz
Tuna (canned, in water)
3 oz
Salmon (cooked)
3 oz
Oysters (fried)
3 oz
Ricotta cheese
1 cup
Pork (pan-fried)
3 oz
Beef (cooked)
3 oz
Shiitake mushrooms (cooked)
1 cup
Turkey (roasted)
3 oz
Enriched spaghetti noodles (cooked)
1 cup
RDA
Women: 55 μg/day
Men: 55 μg/day
0
100
200
300
400
500
600
Micrograms (μg)

Source: USDA Nutrient Database for Standard Reference, Release 17.

Some studies suggest that chromium is required for the hormone insulin to function properly—especially in people with type 2 diabetes.[36] Chromium may also be essential to normal growth and development in children. At the very least, it appears to increase muscle mass and decrease fat mass in laboratory animals.[37] Because of this, a form of chromium called **chromium picolinate** has been widely marketed as an ergogenic aid for athletes. Other types of chromium supplements are promoted as products that help regulate blood glucose.

CHROMIUM DEFICIENCY AND TOXICITY

Chromium deficiency has only been observed in hospitalized patients receiving inadequate nutrition support. Chromium deficiency results in elevated blood glucose levels, decreased sensitivity to insulin, and weight loss. Chromium toxicity is also rare, even when supplemental chromium is consumed. This is probably attributable to chromium's low bioavailability. Toxic levels of chromium have been observed in people exposed to high levels of industrially-released chromium, however. For example, when stainless steel is heated to very high temperatures (such as during welding), chromium is released into the air. Environmental exposure of this kind causes skin irritations and may increase the risk of lung cancer.[38] These effects are never attributed to dietary chromium.

chromium picolinate A form of chromium taken as an ergogenic aid by some athletes.

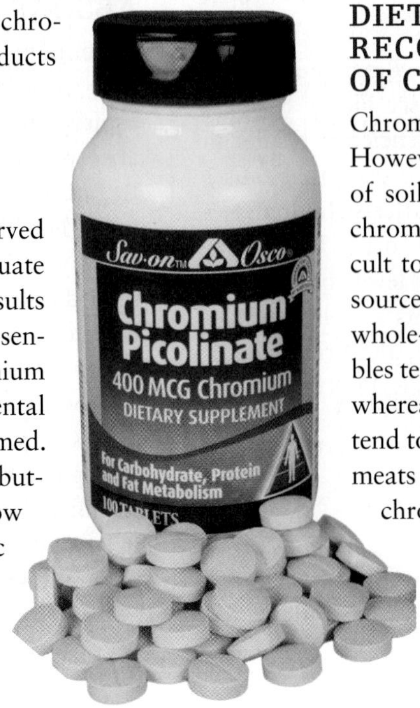

© Custom Medical Stock Photo

DIETARY SOURCES AND RECOMMENDED INTAKE OF CHROMIUM

Chromium is found in a variety of foods. However, because the chromium content of soil can greatly affect the amount of chromium taken up by plants, it is difficult to list foods that are especially good sources of this trace mineral. Nonetheless, whole-grain products, fruits, and vegetables tend to be good sources of chromium, whereas refined cereals and dairy foods tend to contain very little. Some processed meats have relatively high amounts of chromium, as do some beers and wines.

There is insufficient information to establish RDAs for chromium. However, AIs have been set at 25 and 35 μg/day for women and men (19–50 years), respectively. Because high chromium intake poses no known adverse effects, the Institute of Medicine has not established ULs for this mineral.

Manganese

Manganese (Mn) is an essential trace mineral that is a cofactor for metalloenzymes needed for bone formation, glucose production, and energy metabolism. As with chromium, manganese deficiency is rare in humans, and toxicity typically occurs only from excessive environmental exposure.

MANGANESE BIOAVAILABILITY, ABSORPTION, AND FUNCTIONS

Regardless of manganese intake, very little (less than 10 percent) is absorbed. Excess manganese is delivered to the liver, where it is incorporated into bile and excreted in the feces. Manganese is a cofactor for numerous enzymes, such as those involved in glucose production and bone formation. Manganese is also important to several enzymes involved in energy metabolism. Like copper and selenium, manganese is a cofactor for an enzyme that protects cells from free radicals.

DIETARY SOURCES, DEFICIENCY, TOXICITY, AND RECOMMENDED INTAKE OF MANGANESE

Good sources of manganese are whole-grain products, pineapples, nuts, and legumes. Dark green leafy vegetables such as spinach are high in manganese, and water can also contain significant amounts of the dissolved mineral. Because manganese intake is almost always sufficient, deficiency is rare. Manganese toxicity is uncommon, but exposure to high levels of airborne manganese (a result of mining) can cause serious neurological problems. Manganese toxicity can also occur in people with liver disease and those who consume water with high manganese levels. AIs of 2.3 and 1.8 mg/day have been established for men and women, respectively. To prevent nerve damage caused by manganese toxicity, a UL of 11 mg/day has been established for manganese. This UL includes manganese consumed as food, water, and supplements.

Molybdenum

Molybdenum (Mo) is an essential trace mineral that is a cofactor for several important metalloenzymes needed for protein metabolism. Molybdenum deficiency is almost unheard of, and there is little concern among health professionals about dietary inadequacy of this mineral.

manganese (Mn) An essential trace mineral that is a cofactor for metalloenzymes needed for bone formation, glucose production, and energy metabolism.

molybdenum (Mo) An essential trace mineral that is a cofactor for several important metalloenzymes needed for amino acid and purine metabolism.

zinc (Zn) An essential trace mineral involved in gene expression, immune function, and cell growth.

MOLYBDENUM BIOAVAILABILITY, ABSORPTION, AND FUNCTIONS

Molybdenum appears to be absorbed almost completely in the intestine. After circulating in the blood to the liver, molybdenum serves as a cofactor for several enzymes involved in the metabolism of sulfur-containing amino acids (such as methionine and cysteine). Molybdenum is also critical to the structural building blocks of DNA. Finally, molybdenum-containing enzymes are essential to the detoxification of drugs in the liver.

DIETARY SOURCES, DEFICIENCY, TOXICITY, AND RECOMMENDED INTAKE OF MOLYBDENUM

Although foods' molybdenum contents vary greatly depending on soil molybdenum levels, legumes (such as peas and lentils), grains, and nuts tend to be good sources of this mineral. Only one case of primary molybdenum deficiency has been documented. In this case, a hospitalized patient received intravenous nutritional support that lacked molybdenum. The patient's molybdenum deficiency was characterized by abnormal heart rhythms, headache, and visual problems.[39] There are no known adverse effects of high intakes of molybdenum in humans, but toxicity has been shown to harm reproduction capabilities in animals. Molybdenum's RDA is 45 μg/day, and to prevent potential reproductive problems, a UL of 2,000 μg/day has been established for this mineral.

Zinc

Zinc (Zn) is an essential trace mineral involved in gene expression, immune function, and cell growth. Zinc is a cofactor for more than 300 metalloenzymes and is found in almost every cell of the body. Inadequate intake of dietary zinc can disrupt many important physiological functions, such as growth, reproduction, taste, smell, immunity, DNA replication, and protein synthesis. Even

metallothionine An intestinal protein that regulates the amount of zinc released into the blood.

moderate zinc deficiency can cause serious health concerns.

ZINC BIOAVAILABILITY, ABSORPTION, AND FUNCTIONS

Zinc bioavailability is influenced by a variety of dietary factors, many of which are similar to those that influence iron absorption. Zinc's bioavailability is greater in animal-based foods than in plant-based foods. Acidic substances (such as vitamin C) increase its absorption. Because zinc absorption relies on the same transport protein as several other similarly charged minerals (Fe^{2+}, Cu^{2+}, and Ca^{2+}, for example), high intakes of these minerals can decrease zinc bioavailability.

Zinc's absorption is highly regulated, requiring at least two proteins to function properly. After the first protein transports zinc into the intestinal cell, the mineral is bound to the second protein, **metallothionine**, which regulates the amount of zinc released into the blood. The body's synthesis of metallothionine decreases during periods of zinc excess and increases during periods of zinc deficiency. Because intestinal cells are continually shed and replaced, excess zinc bound to metallothionine is excreted in the feces.

Zinc functions primarily as a cofactor. Because enzymes needed for gene expression rely on zinc, this mineral is essential to protein synthesis, cell maturation, growth, and proper immune function. Zinc also acts as a potent antioxidant, and it appears to play a role in the stabilization of cell membranes. Zinc supplementation is often touted as a treatment for the common cold. While most studies do not support this claim, limited data suggest a beneficial effect in children.[40]

DIETARY SOURCES, DEFICIENCY, TOXICITY, AND RECOMMENDED INTAKE OF ZINC

High concentrations of zinc are found in shellfish, meat (especially organ meat), dairy products, legumes, and chocolate (see Figure 8.19). Zinc is also sometimes added to fortified cereals and grain products.

Zinc deficiency was first documented in Egypt and Iran, where people tend to eat plant-based diets high in phytates, which chelate zinc in the intestine. Children with zinc deficiencies exhibit delayed growth, skin irritations, diarrhea, underdeveloped genitals, and delayed sexual maturation. Because zinc supplementation corrects many of these abnormalities, zinc was deemed an essential nutrient.

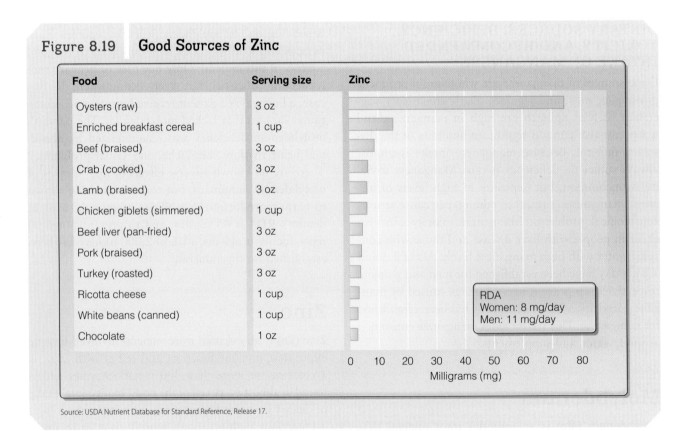

Figure 8.19 Good Sources of Zinc

Food	Serving size
Oysters (raw)	3 oz
Enriched breakfast cereal	1 cup
Beef (braised)	3 oz
Crab (cooked)	3 oz
Lamb (braised)	3 oz
Chicken giblets (simmered)	1 cup
Beef liver (pan-fried)	3 oz
Pork (braised)	3 oz
Turkey (roasted)	3 oz
Ricotta cheese	1 cup
White beans (canned)	1 cup
Chocolate	1 oz

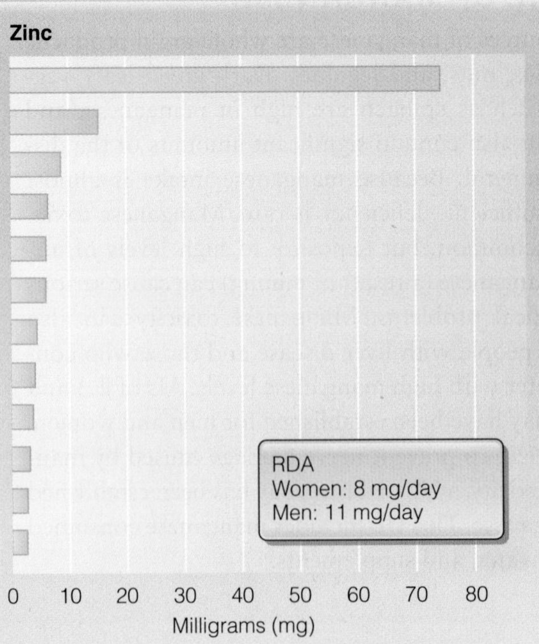

Source: USDA Nutrient Database for Standard Reference, Release 17.

Secondary zinc deficiency can result from chronic conditions such as liver disease, impaired kidney function, and diabetes. A genetic abnormality called **acrodermatitis enteropathica**, is caused by a defect in the protein that transports zinc into intestinal cells. Because zinc is not absorbed, excessive amounts are lost in the feces. Babies born with acrodermatitis enteropathica fail to grow properly and usually develop severely red and scaly skin, especially around the scalp, eyes, and feet. While it is fatal if it goes untreated, acrodermatitis enteropathica can be controlled through lifelong zinc supplementation. When a high enough level of zinc is consumed, adequate amounts can be absorbed—even given the faulty transport protein.

Although dietary zinc toxicity is uncommon, it can be caused by excessive supplementation. Symptoms include poor immune function, depressed levels of high-density lipoprotein (HDL) cholesterol, and impaired copper status. Nausea, vomiting, and loss of appetite have also been reported.

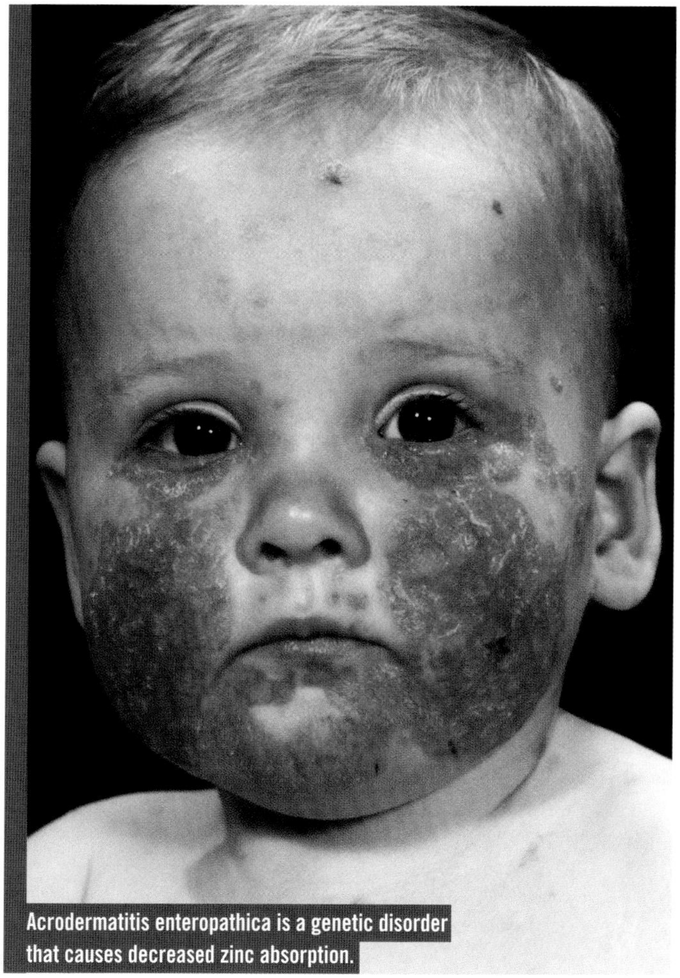

Acrodermatitis enteropathica is a genetic disorder that causes decreased zinc absorption.

RDAs of 11 and 8 mg/day have been set for men and women, respectively. Because zinc bioavailability is lower in plant-based foods, the Institute of Medicine suggests that vegetarians—particularly vegans—consume up to 50 percent more zinc than nonvegetarians. A UL of 40 mg/day has been established for zinc.

Fluoride

Fluoride (F^-) is a trace mineral that strengthens bones and teeth. It is not actually essential to any basic body functions, but because this mineral strengthens bones and teeth when ingested regularly, it is important to know the dietary sources and functions of fluoride within the body.

FLUORIDE BIOAVAILABILITY, ABSORPTION, AND FUNCTIONS

The gastrointestinal tract does not heavily regulate fluoride's absorption. In fact, almost all of the fluoride you consume is absorbed in your small intestine. It then circulates in your blood to your liver and is ultimately integrated into bones and teeth. Excess fluoride is excreted in the urine.

Fluoride affects the health of bones and teeth in several ways. First, it strengthens them by being incorporated into their basic mineral structures. Teeth that contain extra fluoride are especially resistant to bacterial breakdown and cavities (dental caries). Second, topical (not dietary) application of fluoride-containing toothpastes and treatments works directly on the cavity-causing bacteria in the mouth, which are largely responsible for tooth decay. Aside from its beneficial effects on tooth and bone strength, there are no known biological functions of fluoride.

DIETARY SOURCES, DEFICIENCY, TOXICITY, AND RECOMMENDED INTAKE OF FLUORIDE

Very few foods are good sources of fluoride. Potatoes, tea, and legumes contain some fluoride, as do fish with intact bones (such as sardines and some types of canned salmon). In some areas, water naturally contains high amounts of fluoride and is therefore a good

acrodermatitis enteropathica A genetic abnormality that causes secondary zinc deficiency.

fluoride (F^-) A trace mineral that strengthens bones and teeth.

source of this mineral. Because most foods contain little fluoride, many communities add fluoride to their drinking water supplies. You can find out whether your town fluoridates its water by contacting a local dentist or the city water department. Bottled water typically contains little fluoride and is therefore a poor source of this mineral.[41]

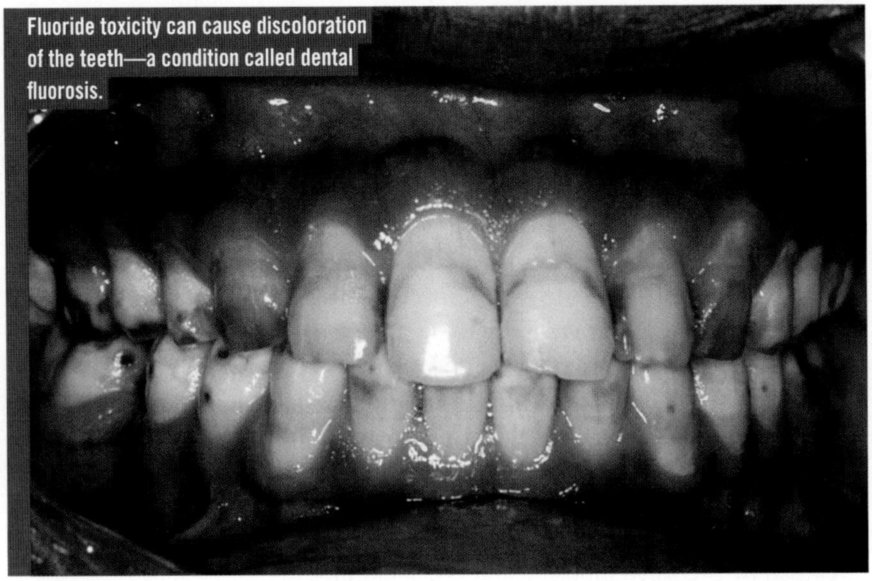

Fluoride toxicity can cause discoloration of the teeth—a condition called dental fluorosis.

Whereas there are no recognized signs or symptoms of fluoride deficiency other than increased risk for dental decay, fluoride toxicity is well documented.[42] Signs and symptoms include gastrointestinal upset, excessive production of saliva, watering eyes, heart problems, and in severe cases, coma. Excessive fluoride intake can cause **dental fluorosis**, pitting and mottling (discoloration) of the teeth, and **skeletal fluorosis**, weakening of the skeleton. Fluorosis is a special concern in small children, who sometimes swallow large amounts of toothpaste on a daily basis. Thus, parents should carefully monitor tooth-brushing routines.

An RDA for fluoride has not been established. However, AIs of 4 and 3 mg/day have been set for men and women, respectively. Fluoride's UL is 10 mg/day.

Are There Other Important Trace Minerals?

In addition to those already covered in this chapter, other minerals may also influence your health. Whether the body requires these minerals has yet to be confirmed or denied through scientific experimentation. Still, each of these minerals deserves comment on possible biological functions and food sources. Similar to fluoride, RDAs have not been established for these minerals. Some have ULs, however.

dental fluorosis
Pitting and mottling of teeth caused by excessive fluoride intake.

skeletal fluorosis
Weakening of the skeleton caused by excessive fluoride intake.

- *Nickel* (Ni) may be critical to protein and lipid metabolism and gene expression. It may also influence bone formation. Nickel is found in legumes, nuts, and chocolate.
- *Aluminum* (Al) may be important to reproduction, bone formation, DNA synthesis, and behavior. It is found in many baked goods, grains, some vegetables, tea, and many antacids.
- *Silicon* (Si) is important to bone growth in laboratory animals. Silicon's dietary sources include whole-grain products, and it is also used as a food additive.
- *Vanadium* (V) influences cell growth and differentiation, and it may affect thyroid hormone metabolism. Dietary sources include mushrooms, shellfish, and black pepper.
- *Arsenic* (As) may be a cofactor for a variety of enzymes, and may be important to DNA synthesis. Seafood and whole grains are good sources of arsenic.
- *Boron* (B) is especially important to reproduction in laboratory animals, and deficiency causes severe fetal malformations. Boron is found in fruits, leafy vegetables, legumes, and nuts.

WHY CHOOSE?
Every 4LTR Press solution comes complete with a visually engaging textbook in addition to an interactive eBook. Go to CourseMate for NUTR to begin using the eBook. Access at www.cengagebrain.com
Complete the Speak Up survey in CourseMate at www.cengagebrain.com
Follow us at www.facebook.com/4ltrpress
©iStockphoto.com/A-Digit | © Cengage Learning 2011

NUTR

9 Energy Balance and Body Weight Regulation

© Alan Thornton/Stone/Getty Images

LEARNING OUTCOMES:

LO1 Understand the significance of stable, negative, and positive energy balance.

LO2 Describe the factors that influence energy intake.

LO3 Define total energy expenditure and detail its components.

LO4 Understand how body weight and composition measurements can be used to assess health.

LO5 Discuss the factors related to and causes of obesity.

LO6 Describe the physiologic mechanisms by which energy balance and body weight are regulated.

LO7 Discuss the components of a successful weight-loss plan.

Chapter 9

LO1 What Is Energy Balance?

Some count them, others curse them—but what role do calories actually play in body weight regulation? Moreover, while most people struggle to lose weight, why do some actually find it difficult to gain weight? If you have ever pondered these questions, you are not alone. In this chapter, you will learn about the relationships among energy intake, energy expenditure, and weight gain—in other words, you will come to understand how the foods you eat and the activities you engage in affect your body weight and composition. You will also learn about body weight regulation, the growing prevalence of obesity in the United States, and the various approaches to weight loss and weight maintenance.

What determines whether the energy in the foods you eat is used to fuel your body or stored for later use? It all comes down to balance—specifically, that between energy intake and energy expenditure. Unless you are attempting to gain or lose weight, the amount of energy you consume should be equal to the amount of energy your body requires. When a person's energy intake equals her energy expenditure, she is in a state of **neutral energy balance**, and her body weight is stable. When she consumes more energy than her body needs, she is in a state of **positive energy balance**. This results in weight gain. When her energy intake is less than her energy expenditure, she is in a state of **negative energy balance**. This results in weight loss. Under most conditions, therefore, a change (or lack thereof) in body weight is a useful indicator of whether you are in neutral, positive, or negative energy balance. The relationship between energy balance and body weight is illustrated in Figure 9.1 on the next page.

Energy Balance and Body Weight

When a person is in positive energy balance, his muscle and/or adipose tissue increases in size and weight. Because increased body weight is primarily associated with increased body fat, positive energy balance is generally considered unhealthy. It is important to remember, however, that positive energy balance can also connote muscular growth. Positive energy balance is also desirable and necessary during periods of growth and development, such as infancy, adolescence, and pregnancy.

Negative energy balance occurs during periods of decreased energy intake, increased energy expenditure, or both. Under this condition, stored energy reserves are broken down, resulting in decreased body weight. Body fat, which exists throughout the body as adipose tissue, serves as the body's primary energy source during negative energy balance. One pound of adipose tissue supplies the equivalent of approximately 3,500 kilocalories of energy. However, it is important to recognize that weight loss does not always result exclusively from loss of body fat during periods of negative energy balance. Water and muscle loss also contribute to total weight loss.

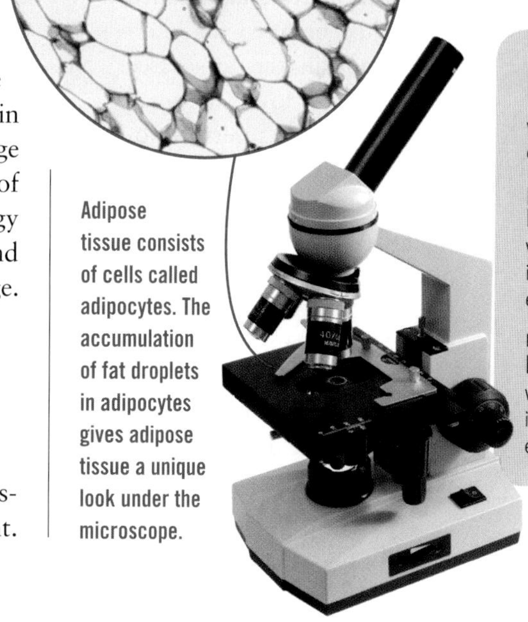

Adipose tissue consists of cells called adipocytes. The accumulation of fat droplets in adipocytes gives adipose tissue a unique look under the microscope.

neutral energy balance A condition whereby energy intake equals energy expenditure.

positive energy balance A condition whereby energy intake is greater than energy expenditure.

negative energy balance A condition whereby energy intake is less than energy expenditure.

A Closer Look at Adipose Tissue

Recall from Chapter 6 that adipose tissue is comprised largely of specialized cells called adipocytes, which contain a central lipid-filled core that consists primarily of triglycerides. The number and size of adipocytes determine the amount of adipose tissue in your body. As an adipocyte fills with lipid, its size increases, a process called **hypertrophic growth.** To accommodate large amounts of lipid, the diameter of an adipocyte can increase 20-fold. When existing cells are full, new adipocytes are formed—a process called **hyperplastic growth.** When a person loses body fat, enlarged adipocytes return to normal size, but the total number of adipocytes remains relatively constant.[1] Figure 9.2 illustrates how adipocyte number and size change during weight gain and loss.

There is a tendency to think of adipose tissue as a sponge that passively soaks up and releases excess lipids. However, adipocytes are anything but passive: they play an active role in regulating energy balance by producing hormones. The discovery that adipocytes produce multiple hormone-like **adipokine** proteins has led many researchers to now think of adipose tissue as an endocrine-like organ. Indeed, adipokines provide important communication between adipocytes and other tissues and organs.

VISCERAL ADIPOSE TISSUE AND SUBCUTANEOUS ADIPOSE TISSUE

Although adipose tissue is found throughout the body, visceral adipose tissue (VAT), also called intra-abdominal fat, is located between the internal organs in the torso (see Figure 9.3). However, much of your body fat reserve is located directly beneath the skin as subcutaneous adipose tissue (SCAT). SCAT is found in most regions of the body, but is most predominant around the thighs, hips, and buttocks. Recall that both VAT and SCAT were introduced in Chapter 6. The distribution of your body fat holds important

hypertrophic growth A process whereby an adipocyte fills with lipid and its size increases.

hyperplastic growth A process whereby new adipocytes are formed.

adipokine A hormone-like protein produced by an adipocyte that plays a communicative role between the adipocyte and other tissues and organs.

Figure 9.1 **The Three States of Energy Balance**

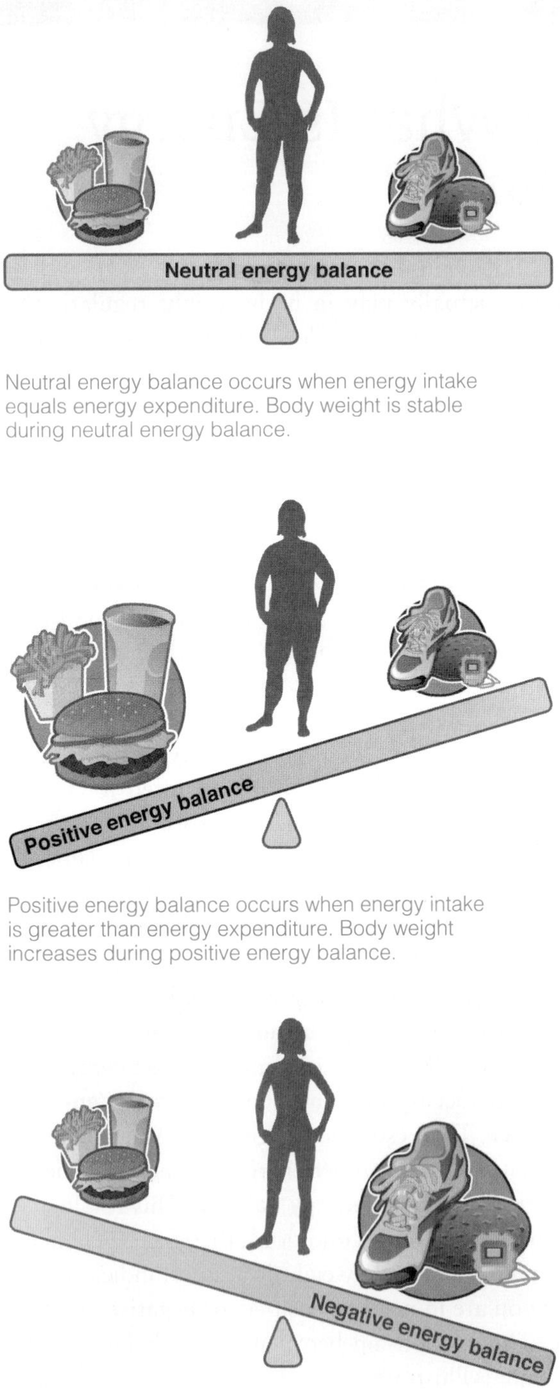

Neutral energy balance occurs when energy intake equals energy expenditure. Body weight is stable during neutral energy balance.

Positive energy balance occurs when energy intake is greater than energy expenditure. Body weight increases during positive energy balance.

Negative energy balance occurs when energy intake is less than energy expenditure. Body weight decreases during negative energy balance.

A change (or lack thereof) in body weight indicates if a person is in neutral, positive, or negative energy balance.

Figure 9.2 Hypertrophic and Hyperplastic Growth of Adipose Tissue

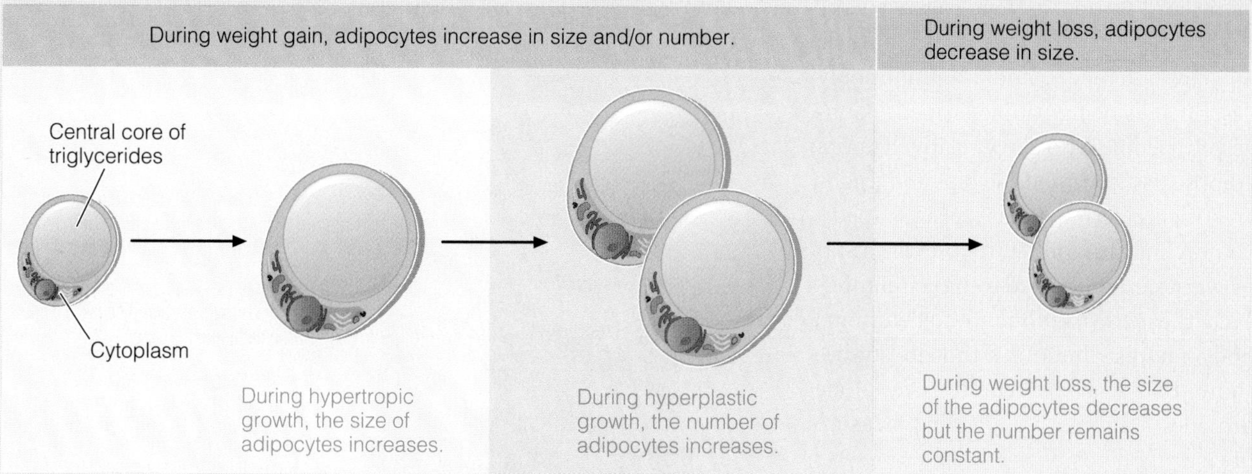
During weight gain, adipocytes increase in size and/or number.
During weight loss, adipocytes decrease in size.
Central core of triglycerides
Cytoplasm
During hypertropic growth, the size of adipocytes increases.
During hyperplastic growth, the number of adipocytes increases.
During weight loss, the size of the adipocytes decreases but the number remains constant.

The amount of a person's adipose tissue depends on the number and size of her adipose cells (adipocytes).

Figure 9.3 Cross-Section of the Upper Body

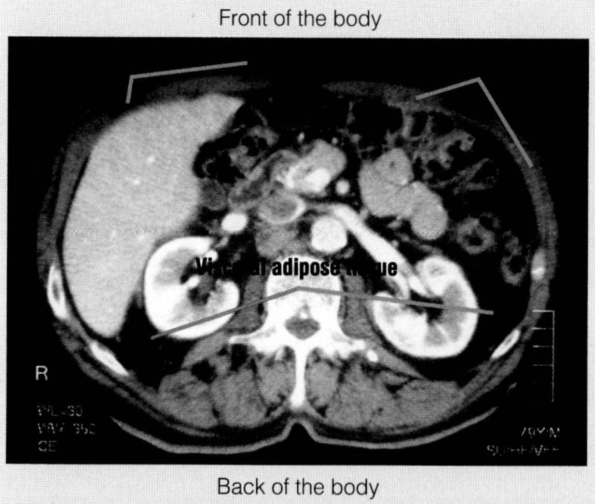
Front of the body
R
Back of the body

Visceral adipose tissue (VAT) is found between the internal organs in the abdomen. Subcutaneous adipose tissue (SCAT) is found directly beneath the skin.

information about your health. The term **central obesity** (or **central adiposity**) describes predominant accumulation of VAT. Although the reasons for why are not clear, people with central obesity are at greater risk for developing weight-related health problems (such as type 2 diabetes, hypertension, and atherosclerosis) than those who store excess adipose tissue elsewhere in the body, such as around the hips.

People with **central obesity** are often described as **apple-shaped**, while those who carry **excess adipose tissue** around their hips and thighs are described as **pear-shaped**.

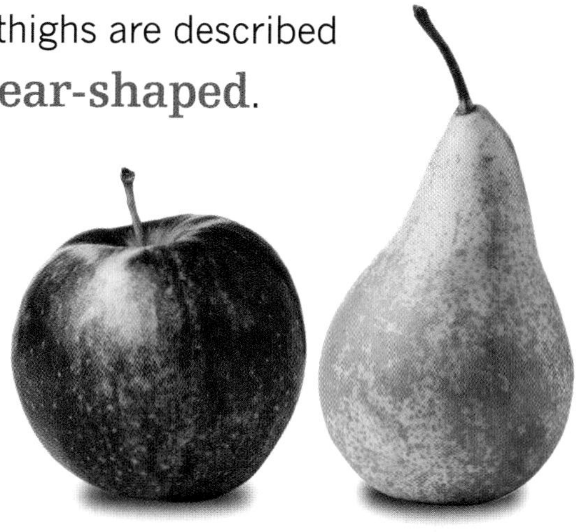

Determining your body fat distribution pattern is easier than you might think—**waist circumference** is a good indicator of central obesity. This measurement is easily obtained by placing a tape measure around the narrowest area of the waist (see Figure 9.4 on the next page). A waist circumference of 40 inches or greater in men

central obesity (or **central adiposity**) A predominant accumulation of adipose tissue between the internal organs in the abdomen.

waist circumference A measurement used as an indicator of central adiposity.

and 35 inches or greater in women generally indicates central adiposity.[2] People with central adiposity tend to have waist circumferences that are larger than their hip circumferences and are commonly described as being "apple-shaped." Conversely, people who carry relatively more adipose tissue around the hips and thighs are often referred to as "pear-shaped." Some clinicians prefer to use the ratio of waist circumference to hip circumference (the *waist-to-hip ratio*) to determine body fat distribution, although studies show that waist circumference alone provides a simple yet effective measure of central adiposity.[3]

Figure 9.4 Waist Circumference Is Used to Assess Central Adiposity

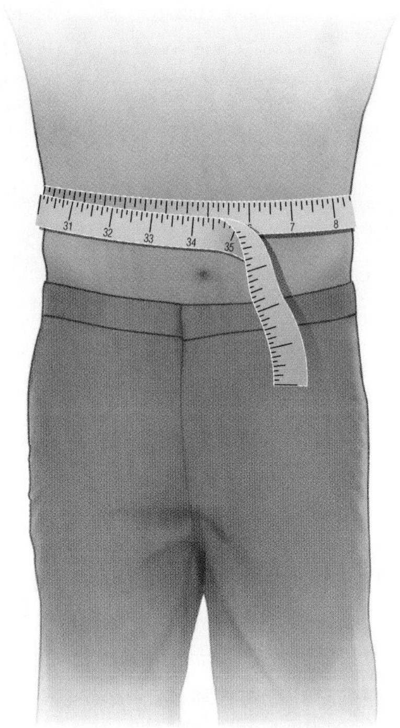

- Person should stand with his feet six to seven inches apart, with his weight evenly distributed to each leg.
- Person should be relaxed, and the measurement should be taken while breathing out.
- The tape measure should be positioned around the narrowest area of the waist.
- The tape should be loose enough to place one finger between the tape and the person's body.
- A waist circumfrence of 40 inches or greater in males and 35 or greater in females indicates central adiposity.

Central adiposity increases a person's risk for weight-related health problems. A person's waist circumference is an important indicator of central adiposity.

LO2 What Factors Influence Energy Intake?

Although it may seem simple at first, understanding energy balance is more complicated than you might think. The energy your body needs comes from the foods you eat, but understanding what you eat, how much you eat, and why you eat is quite complex. Your food choices are related to many facets of your life, and these choices serve numerous purposes beyond providing your body with nourishment. It is also important to recognize that foods people enjoy as adults may be the very foods they avoided as children. Can you think of a food you detested as a child but now enjoy as an adult? Clearly, eating behavior is complicated, and it is important to consider the psychological, physical, social, and cultural forces that influence it.

hunger A basic physiological drive to consume food.

satiety A physiological response whereby a person feels he or she has consumed enough food.

hypothalamus A region of the brain that regulates energy intake by balancing hunger and satiety.

Hunger and Satiety

Hunger is a basic physiological drive to consume food, whereas **satiety** is a physiological response whereby a person feels he has consumed enough food. Humans tend to be periodic eaters, meaning that they usually eat meals at predicable and discrete times throughout the day. Most of the time, people eat to the point of comfort. In other words, they eat until they are satiated. However, people sometimes continue to eat after they are full, even to the point of discomfort. Regardless of one's level of fullness however, a person is usually ready to eat again within three to four hours of when she last ate.

Some people may think that the stomach is the main determinant of hunger and satiety. However, a variety of factors influence these complex physiological sensations. Early experimental work demonstrated that hunger and satiety can be controlled in mice through electrical stimulation of specific regions within the brain. These areas soon became known as the hunger and satiety centers, respectively. Scientists now recognize that a specific, complex region of the brain called the **hypothalamus** regulates energy intake by balancing feelings of hunger and satiety. The hypothalamus receives signals that influence hunger and satiety from different parts of the body, such as the gastrointestinal tract. In response to these signals, the hypothalamus releases hormone-like **neurotransmitter** substances. Certain neurotransmitters promote hunger and the initiation of eating, while others promote satiety and meal termination. Not only do these neurotransmitters regulate short-term food intake, they also play an important role in the long-term regulation of body weight.

THE ROLE OF THE GI TRACT IN HUNGER AND SATIETY

Understanding the precise mechanisms associated with hunger and satiety is of great interest to nutrition researchers. In addition to emotional and social factors, eating behaviors are influenced by a variety of physiological states. One of the most potent physiological satiety triggers is gastric distention (stomach stretching). Gastric distention occurs when food accumulates in the stomach. Stretching of the stomach wall initiates neural signals, which in turn convey information to the brain. The brain responds by releasing neurotransmitters that elicit the sensation of satiety. Gastric stretching may explain why some foods are better at promoting satiety than others. High-volume foods (such as those with large amounts of water and/or fiber) increase gastric stretching, which in turn signals the release of neurotransmitters that help people feel full and satisfied. For example, 473 ml (roughly two cups) of grapes contains approximately 100 kcal, the same amount of energy contained in 59 ml (roughly one-fourth cup) of raisins. However, because the volume of the grapes is greater than that of the raisins, the grapes cause more gastric stretching, which is likely to cause a person to feel more satisfied after eating them.

neurotransmitter A hormone-like substance released by neural tissue such as the hypothalamus.

bariatric surgery Surgical procedure performed to treat obesity.

gastric banding A type of bariatric surgery whereby an adjustable, fluid-filled band is wrapped around the upper portion of the stomach, dividing it into a small upper pouch and a larger lower pouch.

gastric bypass A surgical procedure that involves reducing the size of the stomach and bypassing a segment of the small intestine so that fewer nutrients are absorbed.

How Does Weight-Loss Surgery Work?

Obese people tend to have larger stomachs than do lean people and therefore can accommodate larger volumes of food before gastric stretching triggers satiety.[4] This may be why some severely obese people do not often feel satiated after a meal. To counter this condition, some severely obese people opt to undergo a surgical procedure performed to treat obesity, otherwise known as **bariatric surgery** (see Figure 9.5 on the next page). **Gastric banding** is a type of bariatric surgery whereby an adjustable, fluid-filled band is wrapped around the upper portion of the stomach, dividing it into a small upper pouch and a larger lower pouch. The small stomach pouch fills quickly with food, triggering satiety. Another type of bariatric surgery, called **gastric bypass**, involves reducing the size of the stomach and bypassing a segment of the small intestine so that fewer nutrients are absorbed. These procedures help people lose weight by forcing them to consume less food and by decreasing nutrient digestion and absorption. For many severely obese individuals, weight-loss surgery can improve quality of life both physically and emotionally. However, it is important to remember that any type of bariatric surgery is a life-altering procedure, and its effectiveness depends on a person's willingness to eat a healthy, well-balanced diet and exercise regularly.

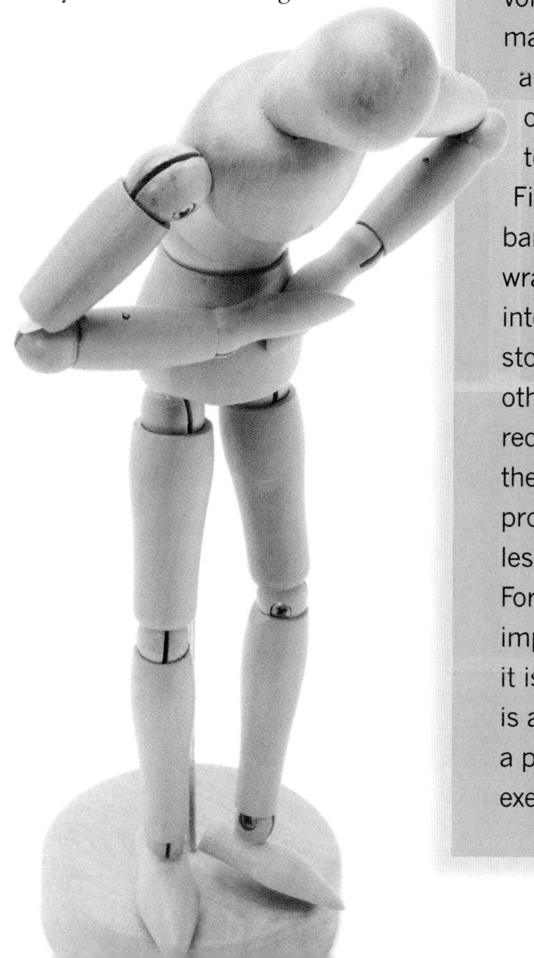

© iStockphoto.com/Kazunori Higuchi

Figure 9.5 **Types of Weight-Loss Surgery**

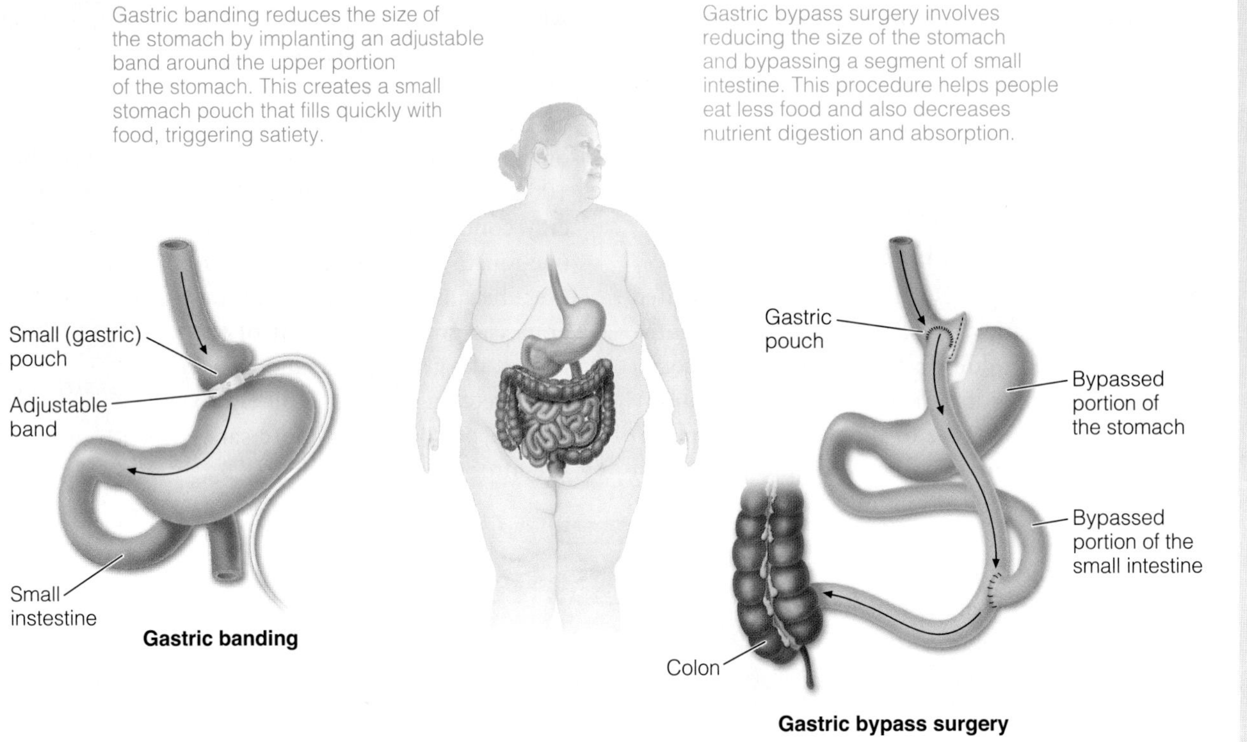

Bariatric surgery is a medical procedure that promotes weight loss. The two most common types of bariatric surgery are gastric banding and gastric bypass surgery.

Although gastric stretching helps relieve the sensation of hunger, it is not the only factor associated with satiety. The sensation of fullness is also triggered by the presence of certain nutrients in the blood following a meal. For example, when food intake causes blood glucose levels to increase, the brain responds by releasing neurotransmitters that stimulate satiety, providing a signal to terminate eating. Some studies also show that high-fat meals suppress feelings of hunger longer than low-fat meals with the same number of calories.[5] The signaling of satiation brought on by the presence of certain nutrients that enter the blood following a meal is an example of a *post-absorptive mechanism*.

A number of hormones produced by the gastrointestinal (GI) tract play a role in regulating hunger and satiety. For example, researchers have recently become interested in the role of **ghrelin**, a newly discovered hormone dubbed the *hunger hormone*. Before a meal is consumed, this potent hunger-stimulating hormone is released by cells in the stomach lining and circulated in the blood. High levels of ghrelin in the blood serve as a pre-meal signal that stimulates the sensation of hunger. Ghrelin levels are highest prior to meals when the stomach is empty, but after food is consumed, ghrelin levels decrease. Recent evidence suggests that some obese people may overproduce ghrelin, which could explain why they do not always experience a feeling of satiety following a meal.[6] While ghrelin serves as a physiological reminder to eat, there is also evidence that other GI hormones released from the small intestine provide equally powerful satiety cues. Some of the factors believed to regulate short-term food intake are summarized in Figure 9.6.

ghrelin A newly discovered hormone secreted by cells in the stomach lining that stimulates hunger.

Psychological Factors

Eating when you are hungry and stopping when you are full are two of the most important keys to maintaining a healthy body weight. Sometimes, however, you may eat for reasons other than hunger, or you may eat past satiety. Whereas hunger and satiety are

Figure 9.6 **Regulation of Short-Term Food Intake**

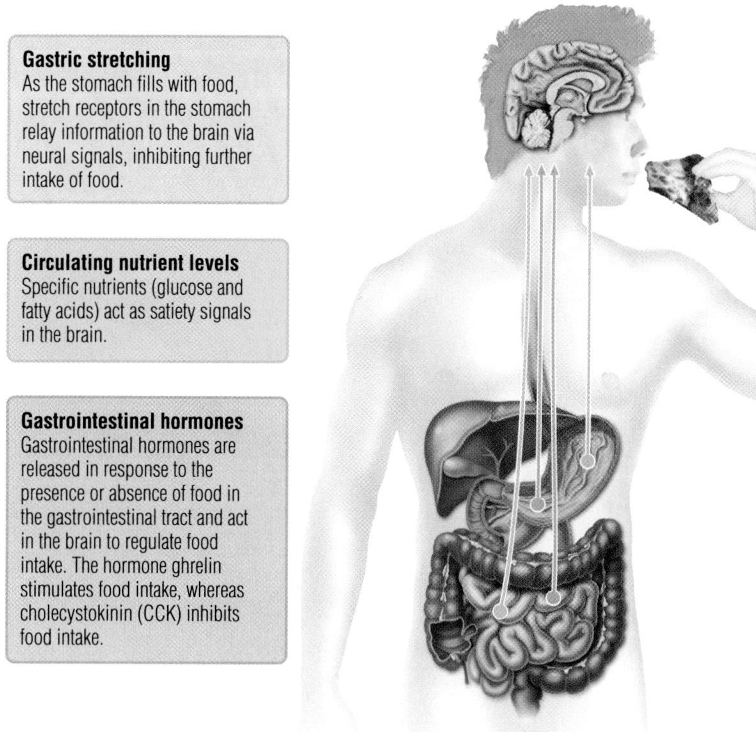

Gastric stretching
As the stomach fills with food, stretch receptors in the stomach relay information to the brain via neural signals, inhibiting further intake of food.
Circulating nutrient levels
Specific nutrients (glucose and fatty acids) act as satiety signals in the brain.
Gastrointestinal hormones
Gastrointestinal hormones are released in response to the presence or absence of food in the gastrointestinal tract and act in the brain to regulate food intake. The hormone ghrelin stimulates food intake, whereas cholecystokinin (CCK) inhibits food intake.

Signals originating from the GI tract stimulate the brain to regulate short-term food intake by releasing neurotransmitters that influence hunger and satiety.

Appetite can easily be aroused by sensory factors such as the appearance, taste, and/or smell of food. For instance, the pleasing smell of baked bread makes many people want to eat, regardless of whether they are hungry. Conversely, unpleasant odors can spoil an appetite, even if a person is hungry. Emotional states can also affect appetite dramatically, but the effect varies from person to person. Some people respond to emotions such as fear, depression, disappointment, excitement, and stress by eating, while others respond by not eating at all. Clearly, emotional states can have a profound influence on appetite, sometimes overriding the normal physiological cues of hunger and satiety.

under physiologic control, food aversions and appetite are influenced primarily by psychological factors. A person with a **food aversion** experiences a strong psychological dislike of a particular food or foods. Sometimes even the thought or smell of certain foods can trigger an adverse physical response. **Appetite** is the psychological longing or desire for food. Appetite is an important determinant of what and when you eat. A strong psychological desire for a particular food or foods is often referred to as a **food craving.** The most commonly craved foods tend to be calorie-rich, such as cookies, cakes, potato chips, and chocolate.[7]

> "**Pica** is a disorder characterized by **cravings for non-food** substances such as dirt, paper, charcoal, or chalk."

LO3 What Determines Energy Expenditure?

As you have learned, a complex interplay of factors influences when and how much you eat. However, energy intake is only one component of neutral energy balance—the other is energy expenditure. The body expends energy to maintain physiological functions, support physical activity, and process food. These components of energy use make up the collective sum of energy used by the body, the **total energy expenditure (TEE).** As illustrated in Figure 9.7 on the next page, TEE has three main components:

(1) Basal metabolism
(2) Physical activity
(3) Thermic effect of food

food aversion A strong psychological dislike of a particular food or foods.

appetite A psychological longing or desire for food.

food craving A strong psychological desire for a particular food or foods.

total energy expenditure (TEE) The collective sum of energy used by the body.

Food Cravings and Food Aversions

Are food aversions the flip side of food cravings? Even the experts are unsure. The reality is that there are no definitive answers when it comes to explaining food cravings and aversions. A food craving is a powerful, irresistible, intense desire for a particular food, whereas food aversions develop when certain foods are viewed as repugnant. Although certain emotional states can provoke food aversions, most are conditioned responses resulting from a paired association between physical discomfort and a particular food. Food aversions tend to be persistent and long lasting, whereas food cravings can come and go. During pregnancy, many women experience cravings for and/or aversions to certain foods. Although some experts believe that food cravings help pregnant women satisfy their needs for nutrients and that food aversions may provide protection from harmful substances, there is very little scientific data to support either of these claims. In other words, a craving for ice cream does not necessarily mean that a woman is calcium deficient.

Total energy expenditure also includes two minor components: **adaptive thermogenesis** and **nonexercise activity thermogenesis.** Adaptive thermogenesis is a temporary expenditure of energy that enables the body to adapt to changes in the environment and physiological conditions such as trauma, low temperatures, and stress. Shivering in response to cold weather is an example of adaptive thermogenesis. Nonexercise activity thermogenesis is expenditure of energy associated with spontaneous movement such as fidgeting and posture maintenance. The contributions of adaptive thermogenesis and nonexercise activity thermogenesis to TEE have yet to be determined, but they are likely significant.

adaptive thermogenesis A temporary expenditure of energy that enables the body to adapt to changes in the environment and physiological conditions.

nonexercise activity thermogenesis An expenditure of energy associated with spontaneous movement such as fidgeting and posture maintenance.

basal metabolism An expenditure of energy to sustain basic, involuntary life functions such as respiration, beating of the heart, nerve function, and muscle tone.

Figure 9.7 **Major Components of Total Energy Expenditure (TEE)***

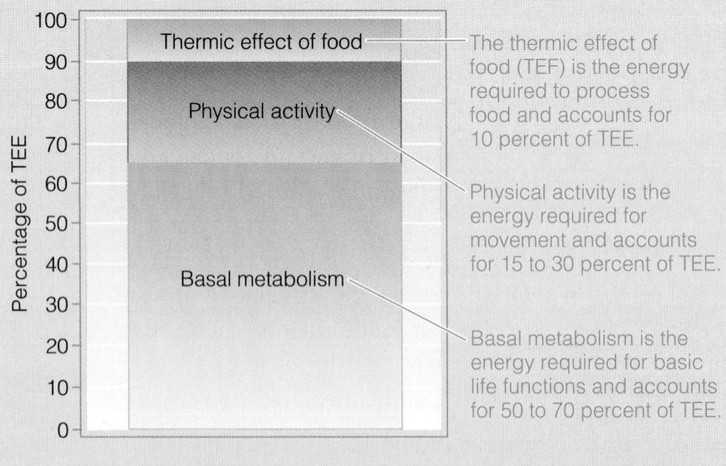
Percentage of TEE
100
90
80
70
60
50
40
30
20
10
0
Thermic effect of food
Physical activity
Basal metabolism
The thermic effect of food (TEF) is the energy required to process food and accounts for 10 percent of TEE.
Physical activity is the energy required for movement and accounts for 15 to 30 percent of TEE.
Basal metabolism is the energy required for basic life functions and accounts for 50 to 70 percent of TEE.

* Energy associated with adaptive thermogenesis and nonexercise activity thermogenesis also contributes slightly to TEE, although its exact contribution is not known.

Adapted from: Levine JA. Non-exercise activity thermogenesis. Proceedings of the Nutrition Society. 2003; 62:667–79.

Basal Metabolism

Basal metabolism is the energy expended to sustain basic, involuntary life functions such as respiration, beating of the heart, nerve function, and muscle tone. **Basal metabolic rate (BMR)**, the amount of energy expended per hour (kcal/hour) so that the body can carry out these functions, accounts for most of TEE—approximately 50 to 70 percent. BMR is measured after an overnight fast and while a person is at complete rest.

basal metabolic rate (BMR) The amount of energy expended per hour (kcal/hour) so that the body can carry out basic, involuntary life functions.

FACTORS INFLUENCING BASAL METABOLIC RATE

Basal metabolic rate is not static. This means that your BMR continually increases or decreases in response to a variety of factors. The major factors influencing BMR include body shape, body composition, age, sex, nutritional status, and genetics.[8] Tall, thin people tend to have higher BMRs than do short, stocky people of equal weight. This is partly because the former have more surface area, resulting in greater loss of body heat. Weight and height being equal, people with high proportions of muscle tend to have higher BMRs than do people with more body fat. This is because it takes more energy to maintain lean tissue (muscle) than body fat. Age can also influence BMR: after the age of 30, BMR decreases by about 2 to 5 percent each decade. Scientists believe that this decrease is caused mainly by loss of muscle. Women tend to have lower BMRs than men do because women tend to be smaller and have proportionately less muscle. Perhaps the most striking factor influencing BMR is dieting.[9] Severe energy restriction can decrease BMR over time because of the loss of muscle inherent to weight loss. The impact of these and other factors on BMR is summarized in Table 9.1.

Physical Activity

After basal metabolism, the energy expended to support physical activity is the second most significant component of TEE. The amount of energy required for physical activity is quite variable, but for most people it accounts for 15 to 30 percent of TEE.[10] Sedentary people are at the lower end of this estimate, whereas physically active people are at the upper end. Some elite athletes require as much as 2,000 to 3,000 extra kilocalories each day to support the demands of physical activity. Many factors affect the amount of energy expended for physical activity. Rigorous activities such as biking, swimming, and running

Table 9.1 Factors Affecting Basal Metabolic Rate (BMR)

Factor	Effect on BMR
Age	After physical maturity, BMR decreases with age.
Sex	Men have higher BMRs than do women of equal size and weight.
Growth	BMR is higher during periods of growth.
Body weight	BMR increases with increased body weight.
Body shape	Tall, thin people have higher BMRs than do short, stocky people of equal weight.
Body composition	Because muscle requires more energy to maintain than does adipose tissue, people with more lean tissue have higher BMRs than do people of equal weight with more adipose tissue.
Body temperature	Increased body temperature causes a transient increase in BMR.
Stress	Stress increases BMR.
Thyroid function	Elevated levels of thyroid hormones increase BMR.
Energy restriction	Loss of body tissue associated with fasting and starvation decreases BMR.
Pregnancy	BMR increases during pregnancy.
Lactation	Milk synthesis increases BMR.

AP Images/Charles Krupa

The amount of energy required for physical activity is quite variable. Some elite athletes may require as many as 2,000 to 3,000 extra kilocalories each day to support the demands of physical activity.

thermic effect of food (TEF, or diet-induced thermogenesis) The expenditure of energy to digest, absorb, and metabolize nutrients following a meal.

overweight Excess weight for a given height.

obese Excess body fat.

body mass index (BMI) A measure of body fat whereby a person's body weight is divided by the square of his or her height.

have higher energy costs than less demanding activities such as walking. Body size also affects energy expended for physical activity. For example, larger people have more body mass to move than smaller people and therefore expend more energy to accomplish the same activity.

Thermic Effect of Food

Another component of TEE is the **thermic effect of food** (**TEF**, or **diet-induced thermogenesis**). TEF is the energy needed to digest, absorb, and metabolize nutrients following a meal. In other words, it is the cost associated with your body's utilization of the foods you eat. The amount of energy associated with TEF depends on the amount of food consumed and the types of nutrients present in that food. In general, the more food consumed, the higher the amount of energy needed, but high-protein foods have higher TEFs than do fatty foods. Considering that meals generally contain a mixture of nutrients, TEF is estimated to be about 5 to 10 percent of total energy intake.[11] In some ways, this component of total energy expenditure is like a caloric sales tax on energy intake. For example, after consuming a 500-kilocalorie meal, a person typically expends 25 to 50 kilocalories (TEF) just to utilize the nutrients it contains.

LO4 How Are Body Weight and Composition Assessed?

You now understand how the balance between energy intake and energy expenditure affects changes in body weight (energy storage). Clearly, trying to keep energy intake and energy expenditure in balance is key to maintaining a stable body weight. However, this can be difficult for many people, resulting in unwanted weight gain. At what point does added weight gain become unhealthy? And where does one draw the line between a few extra pounds and a serious health concern? To answer these questions, it is important first to understand how body weight and composition are defined, measured, and interpreted.

© Robert Daly/Stone/Getty Images

Overweight Is Excess Weight; Obese Is Excess Fat

Although the terms **overweight** and **obese** are often used interchangeably, they have very different meanings. Overweight refers to excess weight for a given height, regardless of whether the extra weight is muscle or adipose tissue. Conversely, an obese person has excess body fat regardless of her weight. It is therefore possible for muscular people (such as athletes) to be considered overweight but not obese, and some inactive people may not be considered overweight, but may still be obese. Most people who are overweight are obese as well, because adult weight gain is usually caused by an increase in adipose tissue rather than an increase in muscle. For this reason, body weight is often used as an indirect indicator of obesity.

BODY WEIGHT ASSESSMENT: HEIGHT–WEIGHT TABLES AND BODY MASS INDEX

Because there is substantial variation in body weight for any given height, defining an *ideal body weight* may not be possible. Consequently, recommended body weights are simply reference values and are not necessarily ideal for all people. The reference standards most commonly used to assess body weight are height–weight tables and **body mass index (BMI)**, whereby a person's body weight is divided by the square of his or her height (kg/m^2).

Recommended weight ranges are based on height–weight tables that were developed to predict weight ranges associated with the longest life expectancies. Although these tables have been revised a number of times, their data may not accurately represent every population group. Even though height–weight tables

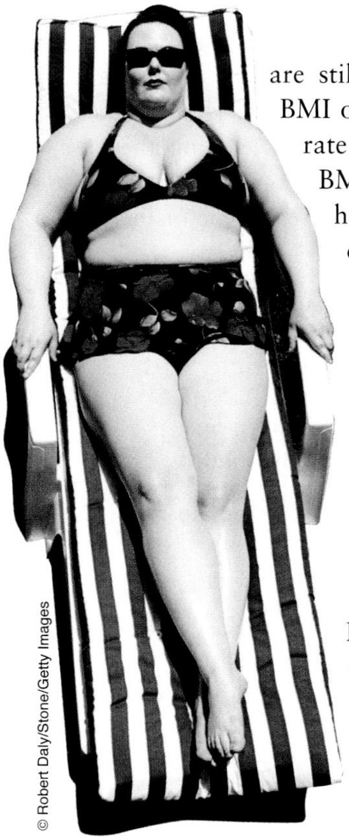

© Robert Daly/Stone/Getty Images

are still used to assess body weight, BMI offers a more reliable and accurate measurement. For this reason, BMI is used more widely than height–weight tables. BMI is calculated using either of the following formulas:

$$\text{BMI (kg/m}^2) = \frac{\text{weight (kg)}}{\text{height (m)}^2}$$

$$\text{BMI (lbs/in}^2) = \frac{\text{weight (lb)} \times 703.1}{\text{height (in)}^2}$$

Based on these formulas, the BMI of a person who weighs 68.2 kg (150 lb) and is 1.65 m (65 in) tall is 25 kg/m². You can use the chart at the back of this book to determine your BMI.

People with low BMIs typically have low amounts of body fat, whereas those with high BMIs tend to have higher amounts of body fat. Because each BMI unit represents six to eight pounds for a given height, an increase in just two BMI units represents a 12- to 16-pound increase in body weight.

BMI is based on the ratio of weight to height, and is therefore a better indicator of obesity than weight alone. The cutoff values for BMI classifications (underweight, healthy weight, overweight, and obese) are based on the association between BMI and weight-related morbidity and mortality (see Figure 9.8). Most medical organizations, including the U.S. Centers for Disease Control and Prevention (CDC), use the following BMI criteria to assess body weight in adults:

- Underweight: <18.5 kg/m²
- Healthy weight: 18.5–24.9 kg/m²
- Overweight: 25.0–29.9 kg/m²
- Obese: ≥ 30 kg/m²

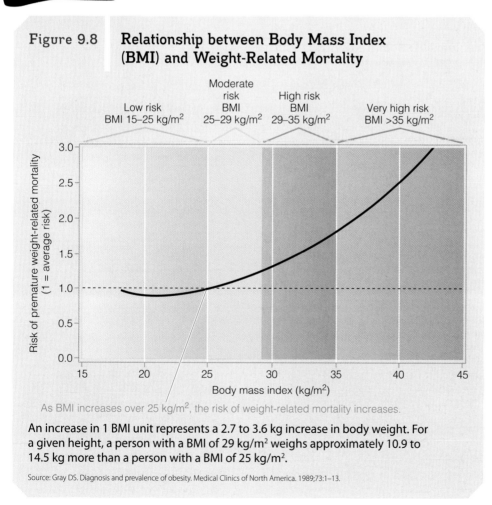

Figure 9.8 Relationship between Body Mass Index (BMI) and Weight-Related Mortality

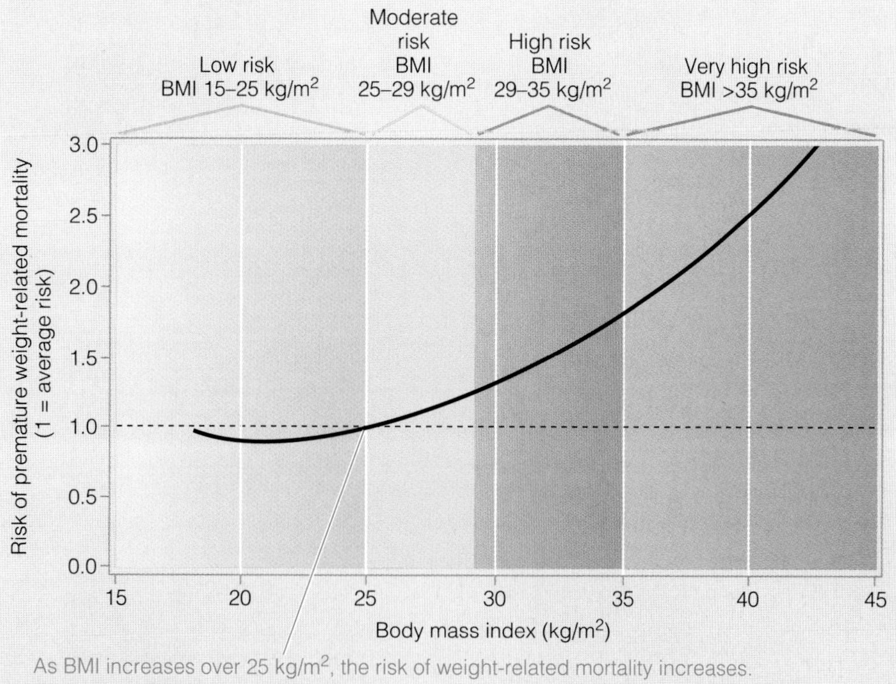

An increase in 1 BMI unit represents a 2.7 to 3.6 kg increase in body weight. For a given height, a person with a BMI of 29 kg/m² weighs approximately 10.9 to 14.5 kg more than a person with a BMI of 25 kg/m².

Source: Gray DS. Diagnosis and prevalence of obesity. Medical Clinics of North America. 1989;73:1–13.

Obesity Is Related to Excess Body Fat

Although BMI is a useful method of interpreting a person's weight given his height, health professionals sometimes want to know a person's body composition. To this end, imagine that the body has two main compartments:

1. The fat compartment
2. The fat-free (lean) compartment

The fat (adipose tissue) compartment consists mostly of stored triglycerides and supporting structures, whereas the fat-free compartment consists mostly of muscle, water, and bone. The amount of fat and fat-free mass a person has is determined by many factors, such as sex, genetics, physical activity, hormones, and diet.

The amount of fat stored in the body changes throughout the life cycle. For example, nearly 30 percent of total body weight in a healthy six-month-old infant is fat, whereas this percentage may be cause for concern in adults. It is recommended that body fat levels be between 12 and 20 percent of total body weight for men and 20 and 30 percent of total body weight for women (see Table 9.2). Body fat over 25 percent in men and over 33 percent in women indicates obesity.[12] Just as too much body fat can cause health problems, too little body fat can also have harmful effects; body fat below 12 percent for men and 20 percent for women is considered too low.

METHODS OF MEASURING BODY COMPOSITION

Obesity increases one's risk for a variety of health problems, including type 2 diabetes, sleep irregularities, joint pain, gout, stroke, heart disease, gallstones, and certain cancers.[13] Some of the health problems associated with obesity are illustrated in Figure 9.9. To assess a person's likelihood of developing these problems, clinicians use a variety of anthropometric methods to estimate body fat. Whereas some of these

Table 9.2 Obesity Classifications Based on Percent Body Fat

	Body Fat (% total body weight)	
Classification	Males	Females
Healthy	12–20	20–30
Borderline obese	21–25	31–33
Obese	>25	>33

Source: Bray G. What is the ideal body weight? Journal of Nutritional Biochemistry. 1998;9:489–92.

Figure 9.9 Health Problems Associated with Obesity

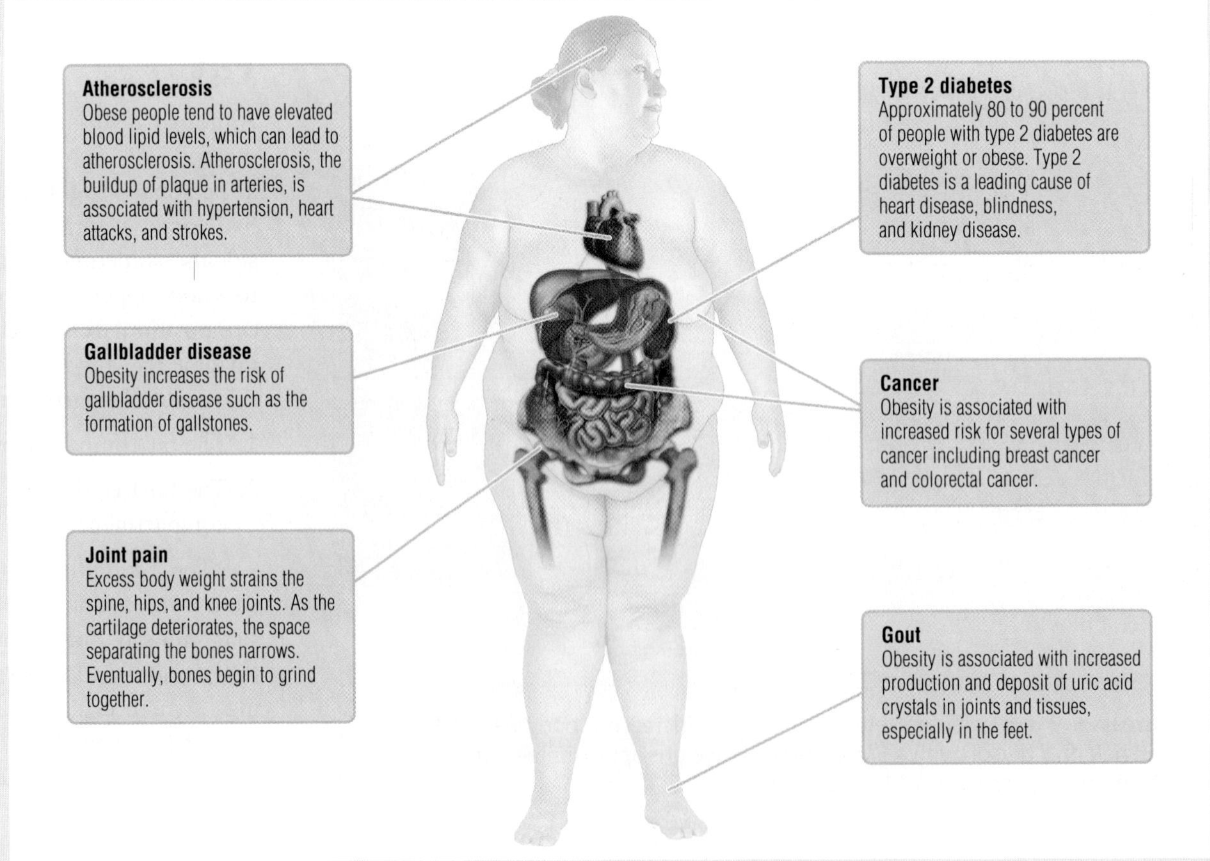

Atherosclerosis
Obese people tend to have elevated blood lipid levels, which can lead to atherosclerosis. Atherosclerosis, the buildup of plaque in arteries, is associated with hypertension, heart attacks, and strokes.
Gallbladder disease
Obesity increases the risk of gallbladder disease such as the formation of gallstones.
Joint pain
Excess body weight strains the spine, hips, and knee joints. As the cartilage deteriorates, the space separating the bones narrows. Eventually, bones begin to grind together.
Type 2 diabetes
Approximately 80 to 90 percent of people with type 2 diabetes are overweight or obese. Type 2 diabetes is a leading cause of heart disease, blindness, and kidney disease.
Cancer
Obesity is associated with increased risk for several types of cancer including breast cancer and colorectal cancer.
Gout
Obesity is associated with increased production and deposit of uric acid crystals in joints and tissues, especially in the feet.

methods are expensive and complex, others are more readily available and easy to use. For illustrations of the common methods used to estimate body composition, see Figure 9.10.

- Often called underwater weighing, **hydrostatic weighing** involves measuring a person's weight both in and out of water. The more fat a person has, the less dense she is, and thus the less she weighs under water. Hydrostatic weighing is usually reserved for clinical evaluations because it requires special equipment and is neither practical nor convenient.
- **Dual-energy x-ray absorptiometry (DEXA)** provides information about the distribution of fat and lean tissue across different areas of the body. During a DEXA measurement, the patient lies still on a table while an x-ray scanning device passes over her body. DEXA is considered the gold standard of body composition analysis and is often used by researchers and clinicians.
- Because lean tissue conducts electrical currents better than adipose tissue, **bioelectrical impedance** can be used to estimate one's body fat percentage by measuring how easily a weak electric current travels through the body. This technique is relatively accurate, simple to use, and is considered an acceptable method for estimating body composition in a clinical setting. Bathroom

hydrostatic weighing A method of estimating body composition in which a person's body weight is measured in and out of water.

dual-energy x-ray absorptiometry (DEXA) A method of estimating body composition in which low-dose x-rays are used to visualize fat and fat-free compartments of the body.

bioelectrical impedance A method of estimating body composition in which a weak electric current is passed through the body.

Figure 9.10 Methods Used to Estimate Body Composition

Hydrostatic weighing
Hydrostatic weighing requires the subject to exhale air from the lungs and be submerged in water. It is important to remain motionless while weight underwater is measured.

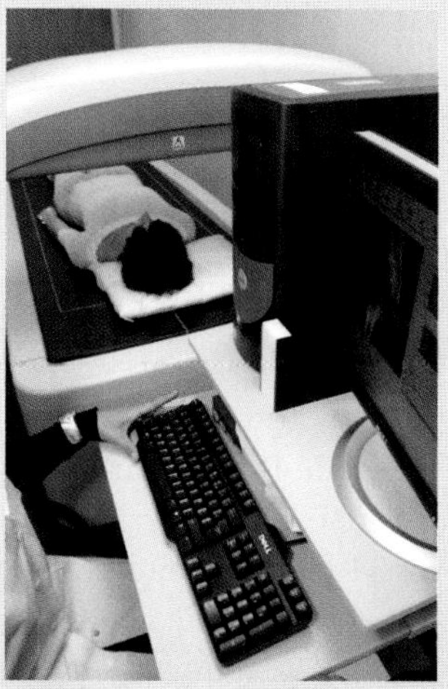

Dual-energy x-ray absorptiometry (DEXA)
While the person is lying on a table, a scanning device passes over the body. The x-ray beams emitted differentiate between the fat mass, lean mass, and skeletal mass. A two-dimentional image of the body is displayed on a computer screen.

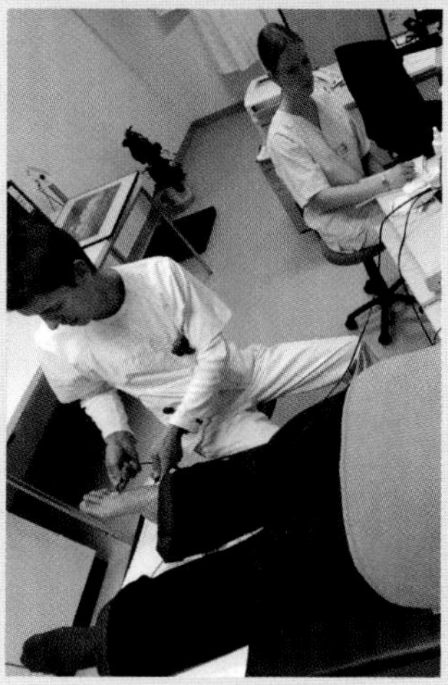

Bioelectrical impedance
Electrodes are placed on a person's hand and foot, and a weak electrical current is passed through the body. The conductivity of the current is measured, which provides an estimate of the fat mass and fat-free mass.

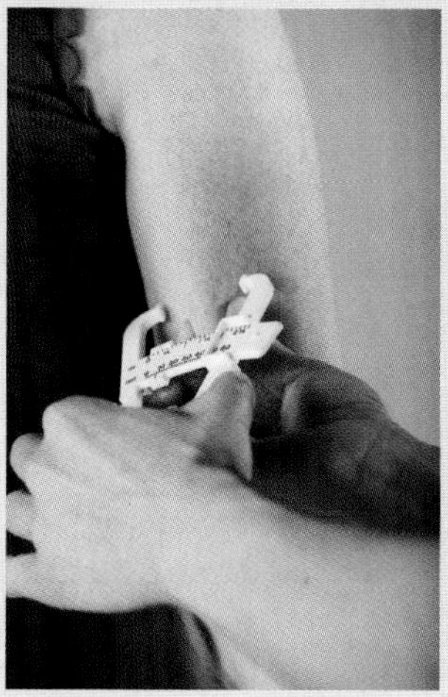

Skinfold thickness
Using a measuring device called a caliper, the heath care worker measures the thickness of a fold of skin with its underlying layer of fat at precise locations on the body.

scales with built-in bioelectrical impedance systems are now available for home use, although little is known about their accuracy.

- The **skinfold thickness method** has been used for many years to estimate body fat. An instrument called a **skinfold caliper** is used to measure the thickness of skin and subcutaneous fat at various locations on the body. Body fat is then estimated using mathematical formulas based on the skinfold thickness measurements.

You are now familiar with some of the methods by which weight and obesity are assessed. But to understand what truly causes weight gain and obesity, you must look beyond the individual and examine the overall influences that affect eating habits.

Table 9.3 Estimated Prevalence of Obesity in the 1970s and 2007–2008

Age group	Prevalence of Obesity (%) 1970s	2007–2008
2–5 years	5	10
6–11 years	4	20
12–19 years	6	18
>19 years	15	34
Number of states with adult obesity rate >25%	0	32

Sources: Flegal KM, Carroll MD, Ogden CL, Curtin LR. Prevalence and trends in obesity among U.S. adults.1999–2008. Journal of the American Medical Association. 2010;303:235-41. Ogden CL, Flegal KM, Carroll MD, Johnson CL. Prevalence and trends in overweight among U.S. children and adolescents. 1999–2000. Journal of the American Medical Association. 2002;288:1728–32. Ogden CL, Carroll MD, Curtin LR, Lamb MM, Flegal KM. Prevalence of high body mass index in U.S. children and adolescents. Journal of the American Medical Association. 2007–2008. 2010;303:242–9. Centers for Disease Control and Prevention. U.S. Obesity Trends. Available from: http://www.cdc.gov/obesity/data/trends.html.

LO5 Genetics versus Environment: What Causes Obesity?

When it comes to one's body, a person is usually his own worst critic. While you might wish that your body looked or was shaped differently, it is important to recognize that the issue of obesity is of concern not because of societal norms but rather because excess weight can be injurious to your health. In the late 19th century, only 3 percent of American adults were overweight and very few were obese. Today, nearly 34 percent of adults are classified as overweight or obese in the United States—a figure that seems to be increasing steadily.[14] Similar trends are evident in children and adolescents: in the last 20 years, the percentage of overweight American children and adolescents has tripled. An overview of obesity trends in the United States is presented in Table 9.3.

Obesity is not only on the rise in the United States, but around the world as well.[15] Recall from Chapter 1 that an increase in obesity that is correlated with industrialization is referred to as the *nutrition transition*. The trend toward obesity is growing worldwide as nations turn from traditional lifestyles to those associated with economic development. Clearly, alarmingly high global obesity rates are cause for concern. But what is causing the global obesity epidemic? To answer this question, it is important to consider both lifestyle and genetic factors, as it is a combination of these factors that causes people to consume more calories and/or expend less energy. Together, these factors can shift energy balance in favor of positive energy balance and weight gain—especially in today's world.

skinfold thickness method A method of estimating body composition in which a skinfold caliper is used to measure the thickness of skin and subcutaneous fat at various locations on the body.

skinfold caliper An instrument used to measure the thickness of skin and subcutaneous fat.

Eating Habits

One of the most important factors that influence body weight is the amount and energy density of the foods consumed. It is likely that an increase in energy intake over the last few decades is a major cause of today's obesity epidemic. Over the past 40 years, the average daily energy intake of American adults has increased by approximately 200 to 300 kilocalories.[16] A variety of societal, cultural, and psychological factors influencing what, how much, when, and where people eat may be contributing to this trend.

SOCIETAL AND CULTURAL INFLUENCES ON EATING HABITS

Food-related societal and cultural norms affect what and how much food people consume. For instance, eating out of the home and consuming increased portion sizes of foods—especially of the energy-dense

> When paired with a **32-ounce soft drink**, a large **McDonald's Quarter Pounder** with Cheese Extra Value Meal contains approximately **1430 kilocalories**.

variety—is becoming more and more common. In the United States, energy-dense, inexpensive, flavorful foods have become readily available and are accepted as a cultural norm. Some estimates indicate that 37 percent of adults and 42 percent of children eat fast food daily.[17] Although the consumption of fast food is not the sole cause of obesity, it may certainly contribute, because many fast food items are high in fat, refined starchy carbohydrates, and calories. As such, a super-sized value meal might provide more than half the calories required in a day. To counter the perception that their food is unhealthy, many fast-food restaurants also offer healthier food choices such as salads, wraps, and sandwiches made with lean meats and whole-grain breads. Selecting these menu options is one way for individuals and families to eat healthier.

Beyond the type of food consumed, studies show that serving size also influences food consumption. That is, when larger food portions are served, people tend to eat more.[18] For example, participants in a study ate 39 percent more M&Ms® when they were given a two-pound bag than when they received a one-pound bag.[19] Why does this happen? It appears that some people depend more on visual cues than on physiological cues such as hunger and satiety to judge how much to eat. This phenomenon is particularly important to consider because many restaurants continue to offer super-sized food portions. For example, Starbucks recently unveiled a new serving size that is 325 ml (seven ounces) larger than the company's previous largest size.

Factors related to food itself can also influence what and how much food a person consumes. When presented with a variety of food choices, people tend to eat more than when they are offered fewer choices.[20] Thus, buffet-style restaurants may be more conducive to overeating than traditional restaurants with more limited menu choices and fixed portions of food. In addition, when people consume or perceive they have consumed fewer calories at one meal, they tend to reward themselves by eating more at subsequent meals.[21] This may explain in part why increased consumption of reduced-calorie foods and beverages does not necessarily lead to a reduction in total calorie intake over time.

SOCIODEMOGRAPHIC AND PSYCHOLOGICAL INFLUENCES ON EATING HABITS

Factors such as economic status, marital status, and education level can also influence a person's risk for obesity. Although sociodemographic factors do not cause obesity directly, some may contribute to the problem indirectly. Beyond sociodemographic factors, some psychological disorders are also related to a person's likelihood of becoming obese. It is not clear, however, whether obesity predisposes individuals to these disorders or if some psychological profiles lead to obesity. All scientists can say for certain is that some personality types appear to be associated with increased risk for obesity. For example, obese individuals are more likely to experience clinical depression and panic attacks than non-obese people.[22] This may stem from the fact that obesity sometimes lowers self-esteem and confidence.

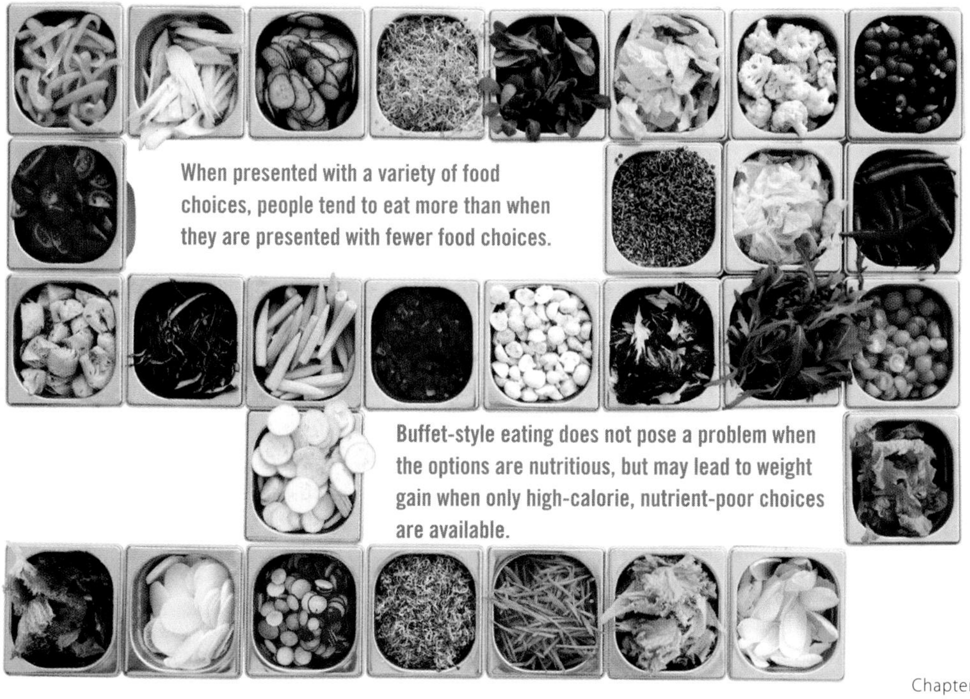

When presented with a variety of food choices, people tend to eat more than when they are presented with fewer food choices.

Buffet-style eating does not pose a problem when the options are nutritious, but may lead to weight gain when only high-calorie, nutrient-poor choices are available.

Conversely, individuals who have difficulty coping with stress may turn to food for emotional comfort and gratification, making them more susceptible to obesity.

Research also suggests that a person's social network may also relate to her risk for obesity.[23] People with friends who have experienced weight gain are more likely to gain weight than those with weight-stable friends. It is not clear how social networks are linked to weight gain, but researchers believe that weight gain among close friends and family members might serve as a permissive cue for others to gain weight as well.

© Gökhan Hotamisligil, Harvard School of Public Health

A **gene mutation** *made the* **mouse** *on the right* **obese.**

Sedentary Lifestyle

Choosing what and how much to eat is not the only lifestyle choice related to unwanted weight gain and obesity. At the same time they are eating more, Americans are becoming less physically active. A decrease in the availability of jobs that require physical work and an increase in the availability of labor-saving devices make daily life less physically demanding for most people. To compound matters, almost 50 percent of adults fail to engage in any kind of leisure-time physical activity that meets the 2008 Physical Activity Guidelines for Americans.[24] Together, these changes have likely had a significant negative impact on the nation's health.

It is important to realize that physical activity is not limited to formal exercise. **Physical activity** is any bodily movement that results in an expenditure of energy, whereas **exercise** is a planned, structured, and repetitive body movement done to improve or maintain physical fitness.[25] Physical activities such as walking up the stairs or mowing the lawn can be just as beneficial as formal exercise. Not surprisingly, the vast majority of studies show that a lack of physical activity increases the risk of being overweight or obese.[26] For most people, walking an extra mile each day—equivalent to taking about 2,000 to 2,500 extra steps—would increase energy expenditure by 100 kcal/day.[27] If all other factors remained unchanged, a person could lose one pound of body weight per month by making this simple change. In addition to helping maintain a healthy body weight, physical activity and exercise also help a person stay healthy and physically fit.

Genetics

Although there is general agreement that lifestyle factors—and not genetics—are the driving force behind the obesity epidemic, both influence body weight. In an environment where energy-dense foods are abundant and physical activity is low, genetic factors make some people more susceptible to weight gain than others.

DISCOVERY OF LEPTIN PROVIDED THE FIRST GENETIC CLUE TO OBESITY

Scientists have long believed that genetic makeup influences body weight. However, direct evidence was lacking until the discovery of a mouse with a gene mutation that made it obese. This mutant animal was referred to as the ***ob/ob*** **(obese) mouse.** *Ob/ob* mice consume large amounts of food, are inactive, and therefore gain weight easily. Soon after the *ob/ob* mouse was discovered, researchers also discovered another mouse with a gene mutation that made it both obese and diabetic. They referred to this mouse as the ***db/db*** **(diabetic) mouse.**[28] In studying these two varieties of mutated mice, scientists learned that the ***ob*** **gene** codes for a hormone called **leptin**, a potent satiety signal produced in many tissues but especially in adipocytes.[29] Because of a mutation in the *ob* gene, *ob/ob* mice do not produce leptin, causing them to eat uncontrollably. Soon thereafter, researchers discovered that the ***db*** **gene** codes for the leptin receptor, found primarily in the hypothalamus. *Db/db* mice do not produce the leptin receptor and therefore cannot appropriately respond to the satiating effects of leptin.

Many people hoped that leptin—touted as the *anti-obesity hormone*—would become the miracle cure for obesity. Yet further research indicated that the vast majority of obese people produce appropriate or even elevated amounts of leptin. In fact, only a few

physical activity Any bodily movement that results in an expenditure of energy.

exercise Planned, structured, and repetitive bodily movement done to improve or maintain physical fitness.

ob/ob **(obese) mouse** A mouse with a gene mutation that makes it obese.

db/db **(diabetic) mouse** A mouse with a gene mutation that makes it both obese and diabetic.

ob **gene** The gene that codes for the hormone leptin.

leptin A potent hormone produced primarily by adipose tissue that signals satiety.

db **gene** The gene that codes for the leptin receptor.

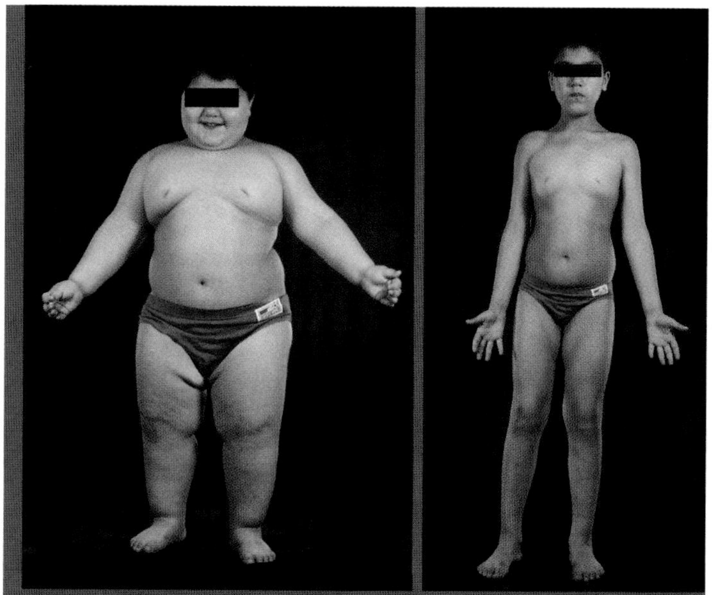

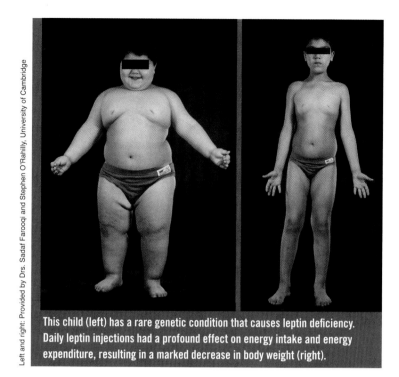

This child (left) has a rare genetic condition that causes leptin deficiency. Daily leptin injections had a profound effect on energy intake and energy expenditure, resulting in a marked decrease in body weight (right).

Left and right: Provided by Drs. Sadaf Farooqi and Stephen O'Rahilly, University of Cambridge

cases of human leptin deficiency have been reported.[30] Still, scientists found that when severely obese, leptin-deficient individuals are given leptin injections, they experience dramatic weight loss. Although leptin has not proved to be effective for treating human obesity in most cases, its discovery led to important insights into body weight regulation and has opened the door to many other exciting discoveries related to the so-called "obesity hormones."

LO6 How Are Energy Balance and Body Weight Regulated?

Although the discovery of leptin has not solved the obesity dilemma, it has deepened scientists' understanding of body weight regulation profoundly. While much remains unknown about body weight regulation and possible defects in key energy-regulating activities, it is now clear that the body plays an active role in influencing how much energy is consumed, how much energy is expended, and how much energy is stored. Physiological systems require homeostatic regulatory mechanisms that maintain checks and balances via hormonal signaling pathways—some of which are described next.

Set Point Theory of Body Weight Regulation

The body's ability to adjust energy intake and energy expenditure on a long-term basis serves an important purpose: body weight regulation. If the body had no means of regulating energy balance, the consequences would be serious. Even a slight imbalance in daily energy intake and energy expenditure could result in substantial weight gain or loss over time. To prevent such imbalances, long-term energy balance regulatory signals communicate the body's energy reserves to the brain, which in turn releases neurotransmitters that influence energy intake and/or expenditure.[31] If this long-term system functions effectively, body weight remains relatively stable.

Scientists have suspected for many years that a complex signaling system regulates body weight by making minor adjustments in energy intake and expenditure.[32] To test this theory, researchers observed weight-gain and weight-loss cycles in food-restricted mice. When food intake was restricted, the mice lost weight. Not surprisingly, when the mice were fed, consumption increased and the mice soon returned to their original weights. However, what was surprising to scientists was the fact that after returning to their original weights, the mice maintained this weight by spontaneously decreasing their intake of food. This phenomenon prompted scientists to propose what they called the **set point theory** of body weight regulation.

Proponents of the set point theory believe that hormones circulating in the blood (such as leptin) regulate body weight by communicating the amount of adipose tissue in the body to the brain. When the amount of adipose tissue increases beyond a certain set point, a signal causes food intake to decrease and/or energy expenditure to increase, favoring weight loss. Conversely, when the amount of adipose tissue decreases below a set point, food intake increases and/or energy expenditure decreases, favoring weight gain. In this way, body weight remains relatively stable on a long-term basis. The discovery of the *ob* gene and leptin provided the first real evidence that hormones produced in adipocytes can communicate adiposity to the brain, which in turn influences hunger and satiety.

set point theory
A scientific concept whereby hormones circulating in the blood are theorized to regulate body weight by communicating the amount of adipose tissue in the body to the brain, which adjusts energy intake and expenditure.

Leptin Communicates the Body's Energy Reserve to the Brain

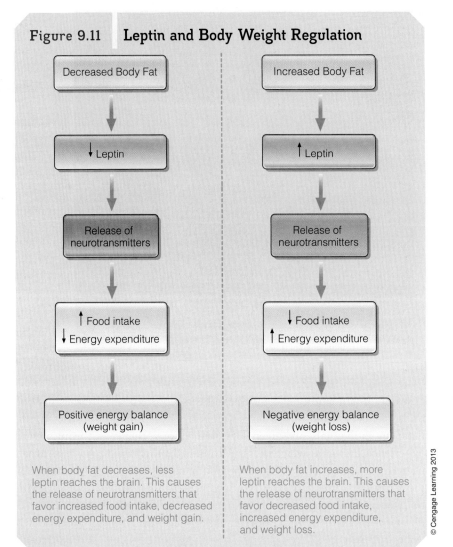

Figure 9.11 **Leptin and Body Weight Regulation**

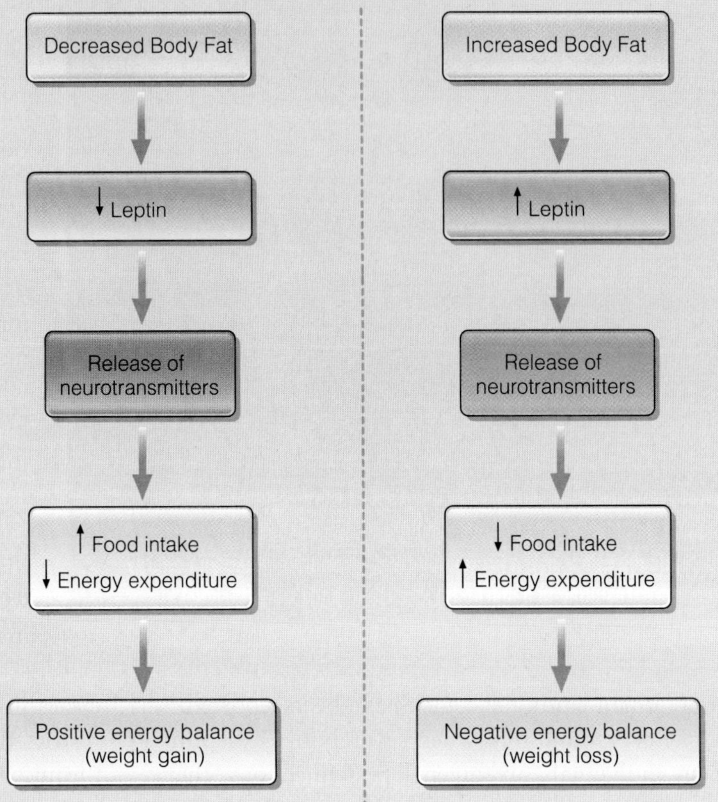

When body fat decreases, less leptin reaches the brain. This causes the release of neurotransmitters that favor increased food intake, decreased energy expenditure, and weight gain.

When body fat increases, more leptin reaches the brain. This causes the release of neurotransmitters that favor decreased food intake, increased energy expenditure, and weight loss.

Although the complex mechanisms that regulate long-term energy balance are not fully understood, leptin appears to play an important role by communicating energy reserves to the brain (see Figure 9.11). As previously mentioned, leptin is produced primarily by adipose tissue. When body fat increases, so does the concentration of leptin circulating in the blood. When body fat decreases, leptin production also decreases. Thus, fluctuations in blood leptin levels reflect changes in the body's primary energy reserve, adipose tissue. Leptin does more than convey the message, however—it also signals a response. Elevated leptin concentrations signal the brain to increase the release of neurotransmitters that prevent further weight gain by prompting a decrease in food intake and/or an increase in energy expenditure. Conversely, decreased blood concentrations of leptin signal the release of neurotransmitters that stimulate food intake and/or decrease energy expenditure, thus protecting the body against further weight loss. In this way, leptin is believed to be part of the complex communication loop that helps maintain a relatively stable body weight over time.

DEFECTS IN LEPTIN RESPONSIVENESS MAY LEAD TO OBESITY

You may be wondering,"If leptin curbs food intake, why are so many people obese? If obese people have elevated leptin levels, why do they continue to gain weight?" These are very good questions. A number of researchers believe that some people may have disconnected leptin-signaling pathways.[33] In other words, the brains of obese people may not be responsive to leptin's appetite suppression signal, regardless of how much leptin is being produced. Clearly, there is more to learn about this regulatory system. As scientists continue to study the role of leptin and other adipokines in long-term energy balance, they will gain a better understanding of the underlying mechanisms associated with body weight regulation.

LO7 What Is the Best Approach to Weight Loss?

Although the diet industry would like you to believe otherwise, the truth is clear: there is no quick or easy way to lose weight and keep it off. Approximately one-third of adults in the United States are on a special diet to lose

The obesity epidemic certainly cannot be attributed to a lack of weight-loss advice. Amazon.com lists over 8,000 books related to weight-loss diets.

weight. There are plenty of diets to choose from—the obesity epidemic certainly cannot be attributed to a lack of weight-loss advice. Beyond advice and meal planning, a variety of non-prescription weight-loss products are available and marketed aggressively to people trying to lose weight. Far from helping, some of these products may be hazardous to a person's health.

Most popular weight-loss plans have specific rules to follow. Some restrict the types of food that can be eaten, others recommend a strict exercise regimen, and still others advocate periods of fasting or cleansing. The real issue that many weight-loss plans fail to address is not what it takes to lose weight but what it takes to keep it off. Rather than succumbing to a fad diet or quick-fix weight-loss approach, consider what actual weight-loss experts have to say.

Healthy Food Choices Promote Overall Health

Although there are many reasons why a person might want to lose weight, the most important is to improve health. Achieving and maintaining weight loss requires lasting lifestyle changes. A person must evaluate, and if necessary, permanently modify the food that he chooses to eat and the amount of physical activity in which he engages. Most health experts suggest that people focus less on weight loss and more on healthy eating patterns and overall fitness. Unfortunately, many misguided weight-loss efforts present food as the enemy rather than as a means to good health. People who decide to lose weight at any cost often sacrifice their health in the process. Simply put, to lose weight and keep it off successfully and in a healthy way, a person must choose to eat a balanced diet of nutrient-dense foods and maintain a moderately high level of physical activity.[34] A healthy weight-loss and weight-maintenance program consists of three components:

(1) Setting reasonable goals
(2) Choosing nutritious foods in moderation
(3) Increasing energy expenditure through daily physical activity

SETTING REASONABLE GOALS

Setting reasonable and attainable goals is an important component of any successful weight-loss program. For instance, an obese person's realistic initial weight-loss goal might be to reduce her current weight by 5 to 10 percent. For someone weighing 180 pounds, this would amount to an initial weight loss of approximately 9 to 18 pounds. Studies show that even a modest reduction in weight can improve overall health.[35] When it comes to weight loss, slow and steady is the way to go, and weekly weight loss should not exceed one to two pounds. This can be achieved by decreasing energy intake by 100 to 200 kilocalories each day or by walking one to two miles each day. Over time, these small changes can result in significant reductions in body weight. Rather than making dramatic dietary changes, small changes such as reducing portion sizes and cutting back on energy-dense snack foods can make a big difference in overall energy intake.

Once your body weight stabilizes and you have maintained the new lower weight for a few months, you can decide whether additional weight loss is needed. Some people benefit from joining weight-maintenance programs such as Weight Watchers® or Take Off Pounds Sensibly®. These types of programs provide long-term support and motivate people to maintain healthy diets and lifestyles.

CHOOSING NUTRITIOUS FOODS IN MODERATION

Weight-loss plans that drastically reduce caloric intake and offer limited food choices often leave people feeling hungry and dissatisfied. Instead, those that allow moderate caloric intake and encourage people to eat foods that are healthy and appealing tend to be more successful. Contrary to popular belief, it is not necessary to avoid foods that contain fat in order to lose

> One glazed **doughnut**, a one-ounce bag of **potato chips**, and a 12-ounce serving of **cola** each contains roughly **150 kilocalories**.

weight. Instead, it is best to limit the intake of foods that contain *trans* fatty acids and saturated fatty acids. Fat intake should emphasize foods that predominantly contain polyunsaturated and monounsaturated fatty acids. Indeed, the 2010 Dietary Guidelines for Americans recommends that a person choose fats as carefully as he chooses carbohydrates.

That dairy products and meats are high-fat foods and therefore should be avoided when trying to lose weight is also a common misconception. Again, the keys to good nutrition and weight maintenance are moderation and making smart choices. Switching from whole to reduced-fat milk is one way to lower caloric intake without losing out on the many vitamins and minerals in milk. Likewise, the types of meat you choose and the methods by which you prepare the meat can greatly affect how many calories you consume. Lean meats prepared by broiling or grilling are both nutritious and satisfying.

Reducing energy intake is best achieved by cutting back on energy-dense foods that have little nutritional value such as potato chips, cookies, and cakes. Aside from their lower energy densities, nutrient-dense foods such as whole-grain breads, legumes, fruits, and vegetables contain many beneficial substances such as micronutrients and fiber. Furthermore, because these foods tend to have greater volumes than do energy-dense foods, they help people feel more satisfied after they eat.

Healthy eating requires a person to pay attention to hunger and satiety cues. However, because visual cues sometimes have a greater influence than do internal cues on the quantity of food consumed, the amount of food served or packaged—rather than hunger and satiety—often determines how much one eats. Some commercially made muffins, for instance, are extremely large and contain as many calories as several slices of bread. Learning to recognize and choose reasonable portions of food is a critical component to successful weight management. In fact, reducing portion sizes by as little as 10 to 15 percent can lower daily energy intake by as much as 300 kilocalories. One way to limit serving size (and incidentally, save money) is to consider sharing a meal the next time you eat at a restaurant. These and other behaviors associated with healthy weight management are listed in Table 9.4.

INCREASING ENERGY EXPENDITURE THROUGH DAILY PHYSICAL ACTIVITY

In addition to choosing nutritious foods in moderation, you can also help tip the energy balance equation toward negative energy balance by engaging in physical activity. Walking one mile each day, which takes people about 15 to 20 minutes, uses about 100 kilocalories. This adds up to 700 kilocalories per week. Beyond going out of your way to exercise, just being physically active throughout the day can make a big difference in promoting weight loss and weight maintenance. When possible, you might consider taking the stairs rather than the elevator, walking or biking rather than driving, and incorporating chores into your routine daily activities. Even if they do not lose weight, individuals who are physically active show improved physical fitness. Regardless of one's weight,

Table 9.4 Behaviors Associated with Healthy Weight Management

• Focusing attention on an overall healthy eating pattern that emphasizes nutrient-dense foods.
• Monitoring food intake, body weight, and physical activity to help make oneself more aware of what and how much is eaten and drunk.
• Selecting small-size and lower-calorie options when eating out.
• Preparing and serving smaller food portions, especially foods and beverages that are high in calories.
• Eating nutrient-dense foods to improve nutrient intake and healthy body weight.
• Reducing screen time to one to two hours each day and increasing physical activity to help with weight loss and weight maintenance.

a lack of exercise may prove the greatest health hazard of all.[36]

It is important to realize that overweight people can still be physically fit, and healthy-weight people can be physically unfit. In fact, studies show that obese individuals who are physically fit have fewer health problems than do healthy-weight individuals who are unfit.[37] Healthy blood pressure levels, blood glucose regulation, and blood lipid levels are important indicators of physical fitness. Beyond improving these measures, exercise also promotes a positive self-image and helps a person take charge of her life. Although some people resist the idea of starting an exercise program, they rarely regret it once they begin.

Physical activity is an effective strategy for preventing unhealthy weight gain in healthy-weight, overweight, and obese individuals. For substantial health benefits, the 2010 Dietary Guidelines for Americans recommends that adults engage in at least 150 minutes a week of moderate-intensity—or 75 minutes a week of vigorous-intensity—aerobic physical activity.[38] People who were formerly obese, however, may require more exercise—60 to 90 minutes each day—to maintain a lower body weight. It is important to remember that after a person loses weight, his total energy requirement is lower than it was before the weight was lost. For example, when a person loses 10 pounds, his basal metabolic rate decreases by 80 kcal/day. Thus, to maintain a 10 pound weight loss, he must further reduce daily energy intake or increase energy expenditure by 80 kilocalories—the amount in a single slice of bread.

Not all obese people are physically **unfit** and/or **unhealthy**. It is the combination of being both **overweight and sedentary** that **increases** a person's **risk** of developing weight-related health problems.

Characteristics of People Who Successfully Lose Weight

One of the best ways to learn what works in terms of weight loss is to study people who have been successful. The National Weight Control Registry is a large, prospective study of individuals who have lost significant amounts of weight and have been successful at keeping it off.[39] To be eligible to participate in the study, individuals must have maintained a weight loss of 30 pounds or more for one year or longer. Researchers interviewed this unique group of successful individuals to learn about their weight loss and weight management practices. However, identifying common characteristics among study participants proved difficult. Whereas some reported counting calories or grams of fat, others used prepackaged diet foods. Some participants preferred losing weight on their own, while others sought assistance from weight-loss programs. Nonetheless, researchers identified one strikingly common thread: almost 89 percent of successful weight-loss participants reported that both diet and physical activity were part of their weight-loss plan. Only 10 percent reported using diet alone, and only 1 percent used exercise alone. On average, study participants engaged in 60 to 90 minutes of physical activity daily, which was equivalent to 2,500 kcal/week for women and 3,300 kcal/week for men. In addition to a high level of physical activity, participants ate breakfast regularly and monitored their weight frequently.

Does Macronutrient Distribution Matter?

With all the weight-loss advice that is available, it can be difficult to sort fact from fiction. Even experts have differing opinions about the best balance of protein, fat, and carbohydrate intake for achieving weight loss. Today, one of the biggest controversies in weight-loss research is the role of dietary carbohydrate and dietary fat in weight loss and weight gain.[40]

Weight-loss diets that are low in fat and high in carbohydrates have long been considered the most effective in terms of weight loss and weight maintenance. Indeed, the 2010 Dietary Guidelines for Americans advocates low-fat food choices with an emphasis on whole grains, fruits, and vegetables. Similarly, recall that the Acceptable Macronutrient Distribution Ranges (AMDRs) suggest that one consume 45 to 65 percent of energy from carbohydrate, 10 to 35 percent from protein, and 20 to 35 percent from fat. However, in 1972 Robert Atkins, one of the first pioneers of the low-carbohydrate diet, turned the nutritional world upside down by proposing that too much carbohydrate—rather than too much fat—actually causes people to gain weight. To see how the AMDR's caloric distribution compares to those of high- and low-carbohydrate weight-loss diets, see Table 9.5. A weight-loss plan may advocate either the AMDR's or Dr. Atkin's caloric distribution, but what scientific evidence exists concerning the effects of macronutrient distribution on weight loss and overall health? Read on to find out.

HIGH-CARBOHYDRATE, LOW-FAT WEIGHT-LOSS DIETS

Many researchers believe that diets high in carbohydrates and low in fat promote weight loss and have an overall beneficial effect on health. For example, Dr. Dean Ornish's popular weight-loss plan outlined in *Eat More, Weigh Less* recommends a relatively low intake of fat (10 to 15 percent of total calories).[41] According to Dr. Ornish's plan, dieters should avoid most meat, dairy foods, oils, and olives, but low-fat meats and dairy products can be eaten in moderation. With an emphasis on fruit, vegetables, and whole grains, this weight-reduction and maintenance plan suggests that about 65 to 75 percent of total calories come from carbohydrates and that protein and fat make up the remainder.

There are several reasons why advocates of low-fat, high-carbohydrate diets believe that such plans help prevent obesity. First, gram for gram, fat has more than twice as many calories as carbohydrate and protein. It is therefore reasonable to assume that consuming less fat leads to lower energy intake, which in turn results in weight loss. Second, fat can make food more flavorful, contributing to overconsumption. Finally, excess calories from fat are stored by the body more efficiently than excess calories from carbohydrate or protein. Converting excess glucose and amino acids into fatty acids requires energy, which contributes to energy expenditure and thus helps promote weight loss.

Long-standing dietary advice aimed at helping people lose weight has consistently focused on reducing dietary fat. Although total energy intake has increased, the percentage of total calories from fat has declined from 45 percent in the 1960s to approximately 33 percent today. On average, 10 to 20 fewer grams of fat are consumed per day.[42] Obviously, decreased fat intake has not resulted in a decreased prevalence of obesity. In fact, obesity rates have increased under the lower-fat regime.[43]

Although it is not clear precisely which dietary factors contributed to this trend, some researchers believe that a failure to choose healthy high-carbohydrate foods may be to blame. While experts hoped that Americans would replace fatty foods with more nutritious high-carbohydrate foods such as whole-grain breads, fruits, and vegetables, the wide availability and low cost of low-fat yet high-calorie and high-carbohydrate snack foods have made such foods staples in many peoples' diets. The ingredient that typically replaces fat in products advertised as "fat-free" is refined carbohydrate (such as white flour or table sugar)—a nutrient-poor, calorie-containing substitute. Because many of these products have the same amount of calories as the original products, the consumption of high-carbohydrate, fat-free snack foods may actually contribute to weight

Table 9.5 Caloric Distribution of High- and Low-Carbohydrate Diets Compared to the Acceptable Macronutrient Distribution Ranges (AMDRs)

Nutrient	AMDR[a]	High-Carbohydrate Diet[b]	Low-Carbohydrate Diet[c]
	Percent of Total Calories		
Fat	20–35	10–15	55–65
Carbohydrate	45–65	65–75	5–20
Protein	10–35	10–25	20–40

[a] Source: Institute of Medicine. Dietary Reference Intakes for energy, carbohydrate, fiber, fat, fatty acids, cholesterol, protein, and amino acids. Washington, DC: National Academies Press; 2005.

[b] Source: Ornish D. Eat more weigh less: Dr. Dean Ornish's life choice diet for losing weight safely while eating abundantly. New York: Harper Collins; 1993.

[c] Source: Atkins RC. Dr. Atkins' new diet revolution, revised. National Book Network; 2003.

gain. Some researchers believe that eating foods high in refined carbohydrate (and especially those low in fat) makes people hungrier, and therefore can make them heavier.[44] In short, the theory that a low-fat diet is the best defense against weight gain is not without debate, and many health experts now believe that there is now enough scientific evidence to lift the ban on dietary fat.[45]

LOW-CARBOHYDRATE WEIGHT-LOSS DIETS

On the opposite end of the weight-loss diet spectrum are low-carbohydrate diets such as the Atkins, Paleolithic, and Zone diet plans. Advocates of these weight-loss plans espouse health claims such as weight loss without hunger and improved cardiovascular health.[46] Indeed, some experts believe that people are more likely to gain weight from excess carbohydrates than from excess fats or proteins because high-carbohydrate foods cause insulin levels to rise, which in turn may lead to weight gain. Thus, limiting one's intake of starch and refined sugars should theoretically help a person lose weight.

The many low-carbohydrate weight-loss diets differ in terms of the types of foods allowed. Some exclude nearly all carbohydrates, while others take a more moderate approach by allowing healthy, carbohydrate-rich foods such as fruits, vegetables, and whole-grain products. Low-carbohydrate diets have become enormously popular in recent years as people have sought out new weight-loss strategies and alternatives to traditional approaches.[47] Studies comparing weight loss associated with low-carbohydrate diets to weight loss associated with low-fat diets show that at six months, greater weight loss is achieved on low-carbohydrate diets. While low-carbohydrate diets appear safe in the short term, some experts have raised doubts about their long-term effectiveness.[48]

Perhaps one of the most legitimate concerns regarding low-carbohydrate diets is the restriction of healthy, high-carbohydrate foods such as fruits, vegetables, and whole-grain products—a valid criticism.[49] Although most low-carbohydrate diet plans recommend that people take dietary supplements, these cannot replace the many other substances (such as fiber and phytochemicals) supplied by these restricted foods. To counter potential nutrient deficiencies and gastrointestinal problems, people following low-carbohydrate diets are strongly encouraged to consume fresh fruits and nonstarchy vegetables. Clearly, moderation is the most important factor when it comes to carbohydrate and fat intake.

TOTAL CALORIES VERSUS MACRONUTRIENT DISTRIBUTION

The reality is that nobody knows for certain the ideal distribution of macronutrients for weight loss. What is known, however, is that the AMDRs provide a range of relative proportions of calories from macronutrients associated with a healthy diet. It is up to individuals, therefore, to determine the eating patterns that best suit their needs within the parameters of the AMDRs. However, research strongly suggests that reducing caloric intake is the single most critical component of successful weight loss.

Most popular diets approach energy balance in a simplistic fashion, as if there is one magic key that can open the door to easy weight control. On the contrary, maintaining neutral or negative energy balance is quite complex. Eating behaviors are shaped by many factors, including genetics, the physiological states of hunger and satiety, and the psychological and social determinants of appetite. In the end, only a combination of healthy eating and exercise can lead to successful long-term weight control. This simple formula can be far from easy to follow, however. Influences ranging from social pressures to genetics to the foods available in a given society all affect one's ability to maintain a healthy weight.

NUTR

10 Life Cycle Nutrition

© Keith Seaman - Camerad Inc/FoodPix/Getty Images

LEARNING OUTCOMES:

- LO1 Understand the physiological changes that take place throughout the life cycle.
- LO2 Describe the major stages of prenatal development.
- LO3 Understand the effects of pregnancy and breastfeeding on maternal nutrient and energy requirements.
- LO4 Explain the importance of breastfeeding during infancy.
- LO5 Discuss the nutritional needs of infants.
- LO6 Discuss the nutritional needs of toddlers and young children.
- LO7 Describe how nutritional requirements change during adolescence.
- LO8 Understand the effect of aging on nutrient and energy requirements.

Chapter 10

LO1 What Physiological Changes Take Place during the Human Life Cycle?

From beginning to end, the human life cycle is a process of continuous change. The continuum of life encompasses infancy, childhood, adolescence, adulthood, and for women, the special life stages of pregnancy and lactation. Throughout life, the body changes in size, proportion, and composition. Because the body changes so tremendously throughout life, so too do its nutritional requirements. For example, nutritional needs during periods of growth differ vastly from those during older adulthood. Regardless of one's current life cycle stage, an appropriate diet is essential to good health. In addition, nutritional status at an early stage can influence health at later stages. For this reason, the food choices you make today may have far greater consequences on your long-term health than you might think.

Physiological Changes during the Life Cycle

Cells form, mature, carry out specific functions, die, and are replaced by new cells. In many ways, the life cycle of a cell mirrors human life itself. That is, after a baby is conceived and born, the next 70 to 90 years are characterized by periods of growth and development, maintenance, reproduction, physical decline, and eventually death. The ability to reproduce enables humans to pass their genetic materials (DNA) on to the next generation.

Aging is an inevitable process, and although individuals grow and develop at different rates, physical changes tend to coincide with various stages of the human life cycle. Age-related physical changes affect body size and composition, which in turn influence nutrient and energy requirements. For this reason, the Dietary Reference Intakes (DRIs) recommend specific nutrient and energy intakes for each life-stage group. Newborn and older infants (0 to 6 months and 7 to 12 months); toddlers (1 to 3 years); young children (4 to 8 years); young and older adolescents (9 to 13 years and 14 to 18 years); young adults, middle age adults, and adults (19 to 30 years, 31 to 50 years, and 51 to 70 years); and older adults (over 70 years) each have unique energy and nutrient needs. Beyond these age-related stages, the DRI recommendations also consider the special stages of pregnancy and lactation. The various DRI life-stage groups are illustrated in Figure 10.1 on the next page. Knowing more about growth and developmental milestones will help you understand why and how your own nutritional needs change throughout life.

GROWTH AND DEVELOPMENT

Growth refers to a physical change resulting from an increase in either cell size (hypertrophy) or number (hyperplasia), whereas **development** is a change in the attainment or complexity of a skill or function. The most rapid rates of growth and development occur during infancy, childhood, adolescence, and pregnancy. In biological terms, human growth and development continue throughout life until physical maturity is reached. The physical changes associated with growth and development generally occur in a predictable and orderly manner throughout infancy and childhood, and increase markedly during adolescence.

The most common and useful ways to assess growth are to measure height and weight. The World Health Organization and the U.S. National Center for Health Statistics (NCHS), a division of the Centers for Disease Control and Prevention (CDC), compiles height and weight reference standards

growth A physical change that results from an increase in either cell size or number.

development A change in the attainment or complexity of a skill or function.

into growth charts that indicate expected growth for well-nourished infants, children, and adolescents.[1] By comparing a child's weight to the average weight of a group of healthy children of similar age and sex, parents and clinicians can evaluate the adequacy of a child's growth. For example, if a five-year-old girl's weight is at the 60th percentile of an NCHS chart, 40 percent of healthy children of similar age and sex weigh more, and 60 percent weigh less. With the aid of these charts, growth can be monitored over time and used as a general indicator of health throughout these important phases of the life cycle.

The assessment of a child's development is as important as the assessment of her growth. Although there is a great deal of variability in terms of human development, developmental patterns are usually predictable. For example, some infants walk as early as 10 months, while others may not take their first steps until several months later. Similarly, infants generally learn to crawl before they learn to walk. Although development varies greatly from child to child, failure to reach major developmental milestones by certain ages is cause for concern and may indicate a problem such as illness or poor nutrition.

Figure 10.1 Stages in the Human Life Cycle and DRI Life-Stage Groups

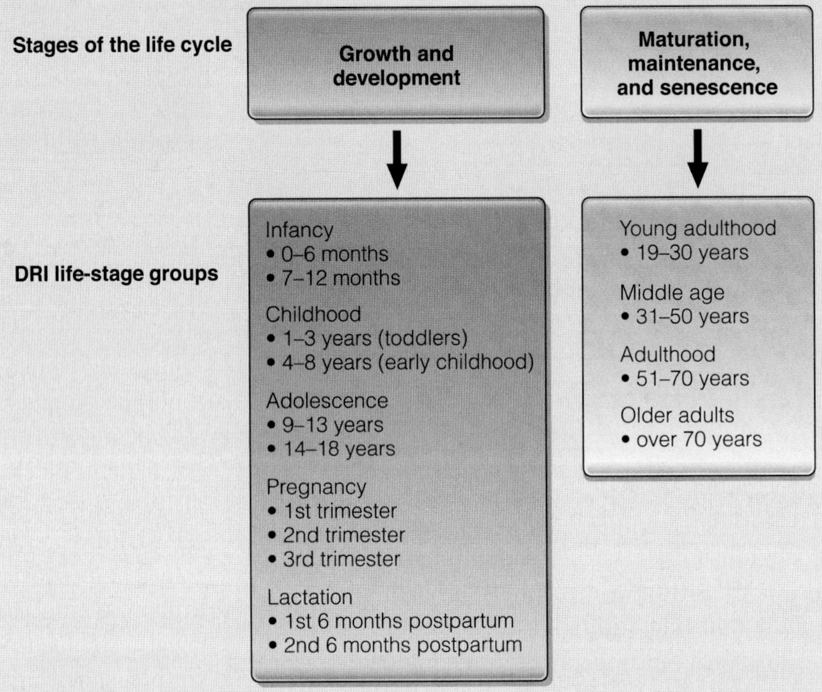

Growth and development affect body size and composition throughout the life cycle. Thus, a person's life stage influences nutrient and energy requirements.

© Cengage Learning 2013

cell turnover The cyclical process by which cells form and break down.

senescence The physical changes characteristically associated with aging.

MATURATION, MATURITY, AND SENESCENCE

As a person approaches physical maturity, his growth and development begin to slow. **Cell turnover**, the cyclical process by which cells form and break down, reaches equilibrium at this time. Once physical maturity is achieved, growth ceases and the body enters a phase of maintenance (during which cell formation equals cell breakdown). As a person ages, the rate at which new cells form decreases, resulting in a loss of some body tissue. Remaining cells become less effective at carrying out their functions and **senescence**, the physical changes characteristically associated with aging, gradually become apparent. Senescence brings about a slow decline in physical function and health, which eventually influences a person's nutrient and calorie requirements.

© Jose Luis Pelaez Inc./Blend Images/Jupiterimages

LO2 What Are the Major Stages of Prenatal Development?

Although it is important for all women to meet their nutritional needs, maintaining adequate nutrition is particularly important during pregnancy. Poor nutritional status before and during pregnancy can have serious long-term effects on an unborn child. For example, the baby of a woman with poor nutritional status is at increased risk for being born too early or too small. Because a mother's nutritional choices have such significant effects on the life of her child, early prenatal care and ongoing health assessment are critical.

Prenatal Development

There may be no other time in a woman's life when her body experiences such extensive changes as during pregnancy. The physiological transformations that a pregnant woman experiences are needed to support the new, emerging life within her. Although every pregnancy is different, much of the growth and development that occur during pregnancy happen in predictable and organized ways. Prenatal development takes places in two periods—the embryonic period and the fetal period. The **embryonic period** comprises the first eight weeks of pregnancy and is subdivided into a pre-embryonic and embryonic phase.

> "**Critical periods** related to sensory perception, psychological development, socialization, and other **areas of development** occur **after birth as well**."

EMBRYONIC PERIOD

Conception takes place when an ovum is fertilized by a sperm, which together form a **zygote**. As the zygote travels toward the uterus, the rapidly dividing cells form a dense cellular sphere called a **blastocyst**. Approximately two weeks after conception, the blastocyst implants itself into the lining of the uterus. This stage of the embryonic period is referred to as the **pre-embryonic phase**. The **embryonic phase** is the stage of prenatal development that spans from the start of the third week to the end of the eighth week after fertilization. During this phase, the developing child, referred to as an **embryo**, grows to about the size of a kidney bean. Cell division continues throughout the embryonic period, eventually forming rudimentary structures that will eventually develop into specific tissues and organs. By the end of the embryonic period, the basic structures of all major body organs are formed. Each organ follows a precise timetable in terms of development. If specific nutrients are lacking when they are needed to construct a certain part of the body, that tissue or organ may not form properly (see Figure 10.2 on the next page).

The term **critical period** is often used to describe the time in prenatal development during which adverse

embryonic period The period of prenatal development that spans from conception through the eighth week of gestation.

zygote An ovum that has been fertilized by a sperm.

blastocyst A dense sphere of cells that implants itself into the lining of the uterus.

pre-embryonic phase The early phase of the embryonic period that begins with fertilization and continues through implantation.

embryonic phase The stage of the embryonic period during which organs and organ systems first begin to form.

embryo A developing human as it exists from the third to the eighth week after fertilization.

critical period A period in prenatal development during which adverse effects on growth and development are irreversible.

© David M. Phillips/Photo Researchers, Inc. / © Andy Walker, Midland Fertility Services/ Photo Researchers, Inc. / © Petit Format/Photo Researchers, Inc. / © Steve Allen/Brand X Pictures/Jupiterimages

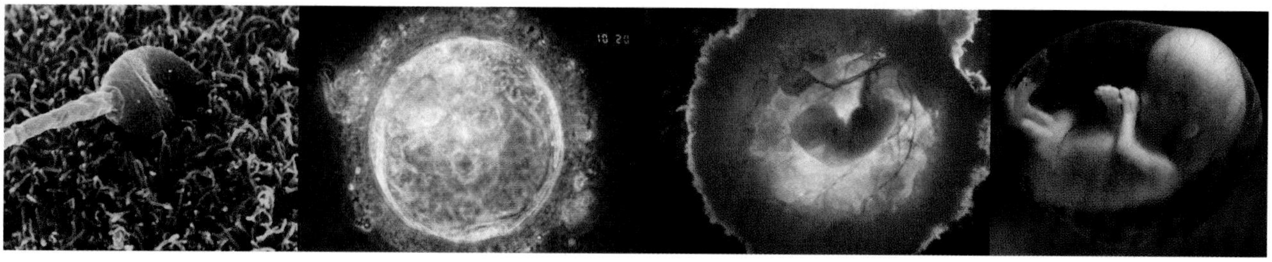

CONCEPTION BLASTOCYST EMBRYO FETUS

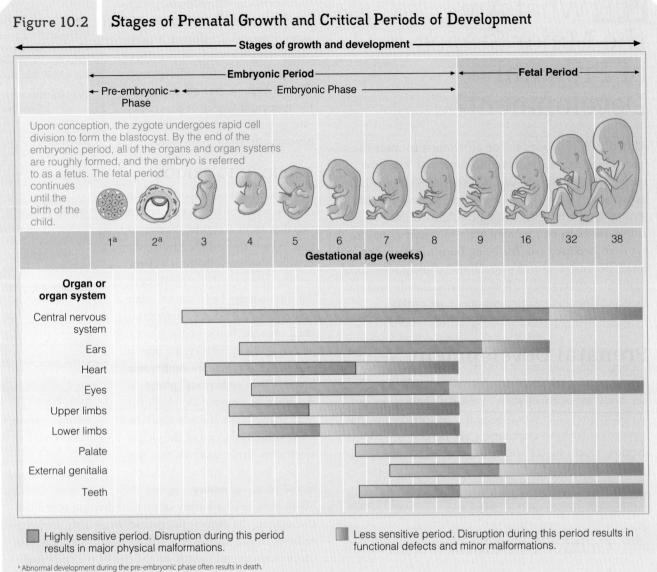

Figure 10.2 Stages of Prenatal Growth and Critical Periods of Development

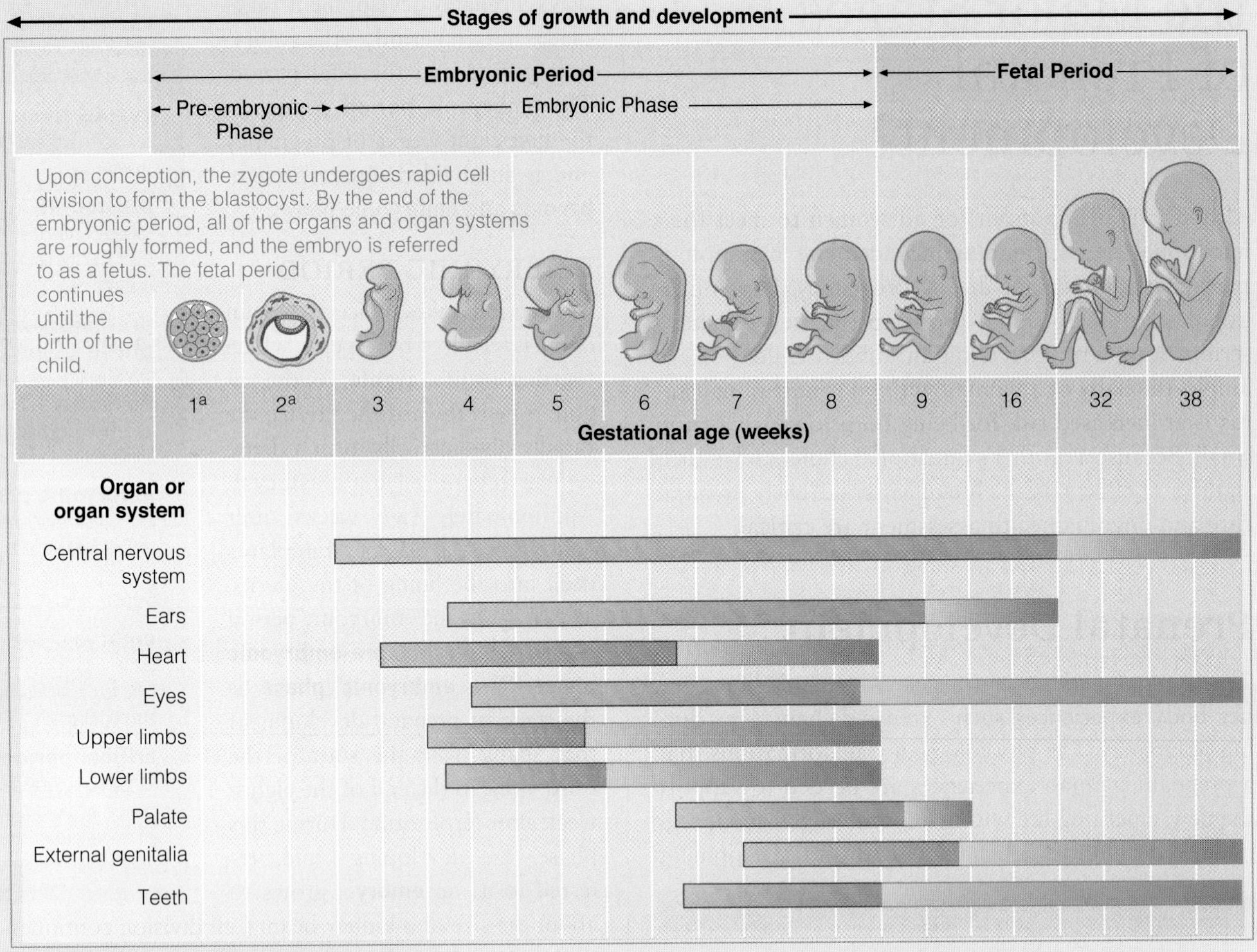

Highly sensitive period. Disruption during this period results in major physical malformations.

Less sensitive period. Disruption during this period results in functional defects and minor malformations.

[a] Abnormal development during the pre-embryonic phase often results in death.

Adapted from: Moore KL, Persaud TVN. The developing human: Clinically oriented embryology, 8th ed. Philadelphia: W.B. Saunders, 2008.

effects on growth and development are irreversible. For example, maternal drug abuse during the critical period of neural development can have irreversible, long-term effects on fetal brain function, sometimes resulting in severe behavioral and learning problems. Although critical periods are most likely to occur during the early stages of pregnancy, they can also occur at later stages.

fetal period The stage of prenatal development that begins at the ninth week of pregnancy and ends at birth.

fetus A developing human as it exists from the ninth week of pregnancy until birth.

FETAL PERIOD

The embryonic period is followed by the **fetal period**, which starts at the beginning of the ninth week of pregnancy and ends at birth. During this second period of pregnancy, the developing child is referred to as a **fetus**. For a fetus to survive outside the womb, a vast amount of additional growth and development is needed. Fetal weight increases by almost 50,000 percent during this period, resulting in a body weight of approximately 3.2 to 3.6 kg (7 to 8 lb) and a length of roughly 51 cm (20 in) at birth. If a mother does not gain an adequate amount of weight or maintain good nutritional health during the fetal period, the fetus's growth may be hindered dramatically. Beyond its direct contribution to the development of a growing child, the food eaten during pregnancy supports dramatic changes in the mother's own body as well.

Birth Defects

While most pregnancies result in the birth of a healthy baby, approximately 3 to 4 percent of babies born in the United States (approximately 150,000 each year) are born with birth defects.[2] A *birth defect* is an abnormality related to a bodily structure or function that can result in mental or physical impairment. Some birth defects are mild, while others are severe or even life threatening. Birth defects are caused by a number of factors such as genetics, environment, lifestyle, or a combination thereof.[3] Although not all birth defects are preventable, a common birth defect, *fetal alcohol syndrome* (FAS), can be prevented if a mother abstains from drinking alcohol during pregnancy. About one to two babies per 1,000 live births have characteristics associated with FAS, such as small head circumference, unusual facial characteristics, and other physical deformities.[4] A less severe form of FAS called *fetal alcohol effect* is associated with learning and behavior problems that are often not apparent until later in life. Scientists do not know when the critical period for FAS development is, and therefore recommend that women completely abstain from alcohol consumption throughout pregnancy.

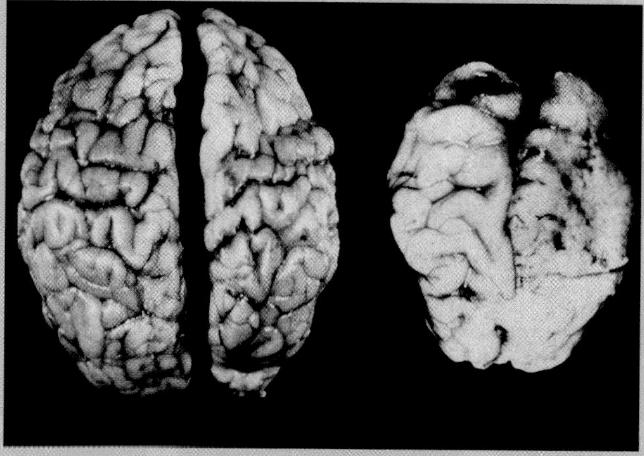

Alcohol exposure during development can induce significant structural changes in the developing brain. The brain on the left is that of a normal six-week-old infant, whereas the brain on the right is that of an infant born with fetal alcohol syndrome.

Courtesy Dr. Sterling Clarren

The Formation of the Placenta

Shortly after the blastocyst implants itself into the lining of the uterus, embryonic and maternal tissues begin to form the **placenta**, an organ that delivers nutrients and oxygen to the developing child. Although the placenta develops early in the pregnancy, it takes several weeks to become fully functional. As illustrated in Figure 10.3 on the next page, the placenta is a highly vascular structure that serves important functions such as the transfer of nutrients and oxygen from the mother's blood to the fetus via the umbilical cord. Potentially harmful substances can also cross from the mother's blood into the fetal circulation through the placenta, so pregnant women must be particularly careful about using medications and other substances that might harm the embryo or fetus. The placenta not only transports substances to the fetus, but also carries waste products away. Waste formed by the fetus passes through the placenta into the mother's blood and is ultimately eliminated from her body.

Gestational Age

How long can a woman expect to be pregnant once she conceives? There are several ways to answer this question. Whereas *embryonic period* and *fetal period* refer to stages of prenatal development, pregnancy is more commonly described in terms of *trimesters*. The first trimester spans the time from conception through week 13. This

placenta An organ made of embryonic and maternal tissues that supplies nutrients and oxygen to the developing child.

gestation length The period of time between conception and birth.

gestational age The number of weeks from the first day of a woman's last normal menstrual cycle to the current date.

full-term A baby born with gestational age between 37 and 42 weeks.

preterm (or **premature**) A baby born with a gestational age less than 37 weeks.

post-term A baby born with a gestational age greater than 42 weeks.

low birth weight (LBW) A baby that weighs less than 2,500 g (5 lb, 8 oz) at birth.

intrauterine growth retardation (IUGR) Slow growth while in the uterus.

small for gestational age (SGA) A baby that has a birth weight below the 10th percentile for gestational age.

appropriate for gestational age (AGA) A baby that has a birth weight between the 10th and 90th percentiles for gestational age.

large for gestational age (LGA) A baby that has a birth weight above the 90th percentile for gestational age.

Figure 10.3 Structure and Functions of the Placenta

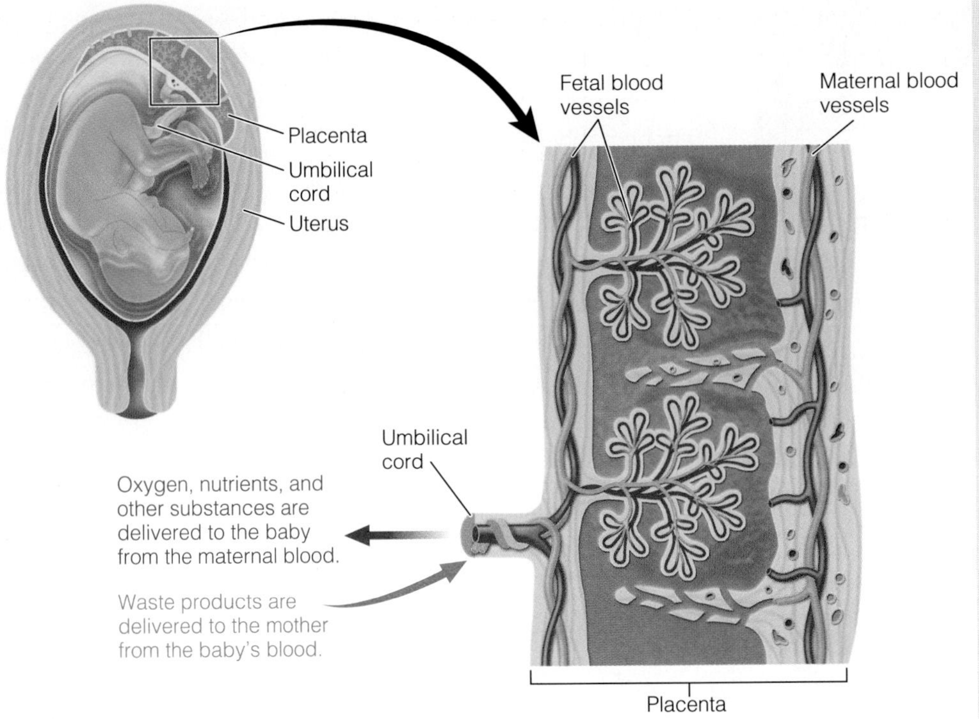
Placenta
Umbilical cord
Uterus
Fetal blood vessels
Maternal blood vessels
Umbilical cord
Oxygen, nutrients, and other substances are delivered to the baby from the maternal blood.
Waste products are delivered to the mother from the baby's blood.
Placenta

The placenta, made of both fetal and maternal tissues, forms early in the pregnancy.

includes the entire embryonic period as well as part of the fetal period. The second trimester lasts from week 14 through week 26, and the third trimester lasts from week 27 to the end of pregnancy.

The duration of pregnancy—the **gestation length**—is the period of time between conception and birth. However, because many pregnant women do not know exactly when they conceived, calculating gestation length can be difficult. A method more commonly used to assess the length of pregnancy is **gestational age**, the number of weeks from the first day of the woman's last menstrual cycle to the current date. Based on gestational age, the average length of pregnancy is about 40 weeks. Babies born with gestational ages between 37 and 42 weeks are considered **full-term** infants; whereas those born with gestational ages less than 37 weeks are considered **preterm** (or **premature**); and those born with gestational ages greater than 42 weeks are considered **post-term**. The earlier a baby is born, the greater the risk for complications that can affect the child's survival and long-term health.

GESTATIONAL AGE AND BIRTH WEIGHT

It is not only important that a baby is born full term, but also that he is born with a healthy weight. As illustrated in Figure 10.4, growth charts are used to classify infants according to birth weight and gestational age. Babies weighing less than 2,500 g (5 lb, 8 oz) at birth are considered to have **low birth weight (LBW)**. LBW infants are small because they are either preterm or have experienced slow growth *in utero*, also known as **intrauterine growth retardation (IUGR)**. Babies who have experienced IUGR are often said to be **small for gestational age (SGA)**, meaning that they have birth weights below the 10th percentile for their gestational age. Infants born with birth weights between the 10th and 90th percentiles for gestational age are said to be **appropriate for gestational age (AGA)**, whereas those with birth weights above the 90th percentile are said to be **large for gestational age (LGA)**.

LBW infants are 40 times more likely to die within the first year of life than AGA infants. In fact, premature birth and LBW are the leading risk factors for infant mortality.[5] Not only does LBW put a baby at risk early in life, it may also have profound long-term effects. Evidence suggests that less than optimal conditions in the womb (uterus) may cause permanent changes in the structure and function of organs and tissues, predisposing individuals to certain chronic diseases later in life.[6] This hypothesis, called the **developmental origins of health and disease hypothesis** (formerly called the fetal origins hypothesis) suggests that less than optimal prenatal and early postnatal conditions may increase a person's risk of developing certain chronic diseases such as cardiovascular disease, stroke, hypertension, type 2 diabetes, and obesity later in life.

developmental origins of health and disease hypothesis A hypothesis that states that less than optimal conditions in the uterus or during early infancy may cause permanent changes in the structure and function of organs and tissues, predisposing individuals to certain chronic diseases later in life.

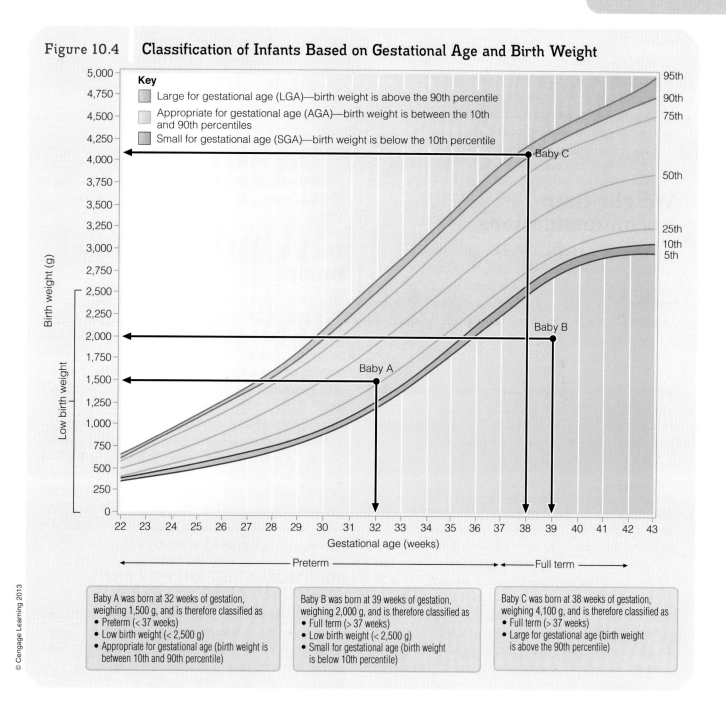

Figure 10.4 Classification of Infants Based on Gestational Age and Birth Weight

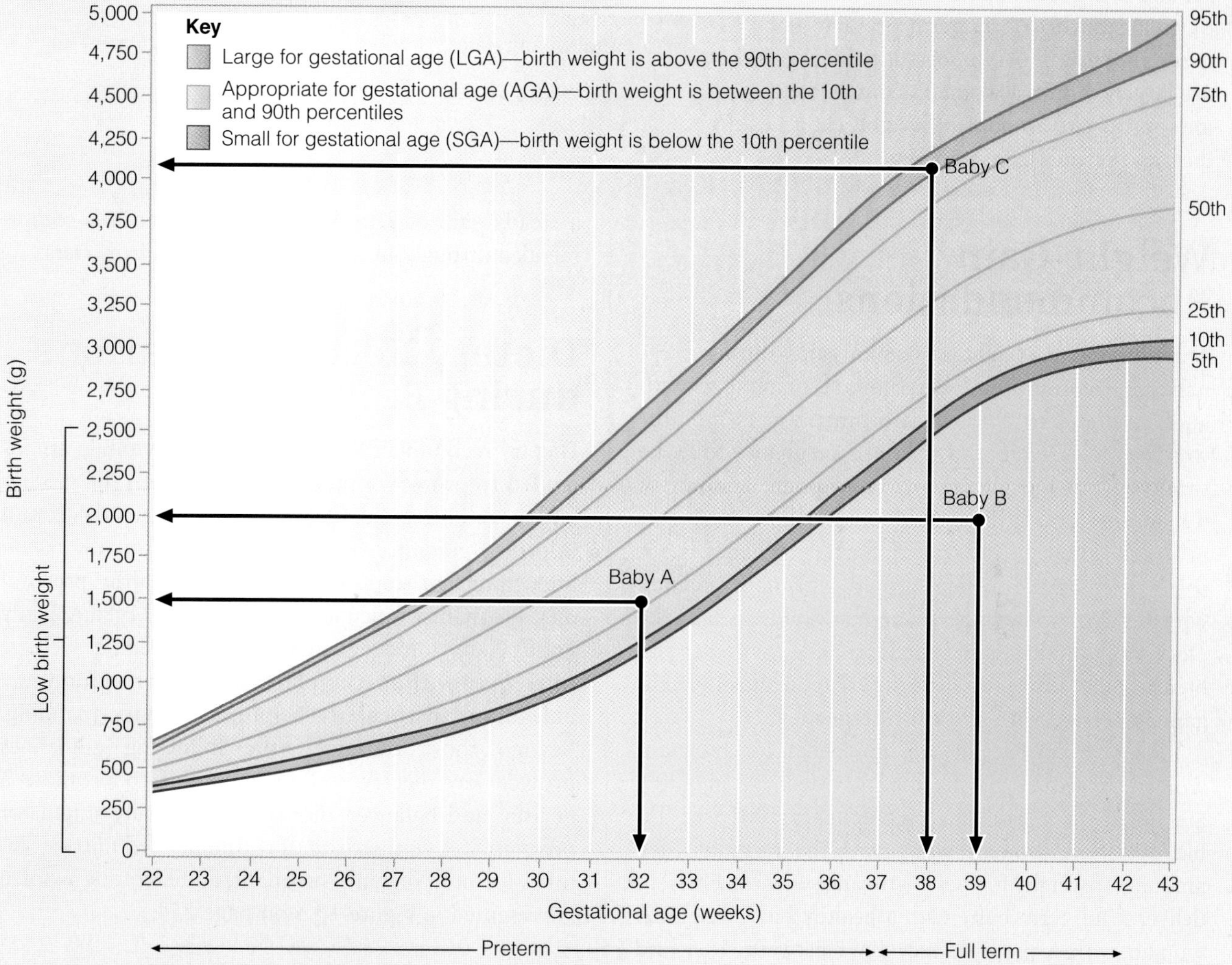

Baby A was born at 32 weeks of gestation, weighing 1,500 g, and is therefore classified as
- Preterm (< 37 weeks)
- Low birth weight (< 2,500 g)
- Appropriate for gestational age (birth weight is between 10th and 90th percentile)

Baby B was born at 39 weeks of gestation, weighing 2,000 g, and is therefore classified as
- Full term (> 37 weeks)
- Low birth weight (< 2,500 g)
- Small for gestational age (birth weight is below 10th percentile)

Baby C was born at 38 weeks of gestation, weighing 4,100 g, and is therefore classified as
- Full term (> 37 weeks)
- Large for gestational age (birth weight is above the 90th percentile)

LO3 What Are the Nutrition Recommendations for a Healthy Pregnancy?

Although unavoidable situations can and do affect pregnancy, a woman can take many precautions to help ensure the birth of a healthy child. The recommendations presented throughout the following sections can help decrease the risk of giving birth to a preterm or LBW baby. Of these, the three most important precautions that are largely within a woman's control are to gain an appropriate amount of weight, eat a healthy diet, and refrain from smoking.

Table 10.1 Components of Weight Gain during Pregnancy

Component	Total Weight Gain
Fetus	3.2 to 3.6 kg (7 to 8 lb)
Placenta	0.7 to 0.9 kg (1½ to 2 lb)
Uterus and supporting structures	1.1 to 1.4 kg (2½ to 3 lb)
Maternal adipose stores	3.2 to 3.6 kg (7 to 8 lb)
Breasts (mammary glands)	0.45 to 0.91 kg (1 to 2 lb)
Body fluids (blood and amniotic fluid)	2.7 to 3.2 kg (6 to 7 lb)
Total weight gain	11.3 to 13.6 kg (25 to 30 lb)

Source: Institute of Medicine. National Research Council. Weight gain during pregnancy: Reexamining the guidelines. National Academies Press. Washington, DC. 2009. Available from: http://www.iom.edu/Reports/2009/Weight-Gain-During-Pregnancy-Reexamining-the-Guidelines.aspx.

Weight-Gain Recommendations

The amount of weight a woman gains during pregnancy is an important determinant of fetal growth and development. Health care practitioners monitor weight gain carefully throughout a pregnancy to make sure that a woman gains the appropriate amount of weight—not too little and not too much. Perhaps surprising to some, weight gained during pregnancy is not solely attributable to fetal growth. Pregnancy-related weight gain also includes tissues and fluids such as the placenta, amniotic fluid that surrounds the fetus, and breast tissue. Table 10.1 lists the components of weight gain associated with a healthy pregnancy.

The current weight-gain guidelines for pregnant women, issued by the Institute of Medicine in 2009, are based on a woman's prepregnancy BMI (see Table 10.2).[7] A woman who gains the recommended amount of weight for her BMI range is most likely to deliver a full-term baby with a healthy birth weight. A woman with a prepregnancy BMI between 18.5 and 24.9 kg/m², a healthy weight range, is advised to gain 25 to 35 pounds, whereas a woman with a prepregnancy BMI between 25.0 to 29.9 kg/m², an overweight range, is encouraged to gain less—between 15 and 25 pounds. In addition to total weight gain, it is also important to monitor the *rate* of weight gain. Whereas little weight gain is necessary during the early stages of pregnancy, a steady gain of 2 to 4 pounds each month is recommended throughout the second and third trimesters.

Dietary Recommendations during Pregnancy

Dietary recommendations for pregnant women are intended to promote optimal health in both the mother and the unborn child. Seeking regular prenatal care and informed guidance from a health care provider is an important first step in establishing a healthy prenatal diet. Pregnant women may also find the MyPlate Daily Food Plans for Pregnancy & Breastfeeding website (http://www.choosemyplate.gov/mypyramidmoms/index.html) particularly helpful in dietary planning. Beyond those resources, the following guidelines provide key dietary recommendations to promote a healthy and balanced diet that is uniquely suited for pregnancy. A comparison of recommended nutrient intakes for nonpregnant, pregnant, and lactating women is presented in Figure 10.5 on page 254.

RECOMMENDED ENERGY AND MACRONUTRIENT INTAKES

Adequate weight gain necessitates adequate energy intake. Therefore, it is important for an expectant mother to satisfy her daily energy requirements as well as those of the growing fetus. Although the energy demands of pregnancy are quite high—about 60,000 kcal over

Table 10.2 Recommended Ranges for Total Weight Gain and Rate of Weight Gain during Pregnancy

Pre-pregnancy BMI (kg/m²)	Recommended Total Weight-Gain Range*	Weekly Rate of Weight Gain during Second and Third Trimesters**
Underweight (<18.5)	12.5 to 18 kg (28 to 40 lb)	0.44 to 0.58 kg (1 to 1.3 lb)
Normal weight (18.5–24.9)	11.5 to 16 kg (25 to 35 lb)	0.35 to 0.50 kg (0.8 to 1 lb)
Overweight (25.0–29.9)	7 to 11.5 kg (15 to 25 lb)	0.23 to 0.33 kg (0.5 to 0.7 lb)
Obese (≥30.0)	5 to 9 kg (11 to 20 lb)	0.17 to 0.27 kg (0.4 to 0.6 lb)

Source: Institute of Medicine. Weight gain during pregnancy: Reexamining the guidelines. National Academies Press. Washington, DC, May 2009. Available from: http://www.iom.edu/Reports/2009/Weight-Gain-During-Pregnancy-Reexamining-the-Guidelines.aspx.

* Weight-gain range for singleton pregnancies.

**Calculations assume a 0.5–2 kg (1.1–4.4 lb) gain in the first trimester.

the course of the pregnancy—very little extra energy is needed during the first trimester. During the second and third trimesters, however, rapid fetal growth requires an energy intake beyond that needed to support the mother's own needs. Pregnant women are generally advised to increase their energy intakes by about 340 and 450 kcal/day for the second and third trimesters, respectively. Thus, a woman with an estimated energy requirement of 2,000 kcal/day (when she is not pregnant) requires approximately 2,340 kcal/day during her second trimester and 2,450 kcal/day during her third trimester of pregnancy.

It is important to get enough carbohydrate, protein, and fat to satisfy energy requirements during pregnancy. If the pregnancy is progressing normally, carbohydrates should remain the primary energy source (45 to 65 percent of total calories). An increase of approximately 45 g/day of additional carbohydrates, equivalent to two to three servings of carbohydrate-rich foods such as whole-grain breads or cereals, is optimal. An additional 25 g protein/day is also needed to support the growth of the baby. This is easily obtained by eating sufficient amounts of high-quality protein sources such as meat, dairy products, and eggs.

Dietary fat should make up approximately 20 to 35 percent of one's total caloric intake during pregnancy. Some essential fatty acids serve other vital roles beyond the provision of energy during pregnancy. For example, linoleic acid and linolenic acid are critical to fetal growth and development. Docosahexaenoic acid (DHA), an ω-3 fatty acid, is essential to fetal brain development and the formation of the retina. To ensure adequate intake of these important fatty acids during pregnancy, women should eat fish and/or ω-3-rich oils (such as canola or flaxseed oil) several times a week. Because some types of fish contain high levels of mercury, the U.S. Food and Drug Administration (FDA) and Environmental Protection Agency (EPA) advise pregnant women to limit their consumption of fish that might contain low levels of mercury (such as salmon, tuna, sardines, and mackerel) and avoid eating fish believed to have high levels of mercury (such as shark, swordfish, king mackerel, and tilefish).[8] Although pregnant women should heed the advisory issued by the FDA and the EPA to limit or avoid certain types of fish, it is important that they consume mercury-free fish and other foods rich in ω-3 fatty acids to ensure optimal fetal and infant development. The 2010 Dietary Guidelines for Americans recommends that pregnant women consume 227 to 340 g (8 to 12 ounces) of seafood per week from a variety of sources while limiting white (albacore) tuna to 170 g (six ounces) per week. Like the EPA, the 2010 Dietary Guidelines advise that

Figure 10.5 **Comparison of Recommended Energy and Nutrient Intakes for Nonpregnant, Pregnant, and Lactating Women**

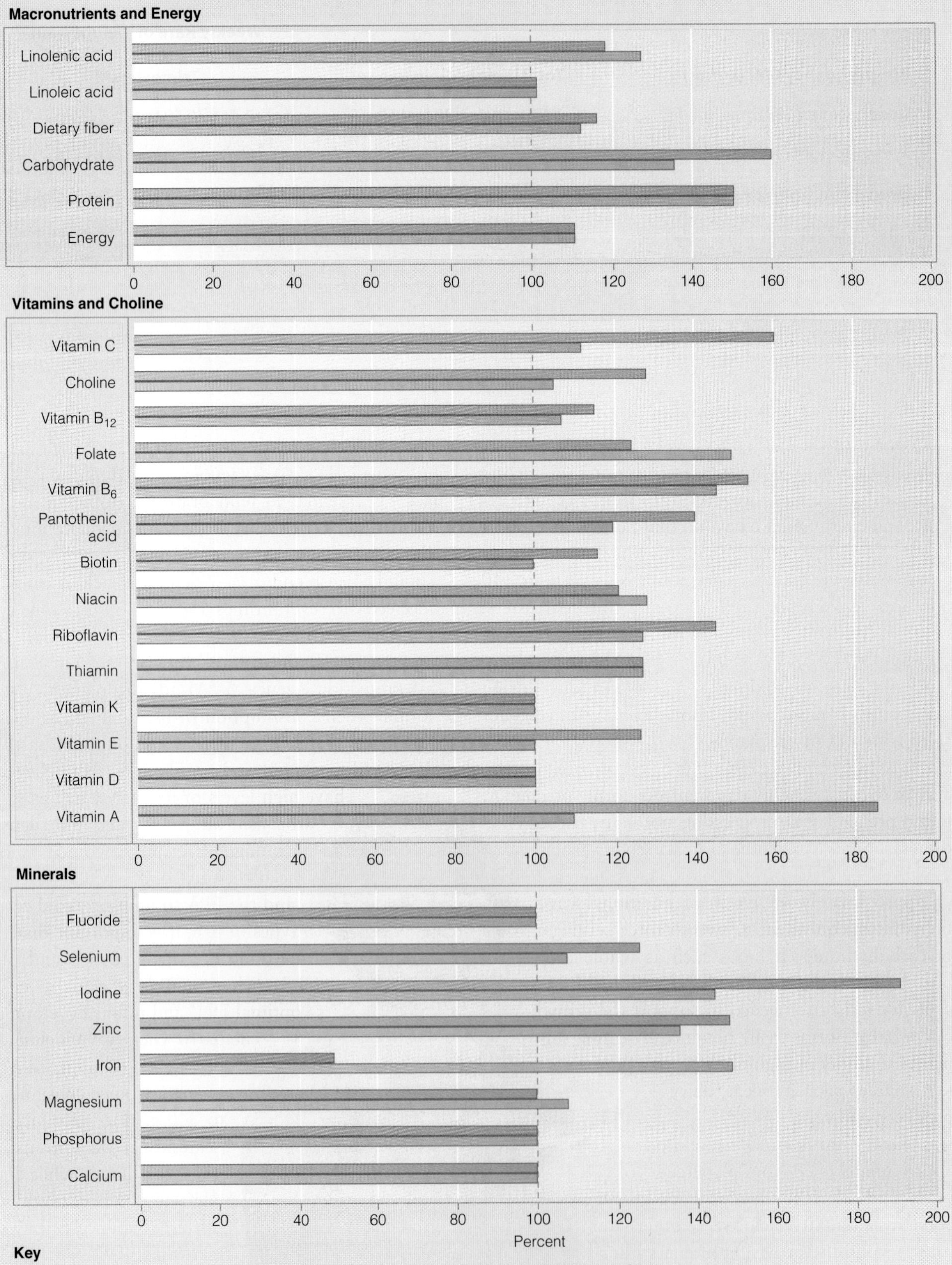

Key

- Recommended intake values for the first six months of lactation for women 19–30 years of age
- Recommended intake values for the second trimester of pregnancy for women 19–30 years of age
- -- Recommended intake values for nonpregnant/nonlactating women 19–30 years of age

pregnant women avoid tilefish, shark, swordfish, and king mackerel.

RECOMMENDED MICRONUTRIENT INTAKE

With few exceptions, the requirements for most vitamins and minerals increase during pregnancy. Perhaps surprisingly, recommended calcium intake does not increase. Although extra calcium is needed for the fetus to grow and develop properly, changes in maternal physiology (such as increased calcium absorption and decreased urinary calcium loss) accommodate these needs without increasing dietary intake. Therefore, the RDA for calcium for pregnant women, 1,000 mg/day, is the same as that for nonpregnant women.

Unlike calcium, the RDA for iron increases substantially during pregnancy: the recommended intake increases from 18 to 27 mg/day. Iron is essential to both the formation of hemoglobin and the growth and development of the fetus and placenta. Most well-planned diets provide women approximately 15 to 18 mg of iron every day, so some pregnant women may have difficulty meeting the recommended intake for iron by diet alone. Therefore, iron supplementation is often encouraged during the second and third trimesters of pregnancy, when iron requirements are the highest.[9]

Adequate folate intake is especially important during pregnancy. Recall from Chapter 7 that folate is critical to cell growth and development, including that of the nervous system. A woman with poor folate status before or in early pregnancy is at an increased risk of having a baby with a neural tube defect, a specific type of birth defect that affects the spinal cord and brain. As the neural tube's critical period occurs early in the pregnancy—21 to 28 days after conception—the neural tube may already be formed before a woman realizes she is pregnant. Because of the importance of folate to neural tube development, an RDA of 600 μgDFE/day has been set for pregnant women. Women capable of becoming pregnant are advised to consume 400 μgDFE/day of folic acid as a supplement or in fortified foods in addition to their regular consumption of folate. Examples of folate-rich foods include dark green leafy vegetables, lentils, orange juice, and enriched cereal grain products.

Staying Healthy during Pregnancy

Every pregnancy is unique. Some women do not experience much unease at all, whereas others experience a wide variety of pregnancy-related discomforts. Pregnancy entails both physical and emotional changes, but implementation of a few simple dietary and lifestyle changes can help minimize many pregnancy-related discomforts.

PREGNANCY-RELATED PHYSICAL COMPLAINTS

Hormonal changes are believed to be the underlying cause of several common physical complaints associated with pregnancy. Morning sickness, fatigue, heartburn, constipation, and food cravings and aversions are often

Not All Neural Tube Defects Are Preventable

Recognizing the importance of folate in the prevention of neural tube defects, the FDA began to require folate fortification of all enriched cereal grain products in 1996. In addition, food manufacturers were granted permission to make health claims on appropriate food labels stating that an adequate intake of dietary folate or folic acid (in supplements and fortified foods) may reduce the risk of neural tube defects. Since these nationwide efforts began, folate status in the United States has improved, and the incidence of neural tube defects has decreased.[10] While folate fortification efforts have been successful at decreasing the occurrence of neural tube defects, not all of these disorders can be prevented by increased folate intake. Some defects are *multifactorial*, meaning that a combination of both environmental and genetic factors contribute to their development. Further investigations into the genetic causes of neural tube defects will hopefully lead to improved methods for early detection and prevention.

pica The urge to consume nonfood items.

gestational diabetes A form of diabetes that develops when pregnancy-related hormonal changes cause cells to become less responsive to insulin.

pregnancy-induced hypertension (or **pre-eclampsia** or **toxemia of pregnancy**) A pregnancy-related condition characterized by high blood pressure, a sudden increase in weight, swelling due to fluid retention, and protein in the urine.

associated with pregnancy-related hormonal changes. Many of these discomforts occur during the early stages of pregnancy, whereas others may persist throughout. In most cases, these discomforts are not serious and can typically be managed with simple diet-related strategies. For example, some women who experience morning sickness, a condition characterized by queasiness, nausea, and vomiting, can find relief by avoiding foods with offensive odors. Other strategies include eating dry toast or crackers, eating small, frequent meals, and eating before getting out of bed in the morning. Heartburn, a common complaint during pregnancy, can often be managed by avoiding spicy or greasy foods, sitting up while eating, and waiting at least two to three hours after eating before lying down. To alleviate constipation, another common pregnancy-related complaint, women are advised to consume adequate amounts of fruits, vegetables, whole grains, and fluids.

Food cravings and food aversions are also common during pregnancy. Powerful urges to consume or avoid certain foods may be caused by hormone-induced heightened senses of taste and smell. While most food cravings and aversions rarely pose serious problems during pregnancy, some expectant mothers develop powerful desires to consume nonfood items such as laundry starch, clay, soil, and burnt matches. The urge to consume nonfood items, called **pica**, has no known cause and can be potentially harmful to the mother and baby.

PREGNANCY-RELATED HEALTH CONCERNS

Although most minor physical discomforts associated with pregnancy are considered normal, it is important for all expectant women to be aware of changes that could indicate a more serious problem. Two common health concerns that can develop during the later stages of pregnancy are gestational diabetes and pregnancy-induced hypertension.

> “The disorder **pica is named** for *Pica*, a **genus of magpie birds**, because magpies are known to **eat many items** not considered to be food.”

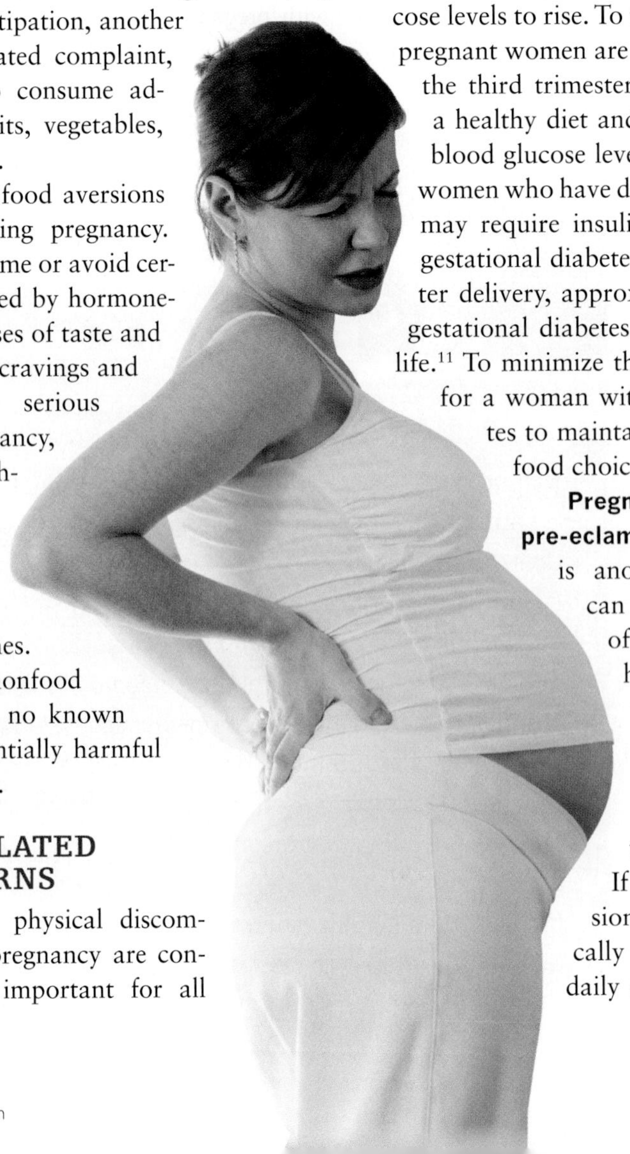

Some women develop a form of diabetes called **gestational diabetes** during pregnancy—usually around 28 weeks or later. Gestational diabetes occurs when pregnancy-related hormonal changes cause cells to become less responsive to insulin, triggering blood glucose levels to rise. To test for gestational diabetes, most pregnant women are given a routine blood test during the third trimester of pregnancy. Once diagnosed, a healthy diet and exercise regimen can help keep blood glucose levels under control. Some pregnant women who have difficulty controlling blood glucose may require insulin injections, however. Although gestational diabetes disappears within six weeks after delivery, approximately half of all women with gestational diabetes develop type 2 diabetes later in life.[11] To minimize this risk, it is especially important for a woman with a history of gestational diabetes to maintain a healthy weight, make sound food choices, and be physically active.

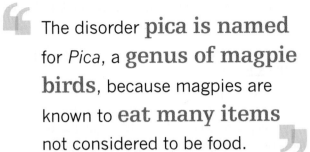

Pregnancy-induced hypertension (or **pre-eclampsia** or **toxemia of pregnancy**) is another serious complication that can develop during the later stages of pregnancy. Pregnancy-induced hypertension, which affects about three percent of pregnancies, is characterized by high blood pressure, sudden swelling and weight gain due to fluid retention, and protein in the urine.[12] If pregnancy-induced hypertension is suspected, women are typically advised to rest and limit their daily activities. Although most women who develop pregnancy-induced hypertension deliver healthy babies, some are not as fortunate. In some cases, a woman's blood pressure can increase to dangerously high levels—a condition referred to as *eclampsia*. When this occurs, the only effective treatment is to induce delivery of the baby.

You now understand the important role that nutrition plays before and during pregnancy. However, because breastfeeding requires careful dietary planning to ensure that energy and nutrient needs are satisfied for both mother and child, maintaining good nutrition is equally important after a pregnancy ends. The physiological changes associated with milk production and the impact this process has on maternal nutrient requirements are just as vital to consider as prenatal nutrition.

LO4 Why Is Breastfeeding Recommended during Infancy?

Women often experience noticeable changes in the size and shape of their breasts during pregnancy. These changes are necessary to prepare the *mammary glands* (commonly known as the breasts) for milk production after the baby is born. Human milk is the ideal food for babies: not only does it support optimal growth and development during infancy and early childhood, the benefits associated with breastfeeding may even extend to later stages of life. Furthermore, because human milk provides immunologic protection against pathogenic viruses and bacteria, it is not surprising that breastfed babies tend to be sick less often than formula-fed babies. Moreover, breastfeeding is beneficial to the mother: it decreases the risks of certain diseases and helps women return to their pre-pregnancy weight more easily.

Lactation

During pregnancy, hormones prepare the mammary glands for milk production. These hormones increase the number of milk-producing cells in **alveolus** clusters and trigger the expansion of ducts that transport milk out of the breast. Upon a baby's delivery, the mammary glands begin to produce and release milk in a process called **lactation**. Women are encouraged to nurse their babies soon after delivery because suckling initiates the process of **lactogenesis**, the onset of milk production.

alveolus A cluster of milk-producing cells.

lactation The production and release of milk.

lactogenesis The onset of milk production.

prolactin A hormone that stimulates alveolar cells to produce milk.

oxytocin A hormone that causes the muscles around the alveoli to contract.

let-down The active process whereby milk is forced out of the alveoli and into the milk ducts.

PROLACTIN AND OXYTOCIN REGULATE MILK PRODUCTION

The hormones **prolactin** and **oxytocin** regulate milk production and the release of milk from alveoli into the milk ducts. When a baby suckles, nerves in the nipple are stimulated, signaling the hypothalamus, which in turn signals the pituitary gland to release both prolactin and oxytocin (see Figure 10.6 on the next page). Prolactin stimulates the alveoli to synthesize milk, whereas oxytocin causes the muscles around the alveoli to contract, forcing the milk out of the alveoli into the milk ducts. This active process of milk release is called milk **let-down**. As the baby suckles, the milk moves through the ducts toward the nipple, and into the baby's mouth. Anxiety, stress, and fatigue can sometimes interfere with the milk let-down reflex, making breastfeeding challenging. For this and other reasons, it is important for women to seek physical and emotional support during the breastfeeding period.

MILK PRODUCTION—A MATTER OF SUPPLY AND DEMAND

Whereas milk production is regulated by many physiological factors, the amount of milk produced is determined largely by how much milk the infant consumes. Women who breastfeed exclusively—meaning that human milk is the sole source of infant feeding—produce more milk than do those who supplement breastfeeding with formula feeding. On average, women produce around 700 ml (three cups) of milk per day during the first six months postpartum, and 600 ml (two and a half cups) per day during the second six months. The reason women produce less milk during the second six months is that most infants are also fed supplemental foods at this age. Because newborns have small stomachs and can only consume small amounts of milk at each feeding, many mothers breastfeed as frequently as every two to three hours. As a baby grows, more milk can be consumed at each feeding, reducing the need to breastfeed as frequently. The American Academy of

Figure 10.6 **Neural and Hormonal Regulation of Lactation**

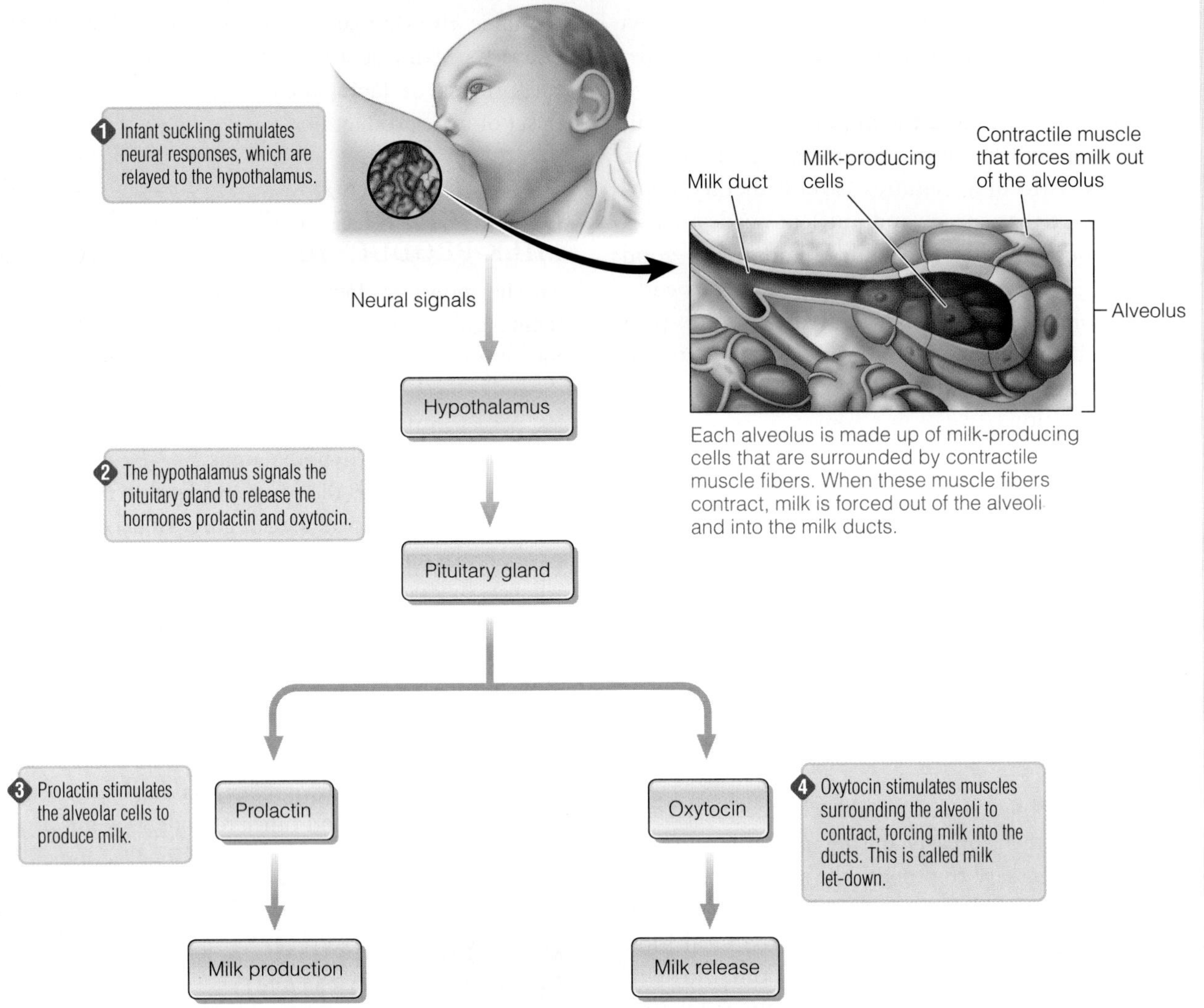

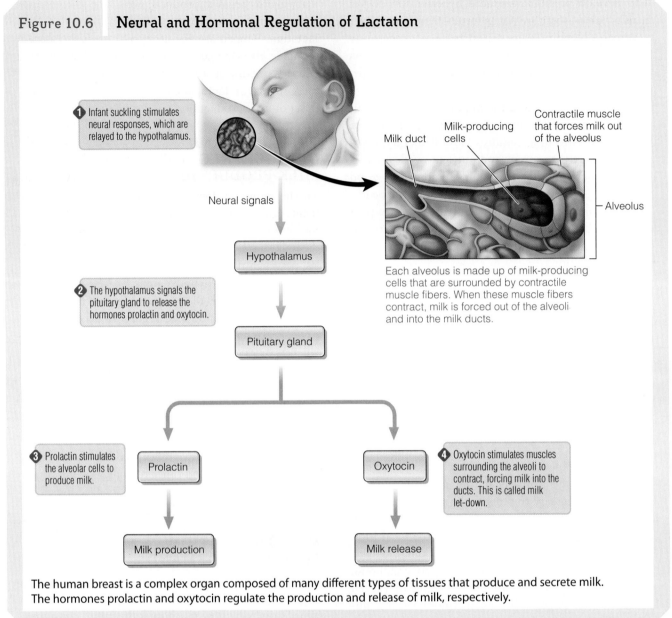

The human breast is a complex organ composed of many different types of tissues that produce and secrete milk. The hormones prolactin and oxytocin regulate the production and release of milk, respectively.

Pediatrics (AAP) recommends that women nurse their newborns at least eight to 12 times each day and feed on demand rather than schedule their babies' feedings.[13]

Breastfeeding requires both proper positioning of the baby at the breast and the ability of the baby to latch onto the nipple. Once these conditions occur, the baby must be able to coordinate sucking and swallowing. Sometimes, parents worry about whether their infant is receiving enough milk. The best indicators of adequate milk production are healthy infant weight gain and appropriate frequency of wet diapers. As a general guideline, parents should expect at least four wet, heavy diapers per day. Most persistent problems associated with breastfeeding can be resolved with the assistance of a medical professional such as a pediatrician or lactation specialist—and all women who experience problems with breastfeeding should seek help. In most cases, a baby who receives adequate amounts of human milk does not need any other sources of nourishment for the first four to six months of life.

Human Milk Is Beneficial to Babies

Over the past 30 years, the number of women who breastfeed has increased steadily in the United States.[14] This trend is in part a response to compelling scientific

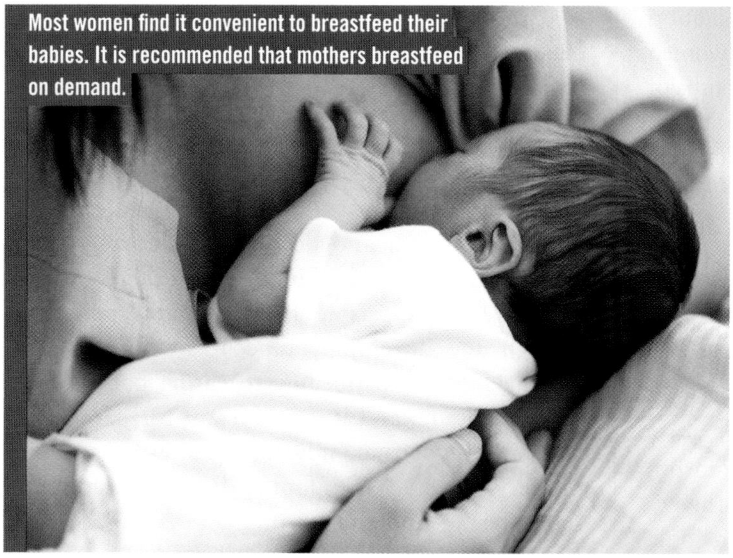
Most women find it convenient to breastfeed their babies. It is recommended that mothers breastfeed on demand.

evidence that human milk is ideally suited to optimal infant growth and development. The AAP recommends exclusive breastfeeding for the first four to six months of life and breastfeeding coupled with complementary foods (excluding cow milk) from the sixth month until at least one year.[15]

Shortly after a mother gives birth, her breasts produce a thick fluid called **colostrum**, which nourishes the newborn and helps prevent disease (as colostrum contains an abundance of substances that fight infections). Over the several days that follow birth, a gradual transition from the production of colostrum to that of mature milk occurs. Like colostrum, mature human milk has many important health benefits. Not only does it reduce the incidence of infectious diseases, but some studies suggest that infants fed human milk are less likely to develop certain types of diseases later in life.[16] Health advantages associated with breastfeeding include reduced risks of asthma, obesity, certain cancers, and type 1 diabetes.

NUTRIENT COMPOSITION OF HUMAN MILK

The nutrient composition of human milk perfectly matches the nutritional needs of the healthy, full-term infant. In addition to nutrients, human milk contains enzymes and other compounds that make certain nutrients easier to digest and absorb. The protein contained in human milk, which is both present in the right amount and easily digested, has an amino acid profile that is uniquely suited to support optimal growth and development during infancy. More than half of an infant's caloric intake comes from lipids. Although there are dozens of fatty acids in human milk, scientists are particularly interested in docosahexaenoic acid (DHA) because of its importance in brain and eye development during infancy. Human milk's high amount of cholesterol, an important component of cell membranes, is also of interest. The primary carbohydrate contained in human milk is lactose. Not only is lactose an important source of energy, it facilitates the absorption of other nutrients. The vitamins and minerals present in human milk are also found in amounts that promote optimal infant growth and development. Given all of its benefits, health care professionals agree that human milk provides the perfect combination of nutrients and disease-fighting substances for infants.

Breastfeeding Is Beneficial to Mothers

Most people are aware that breastfeeding is beneficial to infants, but fewer know that it is also beneficial to mothers. Breastfeeding shortly after giving birth stimulates the uterus to contract, minimizing blood loss and shrinking the uterus to its pre-pregnancy size. Some women also find that breastfeeding helps them return to their pre-pregnancy weights more easily.[17] Beyond its role in stabilizing the body immediately after pregnancy, breastfeeding is associated with several long-term maternal health benefits. It can delay the return of a woman's menstrual cycle, for example. The span of time between birth and the first postpartum menses allows the mother's iron stores to recover and can reduce the likelihood that she will become pregnant again too soon. Because it is possible to ovulate without menstruating, however, women are encouraged to use contraception until they wish to become pregnant again. Finally, breastfeeding reduces a woman's risk of developing breast cancer, ovarian cancer, and possibly osteoporosis later in life.[18]

Maternal Energy and Nutrient Requirements during Lactation

Because milk production requires energy, caloric requirements increase during lactation. The amount of additional energy needed during lactation depends on whether a mother

colostrum A thick fluid produced shortly after birth that nourishes a newborn and helps prevent disease.

is exclusively breastfeeding or feeding by a combination of human milk and infant formula. Because women tend to produce more milk during the first six months of lactation than during the second six months, additional energy required for milk production is approximately 500 and 400 kcal/day, respectively. However, because some of the energy needed for milk production during the first six months should come from stored body fat, total dietary energy intake recommended for the first six months of lactation is lower than that for the second six months. Recommendations for micronutrient intakes during lactation are generally similar to those during pregnancy, though some (such as vitamin A) are somewhat greater and others (such as folate and iron) are somewhat lesser. These values are listed on the Dietary Reference Intakes card at the back of this book. Finally, it is important to note that lactating women should consume sufficient amounts of water and other fluids.

Infant Formula

Although breastfeeding is usually recommended, there are times when breastfeeding is not possible, such as when the mother is taking chemotherapeutic drugs, infected with human immunodeficiency virus (HIV), using illicit drugs, or has untreated tuberculosis. The only acceptable alternative to human milk is commercial infant formula—it is recommended that infants not be fed cow milk at any time during the first year of life. The nutrient content of cow milk is very different from those of human milk and infant formula, and substances found in cow milk might cause other health problems in the baby.

When formula is the best option for nourishing an infant, pediatricians recommend that parents use only formulas fortified with iron to help prevent iron deficiency.[19] Parents should also check infant formula for DHA and arachidonic acid fortification for their positive effects on neural and visual development. Although the important health benefits of these two fatty acids are well recognized, it is not mandatory for manufacturers to add them to infant formula.

© Image Source/Jupiterimages

LO5 What Are the Nutritional Needs of Infants?

Rates of growth and development during the first year of life are astonishing. At no other time in the human lifespan—aside from prenatal periods—do growth and development occur so rapidly. Transitioning from breastfeeding to eating baby food can be challenging, and it is important for parents to be aware of signs that indicate readiness for this next phase. Providing an infant a diet rich in the essential nutrients and an environment that is safe, secure, and engaging helps build a solid foundation for the remainder of life.

Infant Growth and Development

During the first year of life, major developmental milestones such as walking and self-feeding take place, and infant growth follows a fairly predictable pattern. Between four and six months after birth, infant weight doubles and length increases by 20 to 25 percent. Growth rates decrease slightly from six months to one year of life, but the baby continues to grow.

It is important to monitor growth and development throughout infancy. As such, an infant's weight, length, and head circumference are routinely measured during health checkups and recorded on growth charts, as illustrated in Figure 10.7. Growth charts enable pediatricians to monitor and assess infant growth over time. Because infants tend to follow a consistent growth pattern, a dramatic change in length or weight could indicate a problem. Poor breastfeeding technique, for example, can prevent an infant from receiving adequate amounts of energy and nutrients, thus slowing or delaying growth.

In addition to weight and length gains, many developmental changes also occur during infancy. For example, newborns have little control over their bodies. In the first few months of life, however, babies begin to vocalize and are even able to return a friendly smile. Improved muscle control

Figure 10.7 **Growth during the First Three Years of Life**

Birth to 36 months: Girls
Length-for-age and weight-for-age percentiles

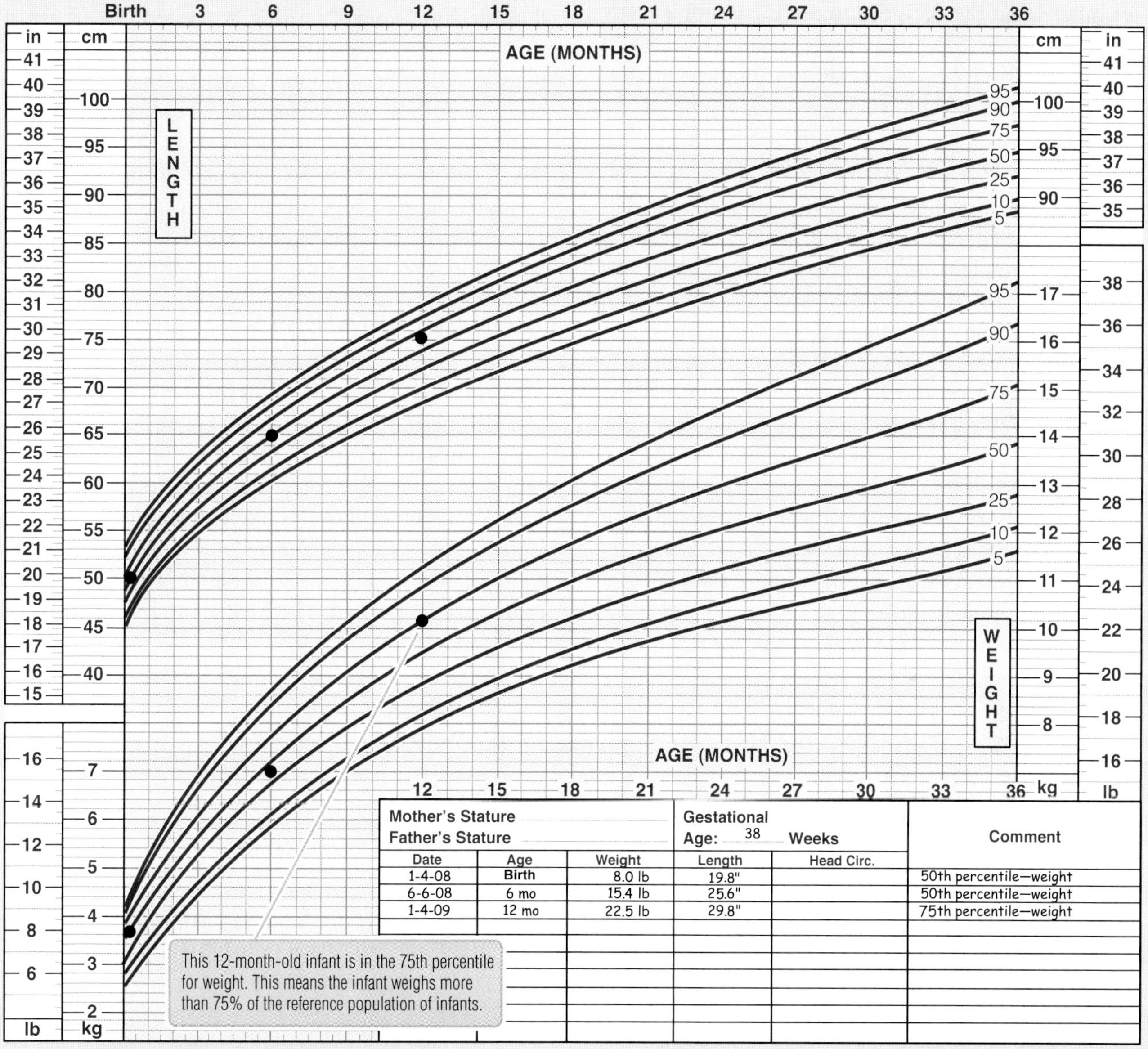

By 4 to 6 months, weight doubles and length increases by approximately 25%.

By 12 months, weight triples and length increases by approximately 50%.

Source: Developed by the National Center for Health Statistics in collaboration with the National Center for Chronic Disease Prevention and Health Promotion (2000). Available from: http://www/cdc.gov/growthcharts.

Weight and length are monitored during infancy and childhood using growth charts.

allows a developing infant to hold his head steady, and by six months, most infants can sit upright with support. These and other developmental milestones affect how and what babies should be fed. Within this first year of life, infants progress from being fed a diet that consists of only human milk and/or infant formula to being able to feed themselves a variety of foods.

Recommended Dietary Supplementation during Infancy

Are human milk, infant formula, and complementary baby foods adequate sources of nutrition during the first year of life, or should a baby also be given nutritional supplements? The answer to this question depends both on whether the infant is breastfed or formula fed and the infant's age. Supplemental vitamin D, fluoride, iron, and fluids are sometimes recommended for infants, but the decisions to use these supplements should be discussed with a health practitioner. Typical recommendations regarding nutrient supplementation are summarized in Table 10.3.

VITAMIN D SUPPLEMENTS

Although scientists have long assumed that breastfed babies receive enough vitamin D from human milk and exposure to sunlight, recent evidence indicates that this is not always so. The AAP recommends that breastfed infants and formula-fed infants who consume less than 16 ounces of infant formula a day receive vitamin D supplementation of 400 IU/day beginning in the first few days of life. Supplementation should continue until a baby can obtain sufficient vitamin D from the diet.[20] Because all infant formulas manufactured in the United States are fortified with adequate amounts of vitamin D, there is little risk of vitamin D deficiency in formula-fed infants who consume more than 473 ml (16 oz) of formula daily.

FLUORIDE SUPPLEMENTS

Because fluoride plays an important role in the formation of teeth and the prevention of dental caries later in life, the AAP recommends that breastfed infants receive fluoride supplements starting at six months of age if the local water source has a fluoride concentration less than 0.3 parts per million.[21] Expectant parents should check with local water departments to find out the fluoride content of the drinking water in their communities. If purified bottled water is used to prepare infant formula, fluoride supplements are recommended. The recommended daily dosage of fluoride is 0.25 mg/day for children between six months and three years of age.

IRON SUPPLEMENTS

Another nutrient sometimes given to infants via supplementation is iron. A full-term infant is born with substantial iron reserves that help meet iron needs during the first six months of life. After six months, an infant consuming iron-fortified formula that contains at least 1 mg/100 kcal iron is likely to maintain adequate iron status. Although human milk contains less iron than infant formula, the iron found in human milk is more easily absorbed. The AAP and CDC recommend

Table 10.3 Recommended Nutrient Supplementation during Infancy

	Infant-Feeding Method	
Nutrient	**Exclusively Breastfed**	**Iron-Fortified Infant Formula**
Vitamin D	• Supplements (400 IU/day) are recommended beginning in the first few days of life and continuing until an infant is consuming at least 16 oz/day of infant formula in addition to or instead of human milk.	• Not needed if an infant is consuming at least 16 oz/day of formula; all formulas in the United States are fortified with vitamin D.
Fluoride	• Supplements (0.25 mg/day) are recommended starting at six months if local water has a fluoride concentration of less than 0.3 ppm.	• Not needed if formula is prepared with water that has at least 0.3 ppm fluoride. Supplements (0.25 mg/day) are recommended starting at six months if the water used to prepare formula has less than 0.3 ppm fluoride.
Iron	• Supplements (1 mg/day per kilogram body weight) are recommended during the second six months of life.	• Not needed; supplements (1 mg/day per kilogram body weight) are recommended for infants not fed iron-fortified infant formula.
Additional fluids	• Not needed unless an infant has excessive fluid loss due to vomiting and/or diarrhea.	• Not needed unless an infant has excessive fluid loss due to vomiting and/or diarrhea.

Source: American Academy of Pediatrics. Pediatric nutrition handbook, 6th ed., Elk Grove Village, IL; 2008.

iron supplementation for infants who are exclusively breastfed during the second six months of life.[22] Infants should continue to receive iron supplementation until complementary iron-rich foods (such as iron-fortified cereals and meat) are introduced.

Iron-fortified cereal is often recommended as a baby's first food.

© Myrleen Ferguson Cate/PhotoEdit

WATER

Because water requirements are likely met throughout the first six months of life if adequate amounts of human milk and/or formula are consumed, infants do not need additional fluid supplementation. However, if vomiting and/or diarrhea cause excessive fluid loss, water replacement becomes necessary. The AAP recommends giving an infant plain water rather than juice when it is necessary to replace fluid loss.[23] Although over-the-counter fluid replacement products with added sugars and/or electrolytes are available, such products are not usually needed. In periods of excessive fluid loss, it is important for parents to contact a health care provider.

Complementary Foods Can Be Introduced between Four and Six Months of Age

The AAP recommends introducing nonmilk complementary foods (excluding cow milk) when an infant is between four and six months of age. Until this time, infants are not physiologically or physically ready for foods other than human milk and infant formula. Many important developmental milestones take place between four and six months, some of which can help parents determine whether an infant is ready to advance to complementary feeding.[24] Signs that an infant is ready for complementary foods include sitting up with support and good head and neck control.

© Larry Williams/Corbis

Because iron status begins to decline at four to six months, pediatricians typically recommend that an infant's first complementary foods be iron-rich, such as iron-fortified cereal or pureed meat. Rice cereal and other single-grain cereals can be mixed with human milk or infant formula to create a smooth, soft consistency. In addition to sufficient amounts of human milk or iron-fortified infant formula, infants require approximately 25 g (one ounce) of iron-fortified cereal (or an equivalent source of iron) every day after six months of age to meet their iron requirements. Some infants find it difficult to consume food from a spoon, but in time, most become quite skilled at eating pureed food. After spoon-feeding is well established, the consistency of foods can be thickened to make eating more challenging. For example, other foods such as pureed vegetables and fruits can be introduced. During this period, complementary foods are considered *extra* because the infant still needs regular feedings of human milk and/or formula.

Although infants should not be given complimentary foods before they reach four to six months of age, there is no evidence that delaying introduction beyond this point is beneficial.[25] Nonetheless, it is important for parents to introduce new foods into their infant's diet gradually. After a new food is first introduced, parents should wait three to four days to make sure the food is tolerated and there are no adverse reactions.[26] Signs and symptoms associated with allergic reactions

Why Are Food Allergies on the Rise?

A swollen face, wheezing, coughing, and hives may not sound life threatening, but when caused by a food allergy, they could mean a trip to the hospital emergency room—or worse. According to the CDC, diagnoses of food allergies have steadily increased in children over the last decade.[27] In 1997, 3.4 percent of U.S. children were diagnosed with a food allergy, but that number has risen to 3.8 percent. Today, more than 3 million children are affected with food allergies. Even more worrisome, the number of children diagnosed with peanut allergies has doubled. Although researchers are unsure why food allergies are on the rise, theories abound. Some data suggest that raising children in an overly clean environment inhibits their immune systems from developing tolerances to normal, everyday proteins (such as those found in common allergenic foods).[28] This theory is called the *hygiene hypothesis*. Another theory suggests that delaying exposure to certain foods (such as nuts and shellfish) until after two or three years of age may make some children more allergy-prone.[29] It is also possible that children are simply being better diagnosed now than in the past. Every theory proposed thus far has certain limitations or problems. Until researchers better understand the immune system's complex responses to certain foods, the best advice for parents is to recognize the early signs and symptoms of food allergies and to help their children avoid offending foods.

or food sensitivities include rashes, diarrhea, runny nose, and in severe cases, difficulty breathing. Foods with high allergic potential include cow milk, eggs, soy, nuts, wheat, fish, and shellfish. While parents should be especially careful when introducing these foods to an infant, there is little scientific evidence that delaying the introduction of these foods until early childhood can prevent allergies later in life.

Infants should be fed human milk and/or iron-fortified formula throughout the first year of life. Because cow milk, goat milk, and soymilk are low in iron, their introduction should be delayed until after a baby's first birthday. Consuming large amounts of milk can displace iron-containing foods, increasing an infant's risk of developing iron deficiency anemia. Similarly, many pediatricians caution parents not to give infants too much fruit juice because it can also displace human milk and iron-fortified formula.[30] Thus, older infants should not be given more than four to six ounces of fruit juice a day. When they are given juice, the AAP and 2010 Dietary Guidelines for Americans recommend that infants and children be given 100 percent pure fruit juice rather than blends or nutrient-sparse fruit drinks.

baby bottle tooth decay A condition whereby dental caries occur in an infant who habitually sleeps with bottles filled with milk, formula, juice, or any other carbohydrate-containing beverage.

Because tooth decay can begin early in life, it is important for parents to establish good dental hygiene practices early on. Infants allowed to fall asleep with bottles filled with milk, formula, juice, or any other carbohydrate-containing beverage are at risk for developing **baby bottle tooth decay.** The pooling of sugars and other compounds found in carbohydrate-rich beverages in an infant's mouth while asleep can damage the newly formed teeth. To avoid baby bottle tooth decay and the dental caries it causes, infants should not be put to bed with bottles that contain anything but water.

As older infants (nine to 10 months of age) become adept at chewing, swallowing, and manipulating food in their hands, they may move toward the next stages of feeding. Certain foods should be avoided during the first few years of life because they pose a risk for choking and are therefore unsafe. These include:

- Popcorn
- Peanuts
- Whole grapes
- Pieces of hot dogs
- Hard candy

In addition, the CDC recommends not feeding honey to children less than one year of age because it can contain spores that cause botulism, a serious foodborne illness. Even very low exposure to these spores can make young children sick.

Baby bottle tooth decay can result when infants are put to sleep with bottles containing carbohydrate-rich liquids such as juice and milk.

LO6 What Are the Nutritional Needs of Toddlers and Young Children?

In terms of nutrient requirements, childhood is divided into two stages: toddlers (ages one to three years) and young children (ages four to eight years). Toddlers and young children grow at a steady rate, but one that is considerably slower than that of infants. Growth charts are used to monitor the adequacy of this growth, but unlike infancy, BMI is used to assess weight in children older than two years of age.

Childhood is a time of growing independence as children gain the ability and confidence to function on their own. Children become more opinionated during this time and often express their likes and dislikes impetuously. (Indeed, feeding toddlers and young children can be challenging.) Childhood is also a time when attitudes about food are formed; parents play an important role in helping toddlers and young children develop healthy relationships with food. Thus, the ways that parents deal with the challenges of feeding children are very important. Regardless of a child's eating habits and tastes, it is important for parents to provide enough nutritious food for optional growth and development.

Feeding Behaviors in Children

"Please, just one bite," begs an anxious parent. This all too familiar plea exemplifies the fact that feeding children is not always easy. Some parents are quite surprised at how quickly mild-mannered infants become willful and opinionated toddlers. The ways that parents respond to feeding challenges can determine whether fussy behaviors persist or fade. Although forcing a child to eat when he does not want to is never recommended, it can be difficult for parents to remain calm when a child refuses to eat. The following six general guidelines regarding common childhood feeding problems can encourage healthy eating:

- *Avoid using food to control behavior.* Most experts agree that using food as a reward or punishment is not a good idea. Although rewarding a child with treats may correct behaviors in the short term, it will likely create serious food issues down the road. For example, giving a child a dessert to reward good behavior may establish a connection between sweet foods and approval. Instead, parents should teach children that food is pleasurable and nourishing—not something to turn to for approval or emotional comfort.
- *Model good eating habits.* Studies show that if parents or siblings enjoy a particular food, a child will be more likely to enjoy it as well.[31] Families that regularly eat meals together are more likely to have children who eat healthier diets than families that do not.[32] Mealtime provides an opportunity for the entire family to be together and share quality time.
- *Be patient.* Childhood food preferences do not always appear rational to adults. Children often judge foods as acceptable or unacceptable based on attributes such as color, texture, and appearance rather than on taste and nutritional value. Also, children likely experience strong flavors such as onions and certain spices more intensely than do adults. Food likes and dislikes change over time. With patience and encouragement, children are often willing to broaden their food preferences. Although challenging for most parents and

caregivers, children usually outgrow their limited food preferences.[33]

- *Introduce new foods.* Making new foods familiar to children is an important first step in food acceptance. Allowing a child to help select and prepare a new food is an effective way to introduce an unfamiliar flavor. Not only does this make the new food more familiar and foster a sense of empowerment, it may also stimulate an interest in healthy food preparation. It is important for parents to remember that accepting new foods takes time. Pressuring children to eat or to try new foods is not recommended.
- *Encourage nutritious snacking.* Because children have small stomachs, they need to eat smaller portions and eat more frequently than do adults. In fact, limiting food consumption to three big meals a day is very difficult for some children. This is why experts recommend that parents and caregivers provide children with nutritious between-meal snacks. Although children who snack frequently are often not hungry at mealtime, this is not usually a problem as long as between-meal snacks are nutrient- and energy-dense.
- *Promote self-regulation.* It is important that children learn to regulate their food intakes based on internal cues of hunger and satiety. For this reason, serving sizes need to be age appropriate, allowing children to ask for more if desired. If a child claims to be too full to eat a certain food, experts do not recommend forcing her to eat all the food on her plate. Instead, parents need to set limits: if a child declines to eat healthy portions of her meal and then asks for dessert, it is reasonable to say no.

> "The Japanese term **betsubara** (meaning "**second stomach**") is used to say that even if one is completely satiated, there is **always room for dessert**."

Feeding young children can be challenging for parents. Children often have strong opinions about what and when they want to eat.

© iStockphoto.com/onebluelight

OVERWEIGHT CHILDREN: A GROWING CONCERN

The percentage of overweight children in the United States is on the rise. Health experts estimate that approximately 17 percent of children age six to 11 are obese (meaning that their BMIs are greater than the 95th percentile).[34] Subsequently, weight-related health conditions such as type 2 diabetes, high blood pressure, and elevated blood lipids are now becoming increasingly prevalent and even commonplace among America's youth. The short- and long-term health and social consequences of childhood obesity are of great concern to parents and health professionals. Because excessive childhood weight gain is likely to continue into adolescence and adulthood, it is important to understand the factors that contribute to this growing trend.

Similar to adults, the behaviors most concretely linked to excessive weight gain in children are unhealthy eating patterns and physical inactivity. Although there is much to learn about the specific meal patterns that promote childhood obesity, easy access to high-fat, energy-dense foods is a primary contributing factor. Changing a child's food environment alone is not sufficient to reverse this growing trend. Efforts to encourage children of all ages to participate in healthful physical activities are also important. For many American children, television and other electronic media have largely replaced physical activity.[35] It is not surprising then that an association between obesity and the amount of time spent watching television has been documented.[36] More than one-third of U.S. children watch three or more hours of television every day—a figure that does not include additional hours spent playing computer and video games.[37] Further compounding the problem, children often consume nutrient-sparse, calorie-dense foods while watching television.

In our modern world, preventing a child from becoming overweight takes considerable effort at home and at school. Regardless of the many pressures and enticements that children encounter every day, it is essential to teach and model the importance of physical activity and good nutrition. After all, there are as many healthy food choices available today as there are unhealthy ones.

Recommended Energy and Nutrient Intakes for Children

Although specific nutrient intake recommendations vary by age, dietary recommendations for children of all ages emphasize the importance of healthy food choices that satisfy nutrient and energy requirements. Regular meals and snacks that include fruits, vegetables, low-fat dairy products, lean meats, whole grains, and legumes should provide all of the nutrients needed for proper growth and development without adding excessive energy to the diet.

Parents and other caregivers are encouraged to use the MyPlate food guidance system (http://www.choosemyplate.gov) to determine the number of servings from each food group needed to meet recommended nutrient and energy intakes for toddlers and young children. MyPlate also provides healthy eating materials designed specifically for this age group. The 2010 Dietary Guidelines for Americans and the 2008 Physical Activity Guidelines for Americans stress the importance of regular physical activity to promote physical health and psychological well-being. Parents are encouraged to make sure that children age six and older engage in at least 60 minutes of moderate- to vigorous-intensity physical activity every day of the week. Younger children are encouraged to play actively several times each day.

CALCIUM

Calcium-rich foods are particularly important to the development of strong, healthy bones throughout childhood. The RDA for calcium increases from 700 to 1,000 mg/day at four years of age. Because milk and other dairy products are good sources of calcium (it would take four cups of broccoli to provide the same amount of calcium as one cup of milk), the 2010 Dietary Guidelines for Americans recommends that children with a daily requirement of 1,000–1,600 kcal/day consume two to three cups of low-fat milk or equivalent milk products every day. In general, one cup of yogurt, one and a half ounces of natural cheese, and two ounces of processed cheese are all equivalent in calcium contents to one cup of milk. Many children do not meet recommended intakes for calcium, which is cause for concern.

IRON

Iron is another nutrient for which meeting the recommended intake is critical to a growing body. Iron deficiency is one of the more common nutritional problems observed in childhood. Toddlers and young children require 7 and 10 mg/day of iron, respectively, to meet their needs. Children who drink large amounts of milk and eat limited varieties of other foods are at greatest risk of iron deficiency.[38] Because over-consumption of cow milk and sugar-sweetened juice can displace iron-rich foods from the diet, the AAP recommends that children age one to five years consume no more than about 700 ml (3 cups) of milk every day, an amount sufficient to meet calcium requirements. To prevent iron deficiency, parents are encouraged to feed their children a varicty of iron-rich foods such as meat, fish, poultry, eggs, legumes, enriched cereal products, whole grains, and other iron-fortified foods.

LO7 How Do Nutritional Requirements Change during Adolescence?

Toward the end of early childhood, hormones trigger changes in height, weight, and body composition that begin to transform a child into an adolescent. This transition marks the beginning of profound physical growth and psychological development. As with other stages of growth, nutrition plays an important role

puberty Maturation of the reproductive system.

menarche The onset of menstruation.

as an adolescent matures from a child into a young adult. Unfortunately, adolescence is a time when unhealthy eating practices often begin to develop. For example, weight dissatisfaction among teens can lead to inappropriate dieting and other destructive weight-loss behaviors.

Growth and Development during Adolescence

Adolescence, the bridge between childhood and adulthood, is signified by the onset of **puberty**. Defined as the maturation of the reproductive system, puberty is initiated by hormonal changes that trigger the physical transformation of a child into an adult. The timing of puberty's onset varies, and adolescents of the same age can differ in terms of physical maturation. Consequently, the nutritional needs of an adolescent may depend more on the stage of physical maturation than chronological age. Nonetheless, females tend to enter puberty at an earlier age (around age 9) than do males (around age 11). The onset of menstruation, known as **menarche**, begins in girls around age 13. For reasons that are not clear, more and more American girls are entering puberty at younger ages. Some researchers believe that this trend may be related in part to the rise in childhood obesity rates as overweight girls tend to develop physically and menstruate earlier than do thinner girls.[39] Consequently, the nutritional needs of an adolescent may depend more on the stage of physical maturation than chronological age.

Considerable physical growth takes place during adolescence. Before the onset of puberty, adolescents attain about 84 percent of their adult height. During the adolescent growth spurt, females and males grow approximately six and eight inches in height, respectively. Because bone mass increases rapidly during adolescence, it is especially important for teens to consume adequate amounts of the nutrients that promote bone health, such as protein, vitamin D, calcium, and phosphorous.

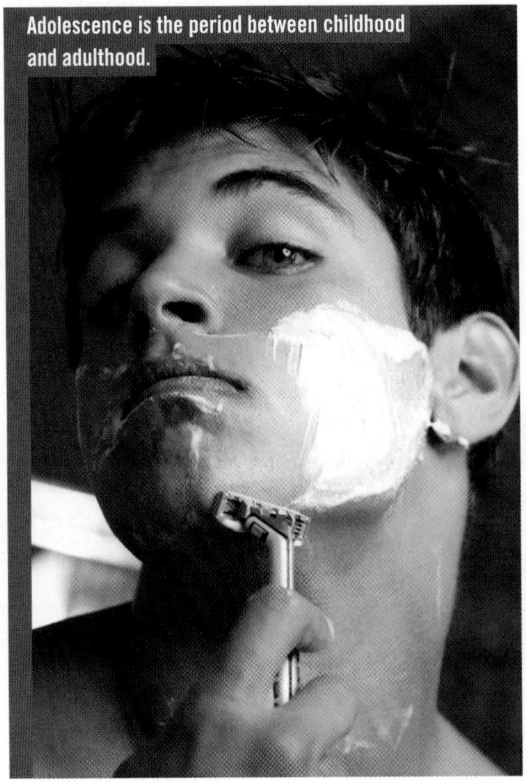
Adolescence is the period between childhood and adulthood.

© Susan Vogel/Upper Cut Images RF/Photolibrary

Changes in linear growth (height) are accompanied by changes in body weight during adolescence. Overall, females and males gain an average of 35 and 45 pounds, respectively. Changes in body composition differ considerably between females and males, however. Whereas a female adolescent experiences a decrease in percentage of lean mass and a relative increase in percentage of fat mass, a male adolescent experiences the opposite. While good nutrition is important throughout the entire life cycle, the accelerated rates of growth and development experienced during adolescence puts this group at particularly high risk for developing diet-related health problems.

Psychological Issues Associated with Adolescent Eating Behaviors

Adolescence is marked not only by rapid physical changes, but also by numerous psychological and developmental changes. For example, adolescents' newfound desire for independence often strains family relationships and leads to rebellious behaviors. Furthermore, a strong urge to fit in and be accepted by peers can lead to social anxiety and further familial tension. For these and many other reasons, healthy eating can become a low priority for adolescents. Peers are often more influential than family members in determining adolescent food preferences, and dieting, skipping meals, and increasing food consumption away from home become common. Although increased body fat is normal and healthy for adolescent females, weight gain can contribute to weight dissatisfaction, which in turn can lead to unhealthy dieting and caloric restriction. Caloric restriction during adolescence can negatively impact growth, development, and reproductive maturation.

Nutritional Concerns and Recommendations during Adolescence

© FoodPix/Jupiterimages/Getty Images

The rapid growth and development associated with adolescence increase the body's need for certain nutrients and energy. Teens can use the MyPlate food guidance system to determine the number of servings from each food group needed to meet recommended nutrient and energy intakes. Using MyPlate's personalized meal planning guide can be particularly advantageous, as an inadequate diet during this stage of life can compromise health and have lasting long-term effects.

FOOD CHOICES OF ADOLESCENTS

As with adults, the majority of calories in an adolescent's diet should come from healthy, nutrient-dense foods such as fruits, vegetables, lean meats, low-fat dairy products, and whole grains. The number of adolescents who meet recommendations for fruit and vegetable consumption is quite low, however. Only 23 percent of adolescent boys and 27 percent of adolescent girls get the recommended number of daily servings of fruit. Similarly, approximately one-third of adolescents sustain adequate intakes of vegetables. It is not surprising then that only a small number of U.S. adolescents (26 to 38 percent) satisfy their requirements for dietary fiber. Sweeteners and added sugars, many of which are associated with the consumption of carbonated beverages, contribute 16 percent of all calories consumed by teens.[40] Between 1977 and 2001, the number of calories consumed from sugar-sweetened soft drinks increased, while the number of calories obtained from milk decreased by 34 percent.[41]

IMPORTANT MICRONUTRIENTS DURING ADOLESCENCE

Because too little dietary calcium can compromise bone health later in life, the RDA for calcium during adolescence is 1,300 mg/day. Getting enough calcium is best achieved by consuming dairy products such as milk, yogurt, and cheese. The average calcium intake for adolescent boys is 1,175 mg/day, which is approximately 91 percent of the RDA for calcium. The mean calcium intake for adolescent girls is 920 mg/day, approximately 71 percent of the RDA.[42] As the mean calcium intake for boys is near the EAR (1,100 mg/day), it is likely that calcium intakes are adequate among this population group.

In addition to calcium, adolescents require iron to support growth. Because of the iron lost in menstruation, the RDA for iron is higher for girls than for boys during later adolescence (15 and 11 mg/day, respectively). Although it appears that most adolescents are well nourished with respect to iron, approximately 8 percent of U.S. teenagers have impaired iron status and 1 to 2 percent have iron deficiency anemia.[43]

While the majority of adolescent boys consume adequate amounts of folate, this may not be the case for girls.[44] Given the pregnancy-related health concerns associated with impaired folate status, this is of great concern for some teens. Because impaired folate status can increase the risk of neural tube defects, it is important for health care professionals to stress the importance of folate-rich foods such as orange juice, green leafy vegetables, and enriched cereals to sexually active girls.

LO8 How Do Age-Related Changes in Adults Influence Nutrient and Energy Requirements?

Adulthood spans the largest segment of the life cycle—approximately 60 to 65 years. Like other stages of life, adulthood carries physical, psychological, and social changes that can affect health. Managing family and career obligations can make adulthood a particularly challenging time in a person's life. A hectic schedule can be stressful, free time can be sparse, and personal health needs often go unattended. It is vital for adults

lifespan The maximum number of years an individual member of a particular species has remained alive.

to recognize the importance of taking care of their own physical and emotional well-being, because a healthy lifestyle can help ensure a full and active life for many years to come.

There is much a person can do to stay physically and mentally fit for as long as possible into older adulthood. Although aging is inevitable, studies show that young and middle age adults who adopt healthy lifestyles (eating nutritious foods, maintaining healthy body weight, participating in regular physical activities, and not smoking) are less likely to develop chronic diseases such as cardiovascular disease, type 2 diabetes, high blood pressure, and certain types of cancer—all of which are leading causes of death.

Adulthood Is Characterized by Physical Maturity and Senescence

Adulthood is the period in the lifespan characterized by physical maturity, which is typically achieved by age 20. Some men continue to grow in their early 20s, and bone mass increases slightly for both men and women until age 30. After achieving physical maturity in young adulthood, adults typically undergo a long period of physical stability (maintenance) before gradually transitioning to senescence. A wide spectrum of health and independence levels is experienced among middle age and older adults. Whereas some people maintain active lifestyles throughout adulthood, others become frail and require increasing amounts of assistance. For this reason, functional status is sometimes a better indicator of nutritional needs than chronological age during this time. Nonetheless, when establishing DRIs, the Institute of Medicine divided adulthood into four groups: young adulthood (19–30 years), middle age adulthood (31–50 years), adulthood (51–70 years), and older adulthood (>70 years).

THE GRAYING OF AMERICA

Life expectancy continues to increase in the United States. In fact, adults over age 85 are the fastest growing segment of the adult population.[45] Note that life expectancy, the expected number of years of life remaining at a given age, is different from **lifespan**, the maximum number of years an individual member of a particular species has remained alive. Both life expectancy and lifespan have risen steadily throughout human history, but the graying of America (first introduced in Chapter 1) is evident now more than ever. Although there is little doubt that a person's genetics play a major role in the aging process, lifestyle choices are vitally important as well.

The graying of America reflects a shift in the age distribution of the United States. Never before have there been so many older adults, and never before has life expectancy been so long. These changes are attributable in part to advances in medical technology and improved health care. They are also partly due to the increased number of births that occurred between the 1940s and early 1960s. People born during this span are often called *baby boomers*. During the baby boom years, many hospitals expanded their obstetric and gynecology units to accommodate the increase in births. Shortly thereafter, many thousands of schools were built throughout the United States to accommodate the growing number of school-age children. Today, there is an increased need for retirement communities and health care practitioners to care for the rising number of older boomers.

Nutritional Concerns and Recommendations during Adulthood

Although adults cannot turn back the hands of time, there is much they can do to keep their bodies strong and healthy. During

Many adults are **physically fit** *and lead* **healthy,** *active lives.*

© Ronnie Kaufman/Blend Images/Jupiterimages

> A **generation of individuals born** between 1980 and the early 2000s known as the ***millennials*** rivals—and perhaps **eclipses**—the baby boomers **in number**.

adulthood, nutrient intake recommendations are intended to both reduce the risk of chronic disease and provide adequate amounts of essential nutrients. The MyPlate food guidance system can assist older adults in planning meals and determining the amount of food from each food group needed to meet recommended nutrient and energy intakes. Because older adults generally have lower energy requirements than do younger adults, food guidelines can help older adults balance energy consumption and select foods that support optimal nutrient intake.

In general, elderly adults comprise an at-risk population for many nutrition-related health problems. Food insecurity, social isolation, depression, illness, hunger, insufficient food availability, and the use of multiple medications can compromise an older person's nutritional status. Physiological changes associated with aging can also influence the development of nutrition-related health problems. A number of these changes are summarized in Table 10.4.

OPTIMIZING BODY COMPOSITION AND BONE HEALTH DURING ADULTHOOD

Most adults experience age-related changes in body composition (such as a decrease in lean mass and an increase in fat mass). Although genetics, physical activity, and nutritional status influence these changes, the average 70-year-old man has approximately 22 more pounds of body fat and 24 fewer pounds of muscle than the average 20-year-old man.[45] Adequate protein intake is important for elderly adults as it helps prevent age-related loss of skeletal muscle. Because it decreases total energy intake requirements, a loss of lean mass leads to age-related weight gain in many middle age adults. In general, all other factors being equal, a 70-year-old woman requires approximately 280 fewer kcal/day than does a 30-year-old woman. Likewise, a

Table 10.4 Physiological Changes Typically Associated with Aging

Cardiovascular System
• Elasticity of blood vessels decreases.
• Cardiac muscles weaken.
• Blood pressure increases.
Endocrine System
• Estrogen, testosterone, and growth hormone levels fall.
• Ability to produce vitamin D diminishes.
Gastrointestinal System
• Saliva and mucus production decreases.
• Loss of teeth.
• Difficulty swallowing.
• Production of gastric juice is reduced.
• Peristalsis decreases.
• Vitamin B_{12} absorption decreases.
Musculoskeletal System
• Bone mass decreases.
• Lean mass decreases.
• Metabolism decreases.
• Strength, flexibility, and agility decrease.
Nervous System
• Appetite regulation is altered.
• Thirst sensation is blunted.
• Ability to smell and taste decreases.
• Sleep patterns change.
• Visual acuity decreases.
Urinary System
• Blood flow to kidneys diminishes.
• Kidney filtration rate decreases.
• Ability to eliminate metabolic wastes decreases.
Respiratory System
• Respiratory rate decreases.

70-year-old man requires 380 fewer kcal/day than does a 30-year-old man.

Of course, an older adult's individual energy requirements depend on many interrelated factors such as physical activity, weight, and changes in the relative amounts of muscle and body fat. Both caloric over- and underconsumption can lead to serious problems in older adults. The 2010 Dietary Guidelines for Americans recommends that all adults prevent gradual weight gain over time by decreasing energy intake and increasing physical activity. Exercise not only decreases fat mass and slows age-related bone loss, it also helps strengthen muscles and improves coordination.

© Philippe Dureuil/es Collection/Glowimages.com

Age-related bone loss can lead to osteoporosis, which in turn can make bones fragile. While men can develop osteoporosis, the condition is by far more common in women. Although many factors influence bone density, maintaining adequate intakes of the nutrients that support bone health (such as protein, calcium, vitamin D, phosphorus, and magnesium) is essential. To promote lasting bone health, the RDA for calcium increases from 1,000 to 1,200 mg/day at age 51 in women and at age 70 in men. However, older adults often find it difficult to meet this RDA. Lactose intolerance increases with age, inhibiting the ability of many to consume calcium-rich dairy products. To counter this problem, health care providers often recommend lactose-reduced dairy products and/or calcium-fortified foods such as soymilk and supplements.

Considerable evidence shows that older adults, especially those who live in northern regions of the United States, are at increased risk of vitamin D deficiency. This condition is likely caused in part by limited exposure to sunlight and in part by a decreased ability to synthesize vitamin D from cholesterol. For these reasons, the RDA for vitamin D increases from 15 to 20 μg/day at age 70. Good sources include fortified milk, eggs, salmon, and tuna.

CHANGES IN THE GASTROINTESTINAL TRACT

Age-related changes in the GI tract affect nutritional status, especially in older adults. Aging muscles become less responsive to neural signals, which can impact digestive functions such as swallowing and peristalsis. In fact, difficulty swallowing is a common cause of choking in the elderly. The threat or experience of this frightening and life-threatening event can compel older adults to avoid eating certain foods altogether. Preparing foods so that they are moist and soft can help prevent choking.

Decreased GI motility can also lead to constipation. Fecal material that remains in the colon for a prolonged period can become hard and compacted, making it difficult to eliminate. Drinking adequate amounts of fluids and eating fiber-rich foods can improve GI function and help prevent this problem. Although the recommended fiber intake for older adults is 20 to 35 g/day, older men and women consume on average only 18 and 14 g/day, respectively. There are many reasons why older adults may be reluctant to consume fiber-rich foods, but it is important that they find ways to incorporate fiber into their diets. Good sources of fiber include whole grains, nuts, beans, fruits, and vegetables. Dietary supplements are also sometimes advised.

With increased age comes a decline in the number of stomach cells that produce gastric juice. This can reduce the bioavailability of nutrients such as calcium, iron, biotin, folate, vitamin B_{12}, and zinc. For example, without intrinsic factor, vitamin B_{12} cannot be absorbed, leading to vitamin B_{12} deficiency. Symptoms associated with vitamin B_{12} deficiency include dementia, memory loss, irritability, delusions, and personality changes—all of which can easily be overlooked or misdiagnosed in older adults. To avoid the development of vitamin B_{12} deficiency, older adults are often advised to take vitamin B_{12} supplements or to consume adequate amounts of vitamin B_{12}-fortified foods. Foods naturally containing vitamin B_{12} include fish, meat, poultry, eggs, and dairy products.

EFFECTS OF MENOPAUSE ON WOMEN'S NUTRITION

As a woman ages, she experiences a natural decline in estrogen production. This can lead to a variety of notable physical changes, such as irregular menstrual cycles and sudden feelings of warmth (hot flashes). The decline in estrogen that occurs during this **perimenopausal** stage of life also accelerates the rate of bone loss. By the time a woman reaches **menopause** around her fifth or sixth decade of life, her ovaries are producing very little estrogen, causing her menstrual cycle to stop completely. Declining levels of estrogen cause bone loss to accelerate further, causing some women's bones to become weak and fragile. Women whose bones were dense before the onset of menopause have fewer problems—another addition to the long list of reasons why adequate intake of nutrients related to bone health is important throughout the lifecycle.

Because the monthly blood loss associated with menstruation ceases, menopause can improve iron status. In fact, the RDA for iron decreases from 18 to 8 mg/day for post-menopausal women—the same amount as recommended for adult men. Regardless of life stage, adequate iron intake remains a concern among older adults who limit their intakes of meat, poultry, and fish.

OTHER NUTRITIONAL ISSUES IN OLDER ADULTS

Other age-related physiological changes can contribute to nutritional deficiencies in older adults. Problems with oral health, missing teeth, or poorly fitting dentures can make food less enjoyable and limit the types of foods a person can eat. Foods that require chewing such as meat, fruits, and vegetables can cause pain, embarrassment, and discomfort for the elderly and therefore may be avoided. It is important for older adults who experience problems with oral health to get proper dental care. Sensory changes in taste and smell can also affect food intake in older adults: the ability to smell diminishes with age, often making food tasteless and unappealing. Furthermore, certain medications can alter taste and diminish appetite. Older adults may find that adding spices to foods makes them more appealing, flavorful, and enjoyable.

The sensation of thirst can also become blunted with age. Because of this, many older adults do not consume enough fluid and are at increased risk of dehydration, which can upset the balance of electrolytes in cells and tissues. A lack of fluid can also disrupt bowel function and exacerbate constipation. Certain medications increase water loss from the body, and some elderly people may intentionally limit fluid intake to avoid embarrassment over loss of bladder control. Symptoms of dehydration, which are often overlooked in the elderly, include headache, dizziness, fatigue, clumsiness, visual disturbances, and confusion. Adequate fluid consumption and early detection of dehydration are very important for older adults. To stay fully hydrated, adults are encouraged to consume 1.9 to 2.4 l (8 to 10 cups) of fluids per day. Fluid can come from both beverages and foods.

perimenopausal A life stage characterized by a natural decline in estrogen production.

menopause A life stage characterized by very little estrogen production, which causes the menstrual cycle to stop.

Finally, the elderly are at particularly high risk of adverse drug–nutrient interactions. Older adults often take multiple medications to treat a variety of chronic diseases. Some drugs can cause loss of appetite, leading to inadequate food intake. Others can alter taste, making food unpleasant to eat. Even nutrient absorption can be affected by certain medications. It is important for older adults to be aware of such problems and seek advice regarding the nutrient-related side effects of their medications.

Assessing Nutritional Risk in Older Adults

Clearly, many interrelated factors put older adults at increased nutritional risk. For this reason, several national health organizations jointly sponsored the Nutritional Screening Initiative (NSI), a collaborative effort that helped identify risk factors closely associated with poor nutritional status in older adults. The risk factors most closely associated with poor nutritional health were compiled and used to develop a screening tool called the NSI DETERMINE checklist, which is illustrated in Figure 10.8 on the next page. As you can see, nutritional risk factors are explained and exemplified by the DETERMINE acronym and a nutritional risk scoring system is outlined. An individual with a score of 6 or more on the NSI DETERMINE checklist is considered to be at high nutritional risk.

Many services are available to help older adults improve nutritional status and overall health. Most communities have congregate meal programs whereby older adults can enjoy nutritious, low-cost meals in the company of others. Most congregate meal programs are subsidized, and total cost is often based on an individual's ability to pay. Another program organized and implemented by the Meals on Wheels® Association

Figure 10.8 Nutritional Screening Initiative DETERMINE Checklist

DETERMINE YOUR NUTRITIONAL HEALTH

Circle the number in the "yes" column for those that apply to you or someone you know. Total your nutritional score.

	YES
I have had an illness or condition that made me change the kind and/or amount of food I eat.	2
I eat fewer than 2 meals per day.	3
I eat few fruits or vegetables or milk products.	2
I have 3 or more drinks of beer, liquor or wine almost every day.	2
I have tooth or mouth problems that make it hard for me to eat.	2
I don't always have enough money to buy the food I need.	4
I eat alone most of the time.	1
I take 3 or more different prescribed or over-the-counter drugs a day.	1
Without wanting to, I have lost or gained 10 pounds in the last 6 months.	2
I am not always physically able to shop, cook and/or feed myself.	2
TOTAL	

Total Your Nutritional Score. If it's–

- 0–2 Good! Recheck your nutritional score in 6 months.
- 3–5 You are at moderate nutritional risk. See what can be done to improve your eating habits and lifestyle.
- 6 or more You are at high nutritional risk. Bring this Checklist the next time you see your doctor, dietitian or other qualified health or social service professional.

DETERMINE: Warning signs of poor nutritional health.

DISEASE
Any disease, illness or chronic condition which causes you to change the way you eat, or makes it hard for you to eat, puts your nutritional health at risk.

EATING POORLY
Eating too little and eating too much both lead to poor health. Eating the same foods day after day or not eating fruit, vegetables, and milk products daily will also cause poor nutritional health.

TOOTH LOSS/MOUTH PAIN
A healthy mouth, teeth and gums are needed to eat. Missing, loose or rotten teeth or dentures which don't fit well, or cause mouth sores, make it hard to eat.

ECONOMIC HARDSHIP
As many as 40% of older Americans have incomes of less than $6,000 per year. Having less—or choosing to spend less—than $25–30 per week for food makes it very hard to get the foods you need to stay healthy.

REDUCING SOCIAL CONTACT
One-third of all older people live alone. Being with people daily has a positive effect on morale, well-being and eating.

MULTIPLE MEDICINES
Many older Americans must take medicines for health reasons. Almost half of older Americans take multiple medicines daily.

INVOLUNTARY WEIGHT LOSS/GAIN
Losing or gaining a lot of weight when you are not trying to do so is an important warning sign that must not be ignored. Being overweight or underweight also increases your chance of poor health.

NEEDS ASSISTANCE IN SELF CARE
Although most older people are able to eat, one of every five have trouble walking, shopping, buying food and cooking food, especially as they get older.

ELDER YEARS ABOVE AGE 80
Most older people lead full and productive lives. But as age increases, risk of frailty and health problems increase. Checking your nutritional health regularly makes good sense.

The Nutrition Screening Initiative, an effort led by several organizations, developed screening tools to assess risk factors associated with poor nutritional health in older adults.

Adapted from Nutritional Screening Initiative. Report on Nutritional Screening: Toward a Common View. Washington DC: Nutritional Screening Initiative; 1991.

of America provides meals to senior citizens. Relying heavily on volunteers, this program works to deliver low-cost meals to homebound elderly and disabled adults. The mission of Meals on Wheels is to nourish and enrich the lives of people who are homebound while promoting dignity and independent living. The federally funded Supplemental Nutrition Assistance Program (SNAP; formerly known as the Food Stamp Program) also assists with the food-related expenses of low-income adults.

It goes without saying that adequate nutrition is an integral and intimate part of life at every stage. Whether a person is receiving her first drops of milk from the breast, eating pureed vegetables, snacking at school, feeding a family, or being fed as an older adult, good nutrition provides the nutrients and energy needed to sustain life. By being aware of your own life stage, nutritional needs, and intake amounts, you can take one of the most profound steps toward ensuring a long, healthy, and happy life.

A Meals on Wheels volunteer delivers a meal to a home-bound older adult.

NUTR

11 Nutrition and Physical Activity

© iStockphoto.com/Frank van den Bergh

LEARNING OUTCOMES:

- LO1 Appreciate the importance of physical activity to overall health.
- LO2 Understand how the body uses nutrients to fuel physical activity.
- LO3 Describe the effects of exercise and athletic training on the body.
- LO4 Understand how various types of physical activity influence dietary requirements.

LO1 What Are the Health Benefits of Physical Activity?

Some people exercise to get or stay in shape, others to improve health and physical fitness, and still others because they enjoy the challenge of training and participating in athletic competitions. Regardless of the reason, exercise offers many health and lifestyle advantages. Because even moderate amounts of physical activity can impart tremendous benefit to a person's health, fitness, and overall quality of life, routine physical activity is simply essential to a long and healthy life. And yet, while the importance of physical activity may be common knowledge, nearly half of all Americans lead sedentary lifestyles.[1] Furthermore, many people are unclear about the types and amounts of physical activity needed to stay healthy and the kinds of foods needed to fuel exercise.

As you learned in previous chapters, obesity and a sedentary lifestyle are risk factors for a multitude of health concerns. In this chapter, you will learn how the body responds to the physical demands of exercise, current public health recommendations regarding physical activity, and the overall importance of nutrition during athletic training and competition.

Physical Activity Improves Health and Fitness

Recall from Chapter 9 that physical activity is any bodily movement that results in increased expenditure of energy. Physical activity includes day-to-day activities such as gardening, household chores, walking, and leisure-time activities. Exercise, also introduced in Chapter 9, is a subcategory of physical activity defined as planned, structured, and repetitive bodily movement done to improve or maintain physical fitness. Whether a person is young, old, sedentary, physically fit, or has some sort of disability, everyone can benefit from physical activity and exercise.

But what does it mean to be physically active? Numerous studies show that as little as 30 minutes of sustained physical activity on most days of the week can substantially improve health and quality of life.[2] Specifically, sustained physical activity can help reduce a person's risk of obesity and certain chronic diseases such as hypertension, stroke, cardiovascular disease, type 2 diabetes, osteoporosis, and some forms of cancer. Unfortunately, while overwhelming evidence shows that physical activity reduces the risk of many debilitating diseases and health conditions, these benefits quickly disappear if a person becomes inactive.[3] Overall benefits of regular physical activity and exercise on health include:

- *Assistance in weight management.* Regular physical activity helps prevent weight gain, and in some cases, facilitates weight loss. This occurs in part because physical activity increases energy expenditure and improves appetite regulation. It also promotes increased muscle mass and decreased body fat.
- *Decreased risk of cardiovascular disease.* Physical activity can improve overall cardiac function, lower resting heart rate, reduce blood pressure, and improve blood lipid regulation. These changes help slow the progression of atherosclerosis and cardiovascular disease.
- *Reduced risk of cancer.* Being overweight increases a person's risk of developing colorectal, esophageal, kidney, breast (postmenopausal), and uterine cancer.
- *Decreased back pain.* Physical activity can increase muscle strength, endurance, flexibility, and overall posture—all of which help prevent injuries and alleviate back pain associated with muscle strain.

- *Decreased risk of type 2 diabetes.* Regular physical activity can decrease blood glucose, insulin resistance, and long-term health complications associated with type 2 diabetes.
- *Optimized bone health.* Weight-bearing physical activity can stimulate bone formation and slow the progression of age-related bone loss.
- *Enhanced self-esteem, stress management, and quality of sleep.* Being physically active provides a sense of achievement and empowerment. This can help reduce depression and anxiety, which in turn fosters improved self-esteem, reduced stress, improved mood, and more restful sleep.

Some people exercise to stay in shape while others exercise to improve overall physical fitness. Regardless of the reason, regular exercise offers many health benefits.

© Masakazu Watanabe/Aflo Foto Agency/Photolibrary

Clearly, the health benefits associated with physical activity are hard to ignore. This is why it is important for you to exercise regularly and make good nutrition a permanent part of your lifestyle.

Physical Activity Recommendations

Because physical *inactivity* is a major public health concern, several health organizations have released specific recommendations as to how much activity a person needs to stay healthy.[4] Although the many different (and sometimes contradictory) guidelines can seem confusing, the 2008 Physical Activity Guidelines for Americans recommends that most people get at least two and a half hours (150 minutes) a week of moderate-intensity aerobic physical activity and two sessions a week of muscle-strengthening activity. Table 11.1 provides a summary of physical activity recommendations for different population groups.

physical fitness Characteristics that a person has or can achieve relating to his or her ability to perform physical activity.

cardiovascular fitness A measure of the circulatory and respiratory systems' ability to supply oxygen and nutrients to working muscles during sustained physical activity.

muscular strength The maximal force exerted by muscles during physical activity.

Five Components of Physical Fitness

The American College of Sports Medicine defines **physical fitness** as a set of characteristics that one has or can achieve relating to her ability to perform physical activity. These characteristics include both health-related and skill-related components, such as cardiovascular endurance and agility, respectively. Regardless of where a person is on the physical fitness continuum, regular physical activity is one of the most important things she can do to maintain health, fitness, and overall well-being. There are countless types of physical activity, most of which are beneficial to health. However, there are five important outcomes of a well-balanced physical fitness regimen:

- **Cardiovascular fitness** refers to the efficiency with which the heart and lungs deliver oxygen and nutrients to working muscles to sustain physical activity. Activities that safely elevate your heart rate (such as walking at a brisk pace, swimming, jogging, and bicycling) can help improve cardiovascular fitness.
- **Muscle strength** refers to the maximal force exerted by muscles during physical activity. Activities such as lifting weights and walking up stairs force muscles to work against resistance, and therefore can help improve muscle and bone strength. Strength-building

Sister Madonna Buder, a nun from Spokane, Washington, has competed in hundreds of triathlons, more than 30 Ironman® races, and two Boston Marathons®. In 2005, she became the first woman over the age of 75 to complete an Ironman distance triathlon. At 82 years of age, she is still competing.

The Canadian Press/Adrian Wyld

Table 11.1 2008 Physical Activity Guidelines for Americans

Population Group	Physical Activity Recommendations
Children and adolescents (age 6–17)	• Children and adolescents should engage in at least one hour of physical activity every day, most of which should consist of either moderate- or vigorous-intensity aerobic physical activity. • Children and adolescents should participate in both muscle- and bone-strengthening activities at least three days per week.
Adults (age 18–64)	• Adults should strive for two and a half hours a week of moderate-intensity, or one and a quarter hours a week of vigorous-intensity aerobic physical activity, or an equivalent combination of moderate- and vigorous-intensity aerobic physical activity. • Aerobic activity should be performed in episodes of at least 10 minutes, spread throughout the week. • Additional health benefits can be obtained by increasing exercise to five hours a week of moderate-intensity aerobic physical activity, or two and a half hours a week of vigorous-intensity physical activity, or an equivalent combination of both. • In addition to aerobic activities, muscle-strengthening activities that involve all major muscle groups should be performed two or more days per week. • If necessary, adults should gradually increase the time spent doing aerobic physical activity and decrease caloric intake to a point at which they can achieve neutral energy balance and a healthy weight.
Older adults (age 65+)	• Older adults should follow the adult guidelines if possible, and those with chronic conditions should be as physically active as their abilities allow. • If they are at risk of falling, older adults should also do exercises that maintain or improve balance.
Adults with disabilities	• Adults with disabilities should follow the adult guidelines if possible and should be as physically active as their abilities allow.
Children and adolescents with disabilities	• When possible, children and adolescents with disabilities should meet the guidelines for all children—or as much activity as conditions allow.
Pregnant and postpartum women	• Pregnant and postpartum women should strive for two and a half hours of moderate-intensity aerobic activity throughout the week. • Women who regularly engage in vigorous-intensity aerobic activity or high amounts of activity can continue their activity provided that their condition remains unchanged and that they consult with their health care providers. • After the first trimester, pregnant women should avoid exercises that involve lying on the back. They should also avoid activities that increase the risk of falling or abdominal trauma, such as contact and/or collision sports.

Adapted from US Department of Health and Human Services. 2008 physical activity guidelines for Americans. Washington, DC: U.S. Department of Health and Human Services; 2008.

activities such as weight lifting and working with elastic bands are referred to as **resistance training** because they involve working against an opposing force.

- **Endurance** refers to a person's ability to exercise for an extended period of time without becoming fatigued. Sustained activities such as skipping rope, swimming, and bicycling can improve endurance.
- **Flexibility** refers to the range of motion around a joint. Activities such as yoga, stretching, and

resistance training Strength-building activities that challenge specific groups of muscles by making them work against an opposing force.

endurance A measure of one's ability to exercise for an extended period of time without becoming fatigued.

flexibility A measure of the range of motion around a joint.

AP Images/Rob Carr

body composition A measure of the relative amounts of muscle, fat, and bone tissue in the body.

FITT principle A collection of four parameters to consider when developing a physical fitness plan: frequency, intensity, type, and time.

swimming can help stretch and lengthen muscles, which improves flexibility.

- **Body composition** refers to the relative amounts of muscle, fat, and bone tissue in the body. Body fat can be decreased through cardiovascular exercise, whereas muscle mass can be increased through resistance-training activities. Weight-bearing activities such as walking and jogging can help increase bone mass.

Getting FITT

To get the most out of a fitness program, it is important to consider four factors collectively referred to as the **FITT principle**—frequency, intensity, type, and time (see Table 11.2). According to the American College of Sports Medicine, the FITT principle should form the basis of any sound physical fitness plan.[5] Physical activity guidelines based on the FITT principle are described next.

FREQUENCY OF PHYSICAL ACTIVITY

In terms of the FITT principle, *frequency* refers to how often a person exercises. To improve cardiovascular and muscular fitness, frequent participation in activities is recommended. Sedentary people should start

Table 11.2 Guidelines and Goal Setting Activities for Achieving FITT Principle

FITT Principle Component	Recommendations for Healthy Adults	Your Goal
Frequency: *The number of times an aerobic exercise or activity is performed; generally expressed in sessions, episodes, or bouts per week.*	3–5 days per week	_________ days/week
Intensity: *Magnitude of the effort required to perform an activity or exercise.*	Moderate-intensity, vigorous-intensity, or a comparable combination of moderate- and vigorous-intensity physical activities as appropriate for personal abilities and goals. • 50–70 percent maximum heart rate for moderate-intensity activities. • 70–85 percent maximum heart rate for vigorous-intensity activities.	Would you prefer participating in only moderate- or vigorous-intensity activities, or a combination of both types?
Type: *The particular type of activity engaged in, for example jogging, walking, playing an organized sport, or yoga.*	Activity type should be chosen based on individual preference, skills, and personal goals.	Name a few activities you would enjoy doing: ______________________ ______________________ ______________________
Time: *The length of time (duration) for which an activity or exercise is performed; generally expressed in minutes per day or hours per week.*	Two and a half hours a week of moderate-intensity, or one and a quarter hours a week of vigorous-intensity activity.	_________ hours/week

Adapted from: U.S. Department of Health and Human Services, Office of Disease Prevention and Health Promotion. 2008 physical activity guidelines for Americans. Washington, DC. 2008. Available from: http://health.gov/PAGuidelines/pdf/paguide.pdf. American College of Sports Medicine (ACSM). ACSM's guidelines for exercise testing and prescription, 8th ed. Philadelphia: Lippincott Williams & Wilkins; 2008. Centers for Disease Control and Prevention. Target heart rate and estimated maximum heart rate. Available from: http://www.cdc.gov/physicalactivity/everyone/measuring/heartrate.html.

a cardiovascular exercise regimen gradually, perhaps with no more than two workouts per week. To achieve muscular strength, two or more workouts per week of resistance activities are typically necessary.

INTENSITY OF PHYSICAL ACTIVITY

The *intensity* of a physical activity refers to the amount of physical effort exerted throughout the activity. Intensity is typically classified as low, moderate, or vigorous, depending on how hard the body is working. Breathing, sweating, and heart rates are often used to judge the intensity of physical activity. It is important for a person to exercise at an intensity level that is appropriate for his particular level of fitness. Training at an intensity level that both benefits health and maintains safety is integral to long-term success. There are several ways to judge the intensity level of a workout:

- *Perceived exertion* is simply how hard one feels her body is working. Based on the physical sensations a person experiences during exercise, perceived exertion is gauged through increased heart rate, breathing, sweating, and muscle fatigue (see Table 11.3).
- The *"talk test"* indicates the relative intensity of an activity by measuring the ease with which someone can converse while exercising. As a rule of thumb, if you are doing moderate-intensity activity, you should be able to talk comfortably. If you are doing vigorous-intensity activity, you should not be able to say more than a few words without pausing for a breath.[6]
- Monitoring *heart rate* is one of the most precise measurements of intensity. For moderate-intensity exercise, a person should maintain a heart rate that is between 50 and 70 percent of his maximum heart rate (estimated by subtracting one's age from 220). For example, because a 20-year-old's maximum heart rate is 200 beats per minute (220–20), his heart rate should be between 100 (50 percent of 200) and 140 (70 percent of 200) beats per minute for moderate-intensity exercise. This age-dependent range is referred to as a *target heart rate range.* Target heart rate ranges for various ages are listed in Table 11.4 on the next page.

indirect calorimetry A method used to estimate energy expenditure based on oxygen consumption and carbon dioxide production.

> "If you experience frequent **nausea, light-headedness**, and/or **blurry vision** during exercise, you are likely working **outside** of a **safe intensity level**."

While individuals can use self-monitoring and tables to estimate their own exertion levels, researchers and clinicians often require a more accurate measure. Such a measure can be obtained using **indirect calorimetry**, which is based on the fundamental principle that energy expended during exercise is proportional to oxygen consumption and carbon dioxide production. As exercise intensity increases, so too

Table 11.3 Signs and Symptoms of Perceived Exertion

Perceived Exertion	Physical Signs and Symptoms	Workout Intensity
No exertion to low exertion	None	Sedentary
Very light to light	Feeling of motion	Low intensity
Somewhat hard	Warmth on cold day, slight sweat on warm day	Moderate intensity
Hard to very hard	Sweating but can still talk without difficulty	High intensity
Extremely hard	Heavy sweating, difficulty talking	
Maximal exertion	Feeling of near exhaustion	

Adapted from Borg, GV. Psychophysical bases of perceived exertion. Medicine & Science in Sports & Exercise. 1982;14:377-81.

does oxygen consumption, until a steady state is reached. By measuring the body's exchange of oxygen and carbon dioxide during exercise, researchers can estimate energy expenditure relative to exercise exertion.

TYPE OF ACTIVITY

The FITT principle notes the importance of the *type* of physical activity performed. To achieve the maximum health benefits, a sound physical fitness program should include several types of exercise. For example, to improve cardiovascular fitness, an exercise regimen should include activities such as walking, cycling, swimming, hiking, cross-country skiing, and dancing. To improve muscle strength, it is important to participate in muscle-strengthening activities such as weight lifting and stair climbing.

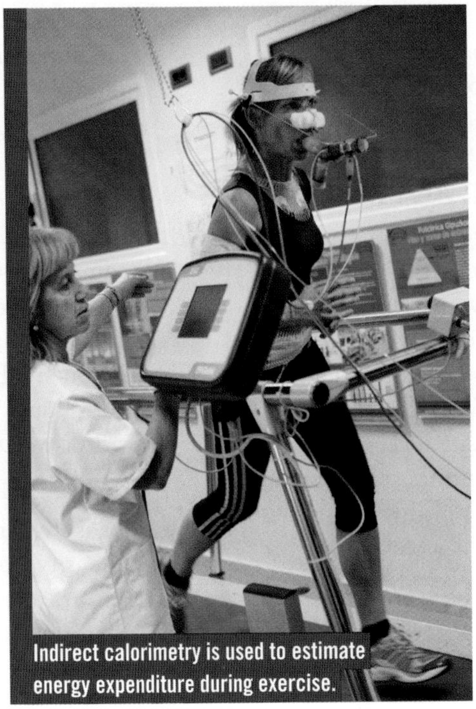

Indirect calorimetry is used to estimate energy expenditure during exercise.

© Javier Larrea/Age fotostock/Photolibrary

TIME OF ACTIVITY

Finally, *time* is an important factor to consider when designing a FITT principle-based fitness regimen. Healthy adults should engage in at least two and a half hours a week of moderate-intensity aerobic activity, or one and a half hours a week of vigorous-intensity aerobic activity.[7] Individuals trying to lose weight or prevent weight regain may need to work out for longer periods, however. A person with a lower fitness level should exercise at a reasonable pace and for a comfortable duration within her target heart rate zone. As fitness level increases, it becomes easier to exercise for longer periods of time. Regardless of one's fitness level, however, exercising too frequently or too intensely can lead to injuries. Exercise can be physically demanding, and the body needs time to recover. To help prevent injuries, some experts recommend warm-up and cool-down periods before and after every workout. Light stretching, walking, and other gentle activities ease the body into and out of rigorous exercise and help prevent injury.

You now have an appreciation for the importance of physical activity and the types and amounts of physical activity needed to gain or maintain a healthy level of physical fitness. Next, you will learn how the body uses energy-yielding nutrients to fuel these activities.

Table 11.4 Maximum and Target Heart Rates

Age (years)	Maximum Heart Rate (beats/min)*	Target Heart Rate Ranges for Moderate-Intensity Activity (beats/min)[b]
20	200	100–140
25	195	98–137
30	190	95–133
35	185	93–130
40	180	90–126
45	175	88–123
50	170	85–119
55	165	83–116
60	160	80–112
65	155	78–109
70	150	75–105

[a]Maximum heart rate is 220 minus age (years).

[b]The CDC recommends 50-70 percent of maximum heart rate for moderate-intensity activity.

Adapted from: Centers for Disease Control and Prevention. Target heart rate and estimated maximum heart rate. Available from: http://www.cdc.gov/physicalactivity/everyone/measuring/heartrate.html.

LO2 How Does Energy Metabolism Change during Physical Activity?

Throughout the day, your body is in a state of perpetual motion. Whether you are standing, sitting, sleeping, walking, or running, your body needs energy. At rest, the body expends approximately 1.0 to 1.5 kcal/minute. During physical exertion, however, energy expenditure can increase to as many as 36 kcal/minute. As illustrated in Table 11.5, physical activity level can contribute substantially to total energy requirements.

ATP Generation

Movement requires motor units within skeletal muscle fibers to contract and relax in response to neural signals from the brain. To perform this complex action, muscle fibers convert the chemical energy in ATP to mechanical energy. In this way, ATP provides the energy muscles need for physical exertion. During exercise, ATP is generated from the metabolic breakdown of glucose, fatty acids, and to a lesser extent, amino acids. The availability of these nutrients is determined by the foods one eats and the nutrients stored in the body.

To fuel physical activity, multiple metabolic pathways work together to provide ATP's energy to muscle cells. Some of these pathways operate in aerobic (oxygen-requiring) conditions, whereas others operate in anaerobic (oxygen-poor) conditions. Although muscles use anaerobic and aerobic metabolic pathways simultaneously, the intensity and duration of activity determines the relative contribution of each. Short, high-intensity exercise (such as a 100-meter sprint) relies heavily on anaerobic pathways, which deliver quick but short-lived bursts of ATP to muscles. By contrast, endurance exercise (such as running a marathon) relies heavily on aerobic pathways, which deliver a more sustained supply of ATP. In total, the body uses three energy systems to generate ATP: the creatine phosphate system, the glycolytic system, and the oxidative system. These three systems work together to ensure adequate energy availability so that the physical demands of exercise can be met (see Figure 11.1 on the next page).

creatine phosphate system The simplest and most rapid energy system; characterized by the use of creatine phosphate to produce ATP.

creatine phosphate A high-energy compound that is broken down to produce ATP.

glycolytic system An energy system characterized by the anaerobic metabolism of glucose (glycolysis).

Table 11.5 The Effects of Physical Activity Level on Total Energy Requirements in Healthy-Weight Adults[a]

Sex and Age (years)	Physical Activity Level (PAL) Sedentary	Low Active	Active	Very Active
Female[b]	kcal/day			
20	1,879	2,085	2,342	2,651
30	1,784	1,989	2,247	2,555
40	1,688	1,894	2,151	2,460
50	1,593	1,799	2,056	2,365
60	1,498	1,703	1,961	2,270
Males				
20	2,542	2,770	3,060	3,536
30	2,447	2,675	2,964	3,441
40	2,352	2,579	2,869	3,345
50	2,256	2,484	2,774	3,250
60	2,161	2,389	2,679	3,155

[a] Calculated using Estimated Energy Requirement (EER) equations for a 5-foot 10-inch, 154-pound reference man and a 5-foot 4-inch, 126-pound reference woman.

[b] Estimates for females do not include women who are pregnant or breastfeeding.

Source: The Institute of Medicine. Dietary Reference Intakes for energy, carbohydrate, fiber, far, fatty acids, cholesterol, protein, and amino acids. Washington, DC: The National Academies Press; 2005.

THE CREATINE PHOSPHATE SYSTEM

At the onset of exercise, small amounts of ATP already present in muscle tissue provide an immediate source of energy. This ATP reserve is quickly depleted, however. The **creatine phosphate system** is the simplest and most rapid means by which active muscles generate ATP (see Figure 11.2 on page 285). Under anaerobic conditions, the high-energy compound **creatine phosphate** can be broken down and combined with adenosine diphosphate (ADP) to produce ATP. Only small amounts of creatine phosphate are stored in muscle, and it takes several minutes to replenish that small supply. Therefore, if physical activity continues, muscles must turn to other systems to generate ATP.

THE GLYCOLYTIC SYSTEM

A continuation of exercise under anaerobic conditions places an even greater demand on muscles to generate ATP. As creatine phosphate stores become depleted, another energy system called the **glycolytic system** takes over. Recall from Chapter 4 that glycolysis is the first step in the metabolic breakdown of glucose, a series of chemical reactions that splits glucose into two three-carbon molecules. Glycolysis can briefly sustain activity after the immediate burst of energy provided by creatine phosphate fades. However, the rate at which muscles take up and use glucose from the blood is not rapid enough to sustain a high rate of ATP production for very long. Moreover, very little ATP is actually produced during glycolysis, limiting the length of time that high-intensity exercise can continue.

During glycolysis, glucose is broken down into two molecules of pyruvate. The fate of pyruvate is determined by oxygen availability: if oxygen is available, pyruvate is funnelled into the next energy system (that is, the oxidative system). If oxygen availability is limited, as is the case during high-intensity exercise, pyruvate is converted to *lactate* (or *lactic acid*). When this occurs, muscles release lactate into the blood, whereby it is delivered to the liver and recycled back into glucose. Once lactate is converted to glucose by the liver, it is released into the blood, ready to re-enter the glycolytic energy system.

Figure 11.1 Relative Contribution of the Body's Energy Systems to Energy Expenditure during Exercise

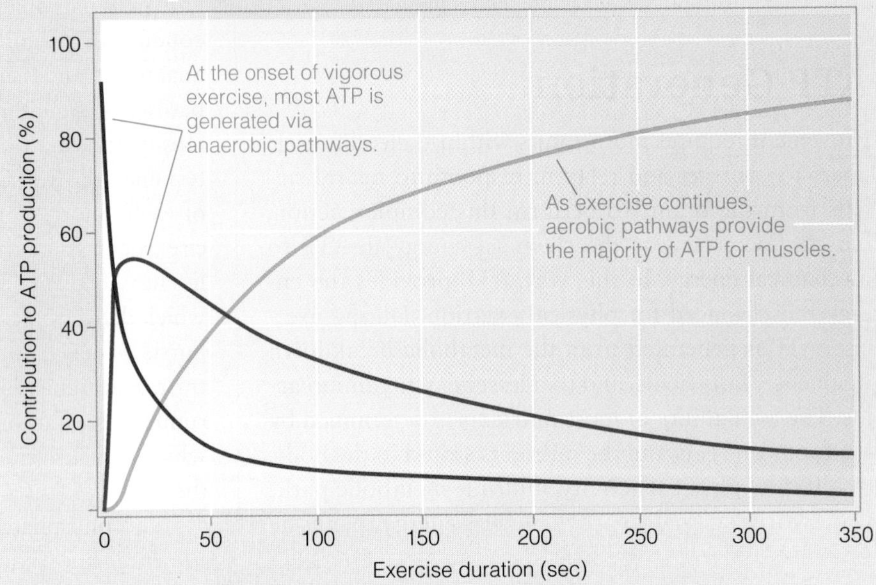

Adapted from Gastin PB. Energy system interaction and relative contribution during maximal exercise. Sports Medicine. 2001;31:725-41.

THE OXIDATIVE SYSTEM

If physical activity continues beyond the capacity of the glycolytic system, muscles must utilize the aerobic **oxidative system** for ATP production. Aerobic pathways are more complex, and therefore generate ATP slower than do anaerobic pathways. After several minutes of low- to moderate-intensity activity, breathing gradually becomes faster and harder. The heart begins to beat more frequently and forcefully. These changes in pulmonary and cardiovascular

oxidative system
An aerobic energy system characterized by steady ATP production in an oxygen-rich environment.

Do Creatine Supplements Enhance Athletic Performance?

Some athletes take creatine supplements to enhance performance and endurance. Widely available and sold over the counter in most grocery stores and pharmacies, creatine-containing products are promoted as beneficial to muscle mass, strength, and recovery time. Although creatine supplementation does increase creatine levels in muscle tissue, whether this translates into improved performance is less clear.

Despite little evidence that they are effective, creatine supplements remain popular among athletes.

Figure 11.2 The Creatine Phosphate System

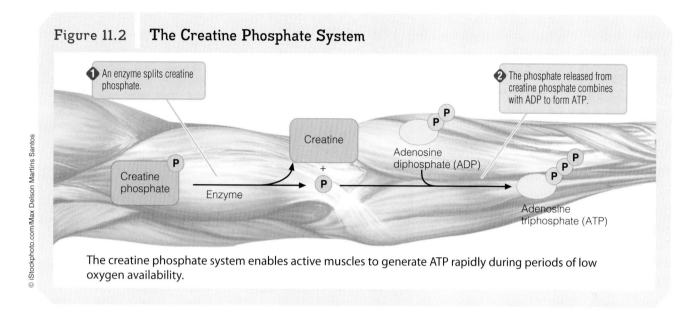

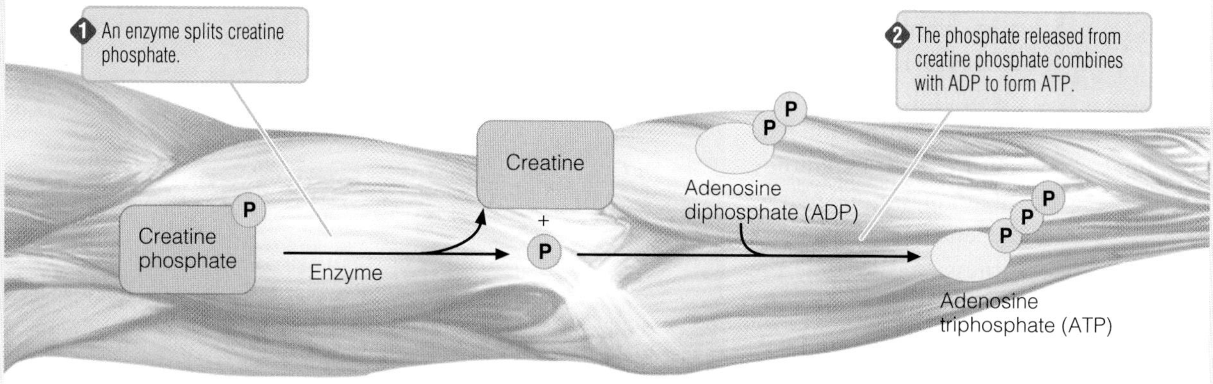

1 An enzyme splits creatine phosphate.
2 The phosphate released from creatine phosphate combines with ADP to form ATP.
Creatine phosphate
P
Enzyme
Creatine
+
P
Adenosine diphosphate (ADP)
Adenosine triphosphate (ATP)

The creatine phosphate system enables active muscles to generate ATP rapidly during periods of low oxygen availability.

Exercise-Related Muscle Soreness

Until recently, the accumulation of lactate in muscles was believed to be the primary cause of muscle soreness after exercise. Although lactate build-up may be associated with acute muscle pain, it does not adequately explain muscle soreness that often lingers for days after a vigorous workout. Scientists now believe that microscopic tears in muscles may be the cause of this long-lasting muscle soreness. This condition, referred to as *delayed onset muscle soreness*, can occur up to 36 hours after exercise—especially if a new or greater-intensity muscle activity is performed.[8] According to researchers, the tiny tears in muscle tissue can lead to inflammation and localized soreness. The good news is that through this process, muscle fibers become stronger and more resistant to tearing. Although this type of muscle damage can be quite painful, it indicates that your muscles will be better equipped to handle the same type of exercise in the future.

function help deliver much-needed oxygen to muscle tissue throughout the body. As sufficient oxygen becomes available, muscle cells are better able to use aerobic metabolism to produce ATP.

Aerobic metabolism utilizes pyruvate (from glycolysis), fatty acids, and to a lesser extent, amino acids to generate ATP. Because most people have ample amounts of stored body fat, it is unlikely that the oxidative system could completely deplete a person's fatty acid energy reserve. For this reason, aerobic metabolic pathways can generate a tremendous amount of ATP over an extended period of time—a critical advantage over anaerobic pathways.

LO3 What Physiologic Adaptations Occur in Response to Athletic Training?

As your body acclimates to the rigors of frequent exercise, a series of physiological changes called **adaptation responses** enable it to perform and recover more efficiently and effectively. The

adaptation response A series of physiological changes that enable the body to perform and recover more efficiently and effectively.

hypertrophy An adaptation response to exercise characterized by enlargement of muscles.

atrophy A process whereby muscles decrease in size.

strength training Exercise that increases skeletal muscle strength, power, endurance, and mass.

anatomical and physiological changes that result from training are very important to the improvement of fitness, athletic condition, and overall health. The nature and magnitude of adaptation responses depend on a person's fitness level and the intensity and duration of training, but everyone who trains regularly experiences some of these physiological changes. Training-induced adaptations can cause muscles to undergo **hypertrophy,** meaning that they become larger. When a person stops training, however, hypertrophic gains quickly diminish. In fact, muscles that are not frequently challenged by physical activity undergo **atrophy,** meaning that they decrease in size. Another example of an adaptation response is the expansion of blood vessels that occurs in response to cardiovascular and strength-building exercise. This expansion of blood vessels increases a person's ability to take in and deliver oxygen to muscles. In addition, training causes muscles to develop a higher tolerance for lactate, allowing them to exercise longer before fatigue sets in. As adaptation responses differ depending on the frequency, intensity, type, and duration of training, most athletic trainers recommend a varied training regimen that alternates between strength- and endurance-focused workouts. Because there are some potential risks associated with doing too much too soon, it is important for beginners to avoid overtraining. Training beyond the body's ability to recover can result in muscle soreness, joint pain, and even serious injury.

The relative contributions of aerobic and anaerobic pathways depend on the intensity of the exercise. Muscles utilize anaerobic pathways when there is a need to generate ATP rapidly, such as during short, high-intensity bursts of activity (left). Activities that entail long periods of endurance rely primarily on aerobic pathways (right).

Hitting the Wall

It can become increasingly difficult during prolonged activity (such as running a marathon) to keep active muscles adequately fueled. When this happens, some athletes lose their stamina, or "hit the wall," a phrase used to describe the feeling of profound fatigue that can occur during an athletic event. Consuming sports drinks, energy bars, or energy gels during exercise can help provide an additional source of glucose; these products provide carbohydrates in a form that is easily digested and rapidly absorbed. Sports drinks are particularly helpful for some athletes because they provide not only glucose, but also water (hydration) and electrolytes. However, because too much fluid can cause stomach cramps, it is important for athletes to find out which products work best for them.

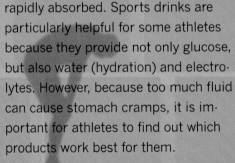

Strength Training and Endurance Training

Different types of exercise result in different types of adaptive responses. **Strength training** (such as weight lifting) challenges muscles with movements that are difficult but brief in duration. These types of activities promote muscle growth (in other words, hypertrophy), which increases muscle size and strength. Conversely, **endurance training** (such as running and swimming) entails steady, low- to moderate-intensity exercise that persists for a longer duration. Endurance training is beneficial to pulmonary and cardiovascular function, and can help improve aerobic capacity (the ability of cells to produce ATP in the presence of an oxygen-rich environment). As the intensity and duration of endurance training increases, muscles require more oxygen for aerobic ATP production. This challenges the cardiovascular system to work even harder.

Cardiovascular adaptations to physical training also include decreased **heart rate** (the number of contractions the heart makes in one minute), decreased **ventilation rate** (breaths expelled per minute), and increased **stroke volume** (the amount of blood pumped per contraction of the heart). As a result, the cardiovascular system pumps blood more efficiently, which facilitates blood flow and the subsequent delivery of nutrients and oxygen to muscles. These adaptive responses increase **maximal oxygen consumption** (or **VO_2 max**), a measure of the cardiovascular system's capacity to deliver oxygen to muscles. A high VO_2 max increases physical stamina and helps delay the onset of fatigue. Endurance training also increases the ability of muscles to use fatty acids for ATP production. This minimizes glucose use, which in turn delays the onset of muscle fatigue. An ability to conserve glucose is particularly important for endurance athletes (such as marathon runners, cyclists, and distance swimmers) who depend on the utilization of glycogen stores as an energy source over an extended period of time.

Some athletic trainers recommend **interval training**, which entails alternating short, fast bursts of intense exercise with slower, less demanding activity. For example, a runner might sprint at full speed for one minute, then jog at a measured pace for five minutes, four times in a row. This type of training helps improve both endurance and **anaerobic capacity**, defined as the maximum amount of work that can be performed under anaerobic conditions. Greater anaerobic capacity means that muscles have a higher tolerance for lactate, and are better able to perform repetitive, high-intensity activity with little or no rest.

While it is important to understand how your body responds to the physical demands of exercise, it is equally important to understand how these demands affect dietary needs and energy requirements. These topics will be addressed in the next section.

> “The **Tabata Method**, a form of **high-intensity interval training**, entails eight cycles of 20 seconds of intense exercise followed by 10 seconds of rest. An entire Tabata workout **lasts just four minutes**.”

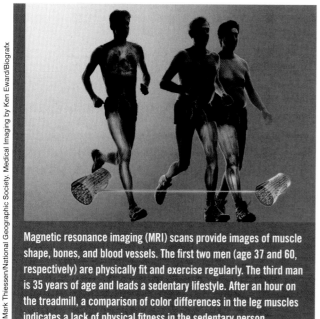

Magnetic resonance imaging (MRI) scans provide images of muscle shape, bones, and blood vessels. The first two men (age 37 and 60, respectively) are physically fit and exercise regularly. The third man is 35 years of age and leads a sedentary lifestyle. After an hour on the treadmill, a comparison of color differences in the leg muscles indicates a lack of physical fitness in the sedentary person.

Ergogenic Aids

Some athletes believe that taking an **ergogenic aid** can enhance physical performance beyond the natural gains obtained through

endurance training Exercise that entails steady, low- to moderate-intensity exercise that persists for a longer duration.

heart rate The number of contractions the heart makes in one minute.

ventilation rate Breaths expelled per minute.

stroke volume The amount of blood pumped per contraction of the heart.

maximal oxygen consumption (or **VO_2 max**) A measure of the cardiovascular system's capacity to deliver oxygen to muscles.

interval training Exercise that entails alternating short, fast bursts of intense exercise with slower, less demanding activity.

anaerobic capacity The maximum amount of work that can be performed under anaerobic conditions.

ergogenic aid A substance taken to enhance physical performance.

physical training alone. Ergogenic aids encompass a wide range of substances, including vitamins, minerals, herbal products, and extracts from organs and glands. Although these products are largely untested, they are widely available and are popular among athletes. Despite their popularity, it is important for athletes to be aware that little effort is made to study the safety and efficacy of ergogenic aids. In fact, the U.S. Food and Drug Administration (FDA) does not require such products to be tested at all before they are placed in the marketplace.

LO4 How Does Physical Activity Influence Dietary Requirements?

In general, nutritional recommendations are similar for all people—regardless of physical activity level. However, people who participate in athletics should take special care to (1) consume enough water to replenish that lost during training and competition, and (2) take in enough energy to meet the demands of physical activity and prevent weight loss. Although exercise may increase micronutrient requirements slightly, these needs can easily be met through an adequate and balanced diet.[9]

Energy Requirements to Support Physical Activity

Athletic individuals must consume enough energy to support normal daily activities as well as those performed as exercise. To satisfy total energy requirements, it is important to consider the frequency, intensity, type, and duration of a workout. Estimated energy expenditures associated with various activities are listed in Table 11.6. Using this information, you can estimate that a sedentary man who weighs 70 kg (154 pounds) needs approximately 2,512 kcal/day simply to maintain his body weight. However, if he decided to run for one hour at a pace of five miles per hour, he would expend about 590 kilocalories in addition to those required to support normal daily activities. As a result, he would need to consume approximately 3,102 kcal to maintain body weight.

An athlete who trains on a regular basis most likely has a very high total energy requirement. In fact, some athletes require 4,000 to 5,000 kcal/day. As discussed in Chapter 2, mathematical formulas developed by the Institute of Medicine as part of the Dietary Reference Intakes (DRIs) can be used to estimate total energy requirements for various physical activity levels.[10]

Table 11.6 Average Energy Expenditure for Selected Physical Activities

	Energy expenditure (kcal/hour)[a]
Moderate-intensity physical activities	
Hiking	330
Dancing	330
Golf (walking and carrying clubs)	330
Bicycling (less than 10 miles/hour)	290
Walking (3½ miles/hour)	280
Weight training (general light workout)	220
Stretching	180
Vigorous-intensity physical activities	
Running/jogging (5 miles/hour)	590
Bicycling (more than 10 miles/hour)	590
Swimming (slow freestyle laps)	510
Aerobic dancing	480
Walking (4½ miles/hour)	460
Heavy yard work (chopping wood)	440
Weight lifting (vigorous effort)	440
Basketball (vigorous)	440

[a] Approximate calories used by a 154-pound man

Adapted from U.S. Department of Agriculture and U.S. Department of Health and Human Services. Dietary Guidelines for Americans, 2010, 7th ed., Washington, DC: US Government Printing Office, December 2010.

Recommended Intakes for Macronutrients

In any diet, it is equally important to consider the ideal distribution of calories in addition to total energy intake. Recall that the Institute of Medicine's Acceptable Macronutrient Distribution Range (AMDR) suggests that 45 to 65 percent of total calories come from carbohydrates, 20 to 35 percent from fat, and 10 to 35 percent from protein. These recommendations likely hold for all adults, including athletes.

CARBOHYDRATES

In order to prevent fatigue and maintain liver and muscle glycogen stores, it is important for athletic individuals to consume adequate amounts of carbohydrates. While the Institute of Medicine does not provide special recommendations regarding carbohydrate intake for athletes, the American Dietetic Association recommends that athletes in training consume 6 to 10 g of carbohydrates per kg of body weight each day.[11] Thus, the recommended daily carbohydrate intake for a 91-kg (200-lb) athlete is approximately 546 to 910 g. (To gain some perspective, one slice of whole-wheat bread contains approximately 17 g of carbohydrates.) While this calculation provides a good estimate for some athletes, total carbohydrate requirements clearly depend on total energy expenditure, sex, and the type of activity performed.

Glycogen is an important energy reserve for athletes participating in distance or endurance sports. To help sustain energy for long periods of time, some athletes try to increase the amount of glycogen stored in their muscles using a technique called **carbohydrate loading.** Advocates of this potentially energy-boosting strategy believe that athletes can stimulate their muscles to store more glycogen by combining the right workout intensity with the right level of carbohydrate intake. Although there is not universal support for the effectiveness of carbohydrate loading, proponents recommend that it be carried out in two stages.[12] During the first stage, an athlete consumes approximately 6 to 8 g carbohydrate per kg of body weight per day. Three to four days prior to competition, the athlete tapers her workout intensity and increases her carbohydrate intake to 10 to 12 g per kg of body weight per day. While athletes participating in high-intensity or endurance competitions often believe that carbohydrate loading gives them a competitive edge, adequate training and a healthy, balanced diet are also important to the improvement of athletic performance.

PROTEIN

Although considerable amounts of protein are needed to maintain, build, and repair muscle, the debate continues as to whether expressly active people (and even competitive athletes) need additional protein.[13] In 2005, the Institute of Medicine concluded that the estimated protein requirement for a healthy adult (0.8 g of protein per kg per day) is equally adequate for a physically active individual.[14] Still, some experts believe that endurance athletes should consume 1.2 to 1.4 g protein per kg per day and that strength-training athletes may require as many as 1.6 to 1.7 g protein per kg per day.[15] Equally important to the amount of protein consumed is the timeframe in which it is consumed. Recent evidence indicates that ingesting protein immediately after exercise may facilitate muscle recovery.[16] Until more information suggests otherwise, most experts agree that it is important for athletes to consume high-quality protein sources (such as meat, dairy, and eggs) to maintain, build, and repair tissue.[17]

LIPIDS

Fatty acids are an important source of energy for physically active individuals. Intake of foods rich in monounsaturated fatty acids (such as olive and canola oils) and polyunsaturated fatty acids (such as vegetable and fish oils) should be emphasized, while intake of saturated and *trans* fatty acids should be limited. Furthermore, because exercise can lead to inflammation, some experts believe that eicosapentaenoic acid- and docosahexaenoic acid-rich ω-3 fatty acids may be especially important for competitive athletes.[18]

Recommended Intakes for Micronutrients

Vitamins and minerals are essential to many functions, including energy metabolism, the repair and maintenance of body structures, protection from oxidative

carbohydrate loading A technique used by some athletes to increase the amount of glycogen stored in their muscles so that they can sustain activity for long periods of time.

sports anemia A temporary type of anemia that occurs at the onset of a training program due to hemodilution.

hemodilution A disproportionate increase in plasma volume as compared to the synthesis of new red blood cells.

damage, and immunity from diseases. As long as an individual consumes a varied diet that supplies adequate energy, vitamin and mineral needs will likely be met as well. Nutrient supplementation to enhance athletic performance is controversial and potentially dangerous. Furthermore, there is little evidence regarding the efficacy of this practice.[19] However, people who consume foods with low nutrient densities and athletes who restrict or avoid animal-based products are at increased risk of inadequate intakes of several micronutrients—especially iron, calcium, and zinc. As these nutrients are especially important in terms of physical activity, deficiencies may foster severe complications for some athletes. For this reason, coaches and trainers need to be aware of the nutritional issues that affect athletes who follow strict vegetarian diets.[20]

IRON

The iron-containing molecules hemoglobin and myoglobin are integral components of red blood cells and muscles, respectively. Recall from Chapter 8 that hemoglobin transports oxygen to cells and removes carbon dioxide, a waste product of energy metabolism. Similarly, myoglobin aids in the delivery of oxygen to muscles. Because iron is so vital to oxygen and carbon dioxide transport, its deficiency can compromise aerobic energy metabolism and thus seriously affect athletic training and performance. Although non-heme iron is available in cereals, grains, and some vegetables, it is not as readily absorbed as the heme iron found in meat. All individuals who do not eat meat—including athletes—are at increased risk of impaired iron status and should have their blood tested periodically.

Endurance athletes, especially runners, may have higher iron requirements than nonathletes. Increased loss of blood in the feces, iron loss associated with excessive sweating, and the rupturing of red blood cells as a result of the feet repeatedly striking hard surfaces contribute to this phenomenon.[21]

© iStockphoto.com/Robert Kirk

Because of the blood loss associated with menstruation, female athletes are also more likely than males to experience impaired iron status. In addition to iron deficiency, some athletes experience **sports anemia**, a temporary type of anemia that occurs at the onset of a training program. Sports anemia is caused by a disproportionate increase in plasma volume as compared to the synthesis of new red blood cells. This physiologic response, referred to as **hemodilution**, can make a person appear iron deficient even though he is not. Because sports anemia is caused by a healthy physiological response to training rather than poor nutritional status, there is no need for athletes with this condition to add additional iron to their diets.

CALCIUM

Although recommended calcium intake is the same for athletes and nonathletes alike, adequate calcium is essential to the growth, maintenance, and repair of bone tissue. Thus, a diet that is low in calcium increases an athlete's risk of low bone density, which in turn increases the likelihood of stress fractures. Stress fractures are small cracks in bones—especially those in the feet—that occur in response to repeated jarring. Women who have menstrual irregularities

Inadequate calcium intake can lead to bone loss, which can lead to stress fractures. It is important for all athletes, especially those who are post-menopausal or have menstrual irregularities, to make sure they meet their recommended calcium intakes.

© Brad Wrobleski/Masterfile

are at particularly high risk of bone loss, and those who exercise excessively while restricting their caloric intakes are at risk of developing the female athlete triad, a syndrome characterized in part by bone loss that will be addressed in Chapter 12.

Athletes who restrict their intakes of dairy products may benefit from foods that are fortified with both calcium and vitamin D, as the latter facilitates absorption of the former. Fortified orange juice, soy products, and breakfast cereals are good sources of these two micronutrients. If an athlete's diet is lacking in calcium-rich foods, he should also consider taking a calcium supplement.[22]

ZINC

Most (70 percent) of the zinc in the American diet comes from animal-based products—primarily meat. This is why athletes who restrict their intakes of meat, dairy, or eggs are at increased risk of not getting enough zinc. Zinc plays an important role in energy metabolism and is needed for the maintenance, repair, and growth of muscles. Clearly, zinc is an especially crucial mineral for athletes. Vegetarian sources of zinc include legumes, hard cheeses, whole-grain products, wheat germ, some fortified cereals, nuts, and soy products. Since supplemental zinc can affect the bioavailability of other nutrients (such as iron and copper), athletes should be careful not to exceed the UL for zinc (40 mg/day).

Fluids, Electrolytes, and Dehydration

While exercise increases body temperature, the body usually dissipates this heat by sweating. However, if the water lost as sweat is not replaced, dehydration can result. To meet the body's need for water, adult men and women should consume approximately 2.6 and 3.8 liters (11 and 16 cups) of water per day, respectively. Athletes, however, often require substantially more than this.[23] Although sweat consists mainly of water, it also contains electrolytes such as sodium, chloride, and potassium, and to a lesser extent, minerals such as iron and calcium. Because extensive sweating can disrupt the body's balance of both fluids and electrolytes, it is important for athletes to stay hydrated.

> "Lucozade Energy™, the **first** commercially available **energy drink**, was originally manufactured **in 1927** to **treat common illnesses** such as influenza."

heat exhaustion A rise in body temperature that occurs when the body has difficulty dissipating heat.

heat stroke A serious condition that can develop if body temperature continues to rise beyond heat exhaustion.

hyponatremia Diminished blood sodium concentration.

DEHYDRATION CAN CAUSE SERIOUS ILLNESS

Excessive sweating causes blood volume to decrease. As blood flow to the skin diminishes, so too does the body's ability to dissipate heat. As a result, core body temperature can rise, increasing the risk of heat exhaustion and heat stroke. **Heat exhaustion** can occur when a person loses as little as 5 percent of her body weight due to sweating.[24] Heat exhaustion is characterized by feelings of illness, muscle spasms, rapid and/or weak pulse, low blood pressure, disorientation, and profuse sweating (see Table 11.7 on the next page). If body heat continues to rise, heat exhaustion can quickly progress to **heat stroke**, which occurs when 7 to 10 percent of body weight is lost due to sweating. This very serious condition is characterized by dry skin, confusion, and loss of consciousness. A person experiencing heat stroke requires immediate medical assistance.

Dehydration can cause a number of problems beyond impaired body temperature regulation. For example, excessive sweating or consumption of large amounts of fluid without adequate amounts of sodium can cause blood sodium concentration to decrease. Diminished blood sodium concentration, called **hyponatremia**, is more common in endurance athletes who participate in activities that last several hours (such as marathon and triathlon contestants). A 2005 study of Boston Marathon runners reported that 13 percent of participants developed hyponatremia.[25] Strength- and endurance-focused athletes alike must be particularly careful when exercising in hot or humid weather, as these conditions can increase sweating and thus the loss of water and electrolytes from the body. Because drinking large amounts of water may actually increase one's risk of hyponatremia, many experts recommend that endurance athletes replenish their bodies' fluids by consuming drinks that contain both carbohydrates and electrolytes, such as sports drinks.[26]

PREVENTING DEHYDRATION AND ELECTROLYTE IMBALANCE

To prevent dehydration, athletes must consume adequate amounts of fluids not only during but also before and after exercise. The National Athletic Trainers' Association recommends that athletes consume 500 to 600 milliliters (two to two and a half cups) of fluid two to three hours before exercising.[27] Similarly, the American College of Sports Medicine stresses the importance of consuming fluids several hours prior to athletic competition.[28]

Because dehydration can compromise performance, athletes should also be sure to consume fluids *during* exercise. The purpose of fluid intake during an athletic event is not to rehydrate, but rather to stay hydrated by replacing the fluid lost through sweating. When a workout lasts less than one hour, plain water is adequate to replenish body fluids. When a workout lasts more than one hour, a fluid that contains small amounts of carbohydrates and electrolytes (such as a sports drink) is likely beneficial.[26] Readily available and well tolerated by most people during exercise, sports drinks provide roughly 10 to 20 g of carbohydrates (40 to 80 kcal) per 237-ml (eight-ounce) serving. It is important to be aware, however, that some sports drinks contain caffeine. Whereas some people enjoy the mild stimulatory effect of caffeine, others find that it causes stomach upset, nervousness, and jitters.

Table 11.7 Signs and Symptoms of Mild to Severe Dehydration

Mild Dehydration	Moderate to Severe Dehydration
• Flushed face	• Low blood pressure
• Extreme thirst	• Fainting
• Dry, warm skin	• Disorientation, confusion, and slurred speech
• Impaired ability to urinate; production of reduced amounts of dark, yellow urine	• Severe muscle contractions in the arms, legs, stomach, and back
• Dizziness made worse when standing	• Convulsions
• Weakness	• Bloated stomach
• Cramping in arms and legs	• Heart failure
• Crying with few or no tears	• Sunken, dry eyes
• Difficulty concentrating	• Skin that has lost its firmness and elasticity
• Sleepiness	• Rapid, shallow breathing
• Irritability	• Fast, weak pulse
• Headache	
• Dry mouth and tongue, with thick saliva	

It is important for athletes to prevent dehydration by consuming adequate amounts of fluids before, during, and after exercise. This is why race organizers often set up water stations along courses.

Nutrition Plays an Important Role in Post-Exercise Recovery

Hard training is physically demanding, and athletes need to consume the right types and amounts of food to make the most of exercise. Adequate nutrition is also an important aspect of successful post-exercise recovery. A well-balanced and complete dietary recovery plan can help the body repair muscle tissue, replace lost nutrients, and fully replete energy reserves such as glycogen.

CARBOHYDRATES REPLENISH GLYCOGEN STORES

It takes approximately 20 hours to replenish liver and muscle glycogen stores after a long, intense athletic workout. Although the rate of glycogen synthesis is variable during the post-exercise period, there are several recommendations to promote optimal muscle glycogen resynthesis. First, because the rate at which muscles synthesize glycogen is highest within the first two hours after exercise, athletes should consume carbohydrate-rich foods as soon as possible after a workout.[29] Consuming 50 to 70 g of carbohydrates within the first hour of post-recovery can jump-start the metabolic process of glycogen synthesis.[30] Ideally, a person should consume several small carbohydrate-rich meals instead of a single large one.[31] This promotes a gradual rise in blood glucose rather than a rapid rise followed by a quick drop.

Second, studies have shown that consuming high-glycemic index foods later in the recovery period may help facilitate glycogen resynthesis.[32] Furthermore, some studies show that the rate of muscle glycogen synthesis may be higher when carbohydrates and proteins are consumed together than when carbohydrates are consumed alone during the post-exercise period.[33] Though an ideal ratio of carbohydrate-to-protein has not been firmly established, some athletic trainers recommend a ratio of 4 g of carbohydrates for every 1 g of protein.[34] This mix of carbohydrates and proteins could promote greater muscle glycogen, but there is no evidence that it improves overall athletic performance.

PROTEIN PROMOTES MUSCLE RECOVERY

Because prolonged and intense exercise increases protein breakdown in muscles, high-quality protein sources should be consumed during the post-exercise recovery phase. The first few hours following exercise are often referred to as an *anabolic window*. This period is an important time for an athlete to both reverse the catabolic state associated with physical activity and increase the rate of protein synthesis. When a person consumes protein-rich foods, the concentration of amino acids in the blood increases. This provides the necessary building blocks for muscle maintenance and recovery during the post-workout period.

As previously discussed, experts disagree as to whether athletes require more protein (relative to body weight) than do nonathletes. Some evidence, however, suggests that certain types of protein may be more advantageous in post-workout muscle recovery than other sources. For example, whey protein from milk has grown in popularity among athletes in recent years. This is because studies comparing the efficacy of whey protein with that of other proteins suggest that whey can speed tissue repair, which in turn accelerates recovery from exercise-induced muscle damage.[35]

WATER AND ELECTROLYTE REPLACEMENT

Even if athletes consume fluids while exercising, they can still lose up to 6 percent of their body weights during strenuous activity. Therefore, fluid replacement is critical during the post-workout recovery phase. Consuming 450 to 675 milliliters (two to three cups) of fluid for each pound of weight lost during exercise adequately replenishes one's fluid balance. A simple way for an athlete to monitor her hydration status is to observe the color of her urine the morning after a hard workout. Light-colored urine is a general indicator that a person is well hydrated, whereas dark-colored urine may indicate that additional fluids are needed. Although sweat contains significant amounts of sodium, the foods consumed after exercise usually contain enough sodium to restore electrolyte balance. Therefore, it is neither necessary nor recommended that athletes take sodium supplements.[36]

NUTR

12 Disordered Eating

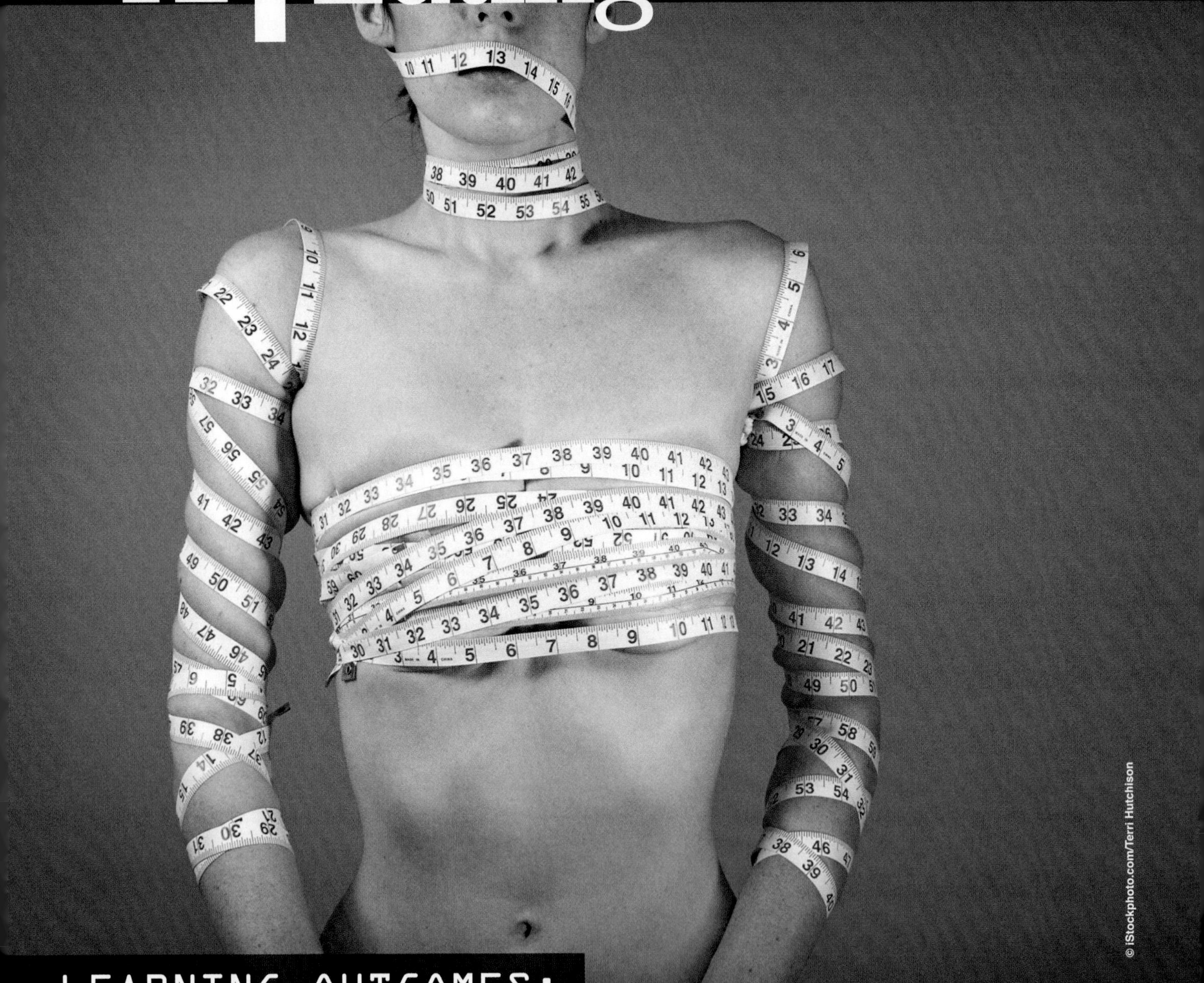

© iStockphoto.com/Terri Hutchison

LEARNING OUTCOMES:

LO1 Understand disordered eating patterns and the major eating disorders.

LO2 Discuss less-common disordered eating behaviors.

LO3 Understand contributing factors related to eating disorders.

LO4 Discuss the connection between eating disorders and athletics.

LO5 Describe strategies to prevent and treat eating disorders.

Chapter 12

LO1 What Is Disordered Eating, and What Are Eating Disorders?

Your body has natural signals that tell you when to eat, what to eat, and how much to eat. Paying attention to and trusting these internal cues helps cultivate a healthy relationship with food—in other words, eating without fear, guilt, or shame. For some people, however, a preoccupation with food and/or weight loss can reach obsessive proportions and disrupt healthy eating behaviors. When this happens, disordered eating sometimes develops. **Disordered eating** is an eating pattern characterized by one or more unhealthy eating behaviors, such as irregular eating, consistent undereating, and consistent overeating.[1] Disordered eating behaviors are common and often arise in response to stress, illness, and dissatisfaction with one's personal appearance. Although disordered eating patterns can be disturbing to others, they typically do not persist long enough to cause serious physical harm. For some people, however, disordered eating progresses into a full-blown **eating disorder**—an extreme disturbance in eating behaviors that can be both physically and psychologically harmful. Eating disorders are complex behaviors that arise from a combination of physical, psychological, and social issues. People with eating disorders often feel isolated, and their relationships with family and friends become strained.

The number of people diagnosed with eating disorders has increased over the last 25 years.[2] Although eating disorders are somewhat rare among the general population, they are relatively common among adolescent girls and young women. The American Psychological Association estimates that in the United States, approximately 8 million girls and women—8 percent of the female population—battle with some form of eating disorder.[3] While eating disorders are most prevalent among adolescent girls and young adult women, a pattern of eating disorders has begun to emerge among middle age and older women as well.[4] This trend may be developing in response to America's youth-oriented culture, which intensifies insecurities associated with aging.

The prevalence of eating disorders in adolescent girls and young women is well documented, but far less is known about eating disorders in adolescent boys and men. Numbers of eating disorders among males may be difficult to estimate because they may be more reluctant to seek help for what is commonly perceived as a female condition. Nevertheless, the prevalence of eating disorders among males has been estimated to be as high as 2 percent.[5] While it is not clear which personal and cultural aspects contribute to the development of eating disorders in adolescent boys and young men, possible factors include a history of obesity, participation in a sport that emphasizes thinness, and a heightened cultural emphasis on physical appearance.

As described in Table 12.1 on the next page, the American Psychiatric Association recognizes three distinct categories of eating disorders. **Anorexia nervosa (AN)** is characterized by an irrational fear of gaining weight or becoming obese; **bulimia nervosa (BN)** is characterized by repeated cycles of bingeing and purging; and **eating disorders not otherwise specified (EDNOS)** includes some, but not all, of the diagnostic criteria for anorexia nervosa and/or bulimia nervosa.[6] Each of these categories is divided into subcategories that depend on the presence or absence of specific behaviors. Because behaviors associated with different eating disorders often overlap,

disordered eating An eating pattern characterized by one or more unhealthy eating behaviors.

eating disorder An extreme disturbance in eating behaviors that can be both physically and psychologically harmful.

anorexia nervosa (AN) An eating disorder characterized by an irrational fear of gaining weight or becoming obese.

bulimia nervosa (BN) An eating disorder characterized by repeated cycles of bingeing and purging.

eating disorders not otherwise specified (EDNOS) A category of eating disorders that includes some, but not all, of the diagnostic criteria for anorexia nervosa and/or bulimia nervosa.

Table 12.1 American Psychiatric Association Classification and Diagnostic Criteria for Eating Disorders

Anorexia Nervosa (AN)
A. Refusal to maintain body weight at or above a minimally healthy weight for age and height.
B. Intense fear of gaining weight or becoming fat, even though underweight.
C. Disturbance in body weight or shape, or denial of the seriousness of the current low body weight.
D. Absence of at least three consecutive menstrual cycles (in postmenarchal, premenopausal females).
Subcategories:
• *Restricting type*: Does not regularly engage in binge eating or purging behavior.
• *Binge-eating/purging type*: Regularly engages in binge eating or purging behavior.
Bulimia Nervosa (BN)
A. Recurrent episodes of binge eating. An episode of binge eating is characterized by both (1) eating, in a discrete period of time, an amount of food that is definitely larger than what most people would eat during a similar period of time and under similar circumstances, and (2) a sense of lack of control over eating during the episode.
B. Recurrent compensatory behaviors to prevent weight gain, such as self-induced vomiting; misuse of laxatives, diuretics, enemas, or other medications; fasting; or excessive exercise.
C. Binge eating and inappropriate compensatory behaviors occurring at least twice a week for three months.
D. Self-evaluation is unduly influenced by body shape and weight.
E. Disturbance does not occur exclusively during episodes of anorexia nervosa.
Subcategories:
• *Purging type*: Regularly engaging in self-induced vomiting or the misuse of laxatives, diuretics, exercise, or enemas.
• *Nonpurging type*: Regularly engaging in inappropriate compensatory behaviors, such as fasting or excessive exercise; but not in self-induced vomiting or the misuse of laxatives, diuretics, or enemas.
Examples of Eating Disorders Not Otherwise Specified (EDNOS)
A. For females, criteria for anorexia nervosa are met, except the individual has regular menses.
B. Criteria for anorexia nervosa are met except that, despite significant weight loss, the individual's current weight is in the healthy range.
C. Criteria for bulimia nervosa are met except that the binge eating and inappropriate compensatory mechanisms occur at a frequency of less than twice a week or a duration of less than three months.
D. Regular use of inappropriate compensatory behavior by an individual of healthy body weight after eating small amounts of food.
E. Repeatedly chewing and spitting out, but not swallowing, large amounts of food.
F. Binge-eating disorder: recurrent episodes of binge eating in the absence of the regular use of inappropriate compensatory behaviors characteristic of bulimia nervosa.

Adapted from Diagnostic and statistical manual of mental disorders, 4th ed. (DSM-IV). Washington, DC: American Psychiatric Association Press, 2004.

it can be difficult to identify the specific disorder that a person has (see Table 12.2).[7]

Anorexia Nervosa (AN)

The American Psychiatric Association recognizes two subcategories of AN. In the first, called **anorexia nervosa, restricting type**, low body weight is maintained through food restriction and/or excessive exercise. In **anorexia nervosa, binge-eating/purging type**, low body weight is maintained through both food restriction and periods of **bingeing** and/or purging. Bingeing is defined as uncontrolled consumption of large quantities of food in a relatively short period of time, whereas **purging** is self-induced vomiting, excessive exercise, and/or misuse of laxatives, diuretics, and/or enemas (see Figure 12.1 on the next page).

When a person with either type of AN looks in the mirror, she tends to be very critical of her body shape and size. Because she experiences a disconnect between actual and perceived body weight and shape, she may believe that she is overweight or obese even though she may be dangerously thin. Because of this distorted self-perception, weight loss becomes an obsession—individuals with AN often take drastic measures to lose weight, spend hours scrutinizing their bodies, and tend to view self-worth in terms of weight and body shape.

Because they tend to perceive food restriction (and often, self-starvation) as an accomplishment rather than a problem, people with AN often have little motivation to change their behaviors. Although the symptoms of AN center around food, the causes are much more complex. AN and other eating disorders usually stem from psychological issues, and both denial of food and the relentless pursuit of thinness become ways of coping with emotions, conflict, and stress.

BEHAVIORS ASSOCIATED WITH ANOREXIA NERVOSA

Typically, people with AN limit their food intakes as well as the varieties of foods consumed. Foods are often categorized as either safe or unsafe depending on whether their consumption is believed to cause weight gain. Foods often perceived as unsafe, and typically the first to be eliminated, include meat, high-sugar foods, and fatty foods. As AN progresses, the diet becomes so restricted that nutritional needs are no longer met. Beyond limiting food intake, individuals with AN may develop unusual food rituals to demonstrate self-restraint and control over the urge to eat. These rituals include chewing food but not swallowing, overchewing, obsessively cooking for others, and cutting food into unusually small pieces.[8]

Obsessive behaviors not related to food are also commonly associated with AN.[9] Many individuals with

Table 12.2 Behaviors Associated with Different Eating Disorders and Disordered Eating Patterns

Type of Eating Disorder and Disordered Eating Pattern	Food Restriction	Bingeing	Compensatory Behavior
Anorexia nervosa, restricting type			Excessive exercise
Anorexia nervosa, binge eating/purging type			Self-induced vomiting and/or misuse of laxatives, diuretics, or enemas
Bulimia nervosa, purging type			Self-induced vomiting and/or misuse of laxatives, diuretics, or enemas
Bulimia nervosa, nonpurging type			Excessive exercise
Binge-eating disorder			
Restrained eating			

Dark blue indicates a primary behavior associated with specific eating disorders and disordered eating, whereas light blue indicates that a behavior occurs to a lesser extent.

anorexia nervosa, restricting type A subcategory of anorexia nervosa characterized by food restriction and/or excessive exercise.

anorexia nervosa, binge-eating/purging type A subcategory of anorexia nervosa characterized by both food restriction and periods of bingeing and purging.

bingeing Uncontrolled consumption of large quantities of food in a relatively short period of time.

purging Self-induced vomiting and/or excessive exercise, misuse of laxatives, diuretics, and/or enemas.

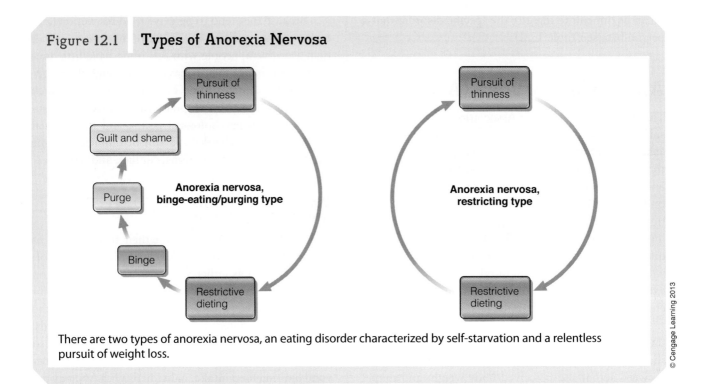

Figure 12.1 **Types of Anorexia Nervosa**

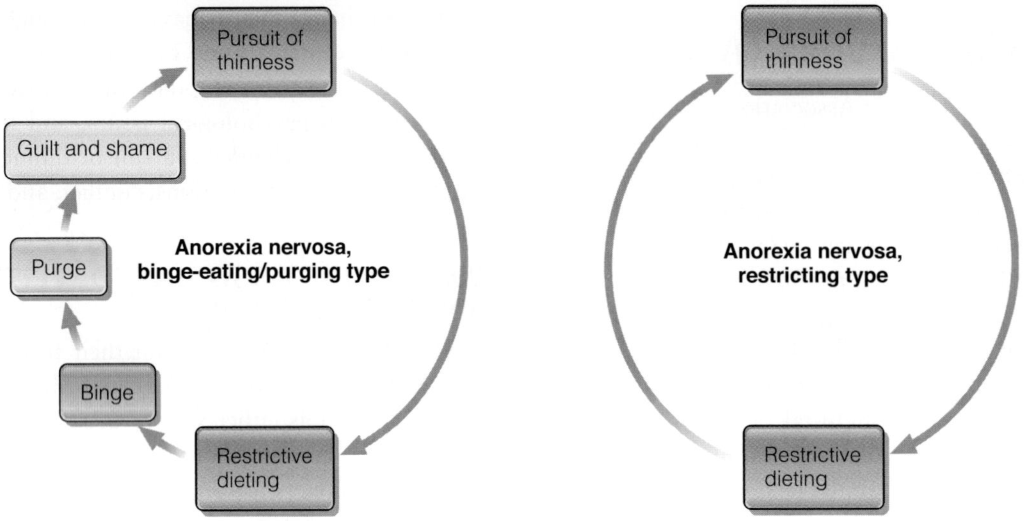

There are two types of anorexia nervosa, an eating disorder characterized by self-starvation and a relentless pursuit of weight loss.

People with distorted body images perceive themselves as fat even if they are dangerously thin.

AN maintain rigid exercise schedules—some work out as much as three to four hours a day. Furthermore, individuals with AN often keep meticulous daily records to track food intake, amount of time spent exercising, and weight loss. Daily monitoring can provide an anorexic person the motivation and mental stamina to sustain a perpetual state of hunger (see Figure 12.2). A page taken from the journal of a person with AN reflects an inability to recognize problematic behaviors (see Figure 12.3 on page 300). Finally, it is not uncommon for people with AN to weigh themselves repeatedly throughout the day. These and other obsessive behaviors help quell concerns about being insufficiently lean and/or excessively fat. Common signs and symptoms, behaviors, and consequences associated with AN are summarized in Table 12.3.

HEALTH CONCERNS ASSOCIATED WITH ANOREXIA NERVOSA

Although AN eventually causes a person to become underweight and appear gaunt, AN-related health problems such as nutrient deficiencies, electrolyte imbalances, and hair loss can develop even before physical signs of AN become apparent. If AN is left untreated, serious health problems—dehydration, cardiac abnormalities, the appearance of lanugo (increased body hair), muscle wasting, and bone loss—can develop. These problems are largely the result of chronic

Figure 12.2 Food, Weight, and Exercise Record of a Person with Anorexia Nervosa

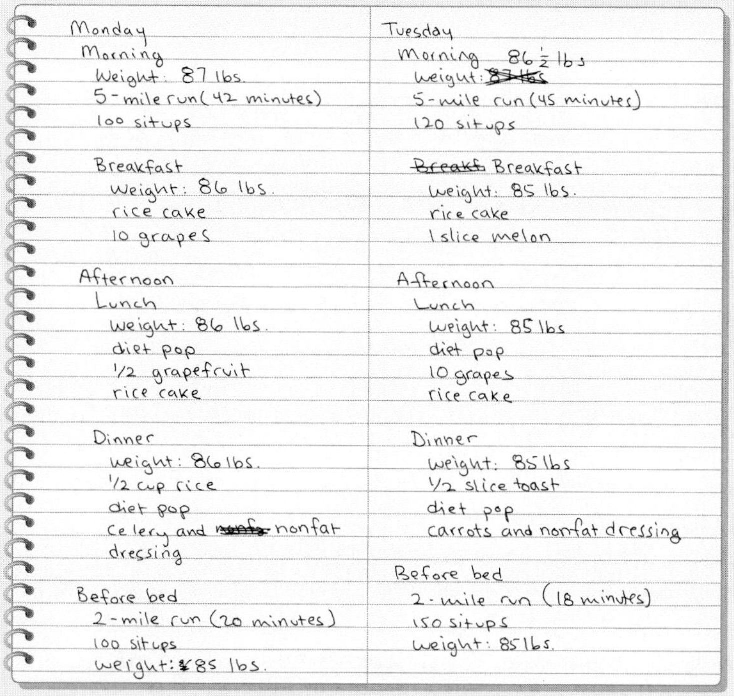
Monday
Morning
Weight: 87 lbs.
5-mile run (42 minutes)
100 situps

Breakfast
Weight: 86 lbs.
rice cake
10 grapes

Afternoon
Lunch
Weight: 86 lbs.
diet pop
1/2 grapefruit
rice cake

Dinner
Weight: 86 lbs.
1/2 cup rice
diet pop
Celery and ~~nonfa~~ nonfat dressing

Before bed
2-mile run (20 minutes)
100 situps
Weight: ~~8~~85 lbs.

Tuesday
Morning 86 1/2 lbs
Weight: ~~87 lbs~~
5-mile run (45 minutes)
120 situps

~~Breakf~~ Breakfast
Weight: 85 lbs.
rice cake
1 slice melon

Afternoon
Lunch
Weight: 85 lbs
diet pop
10 grapes
rice cake

Dinner
Weight: 85 lbs
1/2 slice toast
diet pop
carrots and nonfat dressing

Before bed
2-mile run (18 minutes)
150 situps
Weight: 85 lbs.

Some people with eating disorders keep meticulous records regarding weight loss, exercise and diet.

malnutrition. Once healthy eating habits and body weight are restored, most of these medical concerns can be reversed.

In females, a loss of body fat can lead to a decline in estrogen production, which in turn can lead to disrupted menstruation and poor reproductive function.[10] This physiological response most likely evolved as a means to protect women from pregnancy during periods of low food availability. While it is critical to reproductive function, estrogen also plays an important role in bone health. The longer a woman goes without menstruating—a condition called *amenorrhea*—the greater the loss of bone density. Although estrogen levels return to normal when body weight is restored, it is not likely that bones ever fully recover. For this reason, women who have experienced AN-triggered amenorrhea remain at increased risk for bone disease even after returning to healthy body weights and eating patterns.[11]

Table 12.3 Physical Signs and Symptoms, Behavioral and Emotional Characteristics, and Health Consequences Associated with Anorexia Nervosa

Physical Signs and Symptoms	Behavioral and Emotional Signs and Symptoms	Health Consequences
• At or below the 15th percentile of ideal body weight	• Food rituals such as excessive chewing or cutting food into small pieces	• Reproductive problems
• Thin appearance	• Restriction of the amount and type of food consumed	• Loss of bone mass
• Fainting, dizziness, fatigue, overall weakness	• Adherence to rigid daily schedules and routines	• Electrolyte imbalance
• Intolerance to cold	• Refusal to eat	• Irregular heartbeat
• Significant loss of body fat and lean body mass	• Denial of hunger	• Bruises
• Brittle nails	• Excessive exercise	• Injuries such as stress fractures
• Dry skin, dry hair, thinning hair, hair loss	• Depression or lack of emotion	• Impaired iron status
• Low blood pressure	• Food preoccupation	• Impaired immune status
• Formation of lanugo on the body	• Frequent monitoring of body weight	• Slow heart rate and low blood pressure
• Irregular or absent menstruation	• Tendency toward perfectionism	• Increased risk of heart failure
	• Rigid attitude	• Increased risk of suicide

AN is a serious condition that affects a person's physical and mental well-being. It goes without saying then that AN requires immediate medical, psychological, and nutritional intervention. Although researchers have reported that AN carries a high death rate, this is true only for the most severe cases that require hospitalization.[12] Still, the death rate associated with AN is high: between 5 and 20 percent of those diagnosed with AN die from complications within 10 years of the initial diagnosis.[13] The two primary causes of AN-related death are cardiac arrest and suicide.[14]

Figure 12.3 Thoughts and Behaviors Associated with Anorexia Nervosa

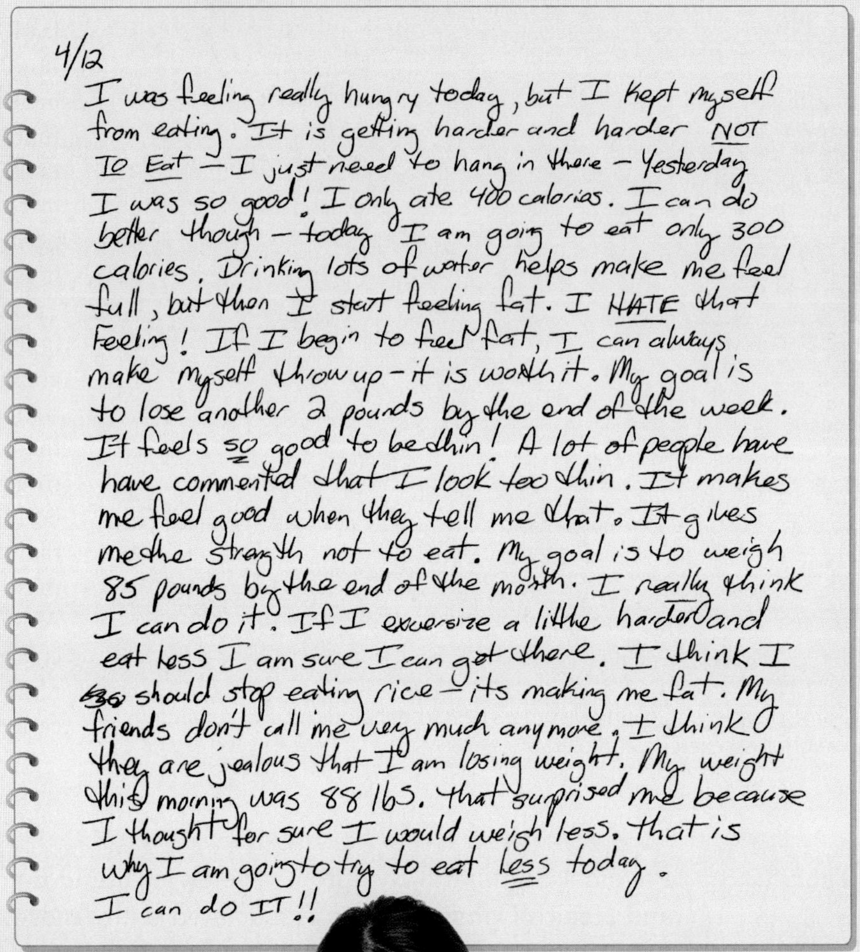
4/12
I was feeling really hungry today, but I kept myself from eating. It is getting harder and harder NOT TO Eat — I just need to hang in there — Yesterday I was so good! I only ate 400 calories. I can do better though — today I am going to eat only 300 calories. Drinking lots of water helps make me feel full, but then I start feeling fat. I HATE that Feeling! If I begin to feel fat, I can always make myself throw up — it is worth it. My goal is to lose another 2 pounds by the end of the week. It feels so good to be thin! A lot of people have have commented that I look too thin. It makes me feel good when they tell me that. It gives me the strength not to eat. My goal is to weigh 85 pounds by the end of the month. I really think I can do it. If I exersize a little harder and eat less I am sure I can get there. I think I should stop eating rice — its making me fat. My friends don't call me very much anymore. I think they are jealous that I am losing weight. My weight this morning was 88 lbs. that surprised me because I thought for sure I would weigh less. that is why I am going to try to eat less today. I can do IT!!

Bulimia Nervosa (BN)

Like people with AN, people with BN use food as a psychological coping mechanism. Unlike those with AN however, people with BN turn *to* food—rather than *away from* it—during periods of stress and emotional conflict. Perhaps surprisingly, those with BN outnumber those with AN by about two to one.[15] The most common type of BN is characterized by repeated cycles of bingeing and purging.[16] Foods consumed during a binge often include sweets such as cookies and ice cream, and can easily add up to thousands of calories. In most cases of BN, the panic associated with bingeing leads to the drastic measure of induced vomiting. However, because some bulimic people compensate for their bingeing with excessive exercise or fasting instead of vomiting, the American Psychiatric Association recognizes *nonpurging* as a specific type of BN.[17] Common signs and symptoms, behaviors, and consequences associated with BN are summarized in Table 12.4.

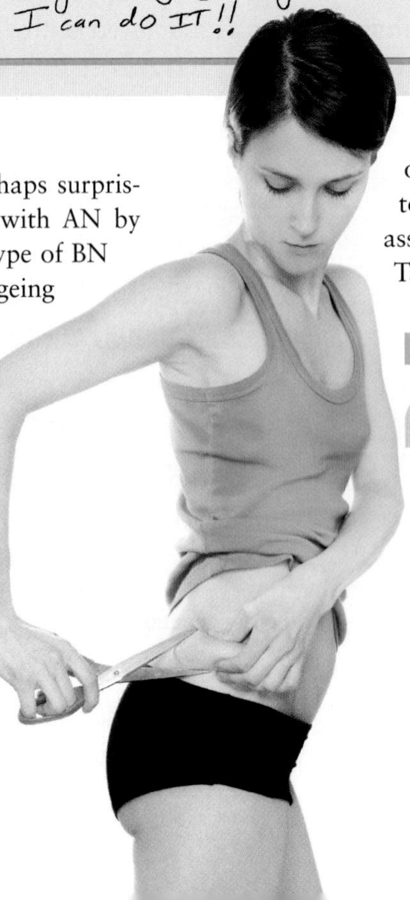

> "The term *bulimia* comes from the Greek word *bulimia*, which means "the hunger of an ox."

Table 12.4 **Physical Signs and Symptoms, Behavioral and Emotional Characteristics, and Health Consequences Associated with Bulimia Nervosa**

Physical Signs and Symptoms	Behavioral and Emotional Signs and Symptoms	Health Consequences
• Fluctuations in body weight	• Feelings of guilt or shame after eating	• Erosion of tooth enamel and tooth decay from exposure to stomach acid
• Swollen or puffy face	• Obsessive concerns about weight	• Electrolyte imbalances that can lead to irregular heart function and possible sudden cardiac arrest
• Odor of vomit on breath or in bathroom	• Repeated attempts at food restriction and dieting	• Inflammation of the salivary glands
• Sores around mouth	• Frequent use of bathroom during and after meals	• Irritation and inflammation of the esophagus that could lead to hemorrhaging and bleeding after vomiting
• Irregular bowel function	• Feelings of being out of control	• Dehydration
• Induces vomiting after eating	• Moodiness and depression	• Weight gain
	• Laxative abuse	• Abdominal pain, bloating
	• Fear of not being able to stop eating voluntarily	• Sore throat, hoarseness
	• Disappearance of food in household	• Broken blood vessels in the eyes
	• Eating to the point of physical discomfort	

RITUALS ASSOCIATED WITH BULIMIA NERVOSA

Because bingeing and purging are often carried out secretly, family and friends may be unaware that someone has BN. Although bulimics may consume several thousand calories during a binge, purging often prevents weight gain. Still, even purging immediately after bingeing cannot completely prevent nutrient absorption, and some weight gain is likely. In fact, people with BN tend to be within or slightly above their recommended weight ranges.[18] Some bulimics alternate between periods of bingeing/purging and periods of food restriction.

Individuals with BN often experience feelings of shame and regret after episodes of bingeing and purging. This can lead to depression, which increases the likelihood of future binge/purge cycles (see Figure 12.4 on the next page). These destructive cycles can be difficult to resist because they often bring short-term relief from complex emotional issues. Many people struggle for years before admitting that they have a problem.

Whereas individuals with AN often feel satisfaction from dieting and weight loss, many individuals with BN feel discontent and are more likely to have low self-esteem. Those with BN also tend to be impulsive, and are prone to other unhealthy behaviors such as substance abuse, self-mutilation, and suicidal tendencies.[19] For these reasons, people with BN are more likely to seek treatment than are those with AN. And sometimes, treatment seeks them: people with BN often only seek help after getting caught in the act of bingeing and purging.

HEALTH CONCERNS ASSOCIATED WITH BULIMIA NERVOSA

Over time, repeated cycles of bingeing and purging damage the body. If a person with BN purges by vomiting frequently, damage is caused when the delicate lining of the esophagus and mouth becomes irritated by repeated exposure to stomach acid. Because the acidity of gastric juice can also damage dental enamel and cause tooth decay, dentists and dental hygienists are often the first to notice the signs of purging. Furthermore, some people with BN induce vomiting by inserting their fingers deep into their mouths, which can cause their hands to become scraped from striking the teeth. Frequent vomiting and/or overuse of laxatives and diuretics can cause dehydration and electrolyte

binge-eating disorder (BED) A disordered eating pattern classified as EDNOS that is characterized by repeated cycles of bingeing without purging.

imbalance. This can lead to irregular heart function, and can even result in sudden cardiac arrest. Fortunately, many people with BN seek help and medical assistance, and most improve with treatment.

Eating Disorders Not Otherwise Specified (EDNOS)

While AN and BN are certainly the most familiar eating disorders, habitual unhealthy eating behaviors that do not meet the precise diagnostic criteria associated with either AN or BN can also be problematic. EDNOS is a general category for eating disorders that are not as well defined or thoroughly researched as AN and BN. Individuals with EDNOS exhibit a spectrum of behaviors and traits associated with AN or BN but do not meet all of the diagnostic criteria. For example, a person who exhibits many of the behaviors associated with AN but has normal weight and/or still menstruates may be classified as having an EDNOS.

BINGE-EATING DISORDER

Only recently has the American Psychiatric Association recognized **binge-eating disorder (BED)** as a disordered eating pattern distinct from bulimia nervosa. There is some uncertainty as to how BED, characterized by repeated cycles of bingeing without purging, should be classified. Currently, BED falls under the broad category of EDNOS. Although many clinical practitioners and researchers believe that BED should be recognized officially as a distinct diagnostic category, others feel that the designation is premature and that the disorder needs further study.[20]

The American Psychiatric Association's provisional diagnostic criteria for BED requires that a person eats excessively large amounts of food in a two-hour period at least twice per week for at least six months; feels a lack of control over the episodes; and experiences feelings of disgust, depression, and/or guilt in response to overeating.[21] Other proposed diagnostic criteria include eating much more rapidly than normal, eating until uncomfortably full, eating large amounts of food even when not physically hungry, and eating alone out of embarrassment at the quantity of food being eaten.[22] As with BN, BED

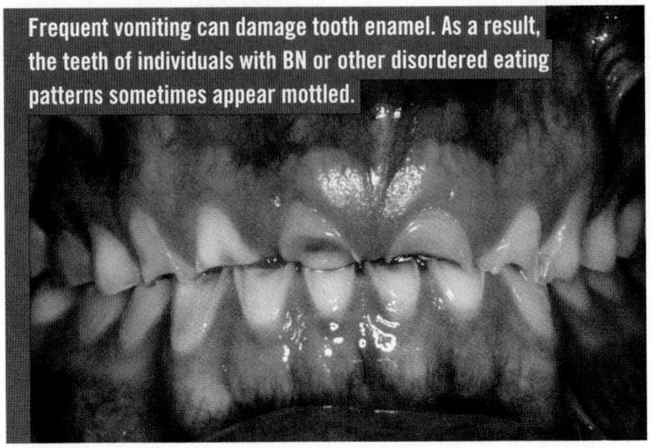

Frequent vomiting can damage tooth enamel. As a result, the teeth of individuals with BN or other disordered eating patterns sometimes appear mottled.

© Edward H. Gill/Custom Medical Stock Photo

Figure 12.4 **Thoughts and Behaviors Associated with Bulimia Nervosa**

-5/9-
I feel so fat and ugly. Yesterday I saw this guy that I like walking with another girl—it made me feel sad. When I got to class, the teacher passed back our exam. I didn't get a very good grade. I WISH I WAS PRETTY AND SMART!!!
Yesterday I felt so bad that I binged and purged most of the afternoon.
I hate when I do that but I can't help myself. I didn't intend for it to happen. I bought all this food thinking that it would last. It lasted for about an hour. Before I knew it the cookies and ice cream were gone. Then I went to the dining hall and ate dinner. !! As soon as I got back to my room I made myself vomit again. I was glad that my roommate wasn't there. She would think I am SO GROSS. That is because I AM GROSS!!! I wish I could stop this. Tomorrow I am going to diet. I am going to get thin. Yeah, right......

© Cengage Learning 2013

binges typically take place in private and are often accompanied by feelings of shame.

Although most people overeat from time to time, the bingeing associated with BED is distinct in that it serves as a habitual coping mechanism. For many people with BED, binges provide an escape from stress and emotional pain by inducing a psychological numbness and a temporary state of emotional well-being. Anger, sadness, anxiety, and other types of emotional distress often trigger binges. Many people with BED struggle with clinical depression, although it is not clear if depression triggers BED or if BED triggers depression.[23] Some studies indicate that people with BED are likely to have been raised in families affected by alcohol abuse.[24] In the case of BED, however, food rather than alcohol becomes the drug of choice.

The number of people with BED is much greater than the combined numbers of people with AN or BN.[25] In total, approximately 1 to 5 percent of the American population has BED.[26] Although people with BED are sometimes of normal weight, most are overweight. In fact, some weight-loss treatment programs estimate that between 20 and 40 percent of obese patients experience BED.[27] Because they are likely to be overweight, people with BED tend to be at increased risk of weight-related health problems such as type 2 diabetes, gallstones, and cardiovascular disease.

An individual who suppresses his desire for food and avoids eating for long periods of time between binges is called a **restrained eater.**[28] While restrained eaters do not meet the specified diagnostic criteria for BN-nonpurging type, there are many similarities. Restrained eaters limit their food intakes and fast to lose weight. After extended periods of food restriction, however, restrained eaters find themselves feeling out of control and respond by bingeing. This cycle of fasting and bingeing can be difficult to stop. Many restrained eaters perceive themselves as overweight, and the consumption of large amounts of food generates further feelings of inadequacy and self-contempt. These feelings can cause restrained eaters to turn back to food for emotional comfort. Like others with disordered eating patterns, restrained eaters find themselves in a vicious cycle that results in poor physical and psychological health.

© Simone van den Berg/Shutterstock.com

restrained eater An individual who suppresses his or her desire for food and avoids eating for long periods of time between binges.

nocturnal sleep-related eating disorder (SRED) A disordered eating pattern characterized by eating while asleep without any recollection of having done so.

LO2 Are There Other Disordered Eating Behaviors?

Many troublesome disordered eating behaviors exist in addition to the eating disorders formally recognized by the American Psychiatric Association. Not eating in public, situational purging, chewing food but spitting it out before swallowing, and obsessive dieting are some examples of such behaviors. Although insufficient information exists for these disordered eating patterns to be classified as distinct eating disorders, such behaviors can be as debilitating and disruptive as AN, BN, and EDNOS. Also, if disordered eating behaviors persist, they can easily progress into a full-blown eating disorder. As for any newly recognized pattern of dysfunctional behavior, further research is needed to cultivate a better understanding of these conditions. Today, some of the most recognized and best understood food-related disturbances are excessive nighttime eating, avoidance of new foods, and obsession with muscularity. Each of these patterns is outlined in the following sections.

Nocturnal Sleep-Related Eating Disorder and Night Eating Syndrome

Nocturnal sleep-related eating disorder (SRED) is a disordered eating pattern characterized by eating while asleep without any recollection of having done so.[29] Reportedly, individuals with SRED leave their beds and walk to their kitchens to prepare and eat food. Individuals may begin to suspect SRED when they notice unexplained missing food and see evidence of late-night kitchen activities. Meals consumed

Night eating syndrome is characterized by an ongoing, persistent pattern of late night binge eating.

© iStockphoto.com/Ryan Lane

during episodes of SRED are typically high in fat and calories and may involve unusual food combinations. Perhaps unsurprisingly, it is common for people with SRED to experience weight gain. Although both men and women develop SRED, the disorder is far more common in women.[30] Some researchers believe that stress, dieting, and depression can trigger episodes of SRED and that food restriction during the day may make some individuals more vulnerable to unconscious binge eating at night.[31] Although rates of occurrence are difficult to estimate, 1 to 3 percent of the American population may experience SRED, and rates as high as 10 to 15 percent have been reported among individuals with eating disorders.[32]

night eating syndrome (NES) A disordered eating pattern characterized by a cycle of daytime food restriction, excessive food intake in the evening, and nighttime insomnia.

food neophobia A disordered eating pattern characterized by an irrational fear or avoidance of new foods.

Night eating syndrome (NES) is a disordered eating pattern closely related to SRED, except individuals are fully aware of their eating. NES is characterized by a cycle of daytime food restriction, excessive food intake in the evening, and nighttime insomnia. Even though most people report that night eating causes them to feel depressed, anxious, and guilty, people with this type of disordered eating typically consume more than half of their daily calories after dinner. Night eating syndrome affects between 1 and 2 percent of American adults and it is more common among women than men.[33] Nine to 15 percent of individuals participating in weight-loss treatment programs report behaviors associated with NES. Signs and symptoms associated with NES include:

- Not feeling hungry for the first several hours after waking
- Overeating in the evening and consuming more than half of daily calories after dinner
- Difficulty falling asleep, accompanied by the urge to eat
- Waking during the night and finding it necessary to eat before falling back to sleep
- Feelings of guilt, shame, moodiness, tension, and agitation—especially at night

Food Neophobia

It is not unusual for parents of young children to complain about picky eating. Many children are reluctant to try new foods, and for some, even slight changes in food routines can be upsetting. In time, most children outgrow these behaviors and begin to accept new foods and eating patterns. Some do not, however. Children who refuse to try new foods as they grow older sometimes develop irrational food-related fears, which can lead to the development of food neophobia later in life.

Food neophobia is a disordered eating pattern characterized by an irrational fear or avoidance of trying new foods. Individuals with food neophobia typically eat very limited ranges of foods and often adhere to well-defined and unusual food rituals and practices.[34] For example, an individual with food neophobia may refuse to eat foods made from two or more food items. He may enjoy certain foods when eaten separately, but may find them disgusting when joined together. In adults, these types of behaviors can be socially restricting and embarrassing. In extreme cases, they can lead to nutritional inadequacies, although some individuals with food neophobia take supplements to compensate for their poor eating habits. People with food neophobia tend to be slightly overweight because they often limit themselves to comfort foods that are high in calories, such as hamburgers, French fries, and macaroni and cheese.[35] Perhaps surprisingly, most adults with food neophobia are not interested in therapy or other treatments. Some researchers believe that food

> As with other **eating disorders** and disordered eating patterns, the **first step in treating** muscle dysmorphia is **admitting that the problem exists**.

neophobia is a type of obsessive-compulsive disorder that requires extensive psychological therapy.[36] Regardless of its classification, researchers are interested in learning more about food neophobia and how the people who are affected by it react to new foods.[37]

Muscle Dysmorphia

Although societal pressures to be thin are directed more intensely toward females than toward males, it is not uncommon for the latter to experience similar body image issues. Instead of being pressured to be thin, boys and men are more often pressured to be strong and muscular. This societal pressure may be reinforced by the media, which often portrays the ideal male body as having well-defined muscles. Once referred to as *reverse anorexia*, **muscle dysmorphia** is a disorder characterized by a preoccupation with increasing muscularity. Muscle dysmorphia is observed primarily in men who have intense fears of being too small, too weak, and/or too skinny.[38] The term *dysmorphia* comes from the Greek words *dys*, which means "bad" or "abnormal," and *morphos*, which means "shape" or "form." Thus, individuals with muscle dysmorphia are preoccupied with perceived defects in their bodies.

Muscle dysmorphia, often described as the opposite of AN, is characterized by a compulsive preoccupation with increasing muscularity.

muscle dysmorphia A disorder characterized by a preoccupation with increasing muscularity.

Behaviors associated with muscle dysmorphia include working out for hours each day, allowing exercise to interfere with family and social life, paying excessive attention to diet, and maintaining unusual food rituals and eating practices. People with muscle dysmorphia may also engage in health-threatening practices such as the use of anabolic steroids to gain muscle mass.[39]

Individuals affected by muscle dysmorphia often have personality traits similar to those of people with recognized eating disorders.[40] In fact, one study reported that one-third of men diagnosed with muscle dysmorphia also had a history of eating disorders.[41] Men with muscle dysmorphia tend to have low self-esteem and typically harbor concerns about masculinity.[42] To compensate for these insecurities, they are driven to achieve overtly strong-looking bodies. Affected by obsessive thoughts of being weak, insecure, and vulnerable, men with muscle dysmorphia can have difficulties maintaining personal and social relationships.

LO3 What Causes Eating Disorders?

Scientists have many theories as to the factors that contribute to eating disorders, but there are no definitive answers. Undoubtedly, a person might develop an eating disorder for a variety of reasons—why some are more vulnerable than others is not clear. Several factors are likely contributors, however. These include sociocultural characteristics, family dynamics, personality traits, and biological (genetic) factors.[43]

Sociocultural Factors

Eating disorders are more prevalent in some cultures than in others; when food is abundant and slimness is valued, eating disorders are most likely to develop. This is one reason why eating disorders are more common in industrialized Western countries like the United States than in regions like Africa, China, and the Middle East.[44] Even within American society, differences in sociocultural status influence the development of eating disorders. However, as multicultural youth acculturate to mainstream American values, distinct cultural norms that once protected them from eating disorders may be eroding. For some, a desire to fit prevailing American beauty and social norms may lead to the development of an eating disorder. The impact of these norms has been felt across socioeconomic, cultural, and racial spectrums. For example, data show that the prevalence of eating disorders among Latina and African-American women is catching up to that of Caucasian women in the United States.[45]

THE ROLE OF THE MEDIA

Based on popular media images, the perfect body is tall, lean, and has well-defined muscles. Whereas the average woman is 1.6 m (5 ft 4 in) tall and weighs 63 kg (140 lb), the average model is 1.8 m (5 ft 11 in) tall and weighs 53 kg (117 lb).[46] This discrepancy has become more pronounced as female fashion models, beauty pageant contestants, and actresses have become increasingly thinner over the years. When the Miss America beauty pageant first began in the 1920s, its contestants' body mass indexes (BMIs) averaged 20 to 25 kg/m^2. Today, nearly all the participants have BMIs below those considered healthy, and nearly half have BMIs consistent with one of the diagnostic criteria for AN (a BMI less than 18.5 kg/m^2).[47] In fact, a recent study reported that women who participated in beauty pageants as children are more likely to experience body dissatisfaction, interpersonal distrust, and impulsiveness than those who did not.[48]

Major media outlets that glamorize unrealistically thin and overly muscular bodies have long been criticized for evoking a sense of inadequacy in impressionable children and young adults. For example, exceptionally slender celebrities who appear on television and in the movies may skew a teenager's standard of thinness, encouraging her to engage in unhealthy eating practices to achieve a thin and supposedly glamorous appearance.[49] Even the body shapes of popular dolls and action figures given to children during their formative years are unrealistic. In short, media likely play an important role influencing how people view themselves, and may even predispose susceptible individuals to eating disorders.

Because dissatisfaction with one's weight and shape is thought to be an essential precursor to the development of an eating disorder, it is important for children to understand that healthy bodies come in many shapes and sizes—despite what the media portrays. It is equally important for older children to be prepared for the physical and emotional changes associated with puberty. The changes in body dimensions and weight gain that occur during adolescence can make girls feel embarrassed about and uncomfortable with their maturing bodies. Because such feelings are not always discussed openly among families and friends, girls may feel as if they are somehow abnormal. In fact, discomfort with the rapid maturation associated with puberty is extremely common.

PRNewsFoto/MGA Entertainment, Inc. / AP Images/Mark Lennihan

SOCIAL NETWORKS

In addition to the media, a person's circle of peers may also contribute to the development of an eating disorder. Attitudes and behaviors about slimness and appearance are often learned from those with whom we associate. A desire to gain

Many **popular dolls marketed to children** *have body proportions that are* **unrealistically thin** *and/or* **muscular.**

acceptance among one's friends can spark body dissatisfaction and dieting in susceptible adolescents. Unfortunately, many teenagers believe that to be liked and to belong they must be thin.

Although eating disorders frequently begin with a desire to lose weight, relatively few people on weight-loss diets actually develop eating disorders. Likewise, not everyone who lives in an affluent society that stigmatizes obesity and advocates extreme slenderness develops an eating disorder. While these factors alone do not cause eating disorders, the sociocultural environments of industrialized countries appear to foster the development of eating disorders in individuals who are already vulnerable for other reasons. One such other reason is an unhealthy family dynamic.

The **BMIs of** beauty contestants have **decreased** over the **last 50 years.**

enmeshment A style of family interaction whereby family members are overly involved with one another and have little autonomy.

chaotic family (or **disengaged family**) A style of family interaction characterized by a lack of cohesiveness and little parental involvement.

Family Dynamics

Aside from sociocultural factors, another important issue related to one's risk of developing an eating disorder is having an unhealthy family dynamic. Because parents influence nearly every aspect of a child's life, it should come as no surprise that family dynamics play a decisive role in the development and perpetuation of an eating disorder. Although no one single family type inescapably causes a child to develop an eating disorder, researchers have found that certain distinguishing behaviors and characteristics increase the likelihood of this happening.[50] Such behaviors and characteristics include overprotectiveness, rigidity, conflict avoidance, abusiveness, chaotic family dynamics, and the presence of a mother with an eating disorder.[51]

ENMESHED FAMILIES

Enmeshment is a style of family interaction whereby family members are overly involved with one another and have little autonomy.[52] An enmeshed family has no clear boundaries among its members. This environment can make it difficult for children to develop independence and individualism. Furthermore, children raised in such families often feel tremendous pressure to please their parents and meet expectations. Rather than doing things for themselves, children with overly involved, protective parents strive to please others. Enmeshed family dynamics promote dependency, which may lay the foundation for the emergence of an eating disorder. Under such circumstances, food often becomes the only component in a child's life over which control can be exerted.

Children who participate in beauty pageants may be more likely to dislike their bodies later in life.

CHAOTIC FAMILIES

In contrast to enmeshed families that have exceedingly tight familial structures, chaotic families have remarkably loose structures. A **chaotic family** (or **disengaged family**) employs a style of family interaction characterized by a lack of cohesiveness and little parental involvement.[53] The roles of family members are loosely defined; children often feel a sense of abandonment; and parents may be depressed, alcoholic, or emotionally absent. A child growing up in this type of home may later develop an eating disorder as a way to fill an emotional emptiness, gain attention, or suppress emotional conflict.

food preoccupation Spending an inordinate amount of time thinking about food.

MOTHERS WITH EATING DISORDERS

The presence of a mother with an eating disorder or body dissatisfaction can negatively influence eating behaviors in children.[54] An inability to demonstrate a healthy relationship with food and model healthy eating for one's children is a serious concern. A mother with an eating disorder is more likely to criticize a daughter's appearance and to encourage her to lose weight, even if doing so would be unhealthy. As a result, children of women with eating disorders are at increased risk of developing eating disorders themselves.[55]

Personality Traits and Emotional Factors

Scientists have long believed that certain personality traits make some people more susceptible to eating disorders than others. Some concerning characteristics include low self-esteem; a lack of self-confidence; and feelings of helplessness, anxiety, and depression.[56] An individual with an eating disorder typically has **food preoccupation**, meaning that an inordinate amount of time is spent thinking about food. Another personality trait commonly associated with eating disorders is *perfectionism*.[57] People who exhibit perfectionism often have difficulty dealing with shortcomings in themselves; an imperfect body is not easily tolerated.

Biological and Genetic Factors

Because certain personality traits and eating behaviors are determined in part by the nervous and endocrine systems, it is logical that brain chemicals and other biological and genetic factors might play a role in the development of an eating disorder. It is also possible, however, that disordered eating may actually disrupt normal neuroendocrine function. For example, studies show that individuals with eating disorders are often clinically depressed.[58] It is difficult to determine whether clinical depression leads to eating disorders or vice versa (or if another factor is at play). In any case, because medication used to treat clinical depression is often effective in the treatment of certain eating disorders, depression is likely a contributing factor.

Scientists are working to identify genes that might influence susceptibility to eating disorders.[59] Studies of identical and fraternal twins have provided evidence suggesting that susceptibility to eating disorders may, in part, be inherited.[60] Although such studies cannot completely differentiate the contribution of genetics from that of environment, some research suggests that the contribution of genetics may actually be greater than that of the environment. How this occurs remains unclear, however.[61]

LO4 Are Athletes at Increased Risk for Eating Disorders?

There are more competitive female athletes today than ever before. Perhaps not coincidentally, the number of female athletes with eating disorders has increased as well. Some studies indicate that the prevalence of eating disorders among female student-athletes and non-athletes does not differ, while other studies indicate otherwise.[62] At any rate, losing weight in an unhealthy way can have serious health consequences above and beyond affecting performance for an athlete. It is of utmost importance for coaches and trainers to recognize and act on the early warning signs and symptoms associated with eating disorders.

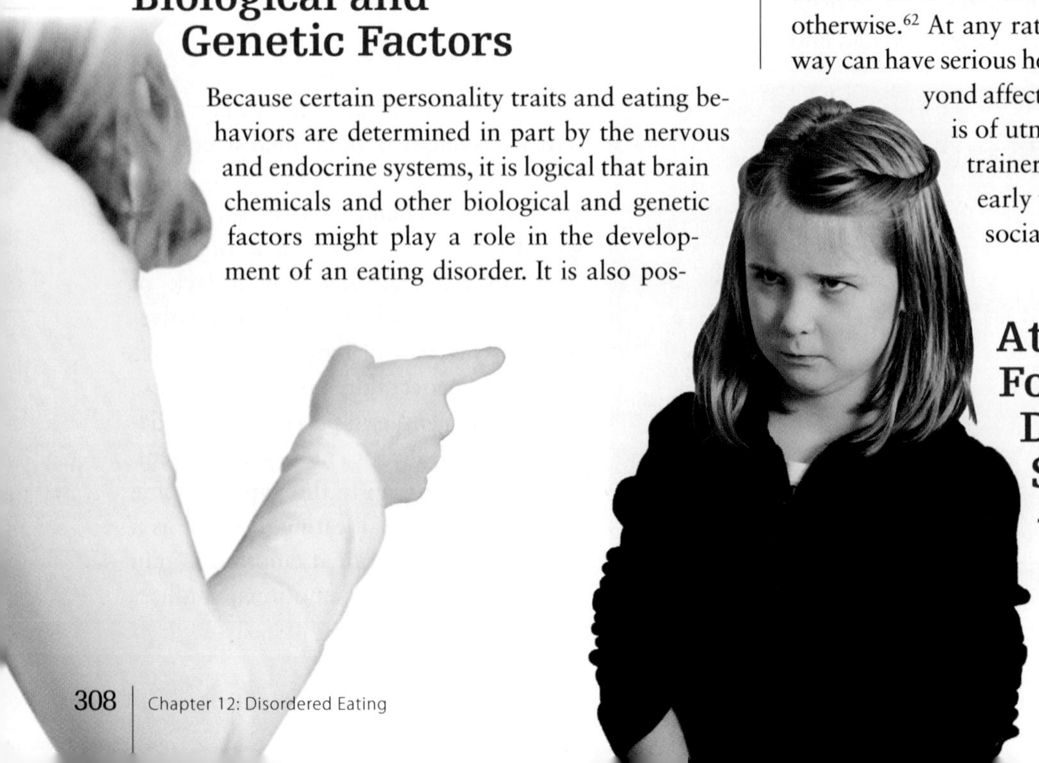

© JacksColdSweat/Shutterstock.com

Athletics May Foster Eating Disorders in Some People

The prevalence of disordered eating and eating disorders among collegiate athletes is estimated to be somewhere

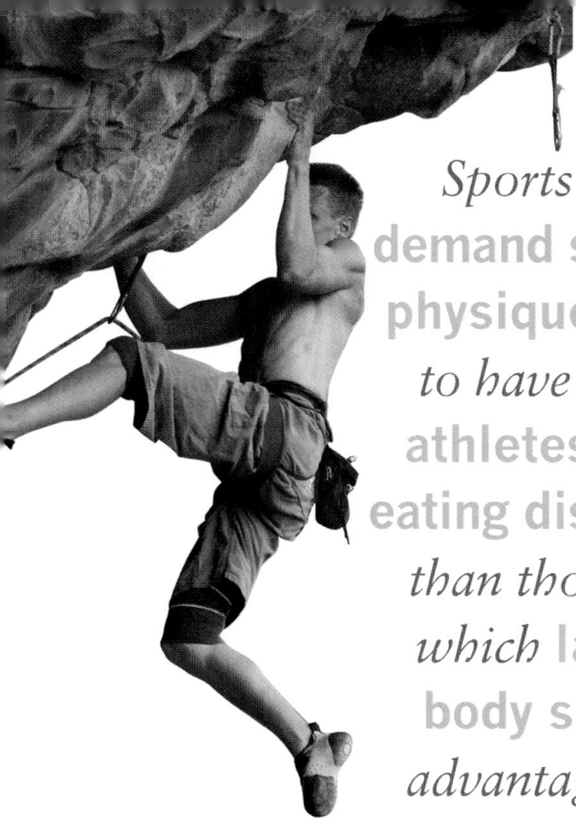

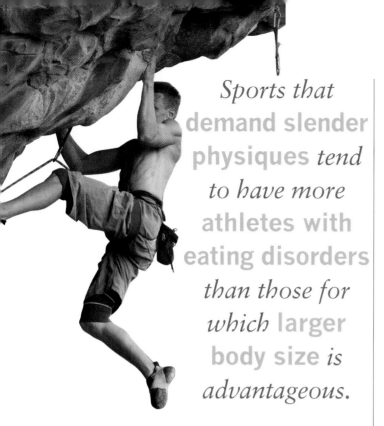

Sports that **demand slender physiques** *tend to have more* **athletes with eating disorders** *than those for which* **larger body size** *is advantageous.*

between 15 and 60 percent.[63] Disagreement and inconsistent estimates may be attributable to the reluctance of some athletes to admit that they experience such problems. In addition, some athletes who exhibit disordered eating behaviors may not satisfy all the criteria needed for diagnosis. Regardless of the exact numbers, athletes—especially female athletes—are considered by many experts to be an at-risk group for developing eating disorders.

For some sports, physical performance is determined not only by motor abilities, strength, and coordination, but also by body weight. Athletes such as ski jumpers, cyclists, rock climbers, and long-distance runners may deliberately try to achieve low body weight to gain a competitive advantage. Athletes in sports that entail strict weight classes (such as boxing and wrestling) often try to fit the lowest possible weight class to avoid being outsized. Sports that demand a thin physical appearance (such as gymnastics) are likely to have more athletes with eating disorders than are sports for which greater size may be beneficial.[64] This may be because judges in such sports often consider size and appearance when rating performance. Athletes who participate in dancing, figure skating, synchronized swimming, gymnastics, and diving have long perceived that body size affects how judges rate their performance.

Because athletes tend to be competitive people and may equate their self-worth with athletic success, they may be especially willing to engage in risky weight-loss practices to achieve that success. Coaches and trainers who believe that excess weight can hinder performance may ignore unhealthy behaviors, and may even encourage them. According to the National Collegiate Athletic Association (NCAA), the sports with the highest numbers of female athletes with eating disorders are cross-country running, gymnastics, swimming, and track and field.[65] Sports with the highest numbers of male athletes with eating disorders are wrestling and cross-country running.

female athlete triad A combination of three interrelated conditions: disordered eating (or eating disorder), menstrual dysfunction, and osteopenia.

The Female Athlete Triad

Because the rigor of athletic training alone is very stressful on the body, athletes with eating disorders are at extremely high risk for developing medical complications. Adequate nourishment is required to meet the physical demands of athletics. This is why serious health problems can arise when an athlete is restricting food intake, bingeing, and/or purging. For example, female athletes are at increased risk for developing a syndrome known as the **female athlete triad**. The female athlete triad is a combination of three interrelated conditions: disordered eating (or eating disorder), menstrual dysfunction, and osteopenia (see Figure 12.5 on the next page).[66]

The risk of stress fractures increases in women when menstrual cycles become irregular or cease.

The interrelation of the three components of the female athlete triad is complex. Disordered eating can lead to very low levels of body fat, which can cause estrogen levels to decrease. Without adequate estrogen levels, menstrual cycles can become irregular, and in some cases, stop completely. A lack of estrogen can cause bone loss, which over time weakens bones, resulting in osteopenia. In severe cases, the entire matrix of the bone can begin to deteriorate, leading to osteoporosis. As illustrated in Figure 12.6 on the next page, a collegiate female runner's risk of stress fractures increases as her menstrual cycle becomes more irregular.[67]

Determining the prevalence of the female athlete triad can be difficult because some women may feel unburdened by the cessation of menstruation and therefore may not be willing to report it. However, these athletes often begin to experience frequent injuries, such as stress fractures, which can draw attention to the fact that they may have a problem. It is important for parents, coaches, and health care providers to be aware of the spectrum of disordered eating patterns and eating disorders so that assistance can be provided to athletes who need it.

Figure 12.5 Female Athlete Triad

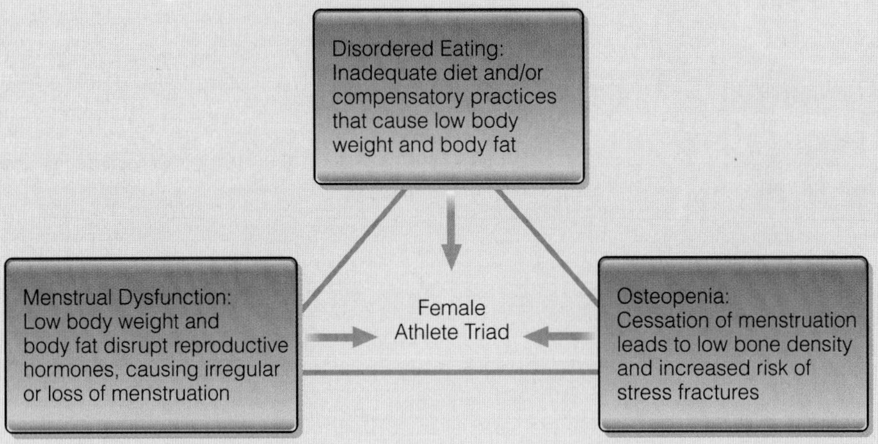

The female athlete triad is a combination of three interrelated conditions: disordered eating (or an eating disorder), menstrual dysfunction, and osteopenia.

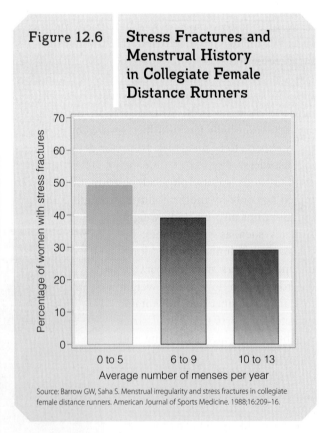

Figure 12.6 Stress Fractures and Menstrual History in Collegiate Female Distance Runners

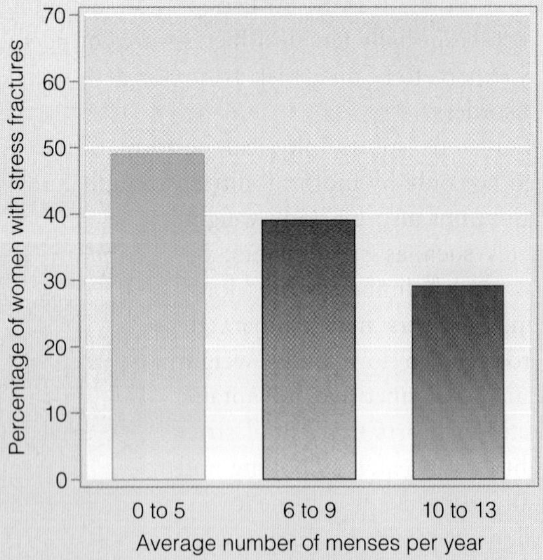

Source: Barrow GW, Saha S. Menstrual irregularity and stress fractures in collegiate female distance runners. American Journal of Sports Medicine. 1988;16:209–16.

LO5 How Can Eating Disorders Be Prevented and Treated?

People with eating disorders are at increased risk for serious medical and/or emotional problems. Thus, they need treatment from qualified health professionals. Typically, a treatment team includes mental health specialists who can help address and treat underlying psychological issues, medical doctors who can treat physiological complications, and a dietitian who can recommend healthy food choices. Important treatment goals for individuals with eating disorders are to learn how to enjoy food without fear and guilt and to respond to the physiological cues of hunger and satiety to regulate food intake. For these goals to become possible, specialists help people with eating disorders to recognize and appreciate their own self-worth. While many treatment options are available for people who have eating disorders, the first step is to recognize that there is a problem and to seek help.

Prevention Programs

Educational programs designed to increase awareness and prevent eating disorders in young girls produce varied results. Although it is important to reach children before eating disorders develop, too many prevention programs focus on deterring dangerous eating-disorder behaviors instead of encouraging healthy attitudes toward food, dieting, and body image. To prevent eating disorders from developing, educational strategies must focus on issues related to overall health and self-esteem. Some suggestions for promoting healthy body images among children and adolescents are listed in Table 12.5.

Table 12.5 Promoting a Healthy Body Image among Children and Adolescents

• Encourage children to focus on positive body features instead of negative ones.
• Help children understand that everyone has a unique body size and shape.
• Be a good role model for children by demonstrating healthy eating behaviors.
• Resist making negative comments about your own weight or body shape.
• Focus on positive, nonphysical traits such as generosity, kindness, and a friendly laugh.
• Do not criticize a child's appearance.
• Never associate self-worth with physical attributes.
• Prepare a child for puberty in advance by discussing physical and emotional changes.
• Enjoy meals together as a family.
• Discuss how the media can negatively affect body image.
• Avoid using food as a way to reward or punish.

Adapted from Story M, Holt K, Sofka D. Bright futures in practice: Nutrition, 2nd ed. Arlington, VA: National Center for Education in Maternal and Child Health, 2002.

Treatment Strategies

A person with an eating disorder may not recognize or cannot admit to having a problem. Concerns expressed by friends and family members often go ignored or dismissed, making loved ones feel confused and frustrated by their inability to help—a completely normal reaction to a very difficult situation. It is important to remember that even though a person with an eating disorder may resist help, family and friends continue to play important supportive roles. Experts recommend that they not focus on the eating disorder *per se*, as this may make the person feel more defensive. Instead, expressing concerns regarding the person's unhappiness and encouraging him to seek help may be of greatest value.

It is important that the medical team chosen to treat a person with an eating disorder has specific training and expertise in the field of eating disorders. Because eating disorders vary greatly, it is critical that appropriate methodologies and treatment goals are implemented. Treatment goals for people with AN include achieving a healthy body weight, resolving psychological issues (such as low self-esteem and distorted body image), and establishing healthy eating patterns. Because some people with AN are severely malnourished, intensive care provided by inpatient facilities may be necessary.[68] Inpatient facilities are staffed by physicians, nurses, social workers, mental health therapists, dietitians, and other professionals to provide care in all aspects of recovery, including medical support, nutritional programming, and psychological counseling. Like those with AN, people with BN often have nutritional, medical, and psychological issues that must be addressed during recovery. In addition to the treatment goals associated with AN, treatment goals for people with BN include the reduction and eventual elimination of bingeing and purging. Establishing a healthy relationship with food, developing strategies to help resist urges to binge and purge, and maintaining healthy body weight without bingeing and purging are all important.

© Custom Medical Stock Photo

The sooner a person with an eating disorder gets help, the better the chance of full recovery is. Regardless of when that happens, recovery can be a long, trying process. Well-meaning advice such as "just eat" only makes matters worse. It is important for families and friends to know that most people with eating disorders eventually do recover. Treatment often involves counseling for the entire family, which helps everyone heal and move forward in life.

NUTR

13 Alcohol, Health, and Disease

© iStockphoto.com/pidjoe

LEARNING OUTCOMES:

LO1 Understand the properties of alcohol, how it is produced, and how it is absorbed in the body.

LO2 Describe the process of alcohol metabolism.

LO3 Discuss the potential benefits of alcohol consumption.

LO4 Explain the health risks associated with alcohol.

LO5 Understand the warning signs and consequences of alcohol abuse.

LO1 What Is Alcohol and How Is It Produced?

The raising of a celebratory toast is an ancient tradition. Despite their rich history of pageantry and ritual, alcoholic beverages have also brought misery and suffering. Although many people who drink do so without harming themselves or others, longtime abusers know all too well that alcohol can lead to psychological and physical dependency. Today, millions of people around the world are seeking help in their efforts to abstain from alcohol use. There is no easy explanation as to why some people can control their drinking while others cannot. Although genetics may predispose some people to alcoholism, cultural factors play major roles as well.

Scientists have long debated whether alcohol is beneficial or detrimental to health. When it is consumed responsibly, alcohol poses little physical, social, or psychological threat. Studies show that moderate alcohol consumption can even reduce the risk of heart disease in middle-aged and older adults, and it may provide some protection against type 2 diabetes and gallstones.[1] When it comes to the benefits of alcohol, however, more is clearly not better.

If you have ever been around someone who has had too much to drink, you know that alcohol is a drug with mind-altering effects. Indeed, most people act and behave differently when under the influence of alcohol. In excess, alcohol can alter judgment; lead to dependency; and damage the liver, pancreas, heart, and brain. Heavy drinking increases the risk of both accidents and some types of cancer, and maternal alcohol consumption can seriously harm an unborn child. Estimated annual costs associated with alcohol abuse in the United States total more than $180 billion. For these and other reasons, the Dietary Guidelines for Americans states clearly that people who choose to drink alcoholic beverages should do so sensibly and in moderation.[2] In this chapter, you will learn how alcohol is produced, absorbed, and circulated. You will also learn about the many effects that alcohol has on health and disease.

Alcohol Is Produced through Fermentation

For all of its profound effects on the body, alcohol is a rather simple molecule. From a chemical perspective, **alcohol** comprises a broad class of organic compounds that have one or more hydroxyl (–OH) groups attached to carbon atoms. There are many types of alcohol, all of which are somewhat similar in terms of chemical structure. Alcohols are quite volatile, and tend to be soluble in water. There are important differences in the types of alcohols, however. Whereas **ethanol** (the type of alcohol found in wine and beer) can be consumed, most forms of alcohol are not safe to drink. Methanol, which is used to make antifreeze, can be lethal if ingested. Although alcohol is not considered a nutrient, it does provide 7 kcal per gram. Table 13.1 on the next page lists the caloric contents of selected alcoholic beverages.

A process called **fermentation**, which was discovered thousands of years ago, produces alcohol. Fermentation occurs when single-cell microorganisms called *yeast* metabolize the sugars found in fruits and grains, ultimately producing ethanol and carbon dioxide. Once the alcohol content of a beverage reaches a certain point,

alcohol An organic compound that has one or more hydroxyl (–OH) groups attached to carbon atoms.

ethanol The type of alcohol found in alcoholic beverages.

fermentation The process by which alcohol is produced. Yeast metabolizes sugars to produce ethanol and carbon dioxide.

Table 13.1 Serving Sizes, Energy Contents, and Alcohol Contents of Selected Alcoholic Beverages

Beverage	Serving Size (ml [oz])	Energy (kcal/serving)[a]	Alcohol (g/serving)
Light beer	355 (12)	103	11
Regular beer	355 (12)	139	13
White wine	148 (5)	98	14
Red wine	148 (5)	98	14
Distilled beverages[b]			
80 proof	44 (1.5)	97	14
90 proof	44 (1.5)	110	16
100 proof	44 (1.5)	124	18
Crème de menthe	44 (1.5)	186	15
Daiquiri	118 (4)	224	14
Whiskey sour	133 (4.5)	226	19
Piña colada	133 (4.5)	245	14

[a] Note that some alcoholic beverages contain energy-yielding nutrients other than alcohol. Therefore, caloric contents cannot always be calculated simply by multiplying grams of alcohol by 7 kcal/g.

[b] Distilled beverages include gin, rum, vodka, whiskey, and other liquors.

Source: U.S. Department of Agriculture, Agricultural Research Service. 2010. USDA National Nutrient Database for Standard Reference, Release 23. Nutrient Data Laboratory. Available from: http://www.ars.usda.gov/ba/bhnrc/ndl.

fermentation stops naturally. Alcohol content is often expressed as **alcohol by volume (ABV)**, the percentage of ethanol in a given volume of liquid. For example, if a beverage has an ABV of 14 percent, 100 ml of that beverage contains 14 ml ethanol. Depending on the strain of yeast used, fermentation usually stops naturally in a barley-based beer when ABV reaches 4 to 8 percent and in a grape-based wine when ABV reaches 11 to 14 percent.

A process called **distillation** can be used to increase the alcohol contents of some alcohol-containing beverages. During distillation, fermented beverages are heated, which causes the alcohol to become a vaporous gas. The alcohol vapors are collected and cooled until they become liquid again. This process of producing pure liquid alcohol concentrate is used to create distilled alcoholic beverages such as gin, vodka, and whiskey. The alcohol content of a distilled alcoholic beverage, also called *hard liquor*, can be determined from its **proof**. A beverage's proof is twice its alcohol content (ABV). For example, a distilled liquor labeled as 80 proof is 40 percent alcohol.

alcohol by volume (ABV) The percentage of ethanol in a given volume of liquid.

distillation The process by which alcohol vapors are condensed and collected to increase alcohol content.

proof A measure of the alcohol content of distilled liquor. Twice a beverage's alcohol content.

Alcohol Absorption

After it is consumed, alcohol does not require digestion. Rather, it is readily absorbed by simple diffusion into the blood. Although some alcohol is absorbed from the stomach, most (80 percent) is absorbed from the small intestine. The actual rate of alcohol absorption is influenced by several factors. For example, alcohol is absorbed more quickly when a person drinks on an empty stomach. The presence of food in the stomach not only dilutes alcohol, it also slows its release into the small intestine. This in turn slows absorption. Although the type of food a person consumes does not have

The **alcohol content** of a distilled alcoholic beverage can be determined from its **proof**, which is **twice the ABV**.

AP Images/Pat Wellenbach

a measurable effect on the rate of alcohol absorption, the alcohol content of a beverage does. In general, the rate of alcohol absorption increases as the alcohol concentration in the beverage increases. However, beverages with high alcohol concentrations (greater than 30 percent ABV) can irritate the stomach lining and increase mucus production. This can delay gastric emptying (the passage of alcohol from the stomach into the small intestine), in turn slowing absorption.[3]

Alcohol Circulation

Once absorbed, alcohol enters the bloodstream and circulates throughout the body. **Blood alcohol concentration (BAC)** is a unit of measure for the amount of alcohol in the blood. BAC is measured in grams per deciliter (g/dl). For example, a person with a BAC of 0.10 has one-tenth of a gram of alcohol per deciliter of blood. Within 20 minutes of consuming one standard drink (12 ounces of beer, 5 ounces of wine, or 1.5 ounces of 80-proof distilled liquor), BAC begins to rise, peaking within 30 to 45 minutes. Because the rate of alcohol metabolism is slower than the rate of alcohol absorption, alcohol accumulates in the blood as more alcoholic beverages are consumed. For this and many other reasons, people who drink should do so slowly and in moderation. In most states, a BAC of 0.08 g/dl is the legal limit for driving (see Figure 13.1).

blood alcohol concentration (BAC) A unit of measure for the amount of alcohol in the blood, measured in grams per deciliter (g/dl).

FACTORS AFFECTING BLOOD ALCOHOL CONCENTRATION

Alcohol readily disperses in water-filled environments both inside and outside of cells. Because tissues in the body vary greatly in their water contents, some take up alcohol more quickly than do others. For example, muscle tissue contains more water than does adipose tissue, and therefore more readily takes up alcohol. In other words, a person's body composition can influence her BAC. If two people of similar body weight ingest the same amount of alcohol, the leaner person will likely have a lower BAC. Body size can also influence BAC. A person with a larger body will likely have more blood and body fluids than will someone with a smaller body. The alcohol consumed by a larger

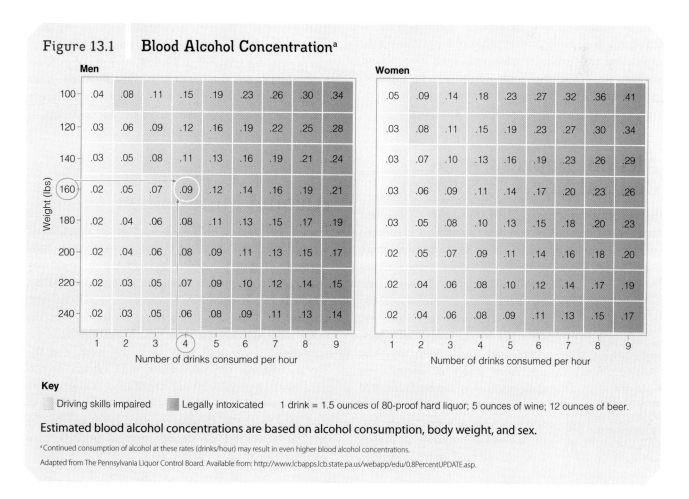

Figure 13.1 Blood Alcohol Concentration[a]

Men

Weight (lbs)	1	2	3	4	5	6	7	8	9
100	.04	.08	.11	.15	.19	.23	.26	.30	.34
120	.03	.06	.09	.12	.16	.19	.22	.25	.28
140	.03	.05	.08	.11	.13	.16	.19	.21	.24
160	.02	.05	.07	.09	.12	.14	.16	.19	.21
180	.02	.04	.06	.08	.11	.13	.15	.17	.19
200	.02	.04	.06	.08	.09	.11	.13	.15	.17
220	.02	.03	.05	.07	.09	.10	.12	.14	.15
240	.02	.03	.05	.06	.08	.09	.11	.13	.14

Number of drinks consumed per hour

Women

Weight (lbs)	1	2	3	4	5	6	7	8	9
100	.05	.09	.14	.18	.23	.27	.32	.36	.41
120	.03	.08	.11	.15	.19	.23	.27	.30	.34
140	.03	.07	.10	.13	.16	.19	.23	.26	.29
160	.03	.06	.09	.11	.14	.17	.20	.23	.26
180	.03	.05	.08	.10	.13	.15	.18	.20	.23
200	.02	.05	.07	.09	.11	.14	.16	.18	.20
220	.02	.04	.06	.08	.10	.12	.14	.17	.19
240	.02	.04	.06	.08	.09	.11	.13	.15	.17

Number of drinks consumed per hour

Key

Driving skills impaired · Legally intoxicated · 1 drink = 1.5 ounces of 80-proof hard liquor; 5 ounces of wine; 12 ounces of beer.

Estimated blood alcohol concentrations are based on alcohol consumption, body weight, and sex.

[a]Continued consumption of alcohol at these rates (drinks/hour) may result in even higher blood alcohol concentrations.

Adapted from The Pennsylvania Liquor Control Board. Available from: http://www.lcbapps.lcb.state.pa.us/webapp/edu/0.8PercentUPDATE.asp.

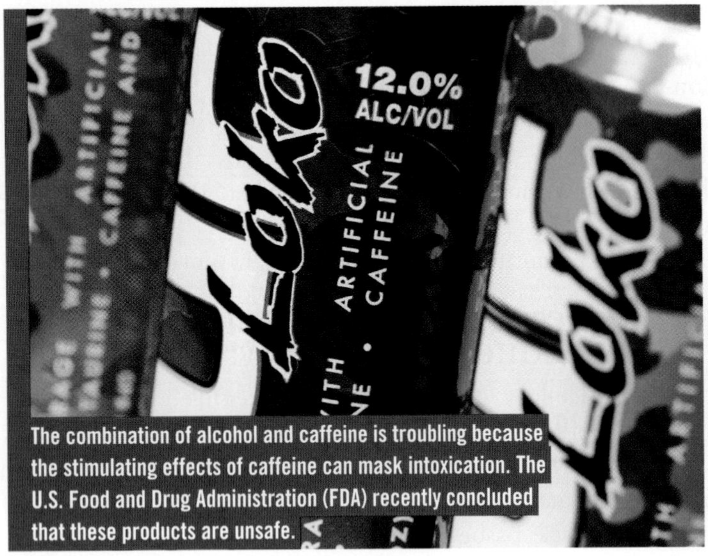

The combination of alcohol and caffeine is troubling because the stimulating effects of caffeine can mask intoxication. The U.S. Food and Drug Administration (FDA) recently concluded that these products are unsafe.

© TRIPPLAAR KRISTOFFER/SIPA/Newscom

person becomes more diluted in the blood, resulting in a lower BAC.

In recent years, caffeinated alcoholic beverages have become very popular. According to many experts, the combination of alcohol and caffeine can be dangerous because the stimulating effects of caffeine can mask intoxication. As a result, an individual may drink more alcohol than intended, which can increase BAC and lead to hazardous and life-threatening behaviors. In 2010, the U.S. Food and Drug Administration (FDA) concluded that products containing both alcohol and caffeine are unsafe and should not be sold. In response, many manufacturers altered their products' formulas or removed them from the marketplace altogether.

Alcohol Depresses the Central Nervous System

Because it sedates brain activities, alcohol is classified as a central nervous system depressant. This is surprising to many, as consuming small amounts of alcohol often makes people feel euphoric. Pleasant feelings arise when alcohol selectively depresses parts of the brain that normally censor social behaviors, aggression, and impulsivity. The inhibitory effect of alcohol on these portions of the brain is called **disinhibition**. In other words, alcohol can cause a temporary loss of inhibition, making a person feel relaxed and more outgoing. However, disinhibition also impairs judgment and reasoning. As BAC increases, areas of the brain that control speech, vision, and voluntary muscular movement also become depressed. If drinking continues, a person may lose consciousness, and his BAC may reach potentially lethal levels. Table 13.2 lists common responses to different levels of alcohol in the blood.

disinhibition A loss of inhibition.

alcohol dehydrogenase (ADH) An enzyme found primarily in the liver that converts ethanol to acetaldehyde.

acetaldehyde dehydrogenase (ALDH) An enzyme that converts acetaldehyde to acetic acid.

LO2 How Is Alcohol Metabolized?

After it is consumed, the body must metabolize alcohol into substances that can safely be eliminated from the body. Contrary to popular belief, there is nothing a person can eat or drink to accelerate this process. Small amounts of unmetabolized alcohol are eliminated from the body by the lungs (in expired air), skin (in sweat), and kidneys (in urine). However, the majority of alcohol is chemically broken down (metabolized) by the liver. There is a limit to how much alcohol the liver can metabolize in any given time, however. The average person can metabolize 14.8 ml (0.5 oz) of pure alcohol per hour. In other words, it takes about one hour for the body to rid itself of the alcohol in a 12-ounce serving of beer. Consequently, if a person continues to drink at a rate greater than her liver can metabolize, her BAC will increase with time.

Alcohol Metabolism

During periods of light to moderate drinking, most alcohol is metabolized through a two-step metabolic pathway. The first step in this metabolic pathway requires the enzyme **alcohol dehydrogenase (ADH)**, which is found primarily in the liver. ADH converts ethanol to acetaldehyde, a toxic molecule that the body must quickly break down. If acetaldehyde accumulates, some will pass from the liver into the blood, causing the unpleasant side effects associated with heavy drinking (headache, nausea, and vomiting)—otherwise known as a hangover. The second step in the metabolic process, catalyzed by the enzyme **acetaldehyde dehydrogenase (ALDH)**, converts acetaldehyde to acetic acid.

Table 13.2 Stages of Alcohol Intoxication

BAC (g/dl)	Stage of Intoxication	Effects
0.01–0.05	Subclinical	• Behavior that is nearly normal by ordinary observation
0.03–0.12	Euphoria	• Increased sociability, talkativeness, and increased self-confidence • Decreased inhibition, diminished attention, and altered judgment • Beginning of sensory-motor impairment, loss of efficiency in finer performance tests
0.09–0.25	Excitement	• Emotional instability, loss of critical judgment • Impaired perception, memory, and comprehension • Decreased sensory response, increased reaction time • Reduced visual acuity, peripheral vision, and recovery from flashes of bright light • Impaired sensory-motor coordination and balance • Drowsiness
0.18–0.30	Confusion	• Disorientation; mental confusion and dizziness • Exaggerated emotional states • Disturbances of vision and perception of color, form, motion, and dimension • Increased pain threshold • Decreased muscle coordination, staggering gait, slurred speech • Apathy, lethargy
0.25–0.40	Stupor	• Extreme lethargy, approaching loss of motor functions • Markedly decreased response to stimuli • Lack of muscular coordination, inability to stand or walk • Vomiting, incontinence • Impaired consciousness, sleep or stupor
0.35–0.50	Coma	• Complete unconsciousness • Depressed or abolished reflexes • Low body temperature • Impaired circulation and respiration • Possible death
0.45+	Death	Death from respiratory arrest

Adapted from Dubowski KM. Stages of acute alcoholic influence/intoxication. 2006. Available from: http://www.borkensteincourse.org/faculty%20documents/dub_stages.pdf.

GENETICS CAN INFLUENCE ALCOHOL METABOLISM

Studies have found that genetic differences (that is, variant forms) in the enzymes ADH and ALDH can affect a person's ability to metabolize alcohol.[4] For example, a high percentage of Asians have a less functional form of ALDH. When such individuals consume alcohol, acetaldehyde levels increase quickly, causing dilation of blood vessels, headaches, and facial flushing. Differences in alcohol-metabolizing enzymes have also been found between men and women.[5] Women tend to have lower ADH activity in stomach cells, which perhaps helps explain why some women have lower tolerances for alcohol. These factors may also account in part for why women are more likely than men to develop alcohol-related health problems.[6]

TOLERANCE AND CROSS-TOLERANCE

Over time, chronic alcohol consumption can activate a group of liver enzymes that assist in alcohol

tolerance A response to chronic drug or alcohol exposure that results in its reduced effectiveness.

metabolism, reducing the amount of time a person remains intoxicated. This is why some chronic drinkers develop a **tolerance** to alcohol, reducing its effects and allowing them to drink excessive amounts before becoming intoxicated. Because these enzymes also aid in the metabolism of other drugs, a heavy drinker may develop a *cross-tolerance* to these substances as well. In other words, when tolerance to alcohol develops, the person also becomes more tolerant (less responsive) to certain other drugs. In fact, in people who drink heavily, the need to metabolize alcohol can out-compete the need to metabolize other drugs, resulting in drug concentrations that reach dangerously high levels. This is why taking certain drugs in combination with alcohol is not only dangerous, but sometimes deadly.

LO3 Does Alcohol Have Any Health Benefits?

Although the detrimental health effects of heavy alcohol consumption have long been known, moderate alcohol consumption may actually benefit health.[7] However, because an amount considered moderate by some might be considered excessive by others, it is important to define exactly what is meant by "moderate alcohol consumption." The 2010 Dietary Guidelines for Americans states that if alcohol is to be consumed, it should only be consumed in moderation, defined as not exceeding one drink per day for women and two drinks per day for men—and it should only be consumed by nonpregnant adults of legal drinking age.[8] In the United States, a standard drink is equivalent to 12 ounces of beer, 5 ounces of wine, or 1.5 ounces of distilled liquor, each of which contains approximately 12 to 14 g of alcohol.[9]

Many people are surprised to learn that consuming alcohol in moderation can have positive health effects. For example, adults who drink in moderation tend to live longer than do nondrinkers and those who drink heavily.[10] In moderation, alcohol consumption appears to be associated with decreased risks of cardiovascular disease, stroke, gallstones, age-related memory loss, and even type 2 diabetes.[11] However, some evidence suggests that other distinct phytochemicals, such as the antioxidants found in red wine, may be especially beneficial.[12]

While moderate alcohol intake may benefit the health of an adult, it provides little (if any) health benefits among young adults. Contrarily, alcohol consumption among young adults is associated with increased risks of injury and death. Studies show that many adults with alcohol-related problems started drinking at a young age; greater frequency of risky drinking behaviors as a young adult is associated with hazardous drinking patterns later in life.[13]

Cross-tolerance *is doubly dangerous.*

Moderate Alcohol Consumption and Cardiovascular Disease

The relationship between moderate alcohol consumption and a reduced risk of cardiovascular disease has been confirmed by hundreds of epidemiological studies.[14] More specifically, adults who consume an average of one to two alcoholic drinks a day have a 30 to 35 percent lower risk of cardiovascular disease than adults who do not consume alcohol (see Figure 13.2).[15] These benefits are lost, however, when alcohol intake becomes excessive. The 2010 Dietary Guidelines for Americans defines excessive drinking as consumption of more than three drinks on any day or more than seven per week for women, more than four drinks on any day or more than 14 per week for men, or consumption within two hours of

Figure 13.2 **Relationship between Average Alcohol Consumption and Coronary Heart Disease**

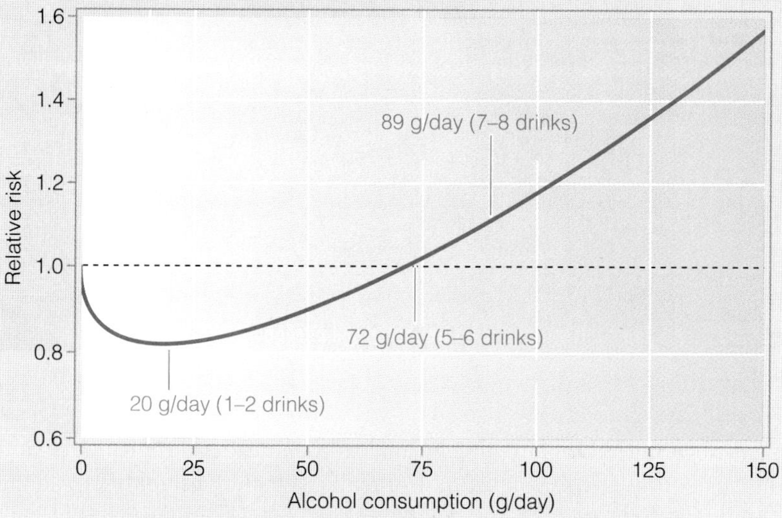

Light to moderate alcohol consumption is associated with reduced risk of coronary heart disease. At high levels (7–8 drinks/day), the risk of coronary heart disease increases.

Source: Corrao G, Rubbiati L, Bagnardi V, Zambon A, Poikolainen K. Alcohol and coronary heart disease: A meta-analysis. Addiction. 2000;95:1505–23. Reprinted with permission of Blackwell Publishing Ltd.

A phytochemical found in grapes and red wine called resveratrol may provide some protection from cardiovascular disease.

four or more drinks for women and five or more drinks for men. Although studies consistently show cardioprotective benefits of light to moderate alcohol consumption, the American Heart Association discourages individuals from beginning or continuing to drink alcohol for this purpose.[16]

HOW MODERATE ALCOHOL CONSUMPTION MAY BENEFIT THE HEART

You may be wondering how alcohol could lower a person's risk of cardiovascular disease. First, recall that high-density lipoproteins (HDLs) help protect against heart disease and stroke. Studies show that a light to moderate daily intake of alcohol is associated with increased HDL-cholesterol levels and therefore may offer some protection from cardiovascular disease.[17] To put this in perspective, one drink per day is associated with a 5 percent increase in HDL, and two to three drinks per day with a 10 percent increase. Light to moderate alcohol consumption is also associated with lower triglyceride levels, and there is evidence that alcohol decreases blood clot formation and helps dissolve existing blood clots.[18] Blood clots can block the flow of blood in arteries, and thus can cause heart attacks and strokes. Finally, alcohol's anti-inflammatory effect helps reduce chronic inflammation, which also lowers cardiovascular disease risk.

ADDITIONAL HEALTH BENEFITS ASSOCIATED WITH RED WINE

In addition to health benefits associated with alcohol, biologically active compounds found in grapes and red wine may provide additional health benefits. There is considerable evidence that the antioxidant **resveratrol**, found in the skin of red grapes, helps reduce inflammation and atherosclerosis, thus providing additional cardiovascular disease protection.[19] In fact, there is some speculation that resveratrol might explain the *French paradox*, a term used to describe the observation that French people have a relatively low occurrence of cardiovascular disease despite indulging in unhealthy behaviors such as eating a great deal of saturated fats and smoking.

LO4 What Serious Health Risks Does Heavy Alcohol Consumption Pose?

In the case of alcohol, more is not better. While numerous studies suggest that moderate alcohol intake may provide health benefits for middle-aged adults, alcohol is clearly hazardous to one's health when consumed in excess.

resveratrol An antioxidant found in the skin of red grapes.

Over time, heavy drinking can lead to impaired nutritional status, liver damage, gout, certain cancers, heart problems, and pancreatic disease (pancreatitis). High-risk drinking also takes a heavy toll on one's family, friends, career, and social life.

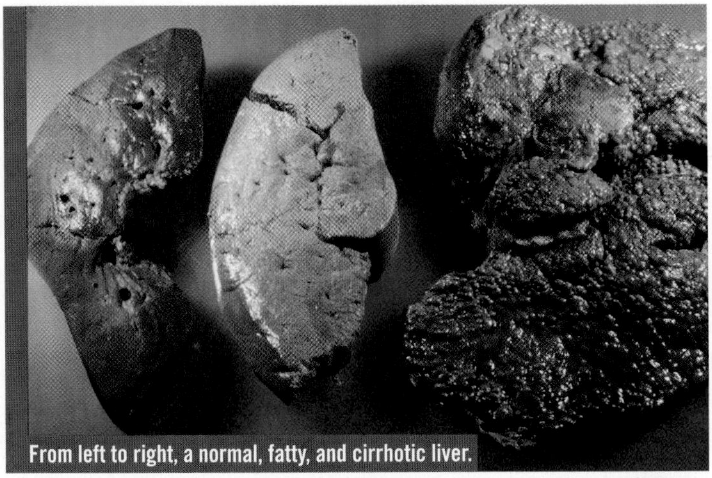

From left to right, a normal, fatty, and cirrhotic liver.

© Arthur Glauberman/Photo Researchers, Inc.

Excessive Alcohol Intake and Nutritional Status

Many studies show that heavy drinking leads to decreased nutrient availability and impaired nutritional status via both primary and secondary malnutrition.[20] Alcoholic beverages have many calories, but very few essential nutrients. In other words, they have low nutrient densities and are sources of empty calories. For many adults, alcohol contributes substantially to total daily energy intake (106 kcal/day), which is why even light to moderate drinking can result in weight gain.[21]

People who abuse alcohol tend to eat very poor diets, which can quickly lead to nutrient deficiencies or primary malnutrition. When alcohol accounts for more than 30 percent of total energy intake, micronutrient intake is likely to be inadequate.[22] Alcohol can also interfere with digestion, absorption, utilization, and excretion of various nutrients, leading to secondary malnutrition.[23] For example, heavy use of alcohol can interfere with the absorption and metabolism of thiamin, resulting in damaged nerves in the brain. The impact of heavy alcohol consumption on selected nutrients is summarized in Table 13.3.

Alcohol Metabolism and the Liver

fatty liver A condition caused by chronic alcohol consumption characterized by the accumulation of triglycerides in the liver.

alcoholic hepatitis Inflammation of the liver caused by obstructed blood flow.

cirrhosis A condition characterized by the presence of scar tissue in the liver.

Chronic alcohol consumption can interfere with normal liver function. For example, it can cause fat to accumulate in and around the liver, a condition referred to as **fatty liver**. This process is reversible if a person stops drinking. In most cases, having a fatty liver does not produce any clinical signs or symptoms. Over time, however, a fatty liver can lead to the development of a more serious condition called **alcoholic hepatitis**. Alcoholic hepatitis is characterized by inflammation and swelling of the liver, which can lead to medical complications. In addition to alcoholic hepatitis, about 10 percent of heavy drinkers develop a condition called **cirrhosis** of the liver. In cirrhosis, liver cells become damaged and are replaced with scar tissue. Cirrhosis can eventually lead to liver failure and even death.

Long-Term Alcohol Abuse and Cancer Risk

The association between alcohol consumption and cancer has been studied extensively. Conclusive evidence shows that heavy drinking (at a rate of 50 g pure alcohol—approximately four drinks—a day) increases a person's risk of developing certain types of cancer, especially those of the mouth, esophagus, colon, liver, and breast.[24] Even a low to moderate intake of alcohol may increase a woman's risk of cancer, particularly breast cancer.[25]

Almost 50 percent of cancers of the mouth and esophagus are associated with heavy drinking.[26] It is not clear how alcohol increases the risk of cancer, although it appears that alcohol may act as both a *carcinogen* and a *co-carcinogen*. Carcinogens are compounds that initiate the formation of cancer, whereas co-carcinogens enhance the carcinogenicity of other cancer-causing chemicals. For example, as cigarettes contain multiple carcinogens, heavy drinkers who smoke are at particularly high risk for developing cancers of the mouth, esophagus, and trachea—it

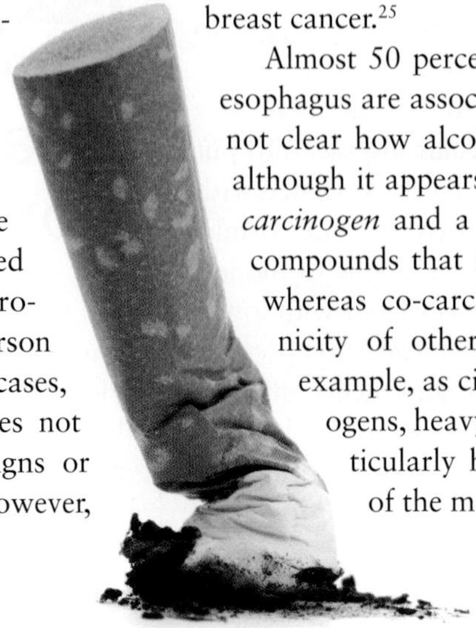

© iStockphoto.com/Milos Luzanin

Table 13.3 **Impact of Heavy Alcohol Consumption on Nutritional Status**

Nutrient	Impact of Alcohol	Health Consequences
Water-Soluble Vitamins		
Thiamin	• Impaired absorption • Increased urinary loss • Altered metabolism • Reduced storage	• Paralysis of eye muscles • Degeneration of nerves with loss of sensation in lower extremities • Loss of balance, abnormal gait • Memory loss, psychosis
Vitamin B_6	• Decreased activation • Increased urinary loss • Displacement from binding protein	• Anemia • Impaired metabolic reactions involving amino acids
Vitamin B_{12}	• Impaired absorption	• Possible neurological damage, numbness and tingling sensations in the arms and legs
Folate	• Decreased absorption • Increased urinary loss	• Anemia • Diarrhea
Riboflavin	• Impaired absorption • Decreased activation	• Deficiency not typical in isolation but occurs in conjunction with other B vitamin deficiencies
Fat-Soluble Vitamins		
Vitamin A	• Decreased conversion to retinal • Reduced absorption	• Impaired vision, night blindness • Liver disease
Vitamin D	• Decreased activation • Impaired absorption	• Bone fractures and osteoporosis
Vitamin E	• Impaired absorption	• Nerve damage • Tunnel vision • Fragility of cell membranes
Vitamin K	• Impaired absorption	• Bruising and prolonged bleeding
Minerals		
Magnesium	• Increased urinary loss	• Muscle rigidity, cramps, and twitching • Irregular cardiac function • Possible hallucinations
Iron	• Increased storage • Impaired absorption • Increased loss in feces	• Both iron deficiency and overload are possible
Zinc	• Decreased absorption • Increased urinary loss • Impaired utilization	• Altered taste, loss of appetite • Impaired wound healing • Night blindness

is possible that alcohol interacts with cigarette smoke to make it more dangerous.

Scientists hypothesize that poor nutritional status associated with heavy drinking may also lead to the development of certain cancers. For example, alcohol interferes with folate availability, which may compromise cell division and DNA repair.[27] Similarly, reduced levels of iron, zinc, vitamin E, certain B vitamins, and vitamin A—all nutrients that are commonly deficient in heavy drinkers—have been experimentally linked to certain types of cancer. Because adequate nutrient intake may offer protection against cancer, an overall dietary inadequacy caused by chronic alcohol intake may weaken a person's natural defense mechanisms.

Alcohol Abuse and the Cardiovascular System

Although moderate alcohol consumption appears to benefit the cardiovascular system, heavy alcohol consumption has been shown to harm it (see Figure 13.3 on the next page). Recall that the Dietary Guidelines for Americans defines excessive drinking as consumption of more than three drinks on any day or more than seven per week for women, more than four drinks on any day or more than 14 per week for men, or consumption within two hours of four or more drinks for women and five or more drinks for men.[28] Cardiovascular consequences associated with long-term heavy drinking include hypertension, stroke, and irregular cardiac function. These detrimental effects can begin to develop in as few as five years.[29]

One of the most serious cardiovascular consequences associated with heavy alcohol consumption is **alcoholic cardiomyopathy**. Over time, chronic exposure to high levels of alcohol causes the heart muscle to weaken. Because the weakened heart cannot contract forcibly, blood flow to vital organs (such as the lungs, liver, kidneys, and brain) is reduced. If drinking continues, alcoholic cardiomyopathy can ultimately result in heart failure. For reasons that are not clear, women are more susceptible to alcoholic cardiomyopathy than are men.

A relationship between heavy drinking and hypertension is also well established: as alcohol intake increases, so does blood pressure.[30] This may be caused in part by alcohol-related liver damage. When a heavy drinker refrains from drinking alcohol, blood pressure often returns to normal. High blood pressure is not the only concern, however. Alcohol can also trigger an irregular heartbeat, a condition called **cardiac arrhythmia**. This condition is sometimes referred to as *holiday heart syndrome* because it is associated with binge drinking on weekends and around holidays.[31] **Binge drinking** is defined as the consumption of four or more drinks for women and five or more drinks for men over a two-hour period. Although its cause is not entirely clear, cardiac arrhythmia may occur as a response to disturbances in electrolyte balance. Whatever its cause, alcohol-induced cardiac arrhythmia can lead to sudden cardiac arrest (heart attack).

alcoholic cardiomyopathy A serious condition whereby the heart muscle weakens in response to chronic alcohol consumption.

cardiac arrhythmia An irregular heartbeat caused by a high intake of alcohol.

binge drinking The consumption of four or more drinks for women and five or more drinks for men over a two-hour period.

pancreatitis A painful condition characterized by inflammation of the pancreas.

Figure 13.3 Effects of Habitual Alcohol Abuse on the Body

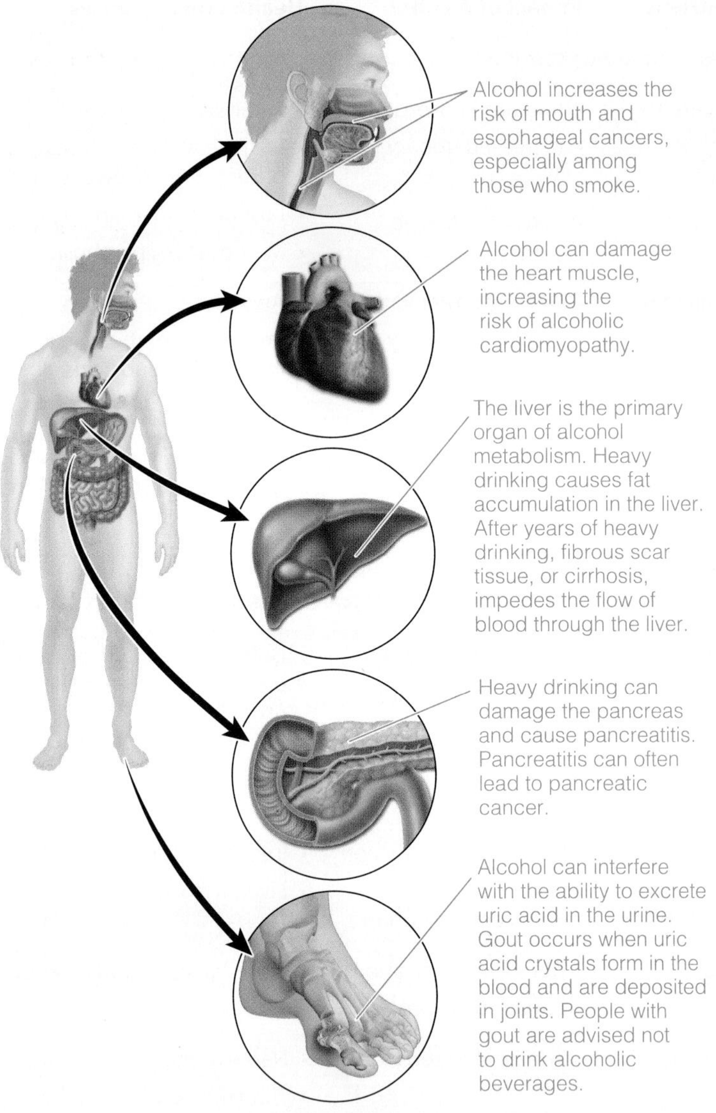
Alcohol increases the risk of mouth and esophageal cancers, especially among those who smoke.
Alcohol can damage the heart muscle, increasing the risk of alcoholic cardiomyopathy.
The liver is the primary organ of alcohol metabolism. Heavy drinking causes fat accumulation in the liver. After years of heavy drinking, fibrous scar tissue, or cirrhosis, impedes the flow of blood through the liver.
Heavy drinking can damage the pancreas and cause pancreatitis. Pancreatitis can often lead to pancreatic cancer.
Alcohol can interfere with the ability to excrete uric acid in the urine. Gout occurs when uric acid crystals form in the blood and are deposited in joints. People with gout are advised not to drink alcoholic beverages.

Chronic alcohol intake can lead to health problems such as cancer, heart disease, liver disease, pancreatic disease, and gout.

Alcohol Abuse and Pancreatic Function

In addition to its effects on the cardiovascular system, excessive drinking can also damage the pancreas. **Pancreatitis** is a painful condition characterized by inflammation of the pancreas. Up to 90 percent of all cases of chronic pancreatitis are associated with heavy drinking. A chronically high alcohol intake (consuming at least seven drinks a day for more than five years) increases a person's risk of developing this condition. Pancreatitis makes it difficult for the pancreas to release pancreatic juice, which in turn impairs digestion and nutrient absorption in the small intestine.[32]

LO5 How Does Alcohol Abuse Contribute to Individual and Societal Problems?

Although the majority of people who drink alcohol do so responsibly and without negative consequences, the line between alcohol use and abuse is often blurry. When alcohol is the cause of significant problems in a person's life, drinking is problematic. **Alcohol use disorders (AUDs)**, a spectrum of disorders that encompasses both recurrent excessive drinking and alcoholism, arises when drinking leads to negative physical, legal, and/or social consequences.[33] AUDs are not discriminatory; they affect men and women, young and old, professional and nonprofessional, and rich and poor.

Though they are frequently misdiagnosed, AUDs occur in great numbers throughout the United States. Annual costs associated with AUD-related hospitalization, premature death, and alcohol-related illness and injury exceed $185 billion.[34] Habitual alcohol abuse affects virtually every aspect of a person's life, including family relationships and performance at work and at school. Arrests related to public intoxication, driving while under the influence of alcohol, and other alcohol-related crimes often result in substantial legal problems. Despite the myriad problems associated with excessive drinking, many individuals with AUDs are not able to stop or limit their alcohol intakes because alcohol use can lead to both physical and emotional dependence.

alcohol use disorders (AUDs) A spectrum of disorders that encompasses both recurrent excessive drinking and alcoholism.

Alcoholics Anonymous® (AA) An organization that offers a 12-step program that provides fellowship and support for individuals who wish to achieve sobriety and stay sober.

Al-Anon® An organization that provides support for family members of alcoholics.

Alateen® An organization that provides support for teenage children of alcoholics.

Treating Alcohol Abuse

Most health professionals recognize alcoholism as a complex and widespread disease that can be treated—but not cured. One of the oldest and most reputable organizations for the treatment of alcoholism is **Alcoholics Anonymous® (AA)**. AA offers a 12-step program that provides fellowship and support for individuals who wish to achieve sobriety and stay sober. More than 2 million people consider themselves lifelong members of AA. Other organizations such as **Al-Anon®** and **Alateen®** provide support for family members of alcoholics. While Al-Anon is open to all affected family members, Alateen caters specifically to teenage children of alcoholics.

© william casey/Shutterstock.com

Alcohol Use on College Campuses

Numerous studies have documented the extent to which alcohol use affects college students.[35] Alcohol abuse, rampant on many college campuses, is associated with a wide range of negative consequences such as vandalism, violence, acquaintance rape, unprotected sex, and death.[36] To give perspective to the magnitude of alcohol-related

Alcoholics Anonymous is a fellowship of men and women who share their experiences to help themselves and others recover from alcoholism.

© mr6/ZUMA Press/Newscom

problems, data from the U.S. Centers for Disease Control and Prevention (CDC) indicate that unintentional alcohol-related injuries among college students increased from nearly 1,600 to more than 1,700 per year between 1998 and 2001. Individuals between 21 and 29 years of age have the highest percentages of fatal alcohol-related traffic accidents.[37] If you find yourself wondering if you or someone you know has a problem with alcohol, you can use the self-assessment tool shown in Table 13.4.

Although 20 percent of college students report that they do not drink alcohol, 25 percent describe themselves as binge drinkers.[38] Indeed, college students are more likely to binge drink than their peers who do not attend college. In fact, most college students who binge drink report that these behaviors started after they entered college.[39] Students who binge drink are more likely than those who do not to miss class, have lower academic rankings, get in trouble with campus law enforcement, and drive while intoxicated.[40] Clearly, more research is needed to understand how the collegiate environment encourages risky drinking behaviors and what can be done to help prevent alcohol-related problems on university campuses.

COLLEGE ADMINISTRATORS DEBATE THE OPTIMAL LEGAL DRINKING AGE

The issue of underage drinking on university campuses continues to perplex academic administrators, who come to little agreement about how to address this problem effectively. While there is convincing evidence that raising the minimum legal drinking age would not be helpful, some college administrators and health professionals believe that keeping the legal drinking age at 21 years is ineffective and does not curtail problem drinking. For this reason, in 2009 a coalition of more than 100 college presidents from across the United States signed a petition asking lawmakers to examine whether the minimum legal drinking age should be lowered.[41] The petition states: "Our experience as college and university presidents convinces us that twenty-one is not working. A culture of dangerous, clandestine 'binge-drinking'—often conducted off-campus—has developed." Still, most public health agencies (including the American Medical Association, the CDC, and the National Highway Traffic Safety Board) support the current drinking age of 21.

TAKING ACTION

Alcohol abuse among college students is a problem that affects the entire campus community. Clearly, traditional educational approaches that focus on distribution of pamphlets and the provision of one-day alcohol awareness programs are not enough—colleges and universities must work together with communities to create a culture that discourages high-risk drinking.[42] To this end, many colleges are developing and implementing comprehensive programs designed to curb the problem of alcohol abuse in unique ways.[43] For example, by increasing campus recreational activities that do not involve drinking, many schools hope to dissuade students from habitually going to bars and clubs. Community-oriented initiatives such as discouraging price discounts on alcohol, two-for-one drink specials, inexpensive beer pitcher sales, and other types of happy hour promotions among campus-adjacent bars are also being implemented.

Binge drinking is a major problem among college students.

© David Young-Wolff/PhotoEdit

Table 13.4 The Johns Hopkins University Hospital Alcohol Screening Quiz

1. Do you lose time from work due to drinking?
2. Is drinking making your home life unhappy?
3. Do you drink because you are shy with other people?
4. Is drinking affecting your reputation?
5. Have you ever felt remorse after drinking?
6. Have you had financial difficulties as a result of drinking?
7. Does your drinking make you careless of your family's welfare?
8. Do you turn to inferior companions and environments when drinking?
9. Has your ambition decreased since you started drinking?
10. Do you crave a drink at a definite time daily?
11. Do you want a drink the next morning?
12. Does drinking cause you to have difficulty sleeping?
13. Has your efficiency decreased since drinking?
14. Is drinking jeopardizing your job or business?
15. Do you drink to escape from worries or trouble?
16. Do you drink alone?
17. Have you ever had a loss of memory as a result of drinking?
18. Has your physician ever treated you for drinking?
19. Do you drink to build up your self-confidence?
20. Have you ever been to a hospital or institution on account of drinking?
If you answered three or more of the questions with a "Yes," there is a strong possibility that your drinking patterns are detrimental to your health and that you may be alcohol dependent. Under these circumstances, you should consider getting an evaluation of your drinking behavior by a health care professional.

Source: Office of Health Care Programs, Johns Hopkins University Hospital. Available from: http://www.alcohol-addiction-info.com/Alcohol_Addiction_Self_Assessment_Tools.html

RECOMMENDATIONS FOR RESPONSIBLE ALCOHOL USE

Both health benefits and risks are associated with alcohol consumption by adults. Although most Americans who drink alcohol do so safely and responsibly, this is certainly not the case for everyone—especially among adolescents and young adults. For this reason, one of the goals identified in Healthy People 2020 is to protect health, safety, and quality of life for all—and especially for children—by reducing alcohol abuse.[44] In 1998, the number of alcohol-related deaths was 5.9 per 100,000. Healthy People 2020 aims to reduce this number to four alcohol-related deaths per 100,000.

The 2010 Dietary Guidelines for Americans also addresses the harmful effects of alcohol when consumed in excess; it states that adults who consume alcohol should do so in moderation.[45] Again, moderation is defined as up to one drink per day for women and up to two drinks per day for men. While these guidelines are generally safe, some people should avoid alcohol completely, such as those younger than the legal drinking age, those who have difficulty restricting their alcohol intakes, those taking medications that can interact with alcohol, and those with specific medical conditions. Further, all should abstain from drinking when driving, operating machinery, or taking part in activities that require attention, skill, or coordination. Pregnant women are advised to abstain from alcohol throughout pregnancy, especially during the first trimester. Finally, women who are breastfeeding should refrain from drinking until their infant is at least three months of age. After this time, women are advised to wait at least four hours after consuming a single alcoholic drink to breastfeed.

14 Keeping Food Safe

© iStockphoto.com/ShyMan

LEARNING OUTCOMES:

- LO1 Discuss the infectious agents that cause foodborne illness.
- LO2 Understand how noninfectious agents can cause foodborne illness.
- LO3 Describe how food manufacturers prevent contamination.
- LO4 Take steps to reduce foodborne illness at home and when eating out.
- LO5 Understand how to reduce foodborne illness while traveling.

LO1 What Causes Foodborne Illness?

Most Americans are fortunate to have an abundant supply of healthful food. Still, there are times when even healthy food can make a person sick. While most people understand the importance of avoiding food that is spoiled or stored improperly, food that appears, smells, and tastes safe to eat can still cause illness. Although every effort is made to ensure that food is healthful and not harmful, food safety remains an important public health concern. The U.S. Centers for Disease Control and Prevention (CDC) estimates that 48 million Americans, or one out of every six people, suffer from foodborne diseases annually. Of them, some 128,000 are hospitalized and 3,000 die.[1] In this chapter, you will learn about common disease-causing agents and how to avoid them so that you can minimize your risk of illness.

You are exposed to thousands of microscopic organisms (microbes) every day. Microbes populate the world you live in, and many even serve useful purposes for you and for your health. For example, microbes in your GI tract assist in food digestion, produce vitamins and other health-promoting substances, and may even help prevent some diseases. Other microbes, however, are pathogenic (disease causing). These microbes can make you sick; consumption of them in foods and beverages is the main cause of foodborne illness. Generally speaking, a **foodborne illness** is a disease caused by the ingestion of unsafe food. Foodborne illnesses are sometimes referred to as *food poisoning*. You can contract a foodborne illness by ingesting either infectious or noninfectious substances. An **infectious agent** (or **pathogen**) is a living microorganism such as a bacterium, virus, mold, fungus, or parasite. A **noninfectious agent** is an inert (nonliving) substance such as a nonbacterial toxin; a chemical residue from processing, pesticides, or antibiotics; or a physical hazard such as glass or plastic.

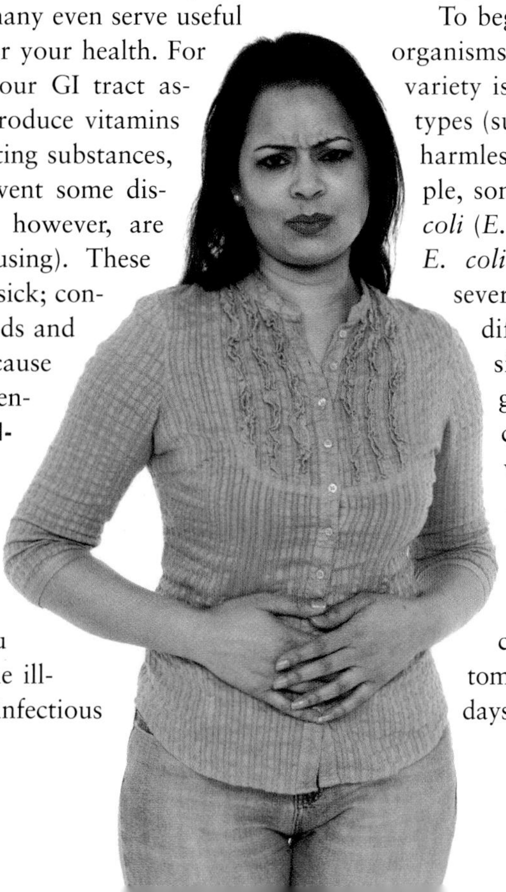

Serotypes

As you have learned, although most foodborne illnesses are caused by microbes, not all microbes are harmful. In fact, even related pathogenic microbes can cause vastly different types of illnesses (or have no effect at all). Therefore, to understand how to prevent foodborne illness, you must first have a basic knowledge of microbiology, and must understand how various organisms can make you sick.

To begin with, a group of closely related microorganisms can have several genetic varieties. Each variety is called a **serotype** (or **strain**). Some serotypes (such as the ones living in your GI tract) are harmless, whereas others cause disease. For example, some serotypes of the bacterium *Escherichia coli* (*E. coli*) live safely in your GI tract, whereas *E. coli* O157:H7 can cause severe illness.[2] Moreover, different serotypes of a single type of pathogenic bacterium can cause illness in different ways. Some pathogenic *E. coli* serotypes cause mild intestinal discomfort within one to three days, while others cause more severe symptoms that take up to eight days to develop. The time

foodborne illness A disease caused by the ingestion of unsafe food.

infectious agent (or **pathogen**) A living microorganism that can cause illness.

noninfectious agent An inert (nonliving) substance that can cause illness.

serotype (or **strain**) A specific genetic variety of an organism.

incubation period The time elapsed between the consumption of a contaminated food and the emergence of sickness.

preformed toxin A toxic substance that already exists in a food before the food is eaten.

methicillin-resistant *Staphylococcus aureus* (MRSA) An antibiotic-resistant strain of *S. aureus* that has received considerable attention from public health officials.

elapsed between the consumption of a contaminated food and the emergence of sickness is called the **incubation period.** Because different pathogens (and serotypes thereof) have different incubation periods, health care providers often consider this variable when trying to determine which foodborne pathogen may have been ingested. Some of the most common infectious agents and their incubation periods are listed in Table 14.1, and the methods by which these agents cause illness are described next.

Preformed Toxins

Some pathogenic organisms produce toxic substances while they are growing in food. Such a substance is called a **preformed toxin** because it already exists in the food before the food is eaten. Preformed toxins typically cause rapid reactions (occurring in one to six hours) such as nausea, vomiting, diarrhea, and sometimes, neurological damage.

STAPHYLOCOCCUS AUREUS (*S. AUREUS*)

One bacterium that produces preformed toxins is *Staphylococcus aureus* (*S. aureus*). This bacterium causes nearly 250,000 foodborne illnesses in the United States every year.[3] Foods commonly contaminated with *S. aureus* include raw and undercooked meat and poultry, cream-filled pastries, and unpasteurized dairy products. Symptoms of *S. aureus* infection include a sudden onset of severe nausea and vomiting, diarrhea, and abdominal cramps. These symptoms typically occur within one to six hours of the contaminated food's consumption.

You may have heard about an antibiotic-resistant strain of *S. aureus* called **methicillin-resistant *Staphylococcus aureus* (MRSA).** This strain has received considerable attention from public health officials. MRSA was identified more than 40 years ago, and at the time, was believed to spread only through direct contact with an infected person—usually in a hospital environment. Scientists now know however that MRSA can spread through the sharing of towels and equipment in athletic and school facilities and that infection can occur via consumption of MRSA-contaminated foods. To date, there has only been one documented outbreak of foodborne illness caused by MRSA in the United States. In this case, researchers believe that an infected food preparer transmitted the bacterium to coleslaw. Three family members consumed the coleslaw and subsequently became ill.[4] Because MRSA cannot be treated effectively by antibiotics, public health officials are carefully monitoring for outbreaks of this bacterium—especially in regard to foodborne illness.

CLOSTRIDIUM BOTULINUM (*C. BOTULINUM*)

Another bacterium that produces preformed toxins, *Clostridium botulinum* (*C. botulinum*), is found

> "In England, ***S. aureus*** is commonly known as ***Golden Staph***, because ***Staphylococcus aureus*** translates literally to 'golden cluster seed.'"

Improperly canned low-acid foods can contain live *Clostridium botulinum* bacteria, which can cause botulism.

© iStockphoto.com/Frank Cutrara

Table 14.1 Infectious Agents of Foodborne Illness, Food Sources, and Symptoms of Infection

Organism	Incubation Period	Duration of Illness	Associated Foods	Signs and Symptoms
Bacteria				
Campylobacter jejuni	2–5 days	2–10 days	Raw and undercooked poultry, untreated water, and unpasteurized milk	Diarrhea (often bloody), abdominal cramping, nausea, vomiting, fever, and fatigue
Clostridium botulinum	12–72 hours	From days to months	Home-canned foods with low acid contents, improperly canned commercial foods, and herb-infused oils	Vomiting, diarrhea, blurred vision, drooping eyelids, slurred speech, dry mouth, difficulty swallowing, and weak muscles
Clostridium perfringes	8–16 hours	24–48 hours	Raw and undercooked meats, gravy, and dried foods	Abdominal pain, watery diarrhea, vomiting, and nausea
Escherichia coli O157:H7	1–8 days	5–10 days	Raw and undercooked meats, raw fruits and vegetables, unpasteurized milk and juice, and contaminated water	Nausea, abdominal cramps, and severe diarrhea (often bloody)
Escherichia coli (entero-toxigenic)	1–3 days	Variable	Water and food contaminated with human feces	Diarrhea, abdominal cramps, and some vomiting
Listeria monocyto-genes	9–48 hours for GI symptoms, 2–6 weeks for invasive disease	Variable	Raw and inadequately pasteurized dairy products and ready-to-eat luncheon meats and frankfurters	Fever, muscle aches, nausea, diarrhea, and premature delivery, miscarriage, or stillbirth
Salmonella	1–3 days	4–7 days	Raw poultry, eggs, and beef; fruit and alfalfa sprouts; and unpasteurized milk	Diarrhea, fever, abdominal pain, and severe headache
Shigella	24–48 hours	4–7 days	Raw and undercooked foods and water contaminated with human fecal material	Fever, fatigue, watery or bloody diarrhea, and abdominal pain
Staphylococcus aureus	1–6 hours	24–48 hours	Improperly refrigerated meats, potato and egg salads, and cream pastries	Severe nausea and vomiting, diarrhea, and abdominal pain
Vibrio cholerae	1–7 days	2–8 days	Contaminated water and many undercooked foods	Diarrhea and vomiting
Viruses				
Hepatitis A virus	15–50 days	2–12 weeks	Mollusks (oysters, clams, mussels, scallops, and cockles)	Jaundice, fatigue, abdominal pain, loss of appetite, nausea, diarrhea, and fever
Norovirus	12–48 hours	12–60 hours	Raw and undercooked shellfish and contaminated water	Nausea, vomiting, diarrhea, ab-dominal pain, headache, and fever
Parasites				
Trichinella (worm)	1–2 days for initial symptoms; others begin 2–8 weeks after infections	Months	Raw and undercooked pork and meats of carnivorous animals	Acute nausea, diarrhea, vomiting, fatigue, fever, and abdominal pain
Giardia intestinalis (protozoan)	1–2 weeks	Days to weeks	Contaminated water and many uncooked foods	Diarrhea, flatulence, and abdominal pain
Molds				
Aspergillus flavus	Days to weeks	Weeks to months	Wheat, flour, peanuts, and soybeans	Liver damage

Adapted from Centers for Disease Control and Prevention. Diagnosis and management of foodborne illnesses: A primer for physicians and other health care professionals. Morbidity and Mortality Weekly Reports. 2004; 53:1–33. Available from: http://www.cdc.gov/mmwr/PDF/rr/rr5304.pdf. Murano P. Understanding food science and technology. Thompson/Wadsworth, 2003.

botulism The foodborne illness caused by *Clostridium botulinum.*

aflatoxin A toxin produced by the *Aspergillus* mold that is found on some agricultural crops.

enteric toxin (or **intestinal toxin**) A toxic substance produced by an organism after it enters the gastrointestinal tract.

mainly in inadequately processed low-acid home-canned foods (such as green beans). A person can also become infected with this microbe by eating improperly canned commercial foods. *C. botulinum* infection causes an illness called **botulism**. Mild cases of botulism cause vomiting and diarrhea, whereas severe cases can cause double vision, blurred vision, drooping eyelids, slurred speech, difficulty swallowing, dry mouth, and muscle weakness. In the most severe cases, botulism can cause paralysis, respiratory failure, and death. In 2007, the CDC reported an outbreak of eight cases of botulism in Indiana, Texas, and Ohio.[5] All infected persons had consumed a particular brand of hot dog chili sauce that was quickly recalled by the manufacturer. Although all of the infected individuals recovered, each endured several days of painful gastrointestinal (GI) symptoms. Because the botulism toxin is destroyed by high temperatures, some experts recommend that home-canned foods be boiled for 10 minutes before they are consumed.

When improperly dried, the *Aspergillus* mold sometimes found in peanuts can produce aflatoxin, a dangerous poison.

ASPERGILLUS AND AFLATOXIN

While most molds that grow on foods such as cheese and bread are not dangerous, some produce unsafe preformed toxins. An example is **aflatoxin**, which is produced by the *Aspergillus* mold found on some agricultural crops (such as peanuts, rice, and wheat). Aflatoxin is occasionally found in the milk of animals fed with contaminated feed. The consumption of aflatoxin is of great concern because this toxin can cause liver damage and cancer, and is often fatal. Although many agricultural practices (such as sufficient drying) are used to minimize the risk of aflatoxin contamination in the United States, consumption of this toxin remains a significant public health issue throughout many regions of the world.

Using Technology to Track Food Safety

Are you interested in new food safety alerts and tips issued by the federal government? If so, you can download a free widget from the U.S. Food and Drug Administration (FDA) website (http://www.foodsafety.gov/widgets/). This low-maintenance application provides instant updates on the latest food safety recalls and offers useful food safety tips appropriate to the season. With the aid of this widget, you might learn that your favorite brand of chicken noodle soup has been recalled because of suspected contamination, or how to cook your turkey to the appropriate internal temperature for Thanksgiving. Keeping abreast of food safety tips will not only help keep you safe but also provide a constant reminder that issues related to food safety can affect you every day.

Enteric (Intestinal) Toxins

In contrast to those that produce preformed toxins before a food is consumed, some organisms produce harmful toxins after they enter the GI tract. Such a toxin is called an **enteric toxin** (or **intestinal toxin**). Enteric toxins draw water into the intestinal lumen, resulting in diarrhea. Although the symptoms of enteric toxin infections vary, incubation periods generally span one to five days—substantially longer than those of most preformed toxins.

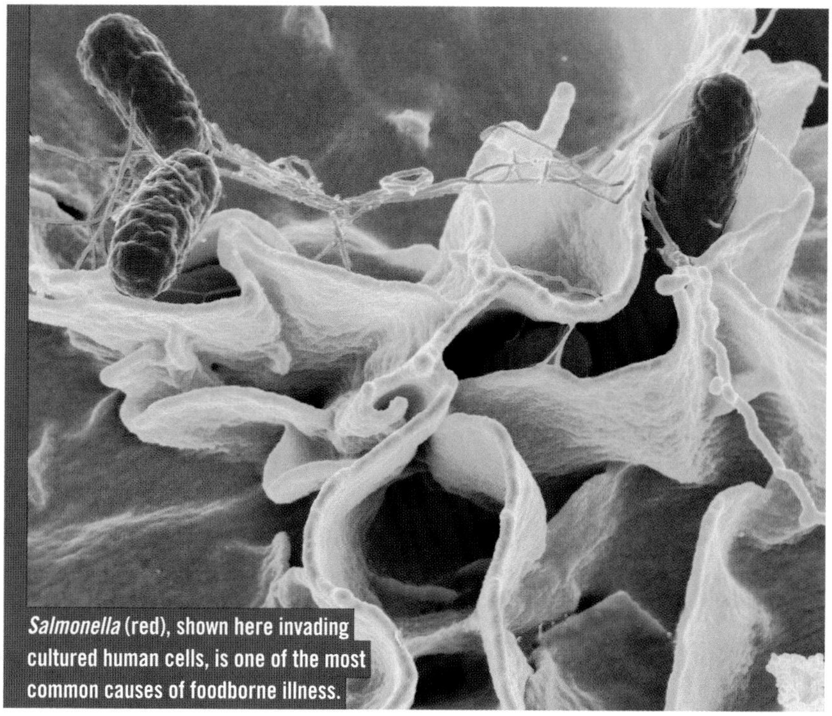
Salmonella (red), shown here invading cultured human cells, is one of the most common causes of foodborne illness.

© Rocky Mountain Laboratories/NIAID/NIH

NOROVIRUSES

A **norovirus**, previously called a Norwalk or Norwalk-like virus, is an example of a pathogen that produces enteric toxins. Symptoms of norovirus infection usually include nausea, vomiting, diarrhea, and abdominal cramping. A person with a norovirus infection may also develop a low-grade fever, chills, headaches, muscle aches, and/or a general sense of fatigue. Symptoms usually begin one to two days after ingestion of the contaminated food. In 2008, a norovirus outbreak occurred on three college campuses. This outbreak resulted in approximately 1,000 cases of reported illness and 10 hospitalizations, prompting closure of one of the three campuses. Although the cause of the related outbreaks (or even if they were traced to a common food) was never determined, the outbreaks eventually subsided. Still, because extensive opportunities for transmission created by shared living and dining areas put college campuses at particularly high risk for norovirus outbreaks, many nutritionists and administrators remain on high alert.[6]

ENTEROTOXIGENIC *E. COLI*

Some forms of *E. coli*, called **enterotoxigenic *E. coli***, produce enteric toxins. The consumption of food or water contaminated with these bacteria can cause severe GI upset; symptoms including diarrhea, abdominal cramps, and nausea tend to occur within 6 to 48 hours after consumption. Foods and beverages typically contaminated with these forms of *E. coli* include uncooked vegetables, fruits, raw and undercooked meats and seafood, unpasteurized dairy products, and untreated tap water. Enterotoxigenic *E. coli* is the primary cause of a condition often referred to as *traveler's diarrhea*, a clinical syndrome resulting from the microbial contamination of food or water during or shortly after travel. If you have ever had severe diarrhea and/or abdominal cramps while traveling, you might have experienced this type of foodborne illness.

Enterohemorrhagic Pathogens

Some pathogens invade the cells of the intestine, seriously irritating the mucosal lining and causing fever, severe abdominal discomfort, and bloody diarrhea. These types of pathogens are called **enterohemorrhagic**, literally meaning that they cause bloody diarrhea and intestinal inflammation. Examples of enterohemorrhagic pathogens include *Salmonella* and an especially dangerous serotype of *E. coli* called *E. coli* O157:H7. Incubation periods for enterohemorrhagic pathogens generally span one to eight days.

SALMONELLA

Salmonella is one of the most common causes of foodborne illness in the United States. It is typically found in raw poultry, eggs, beef, improperly washed fruit, alfalfa sprouts, and unpasteurized milk. The incubation period for this organism is one to three days, and symptoms typically include severe GI upset and headaches. In 2010, over 1,900 *Salmonella*-related illnesses were reported in 11 states throughout the United States. This outbreak, attributed to contaminated eggs,

norovirus An infectious pathogen that produces enteric toxins and often causes foodborne illness.

enterotoxigenic *E. coli* A form of *E. coli* that produces enteric toxins.

enterohemorrhagic Causing bloody diarrhea and intestinal inflammation.

parasite An organism that relies on another organism to survive.

protozoan A single-celled eukaryotic organism. Some protozoa are parasites.

cyst A stage of growth in the life cycle of a protozoan; excreted in the feces.

resulted in a massive nationwide recall of eggs produced by two Iowan distributors.[7] Perhaps surprisingly, numerous *Salmonella* infections have resulted from the consumption of raw alfalfa and mung bean sprouts. In fact, because raw sprouts are particularly prone to a variety of bacterial infections, the FDA recommends that people with compromised immune systems (children, the elderly, and those with autoimmune conditions) avoid them.

E. COLI O157:H7 AND *E. COLI* O104:H4

As mentioned above, *E. coli* O157:H7 is an enterohemorrhagic serotype of *E. coli*. Unpasteurized milk, apple juice, and apple cider sometimes harbor this pathogenic organism, as does improperly prepared meat (including poultry). *E. coli* O157:H7 infection, which has an incubation period of one to eight days, results in nausea, abdominal cramps, and severe diarrhea that is often bloody. In 2006, 183 persons in 26 states were infected with *E. coli* O157:H7.[8] This outbreak rapidly gained national attention as numerous affected individuals had to be hospitalized and at least three died. Fresh spinach was identified as the source of contamination, and the FDA quickly advised consumers not to eat bagged fresh spinach or fresh spinach-containing products unless they were cooked at 71 °C (160 °F) for at least 15 seconds. Partly because of this outbreak, the FDA issued new guidance to the food industry to minimize microbial contamination of fresh-cut fruits and vegetables.[9]

In 2011, an outbreak of a different pathogenic serotype of *E. coli*, *E. coli* O104:H4, resulted in hundreds—if not thousands—of illnesses and numerous deaths throughout the European Union. Several cases were also reported in the United States, but most of those affected had recently traveled to Germany. Authorities identified raw vegetable sprouts grown in Germany as the likely source of contamination.[10]

Parasites: Protozoa and Worms

A **parasite** is an organism that relies on another organism to survive. The consumption of parasite-infested foods can cause foodborne illness. An example of a parasite is the **protozoan**, a single-celled eukaryotic organism. Some protozoa can live as parasites in the intestinal tracts of animals and humans. As part of its reproductive cycle, a protozoan forms **cyst** sacs that are excreted in the feces. If cyst-containing feces come in contact with plants or animals used for food, the resulting food products can be contaminated as well. Thus, if you consume foods or beverages contaminated with protozoan cysts, you can develop a foodborne illness.

One parasitic protozoan, *Giardia intestinalis*, causes diarrhea, abdominal discomfort, and cramping. Symptoms typically begin one to two weeks after infection. *Giardia intestinalis* can be found in untreated swimming pools and hot tubs, and in rivers, ponds, and streams that have been contaminated with the feces of an infected animal or person. This is why chlorine, a substance that kills *Giardia intestinalis*, is typically added to water in public swimming and bathing areas, and why boiling, chemically treating, or filtering water from ponds, streams, and lakes is recommended before drinking.

Consuming foods that contain parasitic worms can also cause foodborne illness. Like protozoa, worms form cysts as part of their life cycles. Once they are ingested, cysts mature into worms that can cross the intestinal lining, travel through the blood, and eventually settle in various locations in the body, including muscles, eyes, and the brain. An example of a parasitic worm is

Consumption of untreated water *is not advised because it may be* **contaminated** *with* Giardia intestinalis.

Trichinella, a roundworm that can invade a variety of animals, including pigs and some fish. Eating undercooked *Trichinella*-contaminated pork and seafood can result in the worm entering the body, causing muscle pain, swollen eyelids, and fever.

Prions

Although **prions** are not living organisms—in fact, they are not organisms at all—they may pose food safety concerns. A prion is an altered protein created when the secondary structure of a normal protein is disrupted. Prions can cause other normal proteins to unravel, setting off a cascade of similar reactions that converts hundreds of normal proteins into abnormal prions. When large numbers of prions build up in a cell, the cell can rupture and release its prions into the surrounding area, which in turn destroys other cells. Eventually, this process kills the surrounding cells and gives the infected tissue a spongy texture. Unfortunately, prions are extremely resilient. They retain the ability to damage other cells even after they have been exposed to extreme heat and/or acid. Therefore, merely cooking food does not destroy prions, and prions are not destroyed by the acidic conditions of the stomach. If a prion-containing food is consumed, the prions can be absorbed into the bloodstream. Though prions are inert, this is why some experts consider some prion-related diseases to be infectious.

© iStockphoto.com/Eric Isselée

Several diseases, such as **bovine spongiform encephalitis** (**BSE**, or **mad cow disease**) are known to be caused by prion ingestion.[11] Mad cow disease is characterized by loss of motor control, confusion, paralysis, wasting, and eventually death. **Creutzfeldt-Jakob disease**, another deadly disease caused by prions, is very rare: it occurs in only one in every million people each year.[12] Creutzfeldt-Jakob disease is usually attributed to direct infection by prions from contaminated medical equipment—during surgery, for example. Simultaneous to a recent BSE outbreak in the United Kingdom, however, researchers discovered a new form of Creutzfeldt-Jakob disease called **variant Creutzfeldt-Jakob disease.** This disease has a relatively long incubation period (years) and is fatal. Considerable evidence links variant Creutzfeldt-Jakob disease to the consumption of BSE-contaminated products.[13] To date, nearly 200 human cases of variant Creutzfeldt-Jakob disease have been reported worldwide, three of which were in the United States. No treatment exists for either form of Creutzfeldt-Jakob disease, and nothing can slow either disease's progression.

prion An altered protein that forms when the secondary structure of a normal protein is disrupted.

bovine spongiform encephalopathy (BSE, or **mad cow disease)** A fatal disease caused by prion ingestion.

Creutzfeldt-Jakob disease A rare but fatal disease caused by a genetic mutation or exposure to prions during surgery.

variant Creutzfeldt-Jakob disease A form of Creutzfeldt-Jakob disease that may be caused by consumption of BSE-contaminated foods.

PRIONS AND PUBLIC POLICY

Because prions are found mainly in nerves and the brain, the World Health Organization (WHO) recommends that all governments prohibit the feeding of highly innervated tissue (such as the brain and spinal cord) from slaughtered cattle to other animals. Similarly, both Canada and the United States have banned the use of such products in human food (including dietary supplements) and in cosmetics. In 2006, Canada also banned the inclusion of cattle tissues capable of transmitting BSE in all animal feeds, pet foods, and fertilizers. Partly in response to concern about BSE, the U.S. Department of Agriculture (USDA) does not allow the slaughter of non-ambulatory (downer) cattle for food purposes.[14] These measures are part of an international effort to unify BSE policy to ensure the safety of human food.

LO2 What Noninfectious Substances Cause Foodborne Illness?

While the consumption of a food that contains an infectious pathogen poses the greatest risk of foodborne illness, the consumption of an inert (nonliving) noninfectious agent can also make you sick. Noninfectious

Red tides occur when marine algae produce brightly colored pigments and toxins that cause shellfish poisoning.

© Don Paulson/Purestock/Glowimages.com

agents include physical contaminants (such as glass and plastic) and other dangerous substances such as toxins, heavy metals, and pesticides.

shellfish poisoning A type of noninfectious foodborne illness caused by the consumption of particular types of contaminated fish and shellfish.

marine toxin A poison produced by ocean algae.

red tide A phenomenon whereby certain ocean algae grow profusely and produce brightly colored pigments that bloom outward and make the surrounding water appear red or brown.

brevetoxin A potent toxin produced by red tide–causing algae that when consumed causes shellfish poisoning.

Algal Toxins

One type of noninfectious foodborne illness, **shellfish poisoning**, can result from the consumption of particular types of contaminated fish and shellfish (for example, clams and oysters). Certain marine animals consume large amounts of ocean algae that sometimes produce poisonous **marine toxin** compounds. Eating marine toxin-contaminated foods can cause tingling, burning, numbness, drowsiness, and difficulty breathing. People with shellfish poisoning may also experience a strange phenomenon called *hot–cold inversion*, whereby cold is perceived as hot and vice versa.

One variety of shellfish poisoning is caused by a phenomenon referred to as **red tide.** This phenomenon occurs when a particular marine alga begins to grow quickly. As it multiplies, it produces brightly colored pigments that bloom outward and make the surrounding water appear red or brown. During this period of rapid growth, the algae produce a potent toxin called **brevetoxin**, which when consumed by humans, causes shellfish poisoning. Approximately 30 cases of shellfish poisoning are reported in the United States every year, with most cases occurring in the coastal regions of the Atlantic Northeast and Pacific Northwest.

Pesticides, Herbicides, Antibiotics, and Hormones

Because no food production system is foolproof, illness-causing pesticides, herbicides, antibiotics, and hormones sometimes come in contact with food products. A number of national and international agencies work together to ensure that the presence of these contaminants in food is negligible or poses no known risk to the consumer or environment. These agencies include the Food and Agriculture Organization (FAO) of the United Nations, the U.S. Environmental Protection Agency (EPA), the FDA, and the USDA.

An example of a dangerous compound addressed by a governmental agency is dichloro-diphenyl-trichloroethane (DDT), a pesticide once used to kill mosquitoes and increase crop yields. The use of DDT was banned in the United States in 1972, when the compound was found to damage wildlife. **Bovine somatotropin (bST)**, otherwise known as bovine growth hormone, has attracted similar public attention in recent years. This hormone is produced naturally by cattle and is used in the dairy industry to increase milk production. Although some people are concerned about its safety, a substantial amount of research suggests that bST is safe for both cows and consumers.[15] Because the evidence stands in the compound's favor, the FDA continues to allow the use of bST, and there are no laws requiring milk produced by cows treated with bST be labeled as such. Even after a compound such as bST has been approved, however, the FDA continues to evaluate its safety.

Food Allergies and Sensitivities

Certain noninfectious compounds can cause illness in small segments of the population, but not because they are toxic or poisonous in the ways already discussed. Instead, these compounds cause illness only for those people who are especially sensitive or allergic to them. Because the percentage of individuals who have adverse reactions to these compounds is so small, however, their presence in food is not prohibited.

Monosodium glutamate (MSG), a flavor enhancer used in a multitude of commercially processed foods, causes severe headaches, facial flushing, and a generalized burning sensation in some people.[16] MSG must be listed on the label of any food to which it is added. Similarly, sulfite compounds are sometimes added to foods such as wine and dried fruits to enhance color and prevent spoilage. Unfortunately, the consumption of sulfites causes breathing difficulties in sulfite-sensitive individuals, especially those with asthma. The FDA requires food manufacturers to label all foods that contain at least 10 parts per million (ppm) sulfites. You can determine if a food has added sulfites by looking for the following terms on its food label:

bovine somatotropin (bST) A growth hormone produced by cattle and used in the dairy industry to increase milk production.

- Sodium bisulfite
- Sodium metabisulfite
- Sodium sulfite
- Potassium bisulfite
- Potassium metabisulfite

Recall from Chapter 5 that exposure to proteins naturally present in some foods causes allergic reactions in certain individuals. People who are allergic to peanuts and other legumes, for example, can experience life-threatening allergic responses if they eat these foods. The most common food allergies are caused by proteins present in eggs, milk, peanuts, soy, and wheat. Researchers estimate that approximately 2 percent of adults and 5 percent of infants and young children have food allergies in the United States.

New Food Safety Concerns

Scientists and public health officials continually endeavor to identify compounds in foods that may cause illness. New research efforts often receive attention in the popular press, causing public concern. This is generally a good thing, as public concern results in scientific scrutiny. Three foodborne substances that have garnered significant public health attention in recent years are acrylamide, melamine, and bisphenol A.

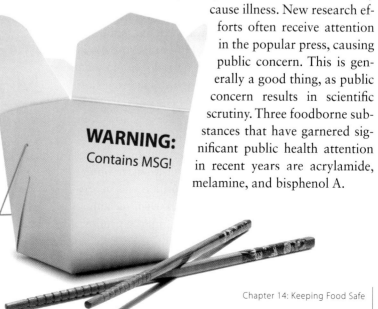

Monosodium glutamate (MSG), a compound that gives some people headaches and causes facial flushing, is commonly used in Asian cuisine.

ACRYLAMIDE

One recently emergent food safety concern involves **acrylamide**, a substance used to make polyacrylamide, which is employed in a number of food-packaging materials. Acrylamide can form in starchy foods exposed to very high temperatures (such as French fries and potato chips), and small amounts of acrylamide have been found in some products packaged in polyacrylamide. Because very high levels of acrylamide cause cancer in laboratory animals, there is concern that dietary acrylamide may be harmful to humans as well.[17] However, most evidence suggests that a person would need to eat a very large amount of fried foods to consume a cancer-causing dose of acrylamide.[18]

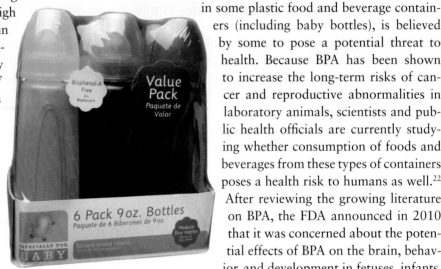

Because of **health concerns** associated with bisphenol A (BPA), BPA-free plastic products are **now widely available**.

MELAMINE

Melamine, a nitrogen-containing chemical, is typically used to make lightweight plastic objects such as dishes. In 2007, the consumption of pet food contaminated with **melamine** caused numerous dogs and cats to become ill and even die.[19] Gluten is a wheat protein commonly used to produce a number of foods, including animal foods. Because melamine is rich in nitrogen, some Chinese gluten manufacturers added melamine to their products to make them appear higher in protein than they actually were. Investigations by the FDA confirmed that melamine-tainted gluten was used to make pet foods sold in the United States. After this story was released, public attention quickly shifted to the possibility that melamine could enter the human food supply as well. In response, the FDA and USDA issued a press release stating that consumption of meat and milk from animals fed melamine-tainted feed posed very low risk to human health.[20] This position was quickly revised, however, when melamine-contaminated milk products caused nearly 60,000 infants and children throughout China to become sick in 2008 and 2009.[21] The FDA now advises consumers worldwide to avoid using infant formula products, milk products, and products with milk-derived ingredients made in China.

acrylamide A substance used to make polyacrylamide that can form in starchy foods exposed to high temperatures.

melamine A nitrogen-containing chemical typically used to make lightweight plastic objects.

bisphenol A (BPA) A chemical found in some plastic food and beverage containers believed by some to pose a potential threat to health.

BISPHENOL A

Early exposure to **bisphenol A (BPA)**, a chemical found in some plastic food and beverage containers (including baby bottles), is believed by some to pose a potential threat to health. Because BPA has been shown to increase the long-term risks of cancer and reproductive abnormalities in laboratory animals, scientists and public health officials are currently studying whether consumption of foods and beverages from these types of containers poses a health risk to humans as well.[22] After reviewing the growing literature on BPA, the FDA announced in 2010 that it was concerned about the potential effects of BPA on the brain, behavior, and development in fetuses, infants, and young children. In response to its findings, the FDA has begun supporting efforts to decrease BPA exposure in the United States.[23] Individuals who want to decrease their own exposure to BPA should:

- Choose glass or metal containers over polycarbonate ones
- Refrain from heating plastics in the microwave
- Wash plastic containers by hand instead of in the dishwasher

LO3 How Do Food Manufacturers Prevent Contamination?

Countless pathogens and inert compounds cause foodborne illness—it is simply beyond the scope of this book to describe each one in detail. However, it is important for you to understand how disease-causing agents of all types are transmitted from one food, surface, or utensil to another (a process known as **cross-contamination**) and from person to person. To do this, you must also understand the techniques that food manufacturers use to keep your food safe.

Handling and Sanitation

To prevent foodborne illness, it is critical that those who handle food do so safely and sanitarily. Pathogens can be transmitted from an infected person to almost any food. Because pathogens spread so easily, food that is otherwise safe can be made unsafe in a matter of seconds. Therefore, it is important that food handlers (including those who harvest, process, and prepare the food) avoid coughing or sneezing on the foods they work with. In addition, because many pathogens pass through the GI tract and are excreted in the feces, handlers must ensure that fecal matter not come in contact with food. People who process and prepare foods are generally required to wear gloves while working and must thoroughly wash their hands after using the toilet.

Food Production, Preservation, and Packaging

Beyond safe food-handling practices, the food-processing industry adheres to additional regulations that keep your food safe. For instance, the USDA requires that meat and poultry be inspected before they are sold and that guidelines for safe handling appear on packages. However, inspection and adherence to guidelines alone cannot guarantee that meat is pathogen free. Food manufacturers also use drying, salting, smoking, fermentation, heating, freezing, and irradiation to help keep food safe.

DRYING, SALTING, SMOKING, AND FERMENTATION

Meat has long been preserved by salting, smoking, and/or drying because these techniques—when done correctly—inhibit bacterial growth. Recall from Chapter 13 that fermentation involves the addition of yeast to a sugary beverage to produce alcohol. The addition of select yeasts and/or bacteria to food is also used to make products such as sauerkraut, yogurt, and pickles. The process of fermentation promotes the growth of nonpathogenic organisms, which in turn minimizes the growth of pathogenic organisms and preserves the food.

cross-contamination The process by which a disease-causing agent is transmitted from one food, surface, or utensil to another.

danger zone The temperature range between 40 and 140 °F, in which most microorganisms prefer to live.

HEAT TREATMENT: COOKING, CANNING, AND PASTEURIZATION

Because most microorganisms prefer living in an environment between 4 and 60 °C (40 and 140 °F), this temperature range is often referred to as the **danger zone**. Manufactures, restaurants, and individuals alike should avoid the danger zone when storing and serving food. Cooling a food to below 4 °C or heating it to above 60 °C inhibits microbial growth and can even kill existing microbes. For this reason, food manufacturers often heat-treat their products. Table 14.2 on the next page lists temperature guidelines for cooking, serving, and reheating foods. As you can see, different foods require different temperatures to maintain safety. The USDA revised its recommended cooking temperatures for pork, steaks, roasts, and chops in 2011. The USDA now recommends cooking all whole cuts of meat (excluding poultry) to 63 °C (145 °F) as measured with a food thermometer placed in the thickest part of the meat and then allowing the meat to rest for three minutes before carving or consuming.

Heating also preserves food for later consumption. Take the canning process, for example. After a food is packaged (or "canned") in a sanitized jar or can, the container—and the food inside—is heated to a high temperature. This kills pathogens and creates a vacuum within the container. The oxygen-free environment within the container helps preserve the food because most foodborne illness–causing organisms require oxygen to grow.

pasteurization A food preservation process whereby food is partially sterilized through brief exposure to a high temperature.

While some organisms (such as *C. botulinum*) can survive in low-acid anaerobic conditions, ensuring that home-canned foods have the proper acidity and are heated sufficiently helps keep them safe. You can learn more about safe canning practices at the USDA website (http://www.fsis.usda.gov/Help/FAQs_Hotline_Preparation/index.asp).

Pasteurization, the process whereby food is partially sterilized through brief exposure to a high temperature, is perhaps the most common form of heat treatment used in food preservation. Foods that are typically pasteurized include milk, juice, spices, ice cream, and cheese. Pasteurization is especially effective for killing some forms of *E. coli*, *Salmonella*, *Campylobacter jejuni*, and *Listeria monocytogenes*. Several outbreaks of *Salmonella* poisoning stemming from unpasteurized apple cider prompted new laws regarding pasteurization to be enacted. Commercially available apple cider must be either pasteurized or labeled as nonpasteurized, and the FDA recommends that homemade cider be heated for 30 minutes at 68 °C (155 °F) or 15 seconds at 82 °C (180 °F).

> **Pasteurization** is named for **French microbiologist Louis Pasteur**, who developed the process to prevent wine and beer from souring.

Table 14.2 Guidelines for Cooking, Serving, and Reheating Foods to Prevent Foodborne Illness

Note that internal temperatures should be measured with a thermometer.	
Cooking	*Pork, Beef, Veal, and Lamb*
	• Cook ground beef, pork, veal, and lamb to at least 71 °C (160 °F).
	• Cook fresh cuts of pork, beef, veal, and lamb to a minimum of 63 °C (145 °F).
	• Allow three minutes stand time for fresh cuts of these meats.
	Poultry
	• Ground turkey and chicken should be cooked to a minimum internal temperature of 74 °C (165 °F).
	• Whole and sectioned chicken, turkey, duck, and goose should also be cooked to a minimum internal temperature of 74 °C (165 °F).
	Eggs
	• Cook eggs until the yolks and whites are firm.
	• Cook egg dishes to at least 71 °C (160 °F).
	Fish
	• Cook fin fish to 63 °C (145 °F) or until the flesh is opaque and separates easily with a fork.
	• Cook shrimp, lobster, and crabs until the flesh is pearly and opaque.
Serving	• Always wash your hands with warm water and soap for 20 seconds before and after handling food.
	• Foods should not sit at room temperature for more than two hours.
Reheating	• Reheat leftovers and casseroles to at least 74 °C (165 °F).
	• Make sure there are no cold spots in food when cooking in a microwave oven.

Adapted from the Partnership for Food Safety Education and FightBAC!® Available from: http://www.fightbac.org.

COLD TREATMENT: COOLING AND FREEZING

As you have likely witnessed, foods spoil less quickly if they are kept cold. To slow or halt the growth of microorganisms, foods should always be refrigerated at 4 °C (40 °F) or colder, or frozen soon after they are prepared. This helps prevent foods from staying in the danger zone for an extended period of time.

© Cordelia Molloy/Photo Researchers, Inc.

The difference in freshness between irradiated strawberries (left) and nonirradiated strawberries (right) picked the same day is quite amazing.

IRRADIATION

In the 1950s, the National Aeronautics and Space Administration (NASA) first used **irradiation** to preserve food for space travel. Shortly thereafter, the FDA approved this food processing practice as a form of food preservation here on Earth. Today, irradiation is approved for meat, poultry, shellfish, eggs, fresh fruits, vegetables, and spices. During irradiation, foods are exposed to radiant energy that damages or destroys bacteria. Irradiation makes foods safer to eat and can dramatically increase shelf life. For example, compared to nonirradiated strawberries (which have a brief shelf life), irradiated strawberries last for several weeks without spoiling. Just as the use of x-rays to inspect luggage neither damages travelers' belongings nor makes them radioactive, irradiation neither damages nutrients nor makes foods radioactive. Still, irradiated foods must be labeled with the *radura symbol* (see Figure 14.1).

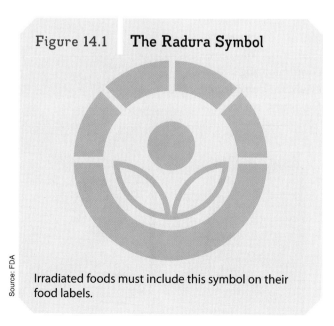

Figure 14.1 The Radura Symbol

Source: FDA

Irradiated foods must include this symbol on their food labels.

LO4 What Steps Can You Take to Reduce Foodborne Illness?

There are many precautions you can take to reduce your risk of foodborne illness. First, you should familiarize yourself with recommendations put forth by regulatory groups and keep abreast of food safety alerts and recalls. Second, you should understand and utilize basic concepts related to safe food handling. In this section, you will learn about a variety of consumer advisory bulletins that are available from the FDA and USDA. You will also learn about a national campaign developed by the U.S. government that provides useful tips for consumers on how to avoid foodborne illness.

Consumer Advisory Bulletins

Both the USDA and FDA maintain user-friendly websites and toll-free phone numbers that provide information about current food safety recommendations. Reliable USDA and FDA resources include:

- *General information*: http://www.foodsafety.gov
- *Recalls and Alerts*: http://www.foodsafety.gov/keep/recalls/

irradiation A food preservation process whereby a food is exposed to radiant energy that damages or destroys bacteria.

FightBAC!® A public education program developed to reduce foodborne bacterial illness. The four major components that comprise FightBAC!® are clean, separate, cook, and chill.

- *To report a foodborne illness*: http://www.fsis.usda.gov/FSIS_Recalls/problems_with_food_products/index.asp
- *USDA's Meat and Poultry Hotline*: 1-888-MPHotline
- *FDA's Food Safety Information Hotline*: 1-888-SAFEFOOD

The FightBac!® Campaign

Although issues related to specific foodborne illnesses may change over time, consumers are encouraged to follow some basic rules when handling, preparing, and storing food. To help consumers avoid the infectious agents that cause foodborne illness, the USDA and the Partnership for Food Safety Education developed a set of food safety guidelines called **FightBAC!®** (see Figure 14.2). The four major components that comprise FightBAC!® are clean, separate, cook, and chill.

CLEAN

To prevent pathogens found in fecal materials from contaminating your food, you should always wash your hands after using the bathroom, changing a diaper, or handling a pet. To do so effectively, wash your hands vigorously with soap for at least 20 seconds and rinse them thoroughly under clean, warm running water. Dry your hands using clean paper or cloth towels. Using an antimicrobial gel can also help ensure that your hands are pathogen free, although you should not consider such a gel a substitute for proper hand washing. A clean cooking environment is also important; you should periodically sanitize counters, equipment, utensils, and cutting boards with a solution of one tablespoon of liquid chlorine bleach in one gallon of water.

> When cleaning, be sure **not to mix liquid bleach and ammonia**. These chemicals react to form a number of toxic (and otherwise **dangerous**) chemicals.

Figure 14.2 FightBac!

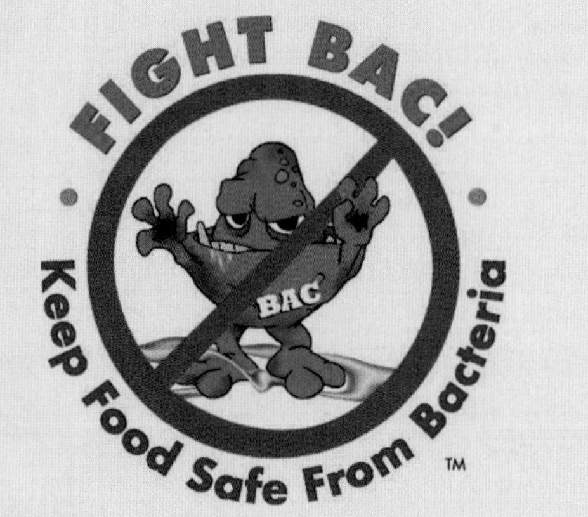

The FightBAC! campaign was created to reduce the incidence of foodborne illness by educating Americans about safe food handling practices at home and at work.

Washing fresh fruits and vegetables can also provide an important protection from foodborne pathogens and noninfectious agents (such as pesticides). Wash produce under clean, cool running water, and if possible, scrub with a clean brush. Fruits and vegetables should be dried with a clean paper or cloth towel. There is some evidence that commercial fruit and vegetable cleaners may help remove *E. coli* O157:H7 and *Salmonella* from produce, but the USDA has not taken a stance on the effectiveness of such products.[24] Note that while you should wash produce, you should not wash raw meat, poultry, or fish, as doing so increases the likelihood of cross-contaminating otherwise noninfected foods and surfaces.

SEPARATE

To prevent cross-contamination, separate raw meat and seafood from other foods in your grocery cart, in your refrigerator, and while preparing a meal. For example, put raw meats in separate, sealed plastic bags in your grocery cart and use separate cutting boards when preparing fresh produce and raw meat. Do not return cooked meat to the same plate that was used to hold the raw meat unless you have thoroughly washed the plate with hot, soapy water.

COOK

As you have learned, heat kills most dangerous microorganisms. Heat can also alter the chemical compositions of some preformed toxins, making them less dangerous. The temperatures and cooking times required to kill pathogens depend on both the particular foods and the organisms or toxins present. To ensure that your cooked food is safe to eat, measure its *internal temperature* using a probe thermometer. (Also, thoroughly clean your thermometer between uses.) Before eating leftovers, reheat your food to its appropriate internal temperature. Just because a food has been previously cooked does not mean it is pathogen-free. In the late 1990s, a *Listeria monocytogenes* outbreak resulted in at least six deaths and two miscarriages. This outbreak prompted the recommendation that lunchmeats and frankfurters be heated before eating.[25] These recommendations may be especially important for pregnant women, infants, older adults, and people with impaired immunity (such as those with HIV or cancer), as they are more likely to become seriously ill when exposed to a pathogen.

CHILL

Because cold temperatures can slow the growth of microorganisms, you should always strive to keep perishable foods chilled. For instance, instead of marinating a piece of meat at room temperature, do so in a refrigerator. Similarly, thaw a frozen food in the refrigerator, submerged in an airtight bag in cold water, or in the microwave instead of at room temperature. After a meal, refrigerate foods as quickly as possible, and separate large amounts of foods into small, shallow containers to allow the food to cool quickly.

Even properly chilled foods can become sources of foodborne illness. For this reason, you should consume leftovers within three to four days. Indeed, per the old saying, "If in doubt, throw it out!" Recommended storage times for refrigerated foods are available at the USDA website (http://www.fsis.usda.gov/PDF/Refrigeration_and_Food_Safety.pdf).

Food Safety in an Emergency

Ensuring that food and water are safe to consume normally requires access to refrigeration, a stove or oven, and clean water. But what should you do when an emergency occurs or electricity is temporarily turned off? The USDA suggests that the best approach to an emergency is to be prepared in advance. A supply of bottled water is absolutely essential, and having three to four days' worth of nonperishable food at an arm's reach can keep you both nourished and safe during an emergency situation. Consider what else you could do to prepare for a likely emergency. For example, if you live in a location that could be affected by a flood, store your emergency food on high shelves that would be safely out of the way of contaminated water. Coolers can keep food cold if the power will be out more than four hours, so having a couple on hand (along with frozen gel packs) is advised. You can learn more about how to prepare for and how to keep food safe during an emergency at the USDA website (http://www.fsis.usda.gov/Fact_Sheets/). Remember the old saying, "An ounce of prevention is worth a pound of cure."

Be Especially Careful When Eating Out

Making sure that your food is safe to eat can be especially difficult at a restaurant, picnic, potluck, or buffet. However, by considering a few simple questions,

you can minimize your risk of foodborne illness in these environments. To gauge the safety of a food, ask yourself whether the four basic FightBac!® guidelines have likely been followed:

- How likely is it that the people handling and preparing this food used sanitary practices?
- Is there evidence that raw ingredients were kept separate from cooked ones?
- Is it likely that this food was cooked properly and kept out of the thermic danger zone?
- Is there evidence that cold ingredients were kept cold?

Unfortunately, the basic rules of food safety often go neglected at picnics and social gatherings, resulting in foodborne illness. Use good judgment and eat only foods you know to be safe. It is better to be safe than sorry, after all!

LO5 What Steps Can You Take to Reduce Foodborne Illness while Traveling?

There are added food safety concerns when traveling abroad or camping. You can keep up to date on current concerns by visiting a special website that the CDC maintains for this very purpose (http://wwwnc.cdc.gov/travel/). While it is not always possible to prevent foodborne illness while traveling, you can substantially lower your risk by being vigilant in your food and beverage choices. Nonetheless, if you do experience traveler's diarrhea while abroad, it is important that you replace lost fluids and electrolytes as soon as symptoms begin to develop. Clear liquids are routinely recommended for adults, and those who develop three or more loose stools in an eight-hour period may benefit from antibiotic therapy.

Drink Only Purified or Treated Water

When traveling outside of the United States or camping in remote regions of the country, it is advisable to drink bottled water and avoid using ice. Before you drink water from a bottle, be sure that it has a fully sealed cap. If the seal is not intact, the container may have been refilled from an unknown source. If bottled water is not available, all water should be boiled for one minute before use. If at a high altitude (greater than 2,000 meters above sea level), boil water for three minutes. If boiled or bottled water is not available, fresh water can be chemically treated to kill pathogens that may be present. Portable water filters can also be used to remove some pathogens.

Avoid or Carefully Wash Fresh Fruit and Vegetables

Although fresh produce that is contaminated with bacteria may not cause illness for local residents, it can cause serious illness for visitors. Thus, when traveling to a foreign country, avoid or carefully wash fresh fruit and vegetables that are eaten without peeling (such as grapes and peppers). Of course, only water known to be clean should be used to clean produce.

Traveling in Areas with Variant Creutzfelt-Jakob Disease

The CDC recommends that when traveling in Europe or another area that has reported cases of BSE in cows, concerned travelers should consider either avoiding beef and beef products or selecting only beef products composed of solid pieces of muscle meat. These considerations, however, should be balanced with the knowledge that the risk of disease transmission is very low. Milk and milk products are not believed to pose any risk for transmission of BSE to consumers—even in areas of the world with emerging incidence of variant Creutzfelt-Jakob disease.

Emerging Issues of Food Biosecurity

The emergence of global terrorism has raised new concerns about food safety both at home and abroad. Indeed, a terrorist or militant faction could do serious widespread harm by contaminating America's food supply. As a result of this threat, **food biosecurity**—measures aimed at preventing the food supply from falling victim to planned contamination—has gained significant national attention in recent years. Of particular interest is the possibility that terrorists could use *C. botulinum* as a widespread disease-causing agent. A supply of botulism antitoxin is maintained by the CDC in the event that such an attack occurs. Congress authorized the **Public Health Security and Bioterrorism Preparedness and Response Act** (or **Bioterrorism Act**) in 2002 to ensure the continued safety of the U.S. food supply. Since its enactment, additional regulations have been approved to help ensure that foods grown and produced in the United States and abroad cannot be tampered with.[26] Clearly, food biosecurity will remain an important issue (for individuals and government agencies alike) for many years to come.

food biosecurity Measures aimed at preventing the food supply from falling victim to planned contamination.

Public Health Security and Bioterrorism Preparedness and Response Act (or **Bioterrorism Act**) Federal legislation aimed to ensure the continued safety of the U.S. food supply.

Outside of willful acts of terrorism, changes in food production and distribution systems may influence food safety and the risk of foodborne illness. For example, a contaminated food produced in a remote country might be distributed through American channels to dozens of states and sold at hundreds of grocery stores. From there, it might appear on your own table. Because production and distribution systems have become global enterprises, the origins of all foods must now be included on food labels. Still, as clearly illustrated by the recent outbreak of melamine contamination in pet food, the U.S. government has little power over whether optimal and honest food production policies are adhered to in other locations. Although the full impact of America's recent reliance on international food production and distribution is not known, some experts warn that fewer restrictions on food importation into the United States may increase the risk of foodborne illness. For such practices to continue, the health benefits imparted by the globalization of America's food supply must greatly outweigh its associated risks.

© Orien Harvey/Lonely Planet Images

Traveling abroad *requires special precautions regarding* **food safety.**

NUTR

15 Food Security, Hunger, and Malnutrition

© iStockphoto.com/Tobias Helbig

LEARNING OUTCOMES:

LO1 Define food security and food insecurity.

LO2 Understand the causes and consequences of food insecurity.

LO3 Discuss the factors that contribute to global food insecurity.

LO4 Take action against food insecurity.

Chapter 15

LO1 What Is Food Security?

You have surely experienced hunger—the physical drive to consume food. While most people ease their hunger by eating, others do not always have this option. When sufficient food is not available or accessible, hunger can lead to serious physical, social, and psychological consequences. The prevalence of persistent hunger in the world is simply astonishing. Although nobody knows for sure the exact number of people who go hungry each day, the United Nations Food and Agriculture Organization (FAO) estimates that approximately 925 million people worldwide experience persistent hunger.[1] This is roughly 14 percent of the world's population, or one in every seven people. Hunger affects people of all ages and in every country in the world—even wealthy countries like the United States.

Food security is defined as the condition whereby a person is able to obtain sufficient amounts of nutritious food to support an active, healthy life. Conversely, **food insecurity** exists when a person does not have adequate physical, social, or economic access to food.[2] Many people are surprised to learn that there is enough food produced in the world to provide every person with at least 2,700 kcal each day.[3] In other words, food insecurity is not caused by insufficient worldwide food production. Rather, it is caused by an inability to obtain sufficient food to feed oneself or one's family. Other constraints, such as limited physical or mental function, can also contribute to food insecurity. In reality, the principal factors related to worldwide food insecurity are poverty, war, and natural disaster.

food security A condition whereby a person is able to access sufficient amounts of nutritious food.

food insecurity A condition whereby a person does not have adequate physical, social, or economic access to food.

There are varying degrees of food insecurity. While temporary hunger may not pose any serious health problems, chronic hunger and undernutrition can have severe consequences. The word *hunger* is often used to describe the physical discomfort experienced by individuals who have consumed insufficient amounts of food. However, it is more commonly used on a global level to describe a chronic shortage of available food. Because many people live with uncertainty as to whether they will have enough to eat, food insecurity and the hunger it causes are major social concerns in the world today.

Responses to Food Insecurity

People living in food-insecure households respond to the threat of hunger in different ways. Whereas some take advantage of charitable organizations that assist people in need, others resort to stealing, begging,

Food insecurity *exists when a household does not have* **sufficient amounts of food.**

low food security A condition characterized by reduced food quality, variety, and/or desirability, but not reduced food intake.

very low food security A condition characterized by disrupted eating patterns and reduced food intake caused by a lack of food access and availability.

and/or scavenging to obtain food.[4] For many people, food insecurity can cause feelings of alienation, deprivation, and distress. It can adversely affect family dynamics and social interactions within the larger context of community, and of course, it can result in hunger. Clearly, the short- and long-term consequences of food insecurity can be devastating for every person in a household.

> "In the **48 contiguous states**, the annual **poverty** threshold is **$22,350** for a **family of four**."

Prevalence of Food Insecurity in the United States

How much money do you spend on food each week? According to the U.S. Department of Labor, American households spend $44 per person on food purchases every week.[5] As you might expect, households experiencing food insecurity spend considerably less. In fact, a recent government report estimated that food-insecure households spend 33 percent less on food than do food-secure households of the same size and composition.[6] Determining the breadth and depth of food insecurity in the United States is a difficult scientific challenge. In order for public health officials to develop effective strategies and target them to the appropriate populations, researchers must fully understand the scope and nature of the problem. Just as there is no single cause of food insecurity, there is no one solution to the problem.

It can be particularly challenging to assess the prevalence of food insecurity in prosperous countries because it is not usually associated with detectable signs of malnutrition. For this reason, clinical measurements of nutritional status (such as weight and height) are not always useful indicators of food insecurity. Instead, the prevalence of food insecurity in U.S. households is typically assessed using data regarding food availability and access. For example, the U.S. Department of Agriculture (USDA) issues a yearly survey that asks individuals questions about their behaviors related to food access and availability.[7] Depending on the responses to this survey, a person or household is classified as having either low food security or very low food security. Households classified as having **low food security** experience reduced food quality, variety, and/or desirability, although there is little indication of reduced food intake. This is not the case with households classified as having **very low food security**. These households are more likely to report disrupted eating patterns and reduced food intake.[8]

Recent survey results indicate that 15 percent of Americans—more than 17 million households—struggled to provide enough food for all of their family members in 2009.[9] Approximately one-third of food-insecure households reported that their eating patterns were greatly affected by a lack of money or other resources. Families with incomes below the poverty line, many of which are headed by single mothers, were at greatest risk of hunger in 2009. Given the current state of the U.S. economy and the rising cost of food (up nearly 3 percent between 2010 and 2011), the number of U.S. households

experiencing food insecurity is expected to increase.[10] Some 23 percent of all children currently living in the United States—nearly 17 million in total—live in households wherein food is scarce at times.[11]

POVERTY IS THE UNDERLYING FACTOR ASSOCIATED WITH FOOD INSECURITY

A person's risk of experiencing food insecurity in the United States is associated with income, ethnicity, family structure, and location of the home. Because many of these risk factors are interrelated, it can be difficult to determine the extent to which each one independently contributes to food insecurity. While many factors may play a role, a link between income—specifically poverty—and food insecurity is indisputable. In fact, poverty is often the common thread that ties the other factors together.

It is important to recognize that poverty is not just a problem among individuals without a source of income—many people who live in poverty maintain steady employment. Based on data provided by the U.S. Census Bureau, 33 percent of working families are officially classified as low income, and 47 million people—16 percent of the population—currently live below the poverty threshold.[12] Experts estimate that 40 percent of adults who request emergency food assistance are employed.[13] In 2009 alone, approximately 5 million households turned to private assistance programs to provide food for their families.[14] To make matters worse, some people who at one time donated money and/or food to such programs now find themselves in need of assistance.[15] It is not surprising then that many assistance programs around the country are experiencing a sharp decline in donations even though the demand for food continues to rise.[16]

When money is limited, people are often forced to reduce food-related expenses to pay for such things as housing, utilities, and health care.[17] Although one in three households living in poverty is food insecure, some households with incomes above the poverty line also experience food insecurity. An unexpected expense such as a medical-related problem or a repair bill can cause some people, at least temporarily, not to have the financial means to purchase sufficient amounts of food.

OTHER FACTORS ASSOCIATED WITH FOOD INSECURITY

food desert An environment that lacks access to affordable and/or nutritious foods.

In addition to income and poverty, many individual and socioeconomic factors can predispose a person or family to food insecurity. For example, in the United States, food insecurity is more prevalent in certain regions of the country and among certain ethnic groups (see Figure 15.1 on the next page).[18] Black and Latino households are at higher risk of food insecurity than most other racial and ethnic groups, and households headed by single women are at even higher risk. Individuals living in urban and rural areas are more likely to experience food insecurity than are those living in suburban regions. It is important to recognize, however, that all three of these factors (ethnicity, head of household, and living location) are strongly associated with income status.[19] Again, poverty is the most telling risk factor of all.

Socioeconomically disadvantaged communities often lack access to affordable and/or nutritious foods. In such an environment, referred to as a **food desert**, residents rely on local food outlets that offer limited and expensive food choices that are low in nutritional value.[20] A lack of supermarkets within disadvantaged communities presents additional hardships. Individuals must either purchase groceries at local convenience stores or navigate public transportation to other communities that have more affordable market choices—a difficult task when bearing a week's worth of groceries. Socioeconomic disparities have adverse health outcomes, and disadvantaged populations tend

Millions of Americans struggle to get enough to eat. Although assistance programs provide food to those in need, many programs report increasing difficulty meeting demand.

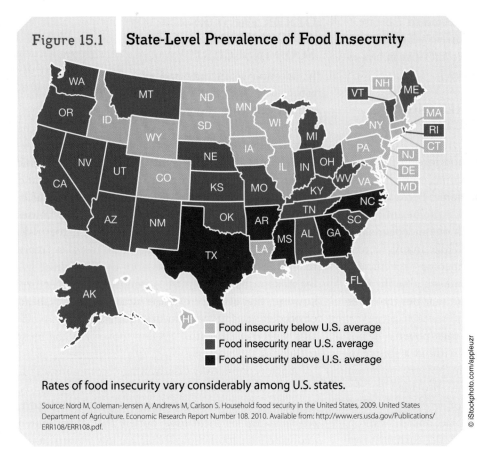

Figure 15.1 State-Level Prevalence of Food Insecurity

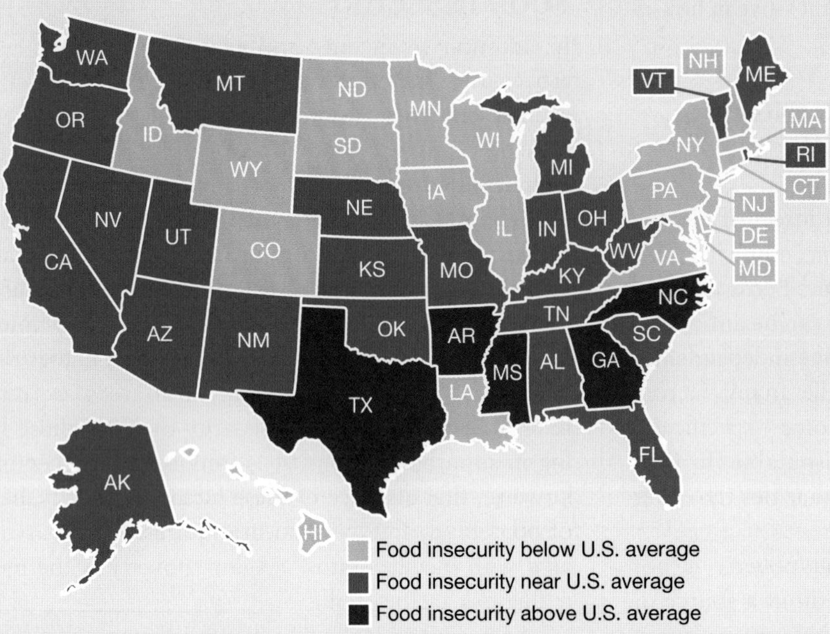

Rates of food insecurity vary considerably among U.S. states.

Source: Nord M, Coleman-Jensen A, Andrews M, Carlson S. Household food security in the United States, 2009. United States Department of Agriculture. Economic Research Report Number 108. 2010. Available from: http://www.ers.usda.gov/Publications/ERR108/ERR108.pdf.

© iStockphoto.com/appleuzr

to have high rates of obesity, type 2 diabetes, and cardiovascular disease.[21] Marketplace incentives designed to attract food retailers into neighborhoods determined to be food deserts are currently underway. Nutritionists and urban developers alike hope that improved access to affordable and healthy foods such as fruits, vegetables, whole grains, and low-fat dairy products will improve health and wellbeing among disadvantaged populations. You can learn more about where food deserts are located in the United States by visiting the USDA's Food Desert Locator (http://www.ers.usda.gov/data/fooddesert/).

LO2 What Are the Consequences of Food Insecurity?

Although food insecurity does not typically lead to starvation or nutrient deficiencies in the United States, it still represents a major public health concern. There are numerous consequences of food insecurity, many of which have been studied most extensively in women and children. For example, some studies have shown that mothers in food-insecure households shield their children from hunger by consuming less food themselves.[22] For this reason, women often experience the negative consequences of food insecurity before their children do.[23] But despite parents' best efforts, it is children who experience the most significant long-term effects of living in a food insecure household.[24] Although parents often try to protect their children from the realities of food insecurity, interviews with children reveal circumstantial awareness that is not always apparent to other family members.[25] Studies show that food-insecure children tend to have difficulties in school, earn lower scores on standardized tests, miss more days of school, exhibit more behavioral problems and depression, and be at increased risk of suicide.[26] Of course, these responses are not direct physiological consequences of food insecurity, but rather repercussions common to households experiencing this problem.

In addition to women and children, older adults are at especially high risk for food insecurity.[27] However, they often experience food insecurity differently than children, young adults, and other population groups. Elders with limited mobility and poor health, for example, may have food available to them, but may experience difficulty or anxiety associated with meal preparation. While older adults face unique challenges, poverty remains a significant indicator of food insecurity. In fact, poverty rates are highest among older women and among older adults who live alone.[28] Many older people have relatively low incomes, which they are unable to supplement with additional employment. Because income is often used to pay for basic living expenses (such as housing and utilities) rather than food, many elderly people go hungry. Perhaps not surprisingly, recent surveys indicate that the prevalence of food insecurity among older adults is on the rise.[29]

Food-Based Assistance in the United States

Fortunately for many, there are a number of programs and services available in the United States to alleviate food insecurity. Some programs are federally funded, whereas others are community efforts staffed by volunteers. Select federally funded food assistance programs are listed in Table 15.1.

SUPPLEMENTAL NUTRITION ASSISTANCE PROGRAM

The **Supplemental Nutrition Assistance Program (SNAP),** formerly known as the Food Stamp Program, is often the first line of defense against hunger for low-income households. Administered and operated by the USDA, SNAP helps more than 28 million people pay for food each month.[30] Individuals who are eligible for SNAP are given *electronic benefit transfer* (*EBT*) cards that can be used like debit cards to make food purchases at grocery stores, convenience stores, and many farmers' markets. A household's monthly monetary allotment depends on the number of people in the household and household members' combined income. The maximum monthly SNAP allotment is $200 per person and about $668 per household.[31] SNAP participants can only use their EBT cards to purchase food items; cards cannot be used to buy tobacco, alcohol, paper products, or other non-food items. Despite the program's benefits, not all individuals eligible for SNAP actually apply to receive assistance.[32] For some, applying for SNAP may be a daunting process, whereas for others, it may feel stigmatizing or demeaning.

Supplemental Nutrition Assistance Program (SNAP)
A federally funded program, formerly known as the Food Stamp Program, that helps low-income households pay for food.

Table 15.1 **Select Federal Food Assistance Programs**

Program	Major Objective	Website
Child and Adult Care Food Program (CACFP)	Provide affordable, quality day care for families, and nutrition for children and elderly adults.	http://www.fns.usda.gov/cnd/CARE/
Expanded Food and Nutrition Education Program (EFNEP)	Assist low-income people in acquiring the knowledge, skills, attitudes, and behaviors necessary to maintain nutritionally balanced diets and contribute to personal development and the improvement of the total family diet and nutritional well-being.	http://www.csrees.usda.gov/nea/food/efnep/efnep.html
Supplemental Nutrition Assistance Program (SNAP)	Provide benefits to low-income people so that they can buy food that improves their diets.	http://www.fns.usda.gov/snap/
Head Start	Promote school readiness and enhance the social and cognitive development of children by providing educational, health, nutritional, social, and other services to enrolled children and families.	http://www.acf.hhs.gov/acf_services.html#hs
Meals on Wheels® Association of America (MOWAA)	Deliver meals to people who are elderly, homebound, disabled, frail, and/or at risk of malnutrition.	http://www.mowaa.org/
National School Lunch and School Breakfast Programs	Provide children with nutritious meals for free or at reduced cost.	http://www.fns.usda.gov/cnd/Default.htm
Special Supplemental Nutrition Program for Women, Infants, and Children (WIC)	Assist in the purchase of nutritious food and provide nutrition education to low-income women, infants, and children who are at nutritional risk.	http://www.fns.usda.gov/wic/

The Special Supplemental Nutrition Program for Women, Infants, and Children (WIC) provides many important services to families in need.

© Jennifer Brown/Star Ledger/Corbis

SPECIAL SUPPLEMENTAL NUTRITION PROGRAM FOR WOMEN, INFANTS, AND CHILDREN

Millions of women, infants, and children in the United States benefit from the Special Supplemental Nutrition Program for Women, Infants, and Children (WIC). This federally funded program, also administered by the USDA, assists pregnant women and families with young children in making nutritious food purchases.[33] Individuals enrolled in WIC receive coupons that can be used to buy a variety of WIC-approved, nutrient-dense foods such as peanut butter, milk, rice, beans, cereal, and canned tuna. Many farmers markets accept WIC coupons, allowing parents to purchase a variety of fresh, locally grown fruits and vegetables. In addition to subsidizing food purchases, WIC provides health assistance and nutrition education to eligible women, young infants, and children. WIC's educational programs encourage exclusive breastfeeding and other optimal infant feeding guidelines set forth by the American Academy of Pediatrics.[34]

National School Lunch Program A federally funded program that provides nutritionally balanced meals either free of charge or at a reduced cost to school-age children at lunch time.

School Breakfast Program A federally funded program that provides nutritionally balanced meals either free of charge or at a reduced cost to school-age children at breakfast time.

food bank An organization that collects donated foods and distributes them to local food pantries, shelters, and soup kitchens.

OTHER FEDERALLY FUNDED FOOD-BASED ASSISTANCE PROGRAMS

Other federally funded food-based assistance programs available in the Unites States include the **National School Lunch Program** and the **School Breakfast Program**. These federally funded programs provide nutritionally balanced meals either free of charge or at a reduced cost to school-age children at lunch and breakfast time, respectively. Administered by the USDA, these programs are available in public schools, nonprofit private schools, residential child-care institutions, and after-school enrichment programs. The National School Lunch Program provided lunch for more than 31 million children each school day in 2009.[35] Since its enactment of in 1946, the National School Lunch Program has served more than 219 billion lunches.[36] In fact, because there is such a need, many schools and child-care programs also provide breakfast, lunch, and snacks to children throughout summer vacation. In 2010, the Institute of Medicine published a comprehensive document that provides guidance as to the optimal types and quantities of foods that should be served in the School Breakfast and National School Lunch Programs.[37] This document was followed by the passage of the Healthy Hunger-Free Kids Act of 2010, legislation that allows the USDA—for the first time in over 30 years—to make real reforms to the school lunch and breakfast programs and thus improve the critical nutritional safety net for millions of at-risk children.[38]

PRIVATELY FUNDED FOOD ASSISTANCE PROGRAMS

In addition to these and other government-funded programs, the private sector provides several services to make food more available to those in need. Private organizations include food recovery programs, food banks and pantries, and food kitchens—many of which are staffed on a volunteer basis by members of the community. A **food bank** is an agency that collects donated foods and distributes them

© Monkey Business Images/Shutterstock.com

AP Images/Grant Halverson

to local food pantries, shelters, and soup kitchens, while a **food pantry** is a program that provides canned, boxed, and sometimes fresh foods directly to individuals in need. Both food banks and food pantries rely on community donations to stock their shelves with nonperishable and perishable items, which are then distributed to people who need them. A **food kitchen** is a program that serves prepared meals to members of the community, but mostly to those who are homeless or living in shelters.

A **food recovery program** such as that run by Feeding® America (formerly called America's Second Harvest) collects and redistributes discarded food that otherwise would have gone to waste. Recovered foods are donated to food pantries, emergency kitchens, and homeless shelters, providing millions of hungry people with food. The USDA estimates that food retailers and producers discard more than 43 million kilograms (96 billion pounds) of edible food annually.[39] There are many types of food recovery efforts, some of which include:

- *Field gleaning* programs gather and distribute agricultural crops that would otherwise not have been harvested.
- *Food rescue* initiatives collect unused perishable foods from retail grocery stores, gardens, restaurants, campus dining facilities, hotels, and/or caterers.
- *Nonperishable food collection* programs gather and donate damaged and/or dated canned and boxed foods from retail sources.

Food distribution programs collect and distribute food that would otherwise be wasted.

> A 2007 National Alliance to End **Homelessness study found** that the **states with the highest rates** of homelessness are Alaska, California, Colorado, Hawaii, Idaho, Nevada, Oregon, Rhode Island, and Washington State.

LO3 What Causes Worldwide Hunger and Malnutrition?

Because poverty is more prevalent in developing countries than in industrialized ones, food insecurity tends to be most ubiquitous and severe in nations with low *per capita* incomes. The Food and Agriculture Organization of the United Nations (FAO) estimates that 16 percent of people living in poor countries do not have enough to eat, and as a result experience persistent hunger.[40] Using the International Food Policy Research Institute's 2010 Global Hunger Index, the FAO reported that the worldwide number of hungry people reached nearly 1 billion in 2010.[41] According to the report, Asia has the largest number of hungry people (578 million), but sub-Saharan Africa has the highest percentage.[42] In this region, 239 million people—33 percent of the population—do not have enough food to eat (see Figure 15.2 on the next page).

food pantry A program that provides canned, boxed, and sometimes fresh foods directly to individuals in need.

food kitchen A program that prepares and serves meals to members of the community who are in need.

food recovery program A program that collects and redistributes discarded food that otherwise would have gone to waste.

Many Factors Contribute to Global Food Insecurity

The causes of global food insecurity are complex, and the factors that contribute to food insecurity in poor countries are often different from those that contribute to food insecurity in the United States. Most experts agree that global food insecurity is not caused by a lack of available food on the international level. Rather, it is caused by diminished local food supplies, which are themselves caused by a variety of socioeconomic conditions. Political instability, a lack of available land for growing crops, population growth, and gender inequalities can all contribute to food instability.

POLITICAL UNREST

Availability of and access to food are often limited by civil strife, war, and political unrest. Political turmoil can displace millions of people from their homes and force them to relocate to crude facilities set up for refugees. In countries with large refugee populations (such as Sudan, Rwanda, and Pakistan), food insecurity and malnutrition are rampant. According to the United Nations (UN), those living in refugee camps have the highest rates of disease and malnutrition of any group worldwide.[43] Because of the danger and logistical problems associated with political unrest, it can be difficult for relief agencies to provide much-needed aid to innocent civilians. Despite recent progress made by repatriation movements, the number of refugees is once again on the rise. Mainly attributable to violence taking place in the Middle East, the worldwide number of forcibly displaced persons is approaching 43 million—15 million of whom are political refugees.[44]

URBANIZATION

The use of land for reasons other than feeding a region's people and supporting local economies can

Although humanitarian aid agencies try to meet the needs of those living in refugee camps, their efforts cannot match the immense needs for food, water, and medicine.

© Reuters/Landov

Figure 15.2 International Prevalence of Food Insecurity

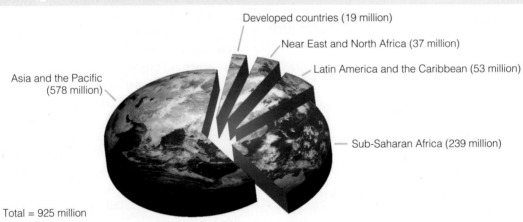

The Food and Agriculture Organization estimates that the overall number of hungry people in the world is approaching 1 billion.

Source: Food and Agriculture Organization of the United Nations. The state of food insecurity in the world: Addressing food security in protracted crises. 2010. Available from http://www.fao.org/docrep/012/al390e/al309e00.pdf.

© iStockphoto.com/Jan Rysavy

contribute significantly to a food shortage.[45] Without the land to grow crops, people cannot produce adequate amounts and varieties of food for themselves and their families. As a result, many people have relocated from rural regions to urban ones with the hope of finding employment opportunities. This shift in America's population, called **urbanization**, is both a consequence and cause of food shortages in many parts of the world. Urbanization and industrialization have had profound impacts on population demographics, transforming food systems and creating new nutritional challenges.[46] For example, the expansion of large supermarket chains in urban areas has greatly impacted small food producers and retailers.[47] Rather than buying their products from small, local farms, supermarket chains are more likely to utilize large consolidated food distribution centers. A shift in food production, procurement, and distribution systems has contributed to the displacement of workers, a decline in traditional food markets, and a fundamental change in local food culture.[48]

© iStockphoto.com/3DStock

Fueled by changes brought on by urbanization, the composition of diets among city-dwellers has shifted away from traditional foods and toward processed foods. Fast food restaurants have largely replaced street vendors who once sold local foods. Recall from Chapter 1 that this dietary shift has led to nutrition transition. The fact that many developing regions face food shortages at the same time that they experience increasing rates of obesity, heart disease, type 2 diabetes, and other diet-related health problems highlights the special challenges related to food insecurity worldwide. The simultaneous occurrence of two distinct nutritional challenges—food shortages and obesity—underscores the importance of addressing the needs of both rural poor and those in urban migration.

POPULATION GROWTH

Population growth across some of the poorest regions of the world has significantly increased the challenge of providing adequate food and water to all those in need. Discouragingly, but perhaps not surprisingly, countries with the fastest growing populations tend to be those already burdened with staggering rates of hunger and malnutrition.[49] As a population proliferates, the ability of a government and assistance programs to provide even the most basic of needs—food and shelter—may be compromised even further.

GENDER INEQUALITY

urbanization A shift in a country's population from rural regions to urban ones.

In many developing nations, a gap exists between the opportunities that are made available to men and women. Some experts believe that promoting gender equality among underprivileged peoples holds the greatest promise for reversing the steadily increasing trend of global hunger.[50] For instance, providing equal educational and employment opportunities to girls and women might increase individual earning capacities, improve maternal nutrition and health, and bolster overall *per capita* income. Clearly, a lack of education and opportunity sustains the vicious cycle of poverty that is passed on from one generation to the next.

Global Food Insecurity and Malnutrition

Although there are many consequences of food insecurity in poor and developing countries, perhaps the most important and devastating consequence is malnutrition. As you have learned, malnutrition is poor nutritional status that results from inadequate or excessive dietary intake. In the case of a food shortage, malnutrition takes the form of undernutrition, which has both short- and long-term effects on the health of individuals, families, and societies. In the case of the nutrition transition, malnutrition takes the form of overnutrition as urbanized individuals shift toward unhealthy, unbalanced diets.

FORMS OF GLOBAL MALNUTRITION

Some malnourished people consume sufficient calories but lack certain nutrients, whereas others lack both calories and nutrients. Iron, iodine, and vitamin A deficiencies, the three most common micronutrient deficiencies, affect billions of people worldwide.[51] Countless infants and children suffer from iron and iodine deficiencies, both of which can impair growth and cognitive development. The lives of millions of preschool-age children are further compromised by vitamin A deficiency, which causes blindness and many other serious consequences.[52]

Simply put, the number of women, infants, and children in the world with micronutrient deficiencies is staggering. According to the United Nations Children's

Fund (UNICEF), the health and welfare of nearly one-third of the world's population are affected by vitamin and mineral deficiencies.[53] Yet the resources exist to rectify these problems: by distributing low-cost, nutrient-rich foods and/or nutrient supplements to those in need, world governments and agencies could eradicate micronutrient deficiencies throughout the world. Some efforts have been made to do just that, but meeting the needs of the hungry is an arduous, time-sensitive process. Although adequate nutrition is important throughout a child's life, the window of opportunity for an intervention to have the greatest impact is between conception and the child's second birthday. The effects of persistent malnutrition on a child's health and development are largely irreversible after the age of two.[54]

INDIVIDUAL CONSEQUENCES OF MALNUTRITION

Like food insecurity, women, infants, and children are especially vulnerable to malnutrition. When experienced during pregnancy, malnutrition can deplete a mother's nutritional stores and increase her risk of having a low-birth-weight (LBW) baby. Poor maternal nutritional status can also increase the risk of neonatal death. In fact, some studies estimate that nearly 60 percent of the deaths of infants and young children in the world are caused in part by malnutrition.[55] Because poor nutrition compromises the immune system, malnutrition can worsen the adverse effects of disease, leading to premature death. When a child is slightly underweight, his risk of death increases to 2.5 times that of a child with a healthy weight.[56] Risk of death

Saving the Lives of Children with Ready-to-Use Therapeutic Food

Malnutrition claims thousands of lives each day, but the burden of persistent hunger is particularly harmful to developing children. For example, almost half of the children in India are underweight and have impaired growth due to malnutrition by the time they are five years old.[57] Although providing children with food seems like the obvious solution, this process can be more complicated than you might think. For example, the children who need food the most often live in remote areas that are difficult to reach. A lack of refrigeration, clean water, and limited cooking amenities also complicates matters. For example, contaminated water and drought-stricken regions render powdered milk useless. Perishable foods spoil if not stored properly, and grains and cereals are not high-quality sources of protein. While it is difficult to provide adequate food to individuals living in these conditions, the availability of ready-to-use therapeutic foods (RUTF) has overcome many such geographic and logistic challenges. RUTF products are prepackaged and require no preparation. Their ingredients are based mainly of peanut butter, vegetable oil, milk powder, sugar, vitamins, and minerals. These nutrient-dense, high-quality protein products require no refrigeration, are easily distributed, and generally have long shelf lives. RUTFs are becoming the standard of care when it comes to refeeding malnourished children worldwide.

A ready-to-use therapeutic food is a nutrient-dense, high-quality protein product that has a long shelf life. The use of RUTFs has been credited with saving the lives of millions of children around the world.

increases even further when a child is severely underweight.

Even if it does not cause premature death, malnutrition can seriously affect a child's growth and development. Children with impaired growth tend to be in the lowest height-for-age percentiles on a growth reference curve. As a consequence of malnutrition, an estimated 24 percent of children younger than five—149 million children—have impaired growth.[57] Africa has the highest percentage of undernourished infants and young children (35 percent), though it is followed closely by Asia, Latin American, and the Caribbean.[58] Even if a malnourished infant survives, impaired growth can compromise her health, well-being, and ability to function later in life.

Even if it does not cause premature death, malnutrition can seriously affect a child's growth and development.

SOCIETAL CONSEQUENCES OF MALNUTRITION

Beyond its effects on the health of the individual, malnutrition can harm whole societies. Extensive food insecurity and malnutrition can result in an entire nation of adults with reduced capacities for physical work, and therefore lower work productivity overall. These consequences can have profound and long-term adverse effects on a country's economic growth and standard of living. Therefore, not only is poverty a cause of hunger, but hunger is a cause of poverty. Addressing both food insecurity and malnutrition is critical to economic progress in underdeveloped nations.

International Organizations Provide Global Food-Based Assistance

Unlike most wealthy countries, impoverished countries often lack stable governments and have few programs in place to assist those in need. Developing countries typically depend on relief efforts provided by international organizations such as the World Health Organization (WHO), the United Nations (UN), U.S. Peace Corps, and Heifer International®. Organizations like these help individuals and entire communities make lasting changes that ultimately improve health and food security across the developing world. International interventions with the greatest impact include efforts to improve maternal nutrition during pregnancy and lactation, reduce societal inequities, provide access to health services, promote self-sufficiency, and reduce illiteracy.[59]

The United Nations is a multinational organization that addresses a multitude of global problems, such as worldwide food insecurity.

THE UNITED NATIONS

Many international organizations are committed to the alleviation of world hunger. One such organization is the United Nations, a multinational organization first established in 1945 to promote peace through international cooperation and collective security. Today, 192 member states comprise the UN, and though the organization serves many purposes, combating international hunger is among its most important efforts. For example, UNICEF, a component of the UN, presented a conceptual framework in 2011 that provides

conditional cash-transfer program An initiative that offers financial reimbursement for individuals or communities that work to improve quality of life.

U.S. Peace Corps A federally funded volunteer program that promotes world peace and friendship worldwide.

communities with incentives for working to improve quality of life (such as by relieving hunger) in the world's poorest countries. Proposed incentives include financial reimbursement for families whose children attend school and for families who start their own small businesses.[60] This type of initiative is called a **conditional cash-transfer program.**

Another UN effort, the Millennium Development Project, has pledged to halve the proportion of people who suffer from hunger in the world by the year 2015.[61] Endorsed by the majority of the countries in the world, this unprecedented effort addresses the needs of the world's poorest countries. To achieve this project's lofty goal, efforts are currently underway to improve education, promote gender equality, reduce infant mortality, improve maternal health, combat AIDS and other infectious diseases, promote sustainable agricultural practices, and develop global economic partnerships. Although on track to meet several of these goals, the Millennium Development Project is an ongoing challenge that requires commitment and dedication from the entire worldwide community. In spite of, and because of its scope, the potential for Millennium Development Project to ease the burden of food insecurity, hunger, and malnutrition throughout the world is substantial.

PEACE CORPS

How can one person make a meaningful difference in the world, especially in regard to world hunger? One such example can be gleaned from President John F. Kennedy, who challenged students in 1961 to serve their country by working to improve quality of life for people in developing countries. This challenge served as inspiration for a federally funded program called the **U.S. Peace Corps.** The mission of the Peace Corps was—and still is—to promote world peace and friendship by:

- Assisting interested countries in meeting their need for trained men and women.
- Bringing a better understanding of Americans to people in other countries.
- Helping promote a better understanding of other peoples on the part of Americans.

Since the Peace Corps was developed, more than 200,000 people have served in more than 139 countries. These volunteers work to improve developing nations by helping farmers grow crops, teaching mothers to better care for their children, and educating entire communities about health and disease prevention. Thus, the Peace Corps offers the opportunity to make a difference in the lives of others by addressing the problems of food insecurity and malnutrition throughout the world.

HEIFER INTERNATIONAL

Heifer International® is a humanitarian effort with a global commitment to foster environmentally sound farming methods that combat both hunger and environmental concerns. This organization recognizes that impoverished peoples often make decisions based on short-term needs rather than cultivating long-term solutions. Heifer International strives to teach families how to restore and manage land in ways that provide food and income for generations to come. Heifer's pragmatic problem-solving approach leads to novel long-term solutions that empower impoverished communities to

© Charles O. Cecil/Alamy

Many **Peace Corps volunteers** *find their work both* **challenging and rewarding.**

provide for themselves. This focus on long-term development (rather than temporary relief) helps restore hope, health, and dignity among those with few resources. Most remarkable, however, is Heifer's *living loans* program, which ensures project sustainability. In a living loan, a community receives a gift of livestock that brings benefits such as food, wool, and nonmechanized power. To repay this loan, the community gifts the offspring of the livestock to another farmer or community. This secondary donation repays the original debt while bringing the hope of prosperity to others. This simple concept of passing on the gift is the founding philosophy on which Heifer International was created, and the means by which it has fostered a living cycle of sustainability for over 65 years.

> "Need a **quick gift idea?** Heifer International allows anyone to make a **gift of livestock** in another person's name at its website, **www.heifer.org**."

LO4 What Can You Do to Alleviate Food Insecurity?

Experts generally agree that there is enough food in the world to feed every living person.[62] Why then are food insecurity and malnutrition so pervasive and devastating? As you have learned, the causes of food insecurity and malnutrition differ by geographic region, gender, political climate, life stage, economic policy, ethnicity, and rate of population growth—among many other factors. Because the causes of food insecurity are so varied and intrinsically interrelated, it is important for nutritionists and policymakers to work both diligently and carefully. It is only by considering the complexity of the issue that the relative importance of each contributing factor can be addressed. And it is only then that effective solutions can be developed.

Leading world health experts agree that improving food availability and access must be a global priority. Although malnutrition is a direct consequence of insufficient dietary intake, the ultimate causes of malnutrition often have more to do with economic and social circumstances. Even if all other factors are on the right track, a high prevalence of certain diseases (such as AIDS), violence, illiteracy, and/or political corruption can devastate food availability and/or access. Thus, to affect a genuine remedy to food insecurity and its related malnutrition, the underlying societal problems must be addressed.

© Jim West/Alamy

Taking Action against Hunger

Although the problem of food insecurity may seem staggering both at home and abroad, it is important to remember that there is much an individual can do to take action against hunger. Despite the scope of the problem, your actions can make a difference in the lives of others. Working alone or collectively toward the elimination of hunger and malnutrition is a worthwhile (and noble) personal and professional priority. The American Dietetic Association (ADA) is one of several professional organizations that work to alleviate world hunger by challenging its members to take action. For instance, the ADA recently created a Dietetic Practice Group to encourage dietitians to work with each other and with other health professionals to reduce poverty and hunger in their communities. The American Society for Nutrition (ASN) is another organization that is committed to the development of international strategies and policies that can help alleviate hunger and poverty around the world.

Endnotes

Chapter 1

1. Vicini J, Etherton T, Kris-Etherton P, Ballam J, Denham S, Staub R, Goldstein D, Cady R, McGrath M, Lucy M. Survey of retail milk composition as affected by label claims regarding farm-management practices. Journal of the American Dietetic Association. 2008;108:1198–203. Dangour AD, Dodhia SK, Hayter A, Allen E, Lock K, Uauy R. Nutritional quality of organic foods: A systematic review. American Journal of Clinical Nutrition. 2009;90:680–5.
2. Ward RE, German JB. Zoonutrients and health. Food Technology. 2003;57:30–36. Pappas E, Schaich KM. Phytochemicals of cranberries and cranberry products: Characterization, potential health effects, and processing stability. Critical Reviews in Food Science and Nutrition. 2009;49:741–81. Gullett NP, Ruhul Amin AR, Bayraktar S, Pezzuto JM, Shin DM, Khuri FR, Aggarwal BB, Surh YJ, Kucuk O. Cancer prevention with natural compounds. Seminars in Oncology. 2010;37:258–81.
3. Gobbetti M, Cagno RD, De Angelis M. Functional microorganisms for functional food quality. Critical Reviews in Food Science and Nutrition. 2010;50:716–27. Mattes RD, Dreher ML. Nuts and healthy body weight maintenance mechanisms. Asia Pacific Journal of Clinical Nutrition. 2010;19:137–41. Rao AV, Snyder DM. Raspberries and human health: A review. Journal of Agricultural Food Chemistry. 2010;58:3871–83. Tulipani S, Mezzetti B, Battino M. Impact of strawberries on human health: Insight into marginally discussed bioactive compounds for the Mediterranean diet. Public Health Nutrition. 2009;12:1656–62.
4. Lieberman HR. Cognitive methods for assessing mental energy. Nutrition and Neuroscience. 2007;10:229–42. Chelben J, Piccone-Sapir A, Ianco I, Shoenfeld N, Kotler M, Strous RD. Effects of amino acid energy drinks leading to hospitalization in individuals with mental illness. General Hospital Psychiatry. 2008;30:187–9. Clauson KA, Shields KM, McQueen CE, Persad N. Safety issues associated with commercially available energy drinks. Journal of the American Pharmacologic Association. 2008;48:e55–63.
5. Carey SS. A beginner's guide to scientific method, 3rd ed. Belmont, CA: Wadsworth/Thomson Learning; 2004.
6. Ibid.
7. National Center for Health Statistics (NCHS). Health, United States, 2010. Available from: http://www.cdc.gov /nchs/data/hus/hus10.pdf; Centers for Disease Control and Prevention. Leading causes of death, 1900–1998. Available from: http://www.cdc.gov/nchs/data/dvs/lead1900_98.pdf.
8. Ibid.
9. Perls T, Terry D. Understanding the determinants of exceptional longevity. Annals of Internal Medicine. 2003; 139:445–9.
10. Anderson RE, Smith RD, Benson ES. The accelerated graying of American pathology. Human Pathology. 1991;22:210–4.
11. Armstrong BL, Conn LA, Pinner RW. Trends in infectious disease mortality in the United States during the 20th century. Journal of the American Medical Association. 1999;281:61–6. Centers for Disease Control and Prevention. Achievements in public health, 1900–1999: Control of infectious diseases. Morbidity and Mortality Weekly Report. 1999;48:621–9.
12. U.S. Department of Health and Human Services. Office of Disease Prevention and Health Promotion. Healthy People 2020. ODPHP Publication No. B0132. November 2010. Available from: http://www.healthypeople.gov.
13. National Center for Health Statistics (NCHS). Health, United States, 2010. Available from: http://www.cdc.gov /nchs/data/hus/hus10.pdf. Centers for Disease Control and Prevention. Leading causes of death, 1900–1998. Available from: http://www.cdc.gov/nchs/data/dvs/lead1900_98.pdf.

Chapter 2

1. Institute of Medicine. Dietary reference intakes for vitamin C, vitamin E, selenium, and carotenoids. Washington, DC: National Academies Press; 2000.
2. Institute of Medicine. Dietary reference intakes: Application in dietary assessment. Washington, DC: National Academies Press; 2000.
3. Institute of Medicine. Dietary Reference Intakes for energy, carbohydrate, fiber, fat, fatty acids, cholesterol, protein, and amino acids. Washington, DC: National Academies Press; 2005.
4. Welsh S, Davis C, Shaw A. A brief history of food guides in the United States—Food Guide Pyramid. Nutrition Today. 1992 (Nov.–Dec.); 6–11. Welsh S, Davis C, Shaw A. Development of the food guide pyramid—food guide pyramid. Nutrition Today. 1992 (Nov.–Dec.);12–15.
5. U.S. Department of Agriculture and U.S. Department of Health and Human Services. Dietary Guidelines for Americans, 2010. 7th Edition, Washington, DC: U.S. Government Printing Office, December 2010.
6. U.S. Department of Agriculture Food Safety and Inspection Service. Nutrition labeling of single-ingredient products and ground or chopped meat and poultry products. Federal Register. Dec. 29, 2010. Available from: http://www.federal register.gov/articles/2010/12/29/2010-32485/nutrition-labeling-of-single-ingredient-products.

Chapter 3

1. Spechler SJ. Clinical manifestations and esophageal complications of GERD. American Journal of Medical Sciences. 2003;326:279–84.
2. DeCross AJ, Marshall BJ. The role of *Helicobacter pylori* in acid-peptic disease. American Journal of Medical Sciences. 1993;306:381–92. Tan VP, Wong BC. *Helicobacter pylori* and gastritis: Untangling a complex relationship 27 years on. Journal of Gastroenterolgy and Hepatology. 2011;26:42–5.
3. Meyer JH. Gastric emptying of ordinary food: effect of antrum on particle size. American Journal of Physiology. 1980;239:G133-5. Hunt JN. Mechanisms and disorders of gastric emptying. Annual Review of Medicine. 1983; 34:219–29.

4. Fasano A, Berti I, Gerarduzzi T, Not T, Colletti RB, Drago S, Elitsur Y, Green P, Guandalini S, Hill ID, Pietzak M, Ventura A, Thorpe M, Kryszak D, Fornaroli F, Wasserman SS, Murray JA, Horvath K. Prevalence of Celiac Disease in at-risk and not-at-risk groups in the United States. A large multicenter study. Archives of Internal Medicine. 2003;163:286–92.
5. Dobrogosz WJ, Peacock TJ, Hassan HM. Evolution of the probiotic concept from conception to validation and acceptance in medical science. Advances in Applied Microbiology. 2010;72:1–41. Kolida S, Saulnier DM, Gibson GR. Gastrointestinal microflora: Probiotics. Advances in Applied Microbiology. 2006;59:187–219. Rastall RA. Bacteria in the gut: Friends and foes and how to alter the balance. Journal of Nutrition. 2004;134:2022S–6S. Pham TT, Shah NP. Biotransformation of isoflavone glycosides by bifidobacterium animalis in soymilk supplemented with skim milk powder. Journal of Food Science. 2007;72:316–24. Guslandi MJ. Probiotic agents in the treatment of irritable bowel syndrome. Journal of International Medical Research. 2007;35:583–9. Hedin C, Whelan K, Lindsay JO. Evidence for the use of probiotics and prebiotics in inflammatory bowel disease: A review of clinical trials. Proceedings of the Nutrition Society. 2007;66:307–15. DiBaise JK, Zhang H, Crowell MD, Krajmalnik-Brown R, Decker GA, Rittmann BE. Gut microbiota and its possible relationship with obesity. Mayo Clinic Proceedings. 2008;83:460–9. Kalliomäki M, Collado MC, Salminen S, Isolauri E. Early differences in fecal microbiota composition in children may predict overweight. American Journal of Clinical Nutrition. 2008;87:534–8. Cani PD, Delzenne NM. Gut microflora as a target for energy and metabolic homeostasis. Current Opinion in Clinical Nutrition and Metabolic Care. 2007;10:729–34.
6. Miller FG, Colloca L, Kaptchuk TJ. The placebo effect: Illness and interpersonal health. Perspectives in Biology and Medicine. 2009;52:518–39. Wampold BE, Minami T, Tierney SC, Baskin TW, Bhati KS. The placebo is powerful: Estimating placebo effects in medicine and psychotherapy from randomized clinical trials. Journal of Clinical Psychology. 2005;6:835–54. Kaptchuk TJ, Kelley JM, Conboy LA, Davis RB, Kerr CE, Jacbson EE, Kirsch I, Schyner RN, Nam BH, Nguyen LT, Park M, Rivers AL, McManus C, Kokkotou E, Drossman DA, Goldman P, Lembo AJ. Components of placebo effect: Randomised controlled trial in patients with irritable bowel syndrome. British Medical Journal. 2008;336:1–8.
7. Xavier RJ, Podolsky DK. Unraveling the pathogenesis of inflammatory bowel disease. Nature. 2007;448:427–34. Lakatos PL. World recent trends in the epidemiology of inflammatory bowel diseases: Up or down? Journal of Gastroenterology. 2006;12:6102–8.

Chapter 4

1. Stanhope KL, Havel PJ. Fructose consumption: recent results and their potential implications. Annals of the New York Academy of Sciences. 2010;1190:15–24.
2. U.S. Department of Agriculture, Economic Research Service. 2011. Table 51—Refined cane and beet sugar: estimated number of per capita calories consumed daily, by calendar year. Table 52—High fructose corn syrup: estimated number of per capita calories consumed daily, by calendar year. Available from: http://www.ers.usda.gov/Briefing/Sugar/Data.htm.
3. Moeller, Fryhofer SA, Osbahr AJ 3rd, Robinowitz CB; Council on Science and Public Health, American Medical Association. The effects of high fructose corn syrup. Journal of the American College of Nutrition. 2009;28:619–26.
4. Johnson R, Appel LJ, Brands M, Howard BV, Lefevre M, Lustig RH, Sacks F, Steffen LM, Wylie-Rosett J. Dietary sugars intake and cardiovascular health: a scientific statement from the American Heart Association. Circulation. 2009;120:1011–20.
5. National Cancer Institute. Sources of added sugars in the diets of the U.S. population ages 2 years and older, NHANES 2005–2006. Risk Factor Monitoring and Methods. Cancer Control and Population Sciences. Available from: http://riskfactor.cancer.gov/diet/foodsources/added_sugars/table5a.html.
6. U.S. Department of Agriculture and U.S. Department of Health and Human Services. Dietary Guidelines for Americans, 2010. 7th Edition, Washington, DC: U.S. Government Printing Office, December 2010.
7. Magnuson BA, Burdock GA, Doull J, Kroes RM, Marsh GM, Pariza MW, Spencer PS, Waddell WJ, Walker R, Williams GM. Aspartame: A safety evaluation based on current use levels, regulations, and toxicological and epidemiological studies. Critical Reviews in Toxicology. 2007;37:629–727.
8. Lomax AR, Calder PC. Prebiotics, immune function, infection and inflammation: A review of the evidence. The British Journal of Nutrition. 2009;101:633–58.
9. Meyer D, Stasse-Wolthuis M. The bifidogenic effect of inulin and oligofructose and its consequences for gut health. European Journal of Clinical Nutrition. 2009;63:1277–89.
10. Anderson JW, Baird P, Davis RH Jr, Ferreri S, Knudtson M, Koraym A, Waters V, Williams CL. Health benefits of dietary fiber. Nutrition Reviews. 2009;67:188–205.
11. Charalampopoulos D, Wang R, Pandiella SS, Webb C. Application of cereals and cereal components in functional foods: A review. International Journal of Food Microbiology. 2002;79:131–41.
12. Venn BJ, Green TJ. Glycemic index and glycemic load: Measurement issues and their effect on diet–disease relationships. European Journal of Clinical Nutrition. 2007;6:S122–31.
13. Astrup A, Meinert Larsen T, Harper A. Atkins and other low-carbohydrate diets: Hoax or an effective tool for weight loss? Lancet. 2004;364:897–9.
14. Ariza MA, Vimalananda VG, Rosenzweig JL. Reviews of endocrine and metabolic disorders. The economic consequences of diabetes and cardiovascular disease in the United States. 2010;11:1–10.
15. Han JC, Lawlor DA, Kimm SY. Childhood obesity. Lancet. 2010;375:1737–48.
16. Groop L, Lyssenko V. Genetics of type 2 diabetes. An overview. Endocrinology and Nutrition. 2009;56:S34–7.
17. Votruba SB, Jensen MD. Regional fat deposition as a factor in FFA metabolism. Annual Review of Nutrition. 2007;27:149–63.
18. Albright A. What is public health practice telling us about diabetes? Journal of the American Dietetic Association. 2008;108:S12–8.
19. Baier LJ, Hanson RL. Genetic studies of the etiology of type 2 diabetes in Pima Indians: hunting for pieces to a complicated puzzle. Diabetes. 2004;53:1181–6.
20. Slavin JL, Jacobs D, Marquart L, Wiemer K. The role of whole grains in disease prevention. Journal of the American Dietetic Association. 2001;101:780–5.

Chapter 5

1. Furst P, Stehle P. What are the essential elements needed for the determination of amino acid requirements in humans? Journal of Nutrition. 2004;134:1558S–65S.
2. Institute of Medicine. Dietary Reference Intakes for energy, carbohydrate, fiber, fat, fatty acids, cholesterol, protein, and amino acids. Washington, DC: National Academies Press; 2005.
3. Furst P, Stehle P. What are the essential elements needed for the determination of amino acid requirements in humans? Journal of Nutrition. 2004;134:1558S–65S.

4. Erlandsen H, Patch MG, Gamez A, Straub M, Stevens RC. Structural studies on phenylalanine hydroxylase and implications toward understanding and treating phenylketonuria. Pediatrics. 2003;112:1557–65. van Spronsen FJ, Enns GM. Future treatment strategies in phenylketonuria. Molecular Genetics and Metabolism. 2010;99:S90–5.
5. American Academy of Pediatrics. Pediatric nutrition handbook, 6th ed. Kleinman RE, editor. Elk Grove Village, IL: American Academy of Pediatrics; 2008.
6. Reeds PJ, Garlick PJ. Protein and amino acid requirements and the composition of complementary foods. Journal of Nutrition. 2003;133:2953S–61S.
7. Schnog JB, Duits AJ, Muskeit FAJ, ten Cate H, Rojer RA, Brandjes DPM. Sickle cell disease: A general overview. Journal of Medicine. 2004;62:364–74. López C, Saravia C, Gomez A, Hoebeke J, Patarroyo MA. Mechanisms of genetically-based resistance to malaria. Gene. 2010;467:1–12.
8. Ibid.
9. Burdge GC, Hanson MA, Slater-Jefferies JL, Lillycrop KA. Epigenetic regulation of transcription: A mechanism for inducing variations in phenotype (fetal programming) by differences in nutrition during early life? British Journal of Nutrition. 2007;97:1036–46. Hanley B, Dijane J, Fewtrell M, Grynberg A, Hummel S, Junien C, Koletzko B, Lewis S, Renz H, Symonds M, Gros M, Harthoorn L, Mace K, Samuels F, van Der Beek EM. Metabolic imprinting, programming and epigenetics—A review of present priorities and future opportunities. British Journal of Nutrition. 2010;104:S1–25. Mathers JC. Early nutrition: Impact on epigenetics. Forum in Nutrition. 2007;60:42–8.
10. Taylor SL, Hefle SL. Food allergy. In: Present knowledge in nutrition, 9th ed. Bowman BA, Russell RM, editors. Washington, DC: ILSI Press; 2006.
11. National Institute of Allergy and Infectious Diseases. Food allergy. Report of the NIH expert panel on food allergy research. 2006. Available from: http://www.niaid.nih.gov/topics/foodallergy/research/pages/reportfoodallergy.aspx.
12. Ibid.
13. Fuller MF, Reeds PJ. Nitrogen cycling in the gut. Annual Review of Nutrition. 1998;18:385–411. Rand WM, Pellet PL, Young VR. Meta-analysis of nitrogen balance studies for estimating protein requirements in healthy adults. American Journal of Clinical Nutrition. 2003;77:109–27.
14. Institute of Medicine. Dietary Reference Intakes for energy, carbohydrate, fiber, fat, fatty acids, cholesterol, protein, and amino acids. Washington, DC: National Academies Press; 2005.
15. Dewey KG. Energy and protein requirements during lactation. Annual Review of Nutrition. 1997;17:19–36.
16. American Dietetic Association, Dietitians of Canada, and the American College of Sports Medicine. Position of the American Dietetic Association, Dietitians of Canada, and the American College of Sports Medicine: Nutrition and athletic performance. Journal of the American Dietetic Association. 2009;109:509–27.
17. Deldicque L, Francaux M. Functional food for exercise performance: Fact or foe? Current Opinions in Clinical Nutrition and Metabolic Care. 2008;11:774–81.Mero A. Leucine supplementation and intensive training. Sports Medicine. 1999;27:347–58.
18. Phillips SM. Protein requirements and supplementation in strength sports. Nutrition. 2004;20:689–95. Wilson J, Wilson GJ. Contemporary issues in protein requirements and consumption for resistance trained athletes. Journal of the International Society for Sports Nutrition. 2006;5:7–27.
19. Bedford JL, Barr SI. Diets and selected lifestyle practices of self-defined adult vegetarians from a population-based sample suggest they are more 'health conscious.' International Journal of Behavior, Nutrition, and Physical Activity. 2005;2:4. Haddad EH, Tanzman JS. What do vegetarians in the United States eat? American Journal of Clinical Nutrition. 2003;78:626S–32S.
20. Antony AC. Vegetarianism and vitamin B-12 (cobalamin) deficiency. American Journal of Clinical Nutrition. 2003;78:3–6. Hunt JR. Bioavailability of iron, zinc, and other trace minerals from vegetarian diets. American Journal of Clinical Nutrition. 2003;78:633S–9S.
21. Kirby M, Danner E. Nutritional deficiencies in children on restricted diets. Pediatric Clinics of North America. 2009;56:1085–103. Mangels AR, Messina V. Considerations in planning vegan diets: Infants. Journal of the American Dietetic Association. 2001;101:670–7. Messina V, Mangels AR. Considerations in planning vegan diets: Children. Journal of the American Dietetic Association. 2001;101:661–9.
22. Pelletier DL, Frongillo EA Jr, Schroeder DG, Habicht JP. The effects of malnutrition on child mortality in developing countries. Bulletin of the World Health Organization. 1995;73:443–8.
23. Jeejeebhoy KN. Protein nutrition in clinical practice. British Medical Bulletin. 1981;37:11–17. Waterlow JC. Classification and definition of protein-calorie malnutrition. British Medical Journal. 1972;3:566–9.
24. Golden M. The development of concepts of malnutrition. Journal of Nutrition. 2002;132:2117S–22S.
25. Hansen RD, Raja C, Allen BJ. Total body protein in chronic diseases and in aging. Annals of the New York Academy of Sciences. 2000;904:345–52. Chao A, Thun MJ, Connell CJ, McCullough ML, Jacobs EJ, Flanders D, Rodriguez C, Sinha R, Calle EE. Meat consumption and risk of colorectal cancer. JAMA (Journal of the American Medical Association). 2005;293:172–82. Miller PE, Lesko SM, Muscat JE, Lazarus P, Hartman TJ. Dietary patterns and colorectal adenoma and cancer risk: A review of the epidemiological evidence. Nutrition and Cancer. 2010;62:413–24.
26. World Cancer Research Fund/American Institute for Cancer Research. Food, nutrition, physical activity, and the prevention of cancer: A global perspective. Washington, DC: American Institute for Cancer Research; 2007.
27. Ibid.

Chapter 6

1. Elias SL, Innis SM. Bakery foods are the major dietary source of trans-fatty acids among pregnant women with diets providing 30 percent energy from fat. Journal of the American Dietetic Association. 2002;102:46–51. Hayes KC, Pronczuk A. Replacing trans fat: The argument for palm oil with a cautionary note on interesterification. Journal of the American College of Nutrition. 2010;253S–84S.
2. Judd JT, Clevidence BA, Muesing RA, Wittes J, Sunkin ME, Podczasy JJ. Dietary *trans* fatty acids: Effects on plasma lipids and lipoproteins of healthy men and women. American Journal of Clinical Nutrition. 1994;59:861–8. Remig V, Franklin B, Margolis S, Kostas G, Nece T, Street JC. *Trans* fats in America: A review of their use, consumption, health implications, and regulation. Journal of the American Dietetic Association. 2010;110:585–92.
3. Hansen SN, Harris WS. New evidence for the cardiovascular benefits of long chain omega-3 fatty acids. Current Atherosclerosis Reports. 2007;9:434–40. Sala A, Folco G, Murphy RC. Transcellular biosynthesis of eicosanoids. Pharmacology Reports. 2010;62:503–10. Singh RK, Gupta S, Dastidar S, Ray A. Cysteinyl leukotrienes and their receptors: Molecular and functional characteristics. Pharmacology. 2010; 85:336–49.
4. Defilippis AP, Blaha MJ, Jacobson TA. Omega-3 fatty acids for cardiovascular disease prevention. Current Treatment Options in Cardiovascular Medicine. 2010;12:365–80. Wood DA, Kotseva K, Connolly S, Jennings C, Mead A, Jones J,

Holden A, De Bacquer D, Collier T, De Backer G, Faergeman O, EUROACTION Study Group. Nurse-coordinated multidisciplinary, family-based cardiovascular disease prevention programme (EUROACTION) for patients with coronary heart disease and asymptomatic individuals at high risk of cardiovascular disease: A paired, cluster-randomised controlled trial. Lancet. 2008;371:1999–2012.
5. Mori TA, Beilin LJ. Omega-3 fatty acids and inflammation. Current Atherosclerosis Reports. 2004;6:461–7. Shahidi F, Miraliakbari H. Omega-3 (n-3) fatty acids in health and disease. Part 1: Cardiovascular disease and cancer. Journal of Medicinal Foods. 2004; 7:387–401. Wijendran V, Hayes KC. Dietary n-6 and n-3 fatty acid balance and cardiovascular health. Annual Review of Nutrition. 2004;24:597–615. Saltiel AR. Fishing out a sensor for anti-inflammatory oils. Cell. 2010;142:672–4.
6. Strumia R. Dermatologic signs in patients with eating disorders. American Journal of Clinical Dermatology. 2005;6:165–73.
7. Foster GD, Wyatt HR, Hill JO, Makris AP, Rosenbaum DL, Brill C, Stein RI, Mohammed BS, Miller B, Rader DJ, Zemel B, Wadden TA, Tenhave T, Newcomb CW, Klein S. Weight and metabolic outcomes after 2 years on a low-carbohydrate versus low-fat diet: A randomized trial. Annals of Internal Medicine. 2010;153:147–57. Vidon C, Boucher P, Cachefo A, Peroni O, Diraison F, Beylot M. Effects of isoenergetic high-carbohydrate compared with high-fat diets on human cholesterol synthesis and expression of key regulatory genes of cholesterol metabolism. American Journal of Clinical Nutrition. 2001;73:878–84.
8. Connor WE, Connor SL. Dietary treatment of familial hypercholesterolemia. Arteriosclerosis. 1989;9:91–105.
9. Allen RR, Carson L, Kwik-Uribe C, Evans EM, Erdman JW Jr. Daily consumption of a dark chocolate containing flavanols and added sterol esters affects cardiovascular risk factors in a normotensive population with elevated cholesterol. Journal of Nutrition. 2008;138:725–31. Klingberg S, Ellegård L, Johansson I, Hallmans G, Weinehall L, Andersson H, Winkvist A. Inverse relation between dietary intake of naturally occurring plant sterols and serum cholesterol in northern Sweden. American Journal of Clinical Nutrition. 2008;87:993–1001. Lin X, Racette SB, Lefevre M, Spearie CA, Most M, Ma L, Ostlund RE Jr. The effects of phytosterols present in natural food matrices on cholesterol metabolism and LDL-cholesterol: A controlled feeding trial. European Journal of Clinical Nutrition. 2010;64:1481–7.
10. Fernandez ML, Webb D. The LDL to HDL cholesterol ratio as a valuable tool to evaluate coronary heart disease risk. Journal of the American College of Nutrition. 2008;27:1–5.
11. Ibid.
12. Esposito K, Ceriello A, Giugliano D. Diet and the metabolic syndrome. Metabolic Syndrome Related Disorders. 2007;5:291–6. Kritchevsky SB, Kritchevsky D, Kromhout D, de Lezenne Coulander C. Diet, prevalence and 10-year mortality from coronary heart disease in 871 middle-aged men. The Zutphen study. American Journal of Epidemiology. 1984;119:733–41. Mottillo S, Filion KB, Genest J, Joseph L, Pilote L, Poirier P, Rinfret S, Schiffrin EL, Eisenberg MJ. The metabolic syndrome and cardiovascular risk a systematic review and meta-analysis. Journal of the American College of Cardiology. 2010;56:1113–32.
13. Lammert F, Wang DQ. New insights into the genetic regulation of intestinal cholesterol absorption. Gastroenterology. 2005;129:718–34. Yang Y, Ruiz-Narvaez E, Kraft P, Campos H. Effect of apolipoprotein E genotype and saturated fat intake on plasma lipids and myocardial infarction in the Central Valley of Costa Rica. Human Biology. 2007;79:637–47. Wu K, Bowman R, Welch AA, Luben RN, Wareham N, Khaw KT, Bingham SA. Apolipoprotein E polymorphisms, dietary fat and fibre, and serum lipids: The EPIC Norfolk study. European Heart Journal. 2007;28:2930–6.
14. Brown MS, Goldstein JL. How LDL receptors influence cholesterol and atherosclerosis. Scientific American. 1984;251:52–60. Dedoussis GV, Schmidt H, Genschel J. LDL-receptor mutations in Europe. Human Mutation. 2004;443–59.
15. Prentice RL. Women's Health Initiative studies of postmenopausal breast cancer. Advances in Experimental Medicine and Biology. 2008;617:151–60. Van Horn L, Manson JE. The Women's Health Initiative: Implications for clinicians. Cleveland Clinic Journal of Medicine. 2008;75:385–90. Wang J, John EM, Horn-Ross PL, Ingles SA. Dietary fat, cooking fat, and breast cancer risk in a multiethnic population. Nutrition and Cancer. 2008;60:492–504. Chan AT, Giovannucci EL. Primary prevention of colorectal cancer. Gastroenterology. 2010;138:2029–43.
16. Al-Serag HB. Obesity and disease of the esophagus and colon. Gastroenterology Clinics of North America. 2005;34:63–82. Key TJ, Schatzkin A, Willett WC, Allen NE, Spencer EA, Travis RC. Diet, nutrition and the prevention of cancer. Public Health Nutrition. 2004;7:187–200. McTiernan A. Obesity and cancer: The risks, science, and potential management strategies. Oncology. 2005;19:871–81. Vrieling A, Kampman E. The role of body mass index, physical activity, and diet in colorectal cancer recurrence and survival: A review of the literature. American Journal of Clinical Nutrition. 2010;92:471–90.
17. World Cancer Research Fund/American Institute for Cancer Research. Food, nutrition, physical activity, and the prevention of cancer: A global perspective. Washington, DC: AICR; 2007.

Chapter 7

1. Carpenter KJ. A short history of nutritional science: Part 1 (1785–1885). Journal of Nutrition. 2003;133:638–45. Carpenter KJ. A short history of nutritional science: Part 2 (1885–1912). Journal of Nutrition. 2003;133:975–84. Carpenter KJ. A short history of nutritional science: Part 3 (1912–1944). Journal of Nutrition. 2003;133:3023–32. Carpenter KJ. A short history of nutritional science: Part 4 (1945–1985). Journal of Nutrition. 2003;133:3331–42.
2. Institute of Medicine. Dietary Reference Intakes for thiamin, riboflavin, niacin, vitamin B_6, folate, vitamin B_{12}, pantothenic acid, biotin, and choline. Washington, DC: National Academy Press; 1998.
3. Ibid.
4. West DW, Owen EC. The urinary excretion of metabolites of riboflavin in man. British Journal of Nutrition. 1963;23:889–98.
5. Cacciapuoti F. Hyper-homocysteinemia: A novel risk factor or a powerful marker for cardiovascular diseases? Pathogenetic and therapeutical uncertainties. Journal of Thrombosis and Thrombolysis. 2011 Jan 14. Scott JM. Homocysteine and cardiovascular risk. American Journal of Clinical Nutrition. 2000;72:33–4.
6. Moyers S, Bailey LB. Fetal malformation and folate metabolism: Review of recent evidence. Nutrition Reviews. 2001;7:215–4.
7. Mitchell LE, Adzick NS, Melchionne J, Pasquariello PS, Sutton LN, Whitehead AS. Spina bifida. Lancet. 2004;364: 1885–95.
8. Pfeiffer CM, Caudill SP, Gunter EW, Osterloh J, Sampson EJ. Biochemical indicators of B vitamin status in the US population after folic acid fortification: Results from the National Health and Nutrition Examination Survey 1999–2000. American Journal of Clinical Nutrition. 2005;82:442–50. Pfeiffer CM, Johnson CL, Jain RB, Yetley EA, Picciano MF,

Rader JI, Fisher KD, Mulinare J, Osterloh JD. Trends in blood folate and vitamin B-12 concentrations in the United States, 1988–2004. American Journal of Clinical Nutrition. 2007;86:718–27.
9. Toh BH, Alderuccio F. Pernicious anaemia. Autoimmunity. 2004;37:357–61.
10. Wojcik M, Burzynska-Pedziwiatr I, Wozniak LA. A review of natural and synthetic antioxidants important for health and longevity. Current Medical Chemistry. 2010;17:3262–88.
11. Jacob RA, Aiello GM, Stephensen CB, Blumberg JB, Milbury PE, Wallock LM, Ames BN. Moderate antioxidant supplementation has no effect on biomarkers of oxidant damage in healthy men with low fruit and vegetable intakes. Journal of Nutrition. 2003;133:740–3. Padayatty SJ, Katz A, Wang Y, Eck P, Kwon O, Lee J-H, Chen S, Corpe C, Dutta A, Dutta SK, Levine M. Vitamin C as an antioxidant: Evaluation of its role in disease prevention. Journal of the American College of Nutrition. 2003;22:18–35.
12. Bhaskaram P. Micronutrient malnutrition, infection, and immunity: An overview. Nutrition Reviews. 2002;60:S60–45.
13. Stephen R, Utecht T. Scurvy identified in the emergency department: a case report. Journal of Emergency Medicine. 2001;21:235–37. Weinstein M, Babyn P, Zlotkin S. An orange a day keeps the doctor away: Scurvy in the year 2000. Pediatrics. 2001;108:E55.
14. Gerster H. Vitamin A—Functions, dietary requirements and safety in humans. International Journal of Vitamins and Nutrition Research. 1997;67:71–90. Hinds TS, West WL, Knight EM. Carotenoids and retinoids: A review of research, clinical, and public health applications. Journal of Clinical Pharmacology. 1997;37:551–8.
15. Campbell JK, Canene-Adams K, Lindshield BL, Boileau TWM, Clinton SK, Erdman JW. Tomato phytochemicals and prostate cancer risk. Journal of Nutrition. 2004;134:3486S–92S. Wertz K, Siler U, Goralczyk R. Lycopene: Modes of action to promote prostate health. Archives of Biochemistry and Biophysics. 2004;430:127–34. Beatty S, Nolan J, Kavanagh H, O'Donovan O. Macular pigment optical density and its relationship with serum and dietary levels of lutein and zeaxanthin. Archives of Biochemistry and Biophysics. 2004;430:70–6. Stringham JM, Hammond BR. Dietary lutein and zeaxanthin: Possible effects on visual function. Nutrition Reviews. 2005;63:59–64.
16. El-Agamey A, Lowe GM, McGarvey DJ, Mortensen A, Phillip DM, Truscott G, Young AJ. Carotenoid radical chemistry and antioxidant/pro-oxidant properties. Archives of Biochemistry and Biophysics. 2004;430:37–48.
17. Stringham JM, Hammond BR. Dietary lutein and zeaxanthin: Possible effects on visual function. Nutrition Reviews. 2005;63:59–64.
18. The Alpha-Tocopherol, Beta Carotene Cancer Prevention Study Group. The effect of vitamin E and beta carotene on the incidence of lung cancer and other cancers in male smokers. New England Journal of Medicine. 1994;330:1029–35. Baron JA, Cole BF, Mott L, Haile R, Grau M, Church TR, Beck GJ, Greenberg ER. Neoplastic and antineoplastic effects of β-carotene on colorectal adenoma recurrence: Results of a randomized trial.Journal of the National Cancer Institute. 2003;95:717–22. Omenn GS, Goodman GE, Thornquist MD, Balmes J, Cullen MR, Glass A, Keogh JP, Meyskens FL, Valanis B, Williams JH, Barnhart S, Hammar S. Effects of a combination of beta carotene and vitamin A on lung cancer and cardiovascular disease. New England Journal of Medicine. 1996;334:1150–55.
19. American Academy of Dermatology. American Academy of Dermatology Association reconfirms need to boost vitamin D intake through diet and nutritional supplements rather than ultraviolet radiation. Available from: http://www.aad.org/aad/Newsroom/Vitamin1D1Consensus1Conf.htm. World Health Organization. Sunbeds, tanning and UV exposure. Available from: http://www.who.int/mediacentre/factsheets/fs287/en/print.html.
20. Bikle DD. Vitamin D and skin cancer. Journal of Nutrition. 2004;134:3472S–8S. Gross MD. Vitamin D and calcium in the prevention of prostate and colon cancer: New approaches for the identification of needs. Journal of Nutrition. 2005;135:326–31. Holick MF. Sunlight and vitamin D for bone health and prevention of autoimmune diseases, cancers, and cardiovascular disease. American Journal of Clinical Nutrition. 2004;6 Suppl:1678S–88S. Welsh J. Vitamin D and breast cancer: Insights from animal models. American Journal of Clinical Nutrition. 2004;80:1721S–4S.
21. Calvo MS, Whiting SJ, Barton CN. Vitamin D intake: A global perspective of current status. Journal of Nutrition. 2005;135:310–6.
22. Kreiter SR, Schwartz RP, Kirkman HN, Charlton PA, Calikoglu AS, Davenport ML. Nutrition rickets in African American breast-fed infants. Journal of Pediatrics. 2000;137:153–7. Pugliese MF, Blumberg DL, Hludzinski J, Kay S. Nutritional rickets in suburbia. Journal of the American College of Nutrition. 1998;17:637–41. Prentice A. Vitamin D deficiency: A global perspective. Nutrition Reviews, 2008;66:S153–64.
23. American Academy of Pediatrics. Pediatric nutrition handbook, 6th ed. Elk Grove Village, IL; 2008. Wagner CL, Greer FR, American Academy of Pediatrics Section on Breastfeeding, American Academy of Pediatrics Committee on Nutrition. Prevention of rickets and vitamin D deficiency in infants, children, and adolescents. Pediatrics. 2008;122:1142–52.
24. Institute of Medicine. Dietary Reference Intakes for calcium and vitamin D. Washington, DC: National Academies Press, 2011.
25. Bostick RM, Potter JD, McKenzie DR, Sellers TA, Kushi LH, Steinmetz KA, Folsom AR. Reduced risk of colon cancer with high intakes of vitamin E: The Iowa Women's Health Study. Cancer Research. 1992;15:4230–7. Kirsh VA, Hayes RB, Mayne ST, Chatterjee N, Subar AF, Dixon LB, Albanes D, Andriole GL, Urban DA, Peters U. PLCO Trial. Supplementation and dietary vitamin E, beta-carotene, and vitamin C intakes and prostate cancer risk. Journal of the National Cancer Institute. 2006;98:245–54. Kline K, Yu W, Sanders BG. Vitamin E and breast cancer. Journal of Nutrition. 2004;134:3458S–62S. Peters U, Littman AJ, Kristal AR, Patterson RE, Potter JD, White E. Vitamin E and selenium supplementation and risk of prostate cancer in the vitamins and lifestyle (VITAL) study cohort. Cancer Causes and Control. 2008;19:75–87. Slatore CG, Littman AJ, Au DH, Satia JA, White E. Long-term use of supplemental multivitamins, vitamin C, vitamin E, and folate does not reduce the risk of lung cancer. American Journal of Respiratory Critical Care Medicine. 2008;177:524–30.
26. Committee on Fetus and Newborn (American Academy of Pediatrics). Controversies concerning vitamin K and the newborn. Pediatrics. 2003;112:191–2.
27. U.S. Food and Drug Administration and Center for Food Safety and Applied Nutrition. Dietary supplement health and education act of 1994. Available from: http://www.cfsan.fda.gov/~dms/dietsupp.html.

Chapter 8

1. Armstrong LE, Pumerantz AC, Roti MW, Judelson DA, Watson G, Dias JC, Sokmen B, Casa DJ, Maresh CM, Lieberman H, Kellogg M. Fluid, electrolyte, and renal indices of hydration during 11 days of controlled caffeine consumption. International Journal of Sport Nutrition and Exercise Metabolism. 2005;15:252–65.

2. Maughan RJ, Griffin J. Caffeine ingestion and fluid balance: A review. Journal of Human Nutrition and Dietetics. 2003;16:411–20.
3. Coris EE, Ramirez AM, Van Durme DJ. Heat illness in athletes: The dangerous combination of heat, humidity and exercise. Sports Medicine. 2004;34:9–16.
4. Bottled water: Pure drink or pure hype? Natural Resources Defense Council, 1999. Available from: http://www.nrdc.org/water/drinking/bw/bwinx.asp.
5. Cook RJ, Barron JC, Papendick JI, Williams GJ. Impact on agriculture of the Mount St. Helens eruptions. Science. 1981;211:16–22.
6. National Osteoporosis Foundation. Fast facts. Available from: http://www.nof.org/osteoporosis/diseasefacts.htm.
7. Syed FA, Ng AC. The pathophysiology of the aging skeleton. Current Osteoporosis Reports. 2010;8:235–40.
8. National Osteoporosis Foundation. Fast facts. Available from: http://www.nof.org/osteoporosis/diseasefacts.htm.
9. Syed FA, Ng AC. The pathology of the aging skeleton. Current Osteoporosis Reports. 2010;8:235–40.
10. Winsloe C, Earl S, Dennison EM, Cooper C, Harvey NC. Early life factors in the pathogenesis of osteoporosis. Current Osteoporosis Reports. 2009;7:140–4.
11. Straub DA. Calcium supplementation in clinical practice: A review of forms, doses, and indications. Nutrition in Clinical Practice. 2007;22:286–96.
12. Jackson RD, Shidham S. The role of hormone therapy and calcium plus vitamin D for reduction of bone loss and risk for fractures: Lessons learned from the Women's Health Initiative. Current Osteoporosis Reports. 2007;5:153–9.
13. Seeman E. Evidence that calcium supplements reduce fracture risk is lacking. Clinical Journal of American Society of Nephrology. 2010;1:S3–11.
14. Heaney RP, Recker RR, Stegman MR, Moy AJ. Calcium absorption in women: Relationships to calcium intake, estrogen, status, and age. Journal of Bone Mineral Research. 1989;4:469–75.
15. Davis JM, Burgess WA, Slentz CA, Bartoli WP. Fluid availability of sports drinks differing in carbohydrate type and concentration. American Journal of Clinical Nutrition. 1990;51:1054–7.
16. Evans GH, Shirreffs SM, Maughan RJ. Postexercise rehydration in man: The effects of osmolality and carbohydrate content of ingested drinks. Nutrition. 2009;25:905–13.
17. Kottke TE, Wu LA. Preventing heart disease and stroke: Messages from the United States. The Keio Journal of Medicine. 2001;50:274–9.
18. Dumler F. Dietary sodium intake and arterial blood pressure. Journal of Renal Nutrition. 2009;19:57–60.
19. Elkhalifa AM, Kinsara AJ, Almadani DA. Prevalence of hypertension in a population of healthy individuals. Medical Principles and Practice. 2011;20:152–5.
20. Reinhold JG, Faradji B, Abadi P, Ismail-Beigi F. Decreased absorption of calcium, magnesium, zinc and phosphorus by humans due to increased fiber and phosphorus consumption as wheat bread. Nutrition Reviews. 1991;49:204–6.
21. Seiner R, Hesse A. Influence of a mixed and a vegetarian diet on urinary magnesium excretion and concentration. British Journal of Nutrition. 1995;73:783–90. Wisker E, Nagel R, Tanudjaja TK, Feldheim W. Calcium, magnesium, zinc, and iron balances in young women: Effects of a low-phytate barley-fiber concentration. American Journal of Clinical Nutrition. 1991;54:553–9.
22. Centers for Disease Control and Prevention. Recommendations to prevent and control iron deficiency in the United States. Mortality and Morbidity Weekly Report. 1998;47: 1–29. Stoltzfus RJ. Defining iron-deficiency anemia in public health terms: Reexamining the nature and magnitude of the public health problem. Journal of Nutrition. 2001;131: 565S–7S.
23. Hallberg L, Rossander-Hultén L. Iron requirements in menstruating women. American Journal of Clinical Nutrition. 1991;54:1047–58.
24. Lopez MAA, Martos FC. Iron availability: An updated review. International Journal of Food Sciences and Nutrition. 2004;55:597–606. Zijp IM, Korver O, Tijburg LBM. Effect of tea and other dietary factors on iron absorption. Critical Reviews in Food Science and Nutrition. 2000;40:371–98.
25. Xiu YM. Trace elements in health and diseases. Biomedical and Environmental Sciences. 1996;9:130–6.
26. Rosenzweig PH, Volpe SL. Iron, thermoregulation, and metabolic rate. Critical Reviews in Food Science and Nutrition. 1999;39:131–48. Cunningham-Rundles S, McNeeley DF, Moon A. Mechanisms of nutrient modulation of the immune response. Journal of Allergy and Clinical Immunology. 2005;115:1119–28. Failla ML. Trace elements and host defense: Recent advances and continuing challenges. Journal of Nutrition. 2003;133:1443S–7S.
27. Bryan J, Osendarp S, Hughes D, Calvaresi E, Baghurst K, van Klinken J-W. Nutrients for cognitive development in school-aged children. Nutrition Reviews. 2004;62:295–306.
28. Gambling L, Danzeisen R, Fosset C, Andersen HS, Dunford S, Srai SKS, McArdle HJ. Iron and copper interactions in development and the effect on pregnancy outcome. Journal of Nutrition. 2003;133:1554S–6S.
29. Institute of Medicine. Dietary Reference Intakes for vitamin A, vitamin K, arsenic, boron, chromium, copper, iodine, iron, manganese, molybdenum, nickel, silicon, vanadium, and zinc. Washington, DC: National Academy Press; 2001.
30. Hoffman R, Benz E, Shattil S, Furie B, Cohen H, Silberstein L, McGlave P. Hematology: Basic principles and practice, 3rd ed. New York: Churchill Livingstone, Harcourt Brace; 2000.
31. Chuttani H, Gupta P, Gulati S, Gupta D. Acute copper sulfate poisoning. American Journal of Medicine. 1965;39:849–54. Bremner I. Manifestations of copper excess. American Journal of Clinical Nutrition. 1998;67:1069S–73S.
32. Zimmermann MB. Assessing iodine status and monitoring progress of iodized salt programs. Journal of Nutrition. 2004;134:1673–7.
33. Valko M, Izakovic M, Mazur M, Rhodes CJ, Telser J. Role of oxygen radicals in DNA damage and cancer incidence. Molecular and Cellular Biochemistry. 2004;266:37–56.
34. Balk EM, Tatsioni A, Lichtenstein AH, Lau J, Pittas AG. Effect of chromium supplementation on glucose metabolism and lipids: A systematic review of randomized controlled trials. Diabetes Care. 2007;30:2154–63.
35. Kozlovsky AS, Moser PB, Reiser S, Anderson RA. Effects of diets high in simple sugars on urinary chromium losses. Metabolism. 1986;35:515–8.
36. Cefalu WT, Hu FB. Role of chromium in human health and in diabetes. Diabetes Care. 2004;27:2741–51. Hopkins Jr. LL, Ransome-Kuti O, Majaj AS. Improvement of impaired carbohydrate metabolism by chromium (III) in malnourished infants. American Journal of Clinical Nutrition. 1968;21:203–11. Mertz W. Interaction of chromium with insulin: A progress report. Nutrition Reviews. 1998;56:174–7.
37. McNamara JP, Valdez F. Adipose tissue metabolism and production responses to calcium proprionate and chromium proprionate. Journal of Dairy Science. 2005;88:2498–507. Page TG, Southern LL, Ward TL, Thompson DLJ. Effect of chromium picolinate on growth and serum and carcass traits of growing-finishing pigs. Journal of Animal Science. 1993;71:656–62.
38. Costa M, Klein CB. Toxicity and carcinogenicity of chromium compounds in humans. Critical Reviews in Toxicology. 2006;36:155–63. Coyle YM, Minahjuddin AT, Hynan LS, Minna JD. An ecological study of the association of metal air pollutants with lung cancer incidence in Texas. Journal of Thoracic Oncology. 2006;1:654–61. Michaels D, Lurie P, Monforton C. Lung cancer mortality in the German

chromate industry, 1958 to 1998. Journal of Occupational and Environmental Medicine. 2006;48:995–7.
39. Abumrad NN, Schneider AJ, Steel D, Rogers LS. Amino acid intolerance during total parenteral nutrition reversed by molybdate therapy. American Journal of Clinical Nutrition. 1981;34:2551–9.
40. Caruso TJ, Prober CG, GwaltneyJr JM. Treatment of naturally acquired common colds with zinc: A structured review. Clinical Infectious Diseases. 2007;45:569–574. Jackson JL, Lesho E, Peterson C. Zinc and the common cold: A meta-analysis revisited. Journal of Nutrition. 2000;130:1512S–5S. Kurugöl Z, Bayram N, Atik T. Effect of zinc sulfate on common cold in children: Randomized, double blind study. Pediatrics International. 2007;49:842–7. Kurugöl Z, Akilli M, Bayram N, Koturoglu G. The prophylactic and therapeutic effectiveness of zinc sulphate on common cold in children. Acta Paediatrica. 2006;95:1175–81.
41. Quock RL, Chan JT. Fluoride content of bottled water and its implications for the general dentist. General Dentistry. 2009;57:29–33.
42. Dhar V, Bhatnagar M. Physiology and toxicity of fluoride. Indian Journal of Dental Research. 2009;20:350–5.

Chapter 9

1. Spalding KL, Arner E, Westermark PO, Bernard S, Buchholz BA, Bergmann O, Blomqvist L, Hoffstedt J, Näslund E, Britton T, Concha H, Hassan M, Rydén M, Frisén J, Arner P. Dynamics of fat cell turnover in humans.Nature. 2008;453:783–7.
2. National Institutes of Health. Clinical guidelines on the identification, evaluation, and treatment of overweight and obesity in adults. National Institutes of Health, National Heart, Lung, and Blood Institute, Obesity Education Initiative. Available from: http://www.nhlbi.nih.gov/guidelines/obesity/practgde.htm.
3. Heinrich KM, Jitnarin N, Suminski RR, Berkel L, Hunter CM, Alvarez L, Brundige AR, Peterson AL, Foreyt JP, Haddock CK, Poston WS. Obesity classification in military personnel: A comparison of body fat, waist circumference, and body mass index measurements. Military Medicine. 2008;17:67–73.
4. Geliebter A. Stomach capacity in obese individuals.Obesity Research. 2001;9:727–8.
5. French S, Robinson T. Fats and food intake. Current Opinion in Clinical Nutrition and Metabolic Care. 2003;6:629–34.
6. Inui A, Asakawa A, Bowers CY, Mantovani G, Laviano A, Meguid MM, Fujimiya M. Ghrelin, appetite, and gastric motility: The emerging role of the stomach as an endocrine organ. Federation of American Societies for Experimental Biology Journal. 2004;18:439–56.
7. Yanovski S. Sugar and fat: Cravings and aversions. Journal of Nutrition. 2003;133:835S–7S.
8. Hulbert AJ, Else PL. Basal metabolic rate: History, composition, regulation, and usefulness. Physiological and Biochemical Zoology. 2004;77:869–76.
9. Luke A, Schoeller DA. Basal metabolic rate, fat-free mass, and body cell mass during energy restriction. Metabolism. 1992;41:450–6.
10. Brooks GA, Butte NF, Rand WM, Flatt JP, Caballero B. Chronicle of the Institute of Medicine physical activity recommendation: How a physical activity recommendation came to be among dietary recommendations. American Journal of Clinical Nutrition. 2004;79:921S–30S.
11. Nair KS, Halliday D, Garrow JS. Thermic response to isoenergetic protein, carbohydrate or fat meals in lean and obese subjects. Clinical Science. 1983;65:307–12.
12. Friedl KE. Can you be large and not obese? The distinction between body weight, body fat, and abdominal fat in occupational standards. Diabetes Technology and Therapeutics. 2004;6:732–49.
13. Katzmarzyk PT, Janssen I, Ardern CI. Physical inactivity, excess adiposity and premature mortality. Obesity Reviews. 2003;4:257–901.
14. National Center for Health Statistics (NCHS). Health, United States, 2010. Available from: http://www.cdc.gov/nchs/data/hus/hus10.pdf.
15. Popkin BM, Gordon-Larsen P. The nutrition transition: Worldwide obesity dynamics and their determinants.International Journal of Obesity and Related Metabolic Disorders. 2004;3:S2–9.
16. Austin GL, Ogden LG, and O Hill J. Trends in carbohydrate, fat, and protein intakes and association with energy intake in normal-weight, overweight, and obese individuals: 1971–2006. American Journal of Clinical Nutrition. 2011; 93:836–43.
17. Briefel RR, Johnson CL. Secular trends in dietary intake in the United States. Annual Review of Nutrition. 2004; 24:401–31.
18. Ello-Martin JA, Ledikwe JH, Rolls BJ. The influence of food portion size and energy density on energy intake: Implications for weight management. American Journal of Clinical Nutrition. 2005;82:236S–41S. Wansink B, Kim J. Bad popcorn in big buckets: Portion size can influence intake as much as taste. Journal of Nutrition Education and Behavior. 2005;37:242–5.
19. Rolls BJ, Roe LS, Kral TVE, Meengs JS, Wall DE. Increasing the portion size of a packaged snack increases energy intake in men and women. Appetite. 2004;42:63–9.
20. Hetherington MM, Foster R, Newman T, Anderson AS, Norton G. Understanding variety: Tasting different foods delays satiation. Physiology and Behavior. 2006;87:263–71.
21. Chandon P, Wansink B. The biasing health halos of fast-food restaurant health claims: Lower calorie estimates and higher side-dish consumption intentions. Journal of Consumer Research. 2007:34;301–14.
22. Kim JY, Oh DJ, Yoon TY, Choi JM, Choe BK. The impacts of obesity on psychological well-being: A cross-sectional study about depressive mood and quality of life. Preventive Medicine and Public Health. 2007;40:191–5.
23. Christakis NA and Fowler JH. The spread of obesity in a large social network over 32 years. New England Journal of Medicine. 2007;57:370–9.
24. National Center for Health Statistics (NCHS). Health, United States, 2010. Available from: http://www.cdc.gov/nchs/data/hus/hus10.pdf.
25. American College of Sports Medicine (ACSM). ACSM's Guidelines for Exercise Testing and Prescription, 8th ed. Philadelphia: Lippincott Williams & Wilkins; 2008.
26. Wareham NJ, van Sluijs EM, Ekelund U. Physical activity and obesity prevention: A review of the current evidence. Proceedings of the Nutrition Society. 2005;64:229–47.
27. Hill JO, Wyatt HR, Reed GW, Peters JC. Obesity and the environment: Where do we go from here? Science. 2003;299:853–5.
28. Ingalls A, Dickie M, Snell GD. Obese, a new mutation in the house mouse. Journal of Heredity. 1950;41:317–8.
29. ColeZhang Y, Proenca R, Maffei M, Leopold L, Friedman JM. Positional cloning of the mouse obese gene and its human homologue. Nature. 1994;372:125–32.
30. FarooqiI S, O'Rahilly S. Monogenic obesity in humans. Annual Review of Medicine. 2005;56:443–58.
31. Marx J. Cellular warriors at the battle of the bulge. Science. 2003;299:846–9.
32. Kennedy AG. The role of the fat depot in the hypothalamic control of food intake in the rat. Proceedings of the Royal Society of London. 1953;140:578–92.
33. Banks WA. Blood-brain barrier as a regulatory interface. Forum in Nutrition. 2010:63:102–10. Tups A. Physiological models of leptin resistance. Journal of Neuroendocrinology. 2009;21:961–71.

34. Wing RR, Phelan S. Long-term weight loss maintenance. American Journal of Clinical Nutrition. 2005;82:222S–5S.
35. National Institutes of Health, and National Heart, Lung and Blood Institute. Clinical guidelines on the identification, evaluation and treatment of overweight and obesity in adults—The evidence report. National Institutes of Health Publication Number 00–4084. Bethesda, MD: National Institutes of Health; October 2000.
36. Fogelholm M. Physical activity, fitness and fatness: Relations to mortality, morbidity, and disease risk factors. A systematic review. Obesity Reviews. 2010;11:202–21.
37. Barlow CE, Kohl HW III, Gibbons LW, Blair SN. Physical fitness, mortality and obesity. International Journal of Obesity and Related Metabolic Disorders. 1995;19:S41–4. McAuley PA, Kokkinos PF, Oliveira RB, Emerson BT, Myers JN. Obesity paradox and cardiorespiratory fitness in 12,417 male veterans aged 40 to 70 years. Mayo Clinic Proceedings. 2010;85:115–21.
38. US Department of Agriculture and US Department of Health and Human Services. Dietary Guidelines for Americans, 2010. 7th Edition, Washington, DC: US Government Printing Office, December 2010.
39. Catenacci VA, Ogden LG, Stuht J, Phelan S, Wing RR, Hill JO, Wyatt HR. Physical activity patterns in the National Weight Control Registry. Obesity. 2008;16:153–61. Butryn ML, Phelan S, Hill JO, Wing RR. Consistent self-monitoring of weight: A key component of successful weight loss maintenance. Obesity. 2007;15:3091–6. Phelan S, Wyatt H, Nassery S, Dibello J, Fava JL, Hill JO, Wing RR. Three-year weight change in successful weight losers who lost weight on a low-carbohydrate diet. Obesity. 2007;15:2470–7.
40. Westman EC, Yancy WS Jr, Vernon MC. Is a low-carb, low-fat diet optimal? Archives of Internal Medicine. 2005;165:1071–2.
41. Center for Nutrition Policy and Promotion and the US Department of Agriculture. Nutrition Insights. Is fat consumption really decreasing? Insight 5 April 1998. Available from: http://www.cnpp.usda.gov/Publications/NutritionInsights/insight5.pdf.
42. Ibid.
43. Willett WC. Dietary fat and body fat: Is there a relationship? Journal of Nutritional Biochemistry. 1998;9:522–4. Willett WC. Is dietary fat a major determinant of body fat? American Journal of Clinical Nutrition. 1998;67:556S–625S.
44. Bray GA, Paeratakul S, Popkin BM. Dietary fat and obesity: A review of animal, clinical and epidemiological studies. Physiology and Behavior. 2004; 83:549–55.
45. Taubes G. The soft science of dietary fat. Science. 2001;291:2536–45.
46. Foster GD, Wyatt HR, Hill JO, McGuckin BG, Brill C, Selma B, Szapary PO, Rader DJ, Edman JS, Klein S. A randomized trial of a low-carbohydrate diet for obesity. New England Journal of Medicine. 2003;248:2082–90.
47. Crowe TC. Safety of low-carbohydrate diets. Obesity Reviews. 2005;6:235–45.
48. Bravata DM, Sanders L, Huang J, Krumholz HM, Olkin I, Gardner CD, Bravata DM. Efficacy and safety of low-carbohydrate diets: A systematic review. Journal of the American Medical Association. 2003;289:1838–49.
49. Schwenke DC. Insulin resistance, low-fat diets, and low- carbohydrate diets: Time to test new menus. Current Opinion in Lipidology. 2005;16:55–60.

Chapter 10

1. Centers for Disease Control and Prevention, National Center for Health Statistics. Clinical growth charts. Available from: http://www.cdc.gov/growthcharts/.
2. National Center for Health Statistics. Data on birth defects. Available from: http://www.cdc.gov/nchs/FASTATS/bdefects.htm.
3. Brent RL. Environmental causes of human congenital malformations: The pediatrician's role in dealing with these complex clinical problems caused by a multiplicity of environmental and genetic factors. Pediatrics. 2004;113:957–68.
4. Eustace LW, Kang DH, Coombs D. Fetal alcohol syndrome: A growing concern for health care professionals. Journal of Obstetrics, Gynecologic, and Neonatal Nursing. 2003;32:215–21.
5. Villar J, Merialdi M, Gulmezoglu AM, Abalos E, Carroli G, Kulier R, de Onis M. Characteristics of randomized controlled trials included in systematic reviews of nutritional interventions reporting maternal morbidity, mortality, preterm delivery, intrauterine growth restriction and small for gestational age and birth weight outcomes. Journal of Nutrition. 2003;133:1632S–9S. US Department of Health and Human Services. Centers for Disease Control and Prevention. National Center for Health Statistics System. National Vital Statistics Reports. Infant mortality statistics from the 2006 period linked birth/infant death data set. 2010. Available from: http://www.cdc.gov/nchs/data/nvsr/nvsr58/nvsr58_17.pdf.
6. Gillman MW. Developmental origins of health and disease. New England Journal of Medicine. 2005;353:1848–50. Barker DJ. The foetal and infant origins of inequalities in health in Britain. Journal of Public Health Medicine. 1991;13:64–8. Barker DJ. Fetal programming of coronary heart disease. Trends in Endocrinology and Metabolism. 2002;13:364–8. Rasmussen KM. The "fetal origins" hypothesis: Challenges and opportunities for maternal and child nutrition. Annual Review of Nutrition. 2000;21:73–95.
7. Institute of Medicine. Weight gain during pregnancy: Reexamining the guidelines. Washington, DC: National Academies Press; 2009. Available from: http://www.iom.edu/pregnancyweightgain.
8. US Department of Health and Human Services, US Environmental Protection Agency. What you need to know about mercury in fish and shellfish. March 2004. Available from: http://www.fda.gov/Food/FoodSafety/Product-Specific Information/Seafood/FoodbornePathogensContaminants/Methylmercury/ucm115662.htm.
9. Beard JL. Effectiveness and strategies of iron supplementation during pregnancy. American Journal of Clinical Nutrition. 2000;71:1288S–94S.
10. Beard JL. Effectiveness and strategies of iron supplementation during pregnancy. American Journal of Clinical Nutrition. 2000;71:1288S–94S. Feinleib M, Beresford SA, Bowman BA, Mills JL, Rader JI, Selhub J, Yetley EA. Folate fortification for the prevention of birth defects: Case study. American Journal of Epidemiology. 2001;154:S60–9.
11. Di Cianni G, Ghio A, Resi V, Volpe L. Gestational diabetes mellitus: An opportunity to prevent type 2 diabetes and cardiovascular disease in young women. Women's Health. 2010;6:97–105.
12. Papageorghiou AT, Campbell S. First trimester screening for preeclampsia. Current Opinion in Obstetrics and Gynecology. 2006;18:594–600. Hutcheon JA, Lisonkova S, Joseph KS. Epidemiology of pre-eclampsia and the other hypertensive disorders of pregnancy. Best Practices & Research Clinical Obstetrics & Gynaecology. 2011;25:391–403.
13. American Academy of Pediatrics. Pediatric nutrition handbook, 6th ed., Elk Grove Village, IL; 2008. American Academy of Pediatrics. 2005 AAP policy statement of breastfeeding and the use of human milk. Pediatrics. 2005;115:496–501.
14. Ryan AS, Wenjun Z, Acosta A. Breastfeeding continues to increase into the new millennium. Pediatrics. 2002;110:1103–9.

15. Grummer-Strawn LM, Scanlon KS, Fein SB. Infant feeding and feeding transitions during the first year of life. Pediatrics. 2008;122:S36–42.
16. Leon DA, Ronalds G. Breast-feeding influences on later life-cardiovascular disease. Advances in Experimental and Medical Biology. 2009;639;153–66.
17. Dewey KG. Impact of breastfeeding on maternal nutritional status. Advances in Experimental Medical Biology. 2004;554:91–100. Winkvist A, Rasmussen KM. Impact of lactation on maternal body weight and body composition. Journal of Mammary Gland Biology and Neoplasia. 1999;4:309–18.
18. Taylor JS, Kacmar JE, Nothnagle M, Lawrence RA. A systematic review of the literature associating breastfeeding with type 2 diabetes and gestational diabetes. Journal of the American College of Nutrition. 2005;24:320–6. Labbok MH. Effects of breasfeeding on the mother. Pediatric Clinics of North America. 2001;48:143–58.
19. Lind T, Hernell O, Lonnerdal B, Stenlund H, Domellof M, Persson LA. Dietary iron intake is positively associated with hemoglobin concentration during infancy but not during the second year of life. Journal of Nutrition. 2004;134:1064–70.
20. Wagner CL, Greer FR. Prevention of rickets and vitamin D deficiency in infants, children, and adolescents. Pediatrics. 2008;122:1142–52.
21. American Academy of Pediatrics. Pediatric nutrition handbook, 6th ed., Elk Grove Village, IL; 2008. American Academy of Pediatrics. 2005. AAP policy statement of breastfeeding and the use of human milk. Pediatrics. 2005;115:496–501.
22. Centers for Disease Control and Prevention. Recommendations to prevent and control iron deficiency anemia in the United States. Morbidity Mortality Weekly Report. 2002;51:897–9. Baker RD, Greer FR. From the American Academy of Pediatrics. Clinical report- Diagnosis and prevention of iron deficiency and iron-deficiency anemia in infants and young children (0–3 years of age). Pediatrics. 2010;126:2010–576.
23. American Academy of Pediatrics Committee on Nutrition. The use and misuse of fruit juice in pediatrics. Pediatrics. 2001;107:1210–3. American Academy of Pediatrics. Pediatric nutrition handbook, 6th ed., Elk Grove Village, IL; 2008.
24. Grummer-Strawn LM, Scanlon KS, Fein SB. Infant feeding and feeding transitions during the first year of life. Pediatrics. 2008;122:S36–42.
25. Thygarajan A, Burks AW. American Academy of Pediatrics recommendations on the effects of early nutritional interventions on the development of atopic disease. Current Opinion in Pediatrics. 2008;20:698–702. Greer FR, Sicherer SH, Burks A, and the Committee on Nutrition and Section on Allergy and Immunology. Effects of early nutritional interventions on the development of atopic disease in infants and children: The role of maternal dietary restriction, breastfeeding, timing of introduction of complementary foods, and hydrolyzed formulas. Pediatrics. 2008;121:183–91.
26. Kleinman RE. American Academy of Pediatrics recommendations for complementary feeding. Pediatrics. 2000;106:1274; American Academy of Pediatrics. Pediatric nutrition handbook, 6th ed., Elk Grove Village, IL; 2008.
27. Centers for Disease Control and Prevention. Food allergy among U.S. children: Trends in prevalence and hospitalizations. 2008. Available from: http://www.cdc.gov/nchs/data/databriefs/db10.htm.
28. Lack G. Epidemiologic risks for food allergy. The Journal of Allergy and Clinical Immunology. 2008;121:1331–6.
29. Prescott SL, Bouygue GR, Videky D, Fiocchi A. Avoidance or exposure to foods in prevention and treatment of food allergy? Current Opinion in Allergy and Clinical Immunology. 2010;10:258–66.
30. American Academy of Pediatrics Committee on Nutrition. The use and misuse of fruit juice in pediatrics. Pediatrics. 2001;107:1210–3. American Academy of Pediatrics. Pediatric nutrition handbook, 6th ed., Elk Grove Village, IL; 2008.
31. Koivisto Hursti UK. Factors influencing children's food choice. Annals of Medicine. 1999;31:26–32.
32. Patrick H, Nicklas TA. A review of family and social determinants of children's eating patterns and diet quality. Journal of American College of Nutrition. 2005;2:83–92.
33. American Academy of Pediatrics. Pediatric nutrition handbook, 6th ed., Elk Grove Village, IL; 2008. American Academy of Pediatrics. 2005. AAP policy statement of breastfeeding and the use of human milk. Pediatrics. 2005;115:496–501.
34. National Center for Health Statistics. Health, United States, 2010: With special feature on death and dying. Hyattsville, MD. 2011.
35. Robinson TN. Television viewing and childhood obesity. Pediatric Clinics of North America. 2001;4:1017–25.
36. Crespo CJ, Smit E, Troiano RP, Bartlett SJ, Macera CA, Andersen RE. Television watching, energy intake, and obesity in US children: Results from the third National Health and Nutrition Examination Survey, 1988–1994. Archives of Pediatric and Adolescent Medicine. 2001;155:360–5.
37. American Academy of Pediatrics Committee on Public Education. Children, adolescents, and television. Pediatrics. 2001;107:423–6.
38. Soemantri AG, Pollitt E, Kim I. Iron deficiency anemia and educational achievement. American Journal of Clinical Nutrition. 1985;42:1221–8; Iannotti LL, Tielsch JM, Black MM, Black RE. Iron supplementation in early childhood: Health benefits and risks. American Journal of Clinical Nutrition. 2006;84:1261–76.
39. Salsberry PJ, Reagan PB, Pajer K. Growth differences by age of menarche in African American and White girls. Nursing Research. 2009;58:382–90.
40. Enns CW. Mickle SH, Goldman JD. Trends in food and nutrient intakes by adolescents in the United States. Family Economics and Nutrition Review. 2003;15:15–27. Muñoz KA, Krebs-Smith SM, Ballard-Barbash R, Cleveland LE. Food intakes of US children and adolescents compared with recommendations. Pediatrics 1997;100:323–9.
41. Nielsen SJ, Popkin BM. Changes in beverage intake between 1977 and 2000. American Journal of Preventive Medicine. 2004;27:205–10.
42. Dwyer J, Picciano FM, Raiten DJ. Estimation of usual intakes: What we eat in American-NHANES. Journal of Nutrition. 2003;133:609S-23S. Available from: http://jn.nutrition.org/content/133/2/609S.full.
43. Centers for Disease Control and Prevention. Recommendations to prevent and control iron deficiency anemia in the United States. Morbidity Mortality Weekly Report. 2005;51;897–9. Baker RD, Greer FR. From the American Academy of Pediatrics. Clinical report—Diagnosis and prevention of iron deficiency and iron-deficiency anemia in infants and young children (0–3 years of age). 2010;126:2010–576.
44. Cavadini C, Siega-Riz AM, Popkin BM. US adolescent food intake trends from 1965 to 1996. Archives of Disease in Children. 2000;83:18–24.
45. Centers for Disease Control and Prevention. The state of aging and health in America. 2004. Available from: http://www.cdc.gov/aging/pdf/state_of_aging_and_health_in_america_2004.pdf.
46. Chernoff R. Geriatric nutrition: The health professional's handbook, 3rd ed. Gaithersburg, MD: Aspen; 2006. Moretti C, Frajese GV, Guccione L, Wannenes F, De Martino MU, Fabbri A, Frajese G. Androgens and body composition in the aging male. Journal of Endocrinological Investigation. 2005;28:56–64.

Chapter 11

1. National Center for Health Statistics. Health, United States, 2010: With special feature on death and dying. Hyattsville, MD; 2010. Available from: http://www.cdc.gov/nchs/data/hus/hus10.pdf.
2. U.S. Department of Health and Human Services, Office of Disease Prevention and Health Promotion. 2008 physical activity guidelines for Americans. Washington, DC. Available from: http://health.gov/PAGuidelines/pdf/paguide.pdf.
3. Vuori IM. Health benefits of physical activity with special reference to interaction with diet. Public Health Nutrition. 2001;4:517–28. Williams MA, Haskell WL, Ades PA, Amsterdam EA, Bittner V, Franklin BA, Gulanick M, Laing ST, Stewart KJ, American Heart Association Council on Clinical Cardiology, American Heart Association Council on Nutrition, Physical Activity, and Metabolism. Resistance exercise in individuals with and without cardiovascular disease: 2007 update: A scientific statement from the American Heart Association Council on Clinical Cardiology and Council on Nutrition, Physical Activity, and Metabolism Circulation. 2007;116:572–84.
4. Blair SN, LaMonte MJ, Nichaman MZ. The evolution of physical activity recommendations: How much is enough? American Journal of Clinical Nutrition. 2004;79:913S–20S.
5. American College of Sports Medicine (ACSM). ACSM's guidelines for exercise testing and prescription, 8th ed. Philadelphia: Lippincott Williams & Wilkins; 2010.
6. Persinger R, Foster C, Gibson M, Fater DCW, Porcari JP. Consistency of the talk test for exercise prescription. Medicine & Science in Sports & Exercise. 2004;36:1632–6. Centers for Disease Control and Prevention. Physical activity for everyone. Available from: http://www.cdc.gov/physicalactivity/everyone/measuring/index.html.
7. U.S. Department of Health and Human Services, Office of Disease Prevention and Health Promotion. 2008 physical activity guidelines for Americans. Washington, DC. Available from: http://health.gov/PAGuidelines/pdf/paguide.pdf.
8. Connolly DA, Sayers SP, McHugh MP. Treatment and prevention of delayed onset muscle soreness. Journal of Strength and Conditioning Research. 2003;17:197–208.
9. Volpe SL. Micronutrient requirements for athletes. Clinics in Sports Medicine. 2007;26:119–30.
10. Institute of Medicine. Dietary Reference Intakes for energy, carbohydrate, fiber, fat, fatty acids, cholesterol, protein, and amino acids. Washington, DC: National Academies Press; 2005.
11. American Dietetic Association, Dietitians of Canada, American College of Sports Medicine, Rodriguez NR, Di Marco NM, Langley S. American College of Sports Medicine position stand. Nutrition and athletic performance. Medicine and Science in Sports and Exercise. 2009;41:709–31.
12. Burke LM. Nutrition strategies for the marathon: Fuel for training and racing. Sports Medicine. 2007;37:344–7.
13. Tipton KD, Witard OC. Protein requirements and recommendations for athletes: Relevance of ivory tower arguments for practical recommendations. Clinics in Sports Medicine. 2007;26:17–36.
14. Institute of Medicine. Dietary Reference Intakes for energy, carbohydrate, fiber, fat, fatty acids, cholesterol, protein, and amino acids. Washington, DC: National Academies Press; 2005.
15. American Dietetic Association, Dietitians of Canada, American College of Sports Medicine, Rodriguez NR, Di Marco NM, Langley S. American College of Sports Medicine position stand. Nutrition and athletic performance. Medicine and Science in Sports and Exercise. 2009;41:709–31.
16. Kreider RB, Campbell B. Protein for exercise and recovery. The Physician and Sportsmedicine. 2009;37:13–21.
17. Tipton KD, Wolfe RR. Protein and amino acids for athletes. Journal of Sports Science. 2004;22:65–79.
18. Simopoulos AP. Omega-3 fatty acids and athletics. Current Sports Medicine Reports. 2007;6:230–6.
19. Maughan RJ, King DS, Lea T. Dietary supplements. Journal of Sports Science. 2004;22:95–113.
20. Barr SI, Rideout CA. Nutritional considerations for vegetarian athletes. Nutrition. 2004;20:696–703; Fuhrman J, Ferreri DM. Fueling the vegetarian (vegan) athlete. Current Sports Medicine Reports. 2010;8:233–41.
21. Suedekum NA, Dimeff RJ. Iron and the athlete. Current Sports Medicine Reports. 2005;4:199–202.
22. Kunstel K. Calcium requirements for the athlete. Current Sports Medicine Reports. 2005;4:203–6.
23. Sawka MN, Cheuvront SN, Carter R 3rd. Human water needs. Nutrition Reviews. 2005;63:S30–9.
24. Shirreffs SM. The importance of good hydration for work and exercise performance. Nutrition Reviews. 2005;63:S14–21.
25. Almond CS, Shin AY, Fortescue EB, Mannix RC, Wypij D, Binstadt BA, Duncan CN, Olson DP, Salerno AE, Newburger JW, Greenes DS. Hyponatremia among runners in the Boston Marathon. New England Journal of Medicine. 2005;352:1550–6.
26. Hsieh M. Recommendations for treatment of hyponatraemia at endurance events. Sports Medicine. 2004;34:231–8.
27. Casa DJ, Armstrong LE, Hillman SK, Montain SJ, Reiff RV, Rich BS, Roberts WO, Stone JA. National Athletic Trainers' Association position statement: Fluid replacement for athletes. Journal of Athletic Training. 2000;35:212–24.
28. Sawka MN, Burke LM, Eichner ER, Maughan RJ, Montain SJ, Stachenfeld NS. American College of Sports Medicine position stand: Exercise and fluid replacement. American College of Sports Medicine. Medicine and Science in Sports and Exercise. 2007;39:377–90.
29. Kerksick C, Harvey T, Stout J, Campbell B, Wilborn C, Kreider R, Kalman D, Ziegenfuss T, Lopez H, Landis J, Ivy JL, Antonio J. International Society of Sports Nutrition position stand: Nutrient timing. Journal of the International Society of Sports Nutrition. 2008;5:17.
30. Sherman WM, Doyle JA, Lamb DR, Strauss RH. Dietary carbohydrate, muscle glycogen, and exercise performance during 7 d of training. American Journal of Clinical Nutrition. 1993;57:27–31.
31. Burke LM, Kiens B, Ivy JL. Carbohydrates and fat for training and recovery. Journal of Sports Science. 2004;22:15–30.
32. Stevenson E, Williams C, Biscoe H. The metabolic responses to high carbohydrate meals with different glycemic indices consumed during recovery from prolonged strenuous exercise. International Journal of Sports Nutrition and Exercise Metabolism. 2005;15:291–307. Walton P, Rhodes EC. Glycaemic index and optimal performance. Sports Medicine. 1997;23:164–72.
33. Baty JJ, Hwang H, Ding Z, Bernard JR, Wang B, Kwon B, Ivy JL. The effect of a carbohydrate and protein supplement on resistance exercise performance, hormonal response, and muscle damage. Journal of Strength and Conditioning Research. 2007;21:321–9.
34. Howarth KR, Moreau NA, Phillips SM, Gibala MJ. Coingestion of protein with carbohydrate during recovery from endurance exercise stimulates skeletal muscle protein synthesis in humans. Journal of Applied Physiology. 2009;106:1394–402.
35. Tang JE, Manolakos JJ, Kujbida GW, Lysecki PJ, Moore DR, Phillips SM. Minimal whey protein with carbohydrate stimulates muscle protein synthesis following resistance exercise in trained young men. Applied Physiology, Nutrition, and Metabolism. 2007;32:1132–8.

36. Speedy DB, Thompson JM, Rodgers I, Collins M, Sharwood K, Noakes TD. Oral salt supplementation during ultradistance exercise. Clinical Journal of Sport Medicine. 2002;12:279–84.

Chapter 12

1. Paxton SJ. Body dissatisfaction and disordered eating. Journal of Psychosomatic Research. 2002;53:961–2.
2. Wilson GT, Shafran R. Eating disorders guidelines from NICE. Lancet. 2005;365:79–1.
3. American Psychiatric Association Work Group on Eating Disorders. Practice guideline for the treatment of patients with eating disorders (revision). American Journal of Psychiatry. 2000;157:1–39.
4. Clarke LH. Older women's perceptions of ideal body weights: The tensions between health and appearance motivations for weight loss. Ageing and Society. 2002;22:751–3.
5. Woodside BD, Garfinkel PE, Lin E, Goering P, Kaplan AS, Goldbloom DS, Kennedy SH. Comparisons of men with full or partial eating disorders, men without eating disorders, and women with eating disorders in the community. American Journal of Psychiatry. 2001;158:570–4.
6. American Psychiatric Association. Diagnostic and statistical manual of mental disorders, 4th ed. (DSM-IV). Washington, DC: American Psychiatric Association; 2004.
7. Eddy KT, Dorer DJ, Franko DL, Tahilani K, Thompson-Brenner H, Herzog DB. Diagnostic crossover in anorexia nervosa and bulimia nervosa: Implications for DSM-V. American Journal of Psychiatry. 2008;165:245–50.
8. Peebles R, Wilson JL, Lock JD. Self-injury in adolescents with eating disorders: Correlates and provider bias. Journal of Adolescent Health. 2011;48:310–3.
9. Halmi KA, Tozzi F, Thornton LM, Crow S, Fichter MM, Kaplan AS, Keel P, Klump KL, Lilenfeld LR, Mitchell JE, Plotnicov KH, Pollice C, Rotondo A, Strober M, Woodside DB, Berrettini WH, Kaye WH, Bulik CM. The relation among perfectionism, obsessive-compulsive personality disorder and obsessive-compulsive disorder in individuals with eating disorders. International Journal of Eating Disorders. 2005;38:371–4.
10. Wolfe BE. Reproductive health in women with eating disorders. Journal of Obstetrics and Gynecology in Neonatal Nursing. 2005;34:255–63.
11. Teng K. Premenopausal osteoporosis, an overlooked consequence of anorexia nervosa. Cleveland Clinic Journal of Medicine. 2011;78:50–8.
12. Birmingham CL, Su J, Hlynsky JA, Goldner EM, Gao M. The mortality rate from anorexia nervosa. International Journal of Eating Disorders. 2005;38:143–6.
13. Patrick L. Eating disorders: A review of the literature with emphasis on medical complications and clinical nutrition. Alternative Medical Review. 2002;7:184–202.
14. Holm-Denoma JM, Witte TK, Gordon KH, Herzog DB, Franko DL, Fichter M, Quadflieg N, Joiner TE Jr. Deaths by suicide among individuals with anorexia as arbiters between competing explanations of the anorexia-suicide link. Journal of Affective Disorders. 2008;107:231–6.
15. Williams PM, Goodie J, Motsinger CD. Treating eating disorders in primary care. American Family Physician. 2008;77:187–95.
16. Hay P, Bacaltchuk J. Bulimia nervosa. Clinical Evidence. 2004;12:1326–47.
17. Cooper Z, Fairburn CG. Refining the definition of binge eating disorder and nonpurging bulimia nervosa. International Journal of Eating Disorders. 2003;34:S89–95.
18. Mitchell JE, Pyle RL, Eckert ED, Hatsukami D, Soll E. Bulimia nervosa in overweight individuals. Journal of Nervous and Mental Disease. 1990;178:324–7.
19. Thompson-Brenner H, Eddy KT, Franko DL, Dorer D, Vashchenko M, Herzog DB. Personality pathology and substance abuse in eating disorders: A longitudinal study. International Journal of Eating Disorders. 2008;41:203–8.
20. Bulik CM, Brownley KA, Shapiro JR. Diagnosis and management of binge eating disorder. World Psychiatry. 2007; 6:142–8.
21. American Psychiatric Association. Diagnostic and statistical manual of mental disorders, 4th ed. (DSM-IV). Washington, DC: American Psychiatric Association; 2004.
22. Latner JD, Clyne C. The diagnostic validity of the criteria for binge eating disorder. International Journal of Eating Disorders. 2008;41:1–14.
23. Vanderlinden J, DalleGrave R, Fernandez F, Vandereycken W, Pieters G, Noorduin C. Which factors do provoke binge eating? An exploratory study in eating disorder patients. Eating and Weight Disorders. 2004;9:300–5.
24. Dansky BS, Brewerton TD, Kilpatrick DG. Comorbidity of bulimia nervosa and alcohol use disorders: Results from the National Women's Study. International Journal of Eating Disorders. 2000;27:180–90.
25. Striegel-Moore RH, Franko DL. Epidemiology of binge eating disorder. International Journal of Eating Disorders. 2003;34:S19–29.
26. Pagoto S, Bodenlos JS, Kantor L, Gitkind M, Curtin C, Ma Y. Association of major depression and binge eating disorder with weight loss in a clinical setting. Obesity. 2007;15:2557–9.
27. Niego SH, Kofman MD, Weiss JJ, Geliebter A. Binge eating in the bariatric surgery population: A review of the literature. International Journal of Eating Disorders. 2007;40:349–59.
28. Masheb RM, Grilo CM. On the relation of attempting to lose weight, restraint, and binge eating in outpatients with binge eating disorder. Obesity Research. 2000;8:638–45.
29. Schenck CH, Mahowald MW. Review of nocturnal sleep-related eating disorders. International Journal of Eating Disorders. 1994;15:343–56.
30. Aronoff NJ, Geliebter A, Zammit G. Gender and body mass index as related to the night-eating syndrome in obese outpatients. Journal of the American Dietetic Association. 2001;101:102–4.
31. Winkelman JW. Sleep-related eating disorder and night eating syndrome: Sleep disorders, eating disorders, or both? Sleep. 2006;29:949–54.
32. The Cleveland Clinic. Sleep-related eating disorders. Available from: http://my.clevelandclinic.org/disorders/sleep_disorders/hic_sleep-related_eating_disorders.aspx.
33. Winkelman JW. Sleep-related eating disorder and night eating syndrome: Sleep disorders. 2006;29:949–54.
34. Marcontell DK, Laster AE, Johnson J. Cognitive-behavioral treatment of food neophobia in adults. Journal of Anxiety Disorders. 2003;17:243–51.
35. Nicklaus S, Boggio V, Chabanet C, Issanchou S. A prospective study of food variety seeking in childhood, adolescence and early adult life. Appetite. 2005;44:289–97.
36. Pelchat ML. Of human bondage: Food craving, obsession, compulsion, and addiction. Physiology and Behavior. 2002;76:347–52.
37. Tuorila H, Mustonen S. Reluctant trying of an unfamiliar food induces negative affection for the food. Appetite. 2010;54:418–21.
38. Pope CG, Pope HG, Menard W, Fay C, Olivardia R, Phillips KA. Clinical features of muscle dysmorphia among males with body dysmorphic disorder. Body Image. 2005;2:395–400.
39. Wroblewska AM. Androgenic-anabolic steroids and body dysmorphia in young men. Journal of Psychosomatic Research. 1997;42:225–34.
40. Grieve FG. A conceptual model of factors contributing to the development of muscle dysmorphia. Eating Disorders. 2007;15:63–80.

41. Olivardia R, Pope HG Jr, Hudson JI. Muscle dysmorphia in male weightlifters: A case-control study. American Journal of Psychiatry. 2000;157:1291–6.
42. Grieve FG. A conceptual model of factors contributing to the development of muscle dysmorphia. Eating Disorders. 2007;15:63–80.
43. Becker AE, Keel P, Anderson-Fye EP, Thomas JJ. Genes and/or jeans? Genetic and socio-cultural contributions to risk for eating disorders. Journal of Addictive Disorders. 2004;23:81–103.
44. Eddy KT, Hennessey M, Thompson-Brenner H. Eating pathology in East African women: The role of media exposure and globalization. Journal of Nervous and Mental Disorders. 2007;195:196–202.
45. George JB, Franko DL. Cultural issues in eating pathology and body image among children and adolescents. Journal of Pediatric Psychology. 2010;35:231–42.
46. National Eating Disorders Association. Statistics: Eating disorders and their precursors. Available from: http://www.nationaleatingdisorders.org.
47. Rubinstein S, Caballero B. Is Miss America an undernourished role model? Journal of the American Medical Association. 2000;283:1569.
48. Wonderlich AL, Ackard DM, Henderson JB. Childhood beauty pageant contestants: Associations with adult disordered eating and mental health. Eating Disorders. 2005;13:291–301.
49. Brown JD, Witherspoon EM. The mass media and American adolescents' health. Journal of Adolescent Health. 2002;31:153–70.
50. Fernández-Aranda F, Krug I, Granero R, Ramón JM, Badia A, Giménez L, Solano R, Collier D, Karwautz A, Treasure J. Individual and family eating patterns during childhood and early adolescence: An analysis of associated eating disorder factors. Appetite. 2007;49:476–85.
51. Coulthard H, Blissett J, Harris G. The relationship between parental eating problems and children's feeding behavior: A selective review of the literature. Eating Behavior. 2004;5:103–15.
52. Humphries LL, Wrobel S, Wiegert HT. Anorexia nervosa. American Family Physician. 1982;26:199–204.
53. Kluck AS. Family factors in the development of disordered eating: Integrating dynamic and behavioral explanations. Eating Behavior. 2008;9:471–83.
54. Mazzeo SE, Zucker NL, Gerke CK, Mitchell KS, Bulik CM. Parenting concerns of women with histories of eating disorders. International Journal of Eating Disorders. 2005;37:S77–9.
55. Agras S, Hammer L, McNicholas F. A prospective study of the influence of eating-disordered mothers on their children. International Journal of Eating Disorders. 1999;25:253–62.
56. Cassin SE, von Ranson KM. Personality and eating disorders: A decade in review. Clinical Psychology Review. 2005;25:895–916.
57. Bulik CM, Tozzi F, Anderson C, Mazzeo SE, Aggen S, Sullivan PF. The relation between eating disorders and components of perfectionism. American Journal of Psychiatry. 2003;160:366–8.
58. George JB, Franko DL. Cultural issues in eating pathology and body image among children and adolescents. Journal of Pediatriac Psychology. 2010;35:231–42.
59. Kaye WH, Bulik CM, Plotnicov K, Thornton L, Devlin B, Fichter MM, Treasure J, Kaplan A, Woodside DB, Johnson CL, Halmi K, Brandt HA, Crawford S, Mitchell JE, Strober M, Berrettini W, Jones I. The genetics of anorexia nervosa collaborative study: Methods and sample description. International Journal of Eating Disorders. 2008;41:289–300.
60. Bulik CM, Slof-Op't Landt MC, van Furth EF, Sullivan PF. The genetics of anorexia nervosa. Annual Review of Nutrition. 2007;27:263–75.
61. Bulik CM, Reba L, Siega-Riz AM, Reichborn-Kjennerud T. Anorexia nervosa: Definition, epidemiology, and cycle of risk. International Journal of Eating Disorders. 2005;37:S2–9.
62. Cox LM, Lantz CD, Mayhew JL. The role of social physique anxiety and other variables in predicting eating behaviors in college students. International Journal of Sport Nutrition. 1997;7:310–7. Reinking MF, Alexander LE. Prevalence of disordered-eating behaviors in undergraduate female collegiate athletes and nonathletes. Journal of Athletic Training. 2005;40:47–51.
63. Sudi K, Ottl K, Payerl D, Baumgartl P, Tauschmann K, Muller W. Anorexia athletica. Nutrition. 2004;20:657–61.
64. Salbach H, Klinkowski N, Pfeiffer E, Lehmkuhl U, Korte A. Body image and attitudinal aspects of eating disorders in rhythmic gymnasts. Psychopathology. 2007;40:388–93.
65. Johnson C, Powers PS, Dick R. Athletes and eating disorders: The National Collegiate Athletic Association study. International Journal of Eating Disorders. 1999;26:179–88.
66. Beals KA, Hill AK. The prevalence of disordered eating, menstrual dysfunction, and low bone mineral density among U.S. collegiate athletes. International Journal of Sport Nutrition and Exercise Metabolism. 2006;16:1–23.
67. Barrow GW, Saha S. Menstrual irregularity and stress fractures in collegiate female distance runners. American Journal of Sports Medicine. 1988;3:209–16.
68. Lock J, Agras WS, Bryson S, Kraemer HC. A comparison of short- and long-term family therapy for adolescent anorexia nervosa. Journal of the American Academy of Child and Adolescent Psychiatry. 2005;44:632–9.

Chapter 13

1. French MT, Zavala SK. The health benefits of moderate drinking revisited: Alcohol use and self-reported health status. American Journal of Health Promotion. 2007;21:484–91.
2. U.S. Department of Agriculture, U.S. Department of Health and Human Services. Dietary Guidelines for Americans, 2010. 7th ed., Washington, DC: U.S. Government Printing Office, December 2010. Available from: http://www.cnpp.usda.gov/Publications/DietaryGuidelines/2010/PolicyDoc/PolicyDoc.pdf.
3. Roberts C, Robinson SP. Alcohol concentration and carbonation of drinks: The effect on blood alcohol levels. Journal of Forensic and Legal Medicine. 2007;14:398–405.
4. Zintzaras E, Stefanidis I, Santos M, Vidal F. Do alcohol-metabolizing enzyme gene polymorphisms increase the risk of alcoholism and alcoholic liver disease? Hepatology. 2006;43:352–61.
5. Sumida KD, Hill JM, Matveyenko AV. Sex differences in hepatic gluconeogenic capacity after chronic alcohol consumption. Clinical Medicine and Research. 2007;5:193–202.
6. Nolen-Hoeksema S. Gender differences in risk factors and consequences for alcohol use and problems. Clinical Psychology Review. 2004;24:981–1010.
7. U.S. Department of Agriculture, U.S. Department of Health and Human Services. Dietary Guidelines for Americans, 2010. 7th ed., Washington, DC: U.S. Government Printing Office, December 2010. Available from: http://www.cnpp.usda.gov/Publications/DietaryGuidelines/2010/PolicyDoc/PolicyDoc.pdf.
8. Ibid.
9. Ibid.
10. Ellison RC. Balancing the risks and benefits of moderate drinking. Annals of the New York Academy of Sciences. 2002;957:1–6.
11. Djoussé L, Gaziano JM. Alcohol consumption and risk of heart failure in the Physicians' Health Study I. Circulation. 2007;115:34–9. Elkind MS, Sciacca R, Boden-Albala B,

Rundek T, Paik MC, Sacco RL. Moderate alcohol consumption reduces risk of ischemic stroke: The Northern Manhattan Study. Stroke. 2006;37(1):13–9.
12. Wollin SD, Jones PJH. Alcohol, red wine and cardiovascular disease. Journal of Nutrition. 2001;131:1401–4.
13. Schmid B, Hohm E, Blomeyer D, Zimmermann US, Schmidt MH, Esser G, Laucht M. Concurrent alcohol and tobacco use during early adolescence characterizes a group at risk. Alcohol and Alcoholism. 2007;42:219–25.
14. Mukamal KJ, Chiuve SE, Rimm EB. Alcohol consumption and risk for coronary heart disease in men with healthy lifestyles. Archives of Internal Medicine. 2006:166:2145–50.
15. Kabagambe EK, Baylin A, Ruiz-Narvaez E, Rimm EB, Campos H. Alcohol intake, drinking patterns, and risk of nonfatal acute myocardial infarction in Costa Rica. American Journal of Clinical Nutrition. 2005;82:1336–45.
16. Goldberg IJ, Mosca L, Piano MR, Fisher EA. AHA Science Advisory. Wine and your heart: A science advisory for healthcare professionals from the Nutrition Committee, Council on Epidemiology and Prevention, and Council on Cardiovascular Nursing of the American Heart Association. Stroke. 2001;32:591–4.
17. Kurth T, Everett BM, Buring JE, Kase CS, Ridker PM, Gaziano JM. Lipid levels and the risk of ischemic stroke in women. Neurology. 2007;68:556–62.
18. Agarwal DP. Cardioprotective effects of light–moderate consumption of alcohol: A review of putative mechanisms. Alcohol and Alcoholism. 2002;37:409–15.
19. Gresele P, Pignatelli P, Guglielmini G, Carnevale R, Mezzasoma AM, Ghiselli A, Momi S, Violi F. Resveratrol, at concentrations attainable with moderate wine consumption, stimulates human platelet nitric oxide production. Journal of Nutrition. 2008;138:1602–08. German JB, Walzem RL. The health benefits of wine. Annual Review of Nutrition. 2000;20:561–93.
20. Manari AP, Preedy VR, Peters TJ. Nutritional intake of hazardous drinkers and dependent alcoholics in the UK. Addiction Biology. 2003;8:201–10. Salaspuro M. Nutrient intake and nutritional status in alcoholics. Alcohol and Alcoholism. 1993;28:85–8.
21. U.S. Department of Agriculture, U.S. Department of Health and Human Services. Dietary Guidelines for Americans, 2010. 7th ed., Washington, DC: U.S. Government Printing Office, December 2010. Available from: http://www.cnpp.usda.gov/Publications/DietaryGuidelines/2010/PolicyDoc/PolicyDoc.pdf.
22. Kesse E, Clavel-Chapelon F, Slimani N, van Liere M, E3N Group. Do eating habits differ according to alcohol consumption? Results of a study of the French cohort of the European Prospective Investigation into Cancer and Nutrition (E3N-EPIC). American Journal of Clinical Nutrition. 2001;74:322–7.
23. González-Reimers E, García-Valdecasas-Campelo E, Santolaria-Fernández F, Milena-Abril A, Rodríguez-Rodríguez E, Martínez-Riera A, Pérez-Ramírez A, Alemán-Valls MR. Rib fractures in chronic alcoholic men: Relationship with feeding habits, social problems, malnutrition, bone alterations, and liver dysfunction. Alcohol. 2005;37:113–7.
24. Baan R, Straif K, Grosse Y, Secretan B, El Ghissassi F, Bouvard V, Altieri A, Cogliano V. WHO International Agency for Research on Cancer Monograph Working Group. Carcinogenicity of alcoholic beverages. Lancet Oncology. 2007;8:292–3.
25. Allen NE, Beral V, Casabonne D, Kan SW, Reeves GK, Brown A, Green J. Moderate alcohol intake and cancer incidence in women. Million Women Study Collaborators. Journal of the National Cancer Institute. 2009;101:296–305.
26. Blot WJ. Alcohol and cancer. Cancer Research. 1992;52:2119S–23S. Seitz HK, Meier P. The role of acetaldehyde in upper digestive tract cancer in alcoholics. Translational Research. 2007;149:293–7.
27. Hamid A, Kaur J. Long-term alcohol ingestion alters the folate-binding kinetics in intestinal brush border membrane in experimental alcoholism. Alcohol. 2007;41:441–6.
28. U.S. Department of Agriculture, U.S. Department of Health and Human Services. Dietary Guidelines for Americans, 2010. 7th ed., Washington, DC: U.S. Government Printing Office, December 2010. Available from: http://www.cnpp.usda.gov/Publications/DietaryGuidelines/2010/PolicyDoc/PolicyDoc.pdf.
29. Piano MR. Alcoholic cardiomyopathy: Incidence, clinical characteristics, and pathophysiology. Chest. 2002;121: 1638–50.
30. Sesso HD, Cook NR, Buring JE, Manson JE, Gaziano JM. Alcohol consumption and the risk of hypertension in women and men. Hypertension. 2008;168:884–90.
31. Djoussé L, Levy D, Benjamin EJ, Blease SJ, Russ A, Larson MG, Massaro JM, D'Agostino RB, Wolf PA, Ellison RC. Long-term alcohol consumption and the risk of atrial fibrillation in the Framingham study. American Journal of Cardiology. 2004;93:710–3.
32. Hanck C, Whitcomb DC. Alcoholic pancreatitis. Gastroenterology Clinics of North America. 2004;33:751–65.
33. American Psychiatric Association. Diagnostic and statistical manual of mental disorders, 4th ed., DSM-IV-TR (Text Revision). Washington, DC: American Psychiatric Publishing, 2000.
34. National Institute on Alcohol Abuse and Alcoholism (NIAA). Updating estimates of the economic costs of alcohol abuse in the United States: Estimates, update methods, and data. Report prepared by the Lewin Group for the National Institute on Alcohol Abuse and Alcoholism. Rockville, MD: NIAAA, National Institutes of Health, Department of Health and Human Services, 2000.
35. Hingson R, Heeren T, Winter M, Wechsler H. Magnitude of alcohol-related mortality and morbidity among U.S. college students ages 18–24: Changes from 1998 to 2001. Annual Review of Public Health. 2005;26:259–79.
36. Turrisi R, Mallett KA, Mastroleo NR, Larimer ME. Heavy drinking in college students: Who is at risk and what is being done about it? Journal of General Psychology. 2006;133:401–20.
37. Yi H, Williams GD, Dufour MC. Surveillance report #65: Trends in alcohol-related traffic crashes, United States, 1977–2001. Bethesda, MD: NIAAA, Division of Biometry and Epidemiology, Alcohol Epidemiologic Data System, August 2003.
38. U.S. Department of Health and Human Services, National Institutes of Health, National Institute on Alcohol Abuse and Alcoholism. What colleges need to know now: An update on college drinking research. NIH Publication No. 07-5010, November 2007.
39. Timberlake DS, Hopfer CJ, Rhee SH, Friedman NP, Haberstick BC, Lessem JM, Hewitt JK. College attendance and its effect on drinking behaviors in a longitudinal study of adolescents. Alcoholism, Clinical and Experimental Research. 2007;31:1020–30.
40. Lange JE, Clapp JD, Turrisi R, Reavy R, Jaccard J, Johnson MB, Voas RB, Larimer M. College binge drinking: What is it? Who does it? Alcoholism, Clinical and Experimental Research. 2002;26:723–30.
41. Ameythyst Initiative. 10 E Street, SE Washington, DC. Available from: http://www.amethystinitiative.org/about/.
42. DeJong W, Larimer ME, Wood MD, Hartman R. NIAAA's Rapid Response to College Drinking problems initiative: Reinforcing the use of evidence-based approaches in college alcohol prevention. Journal of Studies on Alcohol and Drugs. 2009;16S:5–11. Dejong W. Finding common ground for effective campus-based prevention. Psychology of Addictive Behaviors. 2001;15:292–6.

43. Beck KH, Arria AM, Caldeira KM, Vincent KB, O'Grady KE, Wish ED. Social context of drinking and alcohol problems among college students. American Journal of Health Behavior. 2008;32:420–30.
44. U.S. Department of Health and Human Services, Office of Disease Prevention and Health Promotion, Healthy people 2020. Washington D.C., 2011. Available from: http://healthypeople.gov/2020/about/default.aspx.
45. U.S. Department of Agriculture, U.S. Department of Health and Human Services. Dietary Guidelines for Americans, 2010. 7th ed., Washington, DC: U.S. Government Printing Office, December 2010. Available from: http://www.cnpp.usda.gov/Publications/DietaryGuidelines/2010/PolicyDoc/PolicyDoc.pdf.

Chapter 14

1. U.S. Centers for Disease Control and Prevention. Estimates of foodborne illness in the United States. December 15, 2010. Available from: http://www.cdc.gov/foodborneburden/index.html.
2. Trabulsi LR, Keller R, Tardelli Gomes TA. Typical and atypical enteropathogenic *Escherichia coli*. Emerging Infectious Diseases. 2002;8:508–13.
3. U.S. Centers for Disease Control and Prevention. Estimates of foodborne illness in the United States. December 15, 2010. Available from: http://www.cdc.gov/foodborneburden/index.html.
4. Jones TF, Kellum ME, Porter SS, Bell M, Schaffner W. An outbreak of community-acquired foodborne illness caused by methicillin-resistant *Staphylococcus aureus*. Emerging Infectious Diseases. 2002;8:82–4.
5. U.S. Centers for Disease Control and Prevention. Botulism associated with canned chili sauce, July–August 2007. Updated August 24, 2007. Available from: http://www.cdc.gov/botulism/botulism.htm.
6. U.S. Centers for Disease Control and Prevention. Norovirus outbreaks on three college campuses—California, Michigan, and Wisconsin, 2008. Morbidity and Mortality Weekly Reports. 2009;58:1095–100. Available from: http://www.cdc.gov/mmwr/preview/mmwrhtml/mm5839a2.htm.
7. U.S. Centers for Disease Control and Prevention. Investigation update: Multistate outbreak of human *Salmonella enteritidis* infections associated with shell eggs. December 2, 2010. Available from: http://www.cdc.gov/salmonella/enteritidis/.
8. U.S. Centers for Disease Control and Prevention. Ongoing multistate outbreak of *Escherichia coli* serotype O157:H7 infections associated with consumption of fresh spinach—United States, September 2006. Morbidity and Mortality Weekly Reports. 2006;55:1–2. Available from: http://www.cdc.gov/mmwr/preview/mmwrhtml/mm5538a4.htm.
9. U.S. Food and Drug Administration. Guidance for industry: Guide to minimize microbial food safety hazards of fresh-cut fruits and vegetables. February 2008. Available from: http://www.fda.gov/Food/GuidanceComplianceRegulatoryInformation/GuidanceDocuments/ProduceandPlanProducts/ucm064458.htm.
10. Federal Institute for Risk Assessment, Federal Office of Consumer Protection and Food Safety, and Robert Koch Institute. Information update on EHEC outbreak. June 10, 2011. Available from: http://www.rki.de/cln_144/nn_217400/EN/Home/PM082011.html. U.S. Centers for Disease Control and Prevention. Investigation update: Outbreak of shiga toxin-producing *E. coli* O104 (STEC O104:H4) infections associated with travel to Germany. June 15, 2011. Available from: http://www.cdc.gov/print.do?url=http://www.cdc.gov/ecoli/2011/ecoliO104/.
11. Mostl K. Bovine spongiform encephalopathy (BSE): The importance of the food and feed chain. Forum in Nutrition. 2003;56:394–6.
12. Ryou C. Prions and prion diseases: Fundamentals and mechanistic details. Journal of Microbiology and Biotechnology. 2007;17:1059–70.
13. Belay ED, Schonberger LB. The public health impact of prion diseases. Annual Review of Public Health. 2005;26: 191–212. U.S. Centers for Disease Control and Prevention. vCJD (variant Creutzfeldt-Jakob disease). Available from: http://www.cdc.gov/ncidod/dvrd/vcjd/risk_travelers.htm. Roma AA, Prayson RA. Bovine spongiform encephalopathy and variant Creutzfeldt-Jakob disease: How safe is eating beef? Cleveland Clinic Journal of Medicine. 2005;72: 185–94.
14. U.S. Department of Agriculture. FSIS publishes final rule prohibiting processing of "downer" cattle. Available from: http://www.usda.gov.
15. Bovine somatotropin and the safety of cows' milk: National Institutes of Health technology assessment conference statement. Nutrition Reviews. 1991;49:227–32. Etherton TD, Kris-Etherton PM, Mills EW. Recombinant bovine and porcine somatotropin: Safety and benefits of these biotechnologies. Journal of the American Dietetic Association. 1993;93:177–80.
16. Walker R, Lupien JR. The safety evaluation of monosodium glutamate. Journal of Nutrition. 2000;130:1049S–52S.
17. Bull RJ, Robinson M, Laurie RD, Stoner GD, Greisiger E, Meier RJ, Stober J. Carcinogenic effects of acrylamide in Sencar and A/J mice. Cancer Research. 1984;44:107–11.
18. Mucci LA, Dickman PW, Steineck G, Adami H-O, Augustsson K. Dietary acrylamide and cancer of the large bowel, kidney, and bladder: Absence of an association in a population-based study in Sweden. British Journal of Cancer. 2003;88:84–9.
19. U.S. Food and Drug Administration. Melamine pet food recall of 2007. Available from: http://www.fda.gov/animalveterinary/safetyhealth/recallswithdrawals/ucm129575.htm.
20. U.S. Food and Drug Administration. FDA/USDA joint news release: Scientists conclude very low risk to humans from food containing melamine. Available from: http://www.fda.gov/NewsEvents/Newsroom/PressAnnouncements/2007/default.htm.
21. U.S. Food and Drug Administration. Melamine contamination in China. Available from: http://www.fda.gov/NewsEvents/PublicHealthFocus/ucm179005.htm.
22. Durando M, Kass L, Piva J, Sonnenschein C, Soto AM, Luque EH, Muñoz-de-Toro M. Prenatal bisphenol A exposure induces preneoplastic lesions in the mammary gland in Wistar rats. Environmental Health Perspectives. 2007;115:80–6. Newbold RR, Jefferson WN, Padilla-Banks E. Long-term adverse effects of neonatal exposure to bisphenol A on the murine female reproductive tract. Reproductive Toxicology. 2007;24:253–8. U.S. Food and Drug Administration. Update on bisphenol A (BPA) for use in food. January 2010. Available from: http://www.fda.gov/NewsEvents/PublicHealthFocus/ucm064437.htm.
23. U.S. Food and Drug Administration. Update on bisphenol A (BPA) for use in food. January 2010. Available at http://www.fda.gov/NewsEvents/PublicHealthFocus/ucm064437.htm.
24. Kenney SJ, Beuchat LR. Comparison of aqueous commercial cleaners for effectiveness in removing *Escherichia coli* O157:H7 and *Salmonella meunchen* from the surface of apples. International Journal of Food Microbiology. 2002;25:47–55.
25. U.S. Centers for Disease Control and Prevention. Update: Multistate outbreak of listeriosis—United States, 1998–1999. Morbidity and Mortality Weekly Report. 1999;47:1117–8.
26. United States Congress. Public health security and bioterrorism preparedness and response act of 2002. June 12, 2002. Available from: http://www.fda.gov/oc/bioterrorism/Bioact.html.

Chapter 15

1. Food and Agriculture Organization of the United Nations. Global hunger declining, but still unacceptably high. 2010. Available from: http://www.fao.org/docrep/012/al390e/al390e00.pdf.
2. Ibid.
3. Food and Agriculture Organization of the United Nations. The state of food insecurity in the world 2010.Available from: http://www.fao.org/docrep/013/i1683e/i1683e.pdf.
4. Kempson KM, Palmer Keenan D, Sadani PS, Ridlen S, Scotto Rosato N. Food management practices used by people with limited resources to maintain food sufficiency as reported by nutrition educators. Journal of the American Dietetic Association. 2002;102:1795–9.
5. U.S. Department of Agriculture, Economic Research Service. Household food security in the United States, 2008.Available from: http://www.ers.usda.gov/Publications/ERR83/ERR83c.pdf.
6. Nord M, Coleman-Jensen A, Andrews M, and Carlson S. Household food security in the United States, 2009. ERR-108, U.S. Department of Agriculture, Economics Research Service. 2010. Available from: http://www.ers.usda.gov/Publications/ERR108/ERR108.pdf.
7. Ibid.
8. Ibid.
9. Ibid.
10. Ibid.
11. Ibid.
12. DeNavas-Walt C, Proctor BD, Smith JC, U.S. Census Bureau. Income, poverty, and health insurance coverage in the United States: 2009. Current Population Reports, P60-238. Washington, DC: U.S. Government Printing Office, 2010. Available from: http://www.census.gov/prod/2010pubs/p60-238.pdf.
13. DeNavas-Walt C, Proctor BD, Smith JC, U.S. Census Bureau. Income, poverty, and health insurance coverage in the United States: 2009. Current Population Reports, P60-238. Washington, D.C: U.S. Government Printing Office, 2010. Available from: http://www.census.gov/prod/2010pubs/p60-238.pdf.
14. America's Second Harvest. The almanac of hunger and poverty in America 2007. Chicago, IL: 2007. Available from: http://feedingamerica.org/our-network/the-studies/~/media/Files/research/almanac/section1.ashx.
15. Kim M, Ohls J, Cohen R. Hunger in America, 2001. National report prepared for America's Second Harvest. Princeton, NJ: Mathematica Policy Research Inc., 2001. Available from: http://www.mathematica-mpr.com/pdfs/hunger2001.pdf.
16. The United States Conference of Mayors—Sodexho. Hunger and homelessness survey. Available from: http://www.usmayors.org/HHSurvey2007/hhsurvey07.pdf.
17. Olson CM. Nutrition and health outcomes associated with food insecurity and hunger. Journal of Nutrition. 1999;129:521S–4S.
18. Furness BW, Simon PA, Wold CM, Asarian-Anderson J. Prevalence and predictors of food insecurity among low-income households in Los Angeles County. Public Health Nutrition. 2004;7:791–4.
19. Ibid.
20. Beaulac J, Kristjansson E, Cummins S. A systematic review of food deserts, 1966–2007. Preventing Chronic Disease. 2009;6:A105.
21. Whitacre P, Tsai P, Mulligan J. The public health effects of food deserts: Workshop summary. Washington, DC: The National Academies Press, 2009. Available from: http://books.nap.edu/catalog.php?record_id=12623.
22. Kendall A, Olson CM, Frongillo EA Jr. Relationship of hunger and food insecurity to food availability and consumption. Journal of the American Dietetic Association. 1996;96:1019–24.
23. Ibid.
24. Rose-Jacobs R, Black MM, Casey PH, Cook JT, Cutts DB, Chilton M, Heeren T, Levenson SM, Meyers AF, Frank DA. Household food insecurity: Associations with at-risk infant and toddler development. Pediatrics. 2008;121:65–72.
25. Fram MS, Frongillo EA, Jones SJ, Williams RC, Burke MP, Deloach KP, Blake CE. Children are aware of food insecurity and take responsibility for managing food resources. Journal of Nutrition. 2011;141:1114–9.
26. Alaimo K, Olson CM, Frongillo EA. Family food insufficiency, but not low family income, is positively associated with dysthymia and suicide symptoms in adolescents. Journal of Nutrition. 2002;132:719–25.
27. Lee JS, Frongillo EA Jr. Nutritional and health consequences are associated with food insecurity among U.S. elderly persons. Journal of Nutrition. 2001;131:1503–9.
28. Ibid.
29. Ibid.
30. U.S. Department of Agriculture, Food and Nutrition Service. Supplemental Nutrition Assistance Program. Available from: http://www.fns.usda.gov/snap/Default.htm.
31. Ibid.
32. Zedlewski SR. Leaving welfare often severs families' connections to the Food Stamp Program. Journal of the American Medical Women's Association. 2002;57:23–6.
33. Jacknowitz A, Novillo D, Tiehen L. Special Supplemental Nutrition Program for Women, Infants, and Children and infant feeding practices. Pediatrics. 2007;119:281–9.
34. American Academy of Pediatrics. WIC program. Provisional section on breastfeeding. Position statement. Pediatrics. 2001;108:1216–7.
35. U.S. Department of Agriculture Food and Nutrition Service. National school lunch program. Available from: http://www.fns.usda.gov/cnd/lunch/aboutlunch/NSLPFactSheet.pdf.
36. Ibid.
37. Institute of Medicine. School meals: Building blocks for healthy children. Washington, DC: The National Academies Press, 2010.
38. U.S. Public law 111-296. Congress. Healthy hunger-free kids act of 2010. Available from: http://www.gpo.gov/fdsys/pkg/PLAW-111publ296/pdf/PLAW-111publ296.pdf.
39. Kantor LS, Lipton K, Manchester, A, Oliveira V. Estimating and addressing America's food losses. U.S. Department of Agriculture. Economic Research Service. FoodReview, 1996. Available from: http://www.ers.usda.gov/Publications/FoodReview/Jan1997/Jan97a.pdf.
40. Food and Agriculture Organization of the United Nations. Global hunger declining, but still unacceptably high. 2010. Available from: http://www.fao.org/docrep/012/al390e/al390e00.pdf.
41. International Food Policy Research Institute, Concern Worldwide, and Welthungerhilfe. 2010 global hunger index. Available from: http://www.ifpri.org/sites/default/files/publications/ghi10.pdf.
42. Ibid.
43. United Nations High Commissioner for Refugees. 2009 global trends: Refugees, asylum-seekers, returnees, internally displaced and stateless persons. 2010. Available from: http://www.unhcr.org/4c11f0be9.html.
44. Kennedy G, Nantel G, Shetty P. Globalization of food systems in developing countries: impact on food security and nutrition. Food and Agriculture Organization of the United Nations. FAO Food and Nutrition Paper. 2004;83:1–300.
45. Reardon T, Timmer P, Barrett C, Berdegué J. The rise of supermarkets in Africa, Asia and Latin America. American Journal of Agricultural Economics. 2003;85:1140–6.
46. The impact of global change and urbanization on household food security, nutrition, and food safety. Food and Agriculture

Organization of the United Nations. Available from: http://www.fao.org/ag/agn/nutrition/national_urbanization_en.stm.
47. Ibid.
48. Ibid.
49. El-Ghannam AR. The global problems of child malnutrition and mortality in different world regions. Journal of Health and Social Policy. 2003;16:1–26. Horton KD. Bringing attention to global hunger. Journal of the American Dietetic Association. 2008;108:435.
50. Task Force on Education and Gender Equality. United Nations Millennium Project 2005. Taking action: Achieving gender equality and empowering women. Available from: http://www.unmillenniumproject.org/documents/Gender-complete.pdf.
51. Kennedy E, Meyers L. Dietary Reference Intakes: Development and uses for assessment of micronutrient status of women—a global perspective. American Journal of Clinical Nutrition. 2005;81:1194S–7S.
52. Humphrey JH, West KP Jr., Sommer A. Vitamin A deficiency and attributable mortality among under-5-year-olds. Bulletin of the World Health Organization. 1992:70:225–32.
53. United Nations Children's Fund. The state of the world's children 2005. Available from: http://www.unicef.org/publications/files/SOWC_2005_(English).pdf.
54. Population Reference Bureau. 2007 world population data sheet: Malnutrition is a major contributor to child deaths. Available from: http://www.prb.org/Journalists/ Press Releases/2007/2007WPDSBriefing.aspx.
55. Milman A, Frongillo EA, de Onis M, Hwang JY. Differential improvement among countries in child stunting is associated with long-term development and specific interventions. Journal of Nutrition. 2005:135:1415–22.
56. Ibid.
57. Ibid.
58. Ibid.
59. McCall E. Communication for development strengthening the effectiveness of the United Nations. United Nations Development Programme. 2011. Available from: http://www.unicef.org/cbsc/files/Inter-agency_C4D_Book_2011.pdf.
60. Behrman JR, Parker SW, Todd PE. Schooling impacts of conditional cash transfers on young children: evidence from Mexico. Economic Development and Cultural Change. 2009;57:439–77.
61. United Nations Millennium Development Project. United Nations Millennium development goals. Millennium Development Project report 2010. Available from: http://www.un.org/millenniumgoals/pdf/MDG%20Report%202010%20En%20r15%20-low%20res%2020100615%20-.pdf.
62. Nord M, Coleman-Jensen A, Andrews M, and Carlson S. Household food security in the United States, 2009. ERR-108, U.S. Department of Agriculture, Economics Research Service. 2010. Available from: http://www.ers.usda.gov/Publications/ERR108/ERR108.pdf.

Index

A

C

D

E

F

G

H

I

J

K

L

M

N

O

P

Q

R

S

T

U

V

W

X

Y

Z

Review 1

Chapter 1
Why Does Nutrition Matter?

Chapter Summary

LO1: What Is Nutrition?

- The term *nutrition* refers to how living organisms obtain and use food to support all the processes required for their existence. Because this process is complex, the study of nutrition incorporates a wide variety of scientific disciplines.
- A dietitian has the credential *RD*, which stands for *registered dietitian*. Some dietitians are also involved in scientific research.

LO2: What Are Nutrients, and What Do They Do?

- Essential nutrients must be consumed, whereas the body can make sufficient amounts of the nonessential nutrients when needed. When the body cannot make a typically nonessential nutrient in adequate amounts, it becomes conditionally essential.
- The term *organic* is used by chemists to describe most substances that contain carbon and hydrogen atoms, whereas organic foods are those that are produced without using synthetic fertilizers, hormones, or other drugs.
- Because a person must consume more than a gram of them every day, water, carbohydrates, proteins, and lipids are considered macronutrients. Because a person needs only very small amounts of vitamins and minerals, these substances are considered micronutrients.
- Phytonutrients and zoonutrients are compounds found in plant- and animal-based foods respectively. These substances are not considered traditional nutrients but may improve health.
- Foods that contain enhanced amounts of essential nutrients, phytochemicals, or zoonutrients are called functional foods.

LO3: How Are Macronutrients and Micronutrients Classified?

- Carbohydrates, proteins, and lipids all provide energy and are thus referred to as energy-yielding nutrients.
- Carbohydrates, proteins, and lipids have many structural and regulatory functions in the body.
- Water serves as the medium in which all chemical reactions occur, helps eliminate waste products, and regulates body temperature.
- Vitamins, classified as either water- or fat-soluble, serve many purposes (mostly having to do with the regulation of chemical reactions).
- At least 15 minerals, each of which serves a specific purpose, are considered to be essential nutrients. Many of these essential minerals have specific functions regarding structure, regulation, and energy use.

LO4: How Is the Energy in Food Measured?

- Energy is not a nutrient, but in terms of nutrition, the body can use energy found in foods to grow, develop, move, and fuel the many chemical reactions required for life. The body transforms the chemical energy in foods into a usable form called ATP.
- Carbohydrates and proteins contain 4 kcal of energy per gram of substance, while lipids contain 9 kcal per gram.

Key Terms

adenosine triphosphate (ATP) A chemical that provides energy to cells in the body.

calorie The unit of measurement used to express the amount of energy in a food.

cause-and-effect relationship (or **causal relationship**) A relationship whereby an alteration to one variable causes a change in another variable.

chronic degenerative disease A noninfectious disease that develops slowly, persists over a long period of time, and tends to result in progressive breakdown of tissues and loss of function.

conditionally essential nutrient A normally nonessential nutrient that, under certain circumstances, becomes essential.

control group Study participants that do not receive a treatment or intervention.

correlation (or **association**) A relationship whereby an alteration to one variable is related to a change in another variable.

dietitian A nutrition professional who helps people make dietary changes and food choices to support a healthy lifestyle.

disease An abnormal condition of the body or mind that causes discomfort, dysfunction, or distress.

double-blind study A human experiment in which neither the participants nor the scientists know to which group the participants have been assigned.

epidemiologic study A study in which data are collected from a group of people who are not asked to change their behaviors in any way.

energy The capacity to do work.

energy-yielding nutrient A nutrient that the body can use for energy.

essential nutrient A substance that must be obtained from the diet to sustain life.

functional food (or **super food**) A food that likely optimizes human health by providing a high concentration of nutrients, phytochemicals, or zoochemicals.

graying of America A phenomenon occurring in the United States by which an increasing proportion of the population is over the age of 65.

hypothesis A prediction about the relationship between variables.

infant mortality rate The number of infant deaths per 1,000 live births in a given year.

infectious disease An illness that is contagious, caused by a pathogen, and tends to be short-lived.

inorganic compound A substance that does not contain carbon.

intervention study An experiment in which a variable is altered to determine its effect on another variable.

life expectancy A statistical prediction of the average number of years of life remaining for a person at a particular age.

macronutrients A class of nutrients that humans need to consume in relatively large quantities (more than a gram per day).

micronutrients A class of nutrients that humans need to consume in relatively small quantities.

morbidity rate The number of illnesses or diseases in a given period of time.

mortality rate The number of deaths that occur in a certain population group in a given period of time.

nonessential nutrient A substance that sustains life but is not necessarily obtained from the diet.

noninfectious disease An illness that is not contagious, does not involve an infectious agent, and tends to be long-term and chronic.

nutrient A substance found in food that is used by the body for energy, maintenance of body structure, or regulation of chemical processes.

nutrition The science of how living organisms obtain and use food to support processes required for existence.

nutrition transition A shift from undernutrition to overnutrition or unbalanced nutrition that often occurs as a society transitions to a more industrialized economy.

organic compound A substance that contains carbon and hydrogen atoms.

peer-reviewed journal A publication that requires a group of scientists to read and approve a study before it is published.

phytochemical (or **phytonutrient**) A compound found in plants that likely benefits human health beyond the provision of essential nutrients and energy.

placebo An inert treatment given to the control group that cannot be distinguished from the actual treatment.

placebo effect A phenomenon whereby a study participant experiences an apparent effect of the treatment just because the participant believes that the treatment will work.

random assignment A condition by which study participants have equal chance of being assigned to the treatment and control group.

rate A measure of some event, disease, or condition within a specific time span.

researcher bias A phenomenon by which the researcher influences the results of a study.

risk factor A lifestyle, environmental, or genetic factor related to a person's chances of developing a disease.

scientific method A series of steps used by scientists to explain observations.

single-blind study A human experiment in which the participants do not know to which group they have been assigned.

zoochemical (or **zoonutrient**) A compound found in animal-based foods that likely benefits human health beyond the provision of essential nutrients and energy.

- A kilocalorie is sometimes referred to as a Calorie (note the capital *C*) outside of scientific research, as on food labels. Therefore, 1 Calorie is equivalent to 1 kilocalorie, or 1,000 calories.

LO5: How Do Nutritional Scientists Conduct Their Research?

- Most research is conducted using a three-step process called the scientific method, which involves making an observation, generating an explanation (or hypothesis), and testing the explanation by conducting a study.
- Epidemiologic studies are conducted to investigate correlations, whereas intervention studies can test causal relationships.
- Scientists use techniques such as control groups, placebos, blinding, and random assignment to decrease study bias.
- Sometimes it is not possible or practical to test a hypothesis using humans as participants. In these cases, researchers turn to animal models or cell cultures.

LO6: Are All Nutrition Claims Believable?

- Separating fact from fiction can be difficult, but it is largely possible to determine the validity of any nutrition claim.
- Publication in a peer-reviewed journal indicates that the information is probably reliable.
- Although most research is not likely to be influenced by the source of funding, it is possible for a funding agency to have biased the conclusions made by researchers.
- It is important to consider whether the design of the study was appropriate and whether major public health groups support the study's conclusions.

LO7: Nutrition and Health: What Is the Connection?

- Over the past century, the primary public health concerns have shifted from infectious diseases and nutritional deficiencies to chronic diseases and overnutrition.
- These shifts are reflected by changes in morbidity and mortality rates and life expectancy: whereas infant mortality rates and mortality from infectious diseases have decreased, life expectancy and rates of chronic degenerative diseases have increased.
- The shift from undernutrition to overnutrition or unbalanced nutrition as a society becomes more industrialized is called the nutrition transition. This phenomenon is strongly related to many of the chronic diseases facing humankind today.
- Poor dietary practices are associated with a greater risk for chronic disease.

LO8: Why Study Nutrition?

- Consuming a healthy balance of traditional nutrients, phytonutrients, and zoonutrients can decrease your risk of developing obesity, cardiovascular disease, high blood pressure, diabetes, and cancer.
- As the occurrence of chronic diseases increases, it is ever more important to pay attention to what you eat throughout your entire life.

Review 2

Chapter 2
Choosing Foods Wisely

Chapter Summary

LO1: What Is Nutritional Status?

- A person's nutritional status depends on whether sufficient amounts of nutrients and energy are available to support optimal bodily function.
- Both undernutrition and overnutrition are examples of malnutrition, a state of poor nutrition caused by an imbalance between the body's nutrient requirements and nutrient availability.
- Primary malnutrition is due to inadequate diet, whereas secondary malnutrition may be caused by other factors, such as illness.
- Nutrient requirements vary greatly among individuals. They are influenced by genetic, lifestyle, and environmental factors.

LO2: How Is Nutritional Status Assessed?

- Nutritional status can be assessed in several ways, including anthropometric measurements, biochemical measurements, clinical assessment, and dietary assessment (the *ABCD* methods).
- Anthropometric measurements include measures of body dimensions and composition.
- Biochemical analyses of blood and/or urine samples can provide detailed information about nutrient status.
- Clinical assessment involves the conduction of a face-to-face medical history and physical examination to check for signs and symptoms of malnutrition.
- Dietary assessment methods include diet recalls, food frequency questionnaires, and diet records.

LO3: How Much of a Nutrient Is Adequate?

- The Institute of Medicine has formulated a set of nutrient intake standards called the Dietary Reference Intakes (DRIs). These standards include the Estimated Average Requirements (EARs), Recommended Dietary Allowances (RDAs), Adequate Intake (AI) levels, and Tolerable Upper Intake Levels (ULs).
- EARs estimate average nutrient requirements in various population groups; RDAs are based on EARs and can be utilized as nutrient-intake goals for individuals.
- When the Institute of Medicine could not establish EARs and RDAs, AI levels were set as guidelines for intakes.
- ULs are not dietary recommendations but are levels that should not be exceeded.
- Estimated Energy Requirement (EER) equations and Acceptable Macronutrient Distribution Ranges (AMDRs) provide guidance as to total energy intake and distribution of energy intake from macronutrients.

LO4: How Can You Assess and Plan Your Diet?

- The Dietary Guidelines for Americans provide science-based nutritional recommendations to promote optimal health and reduce the risk of chronic disease in the United States.
- The Dietary Guidelines for Americans also incorporates the USDA Food Patterns, which outlines how much of each food group should be consumed to promote optimal health.
- The Dietary Guidelines' Food Patterns forms the basis of MyPlate, a graphic representation of the food groups and recommended intakes.

Key Terms

Acceptable Macronutrient Distribution Range (AMDR) The recommended range of intake for a given energy-yielding nutrient, expressed as a percentage of total daily caloric intake.

Adequate Intake (AI) The daily intake of a nutrient that appears to support adequate nutritional status; established when RDAs cannot be determined.

anthropometric measurement A measurement of a body's physical dimensions or composition.

biochemical measurement Laboratory analysis of a biological sample, such as blood or urine.

body composition The proportions of fat, water, lean tissue, and mineral (bone) mass that make up the body.

Daily Value (DV) A benchmark as to whether a food is a good source of a particular nutrient. May represent a nutrient's recommended daily intake or upper limit.

diet recall A retrospective dietary assessment method by which a person records and analyzes every food and drink consumed over a given time span.

diet record (or **food record**) A prospective dietary assessment method by which a person records and analyzes every food and drink as it is consumed over a given time span.

dietary assessment The evaluation of adequacy of a person's dietary intake.

Dietary Guidelines for Americans A series of recommendations that provide specific nutritional guidance and advice about physical activity, alcohol intake, and food safety.

Dietary Reference Intakes (DRIs) A set of four dietary reference standards used to assess and plan dietary intake: Estimated Average Requirement, Recommended Dietary Allowance, Adequate Intake level, and Tolerable Upper Intake Level.

Estimated Average Requirement (EAR) The daily intake of a nutrient that meets the physiological requirements of half the healthy individuals in a given life-stage and sex.

Estimated Energy Requirement (EER) The average energy intake needed for a healthy person to maintain weight.

food frequency questionnaire A retrospective dietary assessment method by which food selection patterns are assessed over an extended period of time.

Food Tracker A component of the MyPlate website that allows individuals to conduct dietary self-assessments.

health claim An FDA-approved statement that describes a specific health benefit of a food or food component.

malnutrition A state of poor nutritional status caused by an imbalance between the body's nutrient requirements and nutrient availability.

MyPlate A visual food guide developed by the USDA to illustrate the most important food intake pattern recommendations of the 2010 Dietary Guidelines for Americans.

nutrient content claim An FDA-regulated word or phrase that describes how much of a nutrient (or its content) is in a food.

nutrient density The relative ratio of a food's amount of nutrients to its total calories.

nutrient requirement The amount of a nutrient that a person must consume to promote optimal health.

Nutrition Facts panel A required component of most food labels that provides information about the nutrient content of the food.

nutritional adequacy A condition by which a person regularly consumes the required amount of a nutrient to meet physiological needs.

nutritional deficiency A condition caused by inadequate intake of one or more essential nutrients.

nutritional status The extent to which a person's diet meets his or her individual nutrient requirements.

nutritional toxicity Overconsumption of a nutrient that results in dangerous toxic effects.

primary malnutrition A condition by which poor nutritional status is caused strictly by inadequate diet.

qualified health claim A health claim that has less scientific backing and must be accompanied by a disclaimer (or qualifier) statement.

Recommended Dietary Allowance (RDA) The daily intake of a nutrient that meets the physiological requirements of nearly all (roughly 97 percent) healthy individuals in a given life-stage and sex.

regular health claim A health claim that is supported by considerable scientific research.

secondary malnutrition A condition by which poor nutritional status is caused by factors other than diet, such as illness.

sign A physical indicator of disease that can be seen by others, such as pale skin and skin rashes.

symptom A subjective manifestation of disease that generally cannot be observed by other people.

Tolerable Upper Intake Level (UL) The highest level of usual daily nutrient intake likely to be safe.

undernutrition The inadequate intake of one or more nutrients and/or energy.

USDA Food Patterns A USDA publication that categorizes nutritionally similar foods into food groups and makes recommendations regarding the number of servings of each food group that should be consumed daily.

- The MyPlate graphic simply reminds Americans to eat healthful amounts of the five food groups using a familiar mealtime visual: a place setting.
- The MyPlate graphic does not specify the numbers of servings recommended by the Food Patterns. The recommended amount of each food group depends on a person's age, sex, and physical activity level.
- The MyPlate website provides in-depth information about the types and amounts of foods that fit into each food group, assistance with meal planning, and material for special subgroups (such as vegetarians).
- MyPlate is a singular part of a much larger communications initiative developed to help U.S. consumers make better food choices.
- The MyPlate website offers an excellent opportunity for you to conduct a self-assessment of your own dietary intake using the free Food Tracker.

LO5: How Can You Use Food Labels to Plan a Healthy Diet?

- Food labels provide consumers with useful information they can use to make smart food choices.
- Nutrition Facts panels provide information that can help us choose healthful foods. The FDA mandates that several critical elements be listed on every Nutrition Facts panel.
- Daily Values (DVs) give consumers a benchmark as to whether a food is a good source of a particular nutrient. There are two basic types of DVs. The first type, used for select vitamins and minerals, represents a nutrient's recommended daily intake. The second type represents a nutrient's upper intake limit.
- Nutrient content claims and health claims are valuable tools when planning a healthy diet.

LO6: Can You Put These Concepts into Action?

- Though dietary assessment is only one component of a complete nutritional assessment, it is an important first step toward a lifetime of health and nutritional awareness.
- You now have the information needed to assess your own diet and begin to choose the right foods to improve it.
- You might begin by conducting a dietary self-assessment using the MyPlate Food Tracker.
- The food habits you establish now will not only affect your success in college but also influence your eating patterns and health for years to come.

Review 3

Chapter Summary

LO1: Why Learn about Chemistry When Studying Nutrition?

- Chemistry is fundamental to the study of nutrition. Nutrients are chemicals, and the body's utilization of nutrients involves countless chemical reactions.
- Many atoms have an equal number of positively and negatively charged particles, and therefore are neutral. Charged atoms, called ions, serve many vital functions in the body.
- Identical atoms combine with each other to form elements. There are approximately 92 naturally occurring elements, 20 of which are essential to human health.
- When chemical bonds join two or more atoms together, molecules are formed. Molecules can be very small or very large.
- A molecular formula is a representation of the number and type of atoms present in a molecule. When a subscript number follows an element's symbol, it means that there are that many atoms of that type of element present. A number placed before the molecular formula means that that many molecules of the substance are present. For example, the molecular formula for three molecules of water is written $3H_2O$.

LO2: How Are Cells, Tissues, Organs, and Organ Systems Related?

- Molecules make up cells, which make up tissues, which in turn function as building blocks for organs, which work together as components of organ systems.
- Transport mechanisms that do not require energy (such as simple diffusion, facilitated diffusion, and osmosis) are called passive transport systems, whereas mechanisms that require energy (such as carrier-mediated active transport) are called active transport systems.
- Some organelles produce substances necessary to cellular activity, while others function as waste-disposal systems, assisting the recycling of worn-out cellular components.
- There are four tissue types (epithelial, connective, muscle, and neural), which collectively carry out functions such as movement, communication, protection, and structure.
- The nervous and endocrine systems work together to monitor our internal environment, respond to change, and restore balance when necessary. These mechanisms allow us to adapt in an ever-changing, complex environment so that we can maintain homeostasis.

LO3: What Happens during Digestion?

- The digestive system consists of the gastrointestinal (GI) tract and accessory organs, which release a variety of secretions needed for digestion.
- The three functions of the digestive system are the chemical and physical breakdown of food (digestion), the transfer of nutrients into the blood or lymphatic circulatory systems (absorption), and the removal of undigested food residue (elimination).
- After peristalsis propels a bolus down the esophagus, sphincters located throughout the GI tract regulate the flow of the luminal contents from one organ to the next.
- The stomach is uniquely equipped to carry out two important functions: (1) mixing food with the gastric secretions that aid in chemical digestion and (2) temporarily storing food.
- The hormone gastrin is released when food enters the stomach, stimulating the release of gastric juice. Food mixes with gastric juice and turns into chyme.
- After leaving the stomach, the chyme passes into the small intestine. The small intestine is the primary site of chemical digestion and nutrient absorption, although some absorption occurs in the stomach and large intestine.

Key Terms

absorption The transfer of nutrients from the GI tract into the blood or lymph.

accessory organ An organ that is not part of the digestive tract but nonetheless plays an important role in digestion and absorption.

active transport mechanism A transport mechanism that requires energy to move a substance across a cell membrane.

aerobic A metabolic pathway that requires oxygen in order to function.

anabolic pathway A series of metabolic reactions that uses energy to construct a complex molecule from simpler ones.

anaerobic A metabolic pathway that can function under conditions of low oxygen availability.

atom The fundamental unit that makes up the world around us.

bile A fluid produced in the liver and released by the gallbladder that disperses large globules of fat into smaller droplets that are easier to digest.

bioavailability The extent to which a nutrient is absorbed into the circulatory system.

bolus A soft, moist mass of chewed food.

carrier-mediated active transport An active transport mechanism whereby a substance moves from a region of lower concentration to a region of higher concentration with the assistance of a carrier protein and energy.

catabolic pathway A series of metabolic reactions that breaks down a complex molecule into simpler ones, releasing energy in the process.

cell A structural and functional unit that makes up body tissue.

chyme A semi-liquid paste resulting from the mixing of partially digested food with gastric juice in the stomach.

colon The first portion of the large intestine.

connective tissue Tissue that supports, connects, and anchors body structures.

digestion The physical and chemical breakdown of food into a form that allows nutrients to be absorbed.

duodenum The first segment of the small intestine.

electrolyte A molecule that when submerged in water separates into individual ions.

element A pure substance made up of only one type of atom.

elimination The process whereby solid waste is removed from the body.

energy metabolism Chemical reactions that enable cells to use and store energy.

enzyme A biological catalyst that accelerates a chemical reaction.

epithelial tissue Tissue that helps protect the body.

esophagus A narrow muscular tube that begins at the pharynx and ends at the stomach.

excretion The removal of waste products produced in cells.

facilitated diffusion A passive transport mechanism whereby a substance moves from a region of higher concentration to a region of lower concentration with the assistance of a carrier molecule.

feces Solid waste consisting mainly of undigested and unabsorbed matter, dead cells, secretions from the GI tract, water, and bacteria.

gastric emptying The process by which food leaves the stomach and enters the small intestine.

gastric juice Digestive secretions that consist mainly of water, hydrochloric acid, digestive enzymes, mucus, and intrinsic factor.

gastrin A hormone that stimulates release of gastric juice and causes the muscular wall of the stomach to contract vigorously.

gastroesophageal reflux disease (GERD) A condition caused by chronic reflux of the stomach contents into the esophagus, irritating the lining.

gastroesophageal sphincter A circular muscle that regulates the flow of food from the esophagus to the stomach.

homeostasis A state of balance or equilibrium.

hormone A chemical messenger released into the blood by the endocrine system.

inflammatory bowel disease (IBD) A category of chronic conditions characterized by inflammation of the lining of the GI tract.

ion An atom that has a positive or negative electrical charge.

irritable bowel syndrome (IBS) A disorder that typically affects the lower GI tract, causing bouts of cramping, bloating, diarrhea, and constipation.

lacteal A lymphatic vessel found in an intestinal villus into which nutrients are absorbed.

lumen The cavity that spans the entire length of the GI tract.

metabolic pathway A series of interrelated chemical reactions that require the help of enzymes.

metabolism The sum of chemical processes that occur within a living cell to maintain life.

microbiota The natural microbial population that resides in the large intestine.

microvillus A tiny finger-like projection. Microvilli cover the lumenal surfaces of the epithelial cells that line the villi in the small intestine.

molecular formula A representation of the number and type of atoms present in a molecule.

molecule A unit of two or more atoms joined together by chemical bonds.

mucosa The innermost lining of the gastrointestinal tract.

muscle tissue Tissue that is used for movement.

neural tissue Tissue that facilitates communication throughout the body.

organ Two or more different types of tissue functioning together to perform a variety of related tasks.

organelle A structure that is responsible for a specific function within a cell.

osmosis The diffusion of water across a cell membrane.

pancreatic juice A mixture of water, bicarbonate, and various enzymes released by the pancreas.

- Secretions from the pancreas and gallbladder facilitate digestion in the small intestine.

LO4: Nutrient Absorption: What Happens after Digestion?

- Nutrient absorption is the process whereby nutrients are transported from the lumen of the GI tract into either the blood or lymph. The bioavailability of a particular nutrient can be influenced by physiological conditions, other dietary components, and certain medications.
- Materials entering the large intestine consist mostly of undigested remains from plant-based foods, water, bile, and electrolytes. Muscles embedded within the intestinal wall squeeze the undigested food residue, and as material moves through the various regions of the colon, the water and electrolytes are absorbed and returned to the blood for reuse by the body.
- Intestinal microbiota break down undigested food residue, produce nutrients, and inhibit the growth of other disease-causing bacteria.
- Once the remaining material reaches the rectum (the last segment of the large intestine), it is ready to be eliminated from the body.

LO5: How Does the Body Circulate Nutrients and Excrete Waste Products?

- The delivery of nutrients and oxygen to cells is accomplished by the cardiovascular and lymphatic systems.
- The cardiovascular system circulates nutrients and gases, while the lymphatic system circulates fat-soluble nutrients.
- The circulatory systems, liver, kidneys, lungs, and skin work together to excrete cellular waste products and thus prevent the accumulation of toxins in the body.

LO6: What Is Metabolism?

- Chemical energy is transferred into usable ATP through metabolic pathways. Each chemical reaction in a metabolic pathway requires the aid of at least one enzyme.
- Catabolic metabolic pathways (both aerobic and anaerobic) break complex molecules into simpler ones, while anabolic metabolic pathways use energy to construct complex molecules from simpler ones.

passive transport mechanism A transport mechanism that does not require energy to move a substance across a cell membrane.

peristalsis A vigorous, wave-like muscular contraction that propels food from one region of the GI tract to the next.

pharnyx A region toward the back of the mouth that serves as the shared space between the oral and nasal cavities.

prebiotic food A typically fiber-rich food that may stimulate the growth of the microbial population in the large intestine.

probiotic food A food that contains live bacteria, some of which thrive in the colon.

pyloric sphincter A circular muscle that regulates the flow of chyme from the stomach into the small intestine.

rectum The segment of the GI tract that leads to the anal canal.

saliva A secretion released into the mouth by the salivary glands that moistens food and starts the physical process of digestion.

simple diffusion A passive transport mechanism whereby a substance moves from a region of higher concentration to a region of lower concentration without using energy or the assistance of a transport protein.

sphincter A circular band of muscle that regulates the flow of food through the GI tract.

tissue An aggregation of similarly structured and functioning cells that have grouped together to accomplish a common task.

transit time The amount of time it takes for food to travel the entire length of the GI tract.

villus A small finger-like projection. Villi cover the inner surface of the small intestine.

Review 4

Chapter 4
Carbohydrates

Chapter Summary

LO1: What Are Simple Carbohydrates?

- A carbohydrate consisting of a single sugar is called a monosaccharide, and a carbohydrate consisting of two sugars is called a disaccharide. Because of the small sizes of these molecules, monosaccharides and disaccharides are referred to as simple carbohydrates.
- Monosaccharides include glucose (the most abundant monosaccharide in the human body), fructose (a naturally occurring monosaccharide found primarily in honey, fruits, and vegetables), and galactose (which occurs primarily as part of the disaccharide lactose).
- The most common disaccharides are lactose (the most abundant carbohydrate in milk and other dairy products), maltose (a product of the enzymatic breakdown of starches), and sucrose (commonly known as refined table sugar).
- Alternative low-calorie sweeteners such as saccharin, aspartame, and acesulfame K can be added to increase sweetness without increasing the caloric contents of foods.

LO2: What Are Complex Carbohydrates?

- Complex carbohydrates are comprised of many monosaccharides bonded together. The types and arrangements of sugar molecules determine the shape and form of the polysaccharide.
- Glucose molecules are arranged in starch in either an orderly unbranched linear chain or a highly branched configuration. Plants typically contain a mixture of these two types of starch.
- Glycogen consists of glucose molecules bonded together in a highly branched arrangement and is found mainly in liver and skeletal muscle.
- Dietary fibers are a diverse group of carbohydrates found in a variety of foods such as whole grains, legumes, vegetables, and fruits that are not digested or absorbed in the human small intestine.
- Soluble dietary fiber tends to dissolve or swell in water, while insoluble dietary fiber remains relatively unchanged. Consumption of soluble fiber can help lower blood cholesterol levels, promote satiety, and lower blood glucose levels.
- The nutritional value of grain is greatest when all three of its components—bran, germ, and endosperm—are present.

LO3: How Are Carbohydrates Digested, Absorbed, and Circulated?

- Carbohydrates undergo extensive chemical transformations as they move through the GI tract during digestion. Enzymes required for carbohydrate digestion are produced in the salivary glands, pancreas, and small intestine.
- Starch digestion begins in the mouth, but most occurs in the small intestine via pancreatic amylase. Digestion is completed by the enzyme maltase, resulting in free (unbound) glucose molecules.
- The digestion of disaccharides takes place entirely in the small intestine. Each disaccharide has its own specific digestive enzyme.
- Glycemic response is a change in blood glucose following the ingestion of a carbohydrate-rich food.

Key Terms

autoimmune disease An illness that occurs when an abnormal immunological response results in the destruction of one's own bodily tissues.

bran The outer portion of a grain that contains most of the fiber.

carbohydrate An organic compound made up of one or more sugar molecules.

complex carbohydrate (or **polysaccharide**) A category of carbohydrate comprised of many monosaccharides bonded together.

diabetes mellitus A group of metabolic disorders characterized by elevated levels of glucose in the blood.

disaccharide A carbohydrate consisting of two monosaccharides bonded together.

diverticular disease (or **diverticulosis**) A condition whereby straining leads to the formation of pouches called *diverticula* that protrude along the colon wall.

diverticulitis A condition whereby the diverticula become infected or inflamed.

endosperm The portion of a grain that contains mostly starch.

epinephrine A hormone released from the adrenal glands that stimulates glycogenolysis in emergency situations.

fiber (or **dietary fiber**) A diverse group of plant polysaccharides that are not digestible by human enzymes.

fructose A naturally occurring monosaccharide found primarily in honey, fruits, and vegetables.

galactose A monosaccharide that exists primarily as part of a naturally occurring disaccharide found in dairy products.

germ The portion of a grain that contains most of the vitamins and minerals.

glucagon A hormone secreted by the pancreas in response to decreased blood glucose.

glucometer A medical device used to monitor the concentration of glucose in the blood.

gluconeogenesis The synthesis of glucose from noncarbohydrate sources.

glucose The most abundant monosaccharide in the human body; used extensively for energy.

glycemic index (GI) A rating system based on a scale of 0 to 100 used to compare the glycemic responses elicited by different foods.

glycemic load (GL) A rating system used to compare the glycemic responses associated with different foods that takes into account the typical portion of food consumed.

glycemic response The change in blood glucose following the ingestion of a food.

glycogen A polysaccharide found primarily in liver and skeletal muscle that is comprised of glucose molecules.

glycogenolysis The breakdown of glycogen into glucose.

glycolysis An anaerobic metabolic pathway made up of a series of chemical reactions that splits glucose into two three-carbon molecules.

high-fructose corn syrup (HFCS) A widely used sweetener consisting of glucose and fructose that is manufactured from cornstarch.

hyperglycemia A condition characterized by an excess of glucose in the blood.

hypoglycemia A condition characterized by low blood glucose.

insoluble fiber Dietary fiber that remains relatively unchanged in water.

insulin A hormone secreted by the pancreas in response to increased blood glucose.

insulin receptors Specialized proteins located on the outer membranes of certain types of cells that bind insulin.

insulin resistance A condition whereby insulin receptors throughout the body are less responsive to insulin.

ketone An organic compound used as an alternative energy source under conditions of limited glucose availability.

ketosis A condition characterized by excessive ketone accumulation in the blood.

lactase An intestinal enzyme that digests lactose, releasing glucose and galactose molecules.

lactose A disaccharide comprised of galactose joined with glucose. It is the most abundant carbohydrate in milk and many other dairy products.

lactose intolerance A condition whereby the body does not produce enough of the enzyme lactase, making it difficult to digest lactose.

maltase An intestinal enzyme that digests maltose, releasing two glucose molecules.

maltose A disaccharide comprised of glucose joined with glucose. It is not found in many foods.

monosaccharide A carbohydrate consisting of a single sugar molecule.

photosynthesis A process whereby chlorophyll-containing plants produce glucose by combining carbon dioxide (CO_2) and water (H_2O) using energy harvested from sunlight.

pyruvate The end-product of glycolysis; formed by the breakdown of glucose.

salivary amylase An enzyme released from the salivary glands that breaks the chemical bonds in starch.

simple carbohydrate (or **simple sugar**) A category of carbohydrates comprised of monosaccharides and disaccharides.

soluble fiber Dietary fiber that tends to dissolve or swell in water.

sucrase An intestinal enzyme that digests sucrose, releasing glucose and fructose molecules.

sucrose A disaccharide comprised of fructose joined with glucose. It is found in many plants.

type 1 diabetes A form of diabetes whereby the pancreas is no longer able to produce insulin, causing blood glucose levels to become dangerously high.

- Glycemic index (GI) and glycemic load (GL) are rating systems used to assess the glycemic response to specific foods. GL takes into account the typical portion of food consumed while GI does not.

LO4: How Does Your Body Regulate and Use Glucose?

- The pancreatic hormones insulin and glucagon play major roles in blood glucose regulation and energy storage.
- Insulin, released when blood glucose is high, lowers blood glucose levels by (1) enabling cells to take up glucose from the blood; (2) increasing the rate at which glucose is used as an energy source; and (3) promoting the conversion of glucose to body fat, which is stored mostly in adipose tissue.
- Glucagon, released when blood glucose levels are low, increases glucose availability by stimulating the breakdown of liver glycogen. As glycogen stores dwindle, glucagon stimulates gluconeogenesis.
- When carbohydrate intake is limited and glycogen stores are depleted, the body minimizes protein loss by using ketones as an energy source.

LO5: What Is Diabetes?

- Diabetes mellitus is a group of metabolic disorders characterized by elevated levels of glucose in the blood. The two main types of diabetes are referred to as type 1 diabetes and type 2 diabetes.
- Type 1 diabetes, an autoimmune disease, occurs when the pancreas is no longer able to produce insulin. This causes blood glucose levels to become dangerously high.
- Type 2 diabetes is caused by insulin resistance. Because the body's cells do not respond appropriately to insulin's signal, the amount of glucose taken up from the bloodstream is diminished.

LO6: What Are the Recommendations for Carbohydrate Intake?

- The Institute of Medicine's Dietary Reference Intakes (DRIs) are based mainly on ensuring that the brain has adequate glucose for its energy needs.
- Because some carbohydrate-rich foods are more nutrient-dense (and thus nutritional) than others, one must weigh a number of factors before deciding which carbohydrate-containing foods to consume. Following the national health guidelines can help eliminate much of the guesswork when it comes to determining which carbohydrate-rich foods to choose.
- Health experts generally agree that a person should minimize his consumption of foods with high amounts of added sugar (such as cookies, soda, sugary cereals, and heavy syrups).
- The Institute of Medicine's DRIs recommend that adults consume at least 14 g of dietary fiber per 1,000 kcal, which is approximately 21 to 38 grams per day.
- A sudden and/or large increase in fiber intake may cause a gastrointestinal problem such as diarrhea or constipation. Thus, a person should increase her fiber intake gradually and give the body time to adjust.

type 2 diabetes A form of diabetes whereby insulin resistance prevents cells from taking up glucose from the blood.

whole-grain foods Cereal grains that contain bran, endosperm, and the germ in the same relative proportions as exist naturally.

Review 5

Chapter 5
Protein

Chapter Summary

LO1: What Are Proteins?

- Proteins are nitrogen-containing macronutrients made from amino acids linked together via peptide bonds.
- Every amino acid contains a central carbon, a carboxylic acid group, an amino group, and a side chain, or R-group. The body needs 20 amino acids; nine of these must be obtained from the diet and are therefore considered essential nutrients.
- Foods that contain relatively high amounts of all the essential amino acids in the appropriate proportions are considered complete protein sources. Foods lacking or having low amounts of at least one of the essential amino acids are considered incomplete protein sources.
- Combining two or more foods containing incomplete proteins to consume all the essential amino acids is called protein complementation.

LO2: How Do Cells Make Proteins?

- Each gene in a DNA strand contains information about how a protein should be constructed.
- Protein synthesis involves DNA, mRNA, ribosomes, tRNA, and amino acids.
- A series of steps involving cell signaling, transcription, and translation produces hundreds of thousands of different proteins in the body.

LO3: Why Is a Protein's Shape Critical to Its Function?

- The sequence of amino acids making up a protein is called its primary structure. The primary structure folds into secondary and tertiary structures due to the attraction and repulsion of positive and negative charges.
- Peptide chains can join together to form quaternary structures, and these can combine with prosthetic groups to form a protein's final three-dimensional shape.
- Disruption of a protein's shape is called denaturation. This can negatively impact the ability of the protein to function.

LO4: What Is Meant by Genetics and Epigenetics?

- A person's genetic makeup (or genotype) is inherited from his parents.
- Alterations in gene expression that occur without alterations in DNA sequence are collectively called epigenetics.
- Mutations in DNA can influence the cell's ability to produce a functional protein.

LO5: How Are Proteins Digested, Absorbed, and Circulated?

- The process of digestion disassembles food proteins into amino acids that are then absorbed and carried to cells where they are assembled into needed proteins.
- Protein digestion involves hormones such as gastrin, secretin, and CCK, as well as a variety of proteases produced in the stomach, pancreas, and small intestine.
- Protein-digesting enzymes are released as inactive proenzymes and then converted to their active forms (proteases) in the gastrointestinal tract.
- Amino acids are absorbed primarily along the duodenum, where they enter the blood and circulate to the liver for further processing.
- Food allergies and food intolerances are adverse reactions to food proteins and other substances.

Key Terms

α-helix A common secondary structure folding pattern that resembles the shape of a spiral staircase.

α-keto acid A compound similar to an amino acid that does not have an amino group; used to synthesize nonessential amino acids.

albumin A protein present in the blood that plays an important role in regulating fluid balance.

amino acid A nitrogen-containing subunit that combines to form proteins.

amino group The nitrogen-containing component of an amino acid.

anaphylaxis A rapid immune response that causes a sudden drop in blood pressure, rapid pulse, dizziness, and a narrowing of the airways.

antibody (or **immunoglobulin**) A protein, produced by the immune system, that helps fight infection.

ascites A condition characterized by fluid accumulation in the abdominal cavity.

β-folded sheet A common secondary structure folding pattern that resembles the shape of a folded paper fan.

cell signaling The process by which a cell is notified that it should make a particular protein.

cholecystokinin (CCK) A hormone, secreted by intestinal cells, that signals the release of bile from the gallbladder and proenzyme proteases from the pancreas.

chromosome A substance comprised of coiled strands of DNA and special proteins found in a cell's nucleus.

complete protein source A food that supplies an adequate and balanced amount of all essential amino acids.

denaturation The process by which a protein's three-dimensional structure is altered.

deoxyribonucleic acid (DNA) A chemical that provides the instructions for protein synthesis.

edema A condition whereby low levels of albumin in the blood cause fluid to accumulate in body tissues or cavities.

epigenetics Alterations in protein synthesis that do not involve changes in the DNA sequence.

food allergy A condition whereby the body's immune system responds to a food-derived peptide as if it is dangerous.

food intolerance (or **food sensitivity**) A condition whereby the body reacts negatively to a food or food component but does not mount an immune response.

gene A chromosome subunit that tells a cell which amino acids are needed and in what order they must be arranged to synthesize a protein.

genotype The particular DNA inherited from one's parents.

incomplete protein source A food that lacks or supplies low amounts of one or more of the essential amino acids.

kwashiorkor A form of PEM characterized by severe edema in the extremities.

lacto-ovo-vegetarian A vegetarian who consumes dairy products and eggs in an otherwise plant-based diet.

lactovegetarian A vegetarian who consumes dairy products (but not eggs) in an otherwise plant-based diet.

limiting amino acid An essential amino acid that is insufficient or absent in an incomplete protein source.

marasmus A form of PEM characterized by extreme wasting of muscle and loss of adipose tissue.

messenger ribonucleic acid (mRNA) A chemical that carries the instructions contained in DNA outside of the nucleus.
mutation An alteration in a gene that occurs due to a chance genetic modification.
negative nitrogen balance A bodily state whereby nitrogen intake is less than nitrogen loss.
nitrogen balance A bodily state whereby nitrogen intake equals nitrogen loss; sometimes referred to as neutral nitrogen balance.
pepsin An enzyme needed for protein digestion.
pepsinogen The inactive form of the enzyme pepsin.
peptide bond A chemical bond that joins amino acids.
phenotype The observable physical or biochemical characteristics of an organism.
phenylketonuria (PKU) A disease whereby the body does not produce one of the enzymes required to convert the essential amino acid phenylalanine to the normally nonessential amino acid tyrosine.
polypeptide A protein comprised of more than 12 amino acids.
positive nitrogen balance A bodily state whereby nitrogen intake is greater than nitrogen loss.
primary structure (or **primary sequence**) The most basic level of protein structure; determined by the number and sequence of amino acids in a single peptide chain.
proenzyme An inactive precursor of an enzyme.
prosthetic group A nonprotein component of a protein that often contains minerals needed for the protein to carry out its purpose.
protease A type of enzyme that breaks peptide bonds between amino acids.
protein A nitrogen-containing macronutrient made from joined amino acids.
protein complementation The combining of diverse foods with different incomplete proteins to provide adequate amounts of all the essential amino acids.
protein turnover The continual coordinated process of protein breakdown and synthesis.
protein-energy malnutrition (PEM) A condition whereby protein deficiency is accompanied by a deficiency in energy, and usually, one or more micronutrients.
proteolysis The breakdown of proteins.
quaternary structure The most complex level of protein structure; occurs when two or more peptide chains join together.
R-group The side-chain component of an amino acid that distinguishes it from other amino acids.
ribosome A cellular organelle to which mRNA binds and on which proteins are made.
secondary structure Organized and predictable folds that develop in portions of a peptide chain because charged portions of the amino acid backbone attract and repel each other.
secretin A hormone, secreted by intestinal cells, that signals the release of sodium bicarbonate and proteases from the pancreas.
sickle cell anemia (or **sickle cell disease**) A disease whereby a small alteration in the DNA results in the production of defective, misshapen molecules of the protein hemoglobin within red blood cells.
tertiary structure Additional folding of a peptide chain that develops because of interactions between the amino acids' R-groups.
transamination The process by which nonessential amino acids are synthesized.
transcription The process by which mRNA is constructed using DNA as a template.
transfer ribonucleic acid (tRNA) A chemical that carries amino acids to a ribosome to be assembled into a protein.
translation The process by which amino acids are joined via peptide bonds.
vegan A vegetarian who consumes no animal products.
vegetarian A person who does not consume or consumes only some foods and beverages made from animal products.

LO6: Why Do We Need Proteins and Amino Acids?

- Amino acids are used to synthesize proteins needed for structure, catalysis, movement, transport, communication, and protection.
- Proteins can be broken down (a process called proteolysis) and used for ATP production; some amino acids can be converted to glucose.
- Amino acids themselves serve many diverse roles. Some regulate protein synthesis or breakdown, others are involved in cell communication, and still others are converted to neurotransmitters and other signaling molecules.
- During times of energy abundance, amino acids are transformed into fat and stored in adipose tissue.

LO7: How Does the Body Recycle and Reuse Amino Acids?

- Protein turnover represents the balance between protein synthesis and degradation. One's protein status can be assessed by comparing protein intake to the amount of nitrogen lost in body secretions such as urine, sweat, and feces.
- When protein (or nitrogen) intake exceeds loss, the body is in positive nitrogen balance. When nitrogen loss is greater than intake, the body is in negative nitrogen balance.
- Knowing whether a person is in neutral, positive, or negative nitrogen balance can help a clinician diagnose and treat certain disease states and physiologic conditions.

LO8: How Much Protein Do You Need?

- Protein must be consumed to supply both adequate amounts of the essential amino acids and the nitrogen needed to synthesize nonessential amino acids and other nonprotein, nitrogen-containing compounds such as DNA.
- In general, protein requirements are greater for males than females and are highest during infancy, pregnancy, and lactation. A typical college-age male needs 56 g/day of protein, whereas a comparable female needs 46 g/day.
- It is recommended that healthy adults consume approximately 0.8 g of protein daily for each kilogram of body weight. In 2009, the American College of Sports Medicine concluded that protein intakes of 1.2 to 1.7 g/kg/day for endurance and strength-trained athletes may be beneficial.
- The Institute of Medicine's Acceptable Macronutrient Distribution Ranges (AMDRs) recommend that adults obtain 10 to 35 percent of their energy from protein.

LO9: Can Vegetarian Diets Be Healthy?

- Vegans are individuals who eat no animal products, lacto-ovo-vegetarians consume dairy and egg products but not meat, and lactovegetarians consume dairy but not eggs or meat.
- Consuming adequate amounts of protein is typically not difficult for vegetarians, although they may be at greater risk for calcium, zinc, iron, and vitamin B_{12} deficiencies.

LO10: What Are the Consequences of Protein Deficiency and Excess?

- Protein-energy malnutrition (PEM) takes two forms: kwashiorkor and marasmus. Most adults with PEM have signs and symptoms associated with marasmus.
- Treatment of children and adults with PEM is multifaceted, and usually involves a variety of factors including the provision of adequate nutrition.
- For most people, excess protein intake does not result in health complications.
- There is growing evidence that excessive consumption of red or processed meat is associated with increased risk of colorectal cancer.

Chapter Summary

LO1: What Are Lipids?

- A lipid that is liquid at room temperature is called an oil, and one that is solid at room temperature is called a fat.
- Fatty acids, made of carbon, hydrogen, and oxygen atoms, are the most abundant type of lipid in the body. A chain of carbon atoms with alpha (α) and omega (ω) ends forms the fatty acid's backbone. The number of carbon atoms in the backbone determines its chain length.
- A fatty acid with fewer than eight carbon atoms in its backbone is a short-chain fatty acid; one with 8 to 12 carbon atoms is a medium-chain fatty acid; one with more than 12 carbon atoms is a long-chain fatty acid.
- Saturated fatty acids (SFAs) contain only carbon–carbon single bonds, monounsaturated fatty acids (MUFAs) have one carbon–carbon double bond, and polyunsaturated fatty acids (PUFAs) have two or more carbon–carbon double bonds.
- Double bonds can be in either *cis* or *trans* configuration, depending on the placement of the hydrogens around the carbons. In *cis* double bonds, the hydrogen atoms are on the same side. In *trans* double bonds, they are on opposite sides.
- Alpha (α) nomenclature designates the number of carbon atoms, the number and placement of double bonds, and the type of double bonds relative to the alpha end.
- Omega (ω) nomenclature is similar, although the placement of double bonds is determined from the omega end.

LO2: What Are Essential, Conditionally Essential, and Nonessential Fatty Acids?

- The two essential fatty acids are linoleic acid and linolenic acid. Linoleic and linolenic acids are metabolized to longer-chain fatty acids and other compounds such as eicosanoids.
- Because of the extensive amounts of linoleic and linolenic acids stored in adipose tissue, essential fatty acid deficiencies do not generally occur in otherwise healthy people.

LO3: What Is the Difference between Mono-, Di-, and Triglycerides?

- Monoglycerides consist of one fatty acid attached to a glycerol molecule, while diglycerides have two fatty acids, and triglycerides have three.
- The metabolic process by which fatty acids combine with glycerol to form triglycerides is called lipogenesis, and the metabolic process by which fatty acids are separated from their glycerol backbone is called lipolysis.
- Fatty acids not required for energy or other functions are stored as triglycerides in adipose tissue.

LO4: What Are Phospholipids and Sterols?

- A phospholipid consists of a glycerol, two fatty acids, and a phosphate-containing head group. Phospholipids make up cell membranes and circulate lipids throughout the body.
- Sterols are multi-ring compounds that serve many roles in the body. Phytosterols are sterol-like compounds found in plants.
- Cholesterol is an important component of cell membranes and a precursor for the synthesis of steroid hormones such as estrogen and testosterone.

Key Terms

adipocyte A specialized cell, found in adipose tissue, that can accumulate large amounts of triglycerides.

alpha (α) end The end of a fatty acid that contains a carboxylic acid (–COOH) group.

amphipathic A characteristic of a substance that contains both hydrophilic and hydrophobic regions.

aneurysm An outward bulging of a blood vessel due to weakness in the vasculature.

apoprotein (or **apolipoprotein**) A protein embedded within the outer shell of a lipoprotein that enables it to circulate in the blood and interact with the cells that require its contents.

atherosclerosis A narrowing and stiffening of the blood vessels, causing the restriction of blood flow.

β-oxidation The metabolic breakdown of fatty acids to produce ATP.

bile acid An amphipathic substance, consisting of a cholesterol molecule attached to a very hydrophilic subunit, that plays a critical role in digestion and absorption of lipids.

blood clot A small, insoluble particle made of clotted blood and clotting factors.

cardiovascular disease A disease of the heart or vascular system.

chain length The number of carbon atoms in a fatty acid's backbone.

cholesterol An abundant and widely discussed sterol that is used to synthesize bile acids and a variety of hormones such as testosterone.

chylomicron A particle comprised of relatively hydrophobic lipids, cholesterol, and phospholipids that transports dietary lipids in the lymph for circulation.

chylomicron remnant A residual fragment of a chylomicron that remains in the blood until it is taken up by the liver.

***cis* double bond** A carbon–carbon double bond in which the hydrogen atoms are positioned on the same side of the double bond.

diglyceride A lipid comprised of two fatty acids attached to a glycerol backbone.

eicosanoid family A diverse group of hormone-like compounds that help regulate the immune and cardiovascular systems and act as chemical messengers.

emulsification The process by which large lipid globules are broken down and stabilized into smaller lipid droplets.

fat A lipid that is solid at room temperature.

fatty acid The most abundant type of lipid in the body and foods; comprised of a chain of carbons with a methyl (–CH_3) group on one end and a carboxylic acid (–COOH) group on the other.

gastric lipase An enzyme, produced in the stomach, that cleaves fatty acids from glycerol molecules.

head group A phosphate-containing, hydrophilic chemical structure that serves as a component of a phospholipid.
heart disease A slowing or complete obstruction of blood flow to the heart.
high-density lipoprotein (HDL) A lipoprotein, made primarily by the liver, that has the lowest lipid-to-protein ratio. It circulates in the blood to collect excess cholesterol from cells and transport it back to the liver.
hydrophilic Water loving; mixes easily with water.
hydrophobic Water fearing; does not easily mix with water.
intermediate-density lipoprotein (IDL) A lipoprotein that is slightly less dense than a VLDL.
lanugo Very fine hair that grows as a physiological response to insufficient body fat.
LDL receptor A specialized type of protein, located on cell membranes, that binds to the proteins embedded in the surface of an LDL, allowing it to be taken up and broken down by the cell.
lingual lipase An enzyme, produced by the salivary glands, that cleaves fatty acids from glycerol molecules.
linoleic acid An essential ω-6 fatty acid with 18 carbons and two double bonds.
linolenic acid (or **α-linolenic acid**) An essential ω-3 fatty acid with 18 carbons and three double bonds.
lipid An organic macronutrient that is relatively insoluble in water and relatively soluble in organic solvents.
lipogenesis The metabolic process by which fatty acids combine with glycerol to form triglycerides.
lipolysis The metabolic process by which a triglyceride's three fatty acids are removed from the glycerol backbone.
lipoprotein A type of particle that transports lipids throughout the body.
lipoprotein lipase An enzyme that enables chylomicrons and other lipoproteins to deliver fatty acids to cells.
long-chain fatty acid A fatty acid with more than 12 carbon atoms in its backbone.
low-density lipoprotein (LDL) A lipoprotein that delivers cholesterol to the body's cells.
medium-chain fatty acid A fatty acid with eight to 12 carbon atoms in its backbone.
micelle A small droplet of fat formed in the small intestine via emulsification.
monoglyceride A lipid comprised of one fatty acid attached to a glycerol backbone.
monounsaturated fatty acid (MUFA) A fatty acid that contains one carbon–carbon double bond in its backbone.
oil A lipid that is liquid at room temperature.
omega (ω) end The end of a fatty acid that contains a methyl ($-CH_3$) group.
omega-3 (ω-3) fatty acid A fatty acid in which the first double bond is located between the third and fourth carbons from the omega (ω) end.
omega-6 (ω-6) fatty acid A fatty acid in which the first double bond is located between the sixth and seventh carbons from the omega (ω) end.
pancreatic lipase An enzyme, produced in the pancreas, that completes triglyceride digestion by cleaving fatty acids from glycerol molecules.
partial hydrogenation A process by which oil is converted into solid fat by changing many of the carbon–carbon double bonds into carbon–carbon single bonds; some *trans* fatty acids are also produced.

LO5: How Are Triglycerides Digested, Absorbed, and Circulated?

- Lipid digestion occurs in the mouth (lingual lipase), stomach (gastric lipase), and small intestine (pancreatic lipase). Bile emulsifies large lipid globules into smaller droplets in the small intestine.
- Short- and medium-chain fatty acids are transported into intestinal cells unassisted because they are relatively water-soluble.
- More hydrophobic compounds are first repackaged into micelles within the intestinal lumen before moving into the intestinal cells. These lipids are incorporated into chylomicrons and circulated in the lymph.
- Chylomicrons deliver dietary fatty acids to cells via lipoprotein lipase.

LO6: What Are the Types and Functions of Various Lipoproteins?

- The liver produces very low-density lipoproteins (VLDLs) that deliver dietary and other fatty acids to cells.
- The loss of fatty acids from a VLDL results in its conversion to an intermediate-density lipoprotein (IDL) and ultimately a low-density lipoprotein (LDL).
- LDLs deliver cholesterol to cells. The liver produces high-density lipoproteins (HDLs), which pick up cholesterol and return it to the liver.
- Too much LDL can result in the buildup of plaque. High levels of LDL and low levels of HDL are associated with increased risk of cardiovascular disease.

LO7: How Are Dietary Lipids Related to Health?

- Excess energy intake is a major factor in the development of obesity and related complications such as cardiovascular disease.
- Dietary fat intake may play a role in the development of cancer. However, this effect is likely primarily due to the influence of dietary fat on obesity.

LO8: What Are Some Overall Dietary Recommendations for Lipids?

- Consuming 20 to 35 percent of one's calories from lipid, limiting intake of SFA to less than 7 percent of calories, choosing foods low in *trans* fatty acids and cholesterol (200–300 mg/d), and emphasizing foods high in PUFAs and MUFAs are common lipid-specific dietary recommendations.

phospholipid A lipid composed of a glycerol molecule bonded to two hydrophobic fatty acids and a hydrophilic head group.
phytosterol A sterol-like compound made by plants.
plaque A fatty substance that builds up within the walls of blood vessels.
polyunsaturated fatty acid (PUFA) A fatty acid that contains more than one carbon–carbon double bond in its backbone.
saturated fatty acid (SFA) A fatty acid that contains only carbon–carbon single bonds in its backbone.
short-chain fatty acid A fatty acid with fewer than eight carbon atoms in its backbone.
sterol A type of lipid with a multiple-ring chemical structure.
stroke A slowing or complete obstruction of blood flow to the brain.
subcutaneous adipose tissue Adipose tissue located directly under the skin.
***trans* double bond** A carbon–carbon double bond in which the hydrogen atoms are positioned on opposite sides of the double bond.
***trans* fatty acid** A fatty acid containing at least one *trans* double bond.
triglyceride A lipid comprised of three fatty acids attached to a glycerol backbone.
unsaturated fatty acid A fatty acid that contains at least one carbon–carbon double bond in its backbone.
very low-density lipoprotein (VLDL) A lipoprotein, made by the liver, that has a lower lipid-to-protein ratio than do chylomicrons. It delivers fatty acids to cells.
visceral adipose tissue Adipose tissue located around the vital organs in the abdomen.

Review 7

Chapter Summary

LO1: What Do Scientists Know about Vitamins?

- The essential vitamins are classified as water-soluble or fat-soluble, depending on their chemical solubilities in water and lipids.
- Water- and fat-soluble vitamins are often added to foods—this is called fortification. When certain nutrients are added in certain amounts, a fortified food can be labeled as being "enriched."

LO2: What Are Water-Soluble Vitamins?

- Water-soluble vitamins dissolve in water (as opposed to in lipids).
- You can prevent excessive nutrient loss in your foods by properly preparing and storing them.

LO3: What Are the B Vitamins?

- Thiamin is involved in energy metabolism and the synthesis of DNA and RNA.
- Riboflavin is involved in energy metabolism, the synthesis of a variety of vitamins, nerve function, and protection of biological membranes.
- Niacin is involved in energy and vitamin C metabolism and the synthesis of fatty acids and proteins. Niacin deficiency causes pellagra.
- Pantothenic acid is a nitrogen-containing vitamin involved in energy metabolism, hemoglobin synthesis, and phospholipid synthesis.
- Vitamin B_6 is involved in glycogenolysis, metabolism of proteins and amino acids, synthesis of neurotransmitters and hemoglobin, and regulation of steroid hormone function. The vitamin exists in several forms, the most common of which are pyridoxine, pyridoxal, pyridoxamine, and pyridoxal phosphate.
- Biotin is involved in energy metabolism.
- Folate refers to a group of related vitamins involved in single-carbon transfers, amino acid metabolism, and DNA synthesis. The active form of folate in the body is tetrahydrofolate acid (THF).
- Vitamin B_{12} is involved in energy metabolism and methionine production.

LO4: What Is Vitamin C?

- Vitamin C (ascorbic acid) is a water-soluble vitamin that serves antioxidant functions within the body.
- Vitamin C is found in many foods, including a variety of fruits and vegetables. Its deficiency causes scurvy, characterized by bleeding gums and poor wound healing.

LO5: What Are Fat-Soluble Vitamins?

- Fat-soluble vitamins are absorbed mostly in the small intestine, requiring the presence of dietary lipids as well as the action of bile.
- Unlike water-soluble vitamins, your body can store most of the fat-soluble vitamins.

LO6: What Are Vitamin A and the Carotenoids?

- Vitamin A refers to a group of compounds called retinoids, which includes retinol, retinoic acid, and retinal.

Key Terms

α-tocopherol The most biologically active form of vitamin E.

antioxidant A compound that donates electrons or hydrogen ions to other substances, inhibiting oxidation.

ariboflavinosis A condition caused by riboflavin deficiency whereby cheilosis, stomatitis, glossitis, muscle weakness, and confusion occur.

beriberi A life-threatening condition caused by thiamin deficiency. There are four forms: wet, dry, infantile, and cerebral beriberi.

biotin (vitamin B_7) A water-soluble vitamin involved in energy metabolism that is obtained both from the diet and bacteria in the large intestine.

burning feet syndrome A condition believed to be caused by pantothenic acid deficiency whereby tingling in the feet and legs, fatigue, weakness, and nausea may occur.

calcitrol The active form of vitamin D produced in the kidneys. Also known as 1,25-dihydroxyvitamin D (1,25-$[OH]_2$ D_3).

carotenoid A dietary compound with a similar structure to those of the retinoids. Some, but not all, can be converted to vitamin A.

cell differentiation The process by which a nonspecialized, immature cell type becomes a specialized, mature cell type.

cheilosis A condition characterized by sores on the outside and corners of the lips.

cholecalciferol (vitamin D_3) A form of vitamin D found in animal-based foods, fortified foods, and supplements that is also synthesized in the body.

coagulation The process by which blood clots are formed.

coenzyme A vitamin that facilitates the function of an associated enzyme.

dietary folate equivalent (DFE) A unit of measure for the approximate amount of folate in a food that is absorbed by the body.

dietary supplement Product intended to supplement the diet that contains vitamins, minerals, amino acids, herbs or other plant-derived substances, or a multitude of other compounds.

enrichment The fortification of a select group of foods with FDA-specified levels of thiamin, niacin, riboflavin, folate, and iron.

ergocalciferol (vitamin D_2) A form of vitamin D found in plant-based foods, fortified foods, and supplements.

folate (vitamin B_9) A group of related water-soluble vitamins involved in single-carbon transfers, amino acid metabolism, and DNA synthesis.

fortification The addition of nutrients to a food during processing.

free radical A highly reactive molecule with one or more unpaired electrons. Destructive to cell membranes, DNA, and proteins.

glossitis Inflammation of the tongue.

hemolytic anemia A condition caused by vitamin E deficiency whereby red blood cell membranes weaken and rupture, reducing the blood's ability to transport oxygen.
hypercalciuria A condition characterized by elevated urine calcium levels.
hypercarotenodermia A condition whereby carotenoids accumulate in the skin, causing it to become yellow-orange.
hyperkeratosis A condition caused by vitamin A deficiency whereby skin and nail cells overproduce the protein keratin, causing them to become rough and scaly.
hypervitaminosis A A condition caused by vitamin A toxicity whereby blurred vision, liver abnormalities, and reduced bone strength occur.
intrinsic factor A protein produced by the stomach that is needed for vitamin B_{12} absorption.
macular degeneration A chronic disease that causes deterioration of the retina.
megaloblastic, macrocytic anemia A condition caused by folate deficiency whereby cells (including red blood cells) remain large and immature.
menadione A form of vitamin K produced commercially.
menaquinone A form of vitamin K produced by bacteria in the large intestine.
microcytic, hypochromic anemia A condition characterized by low concentrations of hemoglobin, which causes red blood cells to be small and light in color.
neural tube defect A malformation whereby neural tissue does not form properly during fetal development.
niacin (vitamin B_3) An essential water-soluble vitamin involved in energy and vitamin C metabolism and the synthesis of fatty acids and proteins.
niacin equivalent (NE) A unit of measure for the combined amounts of niacin and tryptophan in food.
night blindness A condition characterized by an impaired ability to see in low-light environments.
nonprovitamin A carotenoid A carotenoid that cannot be converted to vitamin A.
osteomalacia An adult condition caused by vitamin D deficiency whereby bones become soft and weak.
osteoporosis A serious disease caused by vitamin D deficiency whereby bones become weak and porous.
pantothenic acid (vitamin B_5) A nitrogen-containing water-soluble vitamin involved in energy metabolism, hemoglobin synthesis, and phospholipid synthesis.
parathyroid hormone (PTH) A hormone released from the parathyroid gland that stimulates the conversion of 25-(OH) D_3 to calcitriol in the kidneys.
pellagra A condition caused by niacin deficiency whereby dermatitis, dementia, diarrhea, and/or death may occur.
pernicious anemia An autoimmune disease caused by vitamin B_{12} deficiency whereby antibodies destroy the stomach cells that produce intrinsic factor.
phylloquinone A form of vitamin K found naturally in plant-based foods.
previtamin D_3 (precalciferol) An intermediate product made in the skin during the conversion of a cholesterol derivative to cholecalciferol.
prohormone A compound converted to an active hormone in the body.
provitamin A carotenoid A carotenoid that can be converted to vitamin A.
reduction-oxidation (redox) reaction A transfer of oxygen or electrons from one molecule to another.
retinoid (or **preformed vitamin A**) A vitamin A compound.

- Vitamin A and the carotenoids are important in regulation of growth, reproduction, vision, immune function, gene expression, and bone formation. The carotenoids are also potent antioxidants and protect proteins, DNA, and cell membranes from free radical damage.
- Good sources of preformed vitamin A are liver, fish, whole milk, and fortified foods. Vitamin A deficiency can result in blindness, infection, and death.

LO7: What Is Vitamin D?

- There are two forms of vitamin D: ergocalciferol (vitamin D_2) is found in plant sources and cholecalciferol (vitamin D_3) is found in animal foods and synthesized in the body.
- In the presence of sunlight, the skin can produce vitamin D from a cholesterol metabolite.
- Vitamin D deficiency can cause rickets in children and osteomalacia and osteoporosis in adults.
- α-Tocopherol, the most biologically active form of vitamin E, is found in oils, nuts, seeds, and some fruits and vegetables. Vitamin E functions mainly as an antioxidant, protecting biological membranes from free radical damage. Vitamin E may also protect the eyes from cataract formation and influence cancer risk by decreasing DNA damage.

LO8: What Are Vitamins E and K?

- Vitamin K refers to three related compounds: phylloquinone, menaquinone, and menadione. Vitamin K is essential to coagulation and proper bone mineralization.
- Dark green vegetables are often good sources of vitamin K, and light and heat can destroy this vitamin in foods. In infants, severe vitamin K deficiency can cause a fatal condition called vitamin K deficiency bleeding.

LO9: Should You Take Dietary Supplements?

- If you have difficulty consuming a good variety and balance of foods in adequate amounts, taking a dietary supplement may help you consume appropriate amounts of the essential nutrients.
- It is important to make informed decisions about which supplements to take and which to avoid.

retinol activity equivalent (RAE) A unit of measure for the combined amounts of preformed vitamin A and provitamin A carotenoids in a food.
riboflavin (vitamin B_2) An essential water-soluble vitamin involved in energy metabolism, the synthesis of a variety of vitamins, nerve function, and protection of biological membranes.
rickets A childhood condition caused by vitamin D deficiency whereby slow growth and bone deformation occur.
scurvy A condition caused by vitamin C deficiency whereby bleeding gums, bruising, poor wound healing, and skin irritations occur.
spina bifida A failure of the neural tube to close properly during the first months of fetal life.
stomatitis Inflammation of the mouth, often caused by dietary deficiencies.
tetrahydrofolate (THF) The active form of folate.
thiamin (vitamin B_1) An essential water-soluble vitamin involved in energy metabolism and the synthesis of DNA and RNA.
vitamin A deficiency disorder (VADD) A spectrum of health-related consequences caused by vitamin A deficiency.
vitamin B_6 A water-soluble vitamin involved in the metabolism of proteins and amino acids, the synthesis of neurotransmitters and hemoglobin, glycogenolysis, and regulation of steroid hormone function.
vitamin B_{12} (cobalamin) A water-soluble vitamin involved in energy metabolism and methionine production.
vitamin C (ascorbic acid) A water-soluble vitamin that serves antioxidant functions within the body.
vitamin D_3 (or **cholecalciferol**) The form of vitamin D that diffuses into the blood and circulates to the liver.
vitamin K deficiency bleeding A disease caused by vitamin K deficiency whereby uncontrollable internal bleeding occurs.
xerophthalmia A condition caused by vitamin A deficiency whereby the cornea and other portions of the eye are damaged, leading to dry eyes, scarring, and even blindness.

Review 8

Chapter Summary

LO1: Why Is Water Essential to Life?

- Water is the most abundant substance in the body. It is critical to many vital bodily functions, such as energy metabolism and temperature regulation.
- Water crosses cell membranes via osmosis. The driving force behind osmosis is the difference in solute concentrations across these membranes.
- Hydrolysis and condensation reactions break and create molecules via the release and addition of water, respectively.
- Dehydration influences cognitive function, ability to engage in aerobic activity, temperature regulation, and risk of urinary tract infections. The body responds to dehydration by releasing antidiuretic hormone (ADH), which decreases the amount of water excreted in the urine.
- Individuals who maintain high levels of physical activity and/or live in hot environments are advised to consume additional water.

LO2: What Are Minerals?

- In nutrition, a mineral is an inorganic substance other than water. Because your body requires them in very small amounts, dietary minerals are considered micronutrients.
- A major mineral is one required in an amount greater than 100 mg/day. A trace mineral is one required only in a minute amount (less than 100 mg/day).
- Minerals are found in both plant and animal sources, but animal-based foods tend to have higher mineral contents than do plant-based foods.
- In general, major minerals are absorbed in the small intestine and travel throughout the body in the blood; excess amounts are excreted in the urine, and toxicities are rare.
- Calcium (Ca) is a major mineral located throughout the skeleton that is essential to structure, coagulation, muscle and nerve function, and energy metabolism.
- An electrolyte is a molecule that produces charged ions when dissolved in water. The three most abundant ions in the body are sodium, chloride, and potassium.
- Electrolytes and their resultant ions are critical to nerve and muscle function, and assist in blood volume and blood pressure regulation.
- Although a high intake of salt has long been associated with increased risk for cardiovascular disease, this is true only in a segment of the population.
- Phosphorus (P) is a major mineral that is essential to cell membranes, bone and tooth structure, DNA, RNA, ATP, lipid transport, and a variety of processes in the body.
- Magnesium (Mg) is a major mineral that is important to many physiological processes, such as energy metabolism and enzyme function.
- The body needs sulfur to synthesize compounds required for healthy connective tissue and nerve function. Sulfur is also an important component of the B vitamins thiamin and biotin, and is therefore essential to energy metabolism.

LO3: What Are Trace Minerals?

- Although they are only needed in minute amounts, the trace minerals are vital to your health.
- The bioavailability of trace minerals is influenced by factors such as nutritional status, genetics, the aging process, and interactions with other food components. Trace

Key Terms

acrodermatitis enteropathica A genetic abnormality that causes secondary zinc deficiency.

aldosterone A hormone produced by the adrenal glands in response to low blood pressure.

antidiuretic hormone (ADH, or **vasopressin)** A hormone released by the pituitary gland during periods of low blood volume that decreases the amount of water excreted in the urine.

bone density The amount of bone tissue in a certain volume of bone.

bone mass The amount of minerals contained in bone.

bone remodeling (or **bone turnover**) The continuous process by which older and damaged bone is replaced by new bone.

calcitonin A hormone released by the thyroid gland that decreases calcium loss from bone, decreases calcium absorption in the small intestine, and increases calcium loss in the urine.

calcium (Ca) A major mineral located throughout the skeleton that is essential to coagulation, muscle and nerve function, and energy metabolism.

ceruloplasmin A transport protein that binds to copper and transports it in the blood.

chelator A compound that binds to nonheme iron in the intestine, making it unavailable for absorption.

chromium (Cr) A trace mineral that may be critical to proper insulin function.

chromium picolinate A form of chromium taken as an ergogenic aid by some athletes.

cofactor A mineral that activates an enzyme by combining with it.

condensation reaction A chemical reaction whereby a chemical bond joins two molecules together, releasing water in the process.

copper (Cu) An essential trace mineral that acts as a cofactor for nine enzymes involved in redox reactions.

cretinism A severe form of iodine deficiency that affects babies born to iodine-deficient mothers.

dehydration A condition whereby the body has an insufficient amount of water.

dental fluorosis Pitting and mottling of teeth caused by excessive fluoride intake.

diuretic A substance or drug that helps the body eliminate water.

evaporative cooling The process by which sweat evaporates from the skin, taking heat with it.

extracellular fluid Fluid located outside of a cell.

ferritin A protein complex that stores iron and releases it only when it is needed.

fluoride (F^-) A trace mineral that strengthens bones and teeth.

goiter A sign of iodine deficiency characterized by an enlarged thyroid gland.

goitrogen A dietary compound, found in some plants, that decreases iodine utilization by the thyroid gland.

hematocrit The percentage of blood volume comprised of red blood cells.

heme iron Iron that is part of hemoglobin and myoglobin and is found only in meat.

hemoglobin A complex protein that transports oxygen from the lungs to cells and carbon dioxide from the cells to the lungs.

hereditary hemochromatosis A genetic abnormality whereby too much iron is absorbed, causing it to accumulate in the body.

hydrolysis reaction A chemical reaction whereby a chemical bond is broken by the addition of a water molecule.

hydroxyapatite A large crystal-like molecule that combines with other minerals to form the structural matrix of bones and teeth.

hypokalemia A condition characterized by low blood potassium concentration.

hyponatremia A condition characterized by low blood sodium concentration.

intercellular fluid Extracellular fluid that fills spaces between or surrounding cells.

intracellular fluid Fluid located inside of a cell.

intravascular fluid Extracellular fluid located in blood and lymph.

iodine (I) An essential trace mineral that serves as a component of the thyroid hormones.

iodine deficiency disorders (IDDs) A broad spectrum of conditions caused by iodine deficiency.

iron (Fe) A trace mineral needed for oxygen and carbon dioxide transport, energy metabolism, stabilization of free radicals, and synthesis of DNA.

Keshan disease A disease caused by severe selenium deficiency that mostly affects children and causes serious heart problems.

kyphosis (or **dowager's hump**) A curvature of the upper spine caused by osteoporosis.

magnesium (Mg) A major mineral that is important to many physiological processes, such as energy metabolism and enzyme function.

major mineral An essential mineral required in an amount greater than 100 mg/day.

manganese (Mn) An essential trace mineral that is a cofactor for metalloenzymes needed for bone formation, glucose production, and energy metabolism.

meat factor A compound found in meat that increases the bioavailability of nonheme iron.

metalloenzyme An enzyme that is activated when it combines with a mineral.

metallothionine An intestinal protein that regulates the amount of zinc released into the blood.

minerals are absorbed primarily in the small intestine. The processes by which blood mineral levels are regulated vary from mineral to mineral.

- Iron (Fe) is a trace mineral that is necessary to oxygen and carbon dioxide transport, energy metabolism, the stabilization of free radicals, and the synthesis of DNA. Two general forms of iron are found in foods: heme iron and nonheme iron. Heme iron is readily absorbed, whereas nonheme iron is not.
- Copper (Cu) is an essential trace mineral that acts as a cofactor for nine enzymes involved in reduction-oxidation (redox) reactions.
- Iodine (I) is an essential trace mineral that serves as a component of the thyroid hormones.
- Selenium (Se) is an essential trace mineral that is critical to redox reactions, thyroid function, and the activation of vitamin C. Selenium is a component of selenoproteins.
- Chromium (Cr) is an essential trace mineral that may be critical to proper insulin function.
- Manganese (Mn) is an essential trace mineral that is a cofactor for metalloenzymes needed for bone formation, glucose production, and energy metabolism.
- Molybdenum (Mo) is an essential trace mineral that is a cofactor for several important metalloenzymes needed for amino acid and purine metabolism.
- Zinc (Zn) is an essential trace mineral involved in gene expression, immune function, and cell growth.
- Fluoride (F^-) is a trace mineral that strengthens bones and teeth.
- In addition to the trace minerals known to be essential to human life, many others may influence health. These include nickel, aluminum, silicon, vanadium, arsenic, and boron. Although scientists do not know whether these minerals are important to human health, there is evidence that they influence the health of other animals.

mineral An inorganic substance other than water that is required by the body in small amounts.

molybdenum (Mo) An essential trace mineral that is a cofactor for several important metalloenzymes needed for amino acid and purine metabolism.

myoglobin An oxygen-storage molecule located within muscles.

nonheme iron Iron that is not part of hemoglobin and myoglobin and is found primarily in plant-based foods.

osteoblast A bone cell that promotes bone formation.

osteoclast A bone cell that promotes the breakdown of older bone.

osteopenia A disease characterized by moderate bone loss in adults.

peak bone mass The greatest total amount of bone that a person has in his or her lifetime.

phosphorus (P) A major mineral that is essential to cell membranes, bone and tooth structure, DNA, RNA, ATP, lipid transport, and a variety of processes in the body.

renin An enzyme, secreted by the kidneys, that converts angiotensinogen to angiotensin I.

selenium (Se) An essential trace mineral that is critical to redox reactions, thyroid function, and the activation of vitamin C.

selenoprotein A protein comprised of selenium-containing amino acids.

skeletal fluorosis Weakening of the skeleton caused by excessive fluoride intake.

solute A substance dissolved in a fluid.

sulfur (S) A major mineral that is a component of certain amino acids (such as cysteine and methionine).

superoxide dismutase A copper-containing enzyme that helps stabilize highly reactive free radical molecules.

thyroid-stimulating hormone (TSH) A hormone, produced by the pituitary gland, that regulates iodine uptake by the thyroid gland.

trace mineral An essential mineral that is required in an amount less than 100 mg/day.

transferrin A protein in the blood that binds to dietary iron once it is released from an intestinal cell.

zinc (Zn) An essential trace mineral involved in gene expression, immune function, and cell growth.

Review 9

Chapter 9
Energy Balance and Body Weight Regulation

Chapter Summary

LO1: What Is Energy Balance?

- The term *energy balance* describes the relationship between energy intake and energy expenditure.
- A person is in positive energy balance when her energy intake exceeds her energy expenditure, negative energy balance when her energy intake is less than her energy expenditure, and neutral energy balance when energy intake equals energy expenditure.
- The number and size of adipocytes determine the amount of adipose tissue in your body. Adipose tissue serves as the body's primary energy reserve.
- When body fat increases, adipocytes can increase in size (hypertrophic growth), number (hyperplasic growth), or both. When body fat decreases, adipocytes decrease in size but not in number.
- A person's distribution of visceral adipose tissue (VAT) and subcutaneous adipose tissue (SCAT) holds important information about his health. Central obesity, a predominant accumulation of VAT, puts a person at greater risk for weight-related health problems.

LO2: What Factors Influence Energy Intake?

- *Hunger* is the basic physiological need for food, whereas satiety is the sensation of having eaten enough.
- A complex region of the brain called the hypothalamus regulates energy intake and thus body weight by balancing feelings of hunger and satiety.
- Signals from the gastrointestinal (GI) tract such as gastric stretching and GI hormones play a role in the regulation of short-term food intake. Circulating concentrations of glucose, fatty acids, and amino acids also influence hunger and satiety.
- The majority of known GI hormones inhibit food intake, with the exception of ghrelin, which stimulates hunger.
- *Appetite* reflects psychological factors that influence hunger and satiety.
- A *food aversion* is a strong psychological dislike of a particular food and a food craving is a strong psychological desire for a particular food.

LO3: What Determines Energy Expenditure?

- The body expends energy to maintain basal metabolism, engage in physical activity, and process food. These three components make up most of a person's total energy expenditure (TEE).
- Smaller components of the TEE include adaptive thermogenesis and nonexercise activity thermogenesis.
- The major factors influencing basal metabolic rate include body shape, body composition, age, sex, nutritional status, and genetics.

LO4: How Are Body Weight and Composition Assessed?

- Overweight is excess weight for a given height, whereas obesity is an abundance of body fat in relation to lean tissue.
- Indices such as height–weight tables and body mass index (BMI) are used to assess body weight.

Key Terms

adaptive thermogenesis A temporary expenditure of energy that enables the body to adapt to changes in the environment and physiological conditions.

adipokine A hormone-like protein produced by an adipocyte that plays a communicative role between the adipocyte and other tissues and organs.

appetite A psychological longing or desire for food.

bariatric surgery Surgical procedure performed to treat obesity.

basal metabolic rate (BMR) The amount of energy expended per hour (kcal/hour) so that the body can carry out basic, involuntary life functions.

basal metabolism An expenditure of energy to sustain basic, involuntary life functions such as respiration, beating of the heart, nerve function, and muscle tone.

bioelectrical impedance A method of estimating body composition in which a weak electric current is passed through the body.

body mass index (BMI) A measure of body fat whereby a person's body weight is divided by the square of his or her height.

central obesity (or **central adiposity**) A predominant accumulation of adipose tissue between the internal organs in the abdomen.

***db* gene** The gene that codes for the leptin receptor.

***db/db* (diabetic) mouse** A mouse with a gene mutation that makes it both obese and diabetic.

dual-energy x-ray absorptiometry (DEXA) A method of estimating body composition in which low-dose x-rays are used to visualize fat and fat-free compartments of the body.

exercise Planned, structured, and repetitive bodily movement done to improve or maintain physical fitness.

food aversion A strong psychological dislike of a particular food or foods.

food craving A strong psychological desire for a particular food or foods.

gastric banding A surgical procedure that involves reducing the size of the stomach and bypassing a segment of the small intestine so that fewer nutrients are absorbed.

gastric bypass A surgical procedure that involves reducing the size of the stomach and bypassing a segment of the small intestine so that fewer nutrients are absorbed.

ghrelin A newly discovered hormone secreted by cells in the stomach lining that stimulates hunger.

hunger A basic physiological drive to consume food.

hydrostatic weighing A method of estimating body composition in which a person's body weight is measured in and out of water.

hyperplastic growth A process whereby new adipocytes are formed.

hypertrophic growth A process whereby an adipocyte fills with lipid and its size increases.

hypothalamus A region of the brain that regulates energy intake by balancing hunger and satiety.

leptin A potent hormone produced primarily by adipose tissue that signals satiety.

negative energy balance A condition whereby energy intake is less than energy expenditure.

neurotransmitter A hormone-like substance released by neural tissue such as the hypothalamus.

neutral energy balance A condition whereby energy intake equals energy expenditure.

nonexercise activity thermogenesis An expenditure of energy associated with spontaneous movement such as fidgeting and posture maintenance.

***ob* gene** The gene that codes for the hormone leptin.

obese Excess body fat.

***ob/ob* (obese) mouse** A mouse with a gene mutation that makes it obese.

overweight Excess weight for a given height.

physical activity Any bodily movement that results in an expenditure of energy.

positive energy balance A condition whereby energy intake is greater than energy expenditure.

satiety A physiological response whereby a person feels he or she has consumed enough food.

set point theory A scientific concept whereby hormones circulating in the blood are theorized to regulate body weight by communicating the amount of adipose tissue in the body to the brain, which adjusts energy intake and expenditure.

skinfold caliper An instrument used to measure the thickness of skin and subcutaneous fat.

skinfold thickness method A method of estimating body composition in which a skinfold caliper is used to measure the thickness of skin and subcutaneous fat at various locations on the body.

thermic effect of food (**TEF**, or **diet-induced thermogenesis**) The expenditure of energy to digest, absorb, and metabolize nutrients following a meal.

total energy expenditure (TEE) The collective sum of energy used by the body.

waist circumference A measurement used as an indicator of central adiposity.

- BMI, based on the ratio of weight to height squared, is also considered a good indicator of body fat.
- BMI is a useful method of interpreting a person's weight given his height, but health professionals sometimes want to know a person's individual body composition.
- Measures of body composition include hydrostatic weighing, dual-energy x-ray absorptiometry (DEXA), bioelectrical impedance, and the skinfold thickness method.

LO5: Genetics versus Environment: What Causes Obesity?

- Many factors resulting in increased energy intake and/or decreased energy expenditure have contributed to the increasing prevalence of obesity in the United States. These included lifestyle, socioeconomic, and cultural factors.
- One of the most important factors that influence body weight is the amount and energy density of the food we consume.
- Although the hormone leptin has not proved effective for treating human obesity in most cases, its discovery has led to important insights into body weight regulation.

LO6: How Are Energy Balance and Body Weight Regulated?

- The set point theory of body weight regulation suggests that body weight is generally stable because of the body's ability to adjust energy intake and expenditure.
- Leptin plays a role in regulating body weight by communicating the body's energy reserve to the brain.
- When body fat increases, so does the concentration of leptin circulating in the blood, typically resulting in overall decreased energy intake and increased energy expenditure. When body fat decreases, leptin production also decreases, resulting in overall increased energy intake and decreased energy expenditure.
- Most obese people produce adequate amounts of leptin, but defects in leptin responsiveness may lead to obesity.

LO7: What Is the Best Approach to Weight Loss?

- Maintaining weight loss requires lasting lifestyle changes such as eating moderate amounts of nutrient-dense foods and engaging in regular exercise.
- The majority of individuals who lose weight and maintain weight loss do so through a combination of healthy diet and physical activity. A healthy weight-loss goal might be to reduce one's current weight by 5 to 10 percent.
- Proponents of low-fat diets believe that reducing one's fat intake leads to the consumption of fewer calories and therefore to greater weight loss, whereas advocates of low-carbohydrate diets believe that increased carbohydrate intake can cause insulin levels to rise, leading to weight gain.
- Limiting carbohydrates can help prevent rises in insulin. Because increases in blood insulin levels may contribute to weight gain, limiting carbohydrates may help a person lose weight (or at least not gain weight)..

Review 10

Chapter Summary

LO1: What Physiological Changes Take Place during the Human Life Cycle?

- Periods of growth, development, maturation, and senescence coincide with specific life stages: infancy, childhood, adolescence, adulthood, and for women, the special life stages of pregnancy and lactation.
- Changes in body size and composition influence nutrient requirements.
- When physical maturity is reached, cell turnover achieves equilibrium. As a person ages, the rate of new cell formation slows, resulting in senescence.

LO2: What Are the Major Stages of Prenatal Development?

- The two stages of prenatal development are the embryonic period and the fetal period.
- The placenta transfers nutrients, oxygen, and other substances from the mother's blood to the fetus; it also transfers waste products from the fetus to the mother.
- Babies born with gestational ages between 37 and 42 weeks are considered full-term infants, whereas those born with gestational ages less than 37 weeks are considered preterm (or premature) and those born with gestational ages greater than 42 weeks are considered post-term.
- Many preterm infants are born with low birth weight (LBW), which can also be caused by intrauterine growth retardation (IUGR). Babies who experience IUGR are often small for gestational age (SGA).
- Gestational age is determined by counting the number of weeks between the first day of a woman's last normal menstrual period and birth.

LO3: What Are the Nutrition Recommendations for a Healthy Pregnancy?

- A pregnant woman who gains an appropriate amount of weight, eats a healthy diet, and refrains from smoking can increase the likelihood of having a full-term baby born with a healthy birth weight.
- During pregnancy, additional energy is needed to support the growth of the fetus, placenta, and maternal tissues. Carbohydrates should remain the primary energy source during this time.
- Additional protein is needed for the formation of fetal and maternal tissues. Dietary fat should provide 20 to 35 percent of total calories during pregnancy.
- Extra calcium is necessary for a fetus to grow and develop, although changes in maternal physiology can accommodate these needs without increasing dietary intake.
- Iron is essential for the formation of hemoglobin and the growth and development of the fetus and the placenta.
- Folate is critical to cell division and development of the nervous system. A woman with poor folate status is at increased risk of having a baby with a neural tube defect.
- Smoking during pregnancy increases the risk of bearing a preterm or LBW baby.

LO4: Why Is Breastfeeding Recommended during Infancy?

- Milk production and the release of milk from the alveoli into the milk ducts are regulated in part by the hormones prolactin and oxytocin.

Key Terms

alveolus A cluster of milk-producing cells.
appropriate for gestational age (AGA) A baby that has a birth weight between the 10th and 90th percentiles for gestational age.
baby bottle tooth decay A condition whereby dental caries occur in an infant who habitually sleeps with bottles filled with milk, formula, juice, or any other carbohydrate-containing beverage.
blastocyst A dense sphere of cells that implants itself into the lining of the uterus.
cell turnover The cyclical process by which cells form and break down.
colostrum A thick fluid produced shortly after birth that nourishes a newborn and helps prevent disease.
critical period A period in prenatal development during which adverse effects on growth and development are irreversible.
development A change in the attainment or complexity of a skill or function.
developmental origins of health and disease hypothesis A hypothesis that states that less than optimal conditions in the uterus or during early infancy may cause permanent changes in the structure and function of organs and tissues, predisposing individuals to certain chronic diseases later in life.
embryo A developing human as it exists from the third to the eighth week after fertilization.
embryonic period The period of prenatal development that spans from conception through the eighth week of gestation.
embryonic phase The stage of the embryonic period during which organs and organ systems first begin to form.
fetal period The stage of prenatal development that begins at the ninth week of pregnancy and ends at birth.
fetus A developing human as it exists from the ninth week of pregnancy until birth.
full-term A baby born with gestational age between 37 and 42 weeks.
gestation length The period of time between conception and birth.
gestational age The number of weeks from the first day of a woman's last normal menstrual cycle to the current date.
gestational diabetes A form of diabetes that develops when pregnancy-related hormonal changes cause cells to become less responsive to insulin.
growth A physical change that results from an increase in either cell size or number.
intrauterine growth retardation (IUGR) Slow growth while in the uterus.
lactation The production and release of milk.
lactogenesis The onset of milk production.

large for gestational age (LGA) A baby that has a birth weight above the 90th percentile for gestational age.
let-down The active process whereby milk is forced out of the alveoli and into the milk ducts.
lifespan The maximum number of years an individual member of a particular species has remained alive.
low birth weight (LBW) A baby that weighs less than 2,500 g (5 lb, 8 oz) at birth.
menarche The onset of menstruation.
menopause A life stage characterized by very little estrogen production, which causes the menstrual cycle to stop.
oxytocin A hormone that causes the muscles around the alveoli to contract.
perimenopausal A life stage characterized by a natural decline in estrogen production.
pica The urge to consume nonfood items.
placenta An organ made of embryonic and maternal tissues that supplies nutrients and oxygen to the developing child.
post-term A baby born with a gestational age greater than 42 weeks.
pre-embryonic phase The early phase of the embryonic period that begins with fertilization and continues through implantation.
pregnancy-induced hypertension (or **pre-eclampsia** or **toxemia of pregnancy**) A pregnancy-related condition characterized by high blood pressure, a sudden increase in weight, swelling due to fluid retention, and protein in the urine.
preterm (or **premature**) A baby born with a gestational age less than 37 weeks.
prolactin A hormone that stimulates alveolar cells to produce milk.
puberty Maturation of the reproductive system.
senescence The physical changes characteristically associated with aging.
small for gestational age (SGA) A baby that has a birth weight below the 10th percentile for gestational age.
zygote An ovum that has been fertilized by a sperm.

- The process by which milk is released from the alveoli into the milk ducts is called milk let-down.
- Breastfeeding is the preferred method of nourishing infants because human milk provides many nutritional and immunological benefits.

LO5: What Are the Nutritional Needs of Infants?

- During the first year of life, a healthy baby's weight almost triples and his length increases by up to 50 percent.
- In communities without fluoridated water, fluoride supplements may be necessary after six months of age. Iron supplements are recommended for infants who exclusively breastfeed during the second six months of life. Vitamin D drops are recommended for breastfed infants beginning in the first few days of life.
- The American Academy of Pediatrics recommends exclusively breastfeeding throughout the first four to six months of life. Human milk and/or infant formula should be the primary source of nutrients and energy throughout the first year of life.
- Nonmilk complementary foods can be added sometime between four and six months of age, depending on an infant's readiness.
- Older infants should be given a wider variety of foods, although they should be chosen carefully to pose minimal risk for choking.

LO6: What Are the Nutritional Needs of Toddlers and Young Children?

- Growth is monitored using growth charts, and BMI is used to assess adiposity in children over 2 years of age.
- Following six general guidelines regarding common childhood feeding problems can encourage healthy eating.
- Nutrients needed for bone health (such as calcium) and to support growth (such as iron) are particularly important during childhood.

LO7: How Do Nutritional Requirements Change during Adolescence?

- Hormonal changes begin the physical transformation from childhood into adolescence, causing shifts in height, weight, and body composition.
- Because the timing of these changes varies, adolescents of the same age can differ in terms of physical maturation and nutritional requirements.
- Linear growth (height) is largely completed at the end of the adolescent growth spurt, although bone mass continues to increase into early adulthood.
- During adolescence, males experience an increase in percentage of lean body mass, whereas the opposite occurs in females.

LO8: How Do Age-Related Changes in Adults Influence Nutrient and Energy Requirements?

- As individuals grow older, they experience a relative loss of lean mass and increase in fat mass. For this reason, energy requirements decrease.
- Many factors can contribute to inadequate food intake in older adults, including poor oral health, altered taste, and decreased ability to smell.
- Age-related changes in the gastrointestinal tract can also affect nutritional status.
- A decrease in production of gastric secretions can impair absorption of iron, calcium, biotin, folate, cobalamin, and zinc.
- Menopause marks the cessation of menstrual cycles.
- Services that provide food to older adults include congregate meal programs and meal delivery programs such as Meals on Wheels.

Review 11

Chapter Summary

LO1: What Are the Health Benefits of Physical Activity?

- Physical activity includes day-to-day activities such as gardening, household chores, walking, and leisure-time activities. Exercise is a subcategory of physical activity defined as planned, structured, and repetitive bodily movement done to improve or maintain physical fitness.
- While the importance of physical activity may be common knowledge, nearly half of all Americans lead sedentary lifestyles.
- Participation in at least two and a half hours of moderate-intensity aerobic activity and two sessions of muscle-strengthening activity per week is recommended for optimal health.
- Physical fitness reflects characteristics that relate to a person's ability to perform physical activity.
- There are five components of physical fitness—cardiovascular fitness, muscular strength, muscular endurance, flexibility, and body composition—each of which contributes to overall health.
- The FITT principle should be the foundation of any physical fitness plan. FITT parameters include frequency, intensity, type, and time.

LO2: How Does Energy Metabolism Change during Physical Activity?

- At rest, the body expends approximately 1.0 to 1.5 kcal/minute. During physical exertion, however, energy expenditure can increase substantially.
- ATP can be generated in both the presence and relative absence of oxygen. These conditions are called aerobic and anaerobic, respectively.
- Aerobic pathways generate large amounts of ATP over an extended period of time, whereas anaerobic pathways generate smaller amounts of ATP for shorter periods of time.
- Until the cardiovascular system is able to increase oxygen delivery to active muscles, the high-energy compound creatine phosphate can be broken down and combined with ADP to produce ATP. This is the simplest and most rapid means by which active muscles generate ATP, although overall yield is low.
- If physical activity continues beyond the capacity of the glycolytic and creatine phosphate systems, muscles must utilize the aerobic (oxidative) system for ATP production. Although aerobic pathways are slower than anaerobic pathways in regard to the rate of ATP formation, the energy yield is rich.

LO3: What Physiologic Adaptations Occur in Response to Athletic Training?

- As your body acclimates to the rigors of frequent exercise, physiological changes called adaptation responses enable it to perform and recover more efficiently and effectively.
- Strength training (such as weight lifting) challenges muscles and helps increase muscle strength, power, endurance, and mass.

Key Terms

adaptation response A series of physiological changes that enable the body to perform and recover more efficiently and effectively.

anaerobic capacity The maximum amount of work that can be performed under anaerobic conditions.

atrophy A process whereby muscles decrease in size.

body composition A measure of the relative amounts of muscle, fat, and bone tissue in the body.

carbohydrate loading A technique used by some athletes to increase the amount of glycogen stored in their muscles so that they can sustain activity for long periods of time.

cardiovascular fitness A measure of the circulatory and respiratory systems' ability to supply oxygen and nutrients to working muscles during sustained physical activity.

creatine phosphate A high-energy compound that is broken down to produce ATP.

creatine phosphate system The simplest and most rapid energy system; characterized by the use of creatine phosphate to produce ATP.

endurance A measure of one's ability to exercise for an extended period of time without becoming fatigued.

endurance training Exercise that entails steady, low- to moderate-intensity exercise that persists for a longer duration.

ergogenic aid A substance taken to enhance physical performance.

FITT principle A collection of four parameters to consider when developing a physical fitness plan: frequency, intensity, type, and time.

flexibility A measure of the range of motion around a joint.

glycolytic system An energy system characterized by the anaerobic metabolism of glucose (glycolysis).

heart rate The number of contractions the heart makes in one minute.

heat exhaustion A rise in body temperature that occurs when the body has difficulty dissipating heat.

heat stroke A serious condition that can develop if body temperature continues to rise beyond heat exhaustion.

hemodilution A disproportionate increase in plasma volume as compared to the synthesis of new red blood cells.

hypertrophy An adaptation response to exercise characterized by enlargement of muscles.

hyponatremia Diminished blood sodium concentration.

indirect calorimetry A method used to estimate energy expenditure based on oxygen consumption and carbon dioxide production.

interval training Exercise that entails alternating short, fast bursts of intense exercise with slower, less demanding activity.

maximal oxygen consumption (or **VO_2 max**) A measure of the cardiovascular system's capacity to deliver oxygen to muscles.

muscular strength The maximal force exerted by muscles during physical activity.

oxidative system An aerobic energy system characterized by steady ATP production in an oxygen-rich environment.

physical fitness Characteristics that a person has or can achieve relating to his or her ability to perform physical activity.

resistance training Strength-building activities that challenge specific groups of muscles by making them work against an opposing force.

sports anemia A temporary type of anemia that occurs at the onset of a training program due to hemodilution.

strength training Exercise that increases skeletal muscle strength, power, endurance, and mass.

stroke volume The amount of blood pumped per contraction of the heart.

ventilation rate Breaths expelled per minute.

- Training-induced adaptations cause muscles to undergo hypertrophy, meaning that they become larger. Muscles that are not frequently challenged undergo atrophy, meaning that they become smaller.
- Endurance training (such as running and swimming) is beneficial to pulmonary and cardiovascular function, and can help improve aerobic capacity (the ability to produce ATP in an oxygen-rich environment). These adaptive responses increase maximal oxygen consumption (or VO_2 max).

LO4: How Does Physical Activity Influence Dietary Requirements?

- Athletic individuals must consume enough energy to support normal daily activities as well as those performed as exercise.
- Carbohydrates maintain glycogen stores, whereas protein builds, maintains, and repairs muscles.
- High-quality protein sources are necessary to maintain, build, and repair tissue.
- Dietary fat provides essential fatty acids, facilitates the use of fat-soluble substances, and is an important source of energy.
- Iron is essential to oxygen and carbon dioxide transport, and calcium is essential to the building, maintenance, and repair of bone tissue.
- Zinc is needed for energy metabolism and for the maintenance, repair, and growth of muscles.
- Excess sweating can disrupt fluid and electrolyte balance, which impairs body temperature and fluid regulation.
- Athletes can prevent dehydration by consuming adequate amounts of fluids before, during, and after exercise.
- To stay fully hydrated, athletes should consume 500 to 600 milliliters (2 to $2^1/_2$ cups) of fluid two to three hours before exercise. 450 to 675 milliliters (2 to 3 cups) should be consumed after exercise for each pound of weight lost.

Review 12

Chapter Summary

LO1: What Is Disordered Eating, and What Are Eating Disorders?

- Disordered eating is characterized by unhealthy eating patterns such as irregular eating, consistent undereating, and/or consistent overeating.
- Although disordered eating patterns can be disturbing to others, they typically do not persist long enough to cause serious physical harm. For some people, however, disordered eating develops into a full-blown eating disorder.
- The three main types of eating disorders are anorexia nervosa (AN), bulimia nervosa (BN), and eating disorders not otherwise specified (EDNOS).
- AN is characterized by a fear of weight gain, distorted body image, and food restriction. People with AN typically limit their food intakes as well as the varieties of foods consumed. Some people with AN also exercise obsessively.
- BN involves cycles of bingeing and purging. Individuals with BN often experience regret and loss of control after bingeing. These feelings can lead to depression, which increases the likelihood of future binge–purge cycles.
- EDNOS is a category for eating disorders that are not as well defined as AN and BN. Individuals with EDNOS exhibit behaviors and traits associated with AN or BN but do not meet all of the diagnostic criteria.

LO2: Are There Other Disordered Eating Behaviors?

- Many disordered eating behaviors exist in addition to the eating disorders formally recognized by the American Psychiatric Association.
- Nocturnal sleep-related eating disorder (SRED) is characterized by eating at night without recollection of having done so.
- Night eating syndrome (NES) is characterized by a cycle of daytime food-restriction, excessive food intake in the evening, and nighttime insomnia.
- Food neophobia is an eating disturbance characterized by an irrational fear or avoidance of trying new foods as well as unusual food rituals and practices.
- Muscle dysmorphia is observed primarily in men who work out excessively to increase muscularity and have intense fears of being too small, too weak, and/or too skinny.

LO3: What Causes Eating Disorders?

- There are many theories as to the factors that contribute to the development of eating disorders, but there are no simple answers.
- Eating disorders are more prevalent in cultures where food is abundant and slimness is valued. However, as multicultural youth acculturate to mainstream American values, distinct cultural norms that once protected them from eating disorders may be eroding.
- Major media outlets that glamorize unrealistically thin and overly muscular bodies have long been criticized for evoking a sense of inadequacy in impressionable children and young adults.
- In addition to the media, a person's circle of peers may also contribute to the development of eating disorders.
- Enmeshed family dynamics promote dependency, which may lay the foundation for the emergence of eating disorders later in life.

Key Terms

anorexia nervosa (AN) An eating disorder characterized by an irrational fear of gaining weight or becoming obese.

anorexia nervosa, binge-eating/purging type A subcategory of anorexia nervosa characterized by both food restriction and periods of bingeing and purging.

anorexia nervosa, restricting type A subcategory of anorexia nervosa characterized by food restriction and/or excessive exercise.

binge-eating disorder (BED) A disordered eating pattern classified as EDNOS that is characterized by repeated cycles of bingeing without purging.

bingeing Uncontrolled consumption of large quantities of food in a relatively short period of time.

bulimia nervosa (BN) An eating disorder characterized by repeated cycles of bingeing and purging.

chaotic family (or **disengaged family**) A style of family interaction characterized by a lack of cohesiveness and little parental involvement.

disordered eating An eating pattern characterized by one or more unhealthy eating behaviors.

eating disorder An extreme disturbance in eating behaviors that can be both physically and psychologically harmful.

eating disorders not otherwise specified (EDNOS) A category of eating disorders that includes some, but not all, of the diagnostic criteria for anorexia nervosa and/or bulimia nervosa.

enmeshment A style of family interaction whereby family members are overly involved with one another and have little autonomy.

female athlete triad A combination of three interrelated conditions: disordered eating (or eating disorder), menstrual dysfunction, and osteopenia.

food neophobia A disordered eating pattern characterized by an irrational fear or avoidance of new foods.

food preoccupation Spending an inordinate amount of time thinking about food.

muscle dysmorphia A disorder characterized by a preoccupation with increasing muscularity.

night eating syndrome (NES) A disordered eating pattern characterized by a cycle of daytime food restriction, excessive food intake in the evening, and nighttime insomnia.

nocturnal sleep-related eating disorder (SRED) A disordered eating pattern characterized by eating while asleep without any recollection of having done so.

purging Self-induced vomiting and/or excessive exercise, misuse of laxatives, diuretics, and/or enemas.

restrained eater An individual who suppresses his or her desire for food and avoids eating for long periods of time between binges.

- Children raised in a chaotic environment may later develop eating disorders as a way to fill emotional emptiness, to gain attention, or to suppress emotional conflict.
- Children of women with eating disorders are at increased risk of developing eating disorders themselves.
- Personality traits associated with eating disorders include low self-esteem, lack of self-confidence, obsessiveness, and feelings of helplessness, anxiety, and depression.
- Brain chemicals and other biological factors might play a role in the development of an eating disorder.

LO4: Are Athletes at Increased Risk for Eating Disorders?

- While some studies indicate that the prevalence of eating disorders among female student-athletes and non-athletes does not differ, other studies indicate otherwise.
- The prevalence of disordered eating and eating disorders among collegiate athletes is estimated to be somewhere between 15 and 60 percent.
- Sports that favor a thin physical appearance are likely to have more athletes with eating disorders than those for which size is not as important.
- Because athletes tend to be competitive people and may equate their self-worth with athletic success, they may be especially willing to engage in risky weight-loss practices to achieve that success.
- Female athletes are at risk for developing the female athlete triad: disordered eating/eating disorder, menstrual irregularities such as amenorrhea, and osteopenia.

LO5: How Can Eating Disorders Be Prevented and Treated?

- People with eating disorders are at risk for serious medical and/or emotional problems. Thus, they need treatment from qualified health professionals.
- Eating disorder prevention involves educational strategies that focus on self-esteem and encourage healthy behaviors instead of simply focusing on the dangers of eating disorders.
- Once an eating disorder has developed, it is important to seek treatment from a qualified team of professionals, which include mental health specialists who can help address and treat underlying psychological issues, medical doctors who can treat physiological complications, and a dietitian who can recommend healthy food choices.
- A person with an eating disorder may not recognize or be unable to admit that he has a problem. Concerns expressed by friends and family members often go ignored or dismissed, making loved ones feel confused and frustrated by their inability to help.
- Treatment goals for people with AN include achieving a healthy body weight, resolving psychological issues (such as low self-esteem and distorted body image), and establishing healthy eating patterns.
- Treatment goals for people with BN include the reduction and eventual elimination of bingeing and purging.
- The sooner a person with an eating disorder gets help, the better the chance of full recovery is. Regardless of when that happens, recovery can be a long, trying process.

Review 13

Chapter 13
Alcohol, Health, and Disease

Chapter Summary

LO1: What Is Alcohol and How Is It Produced?

- Although alcohol has a rich history of pageantry and ritual, adults who consume alcohol should do so safely and in moderation.
- The type of alcohol in beer, wine, and distilled liquor is called ethanol, which is produced through the process of fermentation.
- During fermentation, yeast produces ethanol by metabolizing sugar. To increase the alcohol concentration, fermented beverages can be distilled.
- Alcohol does not require digestion prior to its absorption in the small intestine and stomach.
- Once absorbed, alcohol circulates in the blood. Its concentration is expressed as blood alcohol concentration (BAC), which is affected by a person's body size and composition.
- Alcohol is a central nervous system depressant and acts as a sedative in the brain.
- Alcohol can cause a temporary loss of inhibition (disinhibition), making a person feel relaxed and more outgoing while at the same time impairing judgment and reasoning.

LO2: How Is Alcohol Metabolized?

- Most alcohol is metabolized through a two-step metabolic pathway.
- The first step in this metabolic pathway requires the enzyme alcohol dehydrogenase (ADH), which is found primarily in the liver. The second step in the metabolic process, catalyzed by the enzyme acetaldehyde dehydrogenase (ALDH), converts acetaldehyde to acetic acid.
- Genetic differences in ADH and ALDH activity can affect a person's ability to metabolize alcohol.
- Chronic alcohol consumption can activate a group of liver enzymes that assist in alcohol metabolism, reducing the amount of time a person remains intoxicated, and often leading to alcohol tolerance.
- If a heavy drinker develops a cross-tolerance to a drug, its actions in the body can decrease even though its concentration in the blood may reach dangerously high levels.

LO3: Does Alcohol Have Any Health Benefits?

- The 2010 Dietary Guidelines for Americans states that if alcohol is to be consumed, consumption should not exceed one drink per day for women or two drinks per day for men, and that alcohol should only be consumed by nonpregnant adults of legal drinking age.
- When consumed in moderation, alcohol is associated with decreased risks of cardiovascular disease, gallstones, age-related memory loss, and even type 2 diabetes.
- Adults who consume an average of one to two alcoholic drinks a day have a 30 to 35 percent lower risk of cardiovascular disease than adults who do not consume alcohol; these benefits disappear when alcohol intake becomes excessive.
- There is considerable evidence that the antioxidant resveratrol, found in the skin of red grapes, helps reduce inflammation and atherosclerosis.

Key Terms

acetaldehyde dehydrogenase (ALDH) An enzyme that converts acetaldehyde to acetic acid.

Al-Anon® An organization that provides support for family members of alcoholics.

Alateen® An organization that provides support for teenage children of alcoholics.

alcohol An organic compound that has one or more hydroxyl (–OH) groups attached to carbon atoms.

alcohol by volume (ABV) The percentage of ethanol in a given volume of liquid.

alcohol dehydrogenase (ADH) An enzyme found primarily in the liver that converts ethanol to acetaldehyde.

alcohol use disorders (AUDs) A spectrum of disorders that encompasses both recurrent excessive drinking and alcoholism.

alcoholic cardiomyopathy A serious condition whereby the heart muscle weakens in response to chronic alcohol consumption.

alcoholic hepatitis Inflammation of the liver caused by obstructed blood flow.

Alcoholics Anonymous® (AA) An organization that offers a 12-step program that provides fellowship and support for individuals who wish to achieve sobriety and stay sober.

binge drinking The consumption of four or more drinks for women and five or more drinks for men over a two-hour period.

blood alcohol concentration (BAC) A unit of measure for the amount of alcohol in the blood, measured in grams per deciliter (g/dl).

cardiac arrhythmia An irregular heartbeat caused by a high intake of alcohol.

cirrhosis A condition characterized by the presence of scar tissue in the liver.

disinhibition A loss of inhibition.

distillation The process by which alcohol vapors are condensed and collected to increase alcohol content.

ethanol The type of alcohol found in alcoholic beverages.

fatty liver A condition caused by chronic alcohol consumption characterized by the accumulation of triglycerides in the liver.

fermentation The process by which alcohol is produced. Yeast metabolizes sugars to produce ethanol and carbon dioxide.

pancreatitis A painful condition characterized by inflammation of the pancreas.

proof A measure of the alcohol content of distilled liquor. Twice a beverage's alcohol content.

resveratrol An antioxidant found in the skin of red grapes.

tolerance A response to chronic drug or alcohol exposure that results in its reduced effectiveness.

LO4: What Serious Health Risks Does Heavy Alcohol Consumption Pose?

- While moderate alcohol intake may provide health benefits for middle-aged adults, alcohol is clearly hazardous to one's health when consumed in excess.
- Habitual drinking can adversely affect nutritional status by causing both primary and secondary malnutrition.
- Long-term alcohol abuse can impair liver function, leading to fatty liver, alcoholic hepatitis, and cirrhosis.
- Heavy drinking increases a person's risk of developing certain types of cancer, especially those of the mouth, esophagus, colon, liver, and breast; alcohol may act as both a carcinogen and a co-carcinogen.
- Heavy drinking can cause alcoholic cardiomyopathy, high blood pressure, and cardiac arrhythmia.
- A chronically high alcohol intake (consuming at least seven drinks a day for more than five years) increases a person's risk of developing pancreatitis.

LO5: How Does Alcohol Abuse Contribute to Individual and Societal Problems?

- When alcohol is the cause of significant difficulties in a person's life, drinking is problematic.
- Organizations such as Alcoholics Anonymous®, Al-Anon®, and Alateen® provide support and fellowship for people suffering from the effects of alcohol abuse.
- Alcohol abuse, rampant on many college campuses, is associated with a wide range of risky behaviors and negative consequences such as vandalism, violence, acquaintance rape, unprotected sex, and death.
- Students who binge drink are more likely than those who do not to miss class, have lower academic rankings, experience trouble with campus law enforcement, and drive while intoxicated.
- A coalition of more than 100 college presidents from across the United States signed a petition asking lawmakers to examine whether the minimum legal drinking age should be lowered.
- Colleges and universities must work together with communities and students to create a culture that discourages high-risk drinking.

Chapter Summary

LO1: What Causes Foodborne Illness?

- You are exposed to thousands of microscopic organisms (microbes) every day. Microbes populate the world you live in, and many even serve useful purposes for you and for your health.
- Other microbes, however, are pathogenic (disease causing).
- Foodborne illnesses are sometimes referred to as food poisoning. You can contract a foodborne illness by ingesting an infectious agent such as a bacterium, virus, mold, fungus, or parasite.
- You can also contract a foodborne illness by ingesting a noninfectious agent such as a nonbacterial toxin; a chemical residue from processing, pesticides, or antibiotics; or a physical hazard such as glass or plastic.
- Some bacterial serotypes are harmless, whereas others cause disease.
- Because different pathogens have different incubation periods, health care providers can analyze and characterize different foodborne illnesses.
- Some pathogenic organisms produce toxic substances while they are growing in food. These include methicillin-resistant *S. aureus* (MRSA), *C. botulinum*, and *Aspergillus*.
- Some organisms produce harmful toxins after they enter the GI tract. These include noroviruses and enterotoxigenic *E. coli*.
- Some pathogens invade the cells of the intestine, seriously irritating the mucosal lining and causing fever, severe abdominal discomfort, and bloody diarrhea. These include *Salmonella* and *E. coli* O157:H7.
- A parasite is an organism that relies on another organism to survive. The consumption of parasite-infested foods can cause foodborne illness. An example of a parasite is a protozoan.
- Although prions are not living organisms, they may pose food safety concerns. Several diseases are known to be caused by prion ingestion.

LO2: What Noninfectious Substances Cause Foodborne Illness?

- Inert (noninfectious) compounds in foods can also cause foodborne illness. Noninfectious agents include physical contaminants (such as glass and plastic) and other dangerous substances such as toxins, heavy metals, and pesticides.
- Shellfish poisoning can result from the consumption of particular types of contaminated fish and shellfish (for example, clams and oysters). Eating marine toxin-contaminated foods can cause tingling, burning, numbness, drowsiness, and difficulty breathing.
- Because no food production system is foolproof, illness-causing pesticides, herbicides, antibiotics, and hormones sometimes come in contact with food products.
- Three foodborne substances that have garnered significant public health attention in recent years are acrylamide, melamine, and bisphenol A (BPA).

LO3: How Do Food Manufacturers Prevent Contamination?

- Disease-causing agents of all types can be transmitted from one food, surface, or utensil to another through cross-contamination.
- To prevent foodborne illness, it is critical that those who handle food do so safely and sanitarily.

Key Terms

acrylamide A substance used to make polyacrylamide that can form in starchy foods exposed to high temperatures.

aflatoxin A toxin produced by the *Aspergillus* mold that is found on some agricultural crops.

bisphenol A (BPA) A chemical found in some plastic food and beverage containers believed by some to pose a potential threat to health.

botulism The foodborne illness caused by *Clostridium botulinum*.

bovine somatotropin (bST) A growth hormone produced by cattle and used in the dairy industry to increase milk production.

bovine spongiform encephalopathy (BSE, or **mad cow disease)** A fatal disease caused by prion ingestion.

brevetoxin A potent toxin produced by red tide–causing algae that when consumed causes shellfish poisoning.

Creutzfeldt-Jakob disease A rare but fatal disease caused by a genetic mutation or exposure to by prions during surgery.

cross-contamination The process by which a disease-causing agent is transmitted from one food, surface, or utensil to another.

cyst A stage of growth in the life cycle of a protozoan; excreted in the feces.

danger zone The temperature range between 40 and 140 °F, in which most microorganisms prefer to live.

enteric toxin (or **intestinal toxin**) A toxic substance produced by an organism after it enters the gastrointestinal tract.

enterohemorrhagic Causing bloody diarrhea and intestinal inflammation.

enterotoxigenic *E. coli* A form of *E. coli* that produces enteric toxins.

FightBAC!® A public education program developed to reduce foodborne bacterial illness. The four major components that comprise FightBAC!® are clean, separate, cook, and chill.

food biosecurity Measures aimed at preventing the food supply from falling victim to planned contamination.

foodborne illness A disease caused by the ingestion of unsafe food.

incubation period The time elapsed between the consumption of a contaminated food and the emergence of sickness.

infectious agent (or **pathogen**) A living microorganism that can cause illness.

irradiation A food preservation process whereby a food is exposed to radiant energy that damages or destroys bacteria.

marine toxin A poison produced by ocean algae.

melamine A nitrogen-containing chemical typically used to make lightweight plastic objects.

methicillin-resistant *Staphylococcus aureus* (MRSA) An antibiotic-resistant strain of *S. aureus* that has received considerable attention from public health officials.

noninfectious agent An inert (nonliving) substance that can cause illness.

norovirus An infectious pathogen that produces enteric toxins and often causes foodborne illness.

parasite An organism that relies on another organism to survive.

pasteurization A food preservation process whereby food is partially sterilized through brief exposure to a high temperature.

preformed toxin A toxic substance that already exists in a food before the food is eaten.

prion An altered protein that forms when the secondary structure of a normal protein is disrupted.

protozoan A single-celled eukaryotic organism. Some protozoa are parasites.

Public Health Security and Bioterrorism Preparedness and Response Act (or **Bioterrorism Act**) Federal legislation aimed to ensure the continued safety of the U.S. food supply.

red tide A phenomenon whereby certain ocean algae grow profusely and produce brightly colored pigments that bloom outward and make the surrounding water appear red or brown.

serotype (or **strain**) A specific genetic variety of an organism.

shellfish poisoning A type of noninfectious foodborne illness caused by the consumption of particular types of contaminated fish and shellfish.

variant Creutzfeldt-Jakob disease A form of Creutzfeldt-Jakob disease that may be caused by consumption of BSE-contaminated foods.

- The food-processing industry adheres to additional regulations to keep food safe, but inspection and adherence to guidelines alone cannot guarantee that foods are pathogen free.
- Salting, drying, and smoking remove water from foods, inhibiting pathogenic growth; fermentation involves the addition of nonpathogenic organisms that inhibit the growth of dangerous ones.
- Manufactures, restaurants, and individuals alike should avoid the danger zone (40 to 140 °F) when storing and serving food.
- Methods that slow the growth of or kill pathogens include heating, pasteurization, freezing, and irradiation.

LO4: What Steps Can You Take to Reduce Foodborne Illness?

- There are many precautions you can take to reduce your risk of foodborne illness.
- Both the USDA and FDA maintain user-friendly websites and toll-free phone numbers that provide information about current food safety recommendations.
- The four major components that comprise FightBAC!® are clean, separate, cook, and chill.
- Making sure that your food is safe to eat can be especially difficult at a restaurant, picnic, potluck, or buffet. Food safety recommendations are even more important in these environments.

LO5: What Steps Can You Take to Reduce Foodborne Illness while Traveling?

- While it is not always possible to prevent foodborne illness while traveling, you can substantially lower your risk by being vigilant in your food and beverage choices.
- If you do experience traveler's diarrhea while abroad, it is important that you replace lost fluids and electrolytes as soon as symptoms begin to develop.
- Drink bottled water and avoid using ice. Before you drink water from a bottle, be sure that it has a fully sealed cap.
- Although fresh produce that is contaminated with bacteria may not cause illness for local residents, it can cause serious illness for visitors.
- Concerned travelers visiting regions where variant Creutzfeldt-Jakob disease has been reported might want to consider either avoiding beef and beef products or selecting only beef products composed of solid pieces of muscle meat.
- Food biosecurity and potential bioterrorism have gained significant national attention in recent years.
- Outside of willful acts of terrorism, changes in food production and distribution systems may influence food safety and the risk of foodborne illness.

Review 15

Chapter 15
Food Security, Hunger, and Malnutrition

Chapter Summary

LO1: What Is Food Security?

- Most people are able to ease their hunger, as they have access to sufficient amounts of nutritious food. Others, however, are not. When sufficient food is not available or accessible, hunger can lead to serious physical, social, and psychological consequences.
- The United Nations Food and Agriculture Organization (FAO) estimates that approximately 925 million people worldwide experience persistent hunger.
- Food insecurity exists when people do not have adequate physical, social, or economic access to food.
- Households classified as having low food security experience reduced food quality, variety, and/or desirability, whereas households classified as having very low food security are also likely to report disrupted eating patterns and reduced food intake.
- While the word *hunger* is often used to describe the physical discomfort experienced by individuals who have consumed insufficient amounts of food, it is more commonly used on a global level to describe a shortage of available food.
- Clinical measurements of nutritional status (such as anthropometry) are not always useful indicators of food insecurity. Instead, the prevalence of food insecurity in U.S. households is typically assessed using data regarding food availability and accessibility.
- Many individual and socioeconomic factors can predispose a person or family to food insecurity, but poverty is the most telling risk factor of all.

LO2: What Are the Consequences of Food Insecurity?

- Although food insecurity does not typically lead to starvation or nutrient deficiencies in the United States, it still represents a major public health concern.
- There are numerous consequences of food insecurity, many of which have been studied most extensively in women and children.
- Older adults are at especially high risk for food insecurity, though they often experience it differently than children, young adults, and other population groups.
- Food assistance programs that help at-risk individuals obtain food include the Supplemental Nutrition Assistance Program (SNAP); the Special Supplemental Nutrition Program for Women, Infants, and Children (WIC); the School Breakfast and National School Lunch Programs; food recovery programs; and community food banks and kitchens.
- Some programs are federally funded, whereas others are community efforts staffed by volunteers.

LO3: What Causes Worldwide Hunger and Malnutrition?

- Because poverty is more prevalent in developing countries than in industrialized ones, food insecurity tends to be most ubiquitous and severe in nations with low *per capita* incomes.
- There is sufficient food in the world to feed and nourish all its inhabitants.
- Worldwide food insecurity is usually caused by a combination of poverty, inadequate food distribution, political instability, urbanization, population growth, gender inequities, and other factors.
- Although there are many consequences of food insecurity in poor and developing countries, perhaps the most important and devastating consequence is malnutrition.

Key Terms

conditional cash-transfer program An initiative that offers financial reimbursement for individuals or communities that work to improve quality of life.

food bank An organization that collects donated foods and distributes them to local food pantries, shelters, and soup kitchens.

food desert An environment that lacks access to affordable and/or nutritious foods.

food insecurity A condition whereby a person does not have adequate physical, social, or economic access to food.

food kitchen A program that prepares and serves meals to members of the community who are in need.

food pantry A program that provides canned, boxed, and sometimes fresh foods directly to individuals in need.

food recovery program A program that collects and redistributes discarded food that otherwise would have gone to waste.

food security A condition whereby a person is able to access sufficient amounts of nutritious food.

low food security A condition characterized by reduced food quality, variety, and/or desirability, but not reduced food intake.

National School Lunch Program A federally funded program that provides nutritionally balanced meals either free of charge or at a reduced cost to school-age children at lunch time.

School Breakfast Program A federally-funded program that provides nutritionally balanced meals either free of charge or at a reduced cost to school-age children at breakfast time.

Supplemental Nutrition Assistance Program (SNAP) A federally funded program, formerly known as the Food Stamp Program, that helps low-income households pay for food.

urbanization A shift in a country's population from rural regions to urban ones.

U.S. Peace Corps A federally funded volunteer program that promotes world peace and friendship worldwide.

very low food security A condition characterized by disrupted eating patterns and reduced food intake caused by a lack of food access and availability.

- Consequences of food insecurity in poorer regions of the world include greater incidence of low birth weight, neonatal death, inhibited growth, vitamin A deficiency, iron deficiency, and iodine deficiency.
- Beyond its effects on the health of the individual, malnutrition can harm whole societies.
- Though the United Nations serves many purposes, combating international hunger is among its most important efforts.
- The United Nations Children's Fund (UNICEF) helps combat international hunger by working toward a common goal of improving the quality of life for people in the world's poorest countries.
- Conditional cash-transfer programs may be one way to encourage sound nutritional and educational behaviors and simultaneously alleviate poverty.
- The U.S. Peace Corps offers opportunities to make a difference in the lives of others, helping address the problems of food insecurity and malnutrition in many parts of the world.
- Heifer International® is a humanitarian effort with a global commitment to foster environmentally sound farming methods that combat both hunger and environmental concerns.
- Heifer's living loans program has fostered a living cycle of sustainability for over 65 years.

LO4: What Can You Do to Alleviate Food Insecurity?

- It is only by considering the complexity of food insecurity that the relative importance of each contributing factor can be addressed. It is only then that effective solutions can be developed.
- Leading world health experts agree that improving food availability and access must be a global priority.
- Although malnutrition is a direct consequence of insufficient dietary intake, the ultimate causes of malnutrition often have more to do with economic and societal circumstances.
- Although the problem of food insecurity may seem staggering both at home and abroad, it is important to remember that there is much an individual can do to take action against hunger.
- Working alone or collectively toward the elimination of hunger and malnutrition is a worthwhile and noble personal and/or professional priority.
- The American Dietetic Association (ADA) and American Society for Nutrition (ASN) are two of many professional organizations that work to alleviate world hunger by challenging their members to take action.

Diet Analysis Plus Quick Start Guide

Logging in to *Diet Analysis Plus*

1. Go to the CourseMate for NUTR at **www.cengagebrain.com**.
2. If you have logged in to *Diet Analysis Plus* before, or if you have a CengageBrain account, enter your user name and password and click the **Log In** button to open the **My Home** page. If you have not logged in previously, skip to step 4.
3. If you have logged in before, simply click the **Open** button under *Diet Analysis Plus* on the My Home page to log in to the program. If this is your first time, enter the *Diet Analysis Plus* access code from your access card under **Register another Access Code** and click the **Register** button. Skip steps 4–8.
4. ***CengageBrain Registration*:** Click the **Create an Account** button.
5. Enter the *Diet Analysis Plus* access code from your access card under **Enter Code** and click the **Continue** button.
6. On the **Account Information** page, enter your personal information and preferences, accept the license agreement, and click the **Continue** button.
7. On the **Select Institution** page, use the menus and **Search** button to select the name of your school, and then click the **Continue** button.
8. Click the **Open** button under *Diet Analysis Plus* on the **My Home** page to log in to the program.

Creating a Profile

1. The first time you log in to *Diet Analysis Plus*, you will see the **Create Profile** page. Enter the requested information for your primary profile, and click the **Next** button.

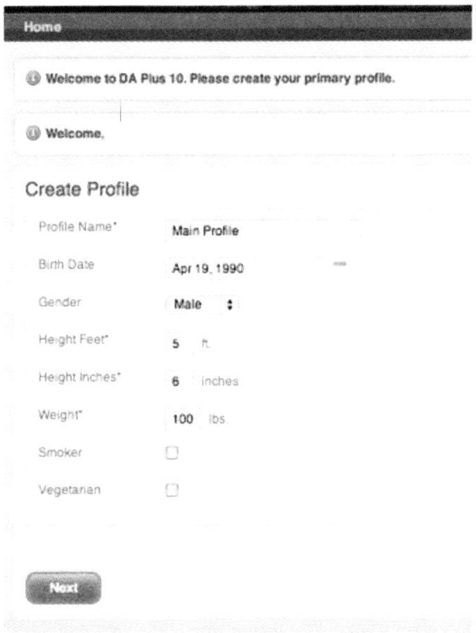

2. You will see the **Activity Questionnaire** page. Answer the questions on this page and click the **Next** button.

3. You will see the **Confirm Profile** page. Review your information. You can click the **Edit** button under your personal or activity information if you need to make corrections. Click the **Save** button.
4. To create an alternate profile, click the **Home** tab, then click the **Create New Profile** link beneath My Profile. Repeat steps 1–3.

Entering Your Course ID

1. Click the **Home** tab.
2. Under My Profile, locate the **Course Identification Number** box and enter the Course Identification Number provided to you by your instructor.
3. Click the **Submit** link. The name of the course for which you are using *Diet Analysis Plus* should now appear beside **Current Course**.

Tracking Your Diet

1. Click the **Track Diet** tab.
2. Click the calendar to the right of the "Select a Date" label and choose the date for which you would like to enter or edit the foods you ate.
3. Next to the "Find Foods" label, enter the name of a food/beverage you ate and click the **Go** button. A food list will appear (with arrows and page numbers if there are more than 15 choices). You may use the Category Filter to narrow your search to a particular group of foods.

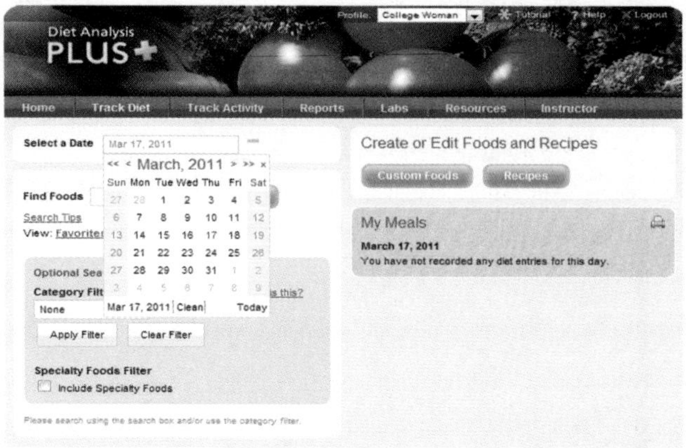

4. Click on the food name closest to what you ate/drank to open the **Serving Size** pop-up screen.
5. Enter the amount that you ate/drank, being careful to choose the correct unit of measurement from the first menu. Select the correct meal or snack from the second menu. Click the **Save** button. Repeat steps 3–5 for each food or drink you consumed that day.

Tracking Your Activity

1. Click the **Track Activity** tab.
2. Click the calendar to the right of the "Select a Date" label and then choose the date for which you would like to enter or edit your activities.
3. Next to the "Find Activities" label, enter the name of an activity you performed and click the **Go** button. An activities list will appear (with arrows and page numbers if there are more than 15 choices).

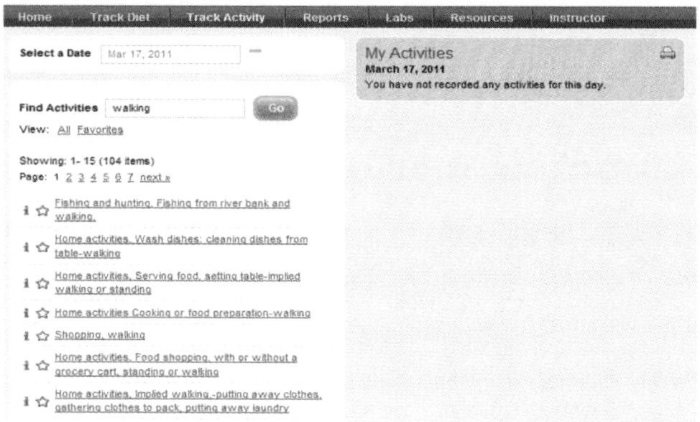

4. Click on the name of the activity that most closely matches what you did to open the **Activity Duration** pop-up screen.

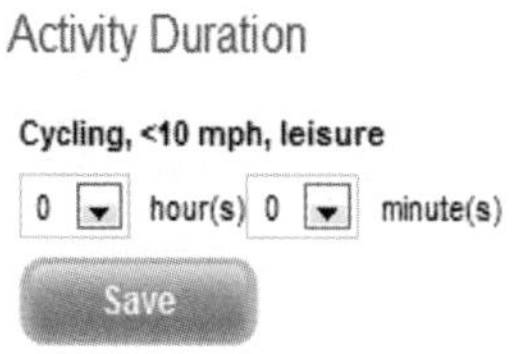

5. Use the menus to enter the amount of time you performed this activity. Click the **Save** button. Repeat steps 3–5 for each activity you completed that day. You must enter activities for all 24 hours of the day (including time spent sleeping, getting dressed, and driving).

Viewing and Printing Reports

1. Click the **Reports** tab.
2. Click on the name of the report you would like to create.
3. Most reports have options for you to adjust at the top of the page. Click the calendars under **Select a Date/Date Range** or **Day 1/2/3** to choose the date(s) you would like to analyze. Under **Choose Meals**, you can uncheck the boxes if you want to exclude meals or snacks from the report. (For the *3 Day Average* report, you will also need to click the **Preview Report** button to view the report.)
4. To generate a version of the report that you can print or e-mail, click the **Print PDF** button to create a PDF of the report, or **Print RTF** to create a Word document.
5. To submit the report to your instructor so that he/she can view it online, click the **Submit Report** button.

Requesting Technical Support

1. Go to **www.cengage.com/support**.
2. You can either click the **Select Product** button under "Self-help" and then choose your version of *Diet Analysis Plus* to access instructions and tutorial videos, or sign in to submit a request for assistance online (click on **create a new case**).
3. Alternatively, you can call technical support at 1-800-354-9706 (option 5, then option 2). Support is available Monday through Thursday, 8:30 a.m.–9:00 p.m. EST, and Friday 8:30 a.m.–6:00 p.m EST.

Dietary Reference Intakes (DRIs)

The Dietary Reference Intakes (DRIs) include four dietary reference standards used to assess and plan dietary intake: Estimated Average Requirements (EARs), Recommended Dietary Allowances (RDAs), Adequate Intake (AI) levels, and Tolerable Upper Intake Levels (ULs). Acceptable Macronutrient Distribution Ranges (AMDRs) and Estimated Energy Requirements (EERs) are also used to calculate and maintain ideal energy and macronutrient intakes. Use the tables below to assess the adequacy of your own intake levels.

Estimated Energy Requirements (EERs), Recommended Dietary Allowances (RDAs), and Adequate Intake (AI) Levels for Energy and the Macronutrients

Age (yr)	Reference BMI (kg/m²)	Reference height, cm (in)	Reference weight, kg (lb)	Water[a] AI (L/day)	Energy EER[b] (kcal/day)	Carbohydrate RDA (g/day)	Total fiber AI (g/day)	Total fat AI (g/day)	Linoleic acid AI (g/day)	Linolenic acid[c] AI (g/day)	Protein RDA (g/day)[d]	Protein RDA (g/kg/day)
Males												
0–0.5	—	62 (24)	6 (13)	0.7[e]	570	60	—	31	4.4	0.5	9.1	1.52
0.5–1	—	71 (28)	9 (20)	0.8[f]	743	95	—	30	4.6	0.5	11	1.2
1–3[g]	—	86 (34)	12 (27)	1.3	1,046	130	19	—	7	0.7	13	1.05
4–8[g]	15.3	115 (45)	20 (44)	1.7	1,742	130	25	—	10	0.9	19	0.95
9–13	17.2	144 (57)	36 (79)	2.4	2,279	130	31	—	12	1.2	34	0.95
14–18	20.5	174 (68)	61 (134)	3.3	3,152	130	38	—	16	1.6	52	0.85
19–30	22.5	177 (70)	70 (154)	3.7	3,067[h]	130	38	—	17	1.6	56	0.8
31–50	22.5	177 (70)	70 (154)	3.7	3,067[h]	130	38	—	17	1.6	56	0.8
>50	22.5	177 (70)	70 (154)	3.7	3,067[h]	130	30	—	14	1.6	56	0.8
Females												
0–0.5	—	62 (24)	6 (13)	0.7[e]	520	60	—	31	4.4	0.5	9.1	1.52
0.5–1	—	71 (28)	9 (20)	0.8[f]	676	95	—	30	4.6	0.5	11	1.2
1–3[g]	—	86 (34)	12 (27)	1.3	992	130	19	—	7	0.7	13	1.05
4–8[g]	15.3	115 (45)	20 (44)	1.7	1,642	130	25	—	10	0.9	19	0.95
9–13	17.4	144 (57)	37 (81)	2.1	2,071	130	26	—	10	1.0	34	0.95
14–18	20.4	163 (64)	54 (119)	2.3	2,368	130	26	—	11	1.1	46	0.85
19–30	21.5	163 (64)	57 (126)	2.7	2,403[i]	130	25	—	12	1.1	46	0.8
31–50	21.5	163 (64)	57 (126)	2.7	2,403[i]	130	25	—	12	1.1	46	0.8
>50	21.5	163 (64)	57 (126)	2.7	2,403[i]	130	21	—	11	1.1	46	0.8
Pregnancy												
1st trimester				3.0	+0	175	28	—	13	1.4	+25	1.1
2nd trimester				3.0	+340	175	28	—	13	1.4	+25	1.1
3rd trimester				3.0	+452	175	28	—	13	1.4	+25	1.1
Lactation												
1st 6 months				3.8	+330	210	29	—	13	1.3	+25	1.3
2nd 6 months				3.8	+400	210	29	—	13	1.3	+25	1.3

Note: For all nutrients, values set for infants are AIs. A dash (—) indicates that a value has not been established.

[a]The water AI includes drinking water, water in beverages, and water in foods. In general, drinking water and other beverages contribute 70 to 80 percent, while foods contribute the remainder.

[b]An EER represents the average dietary energy intake that will maintain energy balance in a healthy person of a given sex, age, weight, height, and physical activity level. Values listed are based on an active person at the reference height and weight and at the midpoint ages for each group until age 19. The values listed for pregnancy and lactation represent energy needed in addition to nonpregnant/nonlactating values.

[c]Linolenic acid referred to in this table and text is α-linolenic acid, an essential omega-3 fatty acid.

[d]Values listed are based on reference body weights.

[e]Assumed to be from human milk.

[f]Assumed to be from human milk and complementary foods and beverages. This includes approximately 0.6 L (3 cups) as total fluid including formula, juices, and drinking water.

[g]For energy, the age groups for young children are 1–2 years and 3–8 years.

[h]For males, subtract 10 kilocalories per day for each year of age above 19.

[i]For females, subtract 7 kilocalories per day for each year of age above 19.

Source: Adapted from the Dietary Reference Intakes series, National Academies Press. Copyright 1997, 1998, 2000, 2001, 2002, 2004, 2005 by the National Academy of Sciences.

Recommended Dietary Allowances (RDAs) and Adequate Intake (AI) Levels for Vitamins

Age (yr)	Thiamin RDA (mg/day)	Riboflavin RDA (mg/day)	Niacin RDA (mg/day)[a]	Biotin AI (μg/day)	Pantothenic acid AI (mg/day)	Vitamin B_6 RDA (mg/day)	Folate RDA (μg/day)[b]	Vitamin B_{12} RDA (μg/day)	Choline AI (mg/day)	Vitamin C RDA (mg/day)	Vitamin A RDA (μg/day)[c]	Vitamin D RDA (μg/day)[d]	Vitamin E RDA (mg/day)[e]	Vitamin K AI (μg/day)
Infants														
0–0.5	0.2	0.3	2	5	1.7	0.1	65	0.4	125	40	400	10	4	2.0
0.5–1	0.3	0.4	4	6	1.8	0.3	80	0.5	150	50	500	10	5	2.5
Children														
1–3	0.5	0.5	6	8	2	0.5	150	0.9	200	15	300	15	6	30
4–8	0.6	0.6	8	12	3	0.6	200	1.2	250	25	400	15	7	55
Males														
9–13	0.9	0.9	12	20	4	1.0	300	1.8	375	45	600	15	11	60
14–18	1.2	1.3	16	25	5	1.3	400	2.4	550	75	900	15	15	75
19–30	1.2	1.3	16	30	5	1.3	400	2.4	550	90	900	15	15	120
31–50	1.2	1.3	16	30	5	1.3	400	2.4	550	90	900	15	15	120
51–70	1.2	1.3	16	30	5	1.7	400	2.4	550	90	900	15	15	120
>70	1.2	1.3	16	30	5	1.7	400	2.4	550	90	900	20	15	120
Females														
9–13	0.9	0.9	12	20	4	1.0	300	1.8	375	45	600	15	11	60
14–18	1.0	1.0	14	25	5	1.2	400	2.4	400	65	700	15	15	75
19–30	1.1	1.1	14	30	5	1.3	400	2.4	425	75	700	15	15	90
31–50	1.1	1.1	14	30	5	1.3	400	2.4	425	75	700	15	15	90
51–70	1.1	1.1	14	30	5	1.5	400	2.4	425	75	700	15	15	90
>70	1.1	1.1	14	30	5	1.5	400	2.4	425	75	700	20	15	90
Pregnancy														
≤18	1.4	1.4	18	30	6	1.9	600	2.6	450	80	750	15	15	75
19–30	1.4	1.4	18	30	6	1.9	600	2.6	450	85	770	15	15	90
31–50	1.4	1.4	18	30	6	1.9	600	2.6	450	85	770	15	15	90
Lactation														
≤18	1.4	1.6	17	35	7	2.0	500	2.8	550	115	1,200	15	19	75
19–30	1.4	1.6	17	35	7	2.0	500	2.8	550	120	1,300	15	19	90
31–50	1.4	1.6	17	35	7	2.0	500	2.8	550	120	1,300	15	19	90

Note: For all nutrients, values for infants are AIs.
[a]Niacin recommendations are expressed as niacin equivalents (NEs), except for recommendations for infants younger than 6 months, which are expressed as preformed niacin.
[b]Folate recommendations are expressed as dietary folate equivalents (DFEs).
[c]Vitamin A recommendations are expressed as retinol activity equivalents (RAEs).
[d]Vitamin D recommendations are expressed as cholecalciferol.
[e]Vitamin E recommendations are expressed as α-tocopherol.

Recommended Dietary Allowances (RDAs) and Adequate Intake (AI) Levels for Minerals

Age (yr)	Sodium AI (mg/day)	Chloride AI (mg/day)	Potassium AI (mg/day)	Calcium RDA (mg/day)	Phosphorus RDA (mg/day)	Magnesium RDA (mg/day)	Iron RDA (mg/day)	Zinc RDA (mg/day)	Iodine RDA (μg/day)	Selenium RDA (μg/day)	Copper RDA (μg/day)	Manganese AI (mg/day)	Fluoride AI (mg/day)	Chromium AI (μg/day)	Molybdenum RDA (μg/day)
Infants															
0–0.5	120	180	400	200	100	30	0.27	2	110	15	200	0.003	0.01	0.2	2
0.5–1	370	570	700	260	275	75	11	3	130	20	220	0.6	0.5	5.5	3
Children															
1–3	1,000	1,500	3,000	700	460	80	7	3	90	20	340	1.2	0.7	11	17
4–8	1,200	1,900	3,800	1,000	500	130	10	5	90	30	440	1.5	1	15	22
Males															
9–13	1,500	2,300	4,500	1,300	1,250	240	8	8	120	40	700	1.9	2	25	34
14–18	1,500	2,300	4,700	1,300	1,250	410	11	11	150	55	890	2.2	3	35	43
19–30	1,500	2,300	4,700	1,000	700	400	8	11	150	55	900	2.3	4	35	45
31–50	1,500	2,300	4,700	1,000	700	420	8	11	150	55	900	2.3	4	35	45
51–70	1,300	2,000	4,700	1,000	700	420	8	11	150	55	900	2.3	4	30	45
>70	1,200	1,800	4,700	1,200	700	420	8	11	150	55	900	2.3	4	30	45

Dietary Reference Intakes (DRIs), Continued

Recommended Dietary Allowances (RDAs) and Adequate Intake (AI) Levels for Minerals

Age (yr)	Sodium AI (mg/day)	Chloride AI (mg/day)	Potassium AI (mg/day)	Calcium RDA (mg/day)	Phosphorus RDA (mg/day)	Magnesium RDA (mg/day)	Iron RDA (mg/day)	Zinc RDA (mg/day)	Iodine RDA (µg/day)	Selenium RDA (µg/day)	Copper RDA (µg/day)	Manganese AI (mg/day)	Fluoride AI (mg/day)	Chromium AI (µg/day)	Molybdenum RDA (µg/day)
Females															
9–13	1,500	2,300	4,500	1,300	1,250	240	8	8	120	40	700	1.6	2	21	34
14–18	1,500	2,300	4,700	1,300	1,250	360	15	9	150	55	890	1.6	3	24	43
19–30	1,500	2,300	4,700	1,000	700	310	18	8	150	55	900	1.8	3	25	45
31–50	1,500	2,300	4,700	1,000	700	320	18	8	150	55	900	1.8	3	25	45
51–70	1,300	2,000	4,700	1,200	700	320	8	8	150	55	900	1.8	3	20	45
>70	1,200	1,800	4,700	1,200	700	320	8	8	150	55	900	1.8	3	20	45
Pregnancy															
≤18	1,500	2,300	4,700	1,300	1,250	400	27	12	220	60	1,000	2.0	3	29	50
19–30	1,500	2,300	4,700	1,000	700	350	27	11	220	60	1,000	2.0	3	30	50
31–50	1,500	2,300	4,700	1,000	700	360	27	11	220	60	1,000	2.0	3	30	50
Lactation															
≤18	1,500	2,300	5,100	1,300	1,250	360	10	13	290	70	1,300	2.6	3	44	50
19–30	1,500	2,300	5,100	1,000	700	310	9	12	290	70	1,300	2.6	3	45	50
31–50	1,500	2,300	5,100	1,000	700	320	9	12	290	70	1,300	2.6	3	45	50

Tolerable Upper Intake Levels (ULs) for Vitamins and Minerals

Age (yr)	Niacin (mg/day)[a]	Vitamin B_6 (mg/day)	Folate (µg/day)[a]	Choline (mg/day)	Vitamin C (mg/day)	Vitamin A (µg/day)[b]	Vitamin D (µg/day)	Vitamin E (mg/day)[c]	Sodium (mg/day)	Chloride (mg/day)	Calcium (mg/day)	Phosphorus (mg/day)
Infants												
0–0.5	—	—	—	—	—	600	25	—	—[e]	—[e]	1,000	—
0.5–1	—	—	—	—	—	600	38	—	—[e]	—[e]	1,500	—
Children												
1–3	10	30	300	1,000	400	600	63	200	1,500	2,300	2,500	3,000
4–8	15	40	400	1,000	650	900	75	300	1,900	2,900	2,500	3,000
Adolescents												
9–13	20	60	600	2,000	1,200	1,700	100	600	2,200	3,400	3,000	4,000
14–18	30	80	800	3,000	1,800	2,800	100	800	2,300	3,600	3,000	4,000
Adults												
19–70	35	100	1,000	3,500	2,000	3,000	100	1,000	2,300	3,600	2,000–2,500	4,000
>70	35	100	1,000	3,500	2,000	3,000	100	1,000	2,300	3,600	2,000	3,000
Pregnancy												
≤18	30	80	800	3,000	1,800	2,800	100	800	2,300	3,600	3,000	3,500
19–50	35	100	1,000	3,500	2,000	3,000	100	1,000	2,300	3,600	2,500	3,500
Lactation												
≤18	30	80	800	3,000	1,800	2,800	100	800	2,300	3,600	3,000	4,000
19–50	35	100	1,000	3,500	2,000	3,000	100	1,000	2,300	3,600	2,500	4,000

Tolerable Upper Intake Levels (ULs) for Vitamins and Minerals

Age (yr)	Magnesium (mg/day)[d]	Iron (mg/day)[b]	Zinc (mg/day)	Iodine (µg/day)	Selenium (µg/day)	Copper (µg/day)	Manganese (mg/day)	Fluoride (mg/day)	Molybdenum (µg/day)	Boron (mg/day)	Nickel (mg/day)
Infants											
0–0.5	—	40	4	—	45	—	—	0.7	—	—	—
0.5–1	—	40	5	—	60	—	—	0.9	—	—	—
Children											
1–3	65	40	7	200	90	1,000	2	1.3	300	3	0.2
4–8	110	40	12	300	150	3,000	3	2.2	600	6	0.3
Adolescents											
9–13	350	40	23	600	280	5,000	6	10	1,100	11	0.6
14–18	350	45	34	900	400	8,000	9	10	1,700	17	1.0
Adults											
19–70	350	45	40	1,100	400	10,000	11	10	2,000	20	1.0
>70	350	45	40	1,100	400	10,000	11	10	2,000	20	1.0
Pregnancy											
≤18	350	45	34	900	400	8,000	9	10	1,700	17	1.0
19–50	350	45	40	1,100	400	10,000	11	10	2,000	20	1.0
Lactation											
≤18	350	45	34	900	400	8,000	9	10	1,700	17	1.0
19–50	350	45	40	1,100	400	10,000	11	10	2,000	20	1.0

NOTE: ULs were not established for vitamins and minerals not listed and for those age groups listed with a dash (—) because of a lack of data, not because these nutrients are safe to consume at any level of intake. All nutrients can have adverse effects when intakes are excessive.

[a]ULs for niacin and folate apply to synthetic forms obtained from supplements, fortified foods, or a combination of the two.

[b]The UL for vitamin A applies to preformed vitamin A only.

[c]The UL for vitamin E applies to any form of supplemental α-tocopherol, fortified foods, or a combination of the two.

[d]The UL for magnesium applies to synthetic forms obtained from supplements or drugs only.

[e]Source of intake should be from human milk (or formula) and food only.

Source: Adapted from the Dietary Reference Intakes series, National Academies Press. Copyright 1997, 1998, 2000, 2001, 2002, 2004, 2005, and 2011 by the National Academy of Sciences.

Acceptable Macronutrient Distribution Ranges (AMDRs)

	Range (percent of energy)		
Macronutrient	Children, 1–3 years	Children, 4–18 years	Adults, 19+ years
Fat	30–40	25–35	20–35
ω-6 polyunsaturated acids[a] (linoleic acid)	5–10	5–10	5–10
ω-3 polyunsaturated fatty acids[a] (linolenic acid)	0.6–1.2	0.6–1.2	0.6–1.2
Carbohydrate	45–65	45–65	45–65
Protein	5–20	10–30	10–35

[a]Approximately 10 percent of the total can come from long-chain ω-3 or ω-6 fatty acids.
Source: Adapted from Institute of Medicine. Dietary reference intakes for energy, carbohydrate, fiber, fat, fatty acids, cholesterol, protein, and amino acids (macronutrients). Washington, DC: National Academies Press, 2005.

Tips for Achieving MyPlate Objectives

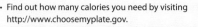

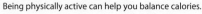

Balance Calories
- Find out how many calories you need by visiting http://www.choosemyplate.gov.
- Being physically active can help you balance calories.

Enjoy Your Food, But Eat Less
- Take the time to fully enjoy your food as you eat it. Eating too fast or when your attention is elsewhere may lead to eating too many calories.
- Pay attention to hunger and fullness cues before, during, and after eating.

Avoid Oversized Portions
- Use a smaller plate, bowl, and glass.
- Portion out foods before you eat them.
- When eating out, choose a smaller size option, share a dish, or take home part of your meal.

Make Half Your Plate Fruits and Vegetables
- Choose red, orange, and dark green vegetables like tomatoes, sweet potatoes, and broccoli, along with other vegetables.
- Add fruit to your meal as part of the main dish or as a dessert.

Switch to Fat-Free or Low-Fat Milk
- Fat-free and low-fat (1%) milk products have the same amount of calcium and other essential nutrients as whole milk, but have fewer calories and less saturated fat.

Make Half Your Grains Whole Grains
- To eat more whole grains, substitute a whole-grain product for a refined product. For example, eat whole-wheat bread instead of white bread and brown rice instead of white rice.

Foods to Eat Less Often
- Cut back on foods high in solid fats, added sugars, and salt. These include cakes, cookies, ice cream, candies, sweetened drinks, pizza, and fatty meats. Consume these foods as occasional treats, not everyday meals.

Compare Sodium in Foods
- Use the Nutrition Facts label to choose lower-sodium versions of foods like soup, bread, and frozen meals.
- Select canned foods labeled "low sodium," "reduced sodium," or "no salt added."

Drink Water Instead of Sugary Drinks
- Cut calories by drinking water and other unsweetened beverages.

Source: Adapted from U.S. Department of Agriculture, Center for Nutrition Policy and Promotion. DG TipSheet No. 1. June 2011. Available from: http://www.choosemyplate.gov/downloads/TenTips/DGTipsheet1ChooseMyPlate.pdf.

Estimated Energy Requirement (EER) Calculations and Physical Activity (PA) Values

An Estimated Energy Requirement (EER) value represents the average energy intake needed for a healthy person to maintain weight. EER values vary by age, sex, weight, height, and physical activity (PA) level. Note that this is different from the other DRI reference values, which only vary by life stage and sex. EERs are calculated using relatively simple mathematical equations. Using the EER equations for adult men and women of healthy weight below, you can calculate your own EER in kilocalories per day (kcal/day). The lower your PA value, the less active you are, and consequently the lower your EER. Note that at every age, active individuals need more energy than do their sedentary counterparts.

Estimated Energy Requirement (EER) Calculations

Age Group	Equations for Estimated Energy Requirement (EER; kcal/d)[a]
0–3 months	[89 × weight (kg) – 100] + 175 kcal
4–6 months	[89 × weight (kg) – 100] + 56 kcal
7–12 months	[89 × weight (kg) – 100] + 22 kcal
13–36 months	[89 × weight (kg) – 100] + 20 kcal
3–8 years (male)	88.5 – [61.9 × age (y)] + PA × [26.7 × weight (kg) + 903 × height (m)] + 20 kcal
3–8 years (female)	135.3 – [30.8 × age (y)] + PA × [10.0 × weight (kg) + 934 × height (m)] + 20 kcal
9–18 years (male)	88.5 – [61.9 × age (y)] + PA × [26.7 × weight (kg) + 903 × height (m)] + 25 kcal
9–18 years (female)	135.3 – [30.8 × age (y)] + PA × [10.0 × weight (kg) + 934 × height (m)] + 25 kcal
19+ years (male)	662 – [9.53 × age (y)] + PA × [15.91 × weight (kg) + 539.6 × height (m)]
19+ years (female)	354 – [6.91 × age (y)] + PA × [9.36 × weight (kg) + 726 × height (m)]
Pregnancy	
14–18 years	
1st trimester	Adolescent EER + 0
2nd trimester	Adolescent EER + 340 kcal
3rd trimester	Adolescent EER + 452 kcal
19–50 years	
1st trimester	Adult EER + 0
2nd trimester	Adult EER + 340 kcal
3rd trimester	Adult EER + 452 kcal
Lactation	
4–18 years	
1st 6 months postpartum	Adolescent EER + 330 kcal
2nd 6 months postpartum	Adolescent EER + 400 kcal
19–50 years	
1st 6 months postpartum	Adult EER + 330 kcal
2nd 6 months postpartum	Adult EER + 400 kcal
Overweight or Obese[b]	
3–18 years (male)	114 – [50.9 × age (y)] + PA × [19.5 × weight (kg) + 1,161.4 × height (m)]
3–18 years (female)	389 – [41.2 × age (y)] + PA × [15.0 × weight (kg) + 701.6 × height (m)]
19+ years (male)	1,086 – [10.1 × age (y)] + PA × [13.7 × weight (kg) + 416 × height (m)]
19+ years (female)	448 – [7.95 × age (y)] + PA × [11.4 × weight (kg) + 619 × height (m)]

[a] "PA" stands for the physical activity value appropriate for the age and physiological state. These can be found in the next table.

[b] Body mass index (BMI) ≥ 25 kg/m^2; values represent estimated total energy expenditure (TEE; kcal/d) for weight maintenance; weight loss can be achieved by a reduction in energy intake and/or an increase in energy expenditure.

Physical Activity (PA) Values

Age Group (sex)	PA Level[a]	PA Value	Age Group (sex)	PA Level[a]	PA Value
3–8 years (male)	Sedentary	1.00	3–8 years (female)	Sedentary	1.00
	Low active	1.13		Low active	1.16
	Active	1.26		Active	1.31
	Very active	1.42		Very active	1.56
3–18 years (overweight male)	Sedentary	1.00	3–18 years (overweight female)	Sedentary	1.00
	Low active	1.12		Low active	1.18
	Active	1.24		Active	1.35
	Very active	1.45		Very active	1.60
9–18 years (male)	Sedentary	1.00	9–18 years (female)	Sedentary	1.00
	Low active	1.13		Low active	1.16
	Active	1.26		Active	1.31
	Very active	1.42		Very active	1.56
19+ years (male)	Sedentary	1.00	19+ years (female)	Sedentary	1.00
	Low active	1.11		Low active	1.12
	Active	1.25		Active	1.27
	Very active	1.48		Very active	1.45
19+ years (overweight/obese male)	Sedentary	1.00	19+ years (overweight/obese female)	Sedentary	1.00
	Low active	1.12		Low active	1.16
	Active	1.39		Active	1.27
	Very active	1.59		Very active	1.44

[a]*Sedentary* activity level is characterized by no physical activity aside from that needed for independent living. *Low active* level is characterized by walking 1.5–3 miles/day at 2–4 mph (or equivalent) in addition to the light activity associated with typical day-to-day life. People who are *active* walk 3–10 miles/day at 2–4 mph (or equivalent) in addition to the light activity associated with typical day-to-day life. *Very active* individuals walk 10 or more miles/day at 2–4 mph (or equivalent) in addition to the light activity associated with typical day-to-day life.

[b]Body mass index (BMI) ≥ 25 kg/m².

Source: Institute of Medicine. Dietary Reference Intakes for energy, carbohydrate, fiber, fat, fatty acids, cholesterol, protein, and amino acids (macronutrients). Washington, DC: National Academies Press, 2005.

Commonly Used Weights, Measures, and Metric Conversion Factors

Length
1 meter (m) = 39 in, 3.28 ft, or 100 cm.
1 centimeter (cm) = 0.39 in, 0.032 ft, or 0.01 m.
1 inch (in) = 2.54 cm, 0.083 ft, or 0.025 m.
1 foot (ft) = 30 cm, 0.30 m, or 12 in.

Temperature

Celsius[a]		Fahrenheit	
Boiling point	100°C	212°F	Boiling point
Body temperature	37°C	98.6°F	Body temperature
Melting point	0°C	32°F	Melting point

- To find degrees Fahrenheit (°F) when you know degrees Celsius (°C), multiply by 1.8 and then add 32.
- To find degrees Celsius (°C) when you know degrees Fahrenheit (°F), subtract 32 and then multiply by 0.56.

Volume
1 liter (l) = 1,000 ml, 0.26 gal, 1.06 qt, 2.11 pt, or 34 oz.
1 milliliter (ml) = 1/1,000 l or 0.03 fluid oz.

Volume (continued)
1 gallon (gal) = 128 oz, 16 c, 3.78 l, 4 qt, or 8 pt.
1 quart (qt) = 32 oz, 4 c, 0.95 l, or 2 pt.
1 pint (pt) = 16 oz, 2 c, 0.47 l, or 0.5 qt.
1 cup (c) = 8 oz, 16 tbsp, 237 ml, or 0.24 l.
1 ounce (oz) = 30 ml, 2 tbsp, or 6 tsp.
1 tablespoon (tbsp) = 3 tsp, 15 ml, or 0.5 oz.
1 teaspoon (tsp) = 5 ml or 0.17 oz.

Weight
1 kilogram (kg) = 1,000 g, 2.2 lb, or 35 oz.
1 gram (g) = 1/1,000 kg, 1,000 mg, or 0.035 oz.
1 milligram (mg) = 1/1,000 g or 1,000 μg.
1 microgram (μg) = 1/1,000 mg.
1 pound (lb) = 16 oz, 454 g, or 0.45 kg.
1 ounce (oz) = 28 g or 0.062 lb.

Energy
1 kilojoule (kJ) = 0.24 kcal, 240 calories, or 0.24 Calories.
1 kilocalorie (kcal) = 4.18 kJ, 1,000 calories, or 1 Calorie.

[a]Also known as *centigrade*.

Summary of the 2010 Dietary Guidelines for Americans

Put forth by the U.S. Department of Agriculture and the U.S. Department of Health and Human Services, the 2010 Dietary Guidelines for Americans encompasses two overarching goals: (1) to help individuals maintain energy balance over time and (2) to help Americans choose nutrient-dense foods and beverages. Underlying these broad goals is the proposition that nutrient needs should be met primarily through the consumption of foods, not supplements. The current version of the Dietary Guidelines (its 7th iteration) goes a step further than previous editions by acknowledging that "everyone has a role in the movement to make American healthy." By working together to enact policies, programs, and partnerships that strengthen America's overall health, organizations and individuals can improve the health of the current generation and ensure better health for generations to come.

Guidelines	Key Recommendations
Balance Calories to Manage Weight	• Prevent and/or reduce risk of becoming overweight and obese through improved eating and physical activity. • Control total calorie intake to manage body weight. For people who are overweight or obese, this will mean consuming fewer calories from foods and beverages. • Increase physical activity and reduce time spent in sedentary behaviors. • Maintain appropriate calorie balance during each stage of life.
Reduce Certain Foods and Food Components	• Reduce sodium intake to less than 2,300 mg/day and further reduce to 1,500 mg/day among persons who are 51 years and older and those of any age who are African American or have hypertension, diabetes, or chronic kidney disease. • Consume less than 300 mg/day dietary cholesterol. • Consume less than 10% of calories from saturated fatty acids. • Keep *trans* fatty acid consumption as low as possible. • Reduce intake of calories from solid fats and added sugars. • Limit consumption of foods that contain refined grains, especially those that contain solid fats, added sugars, and sodium. • If alcohol is consumed, it should be consumed in moderation—up to one drink per day for women and two drinks per day for men—and only by adults of legal drinking age.
Increase Certain Foods and Food Components	• Increase fruit and vegetable intake. • Eat a variety of vegetables, especially dark green, red, and orange vegetables, beans, and peas. • Consume at least half of all grains as whole grains. • Increase intake of fat-free or low-fat milk, milk products, and fortified soy beverages. • Choose a variety of high-protein foods, including seafood, lean meat and poultry, eggs, beans and peas, soy products, and unsalted nuts and seeds. • Use oils to replace solid fats when possible. • Choose foods that provide more potassium, dietary fiber, calcium, and vitamin D. • *Women capable of becoming pregnant*: Choose foods that supply heme iron, additional iron sources, and enhancers of iron absorption such as vitamin C–rich foods. Consume 400 μg/d synthetic folic acid (from fortified foods and/or supplements) in addition to food forms of folate from a varied diet. • *Pregnant or breastfeeding women*: Consume 8 to 12 ounces of seafood per week from a variety of seafood types, while limiting white (albacore) tuna to 6 ounces per week. Due to their high mercury content, tilefish, shark, swordfish, and king mackerel should not be eaten. If pregnant, take an iron supplement as recommended by a health care provider. • *Individuals 50 years and older*: Consume foods fortified with vitamin B_{12}, such as fortified cereals, or dietary supplements.
Building Healthy Eating Patterns	• Select an eating pattern that meets nutrient needs over time at an appropriate calorie level. • Account for all foods and beverages consumed and assess how they fit within a total healthy eating pattern. • Follow food safety recommendations when preparing and eating foods to reduce the risk of foodborne illness.

Source: Adapted from U.S. Departments of Agriculture (USDA) and Health and Human Services (DHHS). Dietary Guidelines for Americans, 2010, 7th ed. Washington, DC: U.S. Government Printing Office, December, 2010.

Based on the ratio of weight to height, body mass index (BMI) is a better indicator of obesity than weight alone. For these reasons, BMI has become the standard for gauging if a person is overweight or obese. This table can help you determine whether your BMI is characteristic of being underweight, healthy weight, overweight, or obese. Commonly used units and abbreviations, also listed below, provide easy reference to selected nutritional conventions.

Weight Classifications Using Body Mass Index (BMI)

Use this chart to calculate your BMI. Locate your weight on the bottom of the chart and your height on the left of the chart. The number located at the intersection of these two values is your BMI.

Height in feet and inches \ Weight in pounds	120	130	140	150	160	170	180	190	200	210	220	230	240	250
4'6"	29	31	34	36	39	41	43	46	48	51	53	56	58	60
4'8"	27	29	31	34	36	38	40	43	45	47	49	52	54	56
4'10"	25	27	29	31	34	36	38	40	42	44	46	48	50	52
5'0"	23	25	27	29	31	33	35	37	39	41	43	45	47	49
5'2"	22	24	26	27	29	31	33	35	37	38	40	42	44	46
5'4"	21	22	24	26	28	29	31	33	34	36	38	40	41	43
5'6"	19	21	23	24	26	27	29	31	32	34	36	37	39	40
5'8"	18	20	21	23	24	26	27	29	30	32	34	35	37	38
5'10"	17	19	20	22	23	24	26	27	29	30	32	33	35	36
6'0"	16	18	19	20	22	23	24	26	27	28	30	31	33	34
6'2"	15	17	18	19	21	22	23	24	26	27	28	30	31	32
6'4"	15	16	17	18	20	21	22	23	24	26	27	28	29	30
6'6"	14	15	16	17	19	20	21	22	23	24	25	27	28	29
6'8"	13	14	15	17	18	19	20	21	22	23	24	25	26	28

Key

BMI (kg/m^2)	Classification
<18.5	Underweight
18.5–24.9	Healthy weight
25.0–29.9	Overweight
≥30	Obese

Source: Centers for Disease Control and Prevention. Overweight and obesity: Defining overweight and obesity. Available from: http://www.cdc.gov/obesity/defining.html.

Commonly Used Nutrition-Related Abbreviations

AI	Adequate Intake
AMDR	Acceptable Macronutrient Distribution Range
ATP	Adenosine Triphosphate
BMI	Body Mass Index
BMR	Basal Metabolic Rate
EAR	Estimated Average Requirement
EER	Estimated Energy Requirement
CDC	U.S. Centers for Disease Control and Prevention
CVD	Cardiovascular Disease
DNA	Deoxyribonucleic Acid
DRI	Dietary Reference Intake
DV	Daily Value
FDA	U.S. Food and Drug Administration
GI	Gastrointestinal
HDL	High-Density Lipoprotein
LDL	Low-Density Lipoprotein
MUFA	Monounsaturated Fatty Acid
PUFA	Polyunsaturated Fatty Acid
RDA	Recommended Dietary Allowance
SFA	Saturated Fatty Acid
TEE	Total Energy Expenditure
UL	Tolerable Upper Intake Level
USDA	U.S. Department of Agriculture